Remunerative Horticulture

Ankan Das • Amit Baran Sharangi
Editors

Remunerative Horticulture

Strategies for Profitable Growth

Springer

Editors
Ankan Das
Department of Horticulture
Institute of Agricultural Science, University of Calcutta
Kolkata, West Bengal, India

Amit Baran Sharangi
Department of Plantation, Spices, Medicinal and Aromatic Crops
Faculty of Horticulture, Bidhan Chandra Krishi Viswavidyalaya
Mohanpur, West Bengal, India

ISBN 978-981-92-1006-0 ISBN 978-981-92-1007-7 (eBook)
https://doi.org/10.1007/978-981-92-1007-7

This Springer imprint is published by the registered company Springer Nature Singapore Pte Ltd.
The registered company address is: 152 Beach Road, #21-01/04 Gateway East, Singapore 189721, Singapore

Foreword

Horticulture has emerged as a vital sector capable of enhancing farm income, generating employment, and promoting sustainable agricultural development. In this context, *Remunerative Horticulture: Strategies for Profitable Growth* is a timely and valuable contribution that highlights the immense potential of this dynamic field.

The book brings together the insights of distinguished scholars and practitioners from across the globe, presenting a comprehensive view of modern horticultural practices. Covering topics from fundamental cultivation techniques to emerging technologies such as robotics, machine learning, and sustainable production systems, the volume offers readers a broad and balanced perspective.

Written in clear and accessible language, this book will benefit students, researchers, entrepreneurs, and professionals alike. By linking scientific knowledge with practical applications—such as value addition, innovative processing, and modem cultivation systems—it encourages readers to explore new opportunities for profitable horticulture.

I am confident that this volume will serve as a useful reference and inspire greater innovation and enterprise in the horticulture sector. I congratulate the editors and contributors for bringing out this meaningful publication.

The Vice-Chancellor of Bidhan Chandra
Krishi Viswavidyalaya
Haringhata, West Bengal, India

Ashok K. Patra

Preface

The editors of *Remunerative Horticulture: Strategies for Profitable Growth* are pleased to present this volume, a collective endeavor that brings together the expertise of distinguished contributors from diverse regions of the world, including India, Afghanistan, China, the United States, Mexico, Sri Lanka, Kenya, Tanzania, Bangladesh, and beyond. The book reflects a shared vision: to highlight horticulture not only as a science of cultivation but also as a powerful instrument for livelihood creation, innovation, and sustainable development.

Across continents, societies are grappling with the challenge of expanding employment opportunities in the face of growing populations and shifting economies. Horticulture, with its versatility and adaptability, offers pathways that can empower individuals and communities to build enterprises, generate income, and strengthen food and nutritional security. This book seeks to provide readers with a comprehensive and professional understanding of the skills, technologies, and practices that make horticulture a vibrant and remunerative field.

The strength of this volume lies in its inclusivity. Each chapter has been authored by experts who bring unique perspectives shaped by their national contexts and professional experiences. Together, these voices form a rich mosaic of ideas—ranging from traditional practices to modern innovations—that underscore the global relevance of horticulture. The integration of such diverse viewpoints ensures that the book resonates with readers across cultures and disciplines, while also contributing to its intellectual depth and impact.

The chapters are carefully organized to provide a holistic view of horticultural practices, covering both pre-harvest and post-harvest dimensions. Readers will encounter discussions that span from fundamental cultivation methods to advanced precision technologies, robotics, and machine learning applications. The book also emphasizes value addition, exploring how simple techniques such as dehydration, drying, or improved packaging can enhance market appeal and profitability.

Equally compelling are the sections on emerging trends such as organic gardening, rooftop and vertical farming, bonsai development, and the use of natural fibers derived from horticultural materials. Waste utilization, microencapsulation, and innovative processing of by-products are presented as opportunities to transform challenges into profitable ventures. By weaving together these diverse themes, the book offers a panoramic view of horticulture as a discipline that is both scientifically rigorous and economically rewarding.

A hallmark of this work is its clarity of language and accessibility. While grounded in scientific knowledge, the chapters are written in a style that appeals to a wide readership, including those without formal training in agriculture. At the same time, the book provides depth and detail that will benefit undergraduate and postgraduate students preparing for advanced studies in horticulture and related fields. Professionals, entrepreneurs, and enthusiasts will also find practical insights that can be applied to real-world contexts.

Beyond its technical content, the book aims to inspire. It demonstrates how horticulture, when approached with creativity and skill, can become a source of empowerment for individuals, families, and societies. By showcasing examples of innovation—from organic produce to urban gardening initiatives—it encourages readers to take proactive steps toward building sustainable enterprises and contributing to community well-being.

We, as editors, believe that this volume will serve as both a reference and a guide for the students, researchers, policymakers, government and nongovernment agencies, amateur readers, and the interested farming communities. It is our hope that readers will not only acquire knowledge but also feel motivated to explore horticulture as a field of opportunity, resilience, and growth.

Kolkata, West Bengal, India — Ankan Das
Mohanpur, West Bengal, India — Amit Baran Sharangi

Contents

Editors and Contributors

About the Editors

Ankan Das is an Associate Professor in the Department of Horticulture and Post Harvest Technology, Palli Siksha Bhavana, Visva-Bharati, India. Previously he has served in the capacity of Assistant Professor in the Department of Horticulture, Institute of Agricultural Science, University of Calcutta, for more than 9 years. He has completed BSc (Hons.) in Horticulture from Uttar Banga Krishi Viswavidyalaya, Cooch Behar and master's and PhD program from Bidhan Chandra Krishi Viswavidyalaya, Nadia, West Bengal. He is a recipient of University Gold Medal for BSc (Hort.) Hons. and University Gold Medal for MSc (Hort.) in Post-Harvest Technology of Horticultural Crop. He has published many research and review papers in journals of national and international repute (including UGC-CARE listed journals) and book chapters with renowned publishers. To his credit, also there are some articles in magazines and popular Bengali newspapers. He has also delivered some lectures in events of academic importance and has presented and participated in many national and international symposiums. He also has gone through academic trainings and workshops and has completed required Faculty Development Programs as of now.

Amit Baran Sharangi is Professor and Former Dean (Faculty of Horticulture) of BCKV (Agricultural University), India, attached with teaching and research in herbs, spices, and MAPs for more than 29 years. He is a Fellow of 10 international and national academies including the Royal Society of Biology (London, UK), National Academy of Biological Sciences, International Society of Noni Sciences, Academy of Science & Technology (WB), etc. and a life member of a good number of societies including the National Academy of Sciences, Indian Academy of Horticultural sciences, etc. with 120 peer-reviewed papers in scientific journals. He has published 40 authored/edited books and numerous book chapters from various international publishers, and is engaged with 75 journals worldwide as editor-in-chief, editorial board member, subject and review editor, and active reviewer (Publons Academy graduate). He has visited abroad extensively on academic mission and obtained several international laurels viz. ENDEAVOUR Post-doctoral Award (Australia), INSA-RSE Visiting Scientist Fellowship & Chartered Biologists (UK), FULBRIGHT Visiting Lecturer Fellowship (USA), Achiever's Award (SADHNA), Marquis Who's Who (USA), Man of the Year (UK), Outstanding Scientist, Bharat Ratna Mother Teresa Gold Medal, Honorary Professorship by the Cypress International Institute University (Texas, USA), etc. He has delivered a couple of invited and keynote lectures at home and abroad (India, the UK, the USA, Australia, New Zealand, Thailand, Israel, Jordan, and Bangladesh) on several aspects of herbs and spices, and chaired/co-chaired and convened on about seven international and five technical sessions. He is associated with a number of research projects as Principal and Co-Principal Investigator having academic and empirical implications. He has served as a member of Selection Committee, Board of Studies, Faculty Council, and Executive Council and presently is a member of BSMA (a specially empowered national body constituted by Director General ICAR, Government of India, to restructure the Post Graduate and PhD Syllabus). He is an active member of several science academies and societies like NASI, NABS, ISNS, IAHS, CHAI, CWSS, and SAH including the New York Academy of Science (NYAS), World Academy of Science, Engineering and Technology (WASET), African Forest Forum, American Biological Society, to name a few.

Contributors

Ashna Acharya CSIR-IHBT Palampur, Palampur, Himachal Pradesh, India

Sumit Acharya Department of Horticulture, Centurion University of Technology and Management, R. Sitapur, Odisha, India

Anjana J. Atapattu Agronomy Division, Coconut Research Institute of Sri Lanka, Lunuwila, Sri Lanka

Priyatharsini Ayyaswamy Department of Floriculture and Landscaping, Horticultural College and Research Institute, Tamil Nadu Agricultural University, Coimbatore, Tamil Nadu, India

Sumana Balo Department of Soil Science and Agricultural Chemistry, Institute of Agricultural Sciences (IAS), Siksha O Anusandhan (Deemed to be University), Bhubaneswar, Khorda, Odisha, India

Solanki Bal School of Smart Agriculture, Adamas University, Kolkata, West Bengal, India

Prerna Baraily College of Agriculture, Uttar Banga Krishi Viswavidyalaya, Majhian, Dakshin Dinajpur, West Bengal, India

Martin Leanardo Battaglia Regenify Ag, Ithaca, NY, USA

Sarthak Bhattacharya Department of Horticulture, Neotia University, South 24 Parganas, Sarisha, West Bengal, India

Urjashi Bhattacharya School of Agriculture and Allied Sciences, The Neotia University, Sarisa, West Bengal, India

C. S. Chaitra Department of Floriculture and Landscaping, Punjab Agricultural University, Ludhaina, Punjab, India

Souvik Chakraborty Department of Post Harvest Technology, Bidhan Chandra Krishi Viswavidyalaya, Haringhata, West Bengal, India

Luciana L. Chappa Department of Agricultural Sciences, College of Agriculture, Mwalimu Nyerere University of Agriculture and Technology, Butiama, Tanzania

Subhrajyoti Chatterjee Department of Horticulture, Centurion University of Technology and Management, R. Sitapur, Odisha, India

Dixit Chaudhary CSIR-IHBT Palampur, Palampur, Himachal Pradesh, India

Rohit Kumar Choudhury Department of Agricultural Chemistry and Soil Science, Faculty of Agriculture, Bidhan Chandra Krishi Viswavidyalaya, Nadia, West Bengal, India

Ankan Das Department of Horticulture, Institute of Agricultural Science, University of Calcutta, Kolkata, West Bengal, India

Anwesha Das Department of Fruit Science, Bidhan Chandra Krishi Viswavidyalaya, Mohanpur, Haringhata, West Bengal, India

Arindam Das Department of Fruit Science, Bidhan Chandra Krishi Viswavidyalaya, Mohanpur, Haringhata, West Bengal, India

Payel Das Department of Horticulture, Institute of Agricultural Science, University of Calcutta, Kolkata, West Bengal, India

Sayani Das Department of Horticulture and Post-Harvest Technology, Palli Siksha bhavana (Institute of Agriculture), Sriniketan, West Bengal, India

Soustav Datta Rathindra Krishi Vigyan Kendra, Palli Siksha Bhavana (Institute of Agriculture), Visva-Bharati, Birbhum, West Bengal, India

Prahlad Deb Department of Horticulture & Postharvest Technology, Institute of Agriculture, Visva-Bharati, Sriniketan, India

S. R. Dhiman Department of Floriculture and Landscaping, Dr. YS Parmar University of Horticulture and Forestry, Solan, Himachal Pradesh, India

Suchandra Dutta Ramakrishna Mission Vivekananda Educational and Research Institute (RKMVERI), Narendrapur, Kolkata, West Bengal, India

Shah Faisal Institute of Food Science and Technology, Chinese Academy of Agricultural Science, Beijing, China

Dinkar Jagannath Gaikwad Research Centre of Excellence for Smart Agriculture, Centurion University of Technology and Management, R. Sitapur, Odisha, India

Sourav Garai Division of Agronomy, Ramakrishna Mission Vivekananda Educational and Research Institute, Narendrapur, Kolkata, West Bengal, India

Abdul Khalil Gardezi Colegio de Posgraduados, Posgrado en Hidrociencias, Montecillos, Mexico

Subhadwip Ghorai Department of Agriculture, Brainware University, Kolkata, West Bengal, India

Arijit Ghosh Department of Agricultural Chemistry and Soil Science, Faculty of Agriculture, Bidhan Chandra Krishi Viswavidyalaya, Nadia, West Bengal, India

Harun I. Gitari Department of Agricultural Science and Technology, School of Agriculture and Environmental Sciences, Kenyatta University, Nairobi, Kenya

R. Gowshika Department of Vegetable Science, Horticultural College and Research Institute, Tamil Nadu Agricultural University, Coimbatore, Tamil Nadu, India

Y. C. Gupta Department of Floriculture and Landscaping, Dr. YS Parmar University of Horticulture and Forestry, Solan, Himachal Pradesh, India

Atal Hameedi Agriculture Faculty, Horticulture Department, Sayed Jamaluddin Afghani University, Asadabad, Afghanistan

Akbar Hossain Soil Science Division, Bangladesh Wheat and Maize Research Institute, Dinajpur, Bangladesh

Chinmaya Jena M. S. Swaminathan School of Agriculture, Centurion University of Technology and Management, Paralakhemundi, Odisha, India

Prakash Kumar Jha Department of Plant and Soil Sciences, Mississippi State University, Starkville, MS, USA

Gomathy Kamalakkannan Department of Fruit and Vegetable Technology, CSIR-Central Food Technological Research Institute, Mysore, Karnataka, India

J. Karthikeyan Department of Vegetable Science, Horticultural College and Research Institute, Tamil Nadu Agricultural University, Coimbatore, Tamil Nadu, India

Bharati Kashyap Department of Floriculture and Landscaping, Dr. YS Parmar University of Horticulture and Forestry, Solan, Himachal Pradesh, India

Arju Ali Khan Department of Fruit Science, Bidhan Chandra Krishi Viswavidyalaya, Mohanpur, Haringhata, West Bengal, India

Anshul Kumar Department of Floriculture and Landscaping, Dr. YS Parmar University of Horticulture and Forestry, Solan, Himachal Pradesh, India

Guddu Kumar Assistant Technical Manager (ATMA), Department of Agriculture, Gopalganj, Bihar, India

Manoj Kumar Division of Vegetable Science, Sher-E-Kashmir University of Agricultural Sciences and Technology, Jammu, Jammu and Kashmir, India

Raj Kumar Department of Vegetable and Spice Crops, Uttar Banga Krishi Viswavidyalaya, Pundibari, West Bengal, India

Nithar Ranjan Madhu Department of Zoology, Acharya Prafulla Chandra College, Kolkata, West Bengal, India

Natanael Magaña-Lira Universidad Autónoma Chapingo, Departamento de Fitotecnia, Texcoco, México

Bibekananda Maiti Department of Plant Physiology, Faculty of Agriculture, Bidhan Chandra Krishi Viswavidyalaya, Nadia, West Bengal, India

Payal Maiti Department of Post Harvest Technology, Bidhan Chandra Krishi Viswavidyalaya, Nadia, West Bengal, India

Sagar Maitra M. S. Swaminathan School of Agriculture, Centurion University of Technology and Management, Paralakhemundi, Odisha, India

Bisruti Maity M. S. Swaminathan School of Agriculture, Centurion University of Technology and Management, Paralakhemundi, Odisha, India

Chinmoy Mandal Department of Pomology and Post Harvest Technology, Faculty of Horticulture, Uttar Banga Krishi Viswavidyalaya, Cooch Behar, West Bengal, India

Joydip Mandal Department of Horticulture and Post-Harvest Technology, Palli Siksha Bhavana (Institute of Agriculture), Sriniketan, West Bengal, India

Kamal Kr. Mandal Department of Fruit Science, Bidhan Chandra Krishi Viswavidyalaya, Nadia, West Bengal, India

Nirmal Mandal Department of Agricultural Biotechnology, Faculty of Agriculture, Bidhan Chandra Krishi Viswavidyalaya, Mohanpur, West Bengal, India

Sergio Roberto Márquez-Berber Universidad Autónoma Chapingo, Departamento de Fitotecnia, Texcoco, México

Rajdeep Mohanta Department of Agriculture, Brainware University, Kolkata, West Bengal, India

Shreya Mondol Department of Horticulture, Institute of Agricultural Science, University of Calcutta, Kolkata, West Bengal, India

Satya Prasad Nanda M. S. Swaminathan School of Agriculture, Centurion University of Technology and Management, Paralakhemundi, Odisha, India

Priti Smita Nayak Department of Horticulture, Centurion University of Technology and Management, R. Sitapur, Odisha, India

Uggumudi Sai Nishitha Department of Floriculture and Landscaping, Horticultural College and Research Institute, Tamil Nadu Agricultural University, Coimbatore, Tamil Nadu, India

Tharindu D. Nuwarapaksha Agronomy Division, Coconut Research Institute of Sri Lanka, Lunuwila, Sri Lanka

Manas Kumar Pandit Department of Vegetable Science, Bidhan Chandra Krishi Viswavidyalaya, Mohanpur, Haringhata, West Bengal, India

Ritoban Pandit Department of Agricultural Meteorology and Physics, Bidhan Chandra Krishi Viswavidyalaya, Mohanpur, Haringhata, West Bengal, India

Somnath Pati Department of Soil and Water Conservation, Faculty of Agriculture, Bidhan Chandra Krishi Viswavidyalaya, Nadia, West Bengal, India

Sarthak Pattanayak Odisha University of Agriculture and Technology, Bhubaneswar, Odisha, India

Keerthivasan Ragupathi Department of Floriculture and Landscaping, Horticultural College and Research Institute, Tamil Nadu Agricultural University, Coimbatore, Tamil Nadu, India

Sudarshan Ramanathan Department of Food Science, School of Life Sciences, St. Aloysius Deemed to be University, Mangaluru, India

Diksha Rani Division of Vegetable Science, Sher-E-Kashmir University of Agricultural Sciences and Technology, Jammu, Jammu and Kashmir, India

Nihal Ravindranath M.S. Swaminathan School of Agriculture, Centurion University of Technology and Management, Paralakhemundi, Odisha, India

Debarpita Datta Ray Department of Genetics and Plant Breeding, Bidhan Chandra Krishi Viswavidyalaya, Mohanpur, Haringhata, West Bengal, India

Sumit Ray M.S. Swaminathan School of Agriculture, Centurion University of Technology and Management, Paralakhemundi, Odisha, India

Diana América Reyna-Izaguirre Universidad Autónoma Chapingo, Departamento de Ingeniería Agroindustrial, Texcoco, Mexico

Shaik Rishitha M. S. Swaminathan School of Agriculture, Centurion University of Technology and Management, Paralakhemundi, Odisha, India

Manideepa Roy Department of Horticulture, Institute of Agricultural Science, University of Calcutta, Kolkata, West Bengal, India

Sourav Roy Department of Agriculture, Brainware University, Kolkata, West Bengal, India

Lalichetti Sagar M. S. Swaminathan School of Agriculture, Centurion University of Technology and Management, Paralakhemundi, Odisha, India

Siddhartha Sahoo Department of Horticulture, Institute of Agricultural Science, University of Calcutta, Kolkata, West Bengal, India

Masina Sairam M. S. Swaminathan School of Agriculture, Centurion University of Technology and Management, Paralakhemundi, Odisha, India

Victry Samanta Department of Horticulture, Institute of Agricultural Science, University of Calcutta, Kolkata, West Bengal, India

D. T. Santosh Research Centre of Excellence for Smart Agriculture, Centurion University of Technology and Management, Paralakhemundi, Odisha, India

Amlanjyoti Sar Department of Fruit Science, Faculty of Horticulture, Bidhan Chandra Krishi Viswavidyalaya, Nadia, West Bengal, India

Debapriya Sarkar Department of Vegetable and Spice Crops, Uttar Banga Krishi Viswavidyalaya, Pundibari, West Bengal, India

Souptik Sarkar Department of Agricultural Chemistry and Soil Science, Faculty of Agriculture, Bidhan Chandra Krishi Viswavidyalaya, Nadia, West Bengal, India

Amit Baran Sharangi Department of Plantation, Spices, Medicinal and Aromatic Crops, Faculty of Horticulture, Bidhan Chandra Krishi Viswavidyalaya, Mohanpur, West Bengal, India

S. S. Vasanthkumar Department of Plant and Environmental Sciences, Clemson University, Clemson, SC, USA

S. P. Yazhini Department of Fruit Science, Horticultural College and Research Institute, Periyakulam, Tamil Nadu Agricultural University, Coimbatore, Tamil Nadu, India

Urban and Peri-Urban Horticulture: Utility, Aesthetics, and Profit

1

Sumit Acharya, Priti Smita Nayak, Atal Hameedi, and Subhrajyoti Chatterjee

Abstract

Urban and peri-urban horticulture has emerged as a dynamic and multifunctional strategy to address food security, environmental sustainability, and quality of life in rapidly urbanizing areas. The integration of horticultural practices, such as rooftop gardens, kitchen gardens, vertical farming, and the cultivation of minor fruits and vegetables, within and around city environments provides a sustainable response to urban challenges. Urban and peri-urban horticulture not only ensures a reliable supply of fresh, nutritious produce at the household and community levels but also enhances urban aesthetics, mitigates the urban heat island effect, and promotes biodiversity. The utility of urban and peri-urban horticulture extends beyond food production. Rooftop and vertical gardens make efficient use of limited space while contributing to thermal insulation, air purification, and water conservation. Kitchen gardens foster food literacy, nutritional awareness, and mental well-being. Minor fruits and vegetables, often underutilized in commercial agriculture, provide dietary diversity and economic opportunities, especially for women and marginalized groups. Additionally, these green spaces can become centers for community interaction, education, and eco-consciousness. From an economic perspective, urban and peri-urban horticulture creates avenues for micro-enterprises, job creation, and income generation, particularly in low-income urban and peri-urban communities. With minimal inputs and innovative approaches such as hydroponics, aquaponics, and organic waste recycling,

S. Acharya · P. S. Nayak · S. Chatterjee (✉)
Department of Horticulture, Centurion University of Technology and Management, R. Sitapur, Odisha, India
e-mail: subhrajyoti.chatterjee@cutm.ac.in

A. Hameedi
Agriculture Faculty, Horticulture Department, Sayed Jamaluddin Afghani University, Asadabad, Afghanistan

A. Das, A. B. Sharangi (eds.), *Remunerative Horticulture*,
https://doi.org/10.1007/978-981-92-1007-7_1

urban and peri-urban horticulture can thrive even in resource-constrained settings. Furthermore, the promotion of aesthetic landscaping through edible greenery enhances property values and urban appeal. In the era of climate change, urban sprawl, and rising food insecurity, urban and peri-urban horticulture represents a holistic, adaptable, and profitable model that intertwines utility, aesthetics, and economic potential. Policymakers, urban planners, and citizens must collaboratively invest in promoting urban and peri-urban horticulture as a pillar of resilient urban development.

Keywords

Urban horticulture · Peri-urban farming · Rooftop gardens · Kitchen gardens · Vertical farming · Minor fruits and vegetables

1.1 Introduction

Urbanization has emerged as one of the most defining features of the twenty-first century, reshaping lifestyles, economies, and landscapes across the globe. With cities expanding rapidly, accompanied by escalating population density and shrinking per capita land resources, the demand for innovative approaches to food production, environmental sustainability, and quality of life has become more urgent than ever. In this context, urban and peri-urban horticulture (UPH) has gained remarkable attention as a sustainable practice that integrates utility, aesthetics, and profitability into the modern urban living. Unlike conventional agriculture, which is largely practiced in rural settings and relies heavily on extensive land resources, UPH offers localized, resource-efficient, and diversified solutions by adapting horticultural practices within or on the fringes of cities. The incorporation of rooftops, balconies, vertical gardens, home backyards, and marginal lands in peri-urban areas has transformed food systems, enhanced urban ecology, and opened up new opportunities for income generation and recreational satisfaction.

The utility of urban and peri-urban horticulture lies primarily in its ability to bring food production closer to consumers, thereby reducing the dependency on long supply chains and minimizing the environmental footprint associated with transportation and storage FAO (2012). Fresh fruits, vegetables, herbs, and ornamentals grown within city spaces not only meet dietary needs but also contribute to nutrition security and resilience against food shortages. In rapidly growing cities, where arable land is limited and perishable commodities are in high demand, UPH provides a feasible alternative to ensure the steady availability of high-quality produce. Rooftop gardens and kitchen gardens, for example, empower households to cultivate minor fruits, leafy greens, and other vegetables for self-consumption, thereby lowering household expenditure while improving dietary diversity. At the same time, community gardens and peri-urban farming initiatives contribute to social bonding, urban employment, and food sovereignty, demonstrating the practical benefits of integrating horticulture into the urban fabric.

Beyond its utility, urban and peri-urban horticulture also enhances the aesthetic dimension of urban environments. The greening of rooftops, balconies, and walls through vertical gardens or potted arrangements not only beautifies otherwise monotonous concrete landscapes but also fosters psychological well-being among urban residents. Numerous studies have highlighted that exposure to greenery, even in small patches, helps reduce stress, improve mental health, and encourage active lifestyles. The visual appeal of flowering plants, climbers, and ornamental foliage has a direct impact on urban aesthetics, creating soothing spaces amidst crowded and polluted cityscapes. Moreover, the integration of horticultural designs into urban planning has proven effective in mitigating the urban heat island effect, lowering ambient temperatures, improving air quality, and contributing to ecological balance (Specht et al. 2014). Thus, the role of horticulture extends beyond food production into shaping healthier, more livable, and environmentally harmonious cities.

Profitability forms yet another cornerstone of urban and peri-urban horticulture. What initially emerged as a household-level or community-based subsistence activity has now evolved into a growing sector with significant commercial prospects. Rooftop farming models, hydroponic setups, vertical farming systems, and container gardens have created business avenues for urban entrepreneurs who capitalize on consumer demand for fresh, organic, and locally grown produce. In particular, the cultivation of minor fruits, exotic vegetables, herbs, and microgreens has proven highly profitable due to their short growing cycles, premium market value, and rising demand among health-conscious urban populations (Orsini et al. 2013). Peri-urban zones, located on the peripheries of cities, often serve as vital supply hubs, providing milk, vegetables, flowers, and fruits that sustain urban markets on a daily basis. With advancements in protected cultivation, hydroponics, and integrated farming systems, peri-urban horticulture has emerged as a bridge connecting urban consumers with fresh, high-quality produce while generating consistent income for growers.

Technological innovation has further strengthened the potential of UPH by optimizing limited resources such as water, space, and soil. Techniques such as drip irrigation, nutrient film technique (NFT), aeroponics, and organic waste composting enable efficient recycling and sustainable crop production in nontraditional spaces. Rooftop gardens equipped with rainwater harvesting systems and solar panels, or vertical gardens supported by automated irrigation, illustrate the synergy between technology and horticultural practices in urban settings. Such models not only enhance productivity but also align with broader goals of climate resilience, the circular economy, and sustainable urban planning. The integration of these methods into housing complexes, educational institutions, office buildings, and community spaces highlights the scalability and adaptability of UPH across diverse contexts.

Equally important is the sociocultural significance of urban horticulture. In many cities, community gardens and rooftop farming collectives have become platforms for knowledge sharing, intergenerational learning, and community engagement. Urban residents, including children and the elderly, find opportunities to reconnect with nature, practice traditional cultivation methods, and engage in eco-friendly lifestyles. These initiatives help cultivate awareness about food origins, nutrition, and environmental stewardship among urban dwellers who are otherwise disconnected from rural farming traditions. Furthermore, urban horticulture has proven instrumental in empowering women, unemployed youth, and marginalized groups by offering livelihood options that require relatively low investment yet ensure consistent returns (Zezza and Tasciotti 2010).

In light of growing urbanization, climate variability, and the pressing need for sustainable development, urban and peri-urban horticulture emerges as an indispensable strategy that blends functionality, beauty, and profitability. It is not merely an urban trend but a resilient response to the challenges of food insecurity, ecological degradation, and urban stress. Rooftop gardens, kitchen gardens, vertical farming systems, and the cultivation of minor fruits and vegetables are not isolated practices but integral components of a holistic vision for sustainable cities. By addressing the nexus of food, health, environment, and livelihood, UPH offers a transformative pathway that supports both individuals and societies in reimagining the future of urban living (Deelstra and Girardet 2000).

1.2 Types and Systems of Urban and Peri-Urban Horticulture

1.2.1 Rooftop Gardens

1.2.1.1 Importance

Rooftop gardens have emerged as one of the most practical and innovative strategies to integrate horticulture into dense urban environments. With the rapid expansion of cities and the limited availability of open land, rooftops provide underutilized spaces that can be transformed into productive green zones. The importance of rooftop gardens lies in their dual contribution to food security and environmental sustainability (Orsini et al. 2014). On one hand, they enable households, institutions, and businesses to produce fresh vegetables, herbs, and fruits directly where they are consumed, reducing dependency on external supply chains (Please refer to Fig. 1.1, which depicts the rooftop cultivation of dragon fruit, which is considered as one of the "Super fruits" in the present time because of its high nutritional value). On the other hand, they contribute to the ecological resilience of cities by mitigating the urban heat island effect, improving air quality, and supporting biodiversity. In many metropolitan areas, rooftop gardening is also seen as a climate adaptation strategy that enhances urban resilience against rising temperatures, flooding, and pollution (Specht et al. 2014). Moreover, the practice aligns with the global push toward sustainable cities and communities, one of the core objectives of the United Nations Sustainable Development Goals (SDG 11).

1.2.1.2 Techniques

A variety of techniques are employed in rooftop gardening, ranging from traditional soil-based cultivation to advanced hydroponic and containerized systems. The

Fig. 1.1 A two-panel image showing a large dragon fruit plant in a blue container on the left, with long, green, spiky leaves, and several red flowers. On the right, a person is tending to the plant, carefully examining a flower. The background features a garden setting with various plants and trees

simplest form involves using pots, grow bags, or raised beds filled with lightweight soil mixes that reduce load stress on building structures. Organic compost derived from household waste is often added to improve soil fertility while ensuring sustainable recycling practices. More sophisticated methods, such as hydroponics and aeroponics, are increasingly adopted in urban areas where soil-based gardening is less feasible. Hydroponic systems allow plants to grow in nutrient-rich water, eliminating the need for soil, while aeroponics uses mist to supply nutrients directly to roots, ensuring efficient water use. Drip irrigation, rainwater harvesting, and modular green roofing systems further enhance productivity and reduce resource waste (Whittinghill and Rowe 2012). In some advanced models, rooftop greenhouses are established, enabling year-round production of vegetables and herbs under controlled conditions. The choice of technique depends largely on the purpose of the garden, available investment, and the structural strength of the rooftop.

1.2.1.3 Benefits

Rooftop gardens provide a wide range of benefits that span food, environment, economics, and social well-being. From a nutritional perspective, they allow households to access pesticide-free, nutrient-dense produce, which contributes to better health and dietary diversity. Environmentally, rooftop gardens act as natural insulation layers, reducing building energy demands by moderating indoor temperatures. This not only lowers electricity costs for cooling in summer and heating in winter but also reduces the carbon footprint of urban buildings. Ecologically, they play a crucial role in filtering air pollutants, absorbing carbon dioxide, and creating microhabitats for pollinators such as bees and butterflies. Economically, rooftop farming can generate supplementary income when surplus produce is sold in local markets or restaurants. Socially, they offer therapeutic and recreational value, providing urban residents with opportunities to engage in gardening activities that reduce stress and enhance community bonding (Thomaier et al. 2015). When implemented in schools and institutions, rooftop gardens also serve as educational platforms for environmental awareness and sustainable practices.

1.2.1.4 Challenges

Despite their many advantages, rooftop gardens face several challenges that can limit their widespread adoption. The first and most critical challenge is structural safety, as not all rooftops are designed to bear the additional load of soil, water, and plants. Waterproofing and proper drainage are essential to prevent leakage and damage to buildings, which may require significant investment. Initial setup costs, particularly for hydroponic and greenhouse systems, can be high and may discourage individual households from adoption. Maintenance is another hurdle, as urban residents with busy lifestyles may not have the time or expertise to manage regular watering, pest control, and harvesting. Climatic extremes such as high winds, intense heat, and heavy rainfall also pose risks, often requiring protective netting, shading, or windbreaks. In developing countries, lack of awareness, limited technical support, and insufficient policy incentives further restrict the scaling up of rooftop gardening practices. Addressing these challenges requires a combination of technical innovation, government support, and community participation.

1.2.1.5 Examples

Several successful examples of rooftop gardens worldwide illustrate their transformative potential. In Singapore, often referred to as a "garden city," government-supported initiatives have encouraged the widespread adoption of rooftop farming in both residential and commercial buildings. The "ComCrop" project, for instance, operates hydroponic farms on rooftops to supply fresh vegetables to local communities. In North America, cities like New York and Toronto have promoted rooftop gardening through green roof policies and incentives, with notable projects such as the Brooklyn Grange Farm, one of the largest rooftop soil-based farms in the world. In India, rooftop gardening has gained momentum in cities like Chennai, Bengaluru, and Delhi, where urban residents cultivate vegetables and herbs for household use, supported by municipal training programs. Similarly, in Nairobi, Kenya, residents have adapted low-cost container gardening methods to produce vegetables such as kale, spinach, and tomatoes on rooftops, contributing to food security in low-income neighborhoods. These examples highlight how rooftop gardens, adapted to local contexts, can simultaneously address food, environmental, and social needs in urban areas.

1.2.2 Kitchen Gardens (Home Gardens)

1.2.2.1 Importance

Kitchen gardens, also known as home gardens or backyard gardens, are among the oldest and most widely practiced forms of urban and peri-urban horticulture. They hold immense importance in improving household nutrition, reducing food expenditure, and fostering self-reliance (Galhena et al. 2013). In densely populated cities where fresh produce is often expensive or adulterated, kitchen gardens empower families to grow a portion of their own vegetables, herbs, and fruits. This practice plays a vital role in addressing micronutrient deficiencies, often referred to as "hidden hunger," by ensuring regular access to fresh, nutrient-rich food items such as leafy greens, carrots, tomatoes, and seasonal fruits. The importance of kitchen gardens extends beyond nutrition; they also help urban families reconnect with nature, a connection that is often weakened due to modern lifestyles. From a sustainability perspective, kitchen gardens represent a model of localized food systems that reduce transportation, packaging, and food waste. They are particularly significant in peri-urban regions, where households can maintain slightly larger garden spaces and integrate small livestock, thus creating diversified and resilient household food systems (Mitchell and Hanstad 2004).

1.2.2.2 Techniques

The techniques used in kitchen gardening vary depending on space availability, household needs, and environmental conditions. In urban settings with limited outdoor areas, container gardening is a common practice, where plants are grown in pots, recycled plastic bottles, or grow bags placed on balconies, terraces, or windowsills. Raised bed gardening, using soil enriched with compost or vermicompost, is another effective technique that allows efficient water use and better crop management. Vertical stacking systems and hanging planters can maximize space utilization in apartments. For households with backyards, traditional soil cultivation with crop rotation and intercropping is widely practiced, often combined with mulching and organic manure to enhance soil fertility. Technological interventions such as drip irrigation, rainwater harvesting, and hydroponic kitchen gardens are increasingly being adopted in urban areas to conserve resources and boost productivity. Moreover, the practice of using kitchen waste such as vegetable peels, eggshells, and coffee grounds to prepare compost helps close the nutrient cycle, ensuring sustainable and cost-effective gardening.

1.2.2.3 Benefits

The benefits of kitchen gardens are multidimensional, for example influencing health, economy, and the environment. From a health perspective, households gain access to fresh, pesticide-free produce that significantly enhances dietary diversity and food safety. Growing vegetables and herbs at home also encourages families to increase their consumption of fresh food, reducing reliance on processed and nutrient-poor alternatives. Economically, kitchen gardens help cut household food expenses by providing a steady supply of daily vegetables and herbs. In peri-urban households, surplus produce can be sold in local markets, generating supplementary income. Environmentally, kitchen gardens promote resource recycling, as organic waste is transformed into compost and greywater can be reused for irrigation. Additionally, small-scale home gardening helps conserve biodiversity by preserving indigenous and traditional crop varieties that may not be commercially available (Marsh 1998). Psychologically, kitchen gardening has therapeutic effects, reducing stress and fostering a sense of accomplishment. For children, home gardens serve as living classrooms that enhance awareness of food systems, ecology, and sustainable living.

1.2.2.4 Challenges

Despite their many advantages, kitchen gardens face challenges that limit their universal adoption. One of the primary constraints is space availability, especially in densely packed urban apartments where access to open soil is nearly impossible. Although container gardening provides a solution, it often restricts the scale and diversity of crops that can be grown. Another major issue is time management, as busy urban residents may find it difficult to maintain gardens regularly, leading to neglected plants or reduced productivity. Pest and disease management in small gardens is also challenging, particularly when chemical pesticides are avoided for safety reasons. Organic methods such as neem oil sprays or biological control require awareness and training, which may not always be available. In peri-urban regions, water scarcity during dry seasons can hinder garden productivity, especially where irrigation facilities are lacking. Furthermore, lack of institutional support, technical guidance, and awareness programs often prevents households from realizing the full potential of kitchen gardens.

1.2.2.5 Examples

Several successful examples highlight the contribution of kitchen gardens to food and nutritional security across the world. In India, the government and nongovernmental organizations have actively promoted kitchen gardens under nutrition programs such as the Poshan Abhiyaan, where families are encouraged to grow vegetables and fruits in small patches of land or containers. States like Karnataka, Kerala, and Odisha have documented success stories where women-led self-help groups have adopted kitchen gardening to improve family nutrition and income (Please refer to Table 1.1, which depicts the cropping scheme of vegetables round the year in the kitchen garden). In Cuba, after the economic crisis of the 1990s, urban households relied heavily on home gardens for fresh produce, giving rise to the "organopónicos" movement, which emphasized organic methods. In African countries such as Kenya and Tanzania, kitchen gardens play a crucial role in combating malnutrition by enabling families to grow nutrient-rich crops like kale, spinach, and sweet potatoes even in small spaces using sack gardens. Similarly, in developed nations like the United States and Germany, the popularity of home gardens has surged due to the rising demand for organic produce and a growing interest in sustainable lifestyles. These examples demonstrate that kitchen gardens, whether in developing or developed contexts, are highly adaptable and provide a resilient model for household-level food systems.

Table 1.1 The cropping scheme with month-wise examples of multiple crops in different columns (Chatterjee 2021)

Cropping scheme	Month	Suitable vegetables
1	June–October	Okra, amaranthus
	October–February	Cabbage, cauliflower, knolkhol, palak, fenugreek
	March–May	Ridge gourd, bottle gourd
2	Round the year	Brinjal and chili
3	July–September	Radish
	October–November	Palak and lettuce
	December–March	Tomato
	April–June	Amaranthus, poi
4	August–November	Tomato
	December–February	Cabbage and cauliflower
	March–July	Bitter gourd
5	July–October	Early cauliflower and cowpea
	November–December	Radish
	January–June	Onion
6	November–February	Potato and French bean
	March–June	Cowpea
	July–October	Mid-early/mid-late cauliflower
7	September–March	Chili
	April–August	Okra

1.2.3 Vertical Gardens/Vertical Farming

1.2.3.1 Importance

Vertical gardens and vertical farming represent one of the most innovative adaptations of horticulture to modern urban realities. As cities expand vertically with high-rise apartments and office complexes, the concept of growing plants on walls, facades, and multilevel structures has gained significant relevance (Despommier 2011). The importance of vertical farming lies in its ability to maximize productivity per unit area, making it particularly valuable in regions where land is scarce or prohibitively expensive. Unlike traditional gardening that requires horizontal plots, vertical systems use vertical surfaces or stacked layers, enabling efficient use of available space. This innovation directly addresses urban food security challenges, providing a consistent supply of fresh produce close to consumers. "Microgreens" can also be grown under a vertical system. Microgreens, also known as "Nutrigreens," are young, edible seedlings derived from diverse vegetables, herbs, and other plants (Please refer to Table 1.2, where suitable vegetables and herbs that can be grown as microgreens are listed). These are typically harvested at a specific developmental stage: when their cotyledonary leaves are fully expanded, and the initial pair of true leaves is either emerging or partially developed (Nayak et al. 2023). Characterized by their concentrated flavor and nutritional value, the stems and leaves of microgreens exhibit significantly elevated levels of vitamins and carotenoids, averaging approximately five times higher than those observed in their mature counterparts

Table 1.2 A list of crops under different families that can be grown as microgreens under a vertical system (Nayak et al. 2023)

Family	Name of the crop
Brassicaceae (previously known as Cruciferae)	Arugula, radish, mizuna, cress, knolkhol, mustard, pakchoi, sprouting broccoli, wasabi
Fabaceae (previously known as Leguminosae)	Lentil, pea, Bengal gram, methi
Amaranthaceae	Green and red amaranthus, spinach
Chenopodiaceae	Swiss chard, palak, quinoa
Apiaceae (previously known as Umbelliferae)	Carrot, coriander, celery
Lamiaceae (previously known as Labiateae)	Basil
Convolvulaceae	Water spinach
Polygonaceae	Sorrel, buckwheat

(Kumari and Kumari 2020). It is the reason for considering microgreens as a potential "super food." Vertical gardens also contribute to improving urban microclimates, reducing heat absorption by buildings, and enhancing aesthetic appeal. Globally, the push toward climate-smart and resource-efficient farming practices has further emphasized the importance of vertical farming as a sustainable solution for future cities (Al-Chalabi 2015).

1.2.3.2 Techniques

Several techniques are employed in vertical gardens and vertical farming, depending on scale, purpose, and available technology. The simplest form is the use of wall-mounted planters or modular panels filled with lightweight growing media such as coco peat, perlite, or vermiculite. These structures are often irrigated through drip systems connected to nutrient reservoirs, ensuring precise water delivery. Hydroponic vertical farming is another widely adopted method, where plants are grown in stacked shelves or towers with roots immersed in nutrient-rich water solutions. Variations such as the nutrient film technique (NFT), aeroponics, and aquaponics have further advanced productivity and resource use efficiency (Refer to Fig. 1.2, which delineates the microgreen setup under a vertical farming system). In nutrient film technique systems, a continuous, shallow stream of nutrient solution is conveyed through specialized channels, consistently irrigating the exposed root systems. Subsequently, at the lower terminus of these channels, the nutrient solution is collected and recirculated to the primary nutrient reservoir (Swain et al. 2021). Aquaponics integrates fish rearing with vertical plant production, recycling fish waste into nutrients for crops. Technological innovations such as automated irrigation, LED grow lights, and climate-controlled greenhouses have elevated vertical farming into a high-tech sector that can function year-round, irrespective of external weather conditions (Kalantari et al. 2017).

Fig. 1.2 The image shows two photographs of a plant growth setup. The left photo displays multiple trays with seedlings under artificial lighting on a two-tiered shelf system. The right photo is a close-up of densely packed seedlings with green leaves and pink stems, illuminated by pink grow lights. The environment appears to be a controlled indoor space for plant cultivation

1.2.3.3 Benefits

The benefits of vertical gardens and vertical farming are diverse, spanning food production, environmental services, and socioeconomic dimensions. First and foremost, vertical farming ensures a steady supply of fresh, pesticide-free vegetables and herbs in urban environments where traditional agriculture is impractical. Crops such as lettuce, spinach, basil, strawberries, and even minor fruits can be cultivated efficiently in vertical setups, often achieving yields many times higher than conventional systems per unit of land. From an environmental perspective, vertical gardens improve urban air quality by absorbing carbon dioxide, filtering dust, and releasing oxygen. They also insulate buildings, reducing indoor temperatures in summer and cutting heating costs in winter. Economically, vertical farming has opened new business opportunities, from small-scale household wall gardens to commercial-scale hydroponic farms supplying supermarkets and restaurants (Banerjee and Adenaeuer 2014). Socially, vertical gardens enhance the aesthetics of cities, transforming barren concrete walls into green, vibrant spaces that reduce stress and improve mental well-being. When adopted in schools, hospitals, and public spaces, they also serve as educational tools to promote sustainable living.

1.2.3.4 Challenges

Despite their potential, vertical gardens and vertical farming face significant challenges that affect adoption and scalability. One of the primary obstacles is the high initial investment, particularly for hydroponic and aeroponic systems that require specialized infrastructure, lighting, and climate control. These costs can be prohibitive for low-income households or small entrepreneurs without external support. Another challenge is the requirement for technical knowledge and skills, as vertical farming involves precise management of nutrient solutions, water quality, and environmental conditions. Any imbalance in pH, nutrient concentration, or temperature can lead to crop losses. Maintenance of wall-mounted gardens also demands regular attention, as uneven irrigation or clogging of drip lines can damage plants. Energy consumption, especially in indoor vertical farms that rely on artificial lighting and climate control, is another concern that can offset environmental benefits unless renewable energy sources are integrated. Additionally, cultural barriers and a lack of awareness about vertical gardening techniques can slow adoption, particularly in developing countries where traditional methods dominate.

1.2.3.5 Examples

Around the world, several inspiring examples demonstrate the success of vertical gardens and vertical farming. In Singapore, where land scarcity is acute, vertical farming has been embraced as part of the nation's food security strategy. Companies like "Sky Greens" operate multitiered vertical farms that use a rotating hydraulic system to grow leafy greens with minimal energy and water input. In Japan, "Mirai," one of the world's largest indoor vertical farms, produces thousands of heads of lettuce daily under LED lights in controlled environments. In the United States, vertical farming companies such as "AeroFarms" and "Plenty" have scaled up to supply supermarkets with fresh greens grown in aeroponic systems without soil or pesticides. In India, vertical gardens have become popular in metropolitan cities like Bengaluru and Delhi, where house walls, balconies, and corporate buildings are being converted into green walls for both aesthetic and functional purposes. Nairobi, Kenya, also showcases low-cost vertical farming systems, where farmers use sack gardens and recycled containers to grow vegetables in slum areas, enhancing food access for low-income communities. These diverse examples illustrate the adaptability of vertical farming, from high-tech commercial ventures to grassroots urban initiatives.

1.2.4 Community and Institutional Gardens

1.2.4.1 Importance

Community and institutional gardens represent collective efforts in urban and peri-urban horticulture, where groups of individuals or institutions come together to cultivate plants for food, aesthetics, education, or social well-being (Guitart et al. 2012). Their importance lies in the way they merge food production with social interaction, cultural expression, and public engagement. In urban settings where individual households may not have sufficient space for gardening, community gardens provide access to shared plots where residents can grow vegetables, fruits, herbs, and flowers. They address urban food insecurity by increasing the availability of fresh produce, particularly for marginalized groups and low-income communities. Institutional gardens, located in schools, hospitals, universities, and workplaces, serve educational and therapeutic functions while also supplying food for cafeterias or local distribution. Beyond food, these gardens play a critical role in environmental stewardship by promoting green spaces, recycling organic waste, and conserving biodiversity within urban landscapes. They are equally significant for building social cohesion, as they bring together people of different ages, cultures, and socioeconomic backgrounds to collaborate toward shared goals (Kingsley and Townsend 2006).

1.2.4.2 Techniques

The techniques used in community and institutional gardening depend largely on available resources, land size, and objectives. In most community gardens, land is divided into individual plots managed by families or members, with shared responsibilities for infrastructure such as composting units, water storage, and tool sheds. Raised beds and container gardening are common in areas with poor soil quality, while organic manures and composting are often emphasized to keep costs low and promote sustainability. Institutional gardens, particularly in schools and universities, may integrate demonstration plots, greenhouses, and hydroponic setups to provide hands-on learning opportunities. Hospitals and rehabilitation centers often design therapeutic gardens with a mix of vegetables, herbs, and ornamentals to stimulate physical and mental health recovery. Techniques such as rainwater harvesting, drip irrigation, and mulching are widely used to conserve water. In urban contexts with very limited space, some community gardens adopt vertical planters, rooftop plots, or hydroponic racks, ensuring that even small land parcels are productive.

1.2.4.3 Benefits

The benefits of community and institutional gardens extend far beyond food production. From a food security perspective, they improve access to fresh, nutrient-rich vegetables and fruits, which is especially critical in urban "food deserts," where fresh produce is scarce (Draper and Freedman 2010). These gardens help reduce household food costs and, in some cases, create opportunities for income generation through the sale of surplus crops. Socially, community gardens serve as spaces for interaction, cooperation, and cultural exchange, fostering a sense of belonging and shared responsibility among urban residents. They provide a platform for intergenerational learning, where elders share traditional knowledge with younger generations, while children gain experiential education about food systems and ecology. In institutional settings, gardens support health and wellness programs. School gardens enhance nutrition education and encourage healthy eating habits, while therapeutic gardens in hospitals aid patients' mental recovery, reduce stress, and improve overall well-being (Armstrong 2000). Environmentally, community gardens contribute to urban greening, reduce heat islands, recycle organic waste into compost, and create habitats for beneficial insects. They also play a role in urban resilience by acting as buffer zones during floods and promoting local adaptation to climate change (Okvat and Zautra 2011).

1.2.4.4 Challenges

Despite their wide-ranging benefits, community and institutional gardens face numerous challenges that can hinder sustainability. Land tenure and security are among the most pressing issues, as many community gardens operate on vacant lots or municipal land that may be reclaimed for development, leaving gardeners vulnerable. Limited access to water and irrigation infrastructure often hampers productivity, especially in cities where water is scarce or costly. Institutional gardens can face challenges related to funding, as schools and hospitals may not allocate sufficient resources for maintenance. Another major constraint is management and organization; community gardens require clear agreements and effective leadership to resolve conflicts over plot allocation, responsibilities, or use of shared resources. Pest and disease management also pose difficulties when large groups cultivate diverse crops in close proximity. In some contexts, lack of awareness, technical expertise, and policy support further limit the growth of community gardening initiatives. Additionally, urban expansion and commercialization of land frequently threaten the survival of such gardens, making long-term sustainability uncertain without legal and institutional backing.

1.2.4.5 Examples

Globally, community and institutional gardens have flourished in diverse forms, adapted to local needs. In the United States, community gardens became widespread during the "Victory Garden" movement of World War II and remain important today, with organizations such as the American Community Gardening Association promoting urban gardening initiatives. New York City's community gardens, particularly in the Bronx and Brooklyn, provide fresh food, green spaces, and cultural hubs for diverse communities. In Canada, Toronto's community gardening program has been integrated into municipal food policy, offering land, training, and support for local residents. In Europe, countries like Germany and the UK have long traditions of "allotment gardens," where citizens rent small plots for cultivation. In developing nations, community gardens are often critical for survival. In Havana, Cuba, urban residents created collective gardens called "organopónicos" to address food shortages, which later evolved into sustainable urban farming systems. In India, institutional gardens in schools and colleges have been promoted under the "School Nutrition Garden" initiative by the Ministry of Human Resource Development, aimed at improving children's nutrition and awareness. Similarly, in Kenya and South Africa, community gardens in townships and peri-urban areas provide both food and livelihoods for low-income families, highlighting their role in poverty alleviation.

1.2.5 Minor Fruits and Vegetables

1.2.5.1 Importance

Minor fruits and vegetables, often referred to as underutilized or neglected crops, play a vital role in diversifying diets, conserving agrobiodiversity, and enhancing resilience in urban and peri-urban horticulture. These crops are typically grown on a small scale and may not dominate mainstream markets, yet they carry significant nutritional, cultural, and ecological value. Examples include amaranth leaves, drumstick (moringa), ridge gourd, ivy gourd, custard apple, guava, ber (Indian jujube), and passion fruit. Their importance includes wider adaptibility to diverse climatic conditions, low input requirements, and rich nutritional profiles, often surpassing those of widely consumed staples. In urban and peri-urban areas, where dietary monotony and micronutrient deficiencies are common, minor fruits and vegetables provide essential vitamins, minerals, antioxidants, and dietary fiber (Grivetti and Ogle 2000). They also serve as climate-smart crops that can thrive in marginal soils, limited spaces, or with minimal water. Moreover, their cultural significance cannot be overlooked, as many minor fruits and vegetables are linked to traditional cuisines, folk medicine, and local heritage, ensuring continuity of food traditions in modern urban lifestyles (Jaenicke and Höschle-Zeledon 2006).

1.2.5.2 Techniques

The cultivation techniques for minor fruits and vegetables vary widely depending on species characteristics, but most are well-suited for integration into urban and peri-urban gardens. Many minor vegetables, such as amaranth, fenugreek, and spinach, can be easily grown in pots, grow bags, or small raised beds using nutrient-rich soils. Perennial vegetables like drumstick and curry leaf trees can be established in kitchen or community gardens, providing year-round harvests. Vertical trellising systems are useful for climbers such as ivy gourd and ridge gourd, making them ideal for rooftops or compact urban plots. For minor fruits, techniques include grafting and budding to improve yield and quality, along with pruning and canopy management to adapt trees for small spaces. In peri-urban contexts, small orchards of guava, papaya, custard apple, and ber can be established using organic manure and drip irrigation for efficient water use. Hydroponic and container-based methods are increasingly applied for leafy minor vegetables, enabling cultivation in balconies and terraces. The integration of organic waste composting, mulching, and rainwater harvesting further enhances sustainability in growing these crops (Padulosi et al. 2013).

1.2.5.3 Benefits

Minor fruits and vegetables provide a wide range of benefits that make them highly valuable in urban and peri-urban horticulture. From a nutritional standpoint, many of these crops are "superfoods." For instance, moringa leaves are rich in vitamin A, vitamin C, calcium, and protein, while amaranth leaves provide iron and folate essential for combating anemia. Minor fruits like guava and ber are loaded with vitamin C and dietary fiber, supporting immunity and digestion. Their short growth cycles, in the case of leafy greens, allow households to harvest fresh produce within weeks, ensuring a continuous supply of nutrients. Economically, minor fruits and vegetables hold niche market potential, as urban consumers increasingly demand exotic, organic, and health-promoting foods. Farmers and entrepreneurs can tap into this demand by cultivating these crops in peri-urban zones and selling them at premium prices. Environmentally, these crops often require fewer chemical inputs, making them suitable for organic cultivation and reducing pollution. Socially, they empower households by diversifying diets, preserving culinary traditions, and promoting local food sovereignty. In addition, they contribute to resilience, as their adaptability to stress conditions such as drought or poor soils makes them reliable even in adverse environments.

1.2.5.4 Challenges

Despite their potential, the cultivation of minor fruits and vegetables faces several challenges. The foremost issue is low awareness among urban consumers, who may prefer familiar, mainstream crops over indigenous or lesser-known crops. This lack of demand discourages farmers from investing in their cultivation. Another challenge is the limited availability of quality planting materials, as seed and sapling distribution systems are often focused on major crops. Postharvest management is also a critical problem; many minor fruits and leafy vegetables are highly perishable and lack established cold chain infrastructure, leading to significant losses. In urban settings, space constraints may limit the number of species that can be grown, while peri-urban farmers may struggle with market access for niche crops. Research and development efforts on improving yields, disease resistance, and shelf life for these crops are also comparatively limited, leaving farmers to rely on traditional practices. Additionally, policy neglect is a significant barrier, as minor fruits and vegetables are often excluded from government horticultural programs, subsidies, and extension services, despite their role in nutrition and sustainability (Bharucha and Pretty 2010).

1.2.5.5 Examples

Numerous examples from around the world highlight the successful integration of minor fruits and vegetables into urban and peri-urban horticulture. In India, cities like Hyderabad and Bengaluru have promoted the cultivation of drumstick, curry leaves, agathi, and amaranth in kitchen gardens as part of nutrition-focused government campaigns (Fig. 1.3 depicts the cultivation of agathi and curry leaves). Guava and papaya orchards in peri-urban belts around Pune and Lucknow supply both local and metropolitan markets, demonstrating their profitability. In Africa, crops like cowpea leaves, amaranth, and African eggplant are commonly grown in

Fig. 1.3 Two-panel image showing plants. The left panel features a close-up of a plant with elongated green leaves and a large, pale yellow flower hanging downward (agathi/heron flower). The right panel displays a top view of a different plant with smaller, darker green leaves, growing in sandy soil (curry leaf). Both images are set outdoors with natural lighting

backyard and community gardens, contributing significantly to household nutrition and income. Kenya, for instance, has popularized sack gardening of kale, amaranth, and spider plant in slum areas, ensuring vegetable access for low-income families (Smith and Eyzaguirre 2007). In Latin America, custard apple and passion fruit are cultivated in peri-urban orchards, while leafy greens such as "chaya" are used in household diets for their medicinal and nutritional properties. In developed countries, there has been renewed interest in "forgotten foods" as part of sustainable diets, with farmers' markets in Europe and the United States increasingly showcasing heritage varieties of vegetables and fruits. These examples emphasize that promoting minor fruits and vegetables in urban and peri-urban horticulture not only strengthens food systems but also safeguards biodiversity and cultural identity.

1.3 Utility and Functional Benefits

Urban and peri-urban horticulture (UPH) provides a wide array of benefits that extend far beyond food production. It addresses the dual challenge of feeding growing urban populations while promoting sustainable and resilient cities. The utility of UPH can be examined through its contribution to food security and nutrition, environmental benefits, and waste recycling. Each of these areas underscores the relevance of horticultural practices as both functional and transformative for cities in the twenty-first century.

1.4 Food Security and Nutrition

One of the most significant benefits of urban and peri-urban horticulture lies in its contribution to food security and nutrition. Rapid urbanization has intensified pressure on global food supply chains, leaving many city dwellers vulnerable to food shortages and price fluctuations. By cultivating fruits, vegetables, and herbs close to consumers, UPH reduces dependency on distant rural producers and creates resilient food systems. Rooftop gardens, community plots, and kitchen gardens ensure year-round access to fresh produce, often at a fraction of the cost of market-purchased goods (Galhena et al. 2013).

Nutritionally, horticultural crops cultivated in urban areas diversify diets by increasing access to micronutrient-rich foods such as leafy greens, tomatoes, beans, and minor fruits. These foods are vital in combating urban malnutrition, particularly among low-income populations that often rely on calorie-dense but nutrient-poor diets. Studies have shown that households maintaining kitchen gardens consume significantly more fruits and vegetables compared to those without access to home-grown produce (Zezza and Tasciotti 2010). Community and institutional gardens also function as educational spaces, where urban residents, especially children, learn about healthy eating habits, thus creating generational improvements in nutritional outcomes.

Moreover, the proximity of production to consumption reduces postharvest losses, ensuring that nutrient value is preserved. Urban agriculture can also respond to cultural and dietary preferences by producing traditional or indigenous crops that may not always be available in commercial supermarkets. Thus, UPH acts as both a buffer against food insecurity and a promoter of healthier, more diverse diets in urban settings (Orsini et al. 2014).

1.5 Environmental Benefits

Beyond food, urban and peri-urban horticulture contributes significantly to environmental quality. Cities face acute challenges related to heat stress, air pollution, and ecological degradation, all of which can be mitigated by integrating horticultural systems into the urban fabric. Rooftop gardens and vertical farms, for example, function as green infrastructure by reducing the urban heat island effect through shading and evapotranspiration (Specht et al. 2014). This cooling effect decreases energy consumption for air conditioning, resulting in both environmental and economic savings.

Vegetation cover from horticultural practices improves air quality by absorbing carbon dioxide and filtering airborne pollutants. Peri-urban horticultural belts serve as ecological buffers, mitigating dust, noise, and vehicular emissions that concentrate in city environments. Moreover, trees and shrubs cultivated in urban landscapes contribute to carbon sequestration, playing a small but meaningful role in global climate mitigation efforts (Whittinghill and Rowe 2012).

Horticultural systems also improve urban hydrology. Green roofs, community gardens, and permeable soil beds reduce stormwater runoff, lowering risks of flooding while improving groundwater recharge. Cities struggling with climate-related extreme rainfall can integrate urban horticulture into stormwater management strategies. Additionally, horticultural diversity enhances biodiversity within urban spaces, creating microhabitats for pollinators such as bees and butterflies, which are essential for ecological balance and agricultural productivity.

These environmental services make UPH an indispensable tool for designing sustainable and resilient cities. Unlike traditional urban infrastructure that often degrades natural systems, horticulture fosters a symbiotic relationship between human settlements and ecological processes (Deelstra and Girardet 2000).

1.6 Waste Recycling

A third functional benefit of urban and peri-urban horticulture is its role in recycling organic waste and closing nutrient cycles. Urban households, restaurants, and food markets generate significant amounts of biodegradable waste, much of which ends up in landfills. By integrating composting and vermicomposting into horticultural systems, these organic residues can be transformed into nutrient-rich fertilizers that enhance soil fertility (Marsh 1998). This not only reduces the burden on municipal waste management systems but also decreases the need for chemical fertilizers, which are costly and environmentally damaging.

Greywater and harvested rainwater can also be utilized in peri-urban horticulture, reducing stress on urban freshwater supplies. In some cities, treated wastewater has been safely applied to peri-urban vegetable cultivation, demonstrating the potential of water recycling for agricultural sustainability. Such practices are especially valuable in water-scarce regions, where agriculture must compete with domestic and industrial demands.

Waste recycling also ties into circular economy principles, where resources are continuously reused within local systems. The transformation of waste into agricultural inputs fosters economic opportunities for urban entrepreneurs engaged in composting services, organic fertilizer production, and community waste-to-food initiatives. Furthermore, recycling food and garden waste reduces greenhouse gas emissions from landfills, contributing to broader climate change mitigation goals.

Through these pathways, UPH positions cities as hubs of innovation in sustainable resource management, turning what was once a liability, urban waste, into a valuable resource for food production and environmental health.

1.7 Aesthetic and Psychological Value

Urbanization is rapidly transforming the global landscape, with more than half of the world's population now residing in cities and peri-urban areas. While this shift offers economic and infrastructural opportunities, it has also intensified challenges such as pollution, overcrowding, stress, and loss of natural spaces. Against this backdrop, urban and peri-urban horticulture (UPH) has emerged as a powerful tool to integrate greenery into cities and their surrounding zones. Beyond the obvious benefits of food security and environmental sustainability, UPH provides significant aesthetic and psychological value, enhancing quality of life in ways that are both visible and deeply felt (Mougeot 2005).

1.8 Aesthetic Value of Urban and Peri-Urban Horticulture

1.8.1 Enhancing Urban Beauty and Livability

Urban horticulture contributes significantly to the beautification of cities by introducing well-planned gardens, rooftop landscapes, vertical green walls, and community parks. These green spaces transform otherwise concrete-dominated surroundings into visually appealing environments. For instance, Singapore's "Garden City" model, which integrates vertical gardens on high-rise buildings, rooftop farms, and landscaped walkways, has made the city internationally admired for its aesthetics while also addressing urban heat (Taylor and Hochuli 2017).

In peri-urban zones, orchards, flower farms, and green belts not only serve as productive landscapes but also act as scenic buffers that enhance the view from urban centers. The sight of well-arranged vegetable patches, blossoming ornamental plants, or terraced fruit gardens adds beauty and visual relief to otherwise monotonous cityscapes.

1.8.2 Cultural and Artistic Expression

Horticulture in urban areas often goes beyond functionality to embody cultural aesthetics. Community gardens, for example, frequently include ornamental plants arranged in patterns inspired by local traditions. In India, urban terrace gardens often feature sacred plants like Tulsi (Holy Basil), which hold cultural and spiritual significance while also beautifying living spaces. Similarly, in Japanese urban areas, Zen gardens and bonsai arrangements serve as forms of living art, offering both aesthetic pleasure and spiritual depth (Beatley 2016).

1.8.3 Green Corridors and Urban Identity

Cities that actively promote urban horticulture often develop a unique identity associated with greenery. For instance, Curitiba in Brazil has developed extensive green corridors and urban parks, transforming its identity into that of a "green city." These green spaces not only increase the city's attractiveness to residents but also draw tourists, thereby linking horticultural aesthetics with economic value (Cabral et al. 2017).

1.9 Psychological Value of Urban and Peri-Urban Horticulture

1.9.1 Stress Reduction and Mental Health

Numerous scientific studies show that spending time in green spaces significantly lowers stress levels, reduces anxiety, and improves mood which is often known as 'Horticulture Therapy'. Exposure to plants, whether in a small balcony garden or a large urban park, stimulates relaxation by lowering cortisol levels and improving heart rate variability (Lee and Maheswaran 2011). The biophilia hypothesis, which suggests that humans have an innate affinity toward nature, explains why horticulture-based green spaces are psychologically restorative (Beatley 2016).

For example, hospital gardens in peri-urban areas are increasingly being designed for "horticultural therapy," where patients recover faster when exposed to plants and natural environments. A rooftop vegetable garden in a hospital setting not only provides fresh produce but also creates calming, restorative views for patients and caregivers alike (Van den Berg et al. 2010).

1.9.2 Enhancing Community Bonding and Social Well-Being

Urban horticulture provides shared spaces for people to engage in collective gardening, thereby fostering social interaction and reducing feelings of isolation. Community gardens in dense urban neighborhoods of New York City, such as those in Harlem and the Bronx, have shown a significant impact in improving mental health by giving residents a sense of belonging and shared purpose (Kuo 2015).

The act of gardening itself can also be therapeutic. Tending to plants, nurturing them from seedlings to maturity, and sharing the harvest instills a sense of accomplishment and purpose. In peri-urban areas, families who manage small orchards or vegetable plots often report enhanced family cohesion and reduced psychological stress due to their shared involvement in horticultural activities (Bratman et al. 2015).

1.9.3 Cognitive Benefits and Productivity

Green spaces in urban environments contribute to improved focus, creativity, and cognitive performance. Research has shown that children exposed to green schoolyards perform better in academic tasks due to improved concentration and reduced attention fatigue (Bratman et al. 2015). Similarly, office buildings that integrate indoor horticulture, such as green walls and potted plants, witness increased employee productivity and job satisfaction (Lee and Maheswaran 2011).

In peri-urban zones, engaging in small-scale horticulture, such as greenhouse vegetable farming, also provides mental stimulation and stress relief for aging populations who find meaning and mental activity in caring for plants (Kuo 2015).

1.10 Practical Examples of Aesthetic and Psychological Impact

- *Singapore's Sky Gardens*: Vertical horticulture on skyscrapers creates breathtaking views while improving the mental well-being of urban residents constantly surrounded by greenery (Taylor and Hochuli 2017).
- *Cuba's Organopónicos (Urban Organic Gardens)*: Established in peri-urban Havana, these gardens are not only productive but also visually appealing and act as community hubs where residents engage in cooperative gardening, reducing urban stress and fostering resilience (Mougeot 2005).
- *Delhi's Terrace Gardens*: Amidst one of the most polluted cities in the world, terrace gardens provide citizens with fresh vegetables and a calming, aesthetic refuge from urban chaos (Cabral et al. 2017).
- *Therapeutic Gardens in the UK*: Designed in hospitals and rehabilitation centers, these gardens serve as stress-relieving and healing environments for patients, showing the psychological value of horticulture (Van den Berg et al. 2010).

- *Tokyo's Rooftop Farms*: Besides producing food, these green rooftops reduce mental fatigue by providing refreshing views and opportunities for urban residents to connect with nature (Bratman et al. 2015).

1.11 The Interconnection of Aesthetic and Psychological Values

The aesthetic and psychological benefits of urban and peri-urban horticulture are deeply interconnected. A visually pleasing landscape enhances emotional well-being, while psychological relaxation makes individuals more appreciative of beauty in their surroundings. The colors, textures, and fragrances of plants evoke sensory pleasure, which directly influences mental states (Kuo 2015). For example, brightly colored flowers like marigolds or petunias can elevate mood, while green foliage fosters calmness and balance.

Furthermore, seasonal changes in horticultural spaces, spring blossoms, summer foliage, and autumn colors introduce variety that continuously refreshes both the eye and the mind, countering the monotony of urban environments. Thus, aesthetic beauty and psychological wellness mutually reinforce one another, creating holistic value (Lee and Maheswaran 2011).

1.12 Challenges and Way Forward

Despite the clear aesthetic and psychological value, urban and peri-urban horticulture faces challenges such as limited space, water scarcity, pollution, and lack of awareness. However, innovative strategies like vertical farming, hydroponic rooftop gardens, and peri-urban green belts are overcoming these hurdles (Cabral et al. 2017). Policy support, public–private partnerships, and community involvement are crucial to scale up horticulture's role in urban design.

Educational campaigns can also highlight the mental health benefits of horticulture, encouraging urban residents to create small-scale gardens on balconies, rooftops, or community plots. When urban planners recognize horticulture as not just a utility but also a cultural and psychological necessity, cities can evolve into healthier, more livable spaces.

1.13 Economic and Entrepreneurial Potential

Urban and peri-urban horticulture (UPH) has emerged as a critical component of sustainable city development, addressing challenges of food security, environmental degradation, and livelihood opportunities in growing urban areas. As urban populations expand, the demand for fresh, locally grown horticultural produce is also increasing. Beyond its environmental and nutritional value, UPH holds immense economic and entrepreneurial potential, creating new avenues for income generation, employment, and business opportunities (Mougeot 2005). The integration of modern technologies, policy support, and innovative business models further strengthens its role in building resilient urban food systems.

1.14 Income Generation

UPH provides direct income opportunities by producing high-value crops such as vegetables, fruits, herbs, flowers, and ornamental plants. Unlike field-based agriculture, horticulture in urban and peri-urban areas relies on small land parcels but offers quick returns due to short cropping cycles (Dubbeling et al. 2016). For example, leafy greens like spinach, lettuce, and microgreens can be harvested within three weeks, allowing multiple harvests annually. Such short-duration crops are profitable for small entrepreneurs, especially when marketed through direct-to-consumer models like farmers' markets, home delivery services, and restaurants.

In peri-urban zones, floriculture and nursery businesses also serve as important income generators. Cut flowers, potted plants, and ornamental species command high prices in urban centers where demand for decorative plants in homes, offices, and events is growing. For instance, peri-urban flower growers around Bengaluru and Pune have established lucrative businesses by supplying roses, gerberas, and orchids to domestic and export markets (Singh and Malhotra 2013).

Value addition is another income stream in UPH. Processing surplus horticultural produce into pickles, juices, herbal teas, or organic fertilizers significantly enhances profitability while reducing postharvest losses (Zezza and Tasciotti 2010). Such micro-enterprises, often run by women and self-help groups, highlight the entrepreneurial dimension of urban horticulture.

1.15 Job Creation

The labor-intensive nature of horticulture provides vast opportunities for job creation in cities and their surrounding areas. UPH generates employment across multiple stages of the value chain—production, harvesting, packaging, distribution, and retail (FAO 2012). Seasonal jobs such as seedling preparation in nurseries, daily work in hydroponic farms, or postharvest handling at collection centers ensure livelihood options for marginalized groups, including migrants and women.

In peri-urban areas, horticulture-based agribusinesses like greenhouse cultivation, aquaponics, and mushroom farming create permanent jobs for technicians, farm workers, and marketing professionals. Additionally, the increasing popularity of landscaping and urban greening projects has boosted demand for 'horticulturists', garden designers, and maintenance staff in metropolitan areas (Cabral et al. 2017). For instance, Delhi's rapid expansion of rooftop gardens and terrace farming initiatives has created employment for installers, trainers, and consultants in the urban green sector.

Indirect jobs are also generated in allied industries such as input supply (seeds, fertilizers, irrigation systems), transportation, cold storage, and digital platforms connecting farmers with consumers. Thus, UPH supports both rural–urban migrants and educated youth seeking alternative livelihoods.

1.16 Urban Agri-preneurship

UPH is fostering a new generation of urban agri-preneurs who combine traditional farming knowledge with modern entrepreneurial skills. These individuals and start-ups are capitalizing on consumer demand for fresh, organic, and locally grown food while using innovative business models (Smit et al. 2001).

For example, rooftop farming enterprises in Mumbai and Chennai not only sell fresh produce but also train households to set up kitchen gardens, creating a dual income stream. Similarly, agri-preneurs in cities like Hyderabad and Bengaluru have launched subscription-based vegetable box services, ensuring regular delivery of farm-fresh produce to urban consumers.

Agri-preneurship also thrives in peri-urban horticulture, where farmers are diversifying into high-value sectors like hydroponic strawberry farming, exotic vegetable cultivation, and herbal plant production. Many young entrepreneurs are leveraging e-commerce platforms to directly connect with consumers, restaurants, and hotels, bypassing middlemen and capturing higher profits (Orsini et al. 2017).

Furthermore, social entrepreneurship in urban horticulture is gaining attention. Community-based models such as cooperative gardens, women-led self-help groups, and school-based gardening programs generate social as well as financial returns, contributing to inclusive growth (Poulsen et al. 2015).

1.17 Technologies and Innovations

Modern technologies are transforming UPH into a high-tech, knowledge-driven enterprise. Some key innovations include the following:

- *Hydroponics and Aeroponics*: Soil-less systems allow cultivation of vegetables in vertical farms and rooftops, maximizing space use in crowded urban areas (Kozai 2018).
- *Aquaponics*: Integration of fish farming with vegetable cultivation provides dual income and resource efficiency.
- *Greenhouses and Polyhouses*: Controlled environments improve the productivity of high-value crops like capsicum, cucumber, and strawberries.
- *Smart Irrigation Systems*: Drip irrigation, sensors, and automated watering reduce labor costs and water use.
- *Digital Platforms and Apps*: E-commerce, supply chain tracking, and precision agriculture tools empower farmers to market efficiently and reduce postharvest losses (FAO 2019).

- *Organic Inputs and Biopesticides*: The demand for chemical-free produce has accelerated innovations in composting, vermiculture, and bio-fertilizer production.

For instance, the rise of *vertical farms in Singapore and Tokyo* demonstrates how urban horticulture can leverage technology to ensure year-round production while generating entrepreneurial opportunities in design, maintenance, and sales (Despommier 2010).

1.18 Challenges and Constraints

Despite its potential, UPH faces significant challenges:

- *Space and Land Constraints*: High land costs and competition for urban space limit the scale of operations (Zezza and Tasciotti 2010).
- *Water Scarcity and Pollution*: Access to clean water is a major barrier, particularly in water-stressed cities.
- *High Initial Investment*: Technologies like hydroponics require substantial capital, deterring small-scale entrepreneurs (Kozai 2018).
- *Policy Gaps*: Lack of clear policies on land use, subsidies, and infrastructure support often hinders growth (FAO 2012).
- *Market Linkages*: Farmers face difficulties in connecting with urban consumers due to inadequate cold chains and logistics.
- *Awareness and Training*: Limited knowledge about advanced horticultural practices reduces productivity and profitability (Orsini et al. 2017).

Addressing these constraints through education, technology transfer, and financial incentives is essential to unlock UPH's full potential.

1.19 Policy, Support Systems, and Future Prospects

Government policies and institutional support play a vital role in scaling up UPH. Initiatives like subsidies for rooftop farming in Delhi, training programs under the National Horticulture Mission in India, and incentives for hydroponic farming in Singapore illustrate the role of supportive frameworks (Singh and Malhotra 2013).

Policies promoting organic certification, urban food markets, and waste-to-compost initiatives can further strengthen UPH's entrepreneurial ecosystem. Partnerships with research institutions and universities provide access to innovations, training, and quality planting material (Dubbeling et al. 2016).

Looking ahead, UPH is expected to play a central role in circular urban economies, where food waste is recycled into compost, rainwater harvesting supports irrigation, and local farms supply fresh produce directly to consumers. With increasing awareness of health and sustainability, the demand for urban-grown produce will only expand, creating new entrepreneurial opportunities (FAO 2019).

1.20 Case Studies and Success Stories

- *Cuba's Organopónicos*: After the 1990s food crisis, Havana transformed vacant lots into organic farms. Today, these peri-urban gardens not only provide food security but also generate employment and revenue for thousands of urban farmers (Premat 2009).
- *AeroFarms, USA*: A Newark-based vertical farming company using aeroponics has become a global leader in indoor farming, supplying high-value greens while creating hundreds of jobs (Despommier 2010).
- *Delhi Rooftop Farming Initiatives*: Supported by NGOs and government schemes, hundreds of households in Delhi have converted terraces into productive gardens, reducing vegetable expenses and generating extra income through neighborhood sales (Singh and Malhotra 2013).
- *Kenya's Urban Farming in Nairobi*: Small-scale entrepreneurs cultivate vegetables and poultry in peri-urban areas, supplying affordable food while creating livelihood opportunities for women and youth (Orsini et al. 2017).
- *Netherlands' Greenhouse Cluster*: The peri-urban horticulture belt near Rotterdam has become a hub for high-tech greenhouse farming, boosting exports and serving as a model of sustainable agri-business (FAO 2012).

1.21 Conclusion

Urban and peri-urban horticulture is no longer a marginal activity but a dynamic sector with strong economic and entrepreneurial potential. It contributes to income generation, job creation, and urban agri-preneurship while leveraging innovations like hydroponics, vertical farming, and digital marketing. Although challenges such as limited land, high investment costs, and policy gaps remain, supportive systems and innovative business models can unlock its vast potential.

Successful case studies worldwide demonstrate that with the right mix of technology, policy, and entrepreneurship, UPH can drive not only food security but also inclusive economic growth. In the future, UPH will be a cornerstone of resilient cities, offering fresh food, green jobs, and sustainable livelihoods for urban populations.

References

Al-Chalabi M (2015) Vertical farming: Skyscraper sustainability? Sustain Cities Soc 18:74–77. https://doi.org/10.1016/j.scs.2015.06.003

Armstrong D (2000) A survey of community gardens in upstate New York: implications for health promotion and community development. Health Place 6(4):319–327. https://doi.org/10.1016/S1353-8292(00)00013-7

Banerjee C, Adenaeuer L (2014) Up, up and away! The economics of vertical farming. J Agric Stud 2(1):40–60. https://doi.org/10.5296/jas.v2i1.4526

Beatley T (2016) Handbook of biophilic city planning & design. Island Press, Washington DC

Bharucha Z, Pretty J (2010) The roles and values of wild foods in agricultural systems. Philos Trans R Soc Lond Ser B Biol Sci 365(1554):2913–2926. https://doi.org/10.1098/rstb.2010.0123

Bratman GN, Hamilton JP, Hahn KS, Daily GC, Gross JJ (2015) Nature experience reduces rumination and subgenual prefrontal cortex activation. Proc Natl Acad Sci 112(28):8567–8572. https://doi.org/10.1073/pnas.1510459112

Cabral I, Keim J, Engelmann R, Kraemer R, Siebert J, Bonn A (2017) Ecosystem services of allotment and community gardens. A Leipzig, Germany case study. Urban For Urban Green 23:44–53. https://doi.org/10.1016/j.ufug.2017.02.008

Chatterjee S (2021) Sohoje Ghoroa Bagan Charcha (Easy maintenance of home garden), 1st edn, Kolkata, pp 15–17

Deelstra T, Girardet H (2000) Urban agriculture and sustainable cities. In: Bakker N, Dubbeling M, Gündel S, Sabel-Koschella U, De Zeeuw H (eds) Growing cities, growing food: urban agriculture on the policy agenda. DSE, Feldafing, pp 43–65

Despommier D (2010) The vertical farm: feeding the world in the 21st century. Thomas Dunne Books/St. Martin's Press, New York

Despommier D (2011) The vertical farm: controlled environment agriculture carried out in tall buildings would create greater food safety and security for large urban populations. J Fur Verbraucherschutz Leb 6(2):233–236. https://doi.org/10.1007/s00003-010-0654-3

Draper C, Freedman D (2010) Review and analysis of the benefits, purposes, and motivations associated with community gardening in the United States. J Community Pract 18(4):458–492. https://doi.org/10.1080/10705422.2010.519682

Dubbeling M, De Zeeuw H, Veenhuizen R (2016) The role of urban agriculture in building resilient cities in developing countries. J Agric Sci 8(12):168–177

FAO (2012) Growing greener cities in Africa: first status report on urban and peri-urban horticulture in Africa. Food and Agriculture Organization of the United Nations

FAO (2019) The state of food and agriculture: moving forward on food loss and waste reduction. Food and Agriculture Organization of the United Nations

Galhena DH, Freed R, Maredia KM (2013) Home gardens: a promising approach to enhance household food security and wellbeing. Agric Food Secur 2(1):8. https://doi.org/10.1186/2048-7010-2-8

Grivetti LE, Ogle BM (2000) Value of traditional foods in meeting macro- and micronutrient needs: the wild plant connection. Nutr Res Rev 13(1):31–46. https://doi.org/10.1079/095442200108728990

Guitart D, Pickering C, Byrne J (2012) Past results and future directions in urban community gardens research. Urban For Urban Green 11(4):364–373. https://doi.org/10.1016/j.ufug.2012.06.007

Jaenicke H, Höschle-Zeledon I (eds) (2006) Strategic framework for underutilized plant species research and development: with special reference to Asia and the Pacific, and to Sub-Saharan Africa. International Centre for Underutilised Crops/Global Facilitation Unit for Underutilized Species, Colombo/Rome, p 33

Kalantari F, Tahir OM, Lahijani AM, Kalantari S (2017) A review of vertical farming technology: a guide for implementation of building integrated agriculture in cities. Adv Eng Forum 24:76–91. https://doi.org/10.4028/www.scientific.net/AEF.24.76

Kingsley JY, Townsend M (2006) 'Dig in' to social capital: community gardens as mechanisms for growing urban social connectedness. Urban Policy Res 24(4):525–537. https://doi.org/10.1080/08111140601035200
Kozai T (ed) (2018) Smart plant factory: the next generation indoor vertical farms. Springer, Singapore
Kumari V, Kumari MSC (2020) Organic microgreens. National Institute of Agricultural Extension Management (MANAGE) and College of Community Science, Acharya N.G. Ranga Agricultural University, Guntur, pp 1–15
Kuo M (2015) How might contact with nature promote human health? Promising mechanisms and a possible central pathway. Front Psychol 6:1093. https://doi.org/10.3389/fpsyg.2015.01093
Lee ACK, Maheswaran R (2011) The health benefits of urban green spaces: a review of the evidence. J Public Health 33(2):212–222. https://doi.org/10.1093/pubmed/fdq068
Marsh R (1998) Building on traditional gardening to improve household food security. FAO, Rome
Mitchell R, Hanstad T (2004) Small home garden plots and sustainable livelihoods for the poor. Livelihood Support Programme Working Paper 11. FAO, Rome
Mougeot LJA (ed) (2005) Agropolis: the social, political and environmental dimensions of urban agriculture. International Development Research Centre and Earthscan, London
Nayak PS, Chatterjee S, Acharya S (2023) Microgreens – a potential game changer for achieving nutritional security. In: Biradar N, Shah RA, Ahmad A (eds) Recent advances in agricultural sciences and technology. ICAR-IGFRI, SRRS/NADCL, Dharwad/Baramulla, pp 231–241
Okvat HA, Zautra AJ (2011) Community gardening: a parsimonious path to individual, community, and environmental resilience. Am J Community Psychol 47(3–4):374–387. https://doi.org/10.1007/s10464-010-9404-z
Orsini F, Kahane R, Nono-Womdim R, Gianquinto G (2013) Urban agriculture in the developing world: a review. Agron Sustain Dev 33(4):695–720. https://doi.org/10.1007/s13593-013-0143-z
Orsini F, Gasperi D, Marchetti L, Piovene C, Draghetti S, Ramazzotti S et al (2014) Exploring the production capacity of rooftop gardens (RTGs) in urban agriculture: the potential impact on food and nutrition security, biodiversity and other ecosystem services in the city of Bologna. Food Secur 6(6):781–792. https://doi.org/10.1007/s12571-014-0389-6
Orsini F, Dubbeling M, de Zeeuw H, Gianquinto G (2017) Rooftop urban agriculture. Springer, Cham. https://doi.org/10.1007/978-3-319-57720-3_1
Padulosi S, Thompson J, Rudebjer P (2013) Fighting poverty, hunger and malnutrition with neglected and underutilized species (NUS): needs, challenges and the way forward. Bioversity International, Rome
Poulsen MN, McNab PR, Clayton ML, Neff RA (2015) A systematic review of urban agriculture and food security impacts in low-income countries. Food Policy 55:131–146. https://doi.org/10.1016/j.foodpol.2015.07.002
Premat A (2009) State power, private plots and the greening of Havana's urban agriculture movement. City Soc 44(1):132–157
Singh HP, Malhotra SK (2013) Urban and peri-urban horticulture for greening the cities, utilising the waste, meeting the needs and servicing the environment. In: Sumangla HP, Malhotra SK, Chowdappa P (eds) Urban and peri-urban horticulture – a perspective. Confederation of Horticulture Associations of India, New Delhi, pp 1–10
Smit J, Nasr J, Ratta A (2001) Urban agriculture: food, jobs and sustainable cities. The Urban Agriculture Network, Inc. (with the financial support and permission of the United Nations Development Programme)
Smith IF, Eyzaguirre PB (2007) African leafy vegetables: their role in the World Health Organization's global fruit and vegetable initiative. Afr J Food Agric Nutr Dev 7(3):1–17. https://doi.org/10.18697/ajfand.14.IPGRI1-1
Specht K, Siebert R, Hartmann I, Freisinger UB, Sawicka M, Werner A, Thomaier S, Henckel D, Walk H, Dierich A (2014) Urban agriculture of the future: an overview of sustainability aspects of food production in and on buildings. Agric Hum Values 31(1):33–51. https://doi.org/10.1007/s10460-013-9448-4

Swain A, Chatterjee S, Viswanath M, Roy A, Biswas A (2021) Hydroponics in vegetable crops: a review. Pharma Innov J 10(6):629–634

Taylor L, Hochuli DF (2017) Defining greenspace: multiple uses across multiple disciplines. Landsc Urban Plan 158:25–38

Thomaier S, Specht K, Henckel D, Dierich A, Siebert R, Freisinger UB, Sawicka M (2015) Farming in and on urban buildings: present practice and specific novelties of zero-acreage farming (ZFarming). Renew Agric Food Syst 30(1):43–54. https://doi.org/10.1017/S1742170514000143

Van den Berg AE, Maas J, Verheij RA, Groenewegen PP (2010) Green space as a buffer between stressful life events and health. Soc Sci Med 70(8):1203–1210. https://doi.org/10.1016/j.socscimed.2010.01.002

Whittinghill LJ, Rowe DB (2012) The role of green roof technology in urban agriculture. Renew Agric Food Syst 27(4):314–322. https://doi.org/10.1017/S174217051100038X

Zezza A, Tasciotti L (2010) Urban agriculture, poverty, and food security: empirical evidence from a sample of developing countries. Food Policy 35(4):265–273. https://doi.org/10.1016/j.foodpol.2010.04.007

Rapid Urbanisation and the Challenges of Socioeconomic Upliftment with Horticulture

2

Sayani Das and Joydip Mandal

Abstract

The twenty-first century is marked by a transformative and permanent shift toward urbanisation, with projections suggesting that more than 65% of the world's population will reside in urban centres by the year 2050. This shift intensifies systemic challenges related to food security, waste management, and malnutrition among underserved communities. Urban and peri-urban horticulture (UPH) offers a strategic response to these pressures by positioning nutrient-rich food production closer to consumers, thereby stabilising availability and reducing reliance on vulnerable, long-distance supply chains. This chapter explores the multifunctional utility of UPH, which transcends basic food production to include socioeconomic upliftment and environmental management. UPH provides vital livelihood opportunities for women, youth, and migrant workers through diverse activities such as nursery management, mushroom cultivation, and microgreen production. Furthermore, it delivers significant aesthetic and psychological benefits, including stress reduction, improved mental well-being, and enhanced community cohesion. The discussion highlights innovative technological models, including vertical farming, hydroponics, rooftop gardening, and protected cultivation, designed to optimise resource use in constrained environments. These systems often utilise sustainable inputs like treated wastewater and organic compost to minimise the urban ecological footprint. Despite these advancements, UPH faces critical constraints such as land scarcity, insecure tenure, and contamination risks from heavy metals or pathogens. In India, initiatives like the Vegetable Initiative for Urban Clusters (VIUC) and the National Horticulture Board's cluster programmes aim to formalise these systems. Integration of smart technologies and precision sensors further enhances the efficiency and profitability of these urban enterprises. Ultimately, this chapter argues

S. Das · J. Mandal (✉)
Department of Horticulture and Post-Harvest Technology, Palli Siksha Bhavana (Institute of Agriculture), Sriniketan, West Bengal, India
e-mail: joydip.mondal@visva-bharati.ac.in

A. Das, A. B. Sharangi (eds.), *Remunerative Horticulture*,
https://doi.org/10.1007/978-981-92-1007-7_2

that embedding horticulture into urban planning and governance is essential for cultivating resilient, climate-adaptive, and inclusive cities. By recognising food production as a formal urban function, cities can leverage UPH to ensure long-term sustainability and human well-being.

Keywords

Urban and peri-urban horticulture (UPH) · Food security · Vertical farming · Socioeconomic upliftment · Sustainable urbanisation · Climate resilience

2.1 Introduction

The rapid and irreversible process of urbanisation has emerged as one of the most transformative demographic trends of the twenty-first century. With the global urban population projected to exceed 65% by the year 2050 (Sumangla et al. 2013), urban centres are becoming focal points of human settlement, economic activity, and cultural transformation. This trend is especially pronounced in developing countries such as India, where urbanisation is driven not only by natural population growth but also by large-scale migration from rural areas in search of better employment opportunities, education, healthcare, and living standards (Suman and Bhatnagar 2019). Rapid urbanisation, particularly in Asia, is reshaping global demographic patterns. It intensifies challenges related to food security, employment, and waste management, especially among underserved urban communities (Bonny et al. 2014). Malnutrition and micronutrient deficiencies continue to affect many urban and peri-urban populations, often due to insufficient consumption of fruits and vegetables (Jindal and Dhaliwal 2017). With urbanisation accelerating and the global population rising, ensuring a stable supply of affordable, nutritious food becomes increasingly urgent. Many expanding cities, especially in developing regions, face mounting pressures from population growth, limited employment opportunities (particularly for migrants from rural areas), and the burden of managing refuse from households, commerce, and industry. These pressures highlight the need for accessible, affordable, and nutrient-rich food for urban dwellers.

One promising solution to strengthen food security is developing home gardens, small-scale urban and peri-urban horticultural production, and small and medium-scale farm enterprises (SMEs) operating within or near urban areas. Urban horticulture, including home gardens, rooftop or balcony gardens, community gardens, and small peri-urban farms, has been proposed as a strategy to improve food availability and nutritional intake, especially among low- and middle-income urban households (Juniawati and Hayuningtyas 2017). Evidence from multiple countries shows that home gardens can significantly contribute to the diversity and quality of diets. For instance, in a recent study of urban residents in Dhaka (Bangladesh), it was found that households trained in home gardening increased their production of diverse fruits and vegetables, improved dietary diversity among women, and raised vegetable consumption (Schreinemachers et al. 2025).

Beyond direct nutritional benefits, urban and peri-urban horticulture offers additional socioeconomic and environmental advantages. Small-scale agriculture in urban zones tends to be labour-intensive, flexible, and capable of producing high-value, perishable crops, such as fruits and vegetables, which require less land and water per unit output than many staple crops (Khan et al. 2020). It can also help create employment opportunities, reduce dependence on long supply chains, and make use of underutilised urban spaces (e.g. rooftops, vacant lots, community land) (Bonny et al. 2014). Some horticultural models are even designed as "nutrition-gardens", scientifically selected to yield crops rich in vitamins and minerals and capable of producing several hundred kilograms of vegetables annually, which are enough to meet a substantial portion of a family's micronutrient requirements (Jindal and Dhaliwal 2017).

Urban and peri-urban horticulture (UPH) represents not merely a survival mechanism but a strategic component of modern urban planning. As noted by Suman and Bhatnagar (2019), horticultural systems, characterised by high turnover, efficient input use, and adaptability to small land parcels, are particularly suited to urban settings. At the same time, persistent micronutrient deficiencies in urban households, fuelled by inadequate intake of fruits and vegetables, underscore the urgency of strengthening horticulture-based interventions that can diversify diets and improve public health (Hamilton et al. 2014).

Moreover, the ecological benefits of UPH extend far beyond food provisioning. Well-designed horticultural landscapes contribute to microclimate regulation, improved air quality, urban biodiversity, and sustainable water management (Bohn and Viljoen 2011; Lovell 2010). As resource constraints intensify due to climate variability and environmental degradation, such multifunctional landscapes become essential assets for creating liveable, climate-adaptive cities. Recent innovations, including soilless cultivation, hydroponics, vertical gardening, aquaponics, and building-integrated agriculture, have further redefined the possibilities of horticulture in spatially constrained environments (Caplow 2009).

Yet the potential of UPH remains undermined by insecure land tenure, inadequate policy support, and unregulated production practices, including unsafe wastewater use and indiscriminate pesticide application (Mbaye and Moustier 2000). Recognising these challenges is critical for integrating horticulture into urban development agendas in ways that promote sustainability, public health, and equitable access to resources. In sum, urban and peri-urban horticulture occupies a pivotal place at the intersection of food security, public health, environmental stewardship, and urban livelihoods. Its growing relevance reflects not only the pressures of urban expansion but also the increasing awareness that resilient cities must produce a significant share of their own nutritious food. The following sections delve deeper into the systems, innovations, opportunities, and constraints that shape the modern landscape of urban and peri-urban horticulture.

2.2 Different Dimensions of Horticulture (Urban and Peri-Urban) Contributing to Urban Sustainability

Urban and peri-urban horticulture (UPH) has emerged as a practical response to the growing challenges of food insecurity, livelihood vulnerability, and environmental stress in rapidly expanding cities. By bringing food production closer to urban consumers, UPH helps ensure access to fresh and nutritious produce while reducing dependence on distant supply chains. At the same time, it creates livelihood opportunities for diverse urban populations and contributes to improving urban environmental quality and social well-being. Owing to its multifunctional role, UPH is increasingly recognised as an important component of sustainable and resilient urban development. The multifaceted dimensions of UPH that contribute to urban sustainability are referred to in Table 2.1.

Urban and peri-urban horticulture (UPH) has gained prominence as an essential element of sustainable urban development because it bridges the widening gap between rising food demand and declining agricultural land near expanding cities. As urban populations continue to swell, ensuring a stable supply of fresh and nutritious produce becomes increasingly difficult through conventional long-distance supply chains. UPH responds to this challenge by positioning food production closer to consumers, enabling quicker harvesting-to-consumption cycles that preserve nutritional quality and reduce losses. This proximity not only enhances dietary diversity among urban households but also stabilises food availability during periods of market fluctuation or transportation disruption. For low-income communities in particular, locally grown fruits and vegetables offer an accessible source of essential micronutrients, making UPH an effective instrument in addressing urban malnutrition and hidden hunger (Korth et al. 2014).

Table 2.1 A detailed view of different dimensions like food and nutrition, livelihood security, economic integration, environmental functions, social and cultural role, long-term resilience, with separate columns mentioning the key aspects and their contribution to urban sustainability

Dimension	Key aspects	Contribution to urban sustainability
Food and nutrition	Local production of fruits and vegetables; short supply chains	Enhances dietary diversity and reduces nutritional deficiencies
Livelihood security	Small-scale farming, nursery raising, microgreens, mushroom cultivation	Provides income opportunities for women, youth, and migrant workers
Economic integration	Local markets, fresh produce vending, small food enterprises	Strengthens urban and peri-urban economies
Environmental functions	Green spaces, composting, efficient water use, biodiversity support	Mitigates heat island effects and improves urban ecology
Social and cultural role	Community gardens, school programmes, therapeutic green spaces	Promotes social interaction, learning, and mental well-being
Long-term resilience	Integration with urban planning and sustainable practices	Supports climate-resilient and inclusive urban development

Alongside its contribution to nutritional security, UPH serves as a vital livelihood resource for a wide spectrum of urban and peri-urban residents. Horticulture's adaptability to limited spaces, whether terraces, balconies, vacant plots, or small peri-urban landholdings, allows individuals and families to participate in income-generating activities without the need for extensive capital. This inclusiveness makes UPH especially relevant for women, youth, and migrant workers, who often face barriers in accessing formal employment. Activities such as nursery management, floriculture, rooftop vegetable cultivation, microgreen production, mushroom farming, and small-scale processing provide diverse opportunities for entrepreneurship. Peri-urban farmers, whose land gradually diminishes under pressure from real estate expansion, rely on high-value horticultural crops to maintain viable livelihoods. Local markets benefit as well, with fresh produce vending, supply-chain linkage, and small food enterprises all contributing to the economic vitality of urban regions (Rammohan et al. 2019).

Beyond food and economic security, UPH plays a crucial role in maintaining and restoring ecological balance within cities. Urban areas increasingly suffer from heat island effects, air pollution, soil degradation, and diminishing green space issues, amplified by dense construction and vehicular emissions. Horticultural landscapes, whether in the form of rooftop gardens, community plots, edible parks, or peri-urban orchards, help mitigate these conditions. Vegetation contributes to cooling through shade and evapotranspiration, reducing local temperatures and easing the burden on urban energy systems. Urban soils enriched through composting and mulching improve carbon storage and support healthier plant growth. Biodiversity also benefits, as functioning horticultural ecosystems host pollinators, beneficial insects, and small fauna that struggle to survive in heavily built environments. Many UPH systems make use of environmentally responsible practices, including drip irrigation, reused grey water, and organic waste recycling, demonstrating how food production can coexist with sustainable resource management (FAO 2022; Schreinemachers et al. 2025).

In addition to these material benefits, UPH strengthens the social and cultural fabric of urban communities. Gardens and horticultural spaces provide residents with opportunities for recreation, learning, and collective engagement. They create inclusive environments where people from diverse backgrounds collaborate, share knowledge, and cultivate a sense of belonging. Community gardens, school horticulture programmes, therapeutic green spaces, and neighbourhood orchards help counteract the alienation often associated with dense urban living. These spaces support mental well-being by offering a connection to nature, fostering relaxation, and encouraging active participation in outdoor activities. For children and youth, exposure to horticulture builds awareness of nutrition, environmental stewardship, and sustainable living practices.

Taken together, the utilities of urban and peri-urban horticulture illustrate its integral role in shaping healthier, more resilient, and inclusive cities. As pressures from rapid urbanisation and climate variability intensify, UPH offers a forward-looking approach that integrates food production with environmental care and socioeconomic development. Its capacity to generate livelihoods, enhance nutrition, improve urban ecology, and promote community cohesion makes it more than a supplemental activity; it becomes a strategic framework for sustainable urban futures. By embedding horticulture into planning processes, providing supportive policies, and fostering technological and community-driven innovations, cities can leverage UPH to ensure long-term resilience and well-being for their populations.

2.3 Aesthetics and Human Well-Being

The aesthetic value of UPH is evident in its capacity to transform grey urban environments into visually and psychologically enriching green spaces. Landscaping in public and private institutions, vertical gardens, and terrace greenspaces improve mental health, reduce stress, and support community cohesion (Wang et al. 2019). Research has demonstrated correlations between access to green spaces and reductions in attention fatigue, depression, and physiological stress. Recreational parks and urban mosaics enhance biodiversity while fulfilling social and ecological functions.

Urban and peri-urban horticulture (UPH) significantly enhances the aesthetic quality of urban environments by transforming grey, concrete-dominated spaces into visually appealing and psychologically beneficial green areas. The incorporation of landscaping in public institutions, private properties, vertical gardens, and terrace green spaces is a proven strategy to improve the urban landscape. These green elements not only add visual appeal but also play a critical role in promoting mental health by reducing stress levels and improving overall psychological well-being (Kuo 2002). Extensive research confirms that regular access to green spaces correlates with measurable decreases in attention fatigue, depression, and physiological stress indicators. Exposure to natural environments replenishes cognitive function impaired by urban stressors, demonstrating a direct link between green space availability and improved mental performance.

Beyond individual health benefits, UPH contributes to stronger community cohesion (Ghallab 2024). Green recreational areas and communal gardens provide venues for social interaction and collective activities, which are essential for fostering a sense of community and shared responsibility among urban residents. From an ecological perspective, well-planned urban green mosaics, including parks, green corridors, and street vegetation, enhance biodiversity within cities. These green networks support diverse flora and fauna, contribute to ecological balance, and deliver ecosystem services such as air purification, microclimate regulation, and habitat provision. It delivers tangible benefits for human well-being, social integration, and ecological health, establishing green infrastructure as a fundamental component of sustainable urban development.

2.4 Models and Technologies

Urban and peri-urban horticulture (UPH) encompasses a diverse set of production models that often overlap and interconnect rather than exist as isolated systems. Many UPH practices, such as rooftop gardening, vertical farming, protected cultivation, hydroponics, indoor farming, and container-based production, share common principles of space optimisation, resource recycling, and climate control, making them adaptable to a wide range of urban constraints (Bonny et al. 2014). These models frequently operate in combinations: for instance, hydroponics is commonly integrated into both rooftop and indoor farming setups, while vertical farming can be housed within protected structures or climate-controlled indoor units. Emerging high-value crops, such as saffron, have also found a niche within UPH through controlled-environment cultivation in polyhouses and indoor vertical systems, demonstrating how traditional crop systems can be reimagined for urban settings. This is illustrated in the form of a chart below (Fig. 2.1). Together, these overlapping and synergistic models illustrate the flexibility and innovation inherent in UPH, enabling cities to enhance food production, diversify livelihoods, and strengthen ecological sustainability.

These models enhance productivity while conserving space and resources. Treated wastewater and nutrient-rich compost from municipal waste offer sustainable input alternatives, reducing the urban ecological footprint.

Urban and peri-urban horticulture (UPH) leverages a range of innovative and traditional models that have been carefully adapted to the unique constraints of urban environments. One of the most common and effective models is rooftop and terrace gardens, which utilise containers, grow bags, and lightweight substrates to convert otherwise unused roof spaces into productive cultivation areas. These

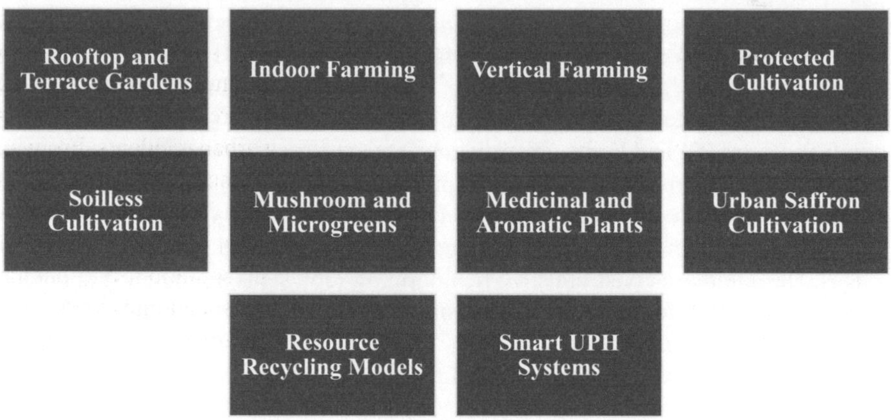

Fig. 2.1 Flow chart depicting various urban agriculture methods. The chart includes rooftop and terrace gardens, indoor farming, vertical farming, protected cultivation, soilless cultivation, mushroom and microgreens, medicinal and aromatic plants, urban saffron cultivation, resource recycling models, and smart UPH systems. Each method is presented in a separate box

lightweight substrates, often made from coconut coir, perlite, vermiculite, and organic compost mixtures, provide the necessary aeration, moisture retention, and nutrient supply while keeping the overall weight manageable for building structures. Rooftop gardens not only optimise space but also contribute to mitigating the urban heat island effect by cooling buildings and improving air quality. The accessibility of such gardens makes them highly attractive for both private residences and public institutions, offering opportunities for urban dwellers to engage directly with food production and greenery (Barreca 2016).

Indoor farming has emerged as a transformative component of urban and peri-urban horticulture, offering possibilities that extend far beyond the constraints of conventional open-field agriculture. Within enclosed spaces such as converted rooms, basements, warehouses, or specially designed growth chambers, crops are cultivated under carefully regulated conditions that govern light, temperature, humidity, and nutrient supply. This high level of environmental control enables uninterrupted, year-round production and ensures uniformity in crop quality, while simultaneously reducing exposure to pests, diseases, and fluctuating climatic stresses. Indoor farming systems frequently integrate soilless methods—most notably hydroponics, aeroponics, and vertically arranged growing racks—which collectively enhance space-use efficiency and allow cultivation in areas where arable land is entirely absent. Although energy requirements for lighting and climate management remain important considerations, ongoing advancements in LED technologies, automation, and renewable energy integration are steadily enhancing the viability and sustainability of these systems. As cities continue to densify, indoor farming offers a promising pathway for producing leafy vegetables, herbs, microgreens, and select high-value crops within close proximity to consumers, thereby strengthening local food security and enriching the functional landscape of UPH (Khan et al. 2020).

Another increasingly important technology is vertical farming, which maximises the use of vertical space by growing plants in stacked layers or on vertical surfaces such as green walls and trellises (Nwosisi and Nandwani 2018). Vertical farming incorporates advanced technologies like LED lighting, automated irrigation, and climate control systems, enabling year-round cultivation of a variety of crops, primarily leafy greens and herbs, regardless of external weather conditions. This system dramatically reduces the land footprint required and can be established in locations without soil access. However, high initial capital costs and significant energy consumption for artificial lighting and environmental controls pose challenges. Despite these constraints, vertical farming represents a promising approach to increasing urban food production density while minimising resource inputs such as water and fertilisers.

Protected cultivation methods, including polyhouses and net houses, have been adapted to urban horticulture to provide controlled environments for growing high-value crops (McCartney and Lefsrud 2018). Polyhouses are structures covered with polyethylene films that regulate temperature, humidity, and light penetration, whereas net houses use insect-proof netting to protect plants from pests and harsh climatic factors. These systems allow for more consistent and higher-quality production of vegetables, flowers, and seedlings by extending growing seasons and reducing dependence on chemical pesticides. They are particularly valuable in urban areas where outdoor cultivation may be limited by pollution and climatic extremes. While their establishment requires investment and technical expertise, protected cultivation offers a sustainable way to intensify production and reduce losses.

Soilless cultivation systems such as hydroponics and aquaponics have revolutionised urban farming by enabling highly efficient, soil-free plant growth (Dos Santos 2016). Hydroponics involves growing plants directly in nutrient-rich water solutions, which can reduce water use by up to 90% compared to traditional farming. Common hydroponic setups include the nutrient film technique (NFT), deep water culture (DWC), and aeroponics, each designed to optimise oxygen and nutrient delivery to the roots. Aquaponics integrates hydroponics with fish farming, creating a closed-loop system where fish waste provides organic nutrients for plants, and plants clean the water for the fish. This symbiotic relationship minimises waste and chemical use while producing both vegetables and fish for consumption. These systems are ideal for rooftop and indoor farming but require constant monitoring and skilled management to maintain nutrient balance and system health.

Certain crops are particularly suited to UPH models due to their space-efficiency and economic value, notably mushroom cultivation and microgreens (Glück-Thaler and Begg 2012; Nethra et al. 2024). Mushrooms can be grown indoors on substrates like straw, sawdust, or coffee grounds, using minimal space and yielding high-value produce. Microgreens, harvested at the seedling stage of various vegetables and herbs, are prized for their intense flavours and concentrated nutrients and can be produced rapidly on vertical racks or small trays. Both mushrooms and microgreens are excellent options for urban growers seeking quick returns with minimal spatial and input demands, making them popular in home gardens and commercial urban farms alike.

A specialised but growing area in UPH is the cultivation of medicinal and aromatic plants (MAPs), which serve dual purposes of primary healthcare and the enhancement of urban therapeutic landscapes (Maknea et al. 2022). Species such as tulsi, aloe vera, mint, lavender, and lemongrass are widely cultivated for their medicinal properties, as well as their ability to attract pollinators and enhance biodiversity. MAPs contribute to local health systems by providing accessible herbal remedies and have commercial potential through essential oils and natural product industries. However, cultivating MAPs requires careful attention to harvest timing, processing, and quality control to preserve their efficacy and value.

Saffron cultivation within urban and peri-urban horticulture represents an emerging frontier that demonstrates how high-value, traditionally field-based crops can be successfully adapted to controlled urban environments. The unique biological requirements of saffron—particularly its need for a defined dry period, moderate temperatures, and well-aerated substrates—make it well suited for cultivation in polyhouses, rooftop units, and indoor growth chambers where environmental parameters can be precisely managed. Urban growers typically employ containerised systems or vertical trays filled with lightweight, sterile media to ensure optimal corm health and efficient use of limited space. Controlled-environment saffron production not only enhances flower induction and stigma yield but also reduces vulnerability to soil-borne diseases and erratic climatic events that increasingly threaten open-field cultivation. Although initial investments in protected structures and quality planting material can be substantial, the exceptionally high market value of saffron, combined with its minimal water requirement and compact growth habit, presents a compelling opportunity for small-scale urban farmers and household producers. As research advances in controlled-environment horticulture, saffron stands out as a promising specialty crop capable of diversifying urban production systems and contributing to niche markets, dietary enrichment, and livelihood enhancement within UPH frameworks (Kour et al. 2022).

Sustainability is central to UPH, and models increasingly incorporate the use of treated wastewater and nutrient-rich compost derived from municipal solid waste to provide sustainable inputs. Treated wastewater irrigation reduces dependence on freshwater supplies and recycles valuable nutrients, although it requires rigorous treatment and monitoring to ensure safety for food production. Similarly, composting of urban organic waste returns essential nutrients to cultivation systems and improves substrate fertility while diverting waste from landfills. Water conservation techniques such as drip irrigation, mulching, fertigation, and rainwater harvesting are widely applied to optimise water use efficiency, ensuring that limited urban water resources are utilised judiciously and sustainably.

The role of digital and smart technologies in enhancing UPH productivity and resource efficiency cannot be overstated. Sensors measuring soil moisture, nutrient content, temperature, and light intensity enable precision management, allowing growers to fine-tune irrigation and fertilisation schedules. Automation systems reduce labour demands and improve consistency, while data analytics and artificial intelligence support predictive crop management, pest detection, and market forecasting. These technological advancements increase the feasibility of large-scale urban farming, making it a more reliable and profitable enterprise. Beyond technical aspects, the implementation of these UPH models produces significant socioeconomic and environmental impacts. By increasing the availability of fresh, nutritious food close to consumers, UPH strengthens urban food security and decreases reliance on complex, long-distance supply chains. Employment opportunities created by UPH span cultivation, processing, technology services, and marketing, often benefiting marginalised urban populations and fostering social inclusion. Environmentally, urban horticulture improves air quality, enhances biodiversity, reduces urban heat islands, and contributes to stormwater management, ultimately helping cities adapt to and mitigate climate change impacts.

2.5 Government Initiatives and Policy Frameworks

In India, UPH has gained traction through national schemes such as the Vegetable Initiative for Urban Clusters (VIUC) under Rashtriya Krishi Vikas Yojana (RKVY), and targeted programmes by the National Horticulture Board (NHB). National dialogues by institutions like Indian Institute of Horticultural Research (IIHR) and National Academy of Agricultural Sciences (NAAS) have stressed integrating UPH into urban planning and governance. The Food and Agriculture Organisation (FAO), International Society for Horticultural Science (ISHS), and CGIAR have also emphasised UPH globally, aligning it with climate action, health, and urban resilience agendas. Supportive measures include policy advocacy for land access, establishment of public sector seed outlets, training in terrace and container gardening, and promotion of safe wastewater use. Urban development bodies, municipal corporations, and state agricultural departments are crucial stakeholders in scaling UPH initiatives.

The Vegetable Initiative for Urban Clusters (VIUC) under the Rashtriya Krishi Vikas Yojana, launched in 2011–2012, is a landmark national programme aimed at promoting the cultivation of vegetables in and around urban centres. It addresses the demand-supply imbalance in fresh produce and facilitates the development of decentralised production systems by supporting rooftop gardens, community gardens, and peri-urban vegetable belts. VIUC encourages aggregation of small growers, promotes group-based extension support, and subsidises inputs such as hybrid seeds, organic manure, and low-cost growing structures. The initiative recognises the spatial constraints and fragmented landholding patterns in urban settings and tailors its interventions accordingly.

The National Horticulture Board (NHB) plays an indirect yet significant role in strengthening urban and peri-urban horticulture (UPH) through its revised Cluster Development Programme (CDP), particularly the component dedicated to Peri-urban Vegetable Clusters. These guidelines emphasise the integration of high-value, technology-driven production systems such as precision horticulture, hydroponics, and vertical farming into zones located near expanding urban markets. By supporting innovative cultivation structures, modern post-harvest handling units, cold-chain infrastructure, and city-linked distribution centres, NHB enhances the efficiency of short supply chains that directly serve urban consumers. This cluster-based approach enables peri-urban growers to adopt protected cultivation, climate-resilient practices, and improved logistics that collectively boost the availability of fresh, high-quality produce in urban areas (NHB 2025).

Apex research institutions such as the Indian Institute of Horticultural Research (IIHR) and the National Academy of Agricultural Sciences (NAAS) have advocated for the formal integration of UPH into city master plans and metropolitan governance systems (NAAS 2013; Singh 2019; Shukla 2024). Their policy papers and expert consultations have emphasised zoning laws that reserve space for food production, guidelines for rooftop cultivation safety, and incentives for real estate developers who incorporate green food-producing spaces in housing and commercial complexes. Recommendations also include decentralising the supply of inputs and subsidies through urban municipal channels and revising building codes to include load-bearing norms for rooftop horticulture.

At the international level, organisations like the Food and Agriculture Organisation (FAO), the International Society for Horticultural Science (ISHS), and the CGIAR network have aligned UPH with global frameworks such as the

Sustainable Development Goals (SDGs), the New Urban Agenda, and climate resilience targets under the Paris Agreement. These institutions support capacity-building, research, and city-level policy reform to integrate UPH into national adaptation plans, urban biodiversity strategies, and food security charters. FAO technical cooperation projects in South Asia have involved pilot UPH models in Indian cities, often integrating them with nutrition-sensitive agriculture and urban poverty alleviation programmes (FAO 2008; King'ori 2004).

One of the critical supportive measures at the policy level is advocacy for secure land access. Governments are drafting provisions for utilising unused urban lands such as railway margins, institutional campuses, and wastelands for horticultural production. These policies aim to grant user rights to organised producer groups, NGOs, and Self Help Groups (SHGs) for time-bound cultivation without affecting land ownership. Municipal bylaws are being amended to recognise food gardens as legitimate land use. Urban development authorities are issuing no-objection certificates for using government school rooftops, bus depots, and flyover spaces for horticulture.

To ensure input accessibility, public sector seed outlets are being established in urban centres through Krishi Vigyan Kendras and agriculture department kiosks. These outlets provide certified seeds suitable for container gardening, along with biofertilisers, neem-based biopesticides, and low-volume irrigation tools like micro-sprinklers and drip sets. Urban seed kits curated for terrace farming are being distributed in cities like Bengaluru, Hyderabad, and Pune under smart city initiatives.

Skill development is being mainstreamed into UPH promotion through training in terrace and container gardening conducted by urban horticulture cells under state agriculture departments. Training modules cover rooftop load management, selection of lightweight media, vertical garden installation, composting kitchen waste, water-efficient irrigation methods, and seasonal crop planning. Special modules for women-led urban farms and youth entrepreneurship in urban horticulture are being scaled under centrally sponsored schemes like the Atmanirbhar Bharat Abhiyan and PM Formalisation of Micro Food Processing Enterprises (PM-FME).

Water use regulations are being revised to encourage the safe reuse of treated wastewater for irrigation in urban and peri-urban agriculture. Municipal sewage treatment plants are increasingly supplying secondary treated water to urban horticulture clusters under guidelines that ensure safe use, in line with standards recommended by the Central Pollution Control Board (CPCB) and the World Health Organisation (WHO). Pilot projects in several Indian cities demonstrate that treated wastewater can be effectively utilised for growing vegetables and leafy greens, provided proper irrigation methods and monitoring are followed. Guidelines are also being developed to manage nutrient supplementation and microbial safety in these water sources (NITI Aayog 2023).

Urban development bodies, municipal corporations, and state agricultural departments are being strategically positioned as implementation anchors. Urban local bodies are embedding UPH in city resilience plans, green infrastructure frameworks, and climate action roadmaps. They are coordinating with smart city SPVs (Special Purpose Vehicles) to allocate budget lines for rooftop gardens, green belts, and edible landscaping in public parks. Municipal waste management units are linking UPH projects with decentralised composting hubs and organic input production centres. State horticulture missions are aligning their urban interventions with the Atal Mission for Rejuvenation and Urban Transformation (AMRUT), Swachh Bharat Mission, and Jal Jeevan Mission for integrated outcomes.

The convergence of urban and peri-urban horticulture (UPH) with food safety, public health, employment generation, and climate adaptation has led to its rising prioritisation across multiple government departments. Policy papers now define UPH not only as an urban farming system but as a key lever for achieving urban ecological balance, social equity, and economic inclusion. The Ministries of Housing and Urban Affairs and NITI Aayog are working together to create rules to standardise and promote UPH in major and growing cities. (NITI Aayog 2023; Ministry of Housing and Urban Affairs 2019; Ministry of Housing and Urban Affairs 2023).

2.6 Challenges and Constraints

Despite its promise, UPH faces constraints including health risks due to heavy metals, microbial contamination, and air pollutants, which necessitate the development and enforcement of safety standards and good agricultural practices (GAPs). These challenges and constraints are illustrated in Fig. 2.2.

Urban and peri-urban horticulture (UPH), while offering multifaceted solutions for sustainable urban living, is hindered by a spectrum of critical challenges that obstruct its widespread adoption and scalability. The most formidable among these is the scarcity and high cost of urban land, which significantly limits the physical expansion of horticultural activities (Lin et al. 2015). In densely populated cities, real estate values are prohibitively high, and land is prioritised for infrastructure, housing, and commercial purposes. Consequently, the availability of open, cultivable spaces is minimal, and urban agriculture often competes with real estate developers and institutional users for even marginal plots. This economic constraint severely affects the feasibility of developing full-scale urban farming enterprises.

Compounding this issue is insecure land tenure and competition from non-agricultural land uses. Most urban growers, particularly those cultivating on vacant government land, railway margins, or peri-urban fringes, operate without formal legal rights. The lack of documented tenure discourages long-term investment in soil improvement, infrastructure, and technology adoption, as growers face the constant threat of eviction. The increasing commercialisation and expansion of urban areas result in agricultural land being rapidly converted into residential or industrial zones, with little policy protection for food-producing spaces. The absence of land-use zoning that formally designates space for UPH further exacerbates this problem (Sumangla et al. 2013).

Water scarcity and competing demands pose another significant challenge for UPH. Urban water systems are primarily designed to meet domestic, industrial, and commercial needs, often neglecting the requirements of agriculture. With

Fig. 2.2 Flow chart illustrating "Challenges and Constraints" at the centre, surrounded by eight factors: "Limited and expensive urban land", "Insecure land tenure", "Water-related constraints", "Environmental contamination risks", "Health and food safety concerns", "Input and management issues", "Institutional fragmentation", and "Awareness and perception gaps". Arrows point from the central theme to each factor, indicating their relationship

groundwater levels depleting and municipal water supply under severe stress, the availability of water for irrigation in urban settings is highly limited and unreliable. This scarcity forces urban farmers to rely on alternative sources, such as untreated wastewater or rainwater harvesting, both of which require infrastructural investment and strict monitoring to ensure safety. Moreover, urban water policies rarely include provisions for horticultural irrigation, leaving growers at a regulatory disadvantage (Kaur et al. 2025).

Environmental and health-related concerns are further amplified by the risks of contamination from untreated waste and agrochemical misuse. Many urban growers, in the absence of formal training or extension support, resort to the use of untreated greywater or sewage, especially in peri-urban areas where wastewater is more accessible than clean water. This practice increases the risk of pathogen transmission and heavy metal accumulation in produce. Additionally, the lack of awareness and monitoring often results in the indiscriminate use of synthetic pesticides and fertilisers, leading to agrochemical runoff, residue build-up, and soil degradation. These risks are especially alarming in densely populated urban areas where food is often grown in close proximity to industrial emissions and vehicular pollution sources (Malone et al. 2023).

A major underlying bottleneck is the lack of institutional coordination and public awareness surrounding UPH. Urban agriculture falls into an administrative grey zone between agriculture and urban development authorities. This fragmentation results in policy ambiguity, poor service delivery, and the absence of holistic planning. Urban municipalities, planning boards, horticulture departments, and environmental agencies often operate in silos, with limited interaction or integrated mandates to support UPH. As a result, issues such as zoning, subsidies, waste reuse, and training are addressed in a disjointed manner. Public awareness is also lacking; many urban residents remain unaware of the benefits of urban farming or perceive it as a backward or informal activity, rather than as a solution to food security, climate resilience, and community well-being.

Environmental and health concerns are amplified by contamination risks. Use of untreated greywater, sewage, and excessive agrochemicals increases the likelihood of heavy metal accumulation, pathogen presence, and soil degradation. Urban soils in industrial zones or near traffic corridors may contain lead, cadmium, and arsenic, while microbial contamination from irrigation water can lead to *Escherichia coli*, *Salmonella*, and other pathogens in produce (Murphy et al. 2025; Varalakshmi and Ganeshamurthy 2012). Air pollutants, including suspended particulates and gaseous emissions, can further degrade the quality of leafy vegetables grown outdoors. These risks highlight the need for robust safety standards, Good Agricultural Practices (GAPs), and monitoring protocols to ensure safe urban-grown food.

Institutional fragmentation and low public awareness also limit UPH's expansion. Urban agriculture often falls between the mandates of agriculture, municipal, and environmental authorities, resulting in disjointed policy support, unclear zoning, and weak service delivery. Many residents are unaware of the benefits of urban farming or perceive it as an informal or low-value activity. A coordinated, cross-sectoral approach is essential to address these challenges and unlock the full potential of UPH (Pradhan et al. 2024).

2.7 Future Perspectives and Strategic Recommendations

UPH has significant potential to enhance food security, ecological sustainability, and social inclusion in cities. However, realising this potential requires coordinated strategies that address technological, environmental, policy, and social dimensions (illustrated in Fig. 2.3). Future perspectives indicate that technological innovations such as vertical farming, hydroponics, and controlled-environment agriculture can improve productivity, resource efficiency, and urban space utilisation (Fei et al. 2025). Integration of UPH into urban planning and governance frameworks is also critical, enabling secure land access, formal zoning for horticultural activities, and alignment with climate resilience and Sustainable Development Goals (Senthamizh 2025).

The long-term viability and scaling of urban and peri-urban horticulture (UPH) depend on a visionary rethinking of urban governance, wherein cities are conceived as integrated socio-ecological systems, and food production is recognised as a

Fig. 2.3 Flow chart with four interconnected boxes. Top left: "Technological Innovation and Productivity Enhancement". Top right: "Integration of UPH into Urban Planning and Governance". Bottom left: "Institutional Support and Capacity Building for Inclusive Growth". Bottom right: "Market Integration, Climate Resilience, and Socio-Ecological Sustainability". The boxes are arranged in a diamond shape

formal urban function. Embedding UPH as a core component of green infrastructure, alongside water management, housing, sanitation, and transport, requires multi-dimensional strategies encompassing urban planning, institutional reform, technological innovation, and market integration (Fei et al. 2025).

A foundational step is the development of city-level horticulture master plans, which map all potential cultivation spaces—rooftops, terraces, institutional campuses, peri-urban agricultural belts, and marginal lands, and designate them through formal land-use zoning. These plans must include technical guidelines on rooftop load-bearing capacity, environmental compliance, crop suitability, water management, and waste recycling. Integrating UPH into building codes and development regulations provides incentives for residents and developers to incorporate terrace gardens, green balconies, vertical farms, and edible landscaping. The process must engage municipal authorities, urban planners, horticultural experts, environmental scientists, and community organisations to ensure inclusivity, technical feasibility, and public participation (Lin et al. 2015).

Equally important is the integration of UPH into nature-based solutions (NbS) for climate resilience. Multifunctional horticultural landscapes, including green roofs, urban orchards, living walls, and edible landscapes in public parks, medians, and institutional grounds, simultaneously provide ecosystem services and produce food. These interventions contribute to carbon sequestration, urban heat island mitigation, microclimate regulation, stormwater management, biodiversity conservation, and improved air quality. Embedding UPH in biodiversity action plans, ecological restoration programmes, and citizen science initiatives fosters community engagement and strengthens local stewardship (Sumangla et al. 2013).

Institutional support remains critical for scaling UPH, particularly in training, technology transfer, and input access. Urban farmers ranging from household gardeners to organised peri-urban growers require ongoing skill development in hydroponics, aquaponics, integrated pest management, composting, water-efficient irrigation, and safe wastewater use. Extension services should incorporate urban-specific modules, and municipal bodies should establish dedicated urban horticulture cells or coordinators. Access to subsidised seeds, organic inputs, lightweight growing structures, and micro-irrigation systems should be streamlined through urban retail networks and online platforms. Financial mechanisms, including interest-free loans, startup grants, and urban agriculture–specific insurance, should target women, youth, and marginalised communities to promote inclusive growth (Senthamizh 2025).

Market integration is essential for the sustainability of UPH. Urban growers need aggregation platforms, cold-chain infrastructure, processing units, and last-mile logistics, including electric delivery vehicles, to reduce post-harvest losses and maintain freshness (Lin et al. 2015). Formation of urban farmer-producer companies (FPCs), self-help group federations, or cooperatives, supported through technical assistance and subsidies, strengthens collective marketing and value addition. Farm-to-table initiatives, community-supported agriculture (CSA) programmes, and urban food hubs can institutionalise urban food chains, offering predictable markets and income security. Fostering innovation requires the establishment of demonstration hubs and urban agriculture incubators. These centres should showcase diverse cultivation models, that is, rooftop and vertical farms, hydroponics, aquaponics, edible landscaping, and urban composting, while providing training, business mentoring, and access to seed capital. Integration of advanced technologies, such as automated fertigation and AI-driven crop diagnostics, can enhance productivity, reduce input use, and improve quality assurance.

Embedding UPH into education, public health, and urban livelihood programmes strengthens both social and nutritional outcomes. Moreover, UPH integration into climate finance frameworks, including green bonds, adaptation funds, and corporate sustainability initiatives, ensures the long-term financial viability of urban horticulture (Khan et al. 2020). A transformative policy environment anchored in political will, multi-stakeholder partnerships, and participatory governance is essential for UPH to transition from a peripheral activity to a strategic urban function. With appropriate planning, investment, and institutional support, cities can evolve into productive, resilient, and inclusive socio-ecological systems, where horticulture contributes meaningfully to food security, climate adaptation, biodiversity, and social well-being (Lin et al. 2015).

2.8 Conclusion

Urban and peri-urban horticulture (UPH) is emerging as a pivotal component in the shaping of resilient, inclusive, and ecologically sustainable urban futures. The urban landscape is undergoing a profound transformation, driven by demographic shifts, climate change, resource limitations, and social aspirations. In this dynamic context, UPH emerges not as a marginal activity but as a core strategy that aligns seamlessly with the ethos of sustainable urbanisation (Lin et al. 2015). Its role transcends traditional food production, extending into realms of environmental stewardship, community engagement, public health, and urban resilience (Suman and Bhatnagar 2019). As cities grapple with multiple stressors, rising temperatures, air and water pollution, loss of biodiversity, and social fragmentation, UPH offers an integrated, nature-based solution that contributes simultaneously to ecological stability and social well-being.

The rising demand for pesticide-free, nutritionally superior food among urban consumers is not merely a market trend; it is a societal shift towards health-conscious living and environmental accountability. This growing awareness, particularly among the urban middle class, has fuelled a surge in interest in locally grown produce that is free from synthetic chemicals and excessive food miles. UPH meets this demand effectively by enabling the production of fresh vegetables, herbs, fruits, and medicinal plants within city boundaries (Opitz et al. 2016). Rooftop gardens, kitchen gardens, and community farms act as decentralised sources of nutrition, especially in cities where supply chains are vulnerable to disruptions. By reducing dependence on long-haul transportation and refrigerated storage, UPH not only lowers the carbon footprint of food but also enhances its nutritional value and freshness.

From an economic standpoint, UPH holds promise as a source of livelihood diversification and local enterprise development (Petts 2012). It supports urban employment through nursery management, compost production, landscape gardening, vertical farming, and sales of organic produce. Women, youth, retirees, and self-help groups particularly benefit from the accessibility and low entry barriers of horticulture-based livelihoods. With the right policy and infrastructural support, urban farming can become a cornerstone of inclusive urban economies that recognise the value of informal and non-traditional employment.

Environmental benefits of UPH are equally substantial (Poot-Pool et al. 2018). Cities, as resource-intensive ecosystems, often operate beyond their regenerative capacity. By integrating productive green infrastructure, UPH closes nutrient and water loops through composting, rainwater harvesting, and treated wastewater reuse. Urban horticultural landscapes also mitigate urban heat islands, improve air quality, buffer extreme weather events, and provide microhabitats for pollinators and other beneficial species, thereby enhancing urban biodiversity.

From a public health perspective, UPH contributes to mental and physical well-being. Access to green spaces and engagement in gardening activities have been correlated with reduced stress, improved mental health, and increased physical activity (Lin et al. 2015). Locally produced, traceable food further strengthens urban food safety, reducing exposure to contaminants and enhancing nutritional security.

Governance and policy integration are essential for the long-term viability of UPH. Mainstreaming urban horticulture requires cross-sectoral coordination across urban development, water, sanitation, housing, health, and climate action plans. Municipal corporations, urban planners, and state governments must recognise UPH as a core urban infrastructure, with dedicated budget allocations, formal policy recognition, and institutional support. In essence, UPH represents more than a food production strategy; it embodies a holistic approach to urban resilience, ecological stewardship, and social cohesion. As we reimagine the role of cities in the twenty-first century, urban and peri-urban horticulture offers a tangible, actionable, and inclusive pathway to resilience. It redefines urban prosperity not merely as economic growth, but as ecological balance, social cohesion, and human well-being (Bonny et al. 2014). The future of UPH is, therefore, not confined to the garden; it is rooted in governance, grows through community, and blossoms in sustainability. As cities look towards climate adaptation, food security, and improved quality of life, UPH will remain an indispensable tool in cultivating urban futures that are green, just, and thriving.

References

Barreca F (2016) Rooftop gardening. A solution for energy saving and landscape enhancement in Mediterranean urban areas. Procedia Soc Behav Sci 223:720–725

Bohn K, Viljoen A (2011) The edible city: envisioning the continuous productive urban landscape (CPUL). Field 4(1)

Bonny B, Bhaskar J, Peter KV (2014) Horticulture in urban and peri-urban agriculture – an analysis. In: The basics of human civilization food, agricutlure and humanity agricutlure, vol 3, p 339

Caplow T (2009) Building integrated agriculture: philosophy and practice. Urban Futures 2030:48–51

Dos Santos MJPL (2016) Smart cities and urban areas – aquaponics as innovative urban agriculture. Urban For Urban Green 20:402–406

Fei S, Wu R, Liu H, Yang F, Wang N (2025) Technological innovations in urban and peri-urban agriculture: pathways to sustainable food systems in metropolises. Horticulturae 11(2):212

Food and Agriculture Organization (FAO) (2008) Urban agriculture: framework for integrating agriculture into urban planning and policy. FAO

Ghallab R (2024) Cultivating connections: exploring the impact of community garden on social cohesion

Glück-Thaler E, Begg C (2012) Oyster mushroom cultivation on coffee: a unique opportunity for urban agriculture. NRSC 374 Special Topics, pp 1–11

Hamilton AJ, Burry K, Mok HF, Barker SF, Grove JR, Williamson VG (2014) Give peas a chance? Urban agriculture in developing countries. A review. Agron Sustain Dev 34(1):45–73

Jindal SK, Dhaliwal MS (2017) Development of vegetable nutrition garden model for diet diversification and improved nutrition security of urban and peri-urban households. Int J Hortic 7

Juniawati J, Hayuningtyas M (2017) Urban agriculture development: a strategy to support food security. KnE Life Sci:701–713

Kaur N, Singh J, Sharma NR, Natt SK, Mohan A, Malik T, Girdhar M (2025) Heavy metal contamination in wastewater-irrigated vegetables: assessing food safety challenges in developing Asian countries. Environ Sci Process Impacts 7:1747–1767

Khan MM, Akram MT, Janke R, Qadri RWK, Al-Sadi AM, Farooque AA (2020) Urban horticulture for food secure cities through and beyond COVID-19. Sustainability 12(22):9592

King'ori P (2004) Assessment of urban and peri-urban agriculture research in the centres of the Consultative Group on International Agricultural Research (CGIAR) in Sub-Saharan Africa, vol 1. International Potato Center

Korth M, Stewart R, Langer L, Madinga N, Rebelo Da Silva N, Zaranyika H, De Wet T (2014) What are the impacts of urban agriculture programs on food security in low and middle-income countries: a systematic review. Environ Evid 3(1):21

Kour K, Gupta D, Gupta K, Dhiman G, Juneja S, Viriyasitavat W, Islam MA (2022) Smart-hydroponic-based framework for saffron cultivation: a precision smart agriculture perspective. Sustainability 14(3):1120

Kuo FE (2002) Horticulture, well-being, and mental health: from intuitions to evidence. In: XXVI international horticultural congress: expanding roles for horticulture in improving human well-being and life quality, vol 639, pp 27–34
Lin BB, Philpott SM, Jha S (2015) The future of urban agriculture and biodiversity-ecosystem services: challenges and next steps. Basic Appl Ecol 16(3):189–201
Lovell ST (2010) Multifunctional urban agriculture for sustainable land use planning in the United States. Sustainability 2(8):2499–2522
Maknea KI, Asănică A, Fabian C, Peticilă A, Tzortzi JN, Popescu D (2022) The use of co-cultivation of aromatic, medicinal plants and vegetables in sustainable urban horticulture. AgroLife Sci J 11(1)
Malone M, Hamlin S, Richard SI (2023) Uprooting urban garden contamination. Environ Sci Pol 142:50–61
Mbaye A, Moustier P (2000) Market-oriented urban agricultural production in Dakar
McCartney L, Lefsrud M (2018) Protected agriculture in extreme environments: a review of controlled environment agriculture in tropical, arid, polar, and urban locations. Appl Eng Agric 34(2):455–473
Ministry of Housing and Urban Affairs (2019) Guidelines for urban water conservation: Jal Shakti Abhiyan. Government of India
Ministry of Housing and Urban Affairs (2023) Annual Report 2022–23. Government of India
Murphy M, Wachira G, Onyango C, Hoffmann V (2025) Heavy metal contamination in urban agriculture: evidence from Nairobi. Environ Sci Pollut Res 32(44):25231–25245
National Academy of Agricultural Sciences (2013) Urban and peri-urban agriculture (Policy Paper No. 67). NAAS, New Delhi
National Horticulture Board (2025) Revised Guidelines for Cluster Development Programme (CDP): Part B – peri-urban vegetable clusters (EOI & Guidelines). National Horticulture Board
Nethra J, Srinivasulu B, Kumar VV, Rao CS (2024) Microgreens: a comprehensive review emphasizing urban agriculture. Int J Environ Climate Chang 14(12):154–168
NITI Aayog (2023) Reuse of treated wastewater in urban/peri-urban agriculture in India. NITI Aayog, New Delhi
Nwosisi S, Nandwani D (2018) Urban horticulture: overview of recent developments. In: Urban horticulture: sustainability for the future, pp 3–29
Opitz I, Berges R, Piorr A, Krikser T (2016) Contributing to food security in urban areas: differences between urban agriculture and peri-urban agriculture in the Global North. Agric Hum Values 33(2):341–358
Petts J (2012) The economics of urban and peri-urban agriculture. In: Continuous productive urban landscapes. Routledge, pp 65–77
Poot-Pool WS, Cetzal-Ix W, Basu SK, Noguera-Savelli E, Noh-Contreras DG (2018) Urban home gardens: a sustainable conservation model for local plants based on Mexican Urban agri-horticultural practices. In: Urban horticulture: sustainability for the future. Springer, Cham, pp 73–88
Pradhan P, Subedi DR, Dahal K, Hu Y, Gurung P, Pokharel S, Joshi A (2024) Urban agriculture matters for sustainable development. Cell Rep Sustain 1(9)
Rammohan A, Pritchard B, Dibley M (2019) Home gardens as a predictor of enhanced dietary diversity and food security in rural Myanmar. BMC Public Health 19(1):1145
Rashtriya Krishi Vikas Yojana (RKVY) (2012) Vegetable Initiative for Urban Clusters (VIUC). Ministry of Agriculture & Farmers Welfare, Government of India
Schreinemachers P, Kumar SS, Uddin NM (2025) Impact of home gardens promoted among urban residents in Dhaka, Bangladesh. Food Secur:1–14
Senthamizh R (2025) Urban agriculture in a changing world: a thematic review of global trends, innovations, governance, and pathways to sustainability. Front Sustain Food Syst 9:1624426
Shukla S (2024) ICAR-IIHR Arka vertical farming technique for urban agriculture. Krishi Jagran
Singh AK (2019) Role of IIHR, Bengaluru, in strengthening horticultural research in the country – way forward. Indian Hortic 62(3)

Suman M, Bhatnagar P (2019) Urban horticulture prospective to secure food provisions in urban and peri-urban environments. Int J Pure Appl Biosci 7:133–140

Sumangla HP, Malhotra SK, Chowdappa P (2013) Urban and peri-urban horticulture – a perspective. Confederation of Horticulture Associations of India, New Delhi, p 34

Urban FAO (2022) Peri-urban agriculture sourcebook – from production to food systems. FAO, Rome

Varalakshmi LR, Ganeshamurthy AN (2012) Heavy metal contamination of water bodies, soils and vegetables in peri-urban areas: a case study in Bangaluru. J Hortic Sci 7(1):62–67

Wang R, Helbich M, Yao Y, Zhang J, Liu P, Yuan Y, Liu Y (2019) Urban greenery and mental wellbeing in adults: cross-sectional mediation analyses on multiple pathways across different greenery measures. Environ Res 176:108535

Organic Horticulture: The Retrospect and Prospects

3

Chinmoy Mandal, Soustav Datta, Sarthak Bhattacharya, Kamal Kr. Mandal, and Prerna Baraily

Abstract

Organic farming is a comprehensive management approach that improves the health of agricultural ecosystems, specifically biodiversity and soil biological activity. Organic approaches have the ability to completely transform horticulture's path towards resilience, longevity, and harmony with the environment by actively reducing the impacts of climate change, slowing soil deterioration, and providing a remedy for chemical contamination. India's top export markets for organic goods are the United States, the European Union, Canada, Switzerland, Southeast Asian nations, South Africa, and Australia. In India, nine states have implemented organic farming–related policies and initiatives. Non-chemical weeding, strip farming, and the production, collection, and distribution of organic supplements are some of the most popular organic practices, yet they all require a significant amount of labour. Organic horticulture is regarded as a vital and

C. Mandal
Department of Pomology and Post Harvest Technology, Faculty of Horticulture, Uttar Banga Krishi Viswavidyalaya, Cooch Behar, West Bengal, India

S. Datta (✉)
Rathindra Krishi Vigyan Kendra, Palli Siksha Bhavana (Institute of Agriculture), Visva-Bharati, Birbhum, West Bengal, India
e-mail: soustav.datta@visva-bharati.ac.in

S. Bhattacharya
Department of Horticulture, Neotia University, South 24 Parganas, Sarisha, West Bengal, India
e-mail: sarthak.bhattacharya@tnu.in

K. Kr. Mandal
Department of Fruit Science, Bidhan Chandra Krishi Viswavidyalaya, Nadia, West Bengal, India

P. Baraily
College of Agriculture, Uttar Banga Krishi Viswavidyalaya, Majhian, Dakshin Dinajpur, West Bengal, India
e-mail: prerna@ubkv.ac.in

A. Das, A. B. Sharangi (eds.), *Remunerative Horticulture*,
https://doi.org/10.1007/978-981-92-1007-7_3

sustainable approach to the issues facing the environment and contemporary horticulture because of its many benefits. The long-term viability of organic farming relies on addressing yield issues, socio-economic obstacles and a lack of effective pest and weed management methods, particularly for horticultural crops. Since 2005, organic horticulture has grown quickly as a sustainable farming method that benefits growers, increases food security in underdeveloped nations, and offers essential ecosystem services. This chapter covers the sustainability of organic horticulture in the present Indian as well as global scenario, evaluating its applicability under the current horticultural conditions and its promise as a sustainable remedy for the economic and environmental challenges.

Keywords

Organic horticulture · Global scenario · Socio-economic status · Future prospects

3.1 Introduction

Organic horticulture is the science of producing vegetables, fruits, or flowers by incorporating the principles of organic farming in soil conservation as well as pest management that sustain soil health, a safe environment and humans. Organic horticulture combines science as well as creativity to boost the environment, promote relationships and improve quality of life (Singh et al. 2022). The goal of modern horticulture is to prioritise environmental sustainability while improving crop quality. In addition to conventional quality characteristics, consumers are requesting vegetables and fruits with reduced pesticide residues and greater nutritional value. Making the switch to organic horticulture presents a viable option because organic fertilisers have shown promise in raising yields, promoting sustainable methods and improving soil health (Khan et al. 2024). The environmental impact of chemical fertilisers is lessened by horticultural innovations, including the incorporation of cover crops and varied crop rotations with organic fertilisers, which also improve soil qualities, microbial diversity and nutrient availability. Investigating a wide variety of products and materials derived from waste is becoming increasingly important to the scientific community. Recent developments include the development of bio-formulations, which combine organic materials with nanomaterials, mineral particles and microbes that stimulate plant growth. The purpose of these 'smart' fertilisers is to increase the efficiency of nutrient utilisation and speed up the shift to more sustainable farming systems (Mora et al. 2021). Algae-based formulations have attracted a lot of interest among these developments because of their affordability and environmental friendliness. Their utilisation has shown improvements in several horticultural crop productivity (Yao et al. 2020).

By encouraging healthy soils (Tal 2018), integrated and applied organic farming increases the amount of nutrients available for plant cultivation (Bender and van der Heijden 2015), providing a more sustainable method. Horticultural organic farming may lessen pest outbreaks by preventing genetic changes and immunological development in insects, which are negative effects linked to pesticide use (Borel 2017).

Because organic farming methods eliminate many of the input expenses associated with traditional farming methods, including the utilisation of synthetic fertilisers, pesticides and herbicides, they are thought to be more affordable and economically competitive (Mie et al. 2017).

Climate is characterised by elements such as temperature, humidity, precipitation, snow, soil and ice cover. Numerous natural variables contribute to the ongoing changes in the climate, but during the past 200 years, human activity has had the most significant and recent influence. According to the greenhouse effect (Alirezaei et al. 2017), its effects are characterised by rising temperature variations brought on by high levels of economic activity and manufacturing, which also raise emissions of greenhouse gases (GHG) such as carbon dioxide and methane (Mikhaylov et al. 2020). Organic horticulture has shown its ability to lower GHG emissions through carbon sequestration and by employing fewer inputs. Accordingly, in the climate change context, a shift from conventional to organic farming methods is becoming recognised as a feasible alternative farming method that has the potential to meet the goals of environmental preservation and climate change mitigation (Syamsiyah et al. 2023).

3.2 Basic Concept Behind Organic Farming in Horticulture

Organic horticulture supports ecological concerns, including soil conservation, and places a high value on environmental preservation. Crop nutrients, or minerals for the crops, are provided by soil microbes that raise the soil's nutrient levels and use insoluble nutrient sources. The basic concepts behind organic farming are as follows (Behera et al. 2011):

(i) It focuses on increasing the soil's biological fertility so that crops can absorb the nutrients they require from a consistent cycle of nutrients created in the soil and released in accordance with plant needs.
(ii) The establishment of ecological balance within the system, the application of bio-pesticides and other cultural practices such as mixed cropping and crop rotation are the main ways by which pests, diseases, as well as weeds are controlled.
(iii) All farm wastes and manures are recycled by organic farmers; nonetheless, the export of farm goods causes a continuous loss of nutrients.
(iv) Improvement of the surroundings to support the growth of wildlife.

3.3 The Evolutionary Process of Organic Principles

The idea of organic farming emerged during the twentieth century as concerns about inadequate food quality, crop variety shortages and soil erosion and depletion grew. It began as an opposition to the industrialisation of farming and a reaction to worries about the utilisation of pesticides and fertilisers made of minerals. The rapid evolution of mechanisation during that era significantly raised crop yields and reduced the cost of farming. The organic farming movement began as a reaction to the ensuing detrimental impacts on the environment. As demand for organic products rises worldwide, organic farming is gaining popularity. Adopting a sustainable farming approach and increasing crop yield without the use of synthetic ingredients are the goals of using organic farming methods (Luttikholt 2007).

Modern Era: Mainstream Acceptance and Global Expansion
Due to the growing demand for organic products from consumers, the organic farming industry has grown significantly in recent decades. A preference for natural products, health concerns about pesticide residues, and increased knowledge of environmental issues are some of the factors driving this movement. More organic foods are now available and come in a wider range at mainstream retail establishments as a result of the organic market's growth. Internationally, the growth of organic trade agreements and standards has made it easier for people to exchange organic products. But discussions concerning the meaning and application of organic farming, the reliability of certification procedures and the difficulties of upholding organic standards on a large scale have also been triggered by this expansion (Yadav et al. 2024).

3.4 The Important Goals of Organic Horticulture (Meena et al. 2013)

(i) Cultivation compatibility with the overall natural cycles of the production system.
(ii) Maintaining and enhancing agro-biodiversity.
(iii) Maximum possible utilisation of renewable resources.
(iv) Establishing a harmonious balance between crops and animal husbandry.
(v) Maintaining and enhancing long-term fertility as well as the biological activity of soil.
(vi) A sufficiently high level of productivity.

3.5 Framework of Organic Horticulture (Padhan et al. 2024)

(i) Ensuring that the farm obtains sufficient produce or crops to fulfil its needs or expectations.
(ii) Making sure that farming techniques align with the cycles and rhythms of nature is the solution.
(iii) Increasing the fertility of the soil and health over time.
(iv) Preserving and growing sustainable variety.
(v) Creating a complete, renewable, secure, and economical cultivating system.
(vi) Fulfilling and receiving the advantages of tradition and wisdom.

3.6 Components of Organic Horticulture

The main components of organic horticulture are crop rotation, including cover crops, fertilisers, mulch, irrigation, weed management as well as pest and disease management. Green manuring, composting, bio-fertilisers, organic cakes, vermicomposting and biodynamics are some of the many organic nutrition sources that are significant. Among the various aspects of organic farming, the application of bio-fertilisers is currently gaining popularity as a low-cost, secure substitute for traditional chemical fertilisers. The bacterial cultures known as bio-fertilisers, which can fix nitrogen from the atmosphere, include Rhizobium from leguminous crops and *Azotobacter* and *Azosprillum* from non-leguminous crops. Likewise, the phosphate-solubilising bacteria (PSB) have demonstrated their usefulness in converting inaccessible soil phosphorus into a form that is accessible. Pioneer plants are vulnerable to deficiencies in organic matter and major nutrients (N, P, K) as well as significant variations in soil temperature and moisture in India's sandy soil ecosystems. *Arbuscular mycorrhizae* are generally acknowledged to be crucial for seedling establishment, growth and survival, especially in marginal habitats where the symbiosis preserves soil structure and increases stress tolerance. Research efforts are mostly focused on the possibility of modifying mycorrhizal relationships to boost plantation, forestry or plant establishment productivity during ecosystem recovery following significant disturbance. There is much interest in their potential utilisation in horticulture (Sharma et al. 2024).

3.6.1 Crop Rotation

A cover crop must be a part of the rotation, along with efforts to preserve or enhance soil organic matter, crop nutrition and soil fertility. It changes the crop ecology, row/drilled crops, shallow/deep roots, leaf density, heavy/light feeders; interferes with insect, weed and disease cycles; and makes pests unable to find hosts when crops are changed (Chandra et al. 2022). For organic farming to be successful, crop rotation and intercropping are essential. Growing host plants as a rotation crop or as an intercrop may increase the number of dangerous pathogens; however, growing leguminous crops alongside them may enhance the amount of nitrogen available, which will ultimately increase fruit yield and quality. In comparison to guava as a solo crop, marigold intercropping has been shown to decrease parasites. Guava wilt can be efficiently controlled by combining two strategies, namely the use of bioagents, ideally in combination with *Aspergillus niger* or *T. harzanium*, and intercropping with crops like marigold or turmeric.

3.6.2 Mulches

Mulches are utilised to conserve moisture, minimise soil disruption; local resources such as straw, newspaper and plastics can be utilised as mulching material. In addition to stabilising soil temperature and lowering weed intensity and irrigation frequency, data show that organic mulches made from sugarcane waste enhance plant growth in an aonla-guava cropping system grown in salt-affected soil (Prasad et al. 2004). The organic mulch materials' breakdown in soil contributes to the soil's increased organic carbon content and nitrogen availability. In Gujarat's semi-arid aonla orchard, organic mulch materials (rice husk, subabool lopping, grasses, maize straw and paddy straw) regulated soil temperature, promoted plant development and decreased moisture evaporation (Singh et al. 2007). Research on the impact of mulches on onions during *kharif* showed that sugarcane trash mulch produced the highest yield, increasing it by 40.70% over the control and the maximum B:C ratio (Sharma et al. 2024).

3.6.3 Fertilization

Apart from the customary cultural methods, organic fruit production requires appropriate nutrient management. The organic concentrates, such as oil cakes and fish or bone meal, will be helpful in providing essential nutrients. Fruit production greatly benefits from the use of bio-fertilisers such as *Azotobacter*, *Azosprillum* and PSB, which are quite useful in providing nutrients that are otherwise unavailable. Elite seedling production in the nursery was facilitated by tamarind seed pelleting with *Azosprillum* and seedling inoculation of VAM at 5 g per poly-pot (Parameswari et al. 2001). Bulky manure application improved sandy soils' ability to retain water and pomegranate water-use efficiency (Sharma et al. 2021). According to Hiwale (2003), pomegranate cv. 'Ganesh' plants treated with a mixture of FYM (50%) + castor cake (25%) + urea (25%) had the highest plant height and fruit yield. In the Chotanagpur region's acidic soil, application of P solubilisers was observed to have a stronger positive impact on guava fruit yield and physicochemical parameters than that of N fixers (Dey et al. 2005).

3.6.4 Biological Pest and Disease Management

Nowadays, many fruit crops use biological management to reduce pests and diseases. The results of orchard management research indicated that some, including *Verticillium lecanii*, have great potential for controlling mealy bugs, thrips, white flies and scales in pomegranates, grapes, guavas, and custard apples (4–6 g l^{-1}). Additionally, NSKE@ 5% spray works well as an alternative to chemical pesticides to manage arid fruit crop pests. The combination of *Trichoderma viridae* + *Paecilomyces* with neem cake can successfully control nematodes and soil-borne diseases. Since disease control is the main factor restricting the yields of organic fruit, choosing resistant rootstocks or cultivars is crucial. In order to practise cleanliness, growers should remove disease vectors, clear up debris, and refrain from incorporating plant material from the same crop that has infected plants.

3.6.5 Organic Manures

By contributing nutrients and organic matter, including nitrogen that is held in reserve by soil microbes, manures help to make the soil more fertile. Generally used organic manures are vermicompost, farmyard manure (FYM), bio-fertilisers and crop residues. Organic manures enhance the soil's tilth and aeration and boost its ability to retain water. Additionally, they promote the growth of microorganisms that make plant food components in the soil easily accessible to crops (Singh et al. 2022). When applied to vegetables, FYM@ 20 t ha^{-1} greatly outperformed the recommended dose of NPK in cabbage in terms of improving head size and production (Chaurusia et al. 2004). Vermicompost significantly enhanced the height of the plant, number of leaves, leaf area index and tuber yield by 24% when compared to the control (Sood and Lal 2004).

3.7 Current Trend in Organic Horticulture

3.7.1 Global Scenario

The leading organic horticulture crops are fruits and coffee, followed by vegetables and cocoa beans. Organic horticulture is expanding steadily. By the end of 2023, nearly 98.9 million hectares of farmland globally were managed under organic systems, operated by 4.3 million producers. Europe maintains ~19.5 million ha under organic management. Latin America holds ~10.3 million ha. Perennial crops (fruits, nuts, coffee, olives, etc.) account for a significant portion of this organic area: in 2022, ~6.4 million hectares of perennial crops were certified organic or under conversion, representing ~3.6% of total perennial crop area. Permanent crops (including fruit trees, coffee, olives, nuts, cocoa, etc.) represented about 6.8% of the global organic farmland in 2023. Specifically, organic olive cultivation in 2023 exceeded ~825,000 hectares globally. The global organic market was valued at 136 billion euros in 2023, driven by rising consumer demand and sustainable practices. This growth is supported by government incentives and increased awareness of health and environmental benefits. (Data retrieved from the sources: https://www.oliveoiltimes.com/world/global-organic-farmland-continues-steady-expansion/137122).

3.7.2 Indian Scenario

India is a top global fruit producer, contributing about 10% of the world's fruit output. In 2023–24, India produced approximately 112.6 MT of fruits and 205 MT of vegetables. The horticulture sector showed strong growth, with overall production rising from 280.7 MT in 2013–14 to 367.72 MT in 2024–25, including 114.51 MT of fruits and 219.67 MT of vegetables. Organic farming is growing rapidly, with India leading globally in the number of organic producers and expanding organic farmland area to about 4.5 million hectares. Key export markets for organic fruits, especially mangoes, include the UK, Netherlands, and Germany, where demand remains high. Fresh fruit and vegetable exports generated around USD 1.8 billion, while processed horticultural exports fetched about USD 2.5 billion in recent years. States like Maharashtra, Madhya Pradesh and Rajasthan are major contributors to organic horticulture production. (Data retrieved from the sources: https://indiadatamap.com/2025/08/21/state-wise-organic-farming-in-india-2025/; https://asqi.in/new-age-organic-farming-what-every-indian-farmer-should-know-in-2025/).

3.7.3 Indian Initiatives

In the past, farming in India used methods similar to organic farming. Even today, a large portion of economically significant forest products, like medicinal plants and herbs, fall into this group. Indian farmers and entrepreneurs are increasingly turning to organic farming, specifically in low-productivity areas, mountainous regions, rain-fed zones and north-eastern states where annual fertiliser usage is <25 kg ha^{-1}. Different schemes as well as programmes have been implemented for the improvement of organic horticulture (Table 3.1). Nine states of India have implemented organic farming–related policies and initiatives. In an effort to boost its mountain farm economy and standard of living, Uttarakhand has pushed for organic farming. Sikkim and Mizoram announced plans to switch to entirely organic farming. Kerala, Maharashtra and Tamil Nadu have backed public–private partnerships to promote organic farming, while Karnataka has created organic policies.

3.7.4 Indian Organic Food Market

India's organic sector is only now starting to attract attention in urban areas, mostly through small stores and supermarkets. However, the recent emergence of hypermarkets in the majority of Indian cities is starting to bring about significant changes to the country's retail scene. Nowadays, there is an organic food section in every supermarket, and there are many organic food shops and restaurants in India's major cities. Customers in India have a preference for organic tea, marmalade, honey, butter, spices, fruits, vegetables and a variety of organic flours.

3.8 Benefits of Organic Horticulture

3.8.1 Crop Productivity and Quality

Numerous studies have examined the impacts of organic farming on crop yield, focusing only on bio-fertilisers and vermicomposting. Cidón et al. (2021) observed a positive correlation between vermicomposting and crop yield as well as revenue. It is the result of vermicomposting's ability to reduce nitrogen concentration, which stabilises as well as improves yield, increases farmer revenue and lowers the cost of

Table 3.1 Different schemes, programmes and institutions for improvement of organic horticulture (Babu and Karunakaran 2021)

Schemes/programme/ institution	Key details
Rashtriya Krishi Vikas Yojana	This central assistance scheme aims to develop more comprehensive plans for the agriculture sectors of several states, integrating cattle, poultry, and fisheries while taking into account agro-climatic conditions, natural resource challenges and technological advancements
National Programme for Organic Production	Launched in 2000 under Government of India's Ministry of Commerce and Industry. The goal of NPOP is to grow organic agriculture and high-quality products in a targeted and well-directed manner
National Project on Organic Farming	This project came into force in 2004 for better implementation of organic farming
National Mission for Sustainable Agriculture	Through the development of suitable adaptation and mitigation strategies, it aims to solve concerns and difficulties pertaining to 'Sustainable Agriculture' in climate change threat context
National Centre for Organic Farming	Promotion of organic farming nationwide through the development of technical capability among all stakeholders, creation of high-quality organic and biological inputs, technological transfer and human resource development
Mission Organic Value Chain Development for North Eastern Region	The scheme targets for improvement of organic value chains in North-eastern regions of India
Mission for Integrated Development for Horticulture	This central scheme for comprehensive development of horticulture sector like vegetables, fruits, flowers, spices and aromatic plants
Capital investment Subsidy Scheme under Soil Health Management Scheme	In order to establish a mechanised fruit and vegetable market waste/agro waste compost production unit, the scheme intends to provide 100% financial help to state government
The Agricultural and Processed Food Products Export Development Authority	This is an apex-export trade promotion government body under Government of India's Ministry of Commerce and Industry

buying and applying nitrogen fertilisers (Gurjar 2022). According to Kumar et al. (2018), vermicomposting increases plants' ability to fully use nutrients necessary for production by reducing nutrient leaching activities in nitrogen fertilisers. Microbial inoculants, also known as bio-fertilisers, are a necessary component of integrated nutrient management that raises crop yields. According to recent research, bio-fertilisers are essential to control the dynamics of organic matter decomposition as well as the availability of plant nutrients (Sneha et al. 2018). The benefit of using bio-fertilisers for crop production is that they sustainably stimulate microbial processes in soil (Kumara et al. 2015). Bio-fertilisers can also increase agricultural productivity by increasing microbial activity in the soil, roots and seeds. This allows nutrients to reach the target plants and increases their production (Yadav and Sarkar 2018).

Crop quality is a crucial factor to take into account when discussing various physiological processes and results in crop plants, including texture, taste, appearance, nutritional and safety characteristics, and texture that affect the value to the environment and consumer. According to a recent study, vermicomposting is linked with better product quality in process changes, increased product durability and enhanced physiological qualities of products- all of which several consumers have been expecting. Because vermicompost is easily absorbed by plants and allows them to bloom efficiently with few flaws, it provides crops with crucial nutrients that are necessary to improve product quality.

3.8.2 Environmental Benefits

Since organic farming does not utilise or release synthetic pesticides into the environment, some of which can damage soil, water and aquatic wildlife, organic farming is less harmful to the environment (Tisdell 2012). Numerous studies have demonstrated the environmental benefits of organic farming, and this is due to (a) enhanced soil fertility, (b) reduced pollution, and (c) improved plant and animal biodiversity (Kasperczyk and Knickel 2006). Therefore, compared to conventional approaches, organic farming clearly has a greater environmental performance per unit area. The main advantages for the environment include lower greenhouse gas emissions, less pollution and more species richness and abundance, which improve plant and animal biodiversity (Fig. 3.1).

3.8.3 Socio-Economic Benefits

For farmers, organic farming offers several advantages like lower input costs, higher prices and coordination within farmer cooperatives. Das et al. (2020) assert that organic farming produces safer, more nutrient-dense food and lowers input costs overall. Additionally, because organic produce is in a specialised market, premium pricing may be achieved for it. According to Mariappan and Zhou (2019), organic farming helps farmers live sustainably, which helps fight climate change in general. Since certified farmer organisations in developing nations often offer services like training, price information and value addition to help farmers meet certification criteria and produce the quality required in global organic markets, in addition to the higher prices that organic produce might fetch, organic certification may also be indirectly associated with additional economic advantages. Since smallholders frequently have limited access to services, the extra services offered by accredited organisations can help smallholder farms improve and generate more income (Meemken and Qaim 2018). Sustainable horticulture can assist in addressing the dual issues of employment development and food security in a developing nation like India. According to Parrott et al. (2006), organic horticulture may significantly

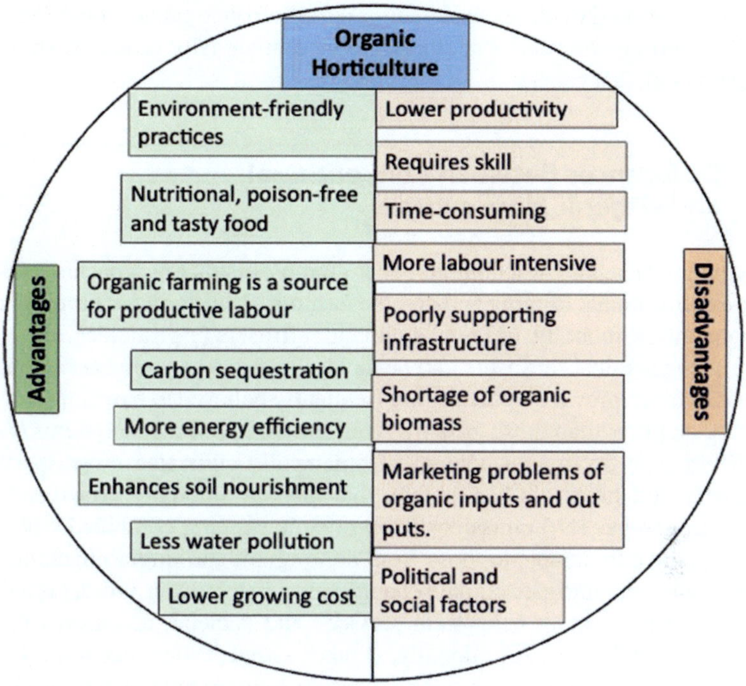

Fig. 3.1 Advantages and disadvantages of organic horticulture

increase farmers' food security and standard of living. Women who usually cannot afford horticultural inputs and have limited access to the traditional credit market can also participate more easily through organic farming.

3.8.4 Health Benefits

Although safety, quality and flavour are major reasons to purchase organic food (Magnusson et al. 2003), consumers' choice for organic products is primarily based on health. Organic farming has several health advantages, including lower occupational health hazards because farmers, as well as farm workers, are less likely to be exposed to harmful pesticides. Vitamins, minerals, and phytonutrients that promote health seem to be more abundant in organic produce. Numerous scientific studies have demonstrated that plant foods with higher concentrations of polyphenols are produced using organic methods and contain phytochemicals with anticarcinogenic, anti-inflammatory, antioxidant and analgesic properties (Aksoy and Öz 2020). Because organic farming produces higher levels of flavonoid expression, eating vegetables and fruits in particular has been shown to help reduce cardiovascular disease (Maldonado-Celis et al. 2019). Moreover, it was shown that strawberry extracts slowed the growth of cancer cells; extracts from organic strawberries inhibited cancer cell growth more than those from commercially produced strawberries (Johansson et al. 2014).

3.9 Differences Between Conventional and Organic Horticulture

Two popular management techniques that employ distinct approaches and procedures are conventional farming and organic farming. While organic farming involves growing plants without the usage of synthetic fertilisers or pesticides, conventional farming uses chemical fertilisers and pesticides to enhance yields and profitability. Plant-based foods grown organically are frequently believed to have superior health-promoting qualities than those grown in conventional production systems (Alsanius et al. 2019). The dry matter content of organically cultivated crops, particularly leafy greens and tubers, is higher than that of conventionally grown greens and tubers. Nitrogen gas (N_2) emissions from organic farming combined with organic fortification are comparable to those from non-organic farming. Enhancing organic farming would help mitigate climate change since, in a broader sense, organic farming is said to offer more ecosystem services and societal advantages than non-organic ones (Table 3.2). Additionally, it takes social justice, economics, and the environment into account, encouraging a comprehensive approach to food production (Skinner et al. 2019).

3.10 Sustainability of Organic Horticulture

3.10.1 Addressing Sustainability Challenges

Horticultural techniques face a number of sustainability issues in an era marked by growing environmental concerns, such as the widespread hazards of chemical contamination, the far-reaching effects of climate change and the increasing degradation of soil. Organic farming stands out as a ray of hope in this context, providing a viable way to ensure the sustainability and durability of horticultural pursuits.

Horticultural systems around the world are under tremendous strain due to climate change, which is characterised by unexpected precipitation, severe temperatures and unpredictable weather patterns. Because organic farming is based on sustainable methods, it naturally aids in mitigating and adapting to climate change. Organic farms can enhance soil carbon storage and reduce GHG emissions by focusing on carbon sequestration through techniques like cover crops. Furthermore, because healthier soils retain moisture better and sustain a wider variety of plant species, organic systems frequently show greater tolerance to climate extremes, reducing the likelihood of crop failures. The climate resilience of organic horticultural systems is further reinforced by their reliance on natural nutrient cycling and decreased reliance on energy-intensive inputs.

Table 3.2 SWOT analysis of organic horticulture

Area	Strength (S)	Weakness (W)	Opportunities (O)	Threats (T)
According to yields	Lower yield with Improved nutritional value (Das et al. 2020)	Lower yield compared to conventional production (Johansson et al. 2014)	Lower yields demand higher prices	Some customers cannot afford to pay more for their purchases (Meemken and Qaim 2018)
Climatological and environmental area	Crops that can withstand drought can be utilised (Altieri and Nicholls 2017)	Decrease in soil organic matter when there is less rainfall (Namboothiripad et al. 2021)	Little rainfall is beneficial for harvesting season	Reduced rainfall makes the soil more acidic
Economic circumstances	It is profitable and enhances employment (Prastiyo and Hardyastuti 2020)	Sometimes a lack of access to financial facilities (Aguilera et al. 2013)	Participation from the government and policymaker can increase the optimism of organic farming (Aghasafari et al. 2020)	A decline in the number of organic farmers in some countries as a result of lower profits in the first year after switching from conventional to organic farming (Cleveland et al. 2017)
Educational scenario	Utilising indigenous knowledge	Low level of farmer literacy	Applying and executing scientific achievements	The weakness of educational and promotional planning
Fertility of the soil	High soil organic matter content, encouraging soil biological activity	Degraded soil can take time to recover (Krause and Machek 2018)	Reduction in environmental degradation	Face environmental problems (losses in biodiversity, water quality and habitat loss)
Method of cultivation	There are less negative effects on the environment when chemical fertilisers and pesticides are not used	Requires more land to produce similar yield to conventional	Positive environmental impacts	Adaptation of organic cultivation at large scale may lead to land use change

(continued)

Table 3.2 (continued)

Area	Strength (S)	Weakness (W)	Opportunities (O)	Threats (T)
Organic fertiliser input	Use of naturally occurring products, like crop residue, manure (Yadav 2017)	High prices for organic fertilisers	Farmers can utilise indigenous knowledge to produce organic fertiliser	The limited supply and variation of organic products in specialised stores
Quality of the produces	More nutritional as well as healthier produce	Lower yield compared to conventional production	Highly marketable produce (Kazimierczak et al. 2019)	Pest and disease damage to produce

Organic methods successfully address soil deterioration, a serious issue impacting global food security. The conventional approach frequently causes soils to lose organic matter, which results in nutritional imbalances, erosion and diminished fertility. On the other hand, organic farming helps to reverse these harmful trends by emphasising the improvement of soil health. By encouraging the accumulation of organic matter, practices including crop rotation, cover crops and composting improve the structure and availability of nutrients in soil. By increasing soil microbial activity and water-holding capacity, these techniques help prevent erosion and promote long-term soil health. By treating the soil as a valuable resource, organic farming tackles one of the fundamental issues endangering the sustainability of horticulture.

Water quality, human health and ecosystems are all seriously threatened by chemical contamination brought on by traditional farming's heavy reliance on synthetic fertilisers and pesticides. This issue is immediately addressed by organic farming's dedication to minimal external interference. Organic methods reduce chemical exposure and encourage biological pest control by using natural substitutes like beneficial insects and bio-pesticides. Additionally, organic farming lowers the possibility of chemical pollutants in the food chain by avoiding synthetic growth regulators and genetically modified organisms (GMOs) (Badu 2023).

In summary, horticulture systems have sustainability issues, and organic farming offers a convincing and workable solution. By actively reducing the effects of climate change, halting soil erosion and providing a remedy for chemical pollution, organic methods show promise in transforming horticulture's path towards durability, resilience and environmental balance.

3.10.2 Implementing Sustainable Practices

Receiving certifications for organic and sustainable practices in the horticultural industry is not only a mark of distinction; it also provides access to a wealth of new markets (Kim et al. 2021). Farmers and horticulturists may now access premium markets with certification, which has become a valuable weapon as ethical consumers look for products that reflect their beliefs and environmental concerns (Sanbagavalli et al. 2020). Third-party confirmation of environmentally conscious and sustainable practices is provided by certification programmes like Global G.A.P. and USDA Organic (Priyadarsini et al. 2019). Customers can feel secure about their purchases because of these certifications, which attest to the products' quality and safety. Additionally, approved horticultural items frequently have preferential access to store shelves and specialist markets. Certifications can serve as a passport to foreign markets for individuals engaged in international trade. Exporting horticulture products frequently requires adherence to certification standards that are specific to several nations and areas. Additionally, attending regional and local farmers' markets provides a chance to establish a direct connection with customers who value certified horticulture products. In conclusion, there is an inherent connection between 'Certification and Market Opportunities' in the horticultural industry. In addition to helping the growers, this also makes the food system more environmentally friendly and sustainable.

3.10.3 Soil Structure, Fertility, Biology and Chemistry Relevance

The physical properties of soil, including air-filled porosity, water-holding capacity, oxygen diffusion rate and infiltration rate, have a significant impact on soil fertility. They have an impact on topsoil erosion, crust formation risk, and plant water-use efficiency (WUE) and nutrient uptake rate. A decrease in bulk density and penetration resistance over time, as well as improvements in germination and root development, are parameters that generally react to organic practices. Among other things, the quantity of earthworms, which are common in organic soils, enhances the physical properties of soil. Horticulture (on-site) and environment (off-site) suffer significant damage as a result of soil erosion. It is now widely known that erosion can be prevented by converting to OH, and in fact, topsoil building eventually resumes (Milgroom et al. 2007).

A stable system with high levels of biological activity that leads to effective nutrient cycling and stress resilience is referred to as healthy soil. The gains in soil biomass and biodiversity that come with the switch to OH are arguably the significant effects. Depending on their place in the food chain, different kinds of soil fauna and flora are impacted by changes in soil management in different ways. Bacteria, protozoa and archaea participate in the early stages of organic matter mineralisation. Consequently, when new organic matter is added, they are the first groups to be impacted. Later, the abundance and activity of other trophic groups, higher in the food chain, like earthworms, mites and predatory nematodes, rise. The presence of mycorrhizal fungi in organically treated soils is a notable difference from conventionally managed soils (Bending et al. 2004). In conventional circumstances, timely nutrient delivery usually satisfies all nutritional requirements. The cost of particular fertiliser and the impact on financial returns, rather than environmental concerns, are what motivate efforts to optimise nutrient use efficiency. The situation is more complicated in OH, where plants primarily absorb nutrients through mineralisation triggered by soil microbes. As an alternative, growers can aid plant nutrition by applying certain soluble organic fertilisers as compost teas or through fertigation.

3.10.4 Humanitarian Aspect of Organic Horticulture

Most people in this society are individualistic; thus, collaboration among farmers and between farmers and customers is quite beneficial to OH. In contrast to the conventional perspective of food as an anonymous commodity, many organic farmers work to create a sense of community around a locally responsive food system. Examples of organisations that use this strategy include community land trusts, farmers' markets and producer-consumer cooperatives. In terms of social impacts, OH seeks to enhance a community's self-reliance and social involvement in order to assist in securing land tenure and boosting rural jobs. Because OH needs more workers, it may improve rural job prospects and slow the rate of urbanisation.

3.10.5 Avoidance of Pollution

Environmental contamination of the air and soil may be caused by CH. Some of the polluting agents can unpredictably cause harm, like eutrophication of streams and lakes far from their application site or the loss of beneficial organisms above and below the ground. In general, OH practitioners are said to use fewer harmful pesticides. Therefore, it is anticipated that OH will release fewer pollutants into the atmosphere. Customers believe organic food is healthier than conventional products because they assume it is almost pesticide-free. Therefore, they are willing to pay the maximum for organically cultivated food.

3.10.6 Maintaining Biodiversity

The measure of the relative diversity of organisms found in a particular ecosystem is called biodiversity. Increased biodiversity boosts farm productivity by facilitating a variety of ecological services like pollination, pest management and soil fertility maintenance. OH seeks to strengthen the autonomy of local ecosystems instead of relying on outside inputs. One of OH's objectives is to preserve and improve natural biodiversity. Both strictly ethical and ecological reasons, as well as occasionally financially motivated plant protection and pollination concerns, contribute to the significant emphasis on preserving biodiversity. Numerous authors have shown that OH has a positive impact on biodiversity, and this effect has been shown for a variety of organism kinds (Bengtsson et al. 2005). Along with imported biological control agents, mating disruption pheromones, and organic growers are also utilising some pesticides that have earned organic certification, which could have a detrimental impact on biodiversity.

3.11 Resources and Application of Organic Fertilizers

3.11.1 Importance of Organic Fertilizers in Horticulture

The goal of modern horticulture is to prioritise environmental sustainability while improving crop quality. Customers are requesting fruits and vegetables that have less pesticide residue and greater nutritional value, in addition to the usual quality attributes. Making the switch to organic horticulture presents a viable option because organic fertilisers have shown promise in raising yields, promoting sustainable agricultural methods and improving soil health, specifically in dry land areas (Fig. 3.2). In addition to vital micronutrients like iron and zinc, organic fertilisers offer a balanced combination of macronutrients. By promoting microbial development and activity and facilitating the slow release of nutrients into the soil, they improve the availability of minerals and water. These procedures help to strengthen the soil's structure, which promotes vegetable crops' healthy root growth (Singh et al. 2020). Furthermore, organic fertilisers decrease pests and diseases, minimise nutrient loss, lessen soil acidity and alleviate heavy metal pollution (Lin et al. 2019). They are a sustainable substitute for synthetic inputs because of their gradual disintegration, which guarantees a consistent supply of nutrients for a long time (Shaji et al. 2021).

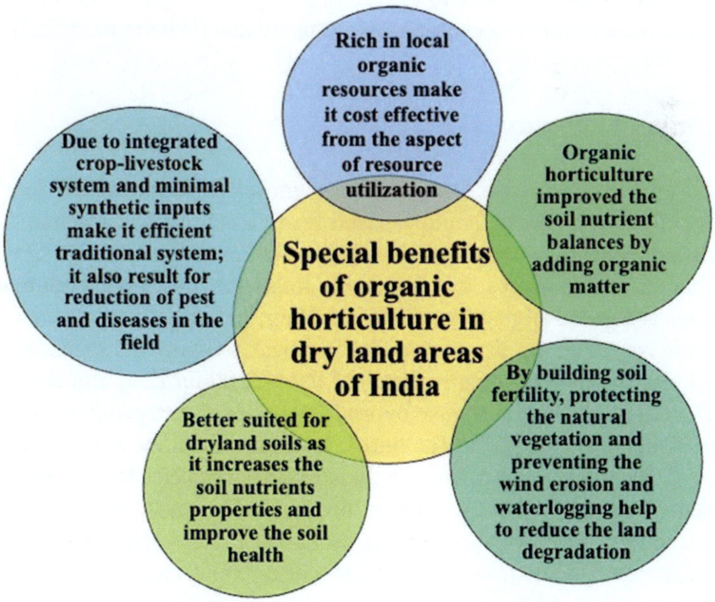

Fig. 3.2 Special benefits of organic horticulture in dry land areas of India

The environmental impact of chemical fertilisers is lessened by horticultural innovations, including the incorporation of cover crops and varied crop rotations with organic fertilisers, which also improve soil qualities, microbial diversity and nutrient availability. In addition to increasing crop yield, this integrated method lessens the negative environmental effects of chemical fertilisers. A wide variety of products and materials derived from waste are being investigated by the scientific community. Recent developments include the development of bio-formulations, which combine organic materials with nanomaterials, mineral particles and plant-growth-promoting microorganisms. Enhancing nutrient utilisation efficiency and easing the shift to more sustainable farming systems are the goals of these 'smart' fertilisers.

3.11.2 Vegetable Resources

Fertilisers are made from a variety of plant products and byproducts. The production of leguminous crops will contribute a net amount of nitrogen to the system. Increased use of nitrogen-fixing crops has been proposed as a potential replacement for industrially fixed nitrogen (Badgley et al. 2007). It is intriguing to utilise seeds as fertilisers because they often have the highest concentration of plant nutrients. Commercially available seed-based fertilisers include ricin (*Ricinus* sp.) cake and lupine (*Lupinus* sp.) meal; lupine meal exhibited a higher rate of mineralisation (Bergstrand 2022). Since ancient times, algae and seaweed have been utilised as plant fertilisers. They are administered to the crop as dried powder, foliar sprays, or liquid extracts, depending on the situation. Because microalgae may be grown for fertilisation, they provide a way to turn diluted organic wastes, such as waste from aquaculture systems, into concentrated plant fertilisers (Wuang et al. 2016).

3.11.3 Animal Resources

Animal-based fertilisers have long been a significant source of nutrients for the cultivation of organic plants. Animal-based fertilisers are often separated into two categories: manures and slaughterhouse waste. To make the material hygienic and easier to handle and transport, it is frequently treated in a variety of ways, including composting, anaerobic digesting, drying, milling, pelleting, etc. Animal type, the feed animals were fed, the use of bedding material and how the manure is handled and stored all affect the manure in terms of its mineral nutrient and organic matter content. A variety of slaughterhouse byproducts are frequently utilised as fertilisers. They are especially well-suited for horticultural applications since they typically have a low C/N ratio and are rich in easily plant-available N. This category includes several items with varying plant nutrient levels, such as meat, horn and bone meal and horn shavings. Fish waste, which is produced in vast quantities by the global fishing industry, can be hydrolysed to create a liquid that is ideal for use as fertiliser. These fish hydrolysates have bio-stimulatory properties and work especially well as foliar sprays (Sahu et al. 2016). One significant aspect of using fish waste as fertiliser is that it allows nutrients to be recovered from eutrophicated oceans. Additionally, fish waste from raising fish on land could be a useful fertiliser (Ekinci et al. 2019).

3.11.4 Organic Fertilizers Application in Potted Crops

Herbs and lettuce, vegetable transplants and ornamentals or bedding plants are examples of potted organic crops grown in greenhouses. The usage of organic fertilisers in potted crops presents difficulties because of the small amount of growing media in the pot, which restricts the volume for the microbial community that mineralises the organic fertilisers, as well as the appropriate overall fertiliser application. Succop and Newman (2004) cultivated basil in pots using liquid organic fertiliser made of kelp, rock phosphate, hydrolysed fish emulsion and poultry compost. When using a peat/perlite substrate or rockwool substrate, there was no significant difference in fresh mass production between the crop produced with organic fertilisation and conventional (mineral) fertilisation. But when perlite was the only substrate, plants fed the organic solution produced more fresh mass than those fed the mineral solution. Using a different strategy, Bergstrand et al. (2019) chose to provide the growing medium with all of the fertilisers the crop needs prior to the crop's initiation. The study comprised basil and *Pelargonium*, which were cultivated in pots using peat-based substrates and fertilised with either dried poultry dung or a product made from lucerne seeds and blood meal. The control treatment was mineral-controlled-release fertilisers, and all treatments were designed to start at an initial concentration of 800 mg N L^{-1}. Regarding fresh weight, *Pelargonium* did not exhibit any variation across treatments.

3.11.5 Organic Fertilizers Application in Greenhouse Crops

In organic systems, the greenhouse crops like sweet pepper, tomato and cucumber may be the most difficult to supply with nutrients. Due to the high nutrient requirements for the entire crop, it is difficult to provide a significant portion of the nutrients prior to crop initiation, and it may be labour-intensive and/or technically difficult to supply additional fertilisers during the cropping period (Burnett et al. 2016). It is challenging for organic fertilisers to meet the high instantaneous nutritional demand during the most intensive growth stage, particularly with regard to nitrogen (Dion et al. 2020). In Bergstrand's (2022) research, the indicated nutrient requirement for organic or conventional tomato is summarised. Fruit-bearing crops, on the other hand, experience a shift in the ratio of the various nutrients, such as the balance between N and K, rather than a continuous relative need for nutrients throughout the crop. Another situation unique to fruit-bearing crops is the growing requirement for calcium during fruiting, which may pose risks because of competition with the uptake of NH_4, which is frequently the major N-form in organic fertilisers. Another potential issue with soil-bound production techniques in greenhouses is the accumulation of Na and SO_4 (Gravel et al. 2012). In an observation by Raviv et al. (2005), composts containing cow manure had the highest yields, and it was proposed that N availability was the primary factor influencing plant development.

3.12 Policies and Regulations

3.12.1 Institution and Certification

In India, there are four types of organisational structures that enable smallholder organic horticulture. These include farmers organised by a company, farmers organised or facilitated by the government, farmers operating under NGO initiatives and farmers who have formed their own organisations like cooperatives, self-help groups (SHGs), etc. For small farmers with few resources to successfully transition to organic farming, some sort of support system is required. This is mostly due to the financial and other limitations that farmers face during the first 'conversion' stage of making the transition from non-organic to organic farming. The interval between the beginning of organic management and the certification of crop or animal husbandry is essentially known as the conversion phase. It is the amount of time needed to neutralise any chemical residues that may have been left in the soil by using certain chemical practices. For annual crops, the conversion phase typically lasts 24 months, although it can last up to 36 months for perennials. Three certification programmes are in place in developing and transitional economies. The first is the well-known and globally recognised third-party certification for individuals. Under an internal control system (ICS), small-scale farmers may receive group certification under the second scheme, which is likewise a third-party certification. The third scheme is the participatory guarantee system (PGS), a participatory certification that focuses on regional or national markets and incorporates small farmers, small business owners, merchants and consumers in the certification process. Every variant is unique to the markets, politics, geographies and communities in which it originated. To further expand organic horticulture in India, there must be significant organisational support. A low-cost, hassle-free certification process, an enabling scenario for small farmers, group certification and an internal control system are among the areas that call for the right institutional support.

3.12.2 Policy Support

For a long time, policies have concentrated on coming up with external solutions to farmers' problems. It must be possible for new policies to establish the circumstances for development that are more reliant on locally accessible resources, as well as on local expertise. In order to support the farmers' own analyses and explain their organised requirements, policymakers will need to figure out how to build relationships and discussions with other actors. Rapid input from conversation and participation would enable alternative policy adaptations. It seems like India is falling far behind in the adoption of organic farming, despite the sincere efforts of several NGOs. Certain challenges need to be addressed at the level of government policy-making in order to establish the groundwork for the growth of organic horticulture throughout the nation. These include the following:

(a) The government must provide significant financial support in order to encourage organic farming.
(b) The growth of the organic product market is essential for boosting domestic sales.
(c) Government assistance to consumer and producer associations for the promotion of organic products.
(d) Lowering the cost of certification and streamlining the certification procedure.

The organic certification process required these four steps (Srivastav et al. 2022):

(i) *Accreditation* is the guarantee that the certification programme is competent to carry out specific work. A governing authority establishes guidelines and criteria and verifies that a certification system is functioning in compliance with them.
(ii) *Standards* are the rules of the production for organic horticulture, or it is defined as production methods, not the product quality, minimum requirements not best practice. It is a dynamic process that is constantly developed and might include certification standards as well as national, regional, or international standards.
(iii) *Inspection* is the on-site visit to verify that the performance of the operation is in accordance with particular standards.
(iv) *Certification* is the written confirmation that the product or process is in compliance with prescribed standards.

List of accredited certifying agencies in India:

(a) Association for Promotion of Organic Farming (APOF)—Bangalore
(b) Indian Society for Certification of Organic Production (ISCOP)—Tamil Nadu
(c) Indian Organic Certification Agency (INDOCRT)—Cochin, Kerala
(d) Skal Inspection and Certification Agency—Bangalore
(e) IMO Control Pvt. Ltd.—Bangalore
(f) Ecocert International- Aurangabad
(g) Bioinspectra—Cochin, Kerala
(h) SGS India Pvt. Ltd.—Gurgaon
(i) International Resources for Fair Trade (IRFD)—Mumbai
(j) National Organic Certification Association (INOCA)—Pune

3.13 Marketing and Economic Scenario

Organic marketing operates on a very different principle from conventional marketing. It is crucial to develop and choose large markets and distribution channels carefully. In addition to higher expenses, such marketing calls for certain knowledge, expertise and experience, all of which unorganised individual farmers are typically unable to acquire (Das 2007). Approximately 85% of the nation's entire organic production is destined for export. As a result, India's domestic organics market remains underdeveloped. Market access for small producers depends on (a) understanding the markets, (b) communication and transport links and (c) an appropriate policy and environment. Small farmers in this evolving environment primarily require improved access to knowledge and capital. Furthermore, in order to meet new standards for products and processes or to prevent exclusion from the supply chain, collaborative action to address scale requirements must be designed. Professional training is necessary for small farmers in both the technical elements of production and marketing.

The adoption and execution of the new sustainable horticulture systems plan has the potential to expand the global organic market from the standpoint of socio-economic and environmental issues. The world's organic markets have grown remarkably, which has inspired many developing nations to increase their attention and expand their organic horticultural industries (Hans 2018). The global organic market size is currently significantly larger than previous estimates. As of 2025, the global organic foods market is valued at approximately USD 186.77 billion and is expected to grow to USD 402.67 billion by 2032, with a CAGR of 11.6% driven by rising consumer preferences and sustainable farming practices. Within this, the herbal and medicinal plants segment is now valued much higher, with the global herbal medicine market size at USD 233.08 billion in 2024, projected to grow to USD 437 billion by 2032, reflecting strong growth with an 8.23% CAGR, fuelled by demand for natural and plant-based therapies.

The total land area under organic cultivation worldwide expanded to nearly 99 million hectares by the end of 2023, with Australia leading at 53 million hectares. India follows with 4.5 million hectares, surpassing Argentina's 4.3 million hectares. Regarding producers, India remains the country with the largest number of organic producers globally, counting approximately 2.36 million certified producers in 2023, representing over 30% of the global organic producers. Uganda and Ethiopia are also significant producers in developing countries, with about 400,000 and 120,000 certified organic producers, respectively. (Data retrieve from the sources: https://www.coherentmarketinsights.com/industry-reports/organic-foods-market; https://www.fortunebusinessinsights.com/herbal-medicine-market-106320; https://www.factmr.com/report/organic-food-ingredients-market; https://www.grandviewresearch.com/industry-analysis/herbal-medicine-market-report).

The potential advantages of OH primarily rely on a number of variables, including dietary benefits from organic foods, economic, environmental and social aspects. Yields are typically a key component of farmers' performance and are therefore closely related to their financial gains. Improving income and the standard of living for the families and communities involved is the ultimate goal of the farming process. In order to forecast the viability of organic technologies in comparison to conventional systems, it is crucial to consider their cost-effectiveness. According to Eyinade et al. (2021), in order to improve rural stability in the majority of developing nations, organic production requires an enhanced labour force. In turn, this increased demand for work presents a chance to generate employment and could boost the economy in many rural areas.

3.14 Remunerative Organic Horticulture

In the global market, organic horticulture will help to provide a competitive, robust and lucrative livelihood. It necessitates switching from traditional commodity horticulture's cost-minimisation tactics to a value-maximisation approach. In order to counteract the biological yield gaps that have historically afflicted organic systems, this method makes use of price premiums, ecosystem service monetisation, direct-to-consumer marketing and cutting-edge technological interventions.

The use of price premiums in the market is the main way that organic horticulture creates financial success. Better-developed markets enable organic products to command higher prices than non-organic ones. Developing countries are unable to reach higher market prices due to the lack of certification processes and poor consumer awareness of organic products. Organic farmers can increase their income through access to particular product markets, international trade opportunities, and added-value supply networks (Pretty 2008). Transitions to organic farming go beyond economic factors because institutional and social factors are crucial to its acceptance (Hossen et al. 2024). Economic impact metrics should incorporate the broader social benefits of organic farming, such as improved public health outcomes, eco-preservation, and a thriving rural social fabric. According to Sakho-Jimbira and Hathie (2020), a more comprehensive evaluation of organic methods shows that they promote resilience and rural development.

3.14.1 Price Premiums and Market Access

Increasing the cost of organic produce is a popular strategy to offset productivity declines in organic farming systems. In developed certification markets, organic foods are valued at 20–50% more than conventional horticultural goods. According to Pretty et al. (2012), farmers encounter several logistical and regulatory obstacles when they are forced to pursue export markets since domestic organic markets are unstable.

3.14.2 Impact of Policy and Institutional Support

Policy support is the essential component required for success. Widespread adoption of low-cost input solutions, coupled with market cooperation and training initiatives, helps developers manage transition risks and premium price acquisitions as a result of appropriate certification procedures (Altieri et al. 2015).

3.14.3 Socio-Economic Benefits and Sustainable Livelihoods

Rural communities benefit from organic farming in ways that go beyond financial rewards. Through cooperative arrangements that improve market access and financial capabilities, organic farming fosters the growth of community-based organisations. These broader social benefits are necessary for the sustainable growth of smallholder livelihoods in developing nations, even though they might not yield substantial returns in short-term financial assessments.

3.15 Challenges and Opportunities for the Development of Organic Horticulture

Despite the widespread acceptance of organic farming in India, there are still certain challenges. Various critical challenges include the following:

(i) Lack of Policy for the Domestic Market and Imports

Pathania (2020) claims that India's existing organic farming policy places a lot of emphasis on export-oriented horticulture while giving little thought to developing organic farming policies for the country's internal market and imports. Because organic production and logo labelling standards are not regulated, it is quite likely that farmers and customers will not be able to distinguish organic products from conventional ones. Since consumers are willing to pay for organic items, fraudulent activities subsequently hinder legitimate parties from realising premiums.

(ii) Declining Crop Yield

According to Elayaraja and Vijai (2020), farmers who switch to organic farming usually see lower yields in early phases of production. However, it is unclear how long it will take for crops grown organically to become sustainable, which makes them economically unprofitable and raises production costs.

(iii) Unavailability of Organic Seeds Upholds Organic Farming

Since indigenous varieties are suited to react to chemicals and fertilisers, they are at risk from concurrent fertiliser-sensitive and genetic planting resources and seeds. As a result, it causes notable disparities in the supply of high-quality organic seeds, which discourages farmers from embracing organic farming.

A coordinated effort from all stakeholders associated with organic farming in India would be necessary to address these issues. This could entail creating regulations that support organic farmers' infrastructure and support services, spreading knowledge and education about the benefits of organic farming and funding research and technology to assist farmers in adjusting to the consequences of climate change. As an example, by collaborating with seed companies and offering subsidies to farmers who use organic seeds, the government may promote the development and distribution of these seeds. Furthermore, encouraging the use of composting and organic fertilisers can assist lessen reliance on chemical fertilisers and enhance soil fertility, both of which over time can raise crop yields. Overcoming these obstacles will ultimately contribute to the sustainability and long-term viability of organic horticulture in India. Depending on crop variety and agro-ecological conditions, a number of studies comparing conventional and organic systems have discovered that the organic systems produce lower yields (Schärer et al. 2022). Numerous factors, including the type of seed used, the plant's environment and many more, may contribute to the problem of reduced yields in organic farming. In many nations, certified seeds and other plant materials appropriate for low-input and organic farming are unavailable, so farmers typically use inorganic seeds before producing their products using the organic approach. As a result, the plants yield less because they naturally need a lot of inputs to be productive. In order to create regulations that effectively support organic farming, it is necessary to identify the variables that encourage the adoption of organic horticultural methods as well as the reasons that cause them to be discontinued.

3.16 Future Prospects

India has a wealth of naturally occurring organic nutrients in many parts of the nation, which will be beneficial for growing crops organically. With its extensive drylands, creative farmers and minimal use of pesticides, India boasts a robust traditional farming system. As a matter of fact, subsistence horticulture has long been practiced in the rain-fed tribal, northeast and hilly regions of the country, which are naturally organic due to the minimal usage of pesticides. In India, only 2% of the country's net sown area is used for organic farming, and only 1.3% of farmers are registered to do so. This figure alone demonstrates the potential for organic farming to grow in India. Therefore, Indian organic farmers will soon be reaffirming their legitimate position in the world's horticultural trade with increased knowledge and capacity building of the producers in accordance with international norms.

The development of solid scientific evidence regarding the advantages of organic farming must also be a priority. In order to encourage more farmers to switch to organic farming, organic and bio-fertilisers should be promoted rather than chemical fertilisers. A cost-effective certification programme that is sympathetic to farmers should be in place. The trend, which began in the developed world, is progressively spreading to developing nations. The demand for organic food is rising locally. India's expanding domestic market is positioned for stronger growth. The expansion of India's own domestic markets is essential to the organic movement's success. Although organic farming has long been practiced in India, the rise of contemporary, scientific, input-intensive horticulture has pushed it to the wall. However, as concerns about food safety as well as quality, the system's long-term viability and the growing evidence of its equal productivity have grown, organic farming has become a viable alternative farming system that not only solves these issues but also guarantees a debt-free, lucrative means of subsistence. The most crucial factor for the future of organic horticulture techniques is the incorporation of contemporary technologies. In order to contribute to sustainable horticulture output and generate information at various levels for use in resource planning, remote sensing monitoring is crucial in organic farming. The employment of GPS technology, remote sensing techniques, and hyperspectral image analysis is the pivotal moment that enables farmers to transition to organic and agro-environmental farming (Padhan et al. 2024). The usage of fully automated drones can be beneficial for increasing production as well as productivity. In organic horticulture, automated drones can be used for soil analysis in field planning, seed pod planting, crop spraying, crop monitoring, weed control, pest and disease control and crop biomass estimation (Pathak et al. 2020). The Internet of Things (IoT) facilitates the development of sensor technology, data storage, data analytics and algorithms for prediction and decision-making. It also helps with irrigation, ensuring that fertilisers and pesticides are distributed as efficiently as possible (Postolache et al. 2022).

Common technologies in organic horticulture in urban areas to maintain soil fertility and produce high-quality product are (i) using compost, (ii) using physical, mechanical and biological mechanisms to control pest and (iii) applying organic methods in the feed and livestock production (Lorenz 2015). When grown in a protected environment, organic horticulture yields high crop productivity. The crops grown in the controlled atmosphere of organic horticulture are highly valued and in high demand since the plant growth is carried out correctly here, and the product quality is enhanced (Dorais and Cull 2016). In organic horticulture, soilless culture can provide higher yields than open field, regular soil conditions. The best substitute for soil-borne disease and pest control is soilless culture (Gautam et al. 2021). Farmers can increase productivity and improve product quality by using hydroponics in organic horticulture practices (Kumari et al. 2018).

3.17 Conclusion

Organic horticulture is a good substitute because it improves the soil, fortifies the base of natural resources, and maintains biological production at levels that match the carrying capacity of the controlled agro-ecosystem. There is a severe lack of data demonstrating a comparison between crop yield and food quality under intensive horticulture and pure organic farming. In Indian farming, introducing and implementing biological and organic plant protection measures on a large scale may prove to be a formidable challenge. The organic food industry has emerged as a lucrative export market for developing nations due to rising customer awareness of food safety and health concerns. Additionally, organic horticulture generates employment, boosts local economies and enhances rural livelihoods. To address problems and optimise the advantages of organic practices for enhanced horticultural production and worldwide food security, successful implementation necessitates ongoing research, extension services and governmental support. It is clear that the production systems for organic horticulture need to expand globally. Owing to its numerous benefits, organic horticulture is seen as a vital and sustainable remedy for the issues facing contemporary farming and the environment. Knowledge of organic horticulture can be applied to research, market development, exports, extension services and the formulation and enhancement of policy programmes.

References

Aghasafari H, Karbasi A, Mohammadi H, Calisti R (2020) Determination of the best strategies for development of organic farming: a SWOT–fuzzy analytic network process approach. J Clean Prod 277:124039. https://doi.org/10.1016/j.jclepro.2020.124039

Aguilera E, Lassaletta L, Gattinger A, Gimeno BS (2013) Managing soil carbon for climate change mitigation and adaptation in Mediterranean cropping systems: a meta-analysis. Agric Ecosyst Environ 168:25–36. https://doi.org/10.1016/j.agee.2013.02.003

Aksoy Z, Öz Ö (2020) Protection of traditional agricultural knowledge and rethinking agricultural research from farmers' perspective: a case from Turkey. J Rural Stud 80:291–301. https://doi.org/10.1016/j.jrurstud.2020.09.017

Alirezaei M, Onat NC, Tatari O, Abdel-Aty M (2017) The climate change-road safety-economy nexus: a system dynamics approach to understanding complex interdependencies. Systems 5(1):6. https://doi.org/10.3390/systems5010006

Alsanius BW, Von Essen E, Hartmann R, Vagsholm I, Doyle O, Schmutz U et al (2019) The "one health"-concept and organic production of vegetables and fruits. Acta Hortic 1242:1–14. https://doi.org/10.17660/ActaHortic.2019.1242.1

Altieri MA, Nicholls CI (2017) The adaptation and mitigation potential of traditional agriculture in a changing climate. Clim Chang 140(1):33–45. https://doi.org/10.1007/s10584-013-0909-y

Altieri MA, Nicholls CI, Henao A, Lana MA (2015) Agroecology and the design of climate change-resilient farming systems. Agron Sustain Dev 35(3):869–890

Babu C, Karunakaran N (2021) Status, benefits and future prospects of organic farming in India: a review. J Manag Res Anal 8(3):103–111. https://doi.org/10.18231/j.jmra.2021.022

Badgley C, Moghtader J, Quintero E, Zakem E, Chappell MJ, Aviles-Vazquez K et al (2007) Organic agriculture and the global food supply. Renew Agric Food Syst 22(2):86–108. https://doi.org/10.1017/S1742170507001640

Badu PR (2023) Organic farming practices for horticultural sustainability. Int J Res Finance Manage 5(4). https://doi.org/10.36948/ijfmr.2023.v05i04.5367

Behera KK, Alam A, Vats S, Sharma HP, Sharma V (2011) Organic farming history and techniques. In: Agroecology and strategies for climate change. Springer Netherlands, Dordrecht, pp 287–328. https://doi.org/10.1007/978-94-007-1905-7_12

Bender SF, van der Heijden MG (2015) Soil biota enhance agricultural sustainability by improving crop yield, nutrient uptake and reducing nitrogen leaching losses. J Appl Ecol 52(1):228–239. https://doi.org/10.1111/1365-2664.12351

Bending GD, Turner MK, Rayns F, Marx MC, Wood M (2004) Microbial and biochemical soil quality indicators and their potential for differentiating areas under contrasting agricultural management regimes. Soil Biol Biochem 36(11):1785–1792. https://doi.org/10.1016/j.soilbio.2004.04.035

Bengtsson J, Ahnström J, Weibull AC (2005) The effects of organic agriculture on biodiversity and abundance: a meta-analysis. J Appl Ecol 42(2):261–269. https://doi.org/10.1111/j.1365-2664.2005.01005.x

Bergstrand KJ (2022) Organic fertilizers in greenhouse production systems–a review. Sci Hortic 295:110855. https://doi.org/10.1016/j.scienta.2021.110855

Bergstrand KJ, Löfkvist K, Asp H (2019) Dynamics of nitrogen availability in pot grown crops with organic fertilization. Biol Agric Hortic 35(3):143–150. https://doi.org/10.1080/01448765.2018.1498389

Borel B (2017) When the pesticides run out. Nature 543(7645):302–304. https://doi.org/10.1038/543302a

Burnett SE, Mattson NS, Williams KA (2016) Substrates and fertilizers for organic container production of herbs, vegetables, and herbaceous ornamental plants grown in greenhouses in the United States. Sci Hortic 208:111–119. https://doi.org/10.1016/j.scienta.2016.01.001

Chandra A, Maurya U, Kumar S (2022) Components of organic horticultural systems. In: Singh AK, Veersain, Yadav S, Srivastav P, Sarfaraz S (eds) Organic horticulture, 1st edn, Chapter 1. Bright Sky Publications, pp 1–6. https://doi.org/10.22271/bs.book.44

Chaurusia SNS, De N, Singh KP (2004) Effect of organic sources of nutrients on the growth and nutrient content on the growth, yield and yield attributing characters of cabbage (*Brassica oleracea* var. *Capitata*). In: Proceeding of National Symposium on organic farming in horticulture, pp 158–161

Cidón CF, Figueiró PS, Schreiber D (2021) Benefits of organic agriculture under the perspective of the bioeconomy: a systematic review. Sustainability 13(12):6852. https://doi.org/10.3390/su13126852

Cleveland DA, Phares N, Nightingale KD, Weatherby RL, Radis W, Ballard J et al (2017) The potential for urban household vegetable gardens to reduce greenhouse gas emissions. Landsc Urban Plan 157:365–374. https://doi.org/10.1016/j.landurbplan.2016.07.008

Das K (2007) Towards a smoother transition to organic farming. Econ Polit Wkly:2243–2245. https://doi.org/10.2307/4419703

Das S, Chatterjee A, Pal TK (2020) Organic farming in India: a vision towards a healthy nation. Food Qual Saf 4(2):69–76. https://doi.org/10.1093/fqsafe/fyaa018

Dey P, Rai M, Kumar S, Nath V, Das B, Reddy NN (2005) Effect of biofertilizer on physicochemical characteristics of guava (*Psidium guajava*) fruit. Indian J Agric Sci 75(2):95

Dion PP, Jeanne T, Thériault M, Hogue R, Pepin S, Dorais M (2020) Nitrogen release from five organic fertilizers commonly used in greenhouse organic horticulture with contrasting effects on bacterial communities. Can J Soil Sci 100(2):120–135. https://doi.org/10.1139/cjss-2019-0056

Dorais M, Cull A (2016) Organic protected horticulture in the world. In: III International symposium on organic greenhouse horticulture 1164, pp 9–22

Ekinci M, Atamanalp M, Turan M, Alak G, Kul R, Kitir N, Yildirim E (2019) Integrated use of nitrogen fertilizer and fish manure: effects on the growth and chemical composition of spinach. Commun Soil Sci Plant Anal 50(13):1580–1590. https://doi.org/10.1080/00103624.2019.1631324

Elayaraja M, Vijai C (2020) Organic farming in India: benefits and challenges. Eur J Mol Clin Med 7(11):3021–3029

Eyinade GA, Mushunje A, Yusuf SFG (2021) A systematic synthesis on the context reliant performance of organic farming. AIMS Agric Food 6(1):142–158. https://doi.org/10.3934/agrfood.2021009

Gautam R, Singh PK, Kumar P, Singh MC, Dhital S, Rani M et al (2021) Advances in soilless cultivation technology of horticultural crops. Indian J Agric Sci 91(4):503–508. https://doi.org/10.56093/ijas.v91i4.112621

Gravel V, Dorais M, Ménard C (2012) Organic fertilization and its effect on development of sweet pepper transplants. HortScience 47(2):198–204. https://doi.org/10.21273/HORTSCI.47.2.198

Gurjar S (2022) Organic farming in Sikkim-a sustainable nexus between crop yield and crop productivity. Indian J Org Farming 1(1):33–65

Hans V (2018) An analysis of sustainable agricultural development in India, pp 1–30. https://doi.org/10.2139/ssrn.3150442

Hiwale SS (2003) Effect of organic and inorganic fertilizers on growth, yield and quality of pomegranate cv. Ganesh. In: National Seminar on organic farming in horticulture for sustainable production, pp 29–30. https://doi.org/10.53550/EEC.2022.v28i02.056

Hossen A, Khan MFI, Arafat Y, Mahmud FU (2024) Machine learning models to predict market movements based on historical Price data and economic indicators. Int J Cent Bank 20(1):497–506

Johansson E, Hussain A, Kuktaite R, Andersson SC, Olsson ME (2014) Contribution of organically grown crops to human health. Int J Environ Res Public Health 11(4):3870–3893. https://doi.org/10.3390/ijerph110403870

Kasperczyk N, Knickel K (2006) Environmental impacts of organic farming. In: Organic agriculture. CSIRO Publishing, p 259. https://doi.org/10.1071/9780643093282.ch16

Kazimierczak R, Średnicka-Tober D, Hallmann E, Kopczyńska K, Zarzyńska K (2019) The impact of organic vs. conventional agricultural practices on selected quality features of eight potato cultivars. Agronomy 9(12):799. https://doi.org/10.3390/agronomy9120799

Khan MT, Aleinikovienė J, Butkevičienė LM (2024) Innovative organic fertilizers and cover crops: perspectives for sustainable agriculture in the era of climate change and organic agriculture. Agronomy 14(12):2871. https://doi.org/10.3390/agronomy14122871

Kim DG, Grieco E, Bombelli A, Hickman JE, Sanz-Cobena A (2021) Challenges and opportunities for enhancing food security and greenhouse gas mitigation in smallholder farming in sub-Saharan Africa. a review. Food Secur 13(2):457–476. https://doi.org/10.1007/s12571-021-01149-9

Krause J, Machek O (2018) A comparative analysis of organic and conventional farmers in The Czech Republic. Agr Econ-Czech 64(1). https://doi.org/10.17221/161/2016-AGRICECON

Kumar A, Prakash CB, Brar NS, Kumar B (2018) Potential of vermicompost for sustainable crop production and soil health improvement in different cropping systems. Int J Curr Microbiol App Sci 7(10):1042–1055. https://doi.org/10.20546/ijcmas.2018.710.116

Kumara K, Singh DR, Praveen KV (2015) Economic benefits from adoption of organic farming in India. Econ Aff 60(3):569–576

Kumari S, Pradhan P, Yadav R, Kumar S (2018) Hydroponic techniques: a soilless cultivation in agriculture. J Pharmacogn Phytochem 1(1):1886–1891

Lin W, Lin M, Zhou H, Wu H, Li Z, Lin W (2019) The effects of chemical and organic fertilizer usage on rhizosphere soil in tea orchards. PLoS One 14(5):e0217018. https://doi.org/10.1371/journal.pone.0217018

Lorenz K (2015) Organic urban agriculture. Soil Sci 180(4/5):146–153. https://doi.org/10.1097/SS.0000000000000109

Luttikholt LW (2007) Principles of organic agriculture as formulated by the International Federation of Organic Agriculture Movements. NJAS Wagening J Life Sci 54(4):347–360. https://doi.org/10.1016/S1573-5214(07)80011-8

Magnusson MK, Arvola A, Hursti UKK, Åberg L, Sjödén PO (2003) Choice of organic foods is related to perceived consequences for human health and to environmentally friendly behaviour. Appetite 40(2):109–117. https://doi.org/10.1016/S0195-6663(03)00002-3

Maldonado-Celis ME, Yahia EM, Bedoya R, Landázuri P, Loango N, Aguillón J et al (2019) Chemical composition of mango (*Mangifera indica* L.) fruit: nutritional and phytochemical compounds. Front Plant Sci 10:1073. https://doi.org/10.3389/fpls.2019.01073

Mariappan K, Zhou D (2019) A threat of farmers' suicide and the opportunity in organic farming for sustainable agricultural development in India. Sustainability 11(8):2400. https://doi.org/10.3390/su11082400

Meemken EM, Qaim M (2018) Organic agriculture, food security, and the environment. Ann Rev Resour Econ 10(1):39–63. https://doi.org/10.1146/annurev-resource-100516-053526

Meena RP, Meena HP, Meena RS (2013) Organic farming: concept and components. Pop Kheti 1(4):5–14

Mie A, Andersen HR, Gunnarsson S, Kahl J, Kesse-Guyot E, Rembiałkowska E et al (2017) Human health implications of organic food and organic agriculture: a comprehensive review. Environ Health 16(1):1–22. https://doi.org/10.1186/s12940-017-0315-4

Mikhaylov A, Moiseev N, Aleshin K, Burkhardt T (2020) Global climate change and greenhouse effect. Entrep Sustain Issues 7(4):2897–2913. https://doi.org/10.9770/jesi.2020.7.4(21)

Milgroom J, Soriano MA, Garrido JM, Gómez JA, Fereres E (2007) The influence of a shift from conventional to organic olive farming on soil management and erosion risk in southern Spain. Renew Agric Food Syst 22(1):1–10. https://doi.org/10.1017/S1742170507001433

Mora MDLL, Calabi-Floody M, Rumpel C (2021) Closing biogeochemical cycles and meeting plant requirements by smart fertilizers and innovative organic amendments. Agronomy 11(6):1158. https://doi.org/10.3390/agronomy11061158

Namboothiripad P, Pushpa J, Mahandrakumar K, Amarnath JS, Prabakaran K (2021) An impact study on organic agriculture-SWOC analysis. Ind J Pure App Biosci 9(1):92–98. https://doi.org/10.18782/2582-2845.8647

Padhan SS, Bhattacharyya S, Deep PR (2024) Organic horticulture techniques for eco-friendly sustainable. Int J Creat Res Thoughts 12(9)

Parameswari K, Srimathi P, Malarkodi K (2001) Influence of biofertilizer pelletization on elite seedling production of tamarind (*Tamarindus indica*) in nursery. Seed Res 29(1):58–62

Parrott N, Olesen JE, Høgh-Jensen H (2006) Certified and non-certified organic farming in the developing world. In: Global development of organic agriculture: challenges and prospects. CABI Publishing, Wallingford, pp 153–179. https://doi.org/10.1079/9781845931812.0153

Pathak H, Kumar G, Mohapatra SD, Gaikwad BB, Rane J (2020) Use of drones in agriculture: potentials, problems and policy needs. ICAR-NIASM 300:4–15

Pathania N (2020) Organic farming: its objectives, principles, types, techniques, favourable circumstances & downsides, vol 1, pp 50–55

Postolache S, Sebastião P, Viegas V, Postolache O, Cercas F (2022) IoT-based systems for soil nutrients assessment in horticulture. Sensors 23(1):403. https://doi.org/10.3390/s23010403

Prasad J, Pathak RA, Pathak RK (2004) Effect of mulching on improvement in soil and plant growth in aonla based cropping system under salt affected soil conditions. In: Organic farming in horticulture, pp 151–154

Prastiyo SE, Hardyastuti S (2020) The impact of regional policy and population growth on environmental Kuznets curve for agricultural sector in Indonesia: a provincial dynamic panel data analysis. Curr World Environ 15(2):260. https://doi.org/10.12944/cwe.15.2.22

Pretty J (2008) Agricultural sustainability: concepts, principles and evidence. Philos Trans R Soc Lond B Biol Sci 363(1491):447–465

Pretty JN, Williams S, Toulmin C (2012) Sustainable intensification: increasing productivity in African food and agricultural systems. Routledge, pp 5–24

Priyadarsini J, Karthick BN, Karthick K, Karthikeyan B, Mohan S (2019) Detection of PH value and pest control for eco-friendly agriculture. In: 2019 5th International conference on advanced computing & communication systems (ICACCS), pp 801–804. https://doi.org/10.1109/ICACCS.2019.8728464

Raviv M, Oka Y, Katan J, Hadar Y, Yogev A, Medina S et al (2005) High-nitrogen compost as a medium for organic container-grown crops. Bioresour Technol 96(4):419–427. https://doi.org/10.1016/j.biortech.2004.06.011

Sahu BB, Barik NK, Paikaray A, Agnibesh A, Mohapatra S, Jayasankar P (2016) Fish waste bio-refinery products: its application in organic farming. Int J Agric Environ Biotechnol 1(4):238605. https://doi.org/10.22161/ijeab/1.4.30

Sakho-Jimbira S, Hathie I (2020) The future of agriculture in sub-Saharan Africa. Policy Brief 2(3):18

Sanbagavalli S, Jeeva M, Somasundaram E (2020) Eco-friendly weed management options for organic farming: a review. Pharma Innov 9:15–18

Schärer ML, Dietrich L, Kundel D, Mäder P, Kahmen A (2022) Reduced plant water use can explain higher soil moisture in organic compared to conventional farming systems. Agric Ecosyst Environ 332:107915. https://doi.org/10.1016/j.agee.2022.107915

Shaji H, Chandran V, Mathew L (2021) Organic fertilizers as a route to controlled release of nutrients. In: Controlled release fertilizers for sustainable agriculture. Academic Press, pp 231–245. https://doi.org/10.1016/B978-0-12-819555-0.00010-4

Sharma BD, Jatav MK, Balai RC, Meena A (2021) Integrated nutrient management for horticultural crops in arid region. In: Dryland horticulture. CRC Press, pp 29–61. https://doi.org/10.1201/9781003180237-3

Sharma BD, Bhargava R, Lal B (2024) Organic farming technologies for horticultural crops. AGBIR 40(1):899–902

Singh AK, Sanjay S, Appa Rao VV, Meshram DT, Bagle BG (2007) Influence of organic mulches on soil, temperature, moisture and growth of aonla cv. NA-7 under rainfed conditions. Hortic J 20(2):71–75

Singh TB, Ali A, Prasad M, Yadav A, Shrivastav P, Goyal D, Dantu PK (2020) Role of organic fertilizers in improving soil fertility. In: Contaminants in agriculture: sources, impacts and management. Springer International Publishing, Cham, pp 61–77. https://doi.org/10.1007/978-3-030-47402-9_4

Singh A, Singh AK, Singh K, Srivastav P (2022) In: Singh AK, Veersain V, Yadav S, Srivastav P, Sarfaraz S (eds) Organic horticulture, 1st edn, Chapter 5. Bright Sky Publications, p 155. https://doi.org/10.22271/bs.book.44

Skinner C, Gattinger A, Krauss M, Krause HM, Mayer J, Van Der Heijden MG, Mäder P (2019) The impact of long-term organic farming on soil-derived greenhouse gas emissions. Sci Rep 9(1):1702. https://doi.org/10.1038/s41598-019-38734-7

Sneha S, Anitha B, Sahair RA, Raghu N, Gopenath TS, Chandrashekrappa GK, Basalingappa MK (2018) Biofertilizer for crop production and soil fertility. Acad J Agric Res 6(8):299–306. https://doi.org/10.15413/ajar.2018.0130

Sood MC, Lal SS (2004) Evaluation of vermicompost as an organic source of plant nutrient in potato. In: Proceeding of national symposium on organic farming in horticulture, pp 212–214

Srivastav P, Singh AK, Singh A, Rai D (2022) Constrains in certification of organic horticulture and export. In: Singh AK, Veersain V, Yadav S, Srivastav P, Sarfaraz S (eds) Organic horticulture. Bright Sky Publications, pp 07–14. https://doi.org/10.22271/bs.book.44

Succop CE, Newman SE (2004) Organic fertilization of fresh market sweet basil in a greenhouse. HortTechnology 14(2):235–239

Syamsiyah J, Ariyanto DP, Herawati A, Herdiansyah G, Dwisetio PK (2023) Rice paddy field on Gentungan, Karanganyar: soil C humic acid, fulvic acid, and stock as affected by period of organic practices. IOP Conf Ser: Earth Environ Sci 1165(1):012013. https://doi.org/10.1088/1755-1315/1165/1/012013

Tal A (2018) Making conventional agriculture environmentally friendly: moving beyond the glorification of organic agriculture and the demonization of conventional agriculture. Sustainability 10(4):1078. https://doi.org/10.3390/su10041078

Tisdell CA (2012) Sustainable agriculture: an update. Working paper no. 183-Economics. University of Queensland, School of Economics

Wuang SC, Khin MC, Chua PQD, Luo YD (2016) Use of Spirulina biomass produced from treatment of aquaculture wastewater as agricultural fertilizers. Algal Res 15:59–64. https://doi.org/10.1016/j.algal.2015.11.026

Yadav M (2017) Towards a healthier nation: organic farming and government policies in India. Int J Adv Res Dev 2(5):153–159

Yadav KK, Sarkar S (2018) Biofertilizers, impact on soil fertility and crop productivity under sustainable agriculture. Environ Ecol 37(1):89–93

Yadav V, Kumar S, Yadav P (2024) Historical background and evolution of the organic agriculture movement. In: Organic farming: cultivating sustainable agriculture. Chapter 2. National Press Associates, pp 9–12

Yao Y, Wang X, Chen B, Zhang M, Ma J (2020) Seaweed extract improved yields, leaf photosynthesis, ripening time, and net returns of tomato (*Solanum lycopersicum* mill.). ACS Omega 5(8):4242–4249. https://doi.org/10.1021/acsomega.9b03941

Scaling up Organic Horticulture for Sustainable Global Growth

4

Keerthivasan Ragupathi, Sudarshan Ramanathan, Uggumudi Sai Nishitha, and Priyatharsini Ayyaswamy

Abstract

Organic horticulture has been recognised as a global movement adhering to the principles of ecology and sustainability, and the world has acknowledged this as the best solution to the environmental, health, and economic issues related to conventional farming. This chapter gives a comprehensive overview of organic horticulture, highlighting its historical background, development, and current state globally and in India. It points out the transition from traditional practices to certified organic production systems and emphasises the role of indigenous knowledge and innovations in shifting the paradigms of organic farming. This chapter details the core principles of organic horticulture, such as soil health management, biodiversity enhancement, natural pest and disease control, and the complete prohibition of synthetic chemicals. The focus of the discussion is primarily on the practices of organic nutrient management that consist of composting, green manuring, biofertilisers, and the application of crop rotations and intercropping to improve the resilience of agroecosystems. The present-day difficulties in organic horticulture are indicated, including yield fluctuations, marketing problems, certification-process delays and insufficient knowledge amongst both growers and buyers. However, this chapter analyzes the potential upsides that can be reflected in both the local and the global markets, the large number of consumers desiring organic fruits and vegetables, and the gradual increase in the involvement of organic horticulture in climate-tolerant and sustainable farming practices. This chapter continues to identify new tools, the influence of policy, and collaboration with different institutions as the key factors and bases to create conditions that are supportive of the organic horticulture industry to grow. Thus,

K. Ragupathi (✉) · U. S. Nishitha · P. Ayyaswamy
Department of Floriculture and Landscape Architecture, Horticultural College and Research Institute, Tamil Nadu Agricultural University, Coimbatore, Tamil Nadu, India

S. Ramanathan
Department of Food Science, School of Life Sciences, St. Aloysius Deemed to be University, Mangaluru, India

A. Das, A. B. Sharangi (eds.), *Remunerative Horticulture*,
https://doi.org/10.1007/978-981-92-1007-7_4

presenting the picture of an enabling environment for organic horticulture to scale up. It further highlights success stories and demonstration farms that represent the best practices and innovations in organic production systems, and it also provides a future perspective on the possibilities of organic horticulture to feed the world, restore and maintain nature's balance, and be a source of a good life in rural areas. Some of the mentioned ways also include the integration of organic practices with such digital tools as the cloud, IoT, and machine learning, alongside value chain development, and climate-smart approaches.

Keywords

Organic horticulture · Certification bottlenecks · Nutrient management · Consumer demand · Composting

4.1 Introduction

Organic horticulture has transitioned from a marginal and utopian practice to a fully accepted method of food and landscape production throughout the world that, in a clear manner, connects ecology, human health, and social equity. This chapter first presents the main concepts and philosophy of organic horticulture, then follows its historical development and global contemporary trends, and lastly locates organic horticulture in the context of the UN Sustainable Development Goals (SDGs), demonstrating that organic practices support several SDG targets.

4.1.1 Concept and Philosophy of Organic Horticulture

Organic horticulture is an all-encompassing and eco-friendly method for producing fruits, vegetables, flowers, and other horticultural plants. The technique helps to keep the soil and the ecosystem healthy and fertile by relying on nature's help and not using synthetic chemicals (IFOAM 2020). The main idea of organic horticulture is based on the principle of "working with nature not against it," where the farm is seen as a living organism made up of interdependent systems of soil, plants, animals, and humans (Edwards et al. 1993). On the other hand, conventional horticultural systems often rely on high external chemical inputs for quick, short-term productivity, while organic horticulture fosters biological cycles, enhances soil organic matter, and promotes biodiversity, leading to long-term sustainability, which eventually attracts a lot of attention from the public and scientists (Reganold and Wachter 2016). The system is based on ecological processes like nutrient cycling, biological pest control, and pollination, which aid in keeping the natural balance within the agroecosystem. The International Federation of Organic Agriculture Movements (IFOAM) indicates that organic horticulture is founded on four essential principles—health, ecology, fairness, and care. The principle of health gives priority to the whole health of the soil, plants, animals, humans, and the planet in an interconnected way. With the Principle of Ecology, it is inevitable that farming in such a way will rely on the ecological systems and cycles that are alive. The Principle of Fairness recognises and supports the existence of fair relations amongst the various parties, whereas the Principle of Care advocates for the diligent and responsible management of nature's resources (IFOAM 2020). Therefore, the very concept of organic horticulture is nothing but a system-dependent and ethical approach. It has also been viewed as a holistic perspective that appreciates ecological balance, resilience, and inter-generational sustainability (Gomiero 2018).

4.1.2 Historical Evolution and Global Trends

The origins of organic horticulture go back to the early twentieth century, when it emerged as a counter-movement to the adoption of industrially produced agriculture. The pioneering work of Sir Albert Howard (1940) in India, which highlighted the importance of soil fertility through composting and organic waste recycling, was the basis for the modern organic movement. In the same way, the contributions of Rudolf Steiner's biodynamic farming (1924) and J.I. Rodale's promotion of organic living in the United States strongly established the principles of ecologically harmonious farming Case (2018).

By the end of the twentieth century, organic farming had already become a so-called good practice which, however, was still confined within a small group of consumers. The establishment of IFOAM in 1972 provided the first step toward support from institutions, which was later followed by implementing national organic standards in Europe, the United States, and Japan. The European Union put into effect a common organic regulation (EEC No. 2092/91) in 1991, whereas the USDA National Organic Programme (NOP) did not come into being until 2002, thereby allowing the same certified products to be (Willer and Lernoud 2019). The total organic land management area has increased noticeably all over the world. As per FiBL & IFOAM (2024), more than 76 million hectares are presently under organic management in 190 nations, with Australia, Argentina, and India being the foremost contributors when measured in land area. The international market for organic products has crossed the USD 130 billion mark as a result of an increase in awareness and the deliberate choice of consumers for sustainable, chemical-free products (FiBL 2024).

In the realm of horticulture, organic methods have been very successful with high-value crops such as fruits, vegetables, spices, herbs, and flower products that are sold in the high-end market and export sector (Scialabba and Müller-Lindenlauf 2010). Amongst the continents, Europe and North America are the main places where organic demand is the strongest, but countries like India, Sri Lanka, Bhutan, and Kenya are quickly developing their organic sectors due to good government policies and increasing domestic markets (Yadav et al. 2020). Thus, the history of organic horticulture has always been about a dynamic dialogue between scientific creativity, environmental awareness, and the market development cycles, which in turn led it to the present-day position of a worldwide sustainable agricultural development cornerstone.

4.1.3 Importance in Sustainable Development Goals (SDGs)

Organic horticulture can be seen as a strategic route that directly leads to several UN Sustainable Development Goals (SDGs); thus, it is a partial solution to global sustainability targets. The practice supports SDG 2 (Zero Hunger) and healthier populations by ensuring the safety, diversity, and localisation of the horticultural products (FAO 2021). It also supports SDG 12 (Responsible Consumption and Production) by using fewer chemical inputs and maintaining the nutrient cycle within the farm Paddock (2023). On top of that, the organic farming method largely increases carbon dioxide capture, which contributes to SDG 13 (Climate Action) via better soil organic matter and less greenhouse gas emissions that are connected with the use of synthetic fertilisers and pesticides (Gattinger et al. 2012). The use of organic practices for soil and biodiversity management also covers SDG 15 (Life on Land) as it helps in conserving pollinators, beneficial organisms, and soil biota.

From the socio-economic angle, the organic horticulture sector supports the UN goals of SDG 8 (decent work and economic growth) and SDG 10 (reduced inequalities) by bringing about a better living and working situation for rural communities, backing up the small farmers, and making the value chains fairer. Besides that, through its focus on health, ethics, and environmental care, organic horticulture is a major participant in the indirect path of SDG 3 (good health and well-being) and SDG 6 (clean water and sanitation) via the chemical cleanup and food safety systems (Fernandez et al. 2019). Global organic production area, number of producers, and market value across regions are summarised in Table 4.1.

4.2 Principles and Practices of Organic Horticulture

The integrated framework of organic horticulture is constructed through three essential components, which include soil health practices, biodiversity enhancement methods, and ecological pest control techniques, along with certification systems that exist in Fig. 4.1.

Table 4.1 Global organic market statistics (2023) and regional distribution

Metric	Global total (2023)	Leading region/ country	Significance
Total organic agricultural area	98.9 million hectares (+2.6% YoY)	Oceania (53.2 M ha)/Australia (53 M ha)	Represents 2.1% of global agricultural land (FiBL & IFOAM 2025)
Total organic producers	4.3 million	India (2.36 million)	Highlights significant reliance on smallholder farmers in emerging economies (FiBL & IFOAM 2025)
Global retail sales value	136.4 billion euros	United States of America (59.0 billion Euros)	Confirms market recovery and dominance of Northern markets (FiBL & IFOAM 2025)

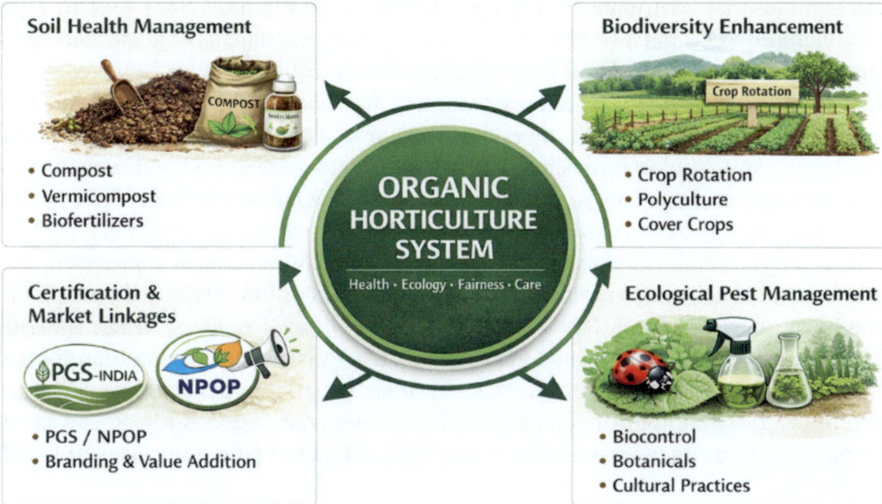

Fig. 4.1 The integrated framework of organic horticulture, consisting of soil health management, biodiversity enhancement, pest control, and certification systems, is illustrated. (Source: Conceptual illustration prepared by authors, 2026)

4.2.1 Soil Health and Organic Matter Management

Soil health is the main factor that supports organic horticulture, and soil organic matter (SOM) lies at the centre. SOM affects almost all soil properties, living, chemical, and physical and decides how much plant growth the soil can support and how long the agroecosystem will be resilient to stress. Soils with high SOM content are healthier and, thus, more capable of storing and supplying water, resisting erosion, cycling nutrients, and supporting diverse soil organism communities (Magdoff and van Es 2000). Soil organic matter (SOM) is a tiny volume of soil, but its impact on the health and productivity of the soil is far-reaching. Its composition includes decaying plant and animal remains, microbial cells, and stable humus fractions (Brady and Weil 1999). Soil globally harbours four to six times more carbon in SOM than in plants, therefore making SOM pivotal in soil fertility and climate moderation (Weil and Brady 2017; Gebreyes 2019).

The organic approach to soil health management aims to enhance the soil's natural strengths and not to control problems by means of external factors. This is mainly done through an increased supply of organic inputs, a diverse range of organic materials, and a reduction in the loss of existing soil organic matter (SOM). The changes in SOM content are indicative of the relationship between inputs and outputs: when the net flow of carbon into the soil through residues, manure, or compost is greater than the flow out through decomposition and erosion, SOM builds up; on the other hand, when the flow of carbon out is greater than that of inflow, SOM diminishes (Magdoff and Amadon 1980). Crop rotations are a very important factor in this process as they alternate between low and high residue crops and include perennials and legumes that contribute considerable underground biomass. Such practices not only increase SOM but also offer more room for microorganisms to grow, thus leading to natural pest control and less soil degradation (Studdert et al. 1997; West and Post 2002). In the same way, efficient crop residue management practices like stubble retention or integration of cover crops result in a significant increase in carbon input to the soil. Studies have shown that root residues sometimes are more effective than aboveground residues in increasing the stable SOM due to their higher resistance and being in a protected position within the soil matrix (Gregorich et al. 2001).

It is very crucial to reduce SOM losses as well. Soil erosion is minimised, and microbial decomposition is slowed down using conservation tillage and cover cropping. Long-term research has shown that the soils under reduced tillage systems continuously hold more organic carbon than the soils under conventional tillage. Therefore, the combination of practices for managing organic matter, such as rotation of crops, retention of residues, addition of compost, reduced tillage and cover cropping, creates a positive feedback loop that will continue to improve soil health gradually.

Soil organic matter (SOM) is one of the most important components of the soil that contributes to the physical properties of the soil, such as soil agglomeration, bulk density, porosity, and water-holding capacity (Bockheim et al. 2003). On the other hand, it also enhances the plant nutrient status by supplying macro- and micronutrients, buffering pH, toxic heavy metal adsorption, and improving cation exchange capacity (Weil and Brady 2017). In addition, SOM positively affects biological activity, since it directly gives energy to the soil organisms, and in turn, supports microbial biomass, nitrogen-fixing bacteria, and beneficial fungi (Acosta-Martinez et al. 2004). Besides, soil management practices such as reduced tillage, crop rotations, cover crops, application of animal manures, composts, and green manures play the most significant roles in maintaining the SOM pools (Gebreyes 2019). The increase in SOM leads not only to the enhancement of crop productivity but also to carbon sequestration in the atmosphere; therefore, it is advantageous for soil health and sustainability in terms of both agriculture and climate change mitigation. The major soil properties improved under organic management practices are summarised in Table 4.2.

Table 4.2 Soil properties improved by organic management

Soil property	Effect of organic practices	References
Organic carbon content	Increase	Gattinger et al. (2012) and Gebreyes (2019)
Microbial biomass	Increase	Acosta-Martinez et al. (2004)
Soil pH	Stabilization	Bernal et al. (2017)
Aggregate stability	Enhanced	Albiach et al. (2001)
Nutrient availability	Improved (N, P, K)	Eghball (2002) and Sanchez et al. (2001)

4.2.2 Organic Nutrient Sources: Compost, Vermicompost, and Biofertilisers

Natural sources of nutrients are the basis of organic horticulture, not only for the growth of plants but also for the sustaining of soil fertility in the long run. The use of manures, compost, vermicompost, and biofertilisers is the main component in the soil and nutrient management in organic farming practices.

Animal manures are, on one hand, great sources of organic matter and nutrients; however, their application must be closely monitored to prevent nutrient imbalance. The high content of phosphorus in manures as compared to plant needs often causes soil accumulation of the nutrient and, eventually, runoff-related pollution of water resources (Magdoff et al. 1997). Therefore, the limitation of the application rates should not only be oriented toward the immediate nitrogen needs of the crop but also take into consideration the future nutrient release in the following seasons. Composting is a process that stabilizes nutrients and increases their efficiency by reducing the easily decomposable fractions. Studies have demonstrated, for instance, that composted manure retained over 4 years 36% of the carbon added as soil organic carbon, while fresh manure only caused a retention of 14–25% (Eghball 2002). Likewise, the repeated application of composted municipal refuse increased the soil organic carbon more than fresh cow manure or sewage sludge (Albiach et al. 2001).

Vermicompost is a well-known organic fertilizer; however, it needs to be applied in larger quantities to achieve effective soil carbon levels. Moreover, the effects of vermicompost on microbial activity, nutrient availability, and crop yield are often short-lived. In addition, biofertilizers consisting of microbial inoculants such as Rhizobium, Azotobacter, and arbuscular mycorrhizal fungi play a significant role in organic nutrient management. Besides these, biofertilizers consisting of microbial inoculants like Rhizobium, Azotobacter, and arbuscular mycorrhizal fungi play a significant role in organic nutrient management. Biofertilizers, through fixing nitrogen from the atmosphere, solubilizing phosphorus, and enhancing nutrient absorption, not only foster plant growth but also promote soil health by maintaining biodiversity and the ecological balance. Compost, vermicompost, manures, and biofertilizers, when combined with crop rotations and residue recycling, create a sustainable way of handling nutrients in organic horticulture.

4.2.3 Pest and Disease Management: Biocontrol, Botanicals, and Cultural Practices

One of the main characteristics of organic horticulture is the non-application of synthetic pesticides and the use of ecological processes, as well as natural products for pest and disease control instead. The strategy is a preventive one, and it concentrates on building up agroecosystems so that they can hold out against pest outbursts rather than combating them with chemical controls.

Soils that are healthy and have a high SOM content naturally support the defence mechanisms of the plants and stress the pests by establishing and nurturing microbial communities that are in competition. As an example, beneficial fungi like Trichoderma not only vie for the spots where they can grow but also produce antibiotics and even attack the pathogens of plants (Tronsmo and Hjeljord 1998). Likewise, fluorescent *Pseudomonas* bacteria not only produce siderophores that take away iron from pathogens like *Fusarium* but also produce antifungal metabolites that prevent and even kill these pathogens (Weller 1988). Crop rotations interrupt the life cycles of soilborne pathogens, nematodes, and insects, and at the same time, they increase the number of different kinds of microbes in the rhizosphere. The practice of reduced tillage and use of organic amendments results in the establishment of communities of predatory beetles, fungi, and other beneficial microorganisms that control the populations of pest insects and also the seeds of weeds by predation (Liebman and Davis 2000). The practice of using cover crops has a further positive effect on resilience by the suppression of weeds, through both shading and allelopathy, the reduction of the nematode population, and even the control of fungal diseases in the crops that are planted afterwards (Davis et al. 2001).

Botanical inputs are other agents that are introduced into organic systems. The incorporation of Brassica cover crops into the soil results in the release of isothiocyanates, which possess fungicidal and nematicidal properties, thus leading to the reduction of pest populations in crops like potatoes and sugar beet (Mojtahedi et al. 1991).

4.2.4 Enhancing Biodiversity Through Crop Rotation and Polyculture

The diversity of life is the foundation of natural systems, and it not only gives them strength and the ability to bounce back but also provides them with the ecological services that are necessary for producing sustainably. The primary methods of attaining this are crop rotation and polyculture. Rotations break the cycles of pests and diseases while at the same time enriching SOM through varied residues. Research indicates that areas of land under diverse crop rotations (including hay and legumes) are able to store 50% more soil organic carbon than regions where continuous monocropping is practiced (Sanchez et al. 2001). Likewise, soil carbon retention for legume-based rotations is higher when compared to monocultures, which is due to increased root biomass and microbial activity (Gregorich et al. 2001).

Polyculture systems, where different species are cultivated at the same time, not only provide shelter for beneficial insects, but they also improve nutrient cycling and soil cover, thus reducing erosion and runoff. The use of cover crops is specifically advantageous: rye, clover, and brassicas not only add organic matter but also keep weeds and pests in check. For instance, rapeseed residues in potato systems led to a decrease in weed biomass by 50% (Boydston and Vaughn 2002), while the use of red clover that is sown along with the main crop has resulted in a significant reduction in weed populations in the following years (Liebman and Davis 2000).

These practices not only add to the proliferation of beneficial soil microbes and the formation of mycorrhizal fungi colonising plant roots but also create a network of ecological interactions that support crop yield. Hence, through the implementation of crop rotations and polycultures, the soil fertility and pest control are not only maintained, but also the ecological integrity of organic horticultural systems is strengthened.

4.3 Certification and Standards

4.3.1 National and International Certification Systems

The certification and standards are the key factors which guarantee the credibility, traceability, and marketability of organic horticultural produce. They act as a seal of approval for consumers that the production and handling systems come up to the organic standards, being free of any synthetic inputs and genetically modified organisms (Willer and Lernoud 2019). The certification is globally regulated by both public and private standards, which are synchronised through the frameworks of organisations like the International Federation of Organic Agriculture Movements (IFOAM), the Codex Alimentarius Commission, and the International Organisation for Standardisation (ISO). The IFOAM Basic Standards are the international reference standards for organic agriculture and specify the rules for production, processing, labelling, and selling (IFOAM 2020). The Codex Alimentarius Guidelines for the Production, Processing, Labelling, and Marketing of Organically Produced Foods (CAC/GL 32–1999) at the intergovernmental level are laid down as the basis for mutual recognition between countries (FAO & WHO 2001). The European Union has also established the EU Regulation (EU) 2018/848, which replaced the previous Council Regulation (EC) No. 834/2007 governing organic certification in the EU. The regulation that covers production rules, inspection systems, and labelling in a uniform manner for the EU member states is one and the same. With the USDA National Organic Programme (NOP) in the United States established in 2002, specific requirements for production, animal raising, and handling are determined and monitored through the use of accredited certifying agencies (USDA 2023). Japan Agricultural Standards (JAS), Canada Organic Regime (COR), and Australian Certified Organic (ACO) are also included as other major systems.

In India, the NPOP (National Programme for Organic Production), a programme conducted by the Agricultural and Processed Food Products Export Development Authority (APEDA), offers the regulatory framework for organic certification and labelling. The India Organic logo is the indicator for organic products that are in compliance with NPOP standards, which are accepted by the EU and Switzerland (APEDA 2023). The Bureau of Indian Standards (BIS) also assists in matters of organic labelling for domestic trade. Generally speaking, certification entails a third-party verification process, which goes through the inspection of records, field visits, residue testing, and regular audits of the farms. The procedure guarantees compliance with organic standards, which involves soil management, input usage, post-harvest treatment, and traceability (Scialabba 2013). On the other hand, the expensive and complicated verification process continues to be a hindrance for small farmers, mostly in developing nations (Kumar et al. 2020). In general, the certification systems are significant in making organic horticultural practices uniform, securing consumer trust, and easing international trade, but they still require the development of inclusive and cost-effective strategies for small and marginal farmers.

4.3.2 Participatory Guarantee Systems (PGS)

The Participatory Guarantee System (PGS) is an optional, community-based method of certification that mainly relies on trust, social networks, and local accountability rather than on strict auditing by formal third parties. It is especially appropriate for small-scale producers and local markets (IFOAM 2019). PGS uses participation, transparency, trust, and shared responsibility as its operating principles that enable farmers, consumers, and stakeholders to jointly confirm organic integrity through peer review and collective decision-making. Third-party certification does not incur the high costs and administrative burdens that PGS does, thus making organic certification more accessible to rural and resource-limited farmers (Nelson et al. 2016). In the IFOAM PGS Guidelines (2008), PGS is characterised as "local quality assurance systems that certify producers based on the active participation of stakeholders and are built on trust, social networks, and knowledge exchange." Farmers in these systems form groups, keep records of their organic practices, and carry out mutual inspections of each other's fields, after which a review committee validates the findings.

The PGS-India program in India, which started in 2015 as a part of Paramparagat Krishi Vikas Yojana (PKVY) scheme, authorises the sale of PGS-certified products in the domestic market. The certification process takes place through the PGS-India Portal, which results in digital traceability and product labelling available with the "PGS-Green" and "PGS-Organic" categories (PGS-India 2022). The whole system takes in more than a million farmers from many states, particularly in organic clusters and farmers' producer organisations (FPOs). The PGS also enables the community to learn and be empowered by the sharing of experiences amongst farmers on composting, pest control, and soil fertility management. Besides, it connects consumers directly with producers, thus enhancing the short food supply chain and the resilience of local markets (Home et al. 2017). Therefore, the Participatory Guarantee System demonstrates a decentralised certification that is non-discriminatory, which is very much in line with the organic philosophy of fairness and care, and at the same time, it ensures the marketability of the certification in the local contexts.

4.3.3 Challenges in Certification and Compliance

With its increasing importance, organic horticulture certification still faces various technical, financial, and institutional challenges that curb its general acceptance, particularly amongst smallholder farmers in third-world countries (Nemes 2009). First, the cost and the complexity of the procedures are very important obstacles to overcome. The costs of documentation, inspection, and renewal for third-party certification can sometimes be as high as 5–15% of the farm revenue, which discourages small producers (Vogl et al. 2005). Moreover, the certification authorities are usually situated in urban areas, thus making it difficult for farmers living far away to have access to these authorities. Second, a lack of proper communication and differences in literacy amongst people lead to noncompliance. A lot of farmers are not aware of the exact technical requirements for the maintenance of the projected certification, which are input documentation, record-keeping, and buffer-zone maintenance (Yadav et al. 2020). Third, disjointed institutional coordination amongst the certification agencies, marketing boards, and government departments causes waste and delays in the approval or export clearance process. Also, the small number of accredited laboratories for residue testing hampers the verification of compliance (APEDA 2023). On the demand side, consumer confusion due to the existence of many organic labels and the inconsistency of standards in different countries is one of the reasons for decreased trust in organic products (Hughner et al. 2007). The lack of equivalence agreements between some areas also negatively impacts the international trade of organic produce, mainly in perishable commodities like fruits and vegetables (Scialabba 2013). Besides, false advertising and labelling errors are still serious issues in both local and foreign markets. All these problems point to the necessity for solid supervision, clear traceability systems, and technological innovations like blockchain-based verification for confirming genuineness (Sharma et al. 2021). That is why, though organic horticulture certification systems are crucial for their credibility, it still requires simplified, participatory, and technology-driven approaches to overcome the compliance burden and to encourage the small-scale producers' equitable participation in the organic value chain.

4.4 Economic and Market Dimensions

4.4.1 Cost–Benefit Analysis of Organic Versus Conventional Horticulture

The economic aspect is the most crucial factor that determines the fate of organic horticulture. On the one hand, organic systems create big management and labour costs, but on the other hand, they open the door to premium prices, low input dependence, and the buildup of soil productivity in the long run, which in total makes it more profitable (Reganold and Wachter 2016). In the case of conventional horticulture, pesticides, synthetic fertilisers, and irrigation are some practices that lead to early yield increases, but at the same time, they result in soaring input prices, environmental pollution, and deterioration of soil health (Gomiero 2018). Conversely, organic horticulture relies on on-farm resources like compost, vermicomposting, and biofertilisers to replace the use of external inputs, thus cutting down the cost of purchased inputs (Pathak and Ram 2013). Nevertheless, the initial years of conversion to organic often see a dip in yields of 10–25% due to the transition period during which the soil biological equilibrium is re-established (Seufert et al. 2012).

Nonetheless, long-term studies show that organic systems, which in some cases yield just a little less, still get the same or even better economic returns because of higher prices and lower costs. The Rodale Institute (2011) did a 30-year comparison and found that due to better water-use efficiency and soil organic matter retention, organic systems under drought conditions yielded 20–30% more net returns than their conventional counterparts. Likewise, in the case of tomato, banana, and citrus, organic management helped in increasing net profit margins by 15–25% in comparison to conventional farming when premium pricing was taken into account (Kumar et al. 2020). Moreover, organic horticulture diversifies farm income through mixed cropping, on-farm input production, and value addition, which protects farmers from market fluctuations. In developing countries, less dependency on inputs makes farmers more resilient to fluctuations in the prices of synthetic fertilisers and pesticides (FAO 2021). To sum up, the transition period may be rocky, but in the end, the economic sustainability of organic horticulture is, in most cases, a win-win situation, as long as the market support, the certification process, and the consumers are there, and the demand is high.

4.4.2 Domestic Market Trends and Export Potential

During the last two decades, the global organic food market has been consistently increasing. It has now reached an astounding USD 135 billion figure in 2023, with fruits and vegetables sharing almost one-third of the total organic sales (FiBL & IFOAM 2024). North America and Europe are the leading continents in this scenario, with their combined consumption being approximately equal to 90% of global retail sales. But on the other hand, regions like Asia, Latin America, and Africa are getting stronger in the market as producers and exporters due to the good agroecological conditions and government support. As per APEDA (2023), the organic market in India is growing at 15–20% on a yearly basis, with the backing of consumer awareness and institutional purchasing. The organic outlets, e-commerce platforms, and farmers' markets in urban areas have made consumers' access to organic fruits, vegetables, and spices more extensive. The National Programme for Organic Production (NPOP) and PGS-India have made it easier for certification and traceability of both domestic and export markets.

India ranks as one of the top three countries producing organic crops across the world, with around 4.4 million hectares certified as organic. A large share of the certified land is used for horticulture (FiBL 2024). The primary export products consist of organic tea, spices, medicinal herbs, mangoes, and processed fruits, and these are mostly sent to the EU, the United States, and Canada. The value of organic exports from India reached more than USD 1.1 billion in 2022–23, which is a clear indication of the strong export potential for high-value horticultural produce (APEDA 2023). The world over, consumers are gradually regarding organic labels as an indicator of health, environmental care, and ethical production, which in turn has encouraged the continuous growth of demand. The introduction of eco-labelling and traceabilty tools, such as QR codes and blockchain-based platforms, is also contributing to the trust of consumers and transparency of the market (Sharma et al. 2021). On the other hand, the uncertainty about the regular supply, the post-harvest handling of the products, and the cold chain infrastructure for the perishable horticultural products are some of the problems that still exist. The smallholders' limited joint marketing and the varying quality standards are sometimes the major reasons for the restricted competitiveness of exports (Kumar et al. 2020). Therefore, it is necessary to strengthen the farmers' cooperatives, create organic clusters, and provide logistics support to make the most of the export potential of organic horticultural products.

4.4.3 Branding, Packaging, and Value Addition

Branding and value addition are the two main factors that contribute to the profitability and the attraction of consumers to organic horticultural products. Because organic produce commands a higher price, it is thus necessary to build consumers' trust through credibility, packaging, and differentiation, which are the main points of consumer acceptance in the market (Hughner et al. 2007). Branding is the process of giving a product a name and identity and communicating its values. Successful organic products often make claims about traceability, local origin, and eco-friendliness. For example, Nature's Basket (India), Alnatura (Germany), and Whole Foods Market (USA) are brands that have created a bond with customers based on the above attributes of their sourcing and the idea of being environmentally conscious (IFOAM 2020). In the case of organic horticulture, packaging must ensure the protection of the product as well as the sustainable development of the environment as its main goals. Biodegradable, compostable, or recyclable materials, such as PLA (polylactic acid) films, paper-based laminates and starch-derived plastics, increasingly replace conventional plastics (Tewary et al. 2025). Packaging also serves as a communication tool, carrying certification logos (e.g. India Organic, EU Leaf, USDA Organic), providing nutritional information, and displaying QR codes that give access to farm traceability data. Value addition through minimal processing, such as drying, dehydration, juicing, or extraction of essential oils, helps reduce post-harvest losses and increases shelf life. Organic processed products like jams, herbal teas, dried fruits, and essential oils often fetch 30–50% higher margins than fresh produce (FAO 2021). Moreover, integrating Geographical Indications (GI) with organic certification (e.g. *Darjeeling Tea*, *Malabar Pepper*) enhances global competitiveness through product authenticity (Willer and Lernoud 2019). Branding and value addition thus extend the organic philosophy beyond production, fostering a holistic value chain that links ecological sustainability with consumer ethics and market innovation.

4.5 Technological Interventions and Innovations

4.5.1 ICT and Digital Tools in Organic Farming

The use of Information and Communication Technologies (ICT) is dramatically changing the organic horticulture sector by making better decisions, enabling the traceability of produce back to the farms, making the use of inputs more efficient, and connecting farmers, researchers, extensionists, and markets more strongly. In organic systems, which are characterised by the absence of synthetic inputs and reliance on ecological processes for management, the use of ICT tools significantly reduces the impacts of the risks, keeps a close monitoring of crop health, gives timely information on pest forecasts, and so on.

Digital agriculture, which is also known as e-agriculture or smart farming, comprises the whole process of data collection, storage, analysis, and sharing aimed at making the entire agricultural operation more efficient, traceable, and sustainable. Digital tools are the backbone of the whole process in organic horticulture, which is knowledge-intensive and ecologically dynamic, since the correctness of inputs can only be checked through monitoring of the soils and climate, disease and pests predicting, and certification and marketing processes being facilitated.

4.5.1.1 Network Programme on Precision Agriculture and ICT Innovations

In 2021, the Indian Council of Agricultural Research (ICAR), Network Programme on Precision Agriculture (NePPA) was established as a national multidisciplinary initiative that connects 16 ICAR institutes from different divisions. The use of precision agriculture technology based on sensors is being encouraged in agricultural practices to ensure input sustainability, environmental safety, and quality of produce. Through the use of digital technology like sensors, IoT, and data analytics, precision agriculture maximises the efficiency of input applications in both organic and conventional systems by monitoring soil health, crop growth, and micro-climate variables at the same time.

Key innovations under NePPA include the following:

- *Variable Rate Technology (VRT)*: Automated, site-specific application of inputs using real-time data. The high-clearance VRT robot developed at ICAR-IARI applies fertilisers precisely as per crop stage.
- *SMART Pond (ICAR-CIFRI)*: An IoT-enabled aquaculture model with mechanical harvesters and sensor-based aeration achieving up to 20 t ha^{-1} $year^{-1}$.
- *riceNxpert (ICAR-NRRI)*: An Android app analysing rice-leaf images to recommend nitrogen doses; trials reported 10–43% higher yields than conventional practice.

Parallel digital marketing and advisory systems include the following:

- *eNAM (National Agriculture Market)*: A pan-India electronic trading portal linking 1389 mandis in 23 states and 4 UTs, ensuring transparent price discovery and rapid digital payments.
- *AGMARKNET* (The Agricultural Marketing Information Network): Connects over 3700 markets to provide real-time arrival and price data for commodity-flow mapping and supply chain integration.
- *Tamil Nadu Agricultural University (TNAU) Expert System*: Delivers crop- and livestock-based advisories (paddy, banana, coconut, etc.) in regional languages under ICAR networks.

Private and Cooperative Initiatives Have also Expanded Digital Agriculture Infrastructure

- *IFFCO-Kisan* integrates IoT-based soil- and moisture-sensors with automated irrigation systems, generating weekly analytics on soil organic carbon, moisture, and vegetation indices.
- *Farmer.CHAT*, an Artificial Intelligence (AI)-driven multilingual assistant by Digital Green, enables conversational, voice-based extension services that overcome digital-literacy barriers.
- *AgNext Technologies* applies AI for instant quality analysis and digital certification of agri-commodities; its chilli-grading assay allows transparent pricing.
- *Farmers' Portal* and *India Observatory* platforms provide integrated information on schemes, soil fertility, crop insurance, and geospatial analytics. Tools like Crop Water Budgeting and GIS-enabled Entitlement Tracking System (GEET) assist in resource-efficient water management.
- A growing ecosystem of mobile applications now supports organic and precision farming: Kisaan Suvidha (market auctions and price alerts), Agrotrace (IoT-based irrigation management), myScheme (government-scheme integration), PM-Surya Ghar (solar-energy adoption), Bijak Vyapaar (grain-trade platform), Kisan Khidmat Ghar (one-stop farmer service hub), and AAKRI (waste-management initiative) are notable examples.
- These initiatives show that digital agriculture provides traceability, transparency, and efficiency from "farm to fork" when matched with the principles of organic and sustainable farming. The combining power of AI, IoT, blockchain, and data analytics, along with the grassroots extension network, opens an excellent opportunity for Indian agriculture to be productive, profitable, and eco-friendly, especially in the organic sector.

4.5.1.2 Major Digital Tools Used in Organic Farming

(a) *Artificial Intelligence (AI)*

The usage of AI, as one of the components of precision organic farming, has a major impact on the quality of the harvest, and it predicts the yield, and the pests and nutrient deficiencies are detected at an early stage. Farmers can tell which plants are sick, count the weeds, and decide on the control measures (even without the use of chemicals). All this with the help of AI-based imaging and sensor data (Javaid et al. 2022). AI technology is also supporting the farmer's decision-making process by advising on bio-inputs, irrigation scheduling, and selecting the right variety according to the specific micro-climates.

(b) *Machine Learning (ML)*

Machine learning is a great tool for data-driven decision-making since it can easily analyse enormous datasets, which are the result of autonomous vehicles, variable rate applicators, GPS-based soil sampling, sensors, cameras, and drones. The ML models make it possible to predict yields, forecast pests, and optimise resources, thus adding efficiency to organic systems. Precision irrigation controllers and automated greenhouse systems are among the tools that use ML to attain the perfect soil moisture and nutrient conditions for organic crops.

(c) *Robotics*

Multi-sensor navigation systems that come with agricultural robots can perform all the tasks, such as inspecting crops, identifying weeds, harvesting, and data collection. In the case of high-value horticultural crops grown under organic cultivation, the use of robotic harvesters is becoming a viable option, especially in places where there is a lack of workforce. Similarly, robots can be programmed to carry out mechanical weeding that is in compliance with organic standards.

(d) *Drones and Remote Sensing*

Drones (unmanned aerial vehicles, UAVs) are utilised more and more for aerial mapping, multispectral imaging, and crop-health evaluation. On organic farms, aerial surveys conducted by drones can discover areas that require specific treatment, such as the lack of irrigation or nutrients, or the presence of pests. The use of both multispectral and thermal pictures can help to make plans for organic nutrient management as well as irrigation scheduling. Data from drones also helps to maintain certification by supplying geo-referenced records of land use and cropping patterns.

(e) *Data Analytics*

Data analytics turns out to be a powerful tool that optimises and makes every stage of the organic value chain—from crop planning to marketing—more efficient and less costly. The use of analytics platforms which combine satellite, sensor, and farm-level data is very helpful for streamlining the growing practices, estimating the number of crops to be harvested, checking the quality of soil, and organising the supply chain. For example, the study of public datasets can be used to make predictions about demand and supply trends in the organic produce market, hence, facilitating decision-making based on market conditions.

(f) *Blockchain Technology*

The organic supply chain made possible by blockchain is characterised by traceability, transparency, and trust. An unalterable distributed ledger could be used to keep track of every transaction, starting with seed origin, going through crop growth, and ending at processing and retail. This technology not only prevents the wrongful use of organic labels but also increases the trust of consumers in organic products.

Origin Trail, for example, showcases how blockchain can tamper-proof data and, at the same time, link the participants, who are farmers, processors, certifiers, and consumers, into an authorised digital ecosystem.

4.5.1.3 Practical Applications and Digital Tools in Organic Farming

The farm, extension and market levels are the fields where the ICT and digital innovations in organic horticulture have to be practically deployed to secure their successful adoption. The applications and tools, in fact, facilitate the farmers getting simultaneous advice, reducing the use of resources to the optimal level, and connecting the various phases of the production and thus certification and market systems through traceability. The digital technologies not only carry the research findings closer to the farmers, but they also enable the organic growers to make decisions that are both environmentally and economically viable based on the evidence.

(a) *Mobile Apps, Chatbots, and Query-Response Systems*

KisanQRS, a deep learning-based query-response system, is a tool that uses the vast database of farmer questions (call centre logs) to provide quick, automated answers on crop management, pest/disease control, fertilisation and so on. Such systems become even more useful when expert extension services are scarce or the number of farmers per expert is very high.

(b) *Extension and Knowledge Dissemination Through Digital Media*

An ICT-enabled extension is a bridge that connects research and farm with no communication gap. The ICAR publication "ICT Tools for Dissemination of Farming System Technologies" (Indian Farming 2025) illustrates how the use of satellite imagery, mobile platforms and AI-driven decision-support systems is reshaping the technology transfer process in India. Such systems send out timely messages about the weather forecasts, pest and disease outbreaks, and organic nutrient application schedules, thus enabling farmers to avoid losses due to adverse climate and at the same time increase their production.

(c) *Remote Sensing, Sensors, and Precision-Farming Tools*

In smallholder organic systems, the use of large-scale precision tools is still a constraint, but their adoption is gradually spreading through cost-sharing models and FPO networks. Climate sensors measuring moisture and nutrients in the soil, as well as monitoring micro-climate parameters, are mainly assisting the scheduling of irrigation and organic fertiliser through their respective efficiencies. Meanwhile, the use of drones and satellites for imaging purposes is helping in the mapping of crop vigour, the identification of nutrient deficiencies, and the detection of pest or disease-stressed zones (Wiley Online Library 2023). Such data make it possible to carry out site-specific, non-chemical interventions that are in accordance with organic standards. The use of IoT-based weather stations together with automated irrigation controllers and GPS-guided farm machines allows farm operations to be precisely timed, thus leading to the reduction of energy and water wastage. Moreover, such technologies play a role in the geo-referencing of farm records, which, in turn, is a reinforcement of the transparency that is needed for organic certification and traceability audits.

4.5.1.4 Benefits and Challenges

One of the major benefits of using digital tools in organic horticulture is the unity of precision agriculture methods through the micro-intervention push of ICT and AI, which acts in favour of the growers by getting higher yields while at the same time reducing and minimising inputs such as chemical residues, thus making the produce safer. Issues of irrigation efficiency and usage of organic nutrients will not be problems anymore as sensors, drones and data analytics are here to help. The introduction of blockchain and digital platforms has helped in making the whole process easier, as they are now working hand in hand to contribute in terms of traceability, certification transparency, and even market access. Smallholders organised in FPOs or cooperatives have been empowered in terms of information, decision-making and market access due to the digital tools.

On the downside of things, there are quite a number of challenges that are making it hard for digital tools to be embraced in organic farming. In rural areas, the internet connections may not be that reliable, and the locals might not be exposed to digital applications, which makes it harder for them to access the tools. The prices of devices and the subscriptions to internet services are both quite high, making it hard for poor farmers to afford. Also, there are very few tools available that are translated into the local dialects, and people still have doubts regarding the security and trustworthiness of the data being processed. On top of that, the slow integration of digital platforms with organic certification protocols can be a bottleneck in complete compliance, and thus, the full benefits of these technologies are not being realised.

4.5.2 Organic Input Production Units and Supply Chains

The organic horticulture practice is substantially dependent on the organic inputs: composts, biofertilisers, biopesticides, green manure, soil amendments, etc., particularly their quality, availability, and consistency. Production of inputs at the local scale plus robust supply chains is essential to reduce cost, ensure quality, foster sustainability, and maintain organic certification standards.

4.5.2.1 On-Farm Input Production Units

4.5.2.1.1 Jaivik Chhabutra/Soil Kitchen (On-Farm Input Production Units)

Tamil Nadu Agricultural University (TNAU) has promoted Jaivik Chhabutra, a "soil kitchen" or on-farm input unit, that is small in size (approx 15 × 15 feet) and uses locally available materials to produce inputs. Components include plastic drums (100, 50, 20–30 L), lid buckets, earthen pots, biodynamic pits, etc. This allows a farmer to prepare soil-nourishing inputs (composts, microbial inoculants, etc.) as needed, reducing dependence on external input suppliers.

Green manure crops (legumes, etc.), FYM, vermicompost and composting are used widely for improving soil organic matter and nutrient cycling. Proper composting not only supplies nutrients but also enhances soil structure, moisture retention, and contributes to pest and disease suppression.

4.5.2.2 Supply Chain and Quality Assurance

4.5.2.2.1 Aggregation, Certification, and Quality Standards

Organic inputs need to meet certain standards (e.g. for microbial purity if biofertilisers, absence of chemical residues in compost, etc.) for certification purposes. Ensuring uniformity, batch testing, labelling and trustworthy supply is important.

4.5.2.2.2 Storage, Logistics, and Market Linkages

Organic inputs are bulky and perishable (e.g. compost). Efficient storage (drying, packaging), transportation, and making them accessible to small and remote farmers are essential. Supply chain issues include cost, gaps in cold chain or moisture protection, and ensuring inputs are affordable.

4.5.3 Integration with Agroecological and Permaculture Designs

Organic horticulture benefits greatly when embedded within agroecological and permaculture frameworks, because these emphasise systems design, diversity, ecological self-regulation, resilience, and long-term soil health. Permaculture designs, agroforestry, guilds, mixed cropping, and habitat provision for biological control agents all contribute to reducing pest/disease outbreaks, improving ecosystem services, and reducing reliance on external inputs.

4.5.3.1 Principles and Design Elements

Plant mixtures such as companion planting, intercropping, alley cropping help to break pest life cycles, provide alternative hosts, confuse/repel pests, attract beneficial insects, stabilise yield and share resources. For example, polycultures can reduce pest infestation rates compared to monocultures. Non-crop plants (flowering strips, hedgerows, beetle-banks), cover crops, mulch, living mulches, ground cover to provide shelter, alternative prey, pollen/nectar for predators/parasitoids. These help maintain natural enemy populations. Use of trees for shade, windbreaks, ponds or water bodies (for dragonflies, etc.), diversity in layer (ground cover, shrubs, trees) to buffer extremes and improve resilience.

4.5.3.2 Pest Management via Design

Rather than relying on chemical pest eradication, design systems that maintain predator/parasitoid communities. Companion plants that repel pests or attract good insects (e.g. marigolds, garlic, borage, etc.) are used in permaculture and organic gardens. For instance, marigolds emit volatiles that deter aphids, and herbs like garlic repel various pests. Trap crops are decoy plants that draw pests away from the main crop. Border plantings of plants that repel pests may also be used. Push–pull is an advanced strategy (used extensively in Africa) where certain plants repel (push) pests from the main crop, while others attract (pull) them towards sacrificial trap crops. Though less documented specifically for small-scale organic horticulture in India, these ideas align with permaculture design principles.

4.6 Policy Support and Institutional Framework

4.6.1 Government Schemes and Incentives (e.g. PKVY, MOVCDNER)

Financial support from the government is essential in promoting organic horticulture via the provision of technical guidance, financial incentives, and the creation of institutional frameworks that make adoption, certification, and marketing easier. India has run numerous major initiatives aimed at the organic sector's development. Amongst them, the Paramparagat Krishi Vikas Yojana (PKVY), which is part of the National Mission for Sustainable Agriculture (NMSA), offers farmers financial assistance for organic cluster formation, certification, training, and learning through observance of the organic practices (DAC & FW 2020). The farmers in these clusters get maximum support of INR 20,000 per hectare for the transition to ecological agriculture that comprises help for compost pits, seeds, and biofertilisers. PKVY stresses the use of participatory certification through PGS-India, which has made it easier for smallholder and disadvantaged farmers to access it. The Mission Organic Value Chain Development for North Eastern Region (MOVCDNER) project is focused on the states of Northeast India that are hilly and ecologically sensitive, and it intends to develop certified organic value chains for fruits, vegetables, spices, and medicinal plants. The programme facilitates cluster-based cultivation, input supply, post-harvest infrastructure, and market linkages, thereby allowing for domestic and export-oriented growth (DAC & FW 2021). Amongst more relevant schemes is the National Programme on Organic Production (NPOP), which delivers certification and quality assurance, and the National Bank for Agriculture and Rural Development (NABARD)-backed organic horticulture credit programmes, which make it easier for small farmers to access credit (APEDA 2023). These government actions together lower the threshold for organic farming, improve market competitiveness, and incorporate sustainability into agricultural policies at the national level. On a global scale, Bhutan, Sikkim, and Cuba have been among the few countries to adopt very ambitious policies for organic agriculture promotion at the state or national level. To illustrate, Sikkim has attained the status of being India's first fully organic state, made possible by subsidies, technical training, and farmer mobilisation programmes, thereby showcasing the efficacy of integrated policy frameworks (Dorji et al. 2020). Major policy initiatives supporting organic horticulture in India are summarised in Table 4.3.

Table 4.3 Key organic horticulture policies and programmes in India

Programme/authority	Objective	Year/reference
Paramparagat Krishi Vikas Yojana (PKVY)	Promote certified organic farming	DAC & FW (2020)
Mission Organic Value Chain Development for North Eastern Region (MOVCDNER)	Develop value chains and market access	DAC & FW (2021)
National Programme for Organic Production (NPOP)	Standardisation & certification of organic produce	APEDA (2023)
Participatory Guarantee System (PGS)	Local certification for smallholders	IFOAM (2019)

4.6.2 Role of NGOs, Cooperatives, and Research Institutions

In organic horticulture, nongovernmental organisations (NGOs), farmer cooperatives, and research institutions play a key role in connecting policy, science, and practical application. Through capacity building, awareness campaigns, and market facilitation, NGOs support smallholder farmers, particularly in inaccessible or neglected areas, providing them with technical assistance as well. Development Alternatives (India), KRIBHCO, and the Organic Farming Association of India are examples of such organisations that partner with farmers in the areas of organic practices and certification (Yadav et al. 2020).

Farmer cooperatives and farmer producer organisations (FPOs) play a crucial role in enhancing the bargaining power of farmers, making the procurement of inputs easier, as well as setting up processing units for value addition. These groups provide a way of doing things through bulk purchases, and thus, they are able to access both the domestic and foreign markets more easily. In the case of India, the government-supported organic clusters are very often in line with the FPOs for the purpose of organising common marketing and thus cutting down the transaction costs. Scientific validation, input development, and the dissemination of knowledge are the areas where institutions like ICAR, IIHR (Bengaluru), CSIR, and state agricultural universities lend a hand through their respective contributions. They are engaged in making organic nutrient formulations, biocontrol agents, disease-resistant varieties, and post-harvest technologies that are organic-friendly (Pathak and Ram 2013). The research outputs help in decision-making at the policy level through the analysis of cost-effectiveness, yield trends, and environmental impacts of organic horticulture, thus making the interventions based on evidence. Global partnerships through FAO, IFOAM, and bilateral research programmes provide means for technology transfer, capacity building, and adoption of the world's best practices, which in turn promote the innovation of organic horticulture (Scialabba and Müller-Lindenlauf 2010).

4.6.3 International Collaborations and Knowledge Exchange

The global community cooperates to the utmost extent possible when it comes to organic horticulture; it allows for the sharing of knowledge, the harmonisation of standards, and the opening up of markets. Various international organisations, such as IFOAM Organics International, FAO Organic Agriculture Programme, and Global Organic Market Access (GOMA), build up research networks and the inter-communication of different policies along with the development of technical capabilities for the producers and the certifying bodies alike (IFOAM 2020). The Knowledge Exchange Programmes comprise training workshops, visits of farmers to one another, joint research projects, and online platforms for the dissemination of information. For instance, Indian organic horticulture clusters have partnered up with academic institutions in Europe and North America in order to bring about improvements in integrated pest management, organic input technologies, and post-harvest handling techniques (Yadav et al. 2020). The recognition of the equivalence of standards between India and the EU, plus USDA-India organic imports, is one example of how bilateral agreements on organic standards can promote trade and eliminate non-tariff barriers, thus ensuring that certified products are in conformity with the requirements of the international market (Willer and Lernoud 2019). Collaborative efforts such as these not only enhance the scientific and technical aspects but also the institutional ecosystem for organic horticulture. As a result, countries can utilise the expertise available worldwide while at the same time adapting the practices to their local agroecological and socio-economic conditions.

4.7 Challenges and Constraints

4.7.1 Yield Gaps and Productivity Concerns

One of the main problems in organic horticulture is the difference in yield compared to conventional systems. Organic methods during the transitional stage to full organic farming can be 10–25% less productive than the conventional ones, especially for high-input horticultural species such as tomato, capsicum, and citrus (Seufert et al. 2012). The reduction in yield is largely due to the slow release of nutrients, the prohibition of synthetic fertilisers, and the improvement of soil quality. Some crops are more vulnerable to nutrient restrictions in organic systems. For instance, the fertility requirements of fruiting vegetables often demand accurate Nitrogen and Potassium management, which is difficult when relying only on organic sources. Furthermore, organic farming practices depend mainly on biocontrol for pests and diseases, which may not always be effective in highly infested situations (Reganold and Wachter 2016). Studies recommend that, over time, soil fertility improvement, integrated organic nutrient management, and enhanced biodiversity can lead to the almost total closing of the yield gap. Sometimes, the organic systems even outdo the conventional ones during periods of environmental stress like drought, owing to the higher soil organic matter content and the consequently better water retention (Gattinger et al. 2012). However, maintaining consistently high productivity is still a major concern for the commercial viability of horticulture, especially for the small-scale farmers.

4.7.2 Limited Awareness and Training Gaps

The successful adoption of organic practices is highly dependent on awareness and technical skills; nevertheless, there are significant training issues that the farmers have to deal with first. The majority of smallholders are not familiar with organic nutrient management, biocontrol techniques, crop rotation strategies, and certification requirements, according to Yadav et al. (2020). Less than perfect training coupled with one's inexperience can be a cause of producing poor quality products and even the unavailability of the whole process. Improper composting, wrong usage of biofertilisers, and not timely pest control can all contribute to reduced yield and poor quality products. Besides, the farmers may not grasp the record-keeping requirements for certification fully, and thus will fall into non-compliance or will be rejected during the audits (APEDA 2023). Governments, NGOs, and research institutions should work together on capacity-building initiatives. The training modules of the PKVY programme, on-farm demonstrations, and farmer field schools are some of the initiatives that have been successful in increasing knowledge and adoption rates. However, the issue of getting to remote areas and marginalised groups still lingers on, thus calling for the implementation of new methods, namely, digital advisory services, mobile apps, and online training platforms (Sharma et al. 2021).

4.7.3 Certification, Infrastructure, and Supply Chain Issues

The growth of organic horticulture is limited by system-level constraints, which include certification, infrastructure, and logistics. The process of third-party certification can be very difficult and costly, especially for smallholders, who are the ones who control the fragmented landholdings. Even if the PGS system is in place, the quality of the crops, the ability to track them, and the consumer trust would still require the strongest monitoring mechanisms (IFOAM 2019). Additional challenges are created through the post-harvest and supply chain infrastructure. The produce from the organic horticulture process is very perishable, yet in most of the producing areas, the cold chain facilities, storage, and transportation are not developed properly (Kumar et al. 2020). If the product is not handled properly, it might deteriorate, get contaminated, or lose its market value. Moreover, the lack of market linkages and the aggregation of smallholder produce will further impede the growth of the organic sector. Erratic supplies, the production of very small lot sizes, and the lack of appropriate distribution networks make it difficult to get into the better domestic and export markets (Scialabba and Müller-Lindenlauf 2010). Strengthening farmer cooperatives, building up warehouse infrastructures, and creating digital market platforms can help overcome these challenges at least to some extent. On top of that, risks of non-compliance with regulations and different standards in various countries complicate international trade, which sometimes results in the rejection of consignments or financial losses. Quality monitoring, the establishment of residue testing labs, and the implementation of digital traceability systems are the necessary activities to cope with these challenges and make the market reliable.

4.8 Case Studies and Success Stories

4.8.1 Organic Horticulture Clusters in India

The Indian government has taken a great leap forward in the promotion of organic horticulture through cluster-based strategies, which provide a full bundle of technical support, certification, and market access, and have been particularly important in the development of these clusters. The Paramparagat Krishi Vikas Yojana (PKVY) has facilitated the establishment of these clusters for high-value crops like vegetables, spices, fruits, and medicinal plants in the first place (DAC & FW 2020). A case in point is the Sikkim organic initiative, where the whole state has become a hundred percent organic agricultural area with the addition of horticultural produce. The availability of marketing and policy support, participatory certification (PGS-India), farmer training, and supply of organic inputs are some of the factors that helped achieve this (Dorji et al. 2020). The project shows that when government support is provided, the community is involved, and the right knowledge is disseminated, then the horticultural production systems can be transformed, and even the ecological resilience might be increased. In Madhya Pradesh and Rajasthan, the vegetable and spice clusters have created a successful scenario of the organic practices, which adoption has led to further soil health improvements, input costs decline, and market returns rise. As the clusters are making use of centralised composting units, community biocontrol nurseries, and coordinated marketing platforms that cut down transaction costs and open up routes to premium markets (Yadav et al. 2020).

4.8.2 Global Best Practices

Innovatively, many countries around the world have gone to the extent of offering models that can be both environmentally friendly and economically smart through their approaches in organic horticulture. Cuba is the most cited case of urban and peri-urban organic agriculture, which has been growing since the economic crisis of the 1990s. The production of Cuban horticulture majorly relies on organic fertilisers, biofertilisers, and integrated pest management, with the activities taking place mostly in small urban gardens and cooperatives. This model is evidence that even in conditions where resources are scarce, organic horticulture can help in food security, farmers' reliance on imported inputs can be cut down, and biodiversity can be increased (Altieri et al. 2012). One of the most notable examples of commitment to organic agriculture is Bhutan's tagline of "100% organic" in horticulture crops, which is included in the country's national policy. The use of organic inputs is covered by a government subsidy, and the state also guarantees the production of organic seeds, and farmers are facilitated in training and getting certified. Bhutan's emphasis on the role of government-backed frameworks and political will in the large-scale adoption of organic horticulture is indeed a valuable lesson (Dorji et al. 2020).

Sikkim, which has been mentioned earlier, is the best state-level organic transformation in India that has taken place. Its success is based on integrated approaches which include the farmer clusters, PGS certification, input supply networks, and value chain development, leading to better farmer incomes and greater environmental sustainability (Dorji et al. 2020). The above world and local instances are testimony to the need for policy integration, capacity building, and participatory models in making organic horticulture a large-scale adoption.

4.8.3 Farmer-Led Innovations and Community Models

Innovation by Farmers leads the way in changing the organic practices according to the local agroecological conditions. In India, communities have come up with low-cost vermicomposting units, botanical pesticide formulations, and intercropping models that increase productivity without disturbing ecological balance (Pathak and Ram 2013). Community-driven cooperatives in Kerala, Himachal Pradesh, and Uttarakhand have developed organic spice and medicinal plant clusters, focusing on joint marketing, PGS certification, and communication between networks of knowledge. These schemes minimise personal risk, grant entry to high-priced markets, and bring about unity amongst people.

Worldwide, innovation networks of farmers have appeared in Latin America and Africa, where small-scale farmers are sharing locally adapted organic practices, pest management solutions, and crop diversification techniques. These top-down models not only make the farm resistant to challenges but also increase its output, and at the same time, the environment is not polluted due to the merging of old knowledge with scientific input (Pretty and Bharucha 2015). Besides, these farmer-led innovations and community-based models have indeed proved that scaling organic horticulture is not entirely a matter of policy or technology; social organisation, local knowledge, and participatory approaches are also the factors that determine the sustainable outcome.

4.9 Future Prospects and Strategic Roadmap

4.9.1 Scaling Up Organic Horticulture: Policy and Practice

Integrated strategies comprising policy support, technological innovation, farmer capacity building, and market development are needed for the scaling of organic horticulture. Policies should offer incentives for converting the land, provide support for the process of certification, create strong market connections, and do all this while reducing barriers for smallholders (Willer and Lernoud 2019).

Strategies for scaling include the following:

- *Cluster-based models* that consolidate resources, inputs, and produce, reducing costs and facilitating access to markets.
- *Farmer Producer Organisations (FPOs)* that enhance collective bargaining, certification management, and value addition.
- *Public–private partnerships* to support infrastructure, research, and digital traceability systems for organic horticultural products.
- *Research and innovation* for developing high-yielding, disease-resistant varieties suitable for organic cultivation, along with sustainable input technologies.

Talents from all over the world, like the example of Bhutan's national organic policy and Cuba's urban horticulture networks, confirm the necessity of political commitment, participatory planning, and community involvement in getting large-scale adoption (Dorji et al. 2020; Altieri et al. 2012). A systems approach is the only way to go to upscale organic horticulture; it has to be done through connecting farm practices with government policies, supply chains, and markets that will guarantee sustainability, profit, and equal nature.

4.9.2 Role in Climate Change Mitigation and Adaptation

Organic horticulture is a big contributor to climate change mitigation and adaptation practices through its focus on soil health, biodiversity, and ecological balance. The combination of these practices, such as organic fertilisation, composting, mulching, and integrating agroforestry, not only enables soil carbon sequestration but also decreases greenhouse gas emissions as well as improves water-use efficiency (Gattinger et al. 2012). Improved soil organic matter gives more resistance to drought and also stabilises yields during extreme weather conditions. Polyculture and crop rotation systems add to the ecosystem's resilience, hence making it less vulnerable to pests and diseases, which are made worse by climate change variability (Altieri et al. 2015). Besides that, organic horticulture does not make use of fossil-fuel-based synthetic inputs, which in turn leads to less indirect carbon emissions associated with conventional production. The organic practices that promote climate-smart horticulture are supportive of both mitigation and adaptation objectives that are in line with SDG 13 (climate action) and wider environmental sustainability goals (FAO 2021).

4.9.3 Linking Organic with Food Security and Rural Development

The usage of organic horticulture as a strategy to increase food security and improve rural living conditions has been suggested, especially in developing countries. Organic horticulture attracts attention by its various ways of production through the nutrient-rich produce and by its connection to local markets, which lead to nutritional security and dietary diversity (Pretty and Bharucha 2015). The smallholder's integration into the organic value chain promotes fair income distribution, labour opportunities in the rural areas, and attracts the young generation into farming. Moreover, the farmer groups and cooperatives facilitate access to knowledge, technology, and financial support, which eventually leads to less rural vulnerability and less migration. In addition, the growing of organic vegetables and fruits is in accordance with environmentally friendly practices, thus preventing soil fertility and water shortages for the coming generations. The interaction of economic viability, ecological sustainability, and social justice has made organic horticulture a major player in the areas of rural development and global food security alike (Scialabba and Müller-Lindenlauf 2010).

4.10 Conclusion

4.10.1 Summary of Lessons Learned

Organic horticulture is an all-encompassing method of sustainable farming that focuses on ecological harmony, soil vitality, diversity of species, and justice for all. In this chapter, we have pointed out a number of major findings:

1. *Foundational Role of Soil and Nutrient Management*: Healthy, biologically active soils are the cornerstone of organic horticulture, supported by organic amendments, compost, vermicompost, and biofertilisers (Doran and Zeiss 2000; Lehmann and Kleber 2015).
2. *Ecological Pest and Disease Management*: Integrated use of biocontrol agents, botanical pesticides, cultural practices, and resistant varieties ensures productivity while safeguarding environmental health (Gurr et al. 2012; Pimentel 2005).
3. *Biodiversity as a Production Strategy*: Crop rotation, polyculture, and agroforestry enhance ecosystem services, increase resilience, and stabilise yields (Altieri 1999; Tuck et al. 2014).
4. *Certification and Market Credibility*: National and international standards, along with participatory systems like PGS, ensure trust, facilitate market access, and create premium opportunities (IFOAM 2019; APEDA 2023).
5. *Economic Viability and Market Potential*: Despite initial yield gaps, organic horticulture offers long-term profitability, market premium opportunities, and export potential, particularly when combined with branding and value addition (Seufert et al. 2012; Willer and Lernoud 2019).

6. *Policy and Institutional Support*: Government schemes, research institutions, NGOs, and global collaborations are essential for scaling, technical guidance, and infrastructure development (DAC & FW 2020; Dorji et al. 2020).
7. *Global Best Practices and Farmer-Led Innovations*: Lessons from Cuba, Bhutan, Sikkim, and farmer-led community models demonstrate the importance of participatory approaches, knowledge exchange, and locally adapted solutions (Altieri et al. 2012; Pretty and Bharucha 2015).

In general, organic horticulture has proven that sustainability, profitability, and ecological integrity are not only compatible but can be achieved together when the technical, institutional, and social aspects are merged.

4.10.2 Way Forward for Research, Policy, and Practice

The fate of organic horticulture is contingent on the unending flow of innovations, strong policy backing, and efficient sharing of knowledge. Amongst the main recommendations are:

1. *Research and Development*: Focus on developing high-yielding, disease-resistant varieties suitable for organic systems, biofertilisers, natural pest control agents, and post-harvest technologies. Long-term studies should assess ecosystem services, soil carbon dynamics, and nutrient cycling.
2. *Policy Integration*: Strengthen incentives, certification systems, and value chain development, particularly for smallholders. Support cluster-based models, FPOs, and technology-enabled traceability systems to improve market access.
3. *Capacity Building*: Expand farmer training programmes, digital advisory services, and participatory learning platforms to bridge knowledge gaps. Emphasise both technical and managerial skills for organic production, certification, and marketing.
4. *Market and Value Chain Development*: Promote branding, value addition, eco-labelling, and export-oriented initiatives to capture premium markets. Develop infrastructure for storage, cold chain, and logistics to reduce post-harvest losses.
5. *Climate and Sustainability Focus*: Position organic horticulture as a climate-smart and socially inclusive strategy, contributing to SDGs, climate mitigation, and resilient food systems.

The application of such strategies in unison can bring about organic horticulture as a revolutionary pathway that connects environmental sustainability, rural areas' development, and global food security, thus providing a win-win situation for both the producers and the consumers over the long term.

References

Acosta-Martinez V, Zobeck TM, Allen V (2004) Soil microbial, chemical and physical properties in continuous cotton and integrated crop–livestock systems. Soil Sci Soc Am J 68(6):1875–1884

Albiach R, Canet R, Pomares F, Ingelmo F (2001) Organic matter components and aggregate stability after the application of different amendments to a horticultural soil. Bioresour Technol 76(2):125–129

Altieri MA (1999) The ecological role of biodiversity in agroecosystems. Agric Ecosyst Environ 74(1–3):19–31

Altieri MA, Funes-Monzote FR, Petersen P (2012) Agroecologically efficient agricultural systems for smallholder farmers: contributions to food sovereignty. Agric Ecosyst Environ 149:1–13

Altieri MA, Nicholls CI, Henao A, Lana MA (2015) Agroecology and the design of climate-resilient farming systems. Agron Sustain Dev 35(3):869–890

APEDA (2023) https://www.pib.gov.in/PressReleaseIframePage.aspx?PRID=1901572&reg=3&lang=2

Bernal MP, Alburquerque JA, Moral R (2017) Composting of animal manures and chemical criteria for compost maturity assessment. A review. Bioresour Technol 100(22):5444–5453

Brady NC, Weil RR (1996) The nature and properties of soils

Bockheim JG, Hinkel KM, Nelson FE (2003) Predicting carbon storage in tundra soils of arctic Alaska. Soil Sci Soc Am J 67(3):948–950

Boydston RA, Vaughn SF (2002) Alternative weed management systems control weeds in potato (Solanum tuberosum). Weed Technol 16(1):23–28

Case AN (2018) The organic profit: Rodale and the making of marketplace environmentalism. University of Washington Press

DAC & FW (2020) Paramparagat Krishi Vikas Yojana (PKVY) guidelines. Department of Agriculture & Farmers Welfare, Ministry of Agriculture and Farmers Welfare, Government of India

DAC & FW (2021) Mission organic value chain development for north eastern region (MOVCDNER) guidelines. Ministry of Agriculture and Farmers Welfare, Government of India

Davis JR, Huisman OC, Westermann DT, Hafez SL, Everson DO, Sorensen LH, Schneider AT (2001) Effects of green manures on Verticillium wilt of potato. Phytopathology 91(6):548–556

Doran JW, Zeiss MR (2000) Soil health and sustainability: managing the biotic component of soil quality. Appl Soil Ecol 15(1):3–11

Dorji T, Chettri N, Sharma E (2020) Organic agriculture in Bhutan: policies, progress, and prospects. Mt Res Dev 40(1):R1–R12

Edwards CA, Grove TL, Harwood RR, Colfer CP (1993) The role of agroecology and integrated farming systems in agricultural sustainability. Agriculture, Ecosystems & Environment 46(1-4):99–121

Eghball B (2002) Soil properties as influenced by phosphorus- and nitrogen-based manure and compost applications. Agron J 94(1):128–135

FAO (2021) The state of food and agriculture: making agrifood systems more resilient to shocks and stresses. Food and Agriculture Organization of the United Nations

FAO & WHO (2001) Guidelines for the production, processing, labeling and marketing of organically produced foods (CAC/GL 32-1999). Codex Alimentarius Commission

Fernandez RM (2019) SDG3 good health and well-being: integration and connection with other SDGs. In Good health and well-being (pp. 629-636). Cham: Springer International Publishing

FiBl (2024). https://www.fibl.org/en/shop-en/1747-organic-world-2024

FiBL & IFOAM (2024) https://www.ifoam.bio/news/global-organic-area-continues-grow-2025

Gattinger A, Müller A, Haeni M, Skinner C, Fliessbach A, Buchmann N et al (2012) Enhanced top soil carbon stocks under organic farming. Proc Natl Acad Sci 109(44):18226–18231

Gebreyes G (2019) Soil organic matter and its role in soil health and crop productivity improvement. Acad Res J Agric Sci Res 7(7):475–483

Gomiero T (2018) Soil degradation, land scarcity and food security: reviewing a complex challenge. Sustainability 10(1):1–36

Gregorich EG, Drury CF, Baldock JA (2001) Changes in soil carbon under long-term maize in monoculture and legume-based rotation. Can J Soil Sci 81(1):21–31

Gurr GM, Wratten SD, Snyder WE (2012) Biodiversity and insect pests: key issues for sustainable management. Wiley-Blackwell

Home P, Kumar RP, Ganguly A, Saha B, Milano-Foster J, Bhattacharya B, Paul S (2017) Genetic redundancy of GATA factors in the extraembryonic trophoblast lineage ensures the progression of preimplantation and postimplantation mammalian development. Development 144(5):876–888

Hughner RS, McDonagh P, Prothero A, Shultz CJ, Stanton J (2007) Who are organic food consumers? A compilation and review of why people purchase organic food. J Consum Behav 6(2–3):94–110

IFOAM (2019) Participatory guarantee systems (PGS): guidelines. International Federation of Organic Agriculture Movements

IFOAM (2020) The world of organic agriculture: statistics and emerging trends 2020. IFOAM & FiBL

Indian Farming (2025) ICT tools for dissemination of farming system technologies. ICAR Publication

Javaid M, Haleem A, Singh R et al (2022) Applications of artificial intelligence in sustainable agriculture. Elsevier

Kumar R, Pathak H, Ram H (2020) Organic vegetable production and its economic viability in India. J Org Syst 15(1):27–39

Lehmann J, Kleber M (2015) The contentious nature of soil organic matter. Nature 528(7580):60–68

Liebman M, Davis AS (2000) Integration of soil, crop and weed management in low-external-input farming systems. Weed Res 40(1):27–47

Magdoff F, Amadon J (1980) Yield trends and soil changes resulting from nitrogen fertilization of continuous corn. Agron J 72(6):913–916

Magdoff F, van Es H (2000) Building soils for better crops, 2nd edn. Sustainable Agriculture Network

Magdoff F, Lanyon L, Liebhardt W (1997) Nutrient cycling, transformations, and flows: implications for a more sustainable agriculture. Adv Agron 60:1–73

Mojtahedi H, Santo GS, Hang AN, Wilson JH (1991) Suppression of root-knot nematode populations with rapeseed as a green manure. Plant Dis 75(3):301–304

Nelson E, Abell R, Goldstein J (2016) Participatory guarantee systems: transforming smallholder organic certification. Ecol Soc 21(2):31

Nemes N (2009) Comparative analysis of organic and non-organic farming systems: A critical assessment of farm profitability. Food and Agriculture Organization of the United Nations, Rome 33

Paddock L (2023) SDG 12: responsible consumption and production. Env't L Rep 53:10133

Pathak H, Ram H (2013) Organic horticulture: principles and practices. J Hortic Sci 8(2):45–60

PGS-India (2022) https://www.pgsindia-ncof.gov.in/

Pimentel D (2005) Environmental and economic costs of the application of pesticides primarily in the United States. Environ Dev Sustain 7(2):229–252

Prashanth P, Reddy MJM, Ray P (2025) Export Opportunity and Challenges in Organic Farming. In Organic Production of Vegetable Crops (pp. 51–72). Apple Academic Press

Pretty J, Bharucha ZP (2015) Integrated pest management for sustainable intensification of agriculture in Asia and Africa. Insects 6(1):152–182

Reganold JP, Wachter JM (2016) Organic agriculture in the twenty-first century. Nat Plants 2:15221

Rodale Institute (2011) https://rodaleinstitute.org/science/articles/

Sanchez JE, Willson TC, Kizilkaya K, Parker E, Harwood RR, Paul EA (2001) Enhancing the quality of soil with alternative and organic amendments: influence on early tomato growth and development. Biol Agric Hortic 18(4):355–370

Scialabba M (2013) Le stagioni delle riforme amministrative: dal New Public Management all'Open Government.

Scialabba NE, Müller-Lindenlauf M (2010) Organic agriculture and climate change. Food and Agriculture Organization of the United Nations
Seufert V, Ramankutty N, Foley JA (2012) Comparing the yields of organic and conventional agriculture. Nature 485(7397):229–232
Sharma A, Kumar V, Kaur R (2021) Blockchain-based traceability for organic supply chains. J Clean Prod 296:126497
Studdert GA, Echeverría HE, Casanovas EM (1997) Crop-pasture rotation for sustaining the quality and productivity of a Typic Argiudoll. Soil Sci Soc Am J 61(5):1466–1472
Tewary A, Mishra A, Sharma I (2025) Green packaging: exploring biodegradable and eco-friendly composite materials. In polymers and composite materials for packaging: smart food packaging and solutions (pp. 247–272). Singapore: Springer Nature Singapore
Tronsmo A, Hjeljord L (1998) Biological control of plant diseases using Trichoderma spp. In: Bay-Petersen J (ed) Proceedings of the regional workshop on biological control, pp 68–75
Tuck SL, Winqvist C, Mota F, Ahnström J, Turnbull LA, Bengtsson J (2014) Land-use intensity and the effects of organic farming on biodiversity: a hierarchical meta-analysis. J Appl Ecol 51(3):746–755
USDA (2023) National Organic Program (NOP). United States Department of Agriculture. https://www.ams.usda.gov/nop
Vogl CR, Kilcher L, Schmidt H (2005) Are standards and regulations of organic farming moving away from small farmers' knowledge?. Journal of Sustainable Agriculture 26(1):5–26
Weil RR, Brady NC (2017) The nature and properties of soils, 15th edn. Pearson
Weller DM (1988) Biological control of soilborne plant pathogens in the rhizosphere with bacteria. Annu Rev Phytopathol 26:379–407
West TO, Post WM (2002) Soil organic carbon sequestration rates by tillage and crop rotation: a global data analysis. Soil Sci Soc Am J 66(6):1930–1946
Wiley Online Library (2023) https://onlinelibrary.wiley.com/
Willer H, Lernoud J (2019) The world of organic agriculture: statistics and emerging trends 2019. Research Institute of Organic Agriculture (FiBL) & IFOAM Organics International
Yadav R, Singh A, Pathak H (2020) Capacity building for organic horticulture in India: challenges and prospects. J Org Syst 15(2):56–69

Hydroponics, Aeroponics, and Aquaponics: The Present and Future

5

Subhadwip Ghorai, Payal Maiti, Sourav Roy, Rajdeep Mohanta, and Shah Faisal

Abstract

Global farming is confronted with increasing threats from climate variability, declining arable land, water shortages, and the challenge of sustaining an increasing population. Soilless farming methods, hydroponics, aeroponics, and aquaponics, pose revolutionary answers by allowing high-yield, resource-conserving, and climate-tolerant crop production. Hydroponics delivers accurate nutrient supply and fast crop cycles in indoor conditions, while aeroponics optimizes oxygenation of roots by means of nutrient misting, offering outstanding water and fertilizer savings. Aquaponics combines fish culture with hydroponics to create closed-loop systems, recycling nutrients and diversifying yields. This chapter explores the science, systems, and world applications of these technologies, from rooftop farms and vertical farm modules in cities to commercial-scale operations. It also reviews economic feasibility, labor and resource efficiencies, and market acceptance, as well as limitations due to high capital costs, technical complexity, and policy gaps. Future directions highlight integration of renewable energy, Artificial Intelligence (AI) and Internet of Things (IoT)-based monitoring, crop-specific breeding for soilless systems and inexpensive modular designs for smallholders. Critically examining opportunities and barriers, this chapter positions hydroponics, aeroponics, and aquaponics as key elements of climate-

S. Ghorai · S. Roy · R. Mohanta (✉)
Department of Agriculture, Brainware University, Kolkata, West Bengal, India

P. Maiti
Department of Post Harvest Technology, Bidhan Chandra Krishi Viswavidyalaya, Nadia, West Bengal, India

S. Faisal
Institute of Food Science and Technology, Chinese Academy of Agricultural Science, Beijing, China

A. Das, A. B. Sharangi (eds.), *Remunerative Horticulture*,
https://doi.org/10.1007/978-981-92-1007-7_5

smart horticulture that can complement traditional farming to improve sustainability, profitability, and food security in the twenty-first century.

Keywords

Hydroponics · Aeroponics · Aquaponics · Sustainability · Soilless cultivation

5.1 Introduction

Global agriculture is faced with unprecedented crises in the twenty-first century. Climate change is impacting weather conditions, leading to irregular rainfalls, intense temperatures, and frequent droughts that undermine traditional agriculture (Zhai and Lee 2024; Lakhiar et al. 2025). Freshwater resources are decreasing because of the impact of growing population and industrial consumption, where agriculture consumes around 70% of freshwater resources worldwide (Mishra 2023). The world population will increase by over 2 billion by 2050, and in turn, demand a revolutionary boost in food production to ensure food security (National Bureau of Statistics of China 2024; United Nations 2018). Moreover, there is rapid urbanization that is reducing available land for cultivation, more than half of the global population living in urban areas, and urban land expansion at the cost of agricultural land (Muchelo et al. 2024). All these require innovative and sustainable horticultural methods to "future-proof" food production (Varzakas and Smaoui 2024).

Soilless cultivation, which encompasses hydroponics, aeroponics, and aquaponics, is one potential strategy. These techniques cultivate crops without traditional soil, utilizing nutrient-dense water solutions or other mediums instead, and provide total control over environmental conditions. Soilless crops yield in smaller areas with much less water than traditional open-field agriculture (Lakhiar et al. 2025). They facilitate cultivation in urban areas (e.g., rooftops, warehouses) and areas with poor soil or extreme climates, underscoring their significance as sustainable horticultural innovations. For instance, hydroponic farms generally consume 90% less water than farms based on soil for the same yield, and are able to produce multiple crops within a year because of conditions that can be controlled (Barbosa et al. 2015). Aeroponics, a more advanced soilless technique involving roots that are sprayed with nutrients, has the potential to lower water consumption by ~98% and fertilizer use by ~60% over soil cultivation while promoting plant growth (Martinez-Mate et al. 2018). Aquaponics combines fish farming and hydroponics, using fish waste as fertilizer for plants and providing both crop and protein crops in an intensified closed-loop system.

This chapter encompasses the science and hydroponics, aeroponics, and aquaponics systems, emphasizing their present uses and future prospects in remunerative horticulture, that is, profitable, sustainable crop production. We present case studies from across the globe (including India) and technologies of integration, such as IoT and AI, that improve system performance. We discuss the economic and environmental feasibility of soilless farming compared to conventional agriculture, and highlight challenges that need to be overcome for increased adoption. Lastly, we consider future trends and research areas that may define the next generation of soilless agricultural systems. By giving a complete overview of the practices, this chapter sets out to show how hydroponics, aeroponics, and aquaponics can support a more resilient and productive horticultural industry in the future and today.

5.2 The Science and Systems of Soilless Cultivation

Soilless culture includes many systems in which plants are grown without the use of soil by providing water, nutrients, and oxygen to the roots in an immediate manner. The main systems include hydroponics, aeroponics, and aquaponics, with differing principles and design specifications (Lakhiar et al. 2025; Kalantari et al. 2018; Eigenbrod and Gruda 2015). All of these systems provide accurate control over the root-zone environment, which can significantly enhance the efficiency of resource utilization and crop yields. This section discusses the scientific basis and common setups for each system, along with their advantages and limitations.

5.2.1 Hydroponics

5.2.1.1 Definition and Principles

Hydroponics is a soilless growth practice where plants are cultivated with their roots suspended in a water-based solution containing nutrients rather than soil (Fussy and Papenbrock 2022). The major concept is to deliver all the mineral nutrients directly to plant roots in soluble form, in addition to sufficient water and oxygen. By removing soil, hydroponics eliminates soilborne diseases and weeds, and enables growers to closely control the root environment (temperature, pH, nutrient levels). This results in more rapid growth and increased yields in ideal circumstances. Hydroponic systems may be installed indoors or in greenhouses and frequently incorporate recirculating nutrient solutions to conserve water (Sundar et al. 2021). Controlled conditions protect the plants from environmental stresses, allowing for year-round cultivation and greater plant density than in fields (Dsouza et al. 2025). In practice, hydroponics has been demonstrated to reduce water use significantly, reports show water savings of 50–90% over soil agriculture, varying with system and crop (NASA 2006; Elsherbiny et al. 2024). For instance, hydroponic Nutrient Film Technique lettuce needed 62% less water than soilborne lettuce in a Spanish experiment (Martinez-Mate et al. 2018). In the Netherlands, high-tech hydroponic greenhouses recorded as much as 50% savings in water and fertilizer usage without reducing yields (Balliu et al. 2021; van Tuyll et al. 2025). These efficiencies underscore why hydroponics is regarded as one of the pillars of climate-smart, resource-efficient horticulture.

5.2.1.2 Common Hydroponic Systems

There are various hydroponic system designs in common usage, with different mechanisms for providing nutrients and oxygen to plant roots. The main systems are:

A very shallow stream (or film) of nutrient solution constantly runs through channels that slope, passing over the roots of plants held in holes in the channel (Graves 1983; Kumar and Dey 2024). The roots absorb nutrients from the thin layer of water and are also exposed to air, thus well-oxygenated. NFT is well-liked by leafy greens and herbs because it is so simple and water-efficient (Niu et al. 2018). Nevertheless, it is dependent upon an active pump; if flow is stopped, roots dry rapidly.

Hydroponics has been investigated and marketed much more widely than either aeroponics or aquaponics, especially under greenhouse and vertical farming conditions. It is for this reason that the next section offers a more elaborate treatment of hydroponics, while later sections on aeroponics and aquaponics are relatively brief to allow for their developing, if promising, status.

5.2.1.2.1 Deep Water Culture (DWC)

Roots of the plants are hung in a deep tank of nutrient solution, usually with an air stone or diffuser bubbling air to aerate the water (Chaudhary and Anand 2024). DWC (alternatively referred to as the raft system when plants rest on a raft) offers continual access to water and nutrients. It is easy to use and appropriate for leafy greens growing very quickly. The answer needs to be well-oxygenated to avoid root rot. DWC systems are simple for novices and cheap to care for, but temperature and pH in the reservoir need to be regulated to maintain conditions at their best (Zamnuri et al. 2024).

5.2.1.2.2 Ebb and Flow (Flood and Drain)

Plant roots (in many cases, in a medium such as gravel or clay pebbles) reside in a grow bed, which is periodically flooded with nutrient solution and then drained. This cycling (usually on a timer) delivers nutrients to roots during floods and air to roots during drains. Ebb and flow systems are flexible and accommodate bigger plants (tomatoes, peppers) with the assistance of a support medium (Jani et al. 2021). Timing needs to be set correctly to prevent waterlogging or drying out roots.

5.2.1.2.3 Drip Irrigation Hydroponics

Nutrient solution is pumped to the roots of each plant through tiny drip emitters, and surplus solution returns to a reservoir for recirculation (Majsztrik and Owen 2020). Drip systems may be combined with an inert growing medium (e.g., coconut coir, perlite, rockwool) that retains water around roots. They are standard in commercial greenhouse cultivation of crops such as tomatoes and cucumbers, allowing for precise regulation of feeding. Drip hydroponics is scalable and reliable, although emitters may clog if not properly filtered (Qamar et al. 2025).

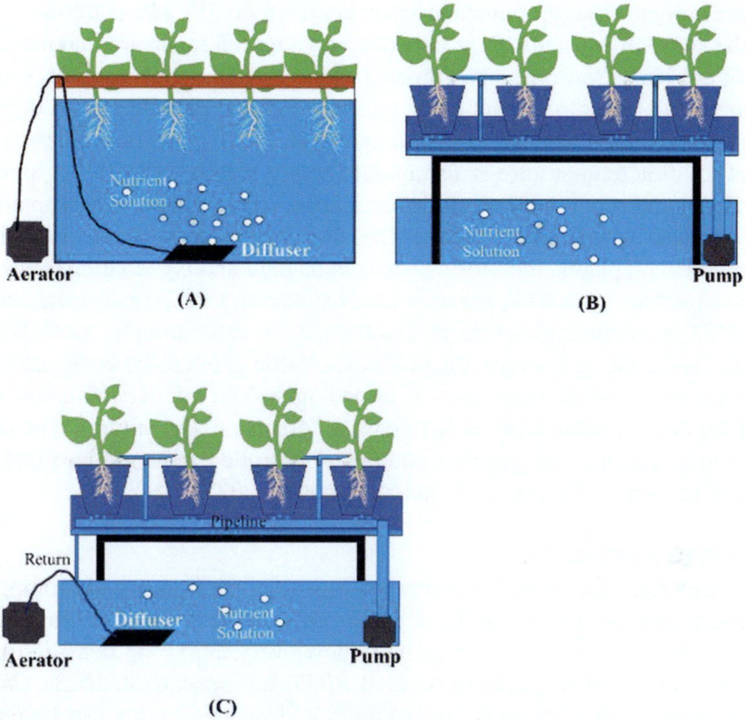

Fig. 5.1 Diagram illustrating three hydroponic systems for plant growth. (**a**) Deep Water Culture: Plants are suspended above a nutrient solution with roots submerged. An aerator and diffuser provide oxygen. (**b**) Nutrient Film Technique: Plants are placed in sloped channels with a thin film of nutrient solution flowing over the roots, circulated by a pump. (**c**) Aeroponics: Plants are suspended with roots exposed to air, periodically misted with nutrient solution. A pump and diffuser circulate the solution

5.2.1.2.4 Wick System

The most basic passive hydroponic system, whereby a wick (usually cloth or felt) extracts nutrient solution from a reservoir into the root zone through capillary action (Schwarz 1995). Wick systems do not require pumps, utilizing the wick's absorption, and are therefore nearly failure-proof but only effective for small, water-efficient plants (herbs, lettuces). They are cheap educational systems but less productive than active hydroponics (Hayati et al. 2024). The diversity of hydroponic designs—ranging from deep water culture to wick and drip systems—can be visualized in Fig. 5.1, which summarizes the most widely adopted approaches.

5.2.1.3 Nutrient Control and Management

An important feature of hydroponics is the control of nutrient solution composition. The plants need the macro-nutrients (N, P, K, Ca, Mg, S) and the micro-nutrients (Fe, Mn, Zn, Cu, B, Mo, etc.) in definite concentrations. In hydroponics, these are provided using formulated nutrient solutions (like the popular Hoagland solution), which are brought to an optimum electrical conductivity (EC) level for the crop. The pH of the solution is usually kept between ~5.5 and 6.5 to ensure maximum availability of nutrients. Sensors and automated dosing systems are commonly employed in recent installations to continuously measure EC and pH and make additions of nutrients or acid/base as necessary (Lakhiar et al. 2018; Sadek and Shehata 2024). Nutrient solution temperature is also maintained (typically ~18–22 °C) for optimal oxygen level and root condition (Rajendran et al. 2024). Because hydroponic roots are not buffered by soil, any imbalance (buildup or deficiency of nutrients) can have a direct impact on plants; monitoring needs to be regular. Oxygenation is also a very important parameter; techniques such as air stones in DWC, or just the cascading flow in NFT, maintain high levels of dissolved O_2 for respiration by roots (Nitu et al. 2024). In recirculating systems, the nutrient solution is filtered and topped off from time to time to avoid the development of pathogens and salt accumulation. In general, hydroponics makes it possible to have a "precision agriculture" type of fertilization and irrigation, quite possibly computer-controlled, to maximize plant growth and use of resources (Ikrang et al. 2022; Reza et al. 2025).

5.2.1.4 Pros and Cons

Hydroponics has many benefits compared to soil cultivation. By separating plants from soil, it prevents soilborne disease and pests, reduces water consumption significantly (through water recycling), and allows very high crop densities and unit area yield (Lakhiar et al. 2025; Korsa et al. 2025; Rajaseger et al. 2023). Crops can more frequently grow quicker; for example, hydroponic lettuce can be ready for harvest in 30 days from transplant, much more quickly than outdoors, enabling several harvests a year (8+ per annum in some hydroponic lettuce NFT farms) (Banboukian et al. 2025). Also, since hydroponics is usually conducted in controlled environments (greenhouses or indoors) (Ragaveena et al. 2021), external conditions such as season or weather have little effect, which allows for constant, year-round production and consistency in quality (Sharma et al. 2025). Numerous hydroponic farms have pesticide-free produce since the closed space and absence of soil decrease pest pressure, which gives value in high-end markets. There are, however, crucial negatives to bear in mind. Hydroponic systems tend to have high initial capital expenditures since they involve infrastructure (pumps, reservoirs, lighting when indoors, buildings) and equipment (Mishra et al. 2024). There can also be high operational expenditures based on energy consumption (for pumps, climate control, and particularly artificial lighting for indoor farms). The systems are technically sophisticated (Sharma et al. 2023) and need expert management, ensuring good nutrient balance, avoiding equipment malfunction, and controlling microbe development in water, all need expertise and watchfulness (Sela Saldinger et al. 2023; Ibrahim et al. 2023). A power or pump failure can kill plants quickly because there is no soil to modulate water availability. Hydroponics also now targets a fairly limited spectrum of crops (leafy greens, herbs, vine crops such as tomatoes) that are profitable; staple cereals are not well adapted to these intensive systems. In conclusion, hydroponics is an efficient, high-control system with enormous potential for productivity and conservation, but with high initial capital costs and management requirements. If executed properly, it can yield more and better quality than traditional farming, but economic viability hinges on attaining enough scale and value in the marketplace to pay for it.

5.2.2 Aeroponics

5.2.2.1 Working Mechanism, Root Suspension and Misting

Aeroponics is a highly sophisticated variety of hydroponics where the roots of the plants are not placed in water or in substrate (Kumar et al. 2023), but are instead suspended in air in an enclosed chamber and sporadically sprayed with a thin mist of nutrient solution. Plants in an aeroponic system are generally suspended with stems/leaves over a misting chamber, with roots hanging loose within the chamber (Lakhiar et al. 2018). A pump-activated misting system (ultrasonic foggers or spray nozzles) introduces a nutrient-containing aerosol to the roots at regular intervals (usually every few minutes). The roots are, therefore, well moistened with the solution and also enjoy great oxygen exposure between rounds of misting. This intense oxygenation of the root-zone is a main benefit of aeroponics; roots are able to absorb oxygen directly from the atmosphere, something that can enhance respiration and nutrient uptake, perhaps speeding up growth rates (Eldridge et al. 2020; Clyde-Smith and Campos 2023). NASA notoriously tested aeroponics for space cultivation and discovered that aeroponic systems can save water by as much as 98% and fertilizer by ~60%, while enabling plants to grow more rapidly (harvest cycles ~45% reduction) compared with soil (Martinez-Mate et al. 2018; NASA 2006). In effect, aeroponics marries the most favorable aspects of both worlds: the accurate feeding of hydroponics with an oxygen-saturated root atmosphere reminiscent of air-pruning and thus capable of promoting robust root growth and fast development.

Practically, aeroponic systems may be high-pressure or low-pressure. High-pressure aeroponics (HPA) employs customized atomizing nozzles to produce extremely fine droplets (< 50 μm), which create a mist that can remain suspended in the air, covering the whole root mass. HPA involves pressurized pumps and highly clean water (to prevent clogging of nozzles) (Lakhiar et al. 2025). Low-pressure aeroponics (LPA), alternatively referred to as "mistponics," employs less-complex sprayers (such as those used in greenhouse misters) to episodically soak the roots with larger droplets; it is cheaper and simpler to install but produces larger droplet sizes that are less likely to penetrate thick root matting (Min et al. 2023; Cai et al. 2023). There are even ultrasonic fogger aeroponic systems that create a nutrient fog (Suganob et al. 2024). These emit nano-scale mist and are mostly utilized in scientific studies or seedling propagation, though scalability is restricted. The basic principle of suspending roots in air and delivering nutrients via fine mist is illustrated in Fig. 5.2, highlighting the distinctive setup of aeroponics compared with hydroponics.

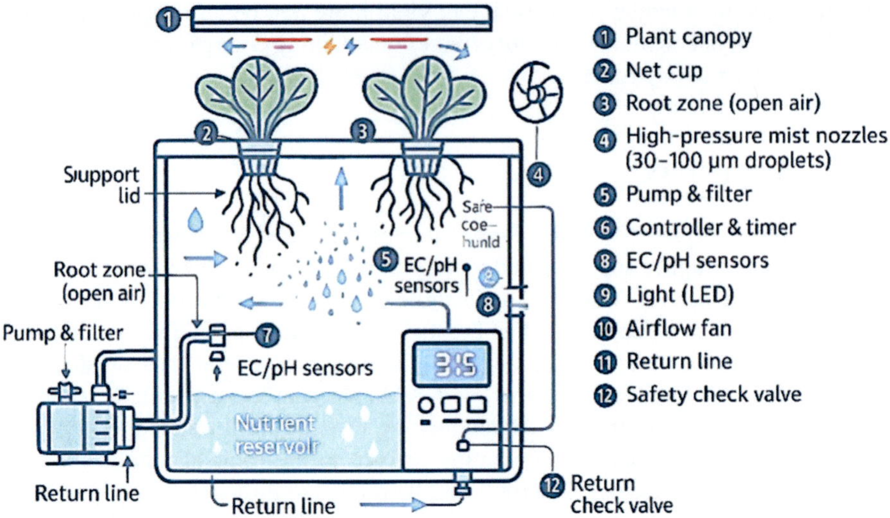

Fig. 5.2 Diagram of an aeroponic system illustrating key components: plant canopy, net cups, open air root zone, high-pressure mist nozzles, pump and filter, controller and timer, EC/pH sensors, LED light, airflow fan, return line, and safety check valve. Nutrient reservoir and support lid are also shown, with arrows indicating water and nutrient flow

5.2.2.2 Advantages Over Traditional Hydroponics

Aeroponics is usually referred to as the most water-efficient and possibly most nutrient-efficient of the soilless systems. Since roots are not immersed in solution all the time, aeroponic systems can utilize much less water, only the small amount that is misted and absorbed (excess can be stored and cycled). NASA research showed that aeroponics might reduce water use by as little as 1–2% of that in standard agriculture for some crops (NASA 2006). Nutrient solution in aeroponics is also utilized sparingly; any excess from misting can be recycled, and nutrient levels can be maintained at lower levels because the delivery to roots is extremely direct (Al-Kodmany 2024). Another important plus is oxygen: roots in aeroponics get access to lots of oxygen 24/7, which can improve root metabolism and health (Lin et al. 2024). This typically means quicker growth and better yields, for example, aeroponically produced lettuces and herbs beat NFT- or DWC-grown ones under optimal misting regimes. Aeroponics also does not need any medium (such as soil, coconut coir), thus eliminating problems of medium expense, disposal, or pathogen carrying in media (Tang et al. 2024). The lack of a medium means transplanting (between aeroponic propagation and another system or field) can be less harsh, as roots are comparatively clean and less entangled. Also, aeroponic systems can be configured vertically (e.g., tower units or wall panels), which can be very high in space-use efficiency. This is utilized in certain vertical farms and "aeroponic towers" for leafy greens (Ptak et al. 2024). Aeroponics is also utilized in some precision farming applications for research, as it facilitates simple sampling of root exudates and root growth observation because the roots are accessible in the chamber (when opened).

5.2.2.3 Technological Requirements and Sensitivity

Aeroponics, with its advantages notwithstanding, is technology-intensive and sensitive to interference (Vettriselvan 2025). These systems rely on pumps, mist nozzles, and frequently sensors/timers, so a power failure or clog will cause roots to dry out rapidly, since there is no water reservoir immediately surrounding them. Aeroponic roots can desiccate in a matter of minutes if misting ceases, particularly in warm temperatures (Lakhiar et al. 2025). System reliability (backup power, clog-resistant nozzles, regular maintenance) is therefore important. High-pressure systems call for water purification (to eliminate particulates and minerals that might clog) and flushing of lines on a routine basis. The equipment and operational expertise required indicate that aeroponics has a greater barrier to entry than simpler hydroponics (Daneshyar 2024). There is limited tolerance to fault as well; even temporary errors in timing or extended nozzle blockage can result in wilting or disease of the plant (Lin et al. 2024) (e.g., roots under drying stress become susceptible to disease). The root chamber microclimate (humidity, temperature) should be controlled to avoid fungal growth or root hypothermia/hyperthermia (Yusuf et al. 2025). Aeroponic systems are usually covered to exclude roots from light (to avoid algae) and prevent drying. That adds complexity, and aeroponics is thus mostly applied in niche applications: in seed potato production (minitubers) in controlled laboratory situations, aeroponics has been extremely effective, producing multiple-fold more potato propagules than traditional approaches (Solankey et al. 2025). But for commercial crop production on a large scale, relatively few aeroponic farms are found because of the expense and risk. It is mostly seen only in research centers, high-tech vertical farms, or small high-tech farms focusing on the greatest efficiency per unit of resources.

5.2.2.4 Case Studies and Precision Farming Applications

Aeroponics has its origins (pun intended) in NASA space farming research, and NASA's discoveries indeed initiated earthly breakthroughs (Badoni et al. 2025). A contemporary example is AeroFarms, a US vertical farm that uses aeroponic misting for leafy greens. In its buildings, plants are cultivated on fabric under LED lights, and their roots are misted with nutrients; AeroFarms claims to use 95% less water than field production and achieve yields up to 390 times higher per square foot per year by layering growth levels (Naresh et al. 2024). Another is in potato seed crop production: aeroponic systems have been utilized by research stations in nations such as China, Turkey, and the USA to produce potato mini-tubers (seed pieces) more productively. Experiments indicated 5 to 6 times more minitubers per plant using aeroponics compared to conventional sand or peat media systems (Tican et al. 2024). This is due to the elevated oxygen environment favoring healthier stolon and tuber development. Aeroponics is also employed to cultivate disease-free plantlets in horticulture, such as the high propagation of strawberry or cassava cuttings, or root development research on several crops (since roots can be reached without soil being dug out) (Tang et al. 2024; Adamek et al. 2024). In precision agriculture research, aeroponic systems enable controlled extraction of experiments on nutrient supply or even on root-zone gas manipulation (such as varying levels of oxygen or CO_2) to study plant reaction.

In short, aeroponics is the state-of-the-art in soilless production. It provides better water and nutrient efficiency and stimulates excellent rates of plant growth, but at the expense of technical sophistication and extreme sensitivity to upkeep. It is not yet common in farm-level commercial agriculture, but it has excellent potential for specialty crops and high-value crops and for applications where efficient use of resources is the key. With further technology development (e.g., more consistent nozzles, improved automation and monitoring), aeroponic farming can become mainstream. Now, it is a niche but potent asset in the toolkit of controlled environment horticulture.

5.2.3 Aquaponics

5.2.3.1 Integration of Aquaculture and Hydroponics

Aquaponics is a soilless system that synergistically integrates hydroponic plant growth with aquaculture (fish culture) in an integrated closed-loop system (Chaudhary and Anand 2024; Nair et al. 2025). Fish (or other aquatic animals such as prawns) are cultured in tanks in an aquaponic system, and their wastewater is pumped to hydroponic grow beds where plants are cultivated (typically in water or substrate). The bacteria and plants subsequently filter and purify the water, which is cycled back to the fish tanks (Rakocy 2012). Essentially, aquaponics produces a mini-ecosystem: fish waste is fertilizer for the plants, and plants (and microbes) clean water for the fish (Rharrhour et al. 2022). The essence of such integration lies in the biological nitrogen cycle and fish–plant symbiosis: Fish produce ammonia (NH_3/NH_4^+) as a metabolic byproduct. At high levels, ammonia is harmful to fish, but naturally occurring nitrifying bacteria (usually colonizing a biofilter within the system) break down ammonia initially into nitrite (NO_2^-) and secondarily into nitrate (NO_3^-) (Ruiz et al. 2020). Nitrate is nitrogen that is easily taken up by plants as a nutrient. Thus, plants serve as biofilters, taking out nitrate (and other nutrients such as phosphate) from the water for nutrition. The outcome is water that is much cleaner (having lower nitrogen compounds) when it is recycled back to the fish tanks, providing a healthier ecosystem for the fish. This is a beneficial cycle for both fish and plants because fish receive freshwater and plants receive fertilized water. Normally, an aquaponics system will also have mechanical filtration parts (to eliminate solid fish waste) and sumps/pumps to pump water between fish and plant parts (Fernández-Cabanás et al. 2020). The integration of aquaculture with hydroponics, mediated by microbial biofilters and nutrient cycling, is schematically represented in Fig. 5.3.

5.2.3.2 System Design and Balancing Requirements

Aquaponics system design involves balancing the requirements of two organisms, fish and plants, which is no easy task. Some of the important parameters are water quality (pH level, temperature, oxygen content, etc.), which must be compatible

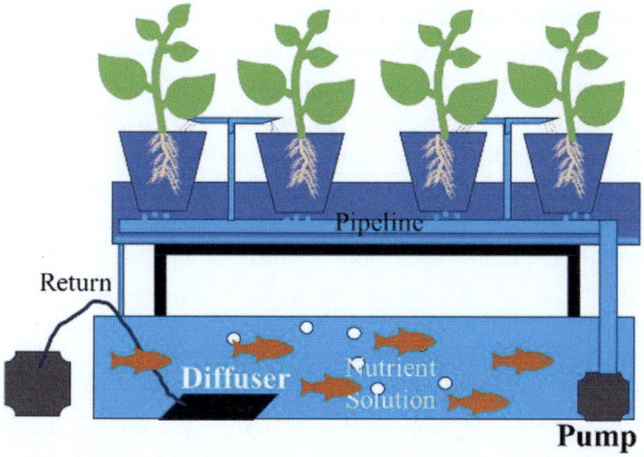

Fig. 5.3 Diagram of an aquaponics system showing plants in pots with visible roots above a water tank. The tank contains fish and is labeled with components such as "Diffuser," "Nutrient Solution," "Pipeline," "Return," and "Pump." The system illustrates the flow of water and nutrients between the fish tank and the plants

with both fish species and plant crops. For instance, most hydroponic crops want to be slightly acidic (Singh et al. 2019) pH of ~6.0–6.5, but nitrifying bacteria work well near neutral pH, and others prefer slightly alkaline water. A compromise pH of ~6.8–7.2 is usually employed to keep everyone relatively happy. Temperature choice will be based on the fish (Vasdravanidis et al. 2022) (tilapia, a popular aquaponics fish, does well in warm water ~25–30 °C, which is also appropriate for tropical vegetables, but trout require cold water). The ratio of components will be important: generally, the fish stocking rate and feed input will be sized so that the waste generated supplies sufficient nutrients to the plants without overloading the system (Yavuzcan Yildiz et al. 2017). A standard rule is 20–40 grams of fish food per day per square meter of plant growing space for leafy greens (Dauda et al. 2019) (this will depend on system details). The system typically has a biofilter (often simply the surfaces in the grow bed or a stand-alone filter unit) to provide healthy bacterial colonies for nitrification. Aquaponic plants may be cultivated in media beds, NFT channels, or deep water rafts, depending on the design (Gupta et al. 2025). Deep water culture beds, also known as deep flow or raft aquaponics, are commonly used for leafy greens. In this system, plants float on rafts placed in a long trough containing fish water, where they absorb nutrients before the water is returned to the fish (Chaudhary and Anand 2024). Media bed aquaponics utilizes gravel or expanded clay on which both bacteria and plant roots live; these are capable of better solids management (the media can trap and mineralize fish waste). Systems may commonly have a settling tank or clarifier to remove excessive solids (fish feces, uneaten feed), which, if not removed, would break down and form anaerobic areas or release ammonia spikes.

5.2.3.3 Applications and Case Studies

Soilless cultivation methods have moved from experimental setups into real-world applications across the globe. In this section, we explore how hydroponics, aeroponics, and aquaponics are being applied in various contexts, from urban rooftop gardens to high-tech vertical farms, and present illustrative case studies. We also look at examples in India and other countries to understand their practical impact on horticulture. These applications underscore the versatility of soilless systems and how they contribute to remunerative horticulture by enabling profitable production in novel settings.

5.2.3.4 Urban Agriculture and Rooftop Farming

Perhaps one of the strongest applications of soilless systems is in urban agriculture (Joshi et al. 2022). Since hydroponics and associated techniques do not involve natural soil, they can be applied on rooftops, in warehouses, or even in transformed shipping containers in city landscapes (Zhou et al. 2024; Gonnella and Renna 2021). Urban hydroponic farms reduce the supply chain by cultivating perishables right where people consume them, keeping the produce fresher and saving on transport costs and emissions. The facilities employ recirculating NFT and DWC systems and have already attained high productivity even in temperate climates using climate control and waste heat from buildings. The second example is in Singapore, where land is limited; the government has encouraged rooftop and vertical farming (Mok et al. 2020). In Singapore, there are many hydroponic rooftop farms situated on shopping mall rooftops or parking garages, yielding leafy greens for local markets. These mini-footprint farms answer a strategic food security objective (Singapore targets "30 by 30"—achieving 30% of its nutritional requirement locally by 2030) (Ministry of the Environment and Water Resources 2019). Even urban dwellers have adopted recreational hydroponics on balconies and rooftops, occasionally employing commercially prepared kits to cultivate vegetables indoors, pesticide-free.

Hydroponic urban agriculture has been economically viable in tapping domestic growers' markets, such as restaurants in large city centers, which tend to pay extra for fresh herbs and green vegetables from domestic growers. In Paris, Agripolis runs a rooftop aquaponic facility producing vegetables and fish, which it sells as high-value "hyper-local" produce. Urban hydroponics is an up-and-coming trend in Indian metros such as Bengaluru, Mumbai, and Delhi. Startups such as UrbanKisaan have installed vertical hydroponic farms in or close to Indian cities and even direct-to-consumer subscription models for fresh greens. UrbanKisaan's vertical farms incorporate stacked hydroponic towers and IoT monitoring; the firm reported consuming 95% less water than traditional farms and raising substantial investment to scale up operations (Barbosa et al. 2015). Even rooftop kitchen gardens are being equipped with compact hydroponic sets for the household, showing increasing acceptance. These urban applications point to the fact that soilless production can be space-saving and profitable by converting concrete jungles into green production areas, in many cases at prime rates.

5.2.3.5 Vertical Farming and Controlled Environment Agriculture (CEA)

Vertical farming involves the growth of crops on vertically inclined surfaces or stacked layers, frequently within fully enclosed indoor facilities using artificial light (a type of controlled environment agriculture). Hydroponics (and aeroponics) form the backbone of vertical farms; as soil would be inconvenient in these systems, soilless methods are employed to supply water and nutrients to vertically aligned plants. Several well-known vertical farms have been developed in the past few years with the goal of providing city dwellers with fresh fruits and vegetables (Oh and Lu 2023). AeroFarms in Newark, USA, cultivates baby greens in a warehouse on stacked beds and uses LED lights. Their practice yields considerably more food than conventional farming and consumes 95% less water (Panotra et al. 2024). Vertical farms represent the high-tech extreme of soilless growing: they incorporate HVAC climate control, computerized irrigation and nutrient feeding, and sometimes AI-based optimization of light spectra and schedules (custom "light recipes" for each plant) (Lakhiar et al. 2025).

The energy requirements are very high (particularly lighting), but vertical farms avoid weather risk and can be located near distribution hubs, saving time for the customer. This can significantly enhance produce quality and shelf-life. Controlled environment agriculture also encompasses greenhouse hydroponics, and in fact, large Dutch and Spanish greenhouse operations employ soilless systems to grow tomatoes, peppers, and cucumbers, on a large scale (Van der Velden et al. 2004). Unlike the majority of Europe, more than 50% of Dutch vegetable production is under glass and utilizes hydroponics or substrate culture (PW Consulting 2024; Tatum 2020). In a part of Spain (Almería), there are approximately 3000 hectares of greenhouses that are soilless, utilizing drip irrigation on substrate to produce high yields in a semi-arid environment (Incrocci et al. 2020). An interesting vertical farming use is in regions with extreme climates (Maheshwari 2021). In the Middle East, National Basket (Saudi Arabia) and Badia Farms (Dubai) have established indoor vertical farms to produce greens under soil less conditions to counter the problem of agriculture in desert climates.

These operate hydroponically under full environmental management and renewable energy incorporation (e.g., solar power) to reduce the high electricity requirements (Pomoni et al. 2023). They facilitate year-round harvest in areas where it is almost impossible to farm outside for most of the year because of heat and lack of water.

The hydroponics system replicates NASA-developed closed environments, recycling water and nutrients with minimal waste. Vertical farming showcases the front edge of controlled environment horticulture.

5.2.3.6 Institutional and Commercial-Scale Operations

Besides urban startups, soilless production has been embraced by research organizations, universities, and big agribusiness firms for diverse reasons (Appicciutoli et al. 2025). Governments and academic organizations typically establish hydroponic or aquaponic facilities as demonstration and test locations to experiment with crop types, nutrient recipes, and integrations of technology. For instance, ICAR–Indian Institute of Horticultural Research (IIHR), Bengaluru, came up with a variation of simplified hydroponics called "cocoponics" based on cocopeat as the substrate, and tried it out on numerous Indian vegetable crops (Ministry of Agriculture and Farmers Welfare 2023). They have educated more than 3000 individuals in soilless cultivation methods in recent years, reflecting increasing interest (Ministry of Agriculture and Farmers Welfare 2023). On the business side, leading greenhouse conglomerates have adopted soilless cultivation to feed mainstream markets. In Europe, Rijk Zwaan and Rooftop Europe run hydroponic greenhouses for commercial-scale production. In North America, BrightFarms constructs supermarket-adjacent greenhouse hydroponic farms to reduce supply chains. And in Asia, particularly Japan, following the 2011 tsunami, numerous plant factories (indoor hydroponic farms) were constructed to provide a secure food source; firms such as Spread Co. operate computer-controlled lettuce factories that yield millions of heads each year.

Notably, commercial usage extends beyond leafy greens. Hydroponic tomato and pepper greenhouses are a fully developed industry. High-wire hydroponic tomatoes grown in rockwool or cocopeat are standard in the Netherlands, Canada, etc., yielding upward of 60–100 kg/m^2 annually of tomatoes, far higher than open-field yields, with precise climate and nutrient control. Almería, Spain, is a case study: while many of its greenhouses continue to utilize some soil or soil blends, many have made the transition to soilless substrate culture in order to eliminate soilborne disease and conserve water (Lakhiar et al. 2025).

This has made Spain a winter vegetable export giant in Europe. In India, adoption on a large scale is in its nascent stages, although there are a few commercial hydroponic farms and agribusinesses concentrating on value-added production (Mohammed et al. 2025). For instance, Bengaluru-based Clover is an agritech firm that assists networks of hydroponic greenhouse growers with an emphasis on a reliable supply of exotic greens and vegetables to urban retailers (Nuthalapati and Nuthalapati 2021). Clover and others have raised venture capital, finding scope for "precision agriculture" in Indian horticulture. Farmers' Hoop and Future Farms are companies establishing hydroponic polyhouse farms in India, reaching out not only to large cities but also to exporting residue-free high-quality produce (Noor et al. 2025).

On the aquaponics front, certain Indian entrepreneurs (e.g., Naik Aquaponics in Baramati) have established demonstration farms to popularize integrated fish-vegetable farming as a rural-friendly model. Technological integration is one of the reasons soilless culture has developed so quickly in the last few years is technological integration, ranging from automation and sensors to data analytics and sophisticated lighting. Hydroponic, aeroponic, and aquaponic systems lend themselves most naturally to high-tech control since they are contained and modular. This can vastly improve efficiency, cut labor, and provide the best conditions for plant growth (Anwarul 2025).

Here, we present a number of critical technological integrations: the function of AI, IoT, and automation, smart nutrient delivery and environmental monitoring innovations; remote sensing and data analytics to optimize the system, and lighting technology breakthroughs (LED spectra tuning).

These advances are turning soilless farms into state-of-the-art production systems commonly referred to as "smart farms" or "precision horticulture" centers. Role of AI, IoT, and Automation: The Internet of Things (IoT), a network of devices and sensors, has emerged as a pillar of contemporary soilless agriculture (Karothia and Chattopadhyay 2024). Farmers can monitor key parameters continuously by installing IoT sensors across a hydroponic/aeroponic system: nutrient solution pH, EC (nutrient level), water temperature, dissolved oxygen, ambient temperature, humidity, light intensity, CO_2 level, and many more (Rahman et al. 2024). These sensors provide inputs to central control systems (usually through cloud platforms), enabling real-time monitoring and alerts.

Automation intervenes through the ability of controllers or algorithms to modify system elements according to sensor readings.

For instance, when pH deviates from the set point, an automatic dosing pump can introduce acid or base to restore it; when the EC of nutrients falls as plants absorb nutrients, the system can introduce fertilizer stock solution to support the desired EC. Climate control (heaters, coolers, humidifiers, etc.) can also be automated to ensure greenhouse or grow-room conditions are held at desired set points (Stasiak and Dixon 2022). Artificial Intelligence (AI) and machine learning are being increasingly employed to process the huge sensor data, as well as visual data from cameras on the farm (Jha et al. 2019). AI can assist in picking up on slight patterns or anticipating problems before they are fully realized. For example, machine learning models can forecast the development of nutrient deficiencies or disease through analysis of sensor trends or leaf color photographs and triggering preemptive changes (Ehsan and Riedl 2024). AI algorithms may also be used to maximize resource utilization: for instance, dynamically adjusting light schedules and intensities to minimize energy consumption but satisfy plant growth requirements (some vertical farms utilize AI to dynamically adjust LED output based on plant growth phase and instantaneous photosynthesis rate). Another use is in robotics: robot-based systems can be applied in seeding, transplanting, and harvesting in some large-scale hydroponic greenhouses (Rodríguez-Nieto et al. 2025).

AeroFarms, for example, employs robotic harvesters and automated conveyors to manage their grow trays, and automation of packing to save on labor expenses (Soofastaei 2025). An example of technological integration is offered by a research study that developed an intelligent NFT and aeroponic greenhouse system employing IoT sensors (Dutta et al. 2025). The IoT system supported real-time monitoring and management of nutrient concentration, temperature, and others, allowing for early anomaly detection and automatic reaction (e.g., adjusting nutrient supply) (Pandey et al. 2025). The result was remarkable: the intelligent-controlled system saved 80–90% water and fertilizer and doubled production per unit of area, reducing the time of the growth cycle by 40% over traditional practices (Sadek and Shehata 2024; Dutta et al. 2025). This shows that automation and accurate control can dramatically improve efficiency.

Table 5.1 An illustration of soilless farming scenario of different countries with key examples, highlighting features and references

Country/region	Technology focus	Key examples	Highlights/outcomes	References
Netherlands	Hydroponics (greenhouses)	High-wire tomato and cucumber production	>50% vegetables under glass; yields 60–100 kg/m^2 annually	Balliu et al. (2021) and van Tuyll et al. (2025)
USA	Aeroponics and vertical farming	AeroFarms (Newark), bowery farming	95% less water; up to 390× yield per area; robotics & AI integration	Naresh et al. (2024) and Kabir et al. (2023)
Singapore	Rooftop and vertical hydroponics	Sky greens, rooftop farms	"30 by 30" food security plan; rooftop food production	Mok et al. (2020)
Spain (Almería)	Substrate hydroponics in greenhouses	~3000 ha soilless greenhouses	Efficient vegetable export system in semi-arid climate	Incrocci et al. (2020)
China	Aquaponics and CEA	Research stations, urban farms	Renewable energy integration; megacity space-saving production	Zhou et al. (2024) and Ghiasi et al. (2023)
India	Hydroponics and aquaponics (emerging)	UrbanKisaan, clover, Naik aquaponics	Urban farms and startups scaling NFT/DWC; subsidies under Mission for integrated development of horticulture (MIDH)/agriculture infrastructure fund (AIF)	Barbosa et al. (2015) and Mohammed et al. (2025)
Middle East (UAE, Saudi Arabia)	Vertical farms, hydroponics	Badia farms, national basket	Desert agriculture solution; solar-powered vertical farms	Pomoni et al. (2023) and Maheshwari (2021)

Bowery Farming has its own farm operating system based on sensors and AI to operate vertical farms with minimal human labor (Kabir et al. 2023). The system continues to learn and adapt to cultivate every crop better with fewer resources. Several countries have successfully implemented soilless systems under diverse contexts, ranging from high-tech vertical farms to rooftop gardens and aquaponic units. Representative global case studies are presented in Table 5.1.

5.2.3.7 Smart Nutrient Delivery and Environmental Monitoring

Soilless system-based nutrient delivery has progressed from manual blending to extremely advanced fertigation systems (Nikolaou et al. 2021). Contemporary hydroponic farms regularly employ fertigation controllers that combine stock solutions of concentrated nutrients with water in precise ratios to attain desired N-P-K and micronutrient concentrations (Fathidarehnijeh et al. 2023; Hasan et al. 2025).

These controllers may tune the mix in real-time through sensor feedback, practicing so-called dynamic dosing. That is, rather than a fixed recipe, the system can decrease the concentration of nitrogen if EC is running too high or if the plants are in a growth phase where less nitrogen is required (such as fruiting stage for tomatoes) (Wahab et al. 2023).

Other systems even tailor nutrient recipes to each zone or crop type on one farm. Environmental monitoring is crucial in soilless cultivation (Fuentes-Peñailillo et al. 2024). Sensors monitor temperature, humidity, and CO_2, with climate control systems maintaining perfect conditions (Mylonas et al. 2019). Remote monitoring allows managers to monitor the farm from phones or computers, see real-time graphs, and receive notifications of issues such as power outages or pump failure (Gupta et al. 2024). It saves manpower and ensures protection, as managers can respond promptly by activating backup pumps or generators if necessary.

Data analysis optimizes agriculture by identifying trends in past data, that is, pH shift or how temperature influences yield (Wen et al. 2025). Smart greenhouses employ predictive analytics, integrating sensor inputs with weather forecasts to control cooling and ventilation. Scientists are also developing digital twins, computer simulations of and optimizations for climate and nutrient approaches, before implementing them in actual systems.

5.2.3.8 Remote Sensing and System Optimization

Big greenhouses increasingly employ remote sensing technologies like infrared cameras, multispectral imaging, and even unmanned aerial vehicles (drones) for keeping an eye on plant health (Jayasuriya et al. 2024). They identify stress, pests, or nutrient stress early on through leaf color or temperature changes, supplementing in-solution sensors by offering a plant-level perspective (Zheng et al. 2023).

Outside data, such as weather forecasts, are also incorporated to maximize operations—such as varying shading or nutrient content on sunny days when transpiration is more elevated or supplementing light on overcast days to ensure light demands for the day.

This method, controlled environment optimization, is based on ongoing, refined adjustments that computerized systems execute far more effectively than people.

The outcome is improved yields, increased quality, and even better nutrition. For example, farms can schedule LED lights to increase intensity or UV exposure in late growth, enhancing antioxidants and flavor or nutritional quality in the ultimate harvest (Loconsole and Santamaria 2021).

5.2.3.9 Innovations in Lighting (LED Spectra Tuning)

Lighting has emerged as one of the most vibrant fields of innovation in controlled environment farming (Neo et al. 2022). With LED technology, growers have the ability to exactly control the spectrum, intensity, and duration of light, as opposed to sunlight's constant spectrum. LEDs initially included mostly red and blue wavelengths, providing a purple hue and focusing on chlorophyll absorption (Stamford et al. 2023; Roosta et al. 2025).

Full-spectrum or adjustable multi-channel LEDs are now used in contemporary systems, enabling growers to make nuances of color to simulate sunlight or induce particular plant reactions.

Various spectra affect not just photosynthesis, but also plant structure and quality (Pashkovskiy et al. 2021). Blue light, for example, yields short, stumpy plants, whereas red light initiates photosynthesis but can lead to stretching if supplied individually. The addition of far-red light induces flowering, and the use of green or white light at or near harvest enhances the appearance for the consumer. In fruiting crops, the red:far-red ratio controls the timing of flowering.

Light cycles are also optimized, such as vertical farms employing cycles such as 18 h of light and 6 h of darkness to moderately enhance productivity (Avgoustaki et al. 2020, 2021; Liu et al. 2022). Energy efficiency is another major concern. LEDs have been continuously improving in photon efficiency (μmol per joule), reducing costs (Liu and van Iersel 2023). Dynamic control systems respond to plant requirements and vary light, providing only the amount needed to achieve a daily target and reducing light when plants are less sensitive (Balasus et al. 2021). This saves energy and minimizes waste.

Interlighting, the placement of LED bars inside tall plant canopies, provides lower leaves with enough light, increasing yields without harming plants, due to LEDs' minimal heat (Bohemen et al. 2025). NASA studies have even influenced the application of LED technology on Earth. Laboratory experiments in closed chambers determined that tailored LED spectra paired with hydroponics could optimize yield and quality (Herrera-Arroyo et al. 2025). Such findings have been implemented by farms such as Green Sense Farms (USA), creating spectral "recipes" for vegetables like basil, increasing productivity as well as flavor through the manipulation of light exposure to change necessary oils.

Such controlled lighting integration and automated nutrient feeding promote quicker growth and cost savings over manual practices. IoT, AI, automation, and smart lighting have revolutionized soilless cultivation, making it possible for small groups to operate extensive, efficient, and data-driven operations. Although resource-intensive, with high skill, capital, and redundancy requirements, such intelligent farms enhance production, quality, and sustainability, representing an important milestone toward scalable, technology-enabled agriculture. Economic and Environmental Viability A critical analysis of hydroponics, aeroponics, and aquaponics needs to evaluate if the aforementioned systems are economically viable and sustainable in the environment against conventional farming (Kumar et al. 2025). In this section, evaluate the input vs. output, labor and resource use efficiency, carbon and water footprint of soilless systems, and their viability to serve organic or premium markets.

We will find that soilless cultivation outperforms in certain aspects (savings in water, efficiency in land use, premium for quality) but underperforms in others (energy-intensive, high capital expenditures) (Fussy and Papenbrock 2022).

Knowledge of these trade-offs is central to ascertaining where and how these systems can be profitable in horticulture. While hydroponics, aeroponics, and aquaponics differ in design and application, they share common goals of improving

Table 5.2 Comparative overview of hydroponics, aeroponics, and aquaponics systems in terms of principles, inputs, efficiency, advantages, and limitations

Parameter	Hydroponics	Aeroponics	Aquaponics
Basic principle	Plants grown in nutrient-rich water (roots partially submerged or supported in inert media)	Roots suspended in air; sprayed/misted with nutrient solution	Integration of aquaculture and hydroponics; fish waste converted into plant nutrients
Water use	70–90% less than soil cultivation	Up to 95–98% less than soil cultivation	~80–90% less than soil; depends on fish–plant balance
Nutrient use	Efficient but requires precise formulation and monitoring	Most efficient; minimal nutrient wastage due to direct misting	Nutrients come largely from fish feed/waste; supplemented when needed
Energy demand	Moderate to high (pumps, climate control, lights if indoor)	High (pressurized misting, strict control, backup systems)	Moderate; energy needed for water circulation, aeration, and climate control
Crop suitability	Leafy greens, herbs, vine crops (tomato, cucumber, capsicum)	High-value crops, propagation material, seed potato minitubers, leafy greens	Leafy greens, herbs, some fruiting vegetables, plus fish protein (tilapia, catfish, trout)
Advantages	Higher yields, year-round production, pesticide reduction, consistent quality	Maximum oxygenation, faster growth, water/nutrient efficiency, clean roots for research/propagation	Dual output (fish + plants), closed-loop sustainability, reduced fertilizer input
Limitations	High initial cost, technical skill required, limited to certain crops	Very high technical sensitivity, nozzle clogging, costly, power-dependent	System balancing is complex, slower crop growth vs. hydroponics, regulatory gaps

input efficiency and ensuring high-quality production. Their key features, advantages, and limitations are summarized in Table 5.2.

5.3 Comparative Analysis of Input Costs Versus Returns

Soilless cultivation, like hydroponics, requires a higher amount of money initially than conventional soil cultivation (Khatri et al. 2024). Farmers need to construct greenhouses or grow rooms, purchase tanks, pumps, pipes, lights, and install climate control and monitoring systems (Ghiasi et al. 2023). Research indicates that hydroponic farms will be 2–10 times more expensive than open-field farms of the same area (Bostancı and Ülger 2022; Banboukian et al. 2025).

A miniature polyhouse with drip hydroponics is more affordable, while a highly technological vertical farm with complete automation is the most costly. This high price is one of the reasons why hydroponics is not widespread among conventional farmers. Maintaining a hydroponic farm also costs more per month. In aquaponics the major costs include electricity to power pumps, lights, and cooling, nutrient solutions, seeds or seedlings, and occasionally fish food. Hydroponics, on the other hand, provides significantly higher yields. For instance, a one-acre soil farm with lettuce will yield X tons in a year.

One acre of a hydroponic greenhouse can yield 3–4 times as much due to multiple harvests within a year. The produce also appears healthier and usually fetches better prices. Most consumers pay 20–50% more for clean, pesticide-free vegetables from greenhouses, particularly during the off-season. Another advantage is a consistent supply. Hydroponic farms do not suffer much from weather, so they can supply crops year-round (Kannan et al. 2022). This simplifies it to sign contracts with hotels and shops and provides a stable income. Hydroponics is most suitable for high-value markets in developing countries such as India. Most farms supply salad greens to hotels and supermarkets, where quality of the supplies are considered more than cheap prices. As land and labor are less expensive in soil farming, hydroponics has to demonstrate its superiority with improved quality and year-round production.

To aid farmers, the government offers subsidies under schemes such as the Mission for Integrated Development of Horticulture (MIDH), for which it subsidizes up to 50% of the installation charge.

5.4 Labor and Resource Efficiency

Soilless agriculture transforms the type of labor by removing repetitive activities like weeding, tilling, and spraying and enabling automation to improve efficiency (Lakhiar et al. 2025; Singh et al. 2025), hence, fewer laborers can handle more areas of production and obtain more yield per hour of work. Meanwhile, labor needs to pivot toward skilled positions—watching systems, controlling sensors, and performing maintenance—which offers scope for higher-paying jobs but is also a problem of retraining old-style farmhands. From a resource perspective, soilless systems are highly efficient in water and nutrient utilization, with hydroponic lettuce requiring only 20–30 liters per kilogram compared to more than 200 liters in the field (Rajaseger et al. 2023), and fertilizers are used more efficiently with no losses.

5.5 Challenges and Limitations

In spite of the promising potential and achievements in hydroponics, aeroponics, and aquaponics, there are several challenges and limitations that, as of now, restrict their general adoption. These obstacles are covered in the principal sections below: high initial capital and complexity, skills and knowledge barriers, regulatory and policy issues, and market acceptance problems. It is necessary to highlight these challenges for the purpose of developing solutions to overcome them, which will be covered in the last section on future prospects.

5.5.1 High Initial Cost and Complexity of Operations

One of the greatest barriers to adoption is the substantial capital requirement for establishing controlled environment structures and hydroponic equipment (Fussy and Papenbrock 2022). Developing controlled environments—from basic shade-net houses to sophisticated indoor farms—needs costly structures, hydroponic equipment, illumination, and control devices. With or without government subsidies, the price per square meter is still high, leaving little room for small-scale farmers or new players. In developing nations, credit access is also an issue, as most farm loans are for traditional needs such as tractors or seeds, not for advanced climate-controlled systems (Embaby 2023). The repayment can take a while unless crops are traded on high margins.

Operational costs create another level of complexity. Energy expenses, particularly for farms with grow lights and cooling systems, are significantly greater than those of traditional farms relying on natural light (Stanghellini and Katzin 2024). Furthermore the soilless systems are more delicate compared to soil. Failure of a pump, nutrient imbalance, or extended power outage can ruin or even kill an entire crop unless redundant systems exist. This involves spare gear, generators, and alarm monitoring, which are more complex and expensive. Soil farming is more forgiving; however, losing a day of irrigation or being late with a fertilizer application seldom destroys a crop (Sever 2022). For most growers, particularly in rural or resource-constrained settings, dealing with these technical risks and dealing with sensitive equipment can prove a significant adoption obstacle.

5.5.2 Skill and Knowledge Barriers

Soilless farming has a steep learning curve. Knowledge applied in conventional soil farming is not entirely transferable, for, in hydroponics, the grower has complete control over all the inputs given to the plant (Fussy and Papenbrock 2022). Plant scientists are what the farmer must become—blending nutrient solutions, measuring pH and EC, and closely working the system. For most, it is new and alien (Rathor et al. 2024). The shortage of trained agronomists or extension agents with experience in hydroponics exacerbates the problem. New growers, lacking training, frequently experience failures, including disease outbreaks in recirculating systems or errors in nutrient balancing. Education and training programs are still not widely available.

Universities only started teaching courses in controlled environment agriculture in the last few years. Most farmers who are interested in hydroponics or aquaponics lack local experts to seek advice from, and they have to fall back on online forums or expensive private consultants (Pattillo et al. 2022). Such knowledge deficiency is particularly severe in low- and middle-income nations. A few institutions like ICAR in India have initiated training courses and issued manuals, but, in relation to the substantial numbers of conventional farmers, those possessing hydroponic expertise are still extremely few.

5.5.3 Regulatory and Policy Gaps

As soilless farming is relatively new in most areas, agricultural policy and regulation are frequently not well-suited to it (Gonnella and Renna 2021). In other nations, zoning regulations fail to categorize indoor farms as agriculture, which causes issues with location permits and taxation. Farmers can also find it difficult to obtain subsidies, crop insurance, or credit since most programs are formulated for conventional field crops (Kramer et al. 2022). Growers in aquaponics may even require both a farming and an aquaculture license, adding bureaucracy for one combined system (Cammies et al. 2021). The other challenge is a lack of definition regarding organic certification.

Hydroponic crops are not permitted by some nations to be marketed as "organic," which restricts market space (Di Gioia and Rosskopf 2021). Encouraging steps have been taken, including India's MIDH scheme, which has now extended support to soilless cultivation, and the agriculture infrastructure fund (AIF), which hydroponic initiatives for concessional finance. Even so, the programs remain limited and are unknown to most farmers or have complicated paperwork to access assistance. Food safety regulations may also be a challenge. Hydroponic produce typically complies with general food safety standards, but aquaponics poses an additional risk factor (Sela Saldinger et al. 2023). Authorities occasionally fear microbial exposure due to fish water, although such systems with good management are safe. Because certification guidelines for aquaponics are not always defined, retailers and authorities might be reluctant to take such produce.

5.5.4 Market Access and Consumer Acceptance

Selling soilless produce profitably remains a challenge. Many consumers still believe soil-grown crops, like tomatoes, taste better, and early hydroponic varieties reinforced this perception by prioritizing yield over flavor (Ţălu 2024). Although quality has since improved, skepticism persists, making consumer education essential. Price sensitivity further limits adoption, especially in developing countries where only niche buyers pay more for hydroponic vegetables (D'Amico et al. 2024). Distribution introduces a further barrier, with wholesale markets preferring bulk supplies and encouraging many growers into direct sales, which limits scale. Cultural resistance is also important—conventional farmers will be reluctant to adopt "farming without soil." Establishing trust through localized demonstrations and success stories is thus essential for acceptance and future growth.

5.5.5 Other Technical Limitations

Soilless systems still have some challenges. Dependence on energy is a significant concern since repeated power outages can easily destroy crops unless expensive backup power is present. Pests and diseases also propagate more rapidly in recirculating systems, so tight sanitation and integrated pest management are crucial. Crop variety is restricted—these approaches are suitable for vegetables, herbs, and minor fruits but not for staples such as wheat, rice, or big fruit trees, so they can complement but not substitute conventional farming. There are also perception problems: a few shoppers think soil-grown produce tastes better or is more "natural," and there is still scientific controversy about flavor and nutrient differences. Generally, soilless cultivation is restrained by technological complexity, high costs, limited knowledge, and acceptance problems. It accounts for less than 1% of world agriculture today, although it is gradually increasing. Overcoming these barriers through innovation, training, favorable policies, and successful demonstrations will be vital in making it more widespread in the future.

5.6 Future Prospects and Research Directions

Hydroponics, aeroponics, and aquaponics have a future based on increased efficiency, reduced costs, and concurrence with sustainability objectives (Ibrahim et al. 2023). Renewable energy integration is one of the areas that need to be addressed, as excessive energy consumption for lights and climate control poses one of the greatest limiting factors. Netherlands and Chinese photovoltaic greenhouses have reported payback times less than 10 years, while Australian ventures such as Sundrop Farms integrate solar-driven desalination with hydroponics to grow vegetables in the desert (Minoofar et al. 2025). To cater to smallholders, scientists are designing low-cost, modular units that incorporate available local materials and minimal designs (Rivera Chavez et al. 2025).

The Kratky system, gravity-fed irrigation, and India's low-cost "Kambala" fodder unit demonstrate how soilless agriculture can be scaled down for rural and resource-constrained environments (Poudel and Di Gioia 2023). For water-deficient and climate-vulnerable areas, these approaches dramatically cut down water consumption relative to conventional farming, which enhances their suitability for climate-vulnerable and water-scarce regions (Khan et al. 2025). Public-private partnerships and policy will be instrumental in scaling adoption (Marx 2019). India's AIF and MIDH schemes and Singapore's 30 × 30 Express grants illustrate the use of subsidies, concessional credit, and regulatory support to stimulate investment. Certification standards, including labels for "clean horticulture," can also enhance consumer willingness to pay and acceptability.

On the research front, focus areas are on developing crop varieties specific to hydroponic conditions, engineering microbial inoculants to replicate soil behavior, and enhancing aeroponic technology (Cichocki et al. 2025).

Hybrid systems integrating aquaponics with biogas or mushroom production demonstrate circular use of resources efforts. Lastly, artificial intelligence and digital innovations will ease management, and space research conducted by NASA underscores the manner in which innovations for extraterrestrial agriculture will be used in Earth-based systems (Yada 2025; Badoni et al. 2025). These paths collectively indicate that soilless culture will shift from niche to mainstream application, particularly in urban and climate-exposed areas.

5.7 Conclusion

Soilless cultivation, such as hydroponics, aeroponics, and aquaponics, is a new method of farming that is able to address most of the issues of conventional agriculture. Such systems provide growers with complete control of plant nutrition and environment. They do not require fertile soil, they are space-efficient, and they have higher yields of better food quality. They consume only a fraction of the water needed in open-field cultivation, ensuring efficient use of limited resources. Hydroponics is already working well for value crops such as leafy greens, cucumbers, tomatoes, and herbs.

Aeroponics, although less commonly utilized, is highly effective in providing water and oxygen, and it is particularly beneficial for seed potato production and even space farming experiments. Aquaponics puts fish and plants together in one system, yielding two types of food with minimal waste. Illustrations from around the globe—rooftop gardens in cities, vertical gardens in North America, greenhouses in Europe, and pilot ventures in India and Africa—demonstrate how these systems can operate in much of the world and enhance yields and efficiency of use. Soilless farming is already restructuring food availability in cities today.

It makes possible year-round production near consumers, lowering spoilage and transport expenses. Most farms are also feeding premium markets, which require clean, fresh, and pesticide-free crops. These systems further benefit the environment by minimizing the use of chemical pesticides, conserving land by growing vertically, and eliminating pollution caused by runoff. Added to renewable energy, they have much reduced carbon footprints, and they can even be established in deserts, on roofs, or on poor lands where regular farming is not viable. Challenges still exist, though. The initial cost is expensive, the payback period is lengthy, and specialized technical expertise is required. It hinges on creating lower-cost and modular systems, educating more qualified laborers, joining cooperatives to exchange knowledge and markets, and establishing stronger government assistance. Governments are already taking action in some cases—for instance, India's MIDH and AIF schemes and Singapore's 30 × 30 Express grants provide subsidies, loans, and policy assistance.

In the future, soilless systems won't completely replace soil farming but will be used alongside it. Hydroponic greenhouses can provide a significant proportion of vegetables, vertical farms can become ubiquitous in the city, and aquaponics can offer safe food in regions that are drought or climate-stressed. Through the use of AI, IoT, and ongoing research in plant breeding, microbes, and system engineering, these farms will be easier to maintain and more efficient.

In short, hydroponics, aeroponics, and aquaponics have transitioned from trials to actual solutions. They already provide apparent advantages for food cultivation, sustainability, and profitability. With more innovation and encouragement, they can assist in feeding the increasing population with nutrients and sustainability while conserving natural resources and making agriculture more climate-change-resilient.

References

Adamek K, Jones AM, Torkamaneh D (2024) Somatic mutation accumulations in micropropagated cannabis are proportional to the number of subcultures. Plants 13(14):1910. https://doi.org/10.3390/plants13141910

Al-Kodmany K (2024) Vertical farms for future cities. In: Priyadarshan PM, Jain SM, Penna S, Al-Khayri JM (eds) Digital agriculture. Springer, Cham. https://doi.org/10.1007/978-3-031-43548-5_6

Anwarul S (2025) Embracing the future. In: Sapra L, Tanwar R, Sapra V (eds) CyberMedics: navigating AI and security in the medical field. CRC Press

Appicciutoli D, Amici AS, Bentivoglio D, Chiaraluce G, Staffolani G, Finco A (2025) Cultivating the future: a bibliometric review of emerging trends in soilless farming. Horticulturae 11(2):140. https://doi.org/10.3390/horticulturae11020140

Avgoustaki DD, Li J, Xydis T (2020) Basil plants grown under intermittent light stress in a small-scale indoor environment: introducing energy demand reduction intelligent technologies. Food Control 118:107389. https://doi.org/10.1016/j.foodcont.2020.107389

Avgoustaki DD, Bartzanas T, Xydis G (2021) Minimising the energy footprint of indoor food production while maintaining a high growth rate: introducing disruptive cultivation protocols. Food Control 130:108290. https://doi.org/10.1016/j.foodcont.2021.108290

Badoni P, Chauhan M, Agrahari A, Hira M (2025) Growing beyond the earth: the potential of extra-terrestrial agriculture from earth to space. In: Prabha C, Hassan MM, Yasmin F, Karim A (eds) Emerging AI applications in earth sciences, Sustainable artificial intelligence-powered applications. Springer, Cham. https://doi.org/10.1007/978-3-031-84583-3_2

Balasus J, Blank J, Babilon S, Hegemann T, Khanh TQ (2021) Energy efficient lighting in plant factories: addressing utilance. Agronomy 11(12):2570. https://doi.org/10.3390/agronomy11122570

Balliu A, Zheng Y, Sallaku G, Fernández JA, Gruda NS, Tuzel Y (2021) Environmental and cultivation factors affect the morphology, architecture and performance of root systems in soilless grown plants. Horticulturae 7(8):243. https://doi.org/10.3390/horticulturae7080243

Banboukian A, Chen Y, Thomas VM (2025) The challenges of controlled environment hydroponic farming: a life cycle assessment of lettuce. Int J Life Cycle Assess 30:1691–1704. https://doi.org/10.1007/s11367-025-02463-6

Barbosa GL, Gadelha FD, Kublik N, Proctor A, Reichelm L, Weissinger E, Wohlleb GM, Halden RU (2015) Comparison of land, water, and energy requirements of lettuce grown using hydroponic vs. conventional agricultural methods. Int J Environ Res Public Health 12(6):6879–6891. https://doi.org/10.3390/ijerph120606879

Bohemen J, Carvalho S, Nicole CC, Smith D, Lefsrud M (2025) LED interlighting: a cost-effective strategy to increase yield and chemical uniformity of Cannabis sativa at commercial scale. Sci Hortic 351:114406. https://doi.org/10.1016/j.scienta.2025.114406

Bostancı KB, Ülger S (2022) Comparison of spinach cultivation in floating hydroponic system and soil in glasshouse and open field conditions. Mediterr Agric Sci 35(1):7–14. https://doi.org/10.29136/mediterranean.1061475

Cai J, Veerappan V, Arildsen K, Sullivan C, Piechowicz M, Frugoli J, Dickstein R (2023) A modified aeroponic system for growing small-seeded legumes and other plants to study root systems. Plant Methods 19(1):21. https://doi.org/10.1186/s13007-023-01000-6

Cammies C, Mytton D, Crichton R (2021) Exploring economic and legal barriers to commercial aquaponics in the EU through the lens of the UK and policy proposals to address them. Aquac Int 29:1245–1263. https://doi.org/10.1007/s10499-021-00690-w

Chaudhary A, Anand S (2024) Soilless cultivation: a distinct vision for sustainable agriculture. In: Pandey K, Kushwaha NL, Pande CB, Singh KG (eds) Artificial intelligence and smart agriculture, Advances in geographical and environmental sciences. Springer, Singapore. https://doi.org/10.1007/978-981-97-0341-8_17

Cichocki J, Trenkner M, Vanicela B, Riethmueller C, Walz M, Chandra S, Pal H (2025) Demystifying the integration of hydroponics cultivation system reinforcing bioeconomy and sustainable agricultural growth. Sci Hortic 341:11397. https://doi.org/10.1016/j.scienta.2025.113973

Clyde-Smith D, Campos LC (2023) Engineering hydroponic systems for sustainable wastewater treatment and plant growth. Appl Sci 13(14):8032. https://doi.org/10.3390/app13148032

D'Amico A, De Boni A, Ottomano Palmisano G, Morea E, Acciani C, Roma R (2024) Consumers' perception and willingness to pay for hydroponic tomatoes: the effects of sustainability and quality attributes. Br Food J 126(13):573–592. https://doi.org/10.1108/BFJ-04-2024-0353

Daneshyar E (2024) Residential rooftop urban agriculture: architectural design recommendations. Sustainability 16(5):1881. https://doi.org/10.3390/su16051881

Dauda AB, Ajadi A, Tola-Fabunmi AS, Akinwole AO (2019) Waste production in aquaculture: sources, components and managements in different culture systems. Aquac Fish 4(3):81–88. https://doi.org/10.1016/j.aaf.2018.10.002

Di Gioia F, Rosskopf EN (2021) Organic hydroponics: a US reality challenging the traditional concept of "organic" and "soilless" cultivation. Acta Hortic 1321:275–282. https://doi.org/10.17660/ActaHortic.2021.1321.36

Dsouza A, Dixon M, Shukla M, Graham T (2025) Harnessing controlled-environment systems for enhanced production of medicinal plants. J Exp Bot 76(1):76–93. https://doi.org/10.1093/jxb/erae248

Dutta M, Gupta D, Tharewal S, Goyal D, Sandhu JK, Kaur M, Alzubi AA, Alanazi JM (2025) Internet of things-based smart precision farming in soilless agriculture: opportunities and challenges for global food security. IEEE Access 13:34238–34268. https://doi.org/10.1109/ACCESS.2025.3540317

Ehsan U, Riedl MO (2024) Explainability pitfalls: beyond dark patterns in explainable AI. Patterns 5(6):1000–1010. https://doi.org/10.1016/j.patter.2024.100971

Eigenbrod C, Gruda N (2015) Urban vegetable for food security in cities: a review. Agron Sustain Dev 35:483–498. https://doi.org/10.1007/s13593-014-0273-y

Eldridge BM, Manzoni LR, Graham CA, Rodgers B, Farmer JR, Dodd AN (2020) Getting to the roots of aeroponic indoor farming. New Phytol 228(4):1183–1192. https://doi.org/10.1111/nph.16780

Elsherbiny O, Gao J, Ma M, Guo Y, Tunio MH, Mosha AH (2024) Advancing lettuce physiological state recognition in IoT aeroponic systems: a meta-learning-driven data fusion approach. Eur J Agron 161:127387. https://doi.org/10.1016/j.eja.2024.127387

Embaby ZA (2023) Innovative greenhouse to improve economic and environmental conditions. In: Abdelhafez AA, Abbas MHH (eds) Climate smart greenhouses—innovations and impacts. IntechOpen, London. https://doi.org/10.5772/intechopen.113335

Fathidarehnijeh E, Nadeem M, Cheema M, Thomas R, Krishnapillai M, Galagedara L (2023) Current perspective on nutrient solution management strategies to improve the nutrient and water use efficiency in hydroponic systems. Can J Plant Sci 104(2):88–102. https://doi.org/10.1139/cjps-2023-0034

Fernández-Cabanás VM, Pérez-Urrestarazu L, Juárez A, Kaufman NT, Gross JA (2020) Comparative analysis of horizontal and vertical decoupled aquaponic systems for basil production and effect of light supplementation by LED. Agronomy 10(9):1414. https://doi.org/10.3390/agronomy10091414

Fuentes-Peñailillo F, Gutter K, Vega R, Silva GC (2024) New generation sustainable technologies for soilless vegetable production. Horticulturae 10(1):49. https://doi.org/10.3390/horticulturae10010049

Fussy A, Papenbrock J (2022) An overview of soil and soilless cultivation techniques—chances, challenges and the neglected question of sustainability. Plants 11(9):1153. https://doi.org/10.3390/plants11091153

Ghiasi M, Wang Z, Mehrandezh M, Paranjape R (2023) A systematic review of optimal and practical methods in design, construction, control, energy management and operation of smart greenhouses. IEEE Access 12:2830–2853. https://doi.org/10.1109/ACCESS.2023.3346436

Gonnella M, Renna M (2021) The evolution of soilless systems towards ecological sustainability in the perspective of a circular economy. Is it really the opposite of organic agriculture? Agronomy 11(5):950. https://doi.org/10.3390/agronomy11050950

Graves CJ (1983) The nutrient film technique. Hortic Rev 5:1–44. https://doi.org/10.1002/9781118060728

Gupta V, Garg A, Agrawal S (2024) Remote monitoring and control systems in agriculture and farming. In: Arya R, Sharma SC, Verma AK, Iyer B (eds) Recent trends in artificial intelligence towards a smart world, Frontiers of artificial intelligence, ethics and multidisciplinary applications. Springer, Singapore. https://doi.org/10.1007/978-981-97-6790-8_10

Gupta A, Sharma T, Sanjaykumar DY, Manojkumar SS, Dhar DW (2025) Amalgamation of hydroponics with the biofilm to enhance plant performance and nutrient management. In: Mathur P, Mukherjee S (eds) Rhizosphere engineering and stress resilience in plants. CRC Press

Hasan M, Bhat AG, S VK, R M, Yadav AK, Saji A, G B, Prakash SA (2025) IoT-based smart fertigation scheduling and wireless microclimate monitoring for a greenhouse Dutch bucket hydroponic system. Irrig Drain *74*:1889–1903. https://doi.org/10.1002/ird.70012

Hayati R, Armadi Y, Yusnaweti Y, Feni R, Asfaruddin A, Ariyani F, Kesumawati N, Suzanna E, Sarina S (2024) Testing alternative hydroponic nutrition of wick system on vegetable growth planted in limited land. J La Lifesci 5(2):94–108. https://doi.org/10.37899/journallalifesci.v5i2.1177

Herrera-Arroyo R, Martínez-Nolasco J, Botello-Álvarez E, Sámano-Ortega V, Martínez-Nolasco C, Moreno-Aguilera C (2025) Smart hydroponic cultivation system for lettuce (*Lactuca sativa* L.) growth under different nutrient solution concentrations in a controlled environment. Appl Syst Innov 8(4):110. https://doi.org/10.3390/asi8040110

Ibrahim LA, Shaghaleh H, El-Kassar GM, Abu-Hashim M, Elsadek EA, Alhaj Hamoud Y (2023) Aquaponics: a sustainable path to food sovereignty and enhanced water use efficiency. Water 15(24):4310. https://doi.org/10.3390/w15244310

Ikrang EG, Ehiomogue PO, Udoumoh UI (2022) Hydroponics in precision agriculture–a review. Ann Fac Eng Hunedoara 20(2):143–148

Incrocci L, Thompson RB, Fernandez-Fernandez MD, De Pascale S, Pardossi A, Stanghellini C, Rouphael Y, Gallardo M (2020) Irrigation management of European greenhouse vegetable crops. Agric Water Manag 242:106393. https://doi.org/10.1016/j.agwat.2020.106393

Jani AD, Meadows TD, Eckman MA, Ferrarezi RS (2021) Automated ebb-and-flow subirrigation conserves water and enhances citrus liner growth compared to capillary mat and overhead irrigation methods. Agric Water Manag 246:106711. https://doi.org/10.1016/j.agwat.2020.106711

Jayasuriya N, Guo Y, Hu W, Ghannoum O (2024) Image-based crop monitoring technologies in protected horticulture: a review. arXiv preprint:arXiv:2401.13928. https://doi.org/10.48550/arXiv.2401.13928

Jha K, Doshi A, Patel P, Shah M (2019) A comprehensive review on automation in agriculture using artificial intelligence. Artif Intell Agric 2:1–2. https://doi.org/10.1016/j.aiia.2019.05.004

Joshi D, Nainabasti A, Awasthi RB, Banjade D, Malla S, Subedi B (2022) A review on soilless cultivation: the hope of urban agriculture. Arc Agric Environ Sci 7(3):473–481. https://doi.org/10.26832/24566632.2022.0703

Kabir MS, Reza MN, Chowdhury M, Ali M, Samsuzzaman AMR, Lee KY, Chung SO (2023) Technological trends and engineering issues on vertical farms: a review. Horticulturae 9(11):1229. https://doi.org/10.3390/horticulturae9111229

Kalantari F, Tahir OM, Joni RA, Fatemi E (2018) Opportunities and challenges in sustainability of vertical farming: a review. J Landscape Ecol 11:35–60. https://doi.org/10.1515/jlecol-2017-0016

Kannan M, Elavarasan G, Balamurugan A, Dhanusiya B, Freedon D (2022) Hydroponic farming—a state of art for the future agriculture. Mater Today Proc 68:2163–2166. https://doi.org/10.1016/j.matpr.2022.08.416

Karothia R, Chattopadhyay MK (2024) Internet of things (IoT) enabled smart sensor node (SSN) to measure the soil and environmental parameters for smart farming. CSI Trans ICT 12(4):119–135. https://doi.org/10.1007/s40012-024-00402-8

Khan MR, Zafar M, Majeed S, Alotaibi NM, Ibadullayeva S, Dinislam K, Biturku J (2025) Nature-based riparian strategies for climate resilience: ecohydrological insights on biodiversity and food security in the Indus Basin. Ecohydrology 18(5):e70091. https://doi.org/10.1002/eco.70091

Khatri L, Kunwar A, Bist DR (2024) Hydroponics: advantages and challenges in soilless farming. Big Data Agric 6:81–88. https://doi.org/10.26480/bda.02.2024.98.105

Korsa G, Ayele A, Haile S, Alemu D (2025) Hydroponic farming: innovative solutions for sustainable and modern cultivation technique. In: Ali B, Iqbal J, Ahmed T (eds) Hydroponic farming—a modern agriculture technique. Intechopen. https://doi.org/10.5772/intechopen.1008336

Kramer B, Hazell P, Alderman H, Ceballos F, Kumar N, Timu AG (2022) Is agricultural insurance fulfilling its promise for the developing world? A review of recent evidence. Annu Rev Resour Econ 14(1):291–311. https://doi.org/10.1146/annurev-resource-111220-014147

Kumar P, Dey SR (2024) Hydroponics phytoremediation: an overview. In: Kumar N (ed) Hydroponics and environmental bioremediation, Springer Water. Springer, Cham. https://doi.org/10.1007/978-3-031-53258-0_13

Kumar P, Sampath B, Kumar S, Babu BH (2023) Hydroponics, aeroponics, and aquaponics technologies in modern agricultural cultivation. In: Mellal MA (ed) Trends, paradigms, and advances in mechatronics engineering. IGI Global. https://doi.org/10.4018/978-1-6684-5887-7.ch012

Kumar R, Manna M, Bhat AG, Satpute AN, Gavhane KP (2025) Intervention of hydroponics and aeroponics for climate-resilient cultivation. In: Rajan R, Ahmad F, Pandey K (eds) Innovations in climate resilient agriculture. Springer, Cham. https://doi.org/10.1007/978-3-031-84802-5_14

Lakhiar IA, Jianmin G, Syed TN, Chandio FA, Buttar NA, Qureshi WA (2018) monitoring and control systems in agriculture using intelligent sensor techniques: a review of the aeroponic system. J Sens 2018(1):8672769. https://doi.org/10.1155/2018/8672769

Lakhiar IA, Yan H, Syed TN, Zhang C, Shaikh SA, Rakibuzzaman M, Vistro RB (2025) Soilless agricultural systems: opportunities, challenges, and applications for enhancing horticultural resilience to climate change and urbanization. Horticulturae 11:568. https://doi.org/10.3390/horticulturae11060568

Lin HA, Coker HR, Park S, Finlayson SA, Tfaily MM, Nagy EM, Hague S, Antony-Babu S, Howe JA, Smith AP (2024) Aeroponic approach for nondestructive root exudate collection and simulation of variable water stress trialed on cotton (*Gossypium hirsutum*). Sci Rep 14(1):28615. https://doi.org/10.1038/s41598-024-79801-5

Liu J, van Iersel MW (2023) Far-red photons increase light capture but have lower photosynthetic capacity than red photons. J Am Soc Hortic Sci 148(6):253–265. https://doi.org/10.21273/JASHS05306-23

Liu X, Xu Y, Wang Y, Yang Q, Li Q (2022) Rerouting artificial light for efficient crops production: a review of lighting strategy in PFALs. Agronomy 12:1021. https://doi.org/10.3390/agronomy12051021

Loconsole D, Santamaria P (2021) UV lighting in horticulture: a sustainable tool for improving production quality and food safety. Horticulturae 7(1):9. https://doi.org/10.3390/horticulturae7010009

Maheshwari S (2021) Vertical farming: resilience towards climate change. In: Kateja A, Jain R (eds) Urban growth and environmental issues in India. Springer, Singapore. https://doi.org/10.1007/978-981-16-4273-9_13

Majsztrik J, Owen JS (2020) Improving nutrient management in the cultivation of ornamental plants in greenhouse, container and field production. In: Reid M (ed) Achieving sustainable cultivation of ornamental plants. Burleigh Dodds Science Publishing, London

Martinez-Mate MA, Martin-Gorriz B, Martínez-Alvarez V, Soto-García M, Maestre-Valero JF (2018) Hydroponic system and desalinated seawater as an alternative farm-productive proposal in water scarcity areas: energy and greenhouse gas emissions analysis of lettuce production in Southeast Spain. J Clean Prod 172:1298–1310. https://doi.org/10.1016/j.jclepro.2017.10.275

Marx A (2019) Public-private partnerships for sustainable development: exploring their design and its impact on effectiveness. Sustainability 11(4):1087. https://doi.org/10.3390/su11041087

Min A, Nguyen N, Howatt L, Tavares M, Seo J (2023) Aeroponic systems design: considerations and challenges. J Agric Eng 54(1):1387. https://doi.org/10.4081/jae.2022.1387

Ministry of Agriculture & Farmers Welfare (2023) Adoption of hydroponics farming. Press Information Bureau

Ministry of the Environment and Water Resources (2019) Committee-of-Supply-2019. Available at: https://www.mewr.gov.sg/about-us/our-organisation/Committee-of-Supply-2019. Accessed 12 Sept 2019

Minoofar A, Zandi A, Gholami A (2025) Precise prediction models for floating and land-based photovoltaics: comparative analysis for advancing sustainable energy and water management solutions. Renew Energy 255:123781. https://doi.org/10.1016/j.renene.2025.123781

Mishra SJ, Rout D, Sahoo D (2024) Analysing the economic viability of hydroponic farming: a comparative cost-benefit analysis. Int J Prog Res Eng Manag Sci 4(6):1806–1811. https://doi.org/10.58257/IJPREMS35128

Mishra RK (2023) Fresh water availability and its global challenge. Br J Multidiscip Adv Stud 4(3):1–78. https://doi.org/10.37745/bjmas.2022.0207

Mohammed SS, Singh IP, Makarfi AM, Karunakaran N, Atapattu JA, Doddanavar IA, Salem YM, Adamu MY, Shamim MZ, Sanyinna SB (2025) Agripreneurship management: the green gold. Cambridge Scholars Publishing

Mok WK, Tan YX, Chen WN (2020) Technology innovations for food security in Singapore: a case study of future food systems for an increasingly natural resource-scarce world. Trends Food Sci Technol 102:155–168. https://doi.org/10.1016/j.tifs.2020.06.013

Muchelo RO, Bishop TF, Ugbaje SU, Akpa SI (2024) Patterns of urban sprawl and agricultural land loss in sub-Saharan Africa: the cases of the Ugandan cities of Kampala and Mbarara. Land 13(7):1056. https://doi.org/10.3390/land13071056

Mylonas A, Kazanci OB, Andersen RK, Olesen BW (2019) Capabilities and limitations of wireless CO_2, temperature and relative humidity sensors. Build Environ 154:362–374. https://doi.org/10.1016/j.buildenv.2019.03.012

Nair CS, Manoharan R, Nishanth D, Subramanian R, Neumann E, Jaleel A (2025) Recent advancements in aquaponics with special emphasis on its sustainability. J World Aquac Soc 56(1):e13116. https://doi.org/10.1111/jwas.13116

Naresh R, Jadav SK, Singh M, Patel A, Singh B, Beese S, Pandey SK (2024) Role of hydroponics in improving water-use efficiency and food security. Int J Environ Clim Change 14(2):608–633. https://doi.org/10.9734/IJECC/2024/v14i23976

NASA Spinoff (2006) Progressive plant growing has business blooming. In: Environmental and agricultural resources. NASASpinoff, New York, pp 64–77

National Bureau of Statistics of China (2024) China sees rising urbanization rate over past 75 years. State Council of the People's Republic of China. Available online: https://english.www.gov.cn/archive/statistics/202409/23/content_WS66f14531c6d0868f4e8eb2c0.html. Accessed on 26 Mar 2025

Neo DC, Ong MM, Lee YY, Teo EJ, Ong Q, Tanoto H, Xu J, Ong KS, Suresh V (2022) Shaping and tuning lighting conditions in controlled environment agriculture: a review. ACS Agric Sci Technol 2(1):3–16. https://doi.org/10.1021/acsagscitech.1c00241

Nikolaou G, Neocleous D, Kitta E, Katsoulas N (2021) Advances in irrigation/fertigation techniques in greenhouse soilless culture systems (SCS). In: Gruda NS (ed) Advances in horticultural soilless culture. Burleigh Dodds Science Publishing, London

Nitu OA, Ivan EŞ, Tronac AS, Arshad A (2024) Optimizing lettuce growth in nutrient film technique hydroponics: evaluating the impact of elevated oxygen concentrations in the root zone under LED illumination. Agronomy 14(9):1896. https://doi.org/10.3390/agronomy14091896

Niu G, Sun Y, Masabni JG (2018) Impact of low and moderate salinity water on plant performance of leafy vegetables in a recirculating NFT system. Horticulturae 4(1):6. https://doi.org/10.3390/horticulturae4010006

Noor A, Haider ST, Akbar A, Ziaf K, Anjum MA (2025) Protected olericulture. In: Iqbal Z, Anjum MA, Ercisli S, Rauf S (eds) Sustainable and innovative vegetable production in times of climate change. Springer, Singapore. https://doi.org/10.1007/978-981-96-6283-8_9

Nuthalapati CS, Nuthalapati C (2021) Has open innovation taken root in India? Evidence from startups working in food value chains. Circ Econ Sustain 1(4):1207–1230. https://doi.org/10.1007/s43615-021-00074-5

Oh S, Lu C (2023) Vertical farming—smart urban agriculture for enhancing resilience and sustainability in food security. J Hortic Sci Biotechnol 98(2):133–140. https://doi.org/10.1080/14620316.2022.2141666

Pandey S, Chaudhary M, Tóth Z (2025) An investigation on real-time insights: enhancing process control with IoT-enabled sensor networks. Discov Internet Things 5(1):29. https://doi.org/10.1007/s43926-025-00124-6

Panotra N, Belagalla N, Mohanty LK, Ramesha NM, Tiwari AK, Abhishek G, Gulaiya S, Yadav K, Pandey SK (2024) Vertical farming: addressing the challenges of 21st century agriculture through innovation. Int J Environ Clim Change 14(4):664–691. https://doi.org/10.9734/IJECC/2024/v14i44150

Pashkovskiy P, Kreslavski VD, Ivanov Y, Ivanova A, Kartashov A, Shmarev A, Strokina V, Kuznetsov VV, Allakhverdiev SI (2021) Influence of light of different spectral compositions on the growth, photosynthesis, and expression of light-dependent genes of scots pine seedlings. Cells 10(12):3284. https://doi.org/10.3390/cells10123284

Pattillo DA, Hager JV, Cline DJ, Roy LA, Hanson TR (2022) System design and production practices of aquaponic stakeholders. PLoS One 17(4):e0266475. https://doi.org/10.1371/journal.pone.0266475

Pomoni DI, Koukou MK, Vrachopoulos MG, Vasiliadis L (2023) A review of hydroponics and conventional agriculture based on energy and water consumption, environmental impact, and land use. Energies 16(4):1690. https://doi.org/10.3390/en16041690

Poudel P, Di Gioia F (2023) Evaluation of alternative soilless growing systems for resource use efficiency, yield and quality performance of multi-leaf lettuce. Acta Hortic 1426:325–332. https://doi.org/10.17660/ActaHortic.2025.1426.45

Ptak M, Wasieńko S, Makuła P (2024) A new approach to vertical plant cultivation maximises crop efficiency. Sustainability 16(16):7189. https://doi.org/10.3390/su16167189

PW Consulting (2024) Worldwide soilless hydroponic vegetables market research report 2025, forecast to 2031. Available online: https://pmarketresearch.com/product/worldwide-soilless-hydroponic-vegetables-market-research-2024-by-type-application-participants-and-countries-forecast-to-2030. Accessed on 5 Apr 2025

Qamar R, Garcia y Garcia A, Shoaib M, Ahmad S (2025) Irrigation management in vegetable crops. In: Iqbal Z, Anjum MA, Ercisli S, Rauf S (eds) Sustainable and innovative vegetable production in times of climate change. Springer, Singapore. https://doi.org/10.1007/978-981-96-6283-8_14

Ragaveena S, Shirly Edward A, Surendran U (2021) Smart controlled environment agriculture methods: a holistic review. Rev Environ Sci Biotechnol 20(4):887–913. https://doi.org/10.1007/s11157-021-09591-z

Rahman MA, Chakraborty NR, Sufiun A, Banshal SK, Tajnin FR (2024) An AIoT-based hydroponic system for crop recommendation and nutrient parameter monitorization. Smart Agric Technol 8:100472. https://doi.org/10.1016/j.atech.2024.100472

Rajaseger G, Chan KL, Tan KY, Ramasamy S, Khin MC, Amaladoss A, Haribhai PK (2023) Hydroponics: current trends in sustainable crop production. Bioinformation 19(9):925. https://doi.org/10.6026/97320630019925

Rajendran S, Domalachenpa T, Arora H, Li P, Sharma A, Rajauria G (2024) Hydroponics: exploring innovative sustainable technologies and applications across crop production, with emphasis on potato mini-tuber cultivation. Heliyon 10(5):e26823. https://doi.org/10.1016/j.heliyon.2024.e26823

Rakocy JE (2012) Aquaponics—integrating fish and plant culture. In: Tidwell JH (ed) Aquaculture production systems. Wiley

Rathor AS, Choudhury S, Sharma A, Nautiyal P, Shah G (2024) Empowering vertical farming through IoT and AI-driven technologies: a comprehensive review. Heliyon 10(15):e34998. https://doi.org/10.1016/j.heliyon.2024.e34998

Reza MN, Lee KH, Karim MR, Haque MA, Bicamumakuba E, Dey PK, Jang YY, Chung SO (2025) Trends of soil and solution nutrient sensing for open field and hydroponic cultivation in facilitated smart agriculture. Sensors 25(2):453. https://doi.org/10.3390/s25020453

Rharrhour H, Wariaghli F, Goddek S, Sadik M, El Moujtahid A, Nhhala H, Yahyaoui A (2022) Towards sustainable food productions in Morocco: aquaponics. E3S Web Conf 337:03004. https://doi.org/10.1051/e3sconf/202233703004

Rivera Chavez ZB, Porcaro A, De Simone MC, Guida D (2025) Improving sustainable viticulture in developing countries: a case study. Sustainability 17(12):5338. https://doi.org/10.3390/su17125338
Rodríguez-Nieto D, Navas E, Fernández R (2025) Automated seeding in hydroponic greenhouse with a dual-arm robotic system. IEEE Access 13:30745–30761. https://doi.org/10.1109/ACCESS.2025.3541954
Roosta HR, Tavakoli MM, Zahedi B, Bikdeloo M (2025) The supplemental LED light spectra and silicon foliar application alleviate the adverse effects of combined salinity-alkalinity stress on Physalis angulata. Sci Rep 15(1):31515. https://doi.org/10.1038/s41598-025-16982-7
Ruiz P, Vidal JM, Sepúlveda D, Torres C, Villouta G, Carrasco C, Aguilera F, Ruiz-Tagle N, Urrutia H (2020) Overview and future perspectives of nitrifying bacteria on biofilters for recirculating aquaculture systems. Rev Aquac 12(3):1478–1494. https://doi.org/10.1111/raq.12392
Sadek N, Shehata D (2024) Internet of things-based smart automated indoor hydroponics and aeroponics greenhouse in Egypt. Ain Shams Eng J 15(2):102341. https://doi.org/10.1016/j.asej.2023.102341
Schwarz M (1995) Culture methods. In: Soilless culture management, Advanced series in agricultural sciences, vol 24. Springer, Berlin/Heidelberg. https://doi.org/10.1007/978-3-642-79093-5_4
Sela Saldinger S, Rodov V, Kenigsbuch D, Bar-Tal A (2023) Hydroponic agriculture and microbial safety of vegetables: promises, challenges, and solutions. Horticulturae 9(1):51. https://doi.org/10.3390/horticulturae9010051
Sever M (2022) Managing for resiliency in the face of drought. Crops Soils 55(6):4–13. https://doi.org/10.1002/crso.20232
Sharma A, Pandey H, Nampoothiri VAS, Kartha BD, Jha R (2023) Production of, factors affecting, gene regulations, and challenges in tissue cultured plant through soilless culture. J Agric Food Chem 71:5804–5811. https://doi.org/10.1021/acs.jafc.2c08162
Sharma T, Ananthakrishnan S, Gawdiya S, Rawat A, Singh A, Suryawanshi Y, Chauhan G, Rana RS (2025) Hydroponics farming: a holistic perspective for crop production. In: Suryawanshi Y, Shahnawaz M, Parihar J, Dar RA, Samyal S, Ade AB (eds) Oilseed crops. Wiley. https://doi.org/10.1002/9781394186426.ch13
Singh H, Dunn B, Payton M (2019) Hydroponic pH modifiers affect plant growth and nutrient content in leafy greens. J Hortic Res 27(1):31–36. https://doi.org/10.2478/johr-2019-0004
Singh G, Bhagat N, Thakur S (2025) Automated production management in horticulture: an industry 4.0 perspective. In: Blockchain-enabled internet of things applications in healthcare: current practices and future directions. Bentham Science Publishers, pp 258–282. https://doi.org/10.2174/97898153052101250101
Solankey SS, Kumari A, Akhtar S, Adarsh A, Sinha A, Kumari M (2025) The present advances in research for potato production: future trends and goals. In: Solankey SS (ed) Advances in research on potato production, Advances in Olericulture. Springer, Cham. https://doi.org/10.1007/978-3-031-82710-5_1
Soofastaei A (2025) Harvesting tomorrow: the future of agriculture in industry 4.0. In: Soofastaei A (ed) Advanced analytics for industry 4.0. CRC Press, Boca Raton
Stamford JD, Stevens J, Mullineaux PM, Lawson T (2023) LED lighting: a grower's guide to light spectra. HortScience 58(2):180–196. https://doi.org/10.21273/HORTSCI16823-22
Stanghellini C, Katzin D (2024) The dark side of lighting: a critical analysis of vertical farms' environmental impact. J Clean Prod 458:142359. https://doi.org/10.1016/j.jclepro.2024.142359
Stasiak M, Dixon M (2022) Growing facilities and environmental control. In: Zheng Y (ed) Handbook of cannabis production in controlled environments. CRC Press, Boca Raton
Suganob NJ, Arroyo CL, Concepcion R (2024) Interplay of fogponics and artificial intelligence for potential application in controlled space farming. AgriEngineering 6(3):2144–2166. https://doi.org/10.3390/agriengineering6030126
Sundar P, Jyothi K, Sundar C (2021) Indoor hydroponics: a potential solution to reuse domestic rinse water. Biosci Biotechnol Res Asia 18(2):373–383. https://doi.org/10.13005/bbra/2924

Ţălu Ş (2024) Insights on hydroponic systems: understanding consumer attitudes in the cultivation of hydroponically grown fruits and vegetables. Hidraulica 1:56–67

Tang L, Syed AU, Otho AR, Junejo AR, Tunio MH, Hao L, Asghar Ali MN, Brohi SA, Otho SA, Channa JA (2024) Intelligent rapid asexual propagation technology—a novel aeroponics propagation approach. Agronomy 14(10):2289. https://doi.org/10.3390/agronomy14102289

Tatum M (2020) Inside Singapore's huge bet on vertical farming. MIT Technol Rev. Available online: https://www.technologyreview.com/2020/10/13/1009497/singapore-vertical-farming-food-security. Accessed 5 Apr 2025

Tican A, Chelmea C, Cioloca M, Popa M, Ştefan M (2024) Preliminary study about comparison of potato minitubers production in conventional culture and modern culture–aeroponic. J Agric Rural Dev Stud 2:48–54. https://doi.org/10.35219/jards.2024.2.05

United Nations, Department of Economic and Social Affairs, Population Division (2018) The 2018 revision of world urbanization prospects. United Nations, New York, NY. Available online: https://www.un.org/en/desa/2018-revision-world-urbanization-prospects. Accessed 26 Mar 2025

Van der Velden NJ, Janse J, Kaarsemaker RC, Maaswinkel RH (2004) Sustainability of greenhouse fruit vegetables: Spain versus The Netherlands; development of a monitoring system. XV international symposium on horticultural economics and management. Acta Hortic 655:275–281. https://doi.org/10.17660/ActaHortic.2004.655.34

van Tuyll A, Lanting M, de Lange E, Boedijn A, van Medevoort J, Koops A, Hemming S, Voogt W, Stanghellini C, van Ittersum MK (2025) A new circular fertiliser framework for greenhouse hydroponic systems using phosphorus as a case study. Nutr Cycl Agroecosyst 130(1):49–68. https://doi.org/10.1007/s10705-024-10376-3

Varzakas T, Smaoui S (2024) Global food security and sustainability issues: the road to 2030 from nutrition and sustainable healthy diets to food systems change. Foods 13(2):306. https://doi.org/10.3390/foods13020306

Vasdravanidis C, Alvanou MV, Lattos A, Papadopoulos DK, Chatzigeorgiou I, Ravani M, Liantas G, Georgoulis I, Feidantsis K, Ntinas GK, Giantsis IA (2022) Aquaponics as a promising strategy to mitigate impacts of climate change on rainbow trout culture. Animals 12(19):2523. https://doi.org/10.3390/ani12192523

Vettriselvan R (2025) Commercial applications of aeroponics: revolutionizing modern agriculture and sustainable food production. In: Utilizing aeroponics techniques for improved farming. IGI Global Scientific Publishing. https://doi.org/10.4018/979-8-3693-2320-5.ch010

Wahab PE, Che-Othman MH, Al-Tawaha AR (2023) Managing nitrogen toxicity in the soilless culture system: a review. J Trop Plant Physiol 15(2):11–26. https://doi.org/10.56999/jtpp.2023.15.2.31

Wen G, Cao Y, Wei X (2025) The data-driven analysis of soil health and crop adaptability: technologies, impacts, and optimization strategies. Adv Res Resour 5(1):350–368. https://doi.org/10.50908/arr.5.1_350

Yada A (2025) Space exploration for earthly benefits: satellite technology—industry, innovation and infrastructure. In: Cutting-edge solutions for advancing sustainable development: exploring technological horizons for sustainability-part 2. Bentham Science Publishers. https://doi.org/10.2174/9789815322408125010

Yavuzcan Yildiz H, Robaina L, Pirhonen J, Mente E, Domínguez D, Parisi G (2017) Fish welfare in aquaponic systems: its relation to water quality with an emphasis on feed and faeces—a review. Water 9(1):13. https://doi.org/10.3390/w9010013

Yusuf AG, Al-Yahya FA, Saleh AA, Abdel-Ghany AM (2025) Optimizing greenhouse microclimate for plant pathology: challenges and cooling solutions for pathogen control in arid regions. Front Plant Sci 16:1492760. https://doi.org/10.3389/fpls.2025.1492760

Zamnuri MA, Qiu S, Rizalmy MA, He W, Yusoff S, Roeroe KA, Du J, Loh KH (2024) Integration of IoT in small-scale aquaponics to enhance efficiency and profitability: a systematic review. Animals 14(17):2555. https://doi.org/10.3390/ani14172555

Zhai L, Lee J-E (2024) Investigating vulnerability, adaptation, and resilience: a comprehensive review within the context of climate change. Atmos 15:474. https://doi.org/10.3390/atmos15040474

Zheng Q, Huang W, Xia Q, Dong Y, Ye H, Jiang H, Chen S, Huang S (2023) Remote sensing monitoring of rice diseases and pests from different data sources: a review. Agronomy 13(7):1851. https://doi.org/10.3390/agronomy13071851

Zhou X, Asioli D, Jürkenbeck K (2024) Investigating consumers' views on foods from soilless farming systems: a review of the literature and discussion of implications and recommendations. Food Qual Prefer 126:105413. https://doi.org/10.1016/j.foodqual.2024.105413

Annunciating the Present Context and Future Directions of Soilless Cultivation of Horticultural Crops for a Sustainable Agriculture

6

D. T. Santosh, Dinkar Jagannath Gaikwad, Masina Sairam, Sumit Ray, Sarthak Pattanayak, Nithar Ranjan Madhu, Lalichetti Sagar, Sourav Garai, Shaik Rishitha, Harun I. Gitari, Martin Leanardo Battaglia, Urjashi Bhattacharya, Anjana J. Atapattu, Akbar Hossain, and Sagar Maitra

Abstract

The world is facing threats such as the shrinkage of farmlands, an ever-increasing population, and the looming vulnerability of the climate, all of which warrant alterating of traditional agricultural practices to ensure food and nutritional security. Crop cultivation, especially of horticultural crops, which are highly dependent on ecology, is mostly affected by climatic factors. Furthermore, rapid

D. T. Santosh · D. J. Gaikwad
Research Centre of Excellence for Smart Agriculture, Centurion University of Technology and Management, R. Sitapur, Odisha, India
e-mail: santosh.dt@cutm.ac.in; gaikwad@cutm.ac.in

M. Sairam · S. Ray · L. Sagar · S. Rishitha · S. Maitra (✉)
M. S. Swaminathan School of Agriculture, Centurion University of Technology and Management, R. Sitapur, Odisha, India
e-mail: sairam.masina@cutm.ac.in; sumit.ray@cutm.ac.in; sagar.maitra@cutm.ac.in; sagar.maitra@cutm.ac.in

S. Pattanayak
Odisha University of Agriculture and Technology, Bhubaneswar, Odisha, India

N. R. Madhu
Department of Zoology, Acharya Prafulla Chandra College, Kolkata, West Bengal, India

S. Garai
Division of Agronomy, Ramakrishna Mission Vivekananda Educational and Research Institute, Narendrapur, Kolkata, West Bengal, India
e-mail: sourav.garai@gm.rkmvu.ac.in

A. Das, A. B. Sharangi (eds.), *Remunerative Horticulture*,
https://doi.org/10.1007/978-981-92-1007-7_6

urbanisation has boosted crop diversification, and several exotic and off-season crops are now available in the market that cannot be grown traditionally. Under these contexts, the age-old method of soil-based cultivation may not fit the growing demands. To address these pressing issues and natural vagaries, soilless cultivation (SLC) practices are gaining pace. Among SLC, hydroponics, aquaponics, aeroponics and vertical farming in soilless media have become popular in recent years. SLC is well-suited for growing green leafy vegetables and herbs, annual and high-value crops, such as strawberries, saffron, etc. Hydroponics is the practice of growing plants in a nutrient-rich water solution that offers control over environmental variables and enhances input use efficiency, yield and quality. Aeroponics is an advanced version of hydroponics where nutrient-rich mist is sprayed intermittently onto roots suspended in the air. Aeroponics reduces water and nutrient usage and enhances crop growth. Aquaponics is a synergistic integration of hydroponics and aquaculture, which combines SLC and fish farming in a closed loop. All SLC techniques offer a paradigm shift from traditional farming to resource-efficient and climate-resilient crop cultivation practices that enable high-quality farm produce. Crops are prone to various pests and diseases in orthodox cultivation practices; whereas SLC reduces the risks of abiotic and biotic stresses and thus reduces the possibility of pesticide residue in the harvested crops. Recent developments in digital agriculture have provided ample scope for the adoption of artificial intelligence, machine learning and the Internet of Things for automation, where a congenial microclimate can be created under controlled environment conditions with precision input management. Under these circumstances, it is observed that current agriculture has welcomed such a transformation of farming techniques in urban and peri-urban areas, aligning with the Development Goals (SDGs), which will further expand its arena. The chapter focuses on the methodology of SLC practices, their pros and cons, and future perspectives.

H. I. Gitari
Department of Agricultural Science and Technology, School of Agriculture and Environmental Sciences, Kenyatta University, Nairobi, Kenya
e-mail: harun.gitari@ku.ac.ke

M. L. Battaglia
Regenify Ag, Ithaca, NY, USA
e-mail: martinb1@vt.edu

U. Bhattacharya
School of Agriculture and Allied Sciences, The Neotia University, Sarisa, West Bengal, India

A. J. Atapattu
Agronomy Division, Coconut Research Institute of Sri Lanka, Lunuwila, Sri Lanka

A. Hossain
Soil Science Division, Bangladesh Wheat and Maize Research Institute, Dinajpur, Bangladesh

Keywords

Agricultural sustainability · Climate change · Climate resilience · Soilless cultivation · Sustainable development goals

Abbreviations

AI	Artificial Intelligence
Ca	Calcium
CEA	Controlled Environment Agriculture
CO_2	Carbon Dioxide
DFT	Deep Flow Technique
DWC	Deep Water Culture
EC	Electrical Conductivity
FAO	Food and Agriculture Organisation
Fe	Iron
GHG	Greenhouse Gases
HVAC	Heating, Ventilation, and Air Conditioning
IoT	Internet of Things
LED	Light-emitting Diode
LED	Light-emitting diode
Mg	Magnesium
ML	Machine Learning
Mn	Manganese
N	Nitrogen
NFT	Nutrient Film Technique
P	Phosphorus
PAR	Photosynthetically Active Radiation
pH	Potential of Hydrogen
PLC	Programmable Logic Controller
PV	Photovoltaic
SDG	Sustainable Development Goals
SLC	Soilless Cultivation
SOP	Standard Operating Procedures
UN	United Nations
US	United States
UV	Ultraviolet Radiation

6.1 Introduction

Food demand is predicted to rise by 30–60% by 2050 compared to baseline levels in the early twenty-first century. The recent trend shows that food demand shifts toward more nutrient-abundant foods, such as fruits, vegetables, meat, and dairy, which will increase the demand for high-value, efficient horticultural production systems (Van Dijk et al. 2021; Huang et al. 2024; Sairam et al. 2025). In the meantime, agricultural yields are already declining, and variability is increasing due to climate change, weather extremes, and soil degradation (Bhadra et al. 2022; Gaikwad et al. 2022). Since agriculture consumes most of the freshwater extraction worldwide, the loss of productive area brought on by soil degradation, erosion, salinisation, and nutrient depletion is further threatened by water scarcity (Hossain et al. 2021). Together, these forces indicate a situation where demand is higher, and supply is less predictable, underscoring the need for mechanisms to decouple production due to water and land resources issues (Ziadat et al. 2025).

The land shortage has reduced the availability of per-capita cropland; however, food demand has increased (Nungula et al. 2023; Zhao et al. 2025). The average area of cultivated land per person worldwide has been declining over the past few decades as population expansion outpaces the growth of cultivated land, and agricultural lands are being repurposed for industry or infrastructure (Sairam et al. 2023). FAO figures depict that per-capita cropland is continuously declining in most regions, suggesting that the land base will not be enough to address future food shortages (Huang et al. 2024). Land erosion, salinity, and organic decay have adversely affected tens to hundreds of millions of hectares of land worldwide, with extreme situations in arid, semi-arid, and low-income regions (Santosh and Gaikwad 2023; Santosh et al. 2024; Rezaei Chiyaneh et al. 2023). Land degradation also limits the amount of usable agricultural land. Growing urban populations warrant fresh, high-quality produce near the centre of consumption, while the expansion of cities and competing land uses further restrict the amount of arable land available for usage near markets. Unfavourable weather conditions, such as temperature extremes, changing rainfall patterns, and biotic factors due to climate change, cause havoc on horticultural crops. There are some fruits, vegetables, and herbs that are physiologically vulnerable; even small changes in temperature, light, or water can have an impact on phenology, yield, and nutritional value. Reduced growing seasons, higher variability in quality, decreased yields, and an increase in pest-disease incidences are the effects that have been recorded and modelled (Ahmad et al. 2023; Ray et al. 2019).

The agriculture sector is extremely worried about seasonal water stress, particularly for crops that require continuous watering during the critical stages. Agriculture is more vulnerable to water competition because it accounts for a large quantity of freshwater withdrawals. Thus, horticulture production encounters substantial economic and environmental expenses due to climate volatility, and systems that ensure a steady output, such as water efficiency and buffering systems against environmental extremes, are paid more. The urgency of radical cultivation systems that disrupt production from soils (and weather) and boost output on a per unit area and per-unit-input basis stems from the two opposing forces of climate change and diminishing arable land. These approaches also provide high-quality produce to markets. They include remarkable water-use efficiency (usually reduced by 70–90% through closed-loop recirculation); nutrient-use efficiency; enhanced production through high density or vertical stacking; and resilience to climate extremes through controlled environment methods. Further, low-carbon energy sources and sensor-based automation can be included to reduce environmental loads (Regmi et al. 2024; Rajendran et al. 2024).

Socio-economic viability should be aligned with the degree of resource efficiency: adoption potential is determined by supply-chain integration, skills, capital and energy costs, and local market conditions. Therefore, transformative practices combine technological advancements with context-sensitive business strategies, bolstering policy and capacity-building to reach a diverse group of producers, from peri-urban commercial farms to urban micro-entrepreneurs (Erekath et al. 2024). In this regard, soilless cultivation (SLC) can establish its magnificent presence by increasing crop productivity and resource-use efficiency per unit area. The following are the main goals of soilless cultivation: hydroponic and vertical systems can produce more high-value vegetables and ornamentals per square metre, require less water, can be planted close to markets, and can produce year-round regardless of the local soil quality (Rajendran et al. 2024; Khatri et al. 2024). Although they are not a panacea, soilless systems provide a strategic toolkit for regions that need horticultural supplies that are robust, as well as those that face high land pressure and continuous soil degradation.

SLC, a broad term encompassing hydroponics, aeroponics, substrate culture and other types of controlled environment production, has now developed beyond a series of experimental research to become commercially viable systems in cities and marginal lands for the production of fresh horticultural products. The ability to produce nutritious food and preserve natural resources is impacted by interconnected environmental, demographic, and socio-economic factors that face global agriculture. Food systems will need to be made more productive and efficient to meet these demands. In the meantime, one of the main causes of environmental externalities is agriculture. Given that modern food systems contribute significantly to greenhouse gas emissions, biodiversity loss, water stress, and nutrient pollution, it is critical to increase food production while reducing environmental impacts per unit of output (Smirnov et al. 2024; Ray et al. 2024).

The commercial adoption of SLC expanded in the mid-1900s due to the introduction of greenhouses, controlled environment technology, and dependable fertilizers, although experiments were conducted decades before. Over the past two decades, high-density and year-round production were standardised by automated fertigation, sensor networks, LED lighting, climate control, and recirculation technologies. With its ability to increase output per square metre of leafy greens and crops with a low turnover time, stacked production has gained popularity in vertical farming and urban SLC systems (Lakhiar et al. 2025). Apart from being close to consumers, mature SLC systems have minimal soil-borne disease pressure, high yields, water and nutrient efficiency, and reduced supply chains and spoilage. Obstacles remain, including the expenses of energy and capital, the technical and administrative skills, pathogen threats to water and systems, and sustainable nutrient sourcing (Erekath et al. 2024; Rajendran et al. 2024).

The chapter focuses on the limitations of conventional crop cultivation and advantages of SLC in the current context of climate change (Fig. 6.1). Further, it

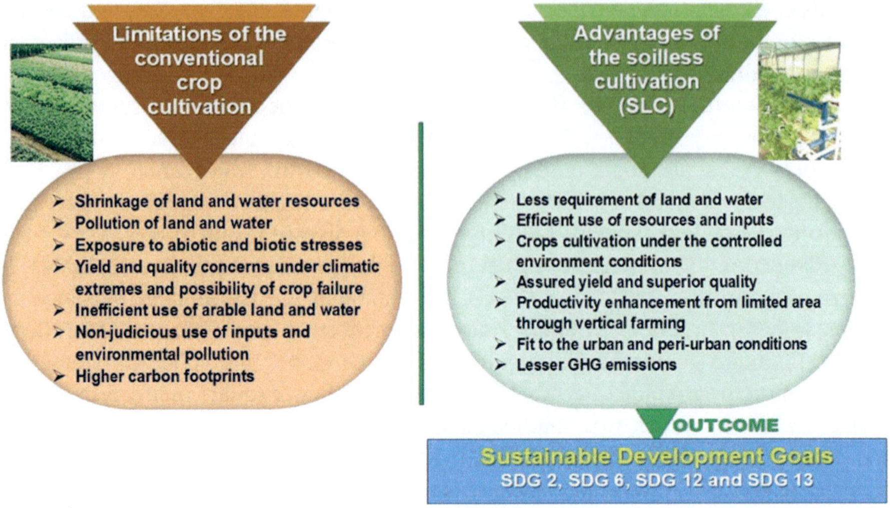

Fig. 6.1 Diagram comparing conventional crop cultivation and soilless cultivation (SLC). On the left, a brown triangle lists limitations of conventional cultivation, including land and water shrinkage, pollution, stress exposure, yield concerns, inefficient resource use, pollution, and high carbon footprints. On the right, a green triangle highlights advantages of SLC, such as reduced land and water needs, efficient resource use, controlled environment cultivation, superior yield, vertical farming productivity, urban adaptability, and lower GHG emissions. The outcome is linked to sustainable development goals 2, 6, 12, and 13

describes several SLC systems commonly used, their prerequisites, methodologies, management, other concerned aspects, and future perspectives targeting food and nutritional security.

6.2 Principles and Types of Soilless Cultivation

Soilless cultivation is a science-based method designed to provide the roots with a precise microclimate and balanced nourishment while attempting to grow plants without the need for natural soil. It is founded on the ideas of sustainable growth of the horticultural sector, the ability to manage the environment, and best use of resources. The acceptance of SLC systems has been extended because of the emerging demand for a superior, year-round, fresh and ready-to-consume goods, the diminution of farmlands and soil deterioration brought on by climate change and anthropogenic activities (Regmi et al. 2024).

The basic idea behind soilless agriculture is to (i) ensure adequate oxygen, moisture, and nutrient delivery at the root-rhizosphere interface, (ii) inject vital nutrients into plant roots in liquid or solid inert media, and (iii) control the chemical (pH, electrical conductivity) and physical (temperature, humidity, and light) conditions to attain ideal development; likewise, recycling or reusing water and nutrient solutions can minimise environmental contamination and losses (Kumar et al. 2024).

Soilless cultivation is at the forefront of sustainable urban and peri-urban crop production because of its high production potential, uniformity, diversity and outstanding quality based on the aforesaid criteria (Erekath et al. 2024).

6.2.1 Basic Concepts of SLC

SLC is a crop production method that replaces a wide range of media for anchoring plant roots and easing nutrient absorption. This medium can be either an inert solid material or a liquid nutrient solution. This concept was developed assuming that soil is a significant physical platform and a nutrient storage factor that can be optimised or redeployed through designed systems. A plant's nutrition can be properly controlled with soilless systems, since all the macro- and micronutrients it needs are dissolved in water and given to it in precisely the right amounts. Soil constraints such as compaction, soil structure, pathogen load, and salinity are not caused by effective nutrient management (Khatri et al. 2024). Also, growing crops in soilless culture makes it easier to control the environment and produce a high yield in saline, desert, or urban locations with limited soil (Rajendran et al. 2024). The technique makes sustainable intensification possible by reducing fertiliser loss (50–60%) and water use (up to 70–90%) compared to open-field farming (Lakhiar et al. 2025). Consequently, SLC can be characterised as a paradigm shift in horticultural production, encompassing precision, sustainability, and flexibility.

6.2.2 Classification of SLC Systems

Depending on the type of growing media and nutrient supply, SLC can be categorised into the following.

- Systems based on liquids, where plant roots are suspended in nutritional solutions and may or may not have mechanical support.
- Systems based on solid media, in which a nutritional solution is supplied via fertigation or irrigation, while inert materials provide structural support.

There are numerous variations of the categories that vary according to the crop choices, availability of resources and the production level.

6.2.2.1 Liquid-Based Systems

- *Hydroponics* is an advanced soilless farming method that uses liquid solutions. Plants grown in water-based systems get the required nutrients from fertiliser solutions that flow either continuously or at regular intervals. Further, it ensures that plants receive proper nutrition under SLC. The most widely adopted hydroponics methods are deep water culture (DWC), ebb and flow, and nutrient film technique (Rajaseger et al. 2023). These systems provide high amounts of nutrients and oxygen, resulting in 20–50% higher productivity of tomato and leafy greens compared to soil-based systems (Vikanksha and Singh 2023; Sahu et al. 2025; Munda et al. 2025).

- *Aeroponics* is an advanced method where plant roots hang in the air and receive nutrients through a fine mist spray. The aeroponics technique allows plants to grow without soil and helps them develop further through direct nutrient absorption. As Fussy and Papenbrock (2022) mentioned, aeroponics cuts the crop duration by 30%, resulting in faster growth rates and an efficient root oxygen supply. They also advocated that aeroponics is a helpful method for growing root crops like potatoes and yams, and it can be used to produce disease-free planting material.
- *Aquaponics* is the practice of cultivating plants in water systems while also raising fish. Fish waste aids in the growth of plants without soil in this process, which undoubtedly employs a closed loop. While plants filter the water for fish, fish faeces give plants natural nutrition. Moreover, it establishes a mutually advantage cycle in which the two living entities support one another's growth. According to Hutchinson et al. (2025), compared to conventional fish farming techniques, the integrated aquaponics system utilises significantly less waste and saves over 90% on water and fertiliser expenditures.

Despite being extremely water-efficient and ecologically controlled, liquid-based systems require careful monitoring of dissolved oxygen, pH, electrical conductivity (EC), and potential pathogen accumulation in recirculating solutions.

6.2.2.2 Solid Media-Based Systems

In SLC systems, solid media are used to provide mechanical support to plants and anchorage for roots. The solid media maintain air and moisture in the rhizosphere. There are several media used in SLC. Of them, perlite, vermiculite, river sand, cocopeat, clay pebbles, oasis cubes and rockwool are common. Perlite is an inflated volcanic mineral with lightweight, sterile properties that facilitate aeration, and it works well when combined with other media (Fig. 6.2a). Vermiculite is a hydrated mineral that enhances nutrient retention and has an excellent cation exchange capacity (Fig. 6.2b). River sand is a common solid medium used in hydroponics due to its excellent drainage and aeration qualities. Because of its roughness and breathable roots, it does not remain totally water saturated, reducing the likelihood of fungal disease. However, because of its low water-holding capacity, it may require frequent irrigation and must be cleaned well to remove silt and organic contaminants before application (Fig. 6.2c). Cocopeat (coir dust), a by-product of coconut husks that is biodegradable, renewable, and has a high capacity to retain water and aerate the soil. It is utilised frequently in tropical horticulture (Fig. 6.2d). Clay pebbles provide solidity to the construction and can be reused after sterilisation (Fig. 6.2e). Oasis cubes are a type of lightweight, stiff foam growing medium which is frequently used in soilless cultivation, seed germination, and seedling development, because of their excellent water retention, uniform moisture distribution, and efficient aeration. The oasis cubes are easy to handle, pre-cut, and transplantable; they are employed commercially in nurseries and controlled environment agriculture (Fig. 6.2f).

Fig. 6.2 Six images of different growing media labelled **a** to **f**. (**a**) Light grey granular material. (**b**) Brown, coarse particles. (**c**) Fine, yellowish-brown powder. (**d**) Dark brown, fibrous material. (**e**) Brown, round pellets. (**f**) Brown, rectangular block with small holes

Solid-media systems provide an option between hydroponics and conventional soil culture due to their ease of use, moderate cost, and lower technical requirements. According to Ali et al. (2024), they can be used by small-scale farmers and commercial greenhouses. However, non-renewable materials like rockwool are hard to recycle, and research is underway to create biodegradable ones (rice husk ash, biochar, and composted organic waste).

6.2.3 Comparative Overview: Soil-Based vs. Soilless Systems

Soil-based and soilless cultivation differ in terms of the way agriculture is physically supported, as well as the way water and nutrients are supplied to the root zone. Different trade-offs, such as productivity, resource use, risk exposure, and risk management complexity, are prominent concerns.

- *Resource-Use Efficiency*: When compared to open-field soil production of high-value horticultural crops, soilless systems (most notably recirculating hydroponics and aeroponics) can reduce crop water use by 50 per cent or more because of root-delivered water and recycling (in closed systems). The efficiency of nutrient usage is improved because leaching losses are reduced and the nutrient solution is precisely designed.
- *Yield and Productivity Per Area*: The high planting densities, regulated microclimate, and ideal root nutrition of intensive soilless systems (greenhouse hydroponics, vertical farms) usually result in higher yields per square metre and faster harvests of leafy greens, herbs, and most fruiting vegetable crops (Pomoni et al. 2023).

- *Environmental Exposure and Resilience*: Local weather, seasonal restrictions, soil-borne illnesses, and degraded soils continue to affect soil systems. When combined with controlled environment technologies, soilless systems improve supply homogeneity by separating the production process from local soils and limiting exposure to a wide range of soil diseases and environmental extremes. However, they emphasise the dangers of waterborne infections in the event of inappropriate recirculation and technical issues (pumps, nutritional dosing, HVAC, energy supply) (Vought et al. 2024).
- *Capital, Energy and Skill Requirements*: Soilless cultivation systems require greater expertise for fertiliser and system control and need higher initial capital (greenhouse structures, lighting, pumps, sensors). Further, provided they are equipped with energy-saving LEDs and renewable energy sources, fully indoor systems (vertical farms) may have a noticeably higher electrical consumption for lighting and climate management (Rathor et al. 2024).
- *Sustainability Trade-Offs*: The SLC systems should be orchestrated with circular approaches, namely, the provision of resource recapture, use of renewable energy and biodegradable media, enabling higher nutrient recovery, embodied energy and material issues to achieve the positive environmental outcomes.

SLC systems are superior to conventional crop production when there is a shortage of land, water, or market accessibility. However, the initial capital investment may be a concern.

6.2.4 Key Components of SLC Systems

The success of SLC relies on some key components, namely, the root zone and growth medium environment, the formulation of the nutrient solution and dosage, pH and water quality management, and environmental control and automation. The system's resilience, resource-use efficiency, and crop output can all be determined by the interactions between these elements (Fathidarehnijeh et al. 2023).

6.2.4.1 Growth Media

The substrate provides anchoring, water retention, aeration, and a slight buffering of nutrient dynamics in solid media systems; in liquid systems, plants can be stabilised using materials (collars, clamps) or additional media. The choice of growing media for SLC depends on the particle size, load of bioburden, cation exchange capacity, water-holding capacity and porosity (Sambo et al. 2019).

6.2.4.2 Root Zone Microclimate

Temperature, oxygenation, and moisture dynamics at the root zone directly affect how well roots respire and absorb nutrients. In stagnant deepwater cultures, aeration is provided by operating air pumps, ensuring intermittent flow, and thin film approaches, because roots cannot perform their physiological activities in environments with low oxygen levels (Samarakoon et al. 2017). The media selection should be in line with the crops' water needs and irrigation schedule to prevent root hypoxia or chronic drought stress.

6.2.4.3 Nutrient Solution Formulation and Management

Formulation and management of nutrient solution depend on the following considerations.

- *Fundamentals*: All macro- and micronutrients are available as soluble forms in nutrient solutions. Target concentrations (in terms of ionic proportions and electrical conductivity, or EC), sufficient but non-toxic concentrations of each element, and the removal of imbalances (such as antagonistic relationships between K, Ca, and Mg) that compromise uptake are the main management goals.
- *Principles of Formulation*: Hoagland solutions are considered the basic recipe at the beginning; however, during later growth stages, crop and stage-specific recipes are provided. Different physiological stages need various nutrients, as fruit set and ripening require changes in Ca, K, and P, while vegetative development requires higher N and K. Temperature, pH, dissolved oxygen, and EC must be measured often. The replenishment mechanism differs depending on the system: recirculating systems measure depleted ions and replenish or substitute solution based on EC and crop demand; however, open (run-to-waste) systems replenish the nutrient solution regularly to prevent the accumulation of pathogens (Fathidarehnijeh et al. 2023).
- *Nutrient Dynamics and Sustainability*: Recirculating systems increase the possibility of ion-selective depletion or accretion; therefore, certain ion concentrations (not just EC) are regularly measured to control them. The creation of nutrient recipes with less environmental impact during disposal or leakage, as well as the recovery of nutrients (in municipal or agricultural waste streams), are receiving increasing attention (Palmitessa et al. 2024).

6.2.4.4 Water Quality and pH Management

- *Water Quality Parameters*: Water quality, which includes salinity/EC, bicarbonate, hardness, sodium adsorption ratio, heavy metals, and microbial load, influences nutrient formulation and plant response to nutrients. The salinity stress due to high EC or alkalinity should be measured and revamped. Such issues can be checked through need-based softening, reverse osmosis or an altered nutrition strategy.
- *pH Control*: Root absorption, microbial activity, and nutrient solubility are all impacted by pH. In commercial systems, routine pH measurement and automated acid/base addition are standard practices because most horticultural SLC crops respond best at pH values of 5.5–6.5. Values below or above this range make certain nutrients (Fe, Mn, and P) less or more available, while moderately buffering or buffering substances in the water or substrate can also cause pH drift (Miller et al. 2020).
- *Control of Pathogens and Contaminants*: The presence of waterborne pathogens (such as *Phytophthora*, *Pythium*, and other microorganisms) can quickly infect

recirculating water. The list of disinfection techniques includes UV, ozonation, microfiltration, and periodic heat or chemical disinfection; however, the cost and the effect on beneficial microbes should be considered (Wallace-Springer et al. 2022).

6.2.4.5 Environmental Control and Automation

- *Climate Control Factors*: The primary above-ground components of photosynthesis and transpiration are ventilation, CO_2 level, relative humidity, light (intensity, spectrum, and photoperiod), and air temperature. Combining all of these factors with root zone control, controlled environment agriculture (CEA) creates ideal growth conditions that can significantly reduce growth time and promote uniformity of growth.
- *Automation and Sensing*: Sensors (EC, pH, temperature, humidity, CO_2, PAR, leaf temperature, soil/root zone moisture), actuators (pumps, valves, HVAC, lighting), and controllers (PLC, microcontrollers) are contributing to the rising automation of SLC systems. Enhancing resilience and reducing labour requirements, predictive control, defect detection, and energy management are made possible when combined with the Internet of Things and machine learning (Ojo and Zahid 2022).
- *Energy and Sustainability Factors*: Energy demand is one of the primary operational costs and sustainability-limiting factors, especially for fully indoor vertical farms. Energy-efficient LEDs, waste-heat recovery, and on-site renewables (solar PV, biogas) can help reduce lifecycle emissions. Energy management must be integrated with crop scheduling and control techniques to strike a balance between cost, carbon footprint and yield.
- *Human Factors and Management*: The need for skilled technical personnel to calibrate, maintain, and interpret data is essential, as automation reduces the need for unskilled workers. Redundancy of critical equipment (pumps, controllers), a well-designed warning system, and clearly specified SOPs are necessary to minimise catastrophic losses from single-point failure (Rathor et al. 2024).

6.3 Major Soilless Cultivation Techniques

Soilless farming refers to innovative methods of growing plants in which nutrients are given in water or a non-nutrient medium rather than using natural soil. These systems enable precise control over ambient factors, root zone aeration, and nutrient availability. The ease of scale, flexibility, and high production in greenhouse and vertical farm buildings make hydroponics the most popular of the several soilless plant cultivation techniques, including hydroponics, aeroponics, and aquaponics (Yang and Kim 2019).

6.3.1 Hydroponics

With hydroponics, plants are grown without soil by submerging their roots in a nutrient-rich water solution. Plant roots anchor to inert materials, such as perlite, cocopeat, vermiculite, rockwool or potting mix. The latest research findings revealed that hydroponics delivers plant nutrients and water directly to plants, reducing input loss and enhancing crop yield. A hydroponics system works based on the proper and efficient nutrient and water management, interacting with the crop root systems on a medium.

6.3.1.1 Hydroponic System Designs for Optimised Nutrient Delivery and Aeration

Various hydroponic systems have been developed over time to enhance fertiliser delivery, oxygen supply, and root health. The most popular of these technologies is the nutrient film technique (NFT) (Fig. 6.3). The roots receive the same oxygen and nutrients as the nutrient solution runs continually in thin layers along slanted channels. Fast-growing crops and herbs are suitable for the NFT, as they have a shallow root system. Only the roots of the plants are submerged in water containing nutrients and air bubbles in the deep water culture (DWC) method (Fig. 6.4). Since it maintains the root temperature and provides the same consistent nutrients, this technique essentially works for spinach and other green vegetables. The idea behind ebb and flow, another hydroponics system, is the same (Fig. 6.5). In this system, a continuous flow of nutrient-rich water is maintained that supports plant roots to absorb nutrients and oxygen. Under the wick system, short-statured herbs can be grown (Kaur et al. 2023).

Fig. 6.3 A composite image showing a hydroponic farming setup. On the left, a diagram illustrates a system of blue pipes and tubes connected to a red container, demonstrating water flow and plant support. On the right, three photographs depict various stages of plant growth in a greenhouse environment. The top image shows rows of leafy greens growing in a controlled setting. The middle image highlights a close-up of plants supported by a blue pipe structure. The bottom image displays a larger view of a greenhouse with different types of plants, including red and green varieties, arranged in rows

Fig. 6.4 Hydroponic setup with leafy greens growing in a yellow tray supported by an orange container. The left image shows the plants from above, highlighting lush green leaves. The right image reveals the underside, displaying roots extending from net pots into the container, which holds water

Fig. 6.5 Vertical hydroponic system with green and orange stacked containers, each holding leafy plants. The left image shows the full structure with plants growing outward. The middle image provides a top-down view, revealing the arrangement of containers and visible roots. The right image focuses on a close-up of the containers, highlighting plant growth and root exposure. The setup is located in a greenhouse environment

6.3.1.2 Nutrient Management in Hydroponic Systems

Nutrient management is vital for crops grown in a hydroponic system. Most systems rely on modified Hoagland solutions that include the nutrient requirements of specific crops and growth phases (Sulaiman et al. 2025). Critical factors include pH (generally maintained between 5.5 and 6.5, where nutrients are best absorbed) and electrical conductivity (EC), which is typically maintained between 1.2 and 3.5 dS m-1. Root health and dissolved oxygen content are directly impacted by the temperature of the nutrient solution, which should ideally be between 18 °C and 24 °C. These ratios are maintained by using precise amounts of micronutrients (Fe, Mn, Zn, Cu, B, Mo) and macronutrients (N, P, K, Ca, Mg, S). To increase the bioavailability of micronutrients, chelating agents can be added to make them more stable. IoT-based monitoring and automation systems are increasingly used to monitor and control contemporary hydroponic farms. A hydroponics system facilitates the highest accuracy and uniformity to determine the pH, dissolved oxygen remaining, and nutrient supply (Samarakoon et al. 2017).

6.3.1.3 Crop Suitability and Yield Potential

The hydroponic system has the potential to raise a wide range of crops. Leafy vegetables such as fenugreek, spinach, pakchoi, Swiss chard and lettuce can be grown in NFT and DFT as they are short-duration crops with a shallow root system, which yield about 20–40 kg/m^2 per year (Gobilik et al. 2021; Sathyanarayana et al. 2022; Kusmiyati et al. 2023; Pastor-Arbulú and Rodríguez-Delfín 2025). Fruit crops need ebb and flow or drip-fed substrate systems, which may yield 20–30% higher than crops cultivated on soil under controlled environment conditions (Gautam et al. 2024; Sharma et al. 2024). The quality and yield are enhanced when they are grown in the hydroponic systems (Xu et al. 2025; Lin et al. 2025). Herbaceous crops with desired aroma like celery, basil, parsley, mint, fenugreek and coriander are suitable to grow under the controlled environment of hydroponic and vertical farming to maintain their quality (Dsouza et al. 2025; Shanila et al. 2024).

6.3.1.4 Advantages and Limitations of Hydroponic Farming

Hydroponics provides several noteworthy benefits over traditional soil-based crop farming. Proper crop management enhances the quality and productivity of crops (Table 6.1). It saves two important inputs, namely water and nutrients, by 90 per cent and 50 per cent, respectively, by ensuring their efficient use (Rajendran et al. 2024). Further, it produces pest-free, disease-free, and healthy crops in a controlled environment, which has made hydroponic cultivation popular among farmers and hobbyists in urban and peri-urban areas. Besides, vertical space is used in the soilless culture, which reduces land requirement and enhances productivity from a unit area (Regmi et al. 2024). However, there are limitations that restrict the spread of hydroponics cultivation of crops. There is a need for knowledge and precise skills to manage crop requirements, the cost involved in the erection of a controlled environment structure and the necessary things to create a microclimate, precise maintenance and power supply requirements (Vought et al. 2024). Further, improper sanitation or lack of biosafety measures may create the possibility of disease infection and/or inadequate maintenance can create problems in the water recirculating and microclimate controlling system (Pomoni et al. 2023).

6.3.2 Aeroponics

Aeroponics is a soilless crop cultivation method that comprises recurrent misting or fogging plants with or without nutrients, though they remain suspended in the air (Fig. 6.5). The aeroponics approach has superiority over conventional crop cultivation with a higher nutrient absorption by plants, augmented aeration for roots, and reduced water needs (Garzón et al. 2023). Aeroponics is an appropriate soilless

Table 6.1 Research evidence supporting the benefits of hydroponic cultivation of crops

Crop	Salient features	References
Spinach	Survival and growth rate were increased due to a proper contact between roots and nutrients The adequate and continuous supply of water, oxygen and nutrients, enhanced the yield of spinach	Gaikwad et al. (2020)
Cherry tomato	A significant enhancement of the fruit size due to appropriate application of nutrients in hydroponics The hydroponics cultivation of tomato improved the fruit quality such as carbohydrates, ascorbic acids, tannins, phenols and flavonoids	Kumar et al. (2024)
Pepper	The nutrient application along with proper oxygen availability in a hydroponic system reported an enhanced chlorophyll content, fresh and dry mass of shoot and root Proper aeration and oxygen supply in the nutrient solution in hydroponics system optimized photosynthesis and promoted growth of pepper	Roosta (2024)
Spinach	Better leaf quality parameters such as leaf area, length and size were observed in NFT hydroponic system Enhanced photosynthetic activity and nitrogen accumulation in the leaves were recorded	Gaikwad et al. (2024)
Lettuce	Lettuce cultivated in NFT resulted in an enhanced root system, leaf length, leaf width, shoot length, and root length in Romain cultivar Cultivation of lettuce in NFT lead to increase in chlorophyll content and nitrogen content	Sahu et al. (2025)

vertical agriculture for urban farming and scientific investigation, as it has a simple and easily manageable compact design under a controlled environment (Mahmoud and Moursy 2025).

6.3.2.1 Principle and Working Mechanism

In aeroponics, the roots get sufficient air while the plants themselves are supported by an air-supporting structure or kept in a tray. The optimal nutritional and oxygen conditions are achieved by introducing a nutrient solution as a fine spray at an average rate using misting nozzles or ultrasonic foggers. As the roots are exposed to oxygen, the automated misting schedules promote faster growth and closed-loop designs avoid wasting water and nutrients.

6.3.2.2 Nutrient Misting and Root Zone Management

In the form of tiny droplets, most of the nutrient mists contain essential macro- and micronutrients. An aerosol that is too small may produce root hair growth without lateral root growth to support growth, whereas droplets that are too big can lower oxygen levels (Tang et al. 2024). To improve misting efficiency, it also features ultrasonic foggers; nonetheless, its operation requires regular maintenance to prevent obstructions from mineral deposits (Garzón et al. 2023).

6.3.2.3 Suitable Crops and Productivity Benefits

Aeroponics supports a wide range of horticultural crops, including fruiting crops like strawberries and tomatoes, herbs like basil and mint, and leafy greens like spinach and lettuce, with higher yields and superior quality (Appicciutoli et al. 2025).

6.3.2.4 Water and Nutrient Efficiency

Aeroponics is an advanced version of limited-water crop cultivation. In this system, nutrient-rich fluid is supplied to plant roots, which saves more than 90% water used in conventional farming. Aeroponics has the potential to supply the precise quantity of nutrients (Salma et al. 2024).

6.3.2.5 Technological Challenges and Maintenance Issues

An aeroponics system requires close supervision as it contains fogging and misting nozzles. Such moisture delivery tools may suffer from clogging, and to circumvent this, periodic maintenance is essential (Joshi et al. 2022). The smooth operation of microclimate, and misting or fogging requires the provision of an uninterrupted energy supply. Automation is an ideal option to manage an aeroponics system efficiently, which is more expensive than a low-cost vertical farming system.

6.3.3 Aquaponics

Aquaponics is an SLC method that combines hydroponic crop cultivation and aquaculture. An aquaponics system facilitates a perfect symbiotic relationship between crops and aquatic animals, including fish, for nutrient cycling (Debroy et al. 2025). The feed leftovers from fish and their faecal materials supply nutrients to crops, which improves the water quality. It is a popular SLC system for hobbyists and urban dwellers, which is potentially important, as it is an ecologically friendly food production system (Somerville et al. 2014).

6.3.3.1 Fish and Plant Selection and Nutrient Cycling

The choice of fish and crop species signifies the sustainability of an aquaponics system. The fish species chosen should be resilient to abiotic and biotic stresses with a faster growth habit, such as tilapia, trout and catfish, etc. Crops are to be low nutrient-requiring and faster nutrient-absorbing, namely, lettuce, basil, aromatic herbs, legumes, and fruiting plants (Steffi et al. 2025). In aquaponics, the role of microbes is vital, as they convert ammonia generated from fish excreta into nitrite and nitrates, which provide nutrients to crops.

6.3.3.2 System Balance and Biofiltration

Managing an aquaponics system is important to avoid stress in fish due to water toxicity and nutrient status. The provision of biofilter substrates enables the proliferation of beneficial microbes that mediate the nitrification process. Regular checking is essential for pH, dissolved oxygen, salinity, temperature, and nutrients. Sanitation in crop and fish environments needs to be maintained carefully to prevent disease infection for fish and crops (Somerville et al. 2014). In this regard, an IoT-enabled automated setup is beneficial (Steffi et al. 2025).

6.3.3.3 Sustainability and Circular Economy Perspectives

The aquaponics system is aligned with the core principles of the circular economy. The symbiosis is an aquaponics system, as the conversion of fish waste into crop nutrients directly and indirectly benefits the environment. It reduces effluent release in the environment and chemical plant nutrient needs. It enhances water use efficiency (Debroy et al. 2025). Moreover, aquaponics ensures local food supply in urban areas, nutritional security, and revenue generation (Tunçelli et al. 2025).

6.3.3.4 Spread of Successful Aquaponic Systems

The commercial and experimental application of aquaponics is well established in the world in urban areas. In the US, the production of fish and vegetables for local markets is common. Singapore witnessed urban rooftop aquaponic farms. In India, the combination of tilapia and fish was proven for higher productivity, efficient nutrient use, and economic viability (Somerville et al. 2014; Debroy et al. 2025).

6.3.4 Vertical Farming

In vertical farming, crops are cultivated in stacks, often set up in buildings, warehouses, or shipping containers. Such farming maintains the principles of CEA, reduces water and land use for crop production, and enhances crop yield per unit area. In urban clusters, vertical farming is practiced under the CEA for vegetable production with artificial lighting systems (Benke and Tomkins 2017; Kabir et al. 2023).

6.3.4.1 Design and Components

In vertical farming systems, crops are grown in a controlled environment; hence, the following rudiments are considered.

- *Cultivating Modules*: Containers or structures, where plants are maintained, often utilising hydroponic or aeroponic systems.
- *Environmental Control Systems*: The optimum conditions for cultivation include advance mechanisms which can regulate temperature, humidity, CO_2 levels and ventilation.
- *Artificial Lighting*: Precise spectral bands through LED are important for growth and development.
- *Automation and Monitoring*: Plant vigour, nutrient profiles and climatic parameters are tracked by the use of sensors and monitoring technology.

The design focuses on space optimisation, energy efficiency, and scalability, which is different from general systems which only focus on scalability.

6.3.4.2 The CEA Merger

CEA is the cultivation of crops in a meticulously regulated environment, especially for factors like light, heat, humidity and nutrients. In vertical farming, CEA plays a vital role, as it guarantees harvest regardless of the external weather scenario. Such a synergy promotes diversification of crops in cityscapes, decreases dependency on synthetic farm inputs and ensures year-round production (Kabir et al. 2023; Despommier 2013).

6.3.4.3 Artificial Illumination Frameworks

Vertical farming is heavily dependent on artificial light, as the indoor natural light is not enough for photosynthesis. LEDs are mostly used these days due to their unique wavelength, which is optimum for absorption and ensures less energy consumption. Recent research shows that the use of enhanced spectral quality and intensity, LEDs boost plant growth and development (Farhangi et al. 2025). The proper light setup to fulfil the needs of crops as per the need, which ultimately increases yield.

6.3.4.4 City and Fringe-Zone Application

The increasing urbanisation leaves less area of arable land, hence making vertical farming more suitable in urban and fringe zones. The unused spaces, such as rooftops, empty lots, and abandoned buildings, can be used for growing fresh produce, decreasing transportation costs and managing carbon emissions. Moreover, this vertical integration in a limited space ensures food and nutritional security for the local people (Despommier 2013; Kalantari et al. 2018).

6.3.4.5 Fiscal Factor and Power Demands

One of the main drawbacks of vertical farming is high costs and energy consumption. The operational cost includes lighting, climate control systems and automation, which requires a significant amount of electricity. However, using energy-efficient technologies like LED lights, renewable energy like solar panels, and local production can decrease the transportation costs, leading to financial success (Kalantari et al. 2018; Ji et al. 2023).

6.4 Advantages of Soilless Cultivation

Soilless cultivation, including hydroponic, aeroponic and aquaponic, is a competitive alternative to conventional farming. The soilless cultivation provides a better resource-use efficiency, enhanced agricultural productivity and potential for urban and peri-urban agriculture (Sathyanarayana et al. 2022).

6.4.1 Efficient Water and Nutrient Utilisation

The enhanced efficiency of nutrient and water use is one of the primary focuses of soilless cultivation. The recirculation of nutrients and water in hydroponics and aeroponic systems reduces the loss by 90%, compared to soil-based farming. Leaching losses are minimised, and nutrient and water uptake are increased due to root zone delivery. In water-scarce areas, soilless farming is ideal as it increases water-use efficiency and conserves resources.

6.4.2 Enhanced Yield and Product Quality

The controlled environmental factors and nutrients lead to enhanced yield and better quality produce. Crops grown in hydroponic and aeroponic systems produce a stable and enhanced yield with higher nutritional quality as compared to soil-grown crops (Despommier 2013). SLC facilitates plants to flourish to their full potential due to an optimum availability of inputs, without soil-borne diseases and enhanced flavour with a longer shelf life.

6.4.3 Year-Round Production and Climate Resilience

Crops are grown under CEA, and SLC are unaffected by outside climatic anomalies, which helps in year-round and off-season production. The precise use of inputs and climatic factors such as temperature, humidity, CO_2, and light decreases the impact of abiotic and biotic stresses and seasonal variability. Hence, increasing food security and climatic resilience.

6.4.4 Reduced Pest and Disease Incidence

SLC reduces the risk of soil-borne pest and disease attacks. The reduced biotic stress also decreases the use of chemicals, hence providing a safer and pesticide-free produce. The production of good quality with a higher yield can be achieved through this system.

6.4.5 Space Optimisation for Urban Agriculture

In the era of urbanisation, arable land is decreasing but leaves ample scope for expansion in vertical space. These vertical spaces, such as rooftops, can be used for vertical farms in urban and semi-urban areas. The use of a hydroponic system can increase production per unit area, which reduces the stress on landless farmers.

6.5 Challenges of Soilless Cultivation

Several issues pose challenges for SLC, related to its implementation and sustainability (Khan et al. 2020). The technical, operational and financial factors should be considered to overcome the challenges.

6.5.1 High Initial Investment and Operational Costs

Some SLC systems, such as hydroponics and aeroponics, have high initial and running costs. The infrastructure costs, fertiliser application methods, controlled climatic conditions, and artificial lighting are higher than those of traditional agriculture. The energy needs for lighting and controlling the climatic conditions are also high for a producer (Al-Chalabi 2015).

6.5.2 Technical Know-How and Skill Requirements

Plant physiology, environmental control, nutritional management, and system maintenance require specific skills and knowledge. Continuous monitoring for water flow, fertiliser solutions, and climatic variables must be be managed by a skilled person. Hence, farmers need training and support to harvest better yields, reducing the chance of crop failure (Ghehsareh et al. 2012).

6.5.3 Energy Consumption and System Sustainability

The use of LEDs and climate control systems requires a continuous energy supply. Even though the soilless system has the capability for year-round production, such systems are energy-dependent and energy-intensive, and the limitation of energy can lead to a complete shutdown. For these systems to be sustainable, renewable energy sources are essential because conventional energy may negate the environmental benefits of using less water and pesticides (Despommier 2013).

6.5.4 Risk of System Failure

Technology plays a key role in soilless farming. Crops can quickly become stressed or perish due to issues with the fertiliser delivery system, pumps, or sensors. Whereas soil can even out minor fluctuations, hydroponic and aeroponic systems instantly transfer changes in water availability or fertiliser content to plants. Preventive maintenance, redundancy, and strict monitoring are therefore necessary to lower the likelihood of system failure.

6.5.5 Market Acceptance and Economic Feasibility

Although soilless culture produces high-quality pesticide-free crops, consumers may not accept this. Demand in the market may fluctuate since some consumers may consider vegetables grown hydroponically or aeroponically to be less natural. The cost of cultivation is rather high under the SLC, which could demand premium pricing or niche markets (Steffi et al. 2025).

6.6 Integration of Digital Technologies and Automation

In the current era of digital innovations, the precision management of crops has become possible, which has shaped farming into Agriculture 5.0. In the CEA, the scope and potential of automation through the adoption of digital technologies are quite applicable, where resource allocation and optimisation can efficiently deliver the target output. Digital agriculture can automate the SLC by employing IoT, AI, and ML for crop management, input cost reduction, advanced monitoring, and quality output (Garzón et al. 2023; Santosh and Gaikwad 2023; Steffi et al. 2025). The IoT-enabled system with sensors supports real-time monitoring of climatic factors, such as temperature, humidity, light intensity, soil moisture, and soil nutrient status, under the CEA. Actuators deliver precise inputs and control the microclimate from a remote location (Jaeger 2024; Kalantari et al. 2018). AI and ML analyse data to predict crop growth and yield. Robotics can support crop management and harvest in an SLC system.

6.7 Crop-Specific Applications

In the SLC, several crops can be grown. The crop management practices, namely, nutrient supply, creation of specific climatic conditions, and adoption of crop-specific interculture, can ensure the target yield and quality during the crop season. As the SLC is preferably practiced under the cation exchange capacity. (CEC), the crop growing season can be manipulated or extended, based on market demand and growers' choice (Garzón et al. 2023; Gaikwad and Maitra 2020).

When lettuce, spinach, basil, and coriander leaves are cultivated in the SLC systems, the hydroponic and aeroponic systems are preferred for faster growth, proper nutrient management, and high-quality produce (Kalantari et al. 2018). Moreover, the CEC prevents disease infection, improves nutrient-use efficiency, and nutritional value, colour, and lustre of produce (Kozai et al. 2019). Fruiting vegetables, such as tomatoes, brinjal, cucumber, and capsicum, are preferably grown under nutrient film technique and deep flow technique in hydroponics, facilitating a continuous flow of water and nutrients (Lakhiar et al. 2020). Off-season cultivation and longer harvest are made possible due to controlled environmental conditions with less abiotic and biotic stresses (Buehler and Junge 2016).

SLC has become popular for ornamental plants, namely, orchids, strawberries, and saffron. These crops are always susceptible to soil-based illnesses since they require particular environmental conditions. Controlling the oxygen, nutrients, and moisture in the root zone through hydroponics and aeroponics can boost yield and enhance product quality (Treftz and Omaye 2016). Vertical farms in North America and Europe can produce year-round, high-quality fruits and flowers, such as orchids and strawberries, in contrast to traditional soil-based agriculture (Dhanasekar 2025).

Comparative empirical studies have consistently demonstrated that soilless farming produces higher quality and higher production than soil-based crops. Fruiting crops have a longer shelf life, better flavour, and uniform size, whereas leafy vegetables provide better biomass with high nutrient content (Khan et al. 2020). Several commercial enterprises have established the efficiency of soilless farming around the world. In Europe, cucumbers, tomatoes, and strawberries are grown in greenhouse systems by using a hydroponic system with the least use of pesticides, whereas in Japan and Singapore, green leafy vegetables and herbs are grown for the city market in vertical farming systems (Kozai et al. 2019; Garzón et al. 2023). Hence, soilless methods are sustainable and economically viable in areas with limited arable land and harsh climatic conditions.

6.8 Environmental and Socio-Economic Implications

Beyond food safety, employment generation, and maximum resource utilization, soilless farming offers environmental and socio-economic benefits. With the increasing demand for high-quality food with limited arable land and water resources, especially in urban and semi-urban areas, the SLC enhances efficiency and cost-effectiveness when digital technology and sustainable practices are integrated.

6.8.1 Resource-Use Efficiency and Carbon Footprint

Enhanced resource-use efficiency is achieved through the precise application of water, nutrients, and energy in SLC techniques. Fertiliser application in the root zone increases its efficient use, which saves 70–90% of water in a hydroponic system compared to traditional farming. Low-carbon emissions and short-distance transport contribute to a reduced carbon footprint.

6.8.2 Reduction of Pesticide Residues and Food Safety Enhancement

The unavailability of soil in SLC decreases the possibility of soil-borne pest and disease incidence, which curtails the use of pesticides, facilitates food security and nutritional quality, with consumer health benefits. Crops grown under a hydroponic system minimise harmful microbial contamination compared to the traditional method. Additionally, the superior quality produce meets consumer preferences and fetches good market value.

6.8.3 Waste Management and Circular Economy Prospects

The efficient recycling of fertiliser and water solutions in the SLC promotes the circular economy. In particular, the use of closed systems for water and nutrient recovery and reuse leads to proper waste management. Additionally, the impact can be decreased by using substrates, such as cocopeat, rockwool, and perlite, which can be recycled and reused. Furthermore, sustainability is enhanced by organic composting of urban waste or aquaponics.

6.8.4 Employment Generation and Urban Livelihood Support

The use of spaces in metropolitan areas for the SLC helps in creating jobs. Experts are needed for the smooth operation, supervision, and management of the soilless system. Moreover, job openings in supporting, marketing, and logistics industries related to soilless cultivation can be created. Moreover, growing fresh vegetables and herbs in urban farms not only provides a livelihood but also increases food security and boosts the local economy.

6.8.5 Policy Support and Economic Viability

The use of sustainable technology, supporting infrastructure costs, ensuring technical support, subsidies, and market arrangements are important for the expansion of the SLC. The proper management of a soilless system can lead to profit booking after two to three years of commercialisation, if marketed to the target urban consumers. Hence, the sustainability and viability of the SLC depend on the proper policy support and policy-making, and creation of demand among urban consumers (Kozai et al. 2019).

6.9 Soilless Cultivation and Sustainable Development Goals (SDGs)

The sustainable production of crops in hydroponics, aeroponics, and aquaponics aligns with the UN's sustainable development goals (SDGs). The potential for enhanced yield and quality with optimised resources, supports environmental sustainability and climatic resilience (Kalantari et al. 2018; Garzón et al. 2023; Ray et al. 2025).

6.9.1 Alignment with Sustainable Development Goals

The production of nutrient-rich food crops under limited available resources aligns with SDG2 (zero hunger). The year-round availability of fresh produce and increased yields per unit area ensure food security and reduce dependency on traditional farming. SLC supports SDG 12 (responsible consumption and production) by reducing resource waste, facilitating sustainable farming practices through closed-loop recycling, judicious nutrient and water management, and limited use of pesticides. SLC conserves freshwater, which directly contributes to SDG 6 (clean water and sanitation). Recirculating fertiliser systems reduce water use by 90% and pollution from nutrient runoff by 90% compared to traditional soil-based farming. Furthermore, these systems guarantee the availability of clean water and protect the ecosystem by preventing pesticide poisoning of soil and water sources.

In urban and peri-urban regions, fresh vegetables and fruits can be raised easily by the SLC, which will reduce greenhouse gas emissions by limiting the need for long-distance transportation through the soilless farming method. Additionally, by reducing the likelihood of severe temperatures, floods, and droughts through using controlled environment systems, climatic variability can be mitigated. SLC can be climate-friendly as there is scope for using renewable energy sources for artificial lighting, which has an impact on SDG 13 (climate action). Further, judicious use of nutrients reduces the burden of GHG emissions associated with the production of fossil fuel-based synthetic fertilisers.

6.9.2 Urban Food Security and Resilience

Soilless agriculture improves food security in cities by providing dependable, high-quality, locally cultivated food. By removing reliance on the rural supply chain, which can be negatively impacted during catastrophes or other climate-related disasters, containerised systems, rooftop hydroponics, and vertical gardens can help urban regions become more self-sufficient.

6.10 Future Perspectives

In the current context of rapid urbanisation and change in food habits, agricultural sustainability, and food and nutritional security are of prime concern, where the SLC can pave a potential horizon. Moreover, the inclusion of cutting-edge technologies such as IoT, AI, and ML, robotics, the use of renewable energy, and the enhancement of demand for crops produced in the SLC in niche markets will be a game-changer (Kalantari et al. 2018; Garzón et al. 2023). There is scope for developing user-friendly, low-cost, and modular systems for small farmers. Today, in developing countries, what exists in the periphery with hobbyists and a few commercial growers will reach large-scale growers' farms with innovative technologies and the enhancement of market demand. However, the expansion of SLC needs further research for the development of user-friendly models, the availability of essential inputs, and trained service providers.

6.11 Conclusion

SLC is an innovative horticultural approach that may face resource constraints; however, in the current context of climate variability, population growth, and urbanisation associated with changes in food demand, it enables global food and nutritional security. Agricultural sustainability is challenged by various abiotic and biotic factors, whereas SLC precisely controls water, nutrient, and climatic variabilities for better quality and quantity of produce. Hence, SLC emerges as a climate-resilient and resource-efficient method. SLC reduces the possibility of soil-borne pest and disease problems, and it ensures high-quality output. Digital agriculture will further shape the SLC through automation and enable a steady supply of fresh and nutritious crops to consumers' doorsteps in the urban areas. The commercialisation of the system provides economic benefits to urban and semi-urban areas. Agricultural sustainability is enhanced by reducing the carbon footprint, pesticide residue, promoting waste recycling, and improving the efficiency of water and nutrient utilisation. Additionally, the jobs generated through SLC have improved food security impact and socio-economic values. These systems not only need educational initiatives but also capacity-building programmes for better adoption. Moreover, SLC aligns with various SDGs such as "SDG 2 (Zero hunger), SDG 3 (Good health and well-being), SDG 11 (Sustainable cities and communities), SDG 12 (Responsible consumption and production), and SDG 13 (Climate action)". The affordable and modular systems can serve small farmers. Renewable energy use and the integration of cutting-edge technologies should be the next step in SLC towards its potential growth in developing countries.

Authors Contribution Literature review & conceptualisation: Santosh D.T., Dinkar Jagannath Gaikwad, Masina Sairam, Sagar Maitra.

Initial Draft: Santosh D.T., Sumit Ray, Sarthak Pattanayak, Nithar Ranjan Madhu, Lalichetti Sagar, Sourav Garai, Shaik Rishitha, Harun I. Gitari, Urjashi Bhattacharya,

Figures/visualisation: Sumit Ray, Dinkar Jagannath Gaikwad, Martin Leanardo Battaglia, Anjana J. Atapattu.

Table: Shaik Rishitha, Masina Sairam.

Final Draft & editing: Akbar Hossain, Harun I. Gitari, Anjana J. Atapattu, Sagar Maitra.

References

Ahmad Z, Waraich EA, Hina A, Ishfaq T, Jameel J (2023) Climate change and global crop production. In: Climate-resilient agriculture, Vol 1: crop responses and Agroecological perspectives. Springer, Cham, pp 27–56. https://doi.org/10.1007/978-3-031-37424-1_2

Al-Chalabi M (2015) Vertical farming: skyscraper sustainability? Sustain Cities Soc 18:74–77. https://doi.org/10.1016/j.scs.2015.06.003

Ali A, Niu G, Masabni J, Ferrante A, Cocetta G (2024) Integrated nutrient management of fruits, vegetables, and crops through the use of biostimulants, soilless cultivation, and traditional and modern approaches—a mini review. Agriculture 14:1330. https://doi.org/10.3390/agriculture14081330

Appicciutoli D, Amici AS, Bentivoglio D, Chiaraluce G, Staffolani G, Finco A (2025) Cultivating the future: a bibliometric review of emerging trends in soilless farming. Horticulturae 11:140. https://doi.org/10.3390/horticulturae11020140

Benke K, Tomkins B (2017) Future food-production systems: vertical farming and controlled-environment agriculture. Sustain Sci Pract Policy 13:13–26. https://doi.org/10.1080/15487733.2017.1394054

Bhadra P, Maitra S, Shankar T, Hossain A, Praharaj S, Aftab T (2022) Climate change impact on plants: plant responses and adaptations. In: Plant perspectives to global climate changes. Academic Press, pp 1–24. https://doi.org/10.1016/B978-0-323-85665-2.00004-2

Buehler D, Junge R (2016) Global trends and current status of commercial urban rooftop farming. Sustainability 8:1108. https://doi.org/10.3390/su8111108

Debroy P, Majumder P, Majumdar P, Das A, Seban L (2025) Analysis of opportunities and challenges of smart aquaponic system: a summary of research trends and future research avenues. Sustain Environ Res 35:18. https://doi.org/10.1186/s42834-025-00255-z

Despommier D (2013) Farming up the city: the rise of urban vertical farms. Trends Biotechnol 31:388–389. https://doi.org/10.1016/j.tibtech.2013.03.008

Dhanasekar S (2025) A comprehensive review on current issues and advancements of internet of things in precision agriculture. Comput Sci Rev 55:100694. https://doi.org/10.1016/j.cosrev.2024.100694

Dsouza A, Dixon M, Shukla M, Graham T (2025) Harnessing controlled-environment systems for enhanced production of medicinal plants. J Exp Bot 76:76–93. https://doi.org/10.1093/jxb/erae248

Erekath S, Seidlitz H, Schreiner M, Dreyer C (2024) Food for future: exploring cutting-edge technology and practices in vertical farm. Sustain Cities Soc 106:105357. https://doi.org/10.1016/j.scs.2024.105357

Farhangi H, Mozafari V, Roosta HR, Shirani H, Farhangi S, Farhangi M (2025) Optimizing LED lighting spectra for enhanced growth in controlled-environment vertical farms. Sci Rep 15:30152. https://doi.org/10.1038/s41598-025-15352-7

Fathidarehnijeh E, Nadeem M, Cheema M, Thomas R, Krishnapillai M, Galagedara L (2023) Current perspective on nutrient solution management strategies to improve the nutrient and water use efficiency in hydroponic systems. Can J Plant Sci 104:88–102. https://doi.org/10.1139/cjps-2023-0034

Fussy A, Papenbrock J (2022) An overview of soil and soilless cultivation techniques—chances, challenges and the neglected question of sustainability. Plants 11:1153. https://doi.org/10.3390/plants11091153

Gaikwad DJ, Maitra S (2020) Hydroponics cultivation of crops. In: Protected cultivation and smart agriculture. New Delhi Publishers, New Delhi, pp 279–287

Gaikwad DJ, Priyadarsini S, Mallick B (2020) Effects of different hydroponics systems and growing media on physiological parameters of spinach. Int J Curr Microbiol Appl Sci 9:1409–1414. https://doi.org/10.20546/ijcmas.2020.905.160

Gaikwad DJ, Ubale NB, Pal A, Singh S, Ali MA, Maitra S (2022) Abiotic stresses impact on major cereals and adaptation options – a review. Crop Res 23:896–915. https://doi.org/10.31830/2348-7542.2022.ROC-913

Gaikwad DJ, Sahu C, Prasanth S (2024) Hydroponics vs soil: an in-depth assessment of morpho-physiological traits in spinach (Spinacia oleracea). Sci Digest 1:6. https://doi.org/10.18805/ag.d-5928

Garzón J, Montes L, Garzón J, Lampropoulos G (2023) Systematic review of technology in aeroponics: introducing the technology adoption and integration in sustainable agriculture model. Agronomy 13:2517. https://doi.org/10.3390/agronomy13102517

Gautam B, Bharti B, Negi N, Mohan J (2024) Trends in the use of hydroponics in horticultural crops. Plant Arch 24:59–64. https://doi.org/10.51470/PLANTARCHIVES.2024.v24.SP-GABELS.010

Ghehsareh AM, Hematian M, Kalbasi M (2012) Comparison of date-palm wastes and perlite as culture substrates on growing indices in greenhouse cucumber. Int J Recycl Org Waste Agric 1:5. https://doi.org/10.1186/2251-7715-1-5

Gobilik J, Rechard CT, Maludin AJ, Alam MA, Benedick S (2021) Efficacy of column hydroponic system for increasing growth and yield of pak-choy (*Brassica rapa* L.) per unit area. Trans Sci Technol 8:7–24

Hossain A, Pramanick B, Bhutia KL, Ahmad Z, Moulick D, Maitra S, Ahmad A, Aftab T (2021) Emerging roles of osmoprotectant glycine betaine against salt-induced oxidative stress in plants: a major outlook of maize (*Zea mays* L.). In: Frontiers in plant-soil interaction: molecular insights into plant adaptation. Academic Press, pp 567–587. https://doi.org/10.1016/B978-0-323-90943-3.00015-8

Huang P, Xie H, Yang Y, Hu X, Liu C, Xu Y, Luo Y (2024) Spatiotemporal variation in rice water requirements and area in the cold rice cultivation region of China: past and future. Agric Water Manag 298:108858. https://doi.org/10.1016/j.agwat.2024.108858

Hutchinson GK, Ames ZR, Nemali K, Ferrarezi RS (2025) Arugula and lettuce responses to greenhouse hydroponic systems: an analysis of yield and resource use efficiencies. Hort Sci 60:601–612. https://doi.org/10.21273/HORTSCI18391-24

Jaeger SR (2024) Vertical farming (plant factory with artificial lighting) and its produce: consumer insights. Curr Opin Food Sci 56:101145. https://doi.org/10.1016/j.cofs.2024.101145

Ji Y, Kusuma P, Marcelis LF (2023) Vertical farming. Curr Biol 33:R471–R473

Joshi D, Nainabasti A, Awasthi RBP, Banjade D, Malla S, Subedi B (2022) A review on soilless cultivation: the hope of urban agriculture. Arch Agric Environ Sci 7:473–481. https://doi.org/10.26832/24566632.2022.0703022

Kabir MSN, Reza MN, Chowdhury M, Ali M, Samsuzzaman, Ali MR, Chung SO (2023) Technological trends and engineering issues on vertical farms: a review. Horticulturae 9:1229. https://doi.org/10.3390/horticulturae9111229

Kalantari F, Tahir OM, Joni RA, Fatemi E (2018) Opportunities and challenges in sustainability of vertical farming: a review. J Landsc Ecol 11:35–60

Kaur G, Upadhyaya P, Chawla P (2023) Comparative analysis of IoT-based controlled environment and uncontrolled environment plant growth monitoring system for hydroponic indoor vertical farm. Environ Res 222:115313. https://doi.org/10.1016/j.envres.2023.115313

Khan S, Purohit A, Vadsaria N (2020) Hydroponics: current and future state of the art in farming. J Plant Nutr 44:1515–1538. https://doi.org/10.1080/01904167.2020.1860217

Khatri L, Kunwar A, Bist DR (2024) Hydroponics: advantages and challenges in soilless farming. Big Data Agric 6:81–88. https://doi.org/10.26480/bda.02.2024.98.105

Kozai T, Niu G, Takagaki M (eds) (2019) Plant factory: an indoor vertical farming system for efficient quality food production. Academic Press

Kumar H, Agarwal A, Yadav PK, Sahay D, Prakash O, Ballabh B, Singh DP (2024) Performance of cherry tomato (*Solanum lycopersicum* var. cerasiformae) for biochemical quality traits under hydroponics. J Crop Weed 20:31–35. https://doi.org/10.22271/09746315.2024.v20.i1.1756

Kusmiyati F, Purbajanti ED, Anwar S, Sas MGA, Safitri D, FaPutra FP (2023) The yield of Packchoi and lettuce under different nutrient formulations with a hydroponic raft system. Int J Sci Res Publ 13:87–91

Lakhiar IA, Gao J, Syed TN, Chandio FA, Tunio MH, Ahmad F, Solangi KA (2020) Overview of the aeroponic agriculture—an emerging technology for global food security. Int J Agric Biol Eng 13:1–10. https://doi.org/10.25165/j.ijabe.20201301.5156

Lakhiar IA, Yan H, Syed TN, Zhang C, Shaikh SA, Rakibuzzaman M, Vistro RB (2025) Soilless agricultural systems: opportunities, challenges, and applications for enhancing horticultural resilience to climate change and urbanization. Horticulturae 11:568. https://doi.org/10.3390/horticulturae11060568

Lin Y, Kim C, Bassil NV, Oliphant JM, Hardigan MA, Kubota C (2025) Characterizing the growth, morphology, productivity, and fruit quality of twenty-three strawberry cultivars in an indoor environment with sole source electric lighting. Front Hortic 4:1621763. https://doi.org/10.3389/fhort.2025.1621763

Mahmoud MA, Moursy MAM (2025) Opportunities and challenges of greenhouses and soilless agricultural systems in combating food insecurity. In: resilient agroecosystems: innovations in cropping systems and climate change mitigation, pp 117–141

Miller A, Adhikari R, Nemali K (2020) Recycling nutrient solution can reduce growth due to nutrient deficiencies in hydroponic production. Front Plant Sci 11:607643. https://doi.org/10.3389/fpls.2020.607643

Munda SC, Maitra S, Gaikwad DJ, Duvvada SK, Sairam M, Santosh DT, Ray S (2025) Influence of growing media and nutrient levels on growth and yield of lettuce (*Lactuca sativa* L.) in hydroponics. Crop Res 60:185–190. https://doi.org/10.31830/2454-1761.2025.cr-1033

Nungula EZ, Mugwe J, Nasar J, Massawe BHJ, Karuma AN, Maitra S, Seleiman MF, Dindaroglu T, Khan N, Gitari HI (2023) Land degradation unmasked as the key constraint in sunflower (*Helianthus annuus*) production: role of GIS in revitalizing this vital sector. Cogent Food Agric 9:2267863. https://doi.org/10.1080/23311932.2023.2267863

Ojo MO, Zahid A (2022) Deep learning in controlled environment agriculture: a review of recent advancements, challenges and prospects. Sensors 22:7965. https://doi.org/10.3390/s22207965

Palmitessa OD, Signore A, Santamaria P (2024) Advancements and future perspectives in nutrient film technique hydroponic system: a comprehensive review and bibliometric analysis. Front Plant Sci 15:1504792. https://doi.org/10.3389/fpls.2024.1504792

Pastor-Arbulú P, Rodríguez-Delfín A (2025) Enhancing the growth, yield and physiological response of two lettuce (*Lactuca sativa* L.) cultivars through NFT system optimization. Front Plant Sci 16:1639002. https://doi.org/10.3389/fpls.2025.1639002

Pomoni DI, Koukou MK, Vrachopoulos MG, Vasiliadis L (2023) A review of hydroponics and conventional agriculture based on energy and water consumption, environmental impact, and land use. Energies 16:1690. https://doi.org/10.3390/en16041690

Rajaseger G, Chan KL, Tan KY, Ramasamy S, Khin MC, Amaladoss A, Haribhai PK (2023) Hydroponics: current trends in sustainable crop production. Bioinformation 19:925. https://doi.org/10.6026/97320630019925

Rajendran S, Domalachenpa T, Arora H, Li P, Sharma A, Rajauria G (2024) Hydroponics: exploring innovative sustainable technologies and applications across crop production, with emphasis on potato mini-tuber cultivation. Heliyon 10:e30736. https://doi.org/10.1016/j.heliyon.2024.e26823

Rathor AS, Choudhury S, Sharma A, Nautiyal P, Shah G (2024) Empowering vertical farming through IoT and AI-driven technologies: a comprehensive review. Heliyon 10:e34998. https://doi.org/10.1016/j.heliyon.2024.e34998

Ray DK, West PC, Clark M, Gerber JS, Prishchepov AV, Chatterjee S (2019) Climate change has likely already affected global food production. PLoS One 14:e0217148. https://doi.org/10.1371/journal.pone.0217148

Ray S, Maitra S, Sairam M, Sravya M, Priyadarshini A, Shubhadarshi S, Padhi DP (2024) An unravelled potential of foliar application of micro and beneficial nutrients in cereals for ensuring food and nutritional security. Int J Exp Res Rev 41:19–42. https://doi.org/10.52756/ijerr.2024.v41spl.003

Ray S, Maitra S, Sairam M, Sameer S, Sagar L, Divya BS, Gitari HI (2025) The nexus between intercropping systems, ecosystem services and sustainable agriculture: a review. Crop Res 26:1–11. https://doi.org/10.31830/2348-7542.2025.roc-1166

Regmi A, Rueda-Kunz D, Liu H, Trevino J, Kathi S, Simpson C (2024) Comparing resource use efficiencies in hydroponic and aeroponic production systems. Technol Hortic 4:e005. https://doi.org/10.48130/tihort-0024-0002

Rezaei Chiyaneh E, Mahdavikia H, Alipour H, Dolatabadian A, Battaglia ML, Maitra S, Harrison MT (2023) Biostimulants alleviate water deficit stress and enhance essential oil productivity: a case study with savory. Sci Rep 13:720. https://doi.org/10.1038/s41598-022-27338-w

Roosta HR (2024) The responses of pepper plants to nitrogen form and dissolved oxygen concentration of nutrient solution in hydroponics. BMC Plant Biol 24:281. https://doi.org/10.1186/s12870-024-04943-7

Sahu C, Gaikwad DJ, Nanda S, Dash GK, Maitra S, Munda SC, Mullapudi S (2025) Comparative performance of lettuce cultivars Batavia and Romain in nutrient film technique under hydroponics system. Crop Res 60:61–67. https://doi.org/10.31830/2454-1761.2025.cr-1009

Sairam M, Maitra S, Praharaj S, Nath S, Shankar T, Sahoo U, Santosh DT, Sagar L, Panda M, Shanthi Priya, Ashwini TR, Gaikwad DJ, Hossain A, Pramanick B, Jatav HS, Gitari HI, Aftab T (2023) An insight into the consequences of emerging contaminants in soil and water and plant responses. In: Emerging contaminants and plants interactions, adaptations and remediation technologies, pp 1–27. https://doi.org/10.1007/978-3-031-22269-6_1

Sairam M, Maitra S, Sagar L, Biswas T, Bárek V, Brestic M, Hossain A (2025) Application of precision nutrient tools for the optimization of fertilizer requirements and assessment of the growth and productivity of maize (*Zea mays* L.) in the northeastern Ghat of India. J Agric Food Res 21:101958. https://doi.org/10.1016/j.jafr.2025.101958

Salma SB, Shubham, Kaushal S (2024) Aeroponics: an emerging food growing system in sustainable agriculture for food security. Int J Res Agron 7:93–97

Samarakoon UC, Fyffe C, Bale J, Ling P, Basnagala S, Donley N, Altland J (2017) Effect of electrical conductivity on the productivity and nutrient uptake of Lactuca sativa L. grown using nutrient film technique (NFT). International symposium on growing media, soilless cultivation, and compost utilization in horticulture 1266:137–144. https://doi.org/10.17660/ActaHortic.2019.1266.19

Sambo P, Nicoletto C, Giro A, Pii Y, Valentinuzzi F, Mimmo T, Cesco S (2019) Hydroponic solutions for soilless production systems: issues and opportunities in a smart agriculture perspective. Front Plant Sci 10:923. https://doi.org/10.3389/fpls.2019.00923

Santosh DT, Gaikwad D (2023) Advances in hydroponic systems: types and management. In: Advances in agricultural technology, pp 16–28

Santosh DT, Debnath S, Maitra S, Sairam M, Sagar L, Hossain A, Moulick D (2024) Alleviation of climate catastrophe in agriculture through adoption of climate-smart technologies. In: Climate crisis: adaptive approaches and sustainability. Springer, pp 307–332

Sathyanarayana SR, Gangadhar WV, Badrinath MG, Ravindra RM, Shriramrao AU (2022) Hydroponics: an intensified agriculture practice to improve food production. Rev Agric Sci 10:101–114. https://doi.org/10.7831/ras.10.0_101

Shanila S, Sreekala GS, Nair DS, Stephen R, Gopinath PP (2024) Hydroponic cultivation of spices and aromatics: techniques, benefits and challenges. Int J Environ Clim Chang 14:234–254. https://doi.org/10.9734/ijecc/2024/v14i114542

Sharma A, Hazarika M, Heisnam P, Pandey H, Devadas VASN, Kesavan AK, Kumar P, Singh D, Vashishth A, Jha R, Misra V, Kumar R (2024) Controlled environment ecosystem: a cutting-edge technology in speed breeding. ACS Omega 9:29114–29138. https://doi.org/10.1021/acsomega.3c09060

Smirnov EN, Karelina EA, Pletnev MG (2024) Markets of agro-industrial complex products in conditions of geoeconomic fragmentation of the international trade. Вестник университета 135. https://doi.org/10.3389/fhort.2025.1621763

Somerville C, Cohen M, Pantanella E, Stankus A, Lovatelli A (2014) Small-scale aquaponic food production: integrated fish and plant farming. FAO Fish Aquac Tech Pap 589

Steffi PF, Thirumalaiyammal B, Thamilmaraiselvi B, Murugan MS, Jenny S, Sathya S, Revathi TR (2025) Cost-effective automated aquaponics systems using Arduino: enhancing legume cultivation, sprout growth, nutrient content and sustainable production: a review. Indian J Agric Res. https://doi.org/10.18805/IJARe.A-6466

Sulaiman H, Yusof AA, Mohamed Nor MK (2025) Automated hydroponic nutrient dosing system: a scoping review of pH and electrical conductivity dosing frameworks. Agri Engg 7:43. https://doi.org/10.3390/agriengineering7020043

Tang L, Syed AUA, Otho AR, Junejo AR, Tunio MH, Hao L, Channa JA (2024) Intelligent rapid asexual propagation technology—a novel aeroponics propagation approach. Agronomy 14:2289. https://doi.org/10.3390/agronomy14102289

Treftz C, Omaye ST (2016) Hydroponics: potential for augmenting sustainable food production in non-arable regions. Nutr Food Sci 46:672–684. https://doi.org/10.1108/NFS-10-2015-0118

Tunçelli G, Ürkü Ç, Memiş D (2025) Isolation and identification of cultivable bacteria in an aquaponic system with koi carp and parsley (*Petroselinum crispum*): implications for water quality and system health. Aquat Sci Eng 40:163–172. https://doi.org/10.26650/ASE.2025.1702567

Van Dijk M, Morley T, Rau ML, Saghai Y (2021) A meta-analysis of projected global food demand and population at risk of hunger for the period 2010–2050. Nat Food 2:494–501. https://doi.org/10.5281/zenodo.4911251

Vikanksha AK, Singh J (2023) Trends in hydroponics practice/technology in horticultural crops: a review. Int J Plant Soil Sci 35:57–65. https://doi.org/10.9734/ijpss/2023/v35i22759

Vought K, Bayabil HK, Pompeo J, Crawford D, Zhang Y, Correll M, Martin-Ryals A (2024) Dynamics of micro and macronutrients in a hydroponic nutrient film technique system under lettuce cultivation. Heliyon 10:e32316

Wallace-Springer N, Wells DE, Pickens JM, Ayipio E, Kemble J (2022) Effects of hydraulic retention time of aquaculture effluent on nutrient film technique lettuce productivity. Agronomy 12:2570. https://doi.org/10.3390/agronomy12102570

Xu S, Shi D, Chen H, Ma F, Tao G, Wu W, Lin D, Wu S, Fei Q, Hu Y, Meng L (2025) Substrate cultivation system improved the quality of 'Hongyan' strawberry fruits compared with the soil cultivation system. Food Chem:144430. https://doi.org/10.1016/j.foodchem.2025.144430

Yang T, Kim HJ (2019) Nutrient management regime affects water quality, crop growth, and nitrogen use efficiency of aquaponic systems. Sci Hortic 256:108619. https://doi.org/10.1016/j.scienta.2019.108619

Zhao T, Zhang X, Liu W, Wang J, Li Z, Liu L (2025) Increase in global per capita cropland imbalance across countries from 1985 to 2022: a threat to achieving sustainable development goals. Geogr Sustain 6:100239. https://doi.org/10.1016/j.geosus.2024.09.005

Ziadat F, Conchedda G, Haddad F, Njeru J, Brès A, Dawelbait M, Li L (2025) Desertification and agrifood systems: restoration of degraded agricultural lands in the Arab region. Agriculture 15:1249. https://doi.org/10.3390/agriculture15121249

Application of ITK in Modern Horticulture: Redefining Tradition

7

S. S. Vasanthkumar, J. Karthikeyan, S. P. Yazhini, and R. Gowshika

Abstract

Indigenous technical knowledge (ITK) in horticulture represents a repository of sustainable, eco-friendly practices rooted in local traditions, cultural heritage, and farmer ingenuity. Orally transmitted across generations, ITK embodies centuries of experiential learning and adaptation to diverse agroecological conditions. Recent interest in integrating ITK with modern scientific and technological advances has opened promising pathways for sustainable, resilient horticultural systems. This chapter examines the pivotal role of ITK in redefining modern horticulture by enhancing soil fertility, pest and disease management, water conservation, seed preservation, post-harvest processing, and crop diversification. Case studies from India's tribal, arid, and hill regions illustrate the successful integration of locally adapted ITK practices with innovations such as drip fertigation, protected cultivation, biopesticide formulations, and organic certification systems. This synergy improves productivity, profitability, and strengthens local food systems and climate resilience. The chapter also highlights efforts to document, validate, and disseminate ITK through farmer field schools, participatory rural appraisal, and open-access digital platforms. The roles of farmers, researchers, and institutions in promoting ITK via inclusive policy frameworks are critically analyzed. Ultimately, ITK emerges as a dynamic, adaptable, and culturally sensitive tool for advancing climate-smart horticulture and rural development.

S. S. Vasanthkumar (✉)
Department of Plant and Environmental Sciences, Clemson University, Clemson, SC, USA

J. Karthikeyan · R. Gowshika
Department of Vegetable Science, Horticultural College and Research Institute, Tamil Nadu Agricultural University, Coimbatore, Tamil Nadu, India

S. P. Yazhini
Department of Fruit Science, Horticultural College and Research Institute, Periyakulam, Tamil Nadu Agricultural University, Coimbatore, Tamil Nadu, India

A. Das, A. B. Sharangi (eds.), *Remunerative Horticulture*,
https://doi.org/10.1007/978-981-92-1007-7_7

Keywords

Indigenous technical knowledge · Sustainable horticulture · Traditional practices · Climate resilience · Integrated farming · Participatory innovation

7.1 Introduction

7.1.1 Definition and Scope of Indigenous Technical Knowledge in Horticulture

Indigenous technical knowledge (ITK) in horticulture refers to a comprehensive body of empirical wisdom, practices, and innovations developed and refined across generations through observation, experimentation, and adaptation to local agro-ecological conditions. This knowledge system encompasses a wide range of horticultural practices, including traditional propagation techniques (e.g., grafting, budding, layering), organic pest and disease management, composting, green manuring, farmyard manure application, and water conservation methods such as mulching and rainwater harvesting. Beyond cultivation, ITK includes post-harvest processing, storage, seed selection and preservation, pest and disease prediction using environmental indicators, and integrated crop–livestock systems. It also incorporates ethnobotanical knowledge of medicinal and aromatic plants, local food processing, and the development of climate-resilient crop varieties. Characterized as site-specific, culturally embedded, cost-effective, and environmentally sustainable, ITK is particularly valuable for small-scale farmers operating in resource-constrained environments. Recent studies have validated the scientific basis of many ITK practices, demonstrating their potential for broader application and integration with modern horticultural sciences (Ficiciyan et al. 2018).

7.1.2 Historical and Cultural Roots of ITK in Indian Agriculture

The origins of ITK in Indian agriculture date back to the Vedic era (1500–500 BCE), as documented in ancient texts such as *the Arthashastra by Kautilya, the Kashyapiya Krishisukti, and the Brihat-Samhita* by Varahamihira. These texts provide detailed insights into agricultural implements, crop selection, soil management, and weather prediction methods, reflecting an advanced understanding of monsoon cycles, crop calendars aligned with astronomical cycles, and region-specific farming practices (Giri et al. 2020). Cultural and religious festivals, such as Ugadi (initiating deep plowing), Nagapanchami (promoting soil health through termite control), and Deepavali (encouraging pollinator attraction via lighting), played a pivotal role in preserving and transmitting agricultural knowledge (Panda et al. 2024). Region-specific innovations emerged from India's diverse ecosystems, including Nagaland's Zabo system, which integrates water conservation with agroforestry, Bihar's Ahar-Pyne system for capturing floodwater in terraced structures, and advanced terrace cultivation in the Himalayas. These practices were embedded within social, governance, and spiritual frameworks, emphasizing the interconnectedness of humans, land, and nature to ensure sustainable resource management across generations (Mondal 2024; Murry 2021).

7.1.3 Significance of ITK in the Modern Horticultural Landscape

ITK offers sustainable solutions to contemporary horticultural challenges, including climate change, resource scarcity, and biodiversity loss. Practices such as vermicomposting, organic pest control using botanical extracts, and traditional water harvesting systems like Johads and Taanks enhance ecosystem services while maintaining productivity (Das et al. 2019; Yadav 2023). Johads, community-managed crescent-shaped earthen dams prevalent in North India, collect monsoon rainwater for controlled percolation and groundwater recharge, achieving water use efficiency comparable to that of modern irrigation systems using locally available materials (Kolarkar et al. 1983). ITK interventions, such as green manuring with leguminous crops, application of organic manures (e.g., castor and neem cakes) to control soil-borne diseases, and traditional intercropping, significantly improve soil health, biodiversity, and natural pest management (Salgado et al. 2021). Additionally, ITK contributes to climate resilience through drought-resistant crop varieties developed via traditional selection, phenological weather forecasting, and adaptive farming practices (Sharma et al. 2025). Integrating ITK with modern horticultural technologies creates culturally, economically, and environmentally viable solutions that enhance food security, livelihoods, and agricultural biodiversity. For example, traditional bamboo drip irrigation systems demonstrate resource efficiency comparable to modern systems, require minimal maintenance, and use locally sourced materials (Murry 2021). The documentation and scientific validation of ITK practices facilitate innovation and knowledge exchange, enabling farmers to address modern challenges while preserving cultural heritage.

7.1.4 Objectives and Structure of the Chapter

This chapter pursues four primary objectives to harmonize traditional wisdom with modern horticultural science: (1) to systematically document and scientifically validate ITK practices for broader application across diverse agro-climatic zones; (2) to elucidate the historical development, cultural significance, and regional variations of ITK in Indian horticulture; (3) to evaluate the integration of ITK with modern technologies to enhance sustainability, productivity, and climate resilience while maintaining cultural authenticity; and (4) to propose policy frameworks for the conservation, documentation, validation, and dissemination of ITK, ensuring the protection of intellectual property rights for local communities and facilitating knowledge transfer.

7.2 Methodology

7.2.1 ITK Selection and Documentation Methodology

The documentation of indigenous technical knowledge (ITK) practices employs participatory research methods to ensure community engagement and scientific authenticity. Data collection involves ethnographic techniques, including in-depth interviews with traditional practitioners, focus group discussions with farming communities, and direct field observations conducted across cropping seasons. Methods such as oral histories, photographic and video documentation, and detailed field notes capture both explicit and tacit knowledge embedded in ITK practices. Semi-structured questionnaires systematically collect data across nine key horticultural areas: field preparation, sowing timing, cropping patterns, soil fertility management, plant protection, weed management, post-harvest processing, seed selection and preservation, and water management. Triangulation, achieved through consultations with diverse knowledge holders, traditional leaders, and experienced practitioners, ensures robustness and reliability of the data. Purposive sampling across various agro-climatic zones guarantees geographic representation and captures region-specific adaptations and variations in ITK practices (Panda et al. 2024).

7.2.2 Validation and Scientific Analysis Framework

The quantification of indigenous knowledge (QuIK) approach is utilized to transform qualitative ITK data into measurable parameters for scientific analysis. This method employs inductive coding to develop analytical categories from field data, followed by the collection of biometric and yield data to evaluate the effectiveness of documented practices. Participatory on-farm trials, co-designed and implemented with farmers, assess key performance indicators, including yield, resource use efficiency, environmental sustainability, economic viability, and social acceptability. These trials involve farmers in monitoring and evaluation to ensure practical relevance. Cross-sectoral collaboration among research institutions, extension services, and farming communities ensures scientific rigor while maintaining cultural sensitivity, thus preserving the integrity of ITK practices (Salgado et al. 2021).

7.2.3 Integration Assessment and Policy Analysis

The integration of ITK with modern horticultural technologies is evaluated through a comparative analysis of traditional practices, conventional modern methods, and hybrid approaches combining both systems. Case studies document successful integration models across diverse regions and cropping systems, detailing implementation strategies, outcomes, and lessons learned. Policy analysis examines existing intellectual property frameworks, benefit-sharing mechanisms, and institutional support systems that either facilitate or hinder ITK integration. Stakeholder

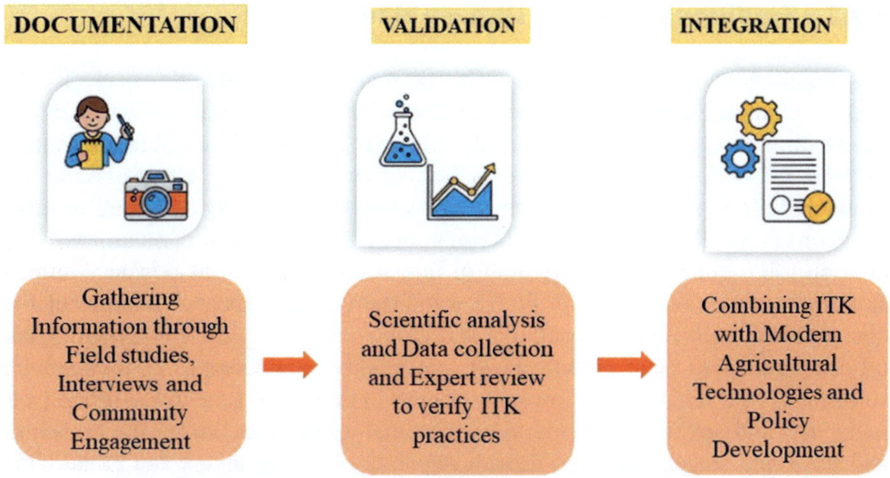

Fig. 7.1 Methodology involved in indigenous technical knowledge

consultations with farmers, researchers, extension officers, policymakers, and community leaders identify barriers and opportunities for scaling ITK applications. Economic analyses assess cost-benefit ratios, return on investment, and market potential for ITK-based products and services compared to conventional alternatives. Environmental impact assessments measure resource use efficiency, carbon footprint, biodiversity conservation, and ecosystem service provision (Fig. 7.1). Participatory evaluation techniques engage communities to assess the effectiveness and cultural acceptability of integration approaches, ensuring that scaling strategies align with local values and agricultural goals (Absanto et al. 2025; More et al. 2025).

7.3 Role of Indigenous Technical Knowledge in Sustainable Horticulture

7.3.1 Soil Health and Fertility Management

Indigenous technical knowledge (ITK) promotes sustainable soil fertility through organic amendments. Farmyard manure (FYM), composed of cattle dung, crop residues, and ash, provides essential nutrients (nitrogen, phosphorus, potassium, and micronutrients) while improving soil water retention. Traditional composting pits, layered with crop residues, tobacco stalks, and ash, facilitate anaerobic decomposition to produce humus-rich compost, enhancing microbial diversity and soil aeration. Green manuring with leguminous crops such as *Crotalaria juncea* (Sun hemp) and *Sesbania aculeata* (Daincha) fixes atmospheric nitrogen, enriching soil fertility and supporting sustainable soil ecosystems for long-term horticultural productivity (Salgado et al. 2021).

7.3.2 Traditional Pest and Disease Management

ITK employs eco-friendly pest and disease management strategies using locally sourced botanical and animal-derived products. Neem seed kernel extracts, containing azadirachtin, reduce populations of sap-sucking pests such as aphids and whiteflies by 60–80% through antifeedant and growth-inhibitory effects (Kaur and Kocher 2023). Fermented cow urine, mixed with 10% cow dung and aged for seven days, serves as a biopesticide rich in ammonia and phenolic compounds, with 5% dilution foliar sprays effectively control fungal diseases such as powdery mildew and anthracnose (Wahedi et al. 2017). Panchagavya, a fermented mixture of five cow products (milk, curd, ghee, dung, and urine) combined with fruits, jaggery, and oil cake, enhances plant growth, yield, and quality parameters (e.g., total soluble solids, ascorbic acid, and chlorophyll content) in crops like tomato (*Solanum lycopersicum*) and soybean (*Glycine max*). Optimal application during the branching and flowering stages leverages growth regulators such as auxins and gibberellins (Wahedi et al. 2017). Wood ash, applied as a dusting powder or soil amendment, controls soft-bodied pests (e.g., slugs and snails) and supplies silica and potassium to strengthen plant cell walls (Bang-Andreasen et al. 2017). Validated by the Scientific Validation and Research on Panchagavya (SVAROP) program, these methods ensure effective pest management while minimizing chemical toxicity and maintaining ecological balance.

7.3.3 Indigenous Seed Storage and Preservation

ITK systems prioritize seed viability and genetic purity to sustain climate-adapted and disease-resistant cultivars. Seeds are sun-dried to 8–10% moisture content and stored in airtight earthen pots sealed with cow dung paste, which leverages low oxygen levels and antimicrobial properties to deter storage pests (Wambugu et al. 2009). Botanical protectants, including neem (*Azadirachta indica*) leaves, *Pongamia pinnata* seeds, and turmeric (*Curcuma longa*) powder, provide fumigant action, reducing weevil infestations in stored legumes and cereals by over 70% (Lucy et al. 2025). Smoke treatment with teak (*Tectona grandis*) or neem for two hours imparts phenolic compounds that suppress fungal spores and insect eggs. Community seed banks, often managed by women's self-help groups, support periodic regeneration cycles, on-farm experimentation, and knowledge exchange, preserving agrobiodiversity and facilitating the dissemination of resilient genotypes among local horticulturists (Sharma et al. 2025).

7.3.4 Water Management

ITK water management practices optimize resource use in water-scarce regions. The Ahar-Pyne system in Bihar captures floodwater in ponds for controlled release during dry seasons, supporting vegetable and fruit production (Mondal 2024). In

Rajasthan, Khadin bunds and taankas collect monsoon runoff and rooftop water for orchard irrigation, minimizing evaporation and recharging groundwater. Johads, crescent-shaped earthen check dams, enhance percolation and groundwater recharge, with over 4500 revitalized in Alwar district, demonstrating water use efficiency comparable to modern drip irrigation systems while using low-cost, locally sourced materials (Kolarkar et al. 1983). Organic mulching with materials such as rice straw, groundnut haulms, and sugarcane trash, applied at a depth of 5 cm, reduces surface evaporation by 35–45% and moderates soil temperature extremes (Li et al. 2020). Zai pits, small earthen basins filled with compost and silt, concentrate moisture and nutrients around plant roots, enhancing sapling survival and productivity in semi-arid landscapes (Vinayak et al. 2022). These low-tech, community-based ITK solutions rival the efficiency of modern irrigation technologies.

7.3.5 Use of Local Biodiversity and Companion Planting

ITK leverages local biodiversity and plant interactions to enhance horticultural resilience. Intercropping systems, such as *Tagetes* spp. (marigold) with tomatoes (*Solanum lycopersicum*) to reduce nematode infestations or *Ocimum tenuiflorum* (basil) with eggplants (*Solanum melongena*) to deter whiteflies, achieve up to 50% pest reduction and 10–20% yield increases (Risch 1983). Ethnobotanical knowledge guides the use of aromatic plants (e.g., lemongrass [*Cymbopogon* spp.], citronella [*Cymbopogon nardus*]) and flowering plants (e.g., sunflower [*Helianthus annuus*], cosmos [*Cosmos bipinnatus*]) to attract pollinators, improving fruit set in orchards of mango (*Mangifera indica*), guava (*Psidium guajava*), and citrus (*Citrus* spp.). Agroforestry systems integrating native fruit trees (*Ziziphus* spp., *Emblica officinalis*) with annual vegetables provide year-round harvests, diversify income, and enhance soil carbon sequestration. These practices foster resilient, multifunctional horticultural landscapes that support ecosystem services and rural livelihoods (Sharma et al. 2025). The comparative performance of Indigenous technical knowledge (ITK) and modern conventional methods across different agricultural aspects is summarized in Table 7.1.

7.4 Successful Case Studies

7.4.1 Tribal Communities (Odisha, North East India)

Tribal communities in India leverage Indigenous technical knowledge (ITK) to sustain horticultural and agricultural practices that enhance biodiversity and livelihoods. In Odisha, the Kutia Kondh tribe practices shifting cultivation of millets, such as sorghum (*Sorghum bicolor*) and finger millet (*Eleusine coracana*), maintaining soil fertility through crop rotation and fallow periods. Cultural festivals like the Burlang Yatra reinforce these practices, reflecting deep cosmological

Table 7.1 Comparative analysis of ITK and modern methods

Aspect	Traditional/ITK performance	Modern/ conventional performance	Performance difference	Reference
Crop yields	70–77% of modern yields (mean: 14.31)	100% baseline (mean: 18.95)	24.5% lower (traditional)	Ficiciyan et al. (2018)
Intercropping benefits	LER 1.67 (sorghum/ pigeon pea)	LER 1.0 (monoculture)	67% yield advantage	Willey (1990)
Water use efficiency	1,979,033 gal/ac (flood)	1,166,313 gal/ac (drip)	41% water savings (drip)	Absanto et al. (2025)
Pest reduction	50% pest reduction	Baseline pest levels	50% pest reduction benefit	Ficiciyan et al. (2018)
Economic returns	Rs 112,374 (drip)	Rs 88,963 (flood)	20.8% higher (drip)	More et al. (2025)
Technical efficiency	58.9% (furrow)	92.8% (sprinkler)	57% higher (modern)	Absanto et al. (2025)

connections with nature and promoting eco-friendly farming (Joshi 2021). Sacred groves in Odisha preserve over 2000 sites, conserving 39 plant species used in Shraddha rituals and 63 species for various purposes, supporting agro-biodiversity (Panda et al. 2024). In Arunachal Pradesh, the Aadi tribe cultivates pest-resistant Dheka rice, using *Citrus maxima* (pummelo) bark and wood ash to deter housefly infestations, demonstrating effective traditional pest management (Giri et al. 2020). In Meghalaya's Ri-Bhoi district, 315 farmers across 110 hectares adopted integrated organic farming systems (IOFS), integrating rainwater harvesting, vermicomposting, and livestock management, achieving 20–50% yield increases in crops such as maize (*Zea mays*), French bean (*Phaseolus vulgaris*), potato (*Solanum tuberosum*), ginger (*Zingiber officinale*), tomato (*Solanum lycopersicum*), carrot (*Daucus carota*), and chili (*Capsicum* spp.) (Das et al. 2019). In the Nilgiris, tribal women documented 67 ITK practices, including frost protection for tea (*Camellia sinensis*) using fern wraps and shade trees like silver oak (*Grevillea robusta*) and *Moringa oleifera* in coffee (*Coffea* spp.) plantations (Pooja et al. 2025). In North East India, tribal communities employ traditional ecological knowledge for sustainable resource use, with systems like Nagaland's Zabo, which integrates forestry, agriculture, and animal rearing with water harvesting, and Meghalaya's bamboo drip irrigation, matching modern systems' efficiency using locally sourced materials (Murry 2021; Tynsong et al. 2020).

7.4.2 Arid and Semi-Arid Regions (Rajasthan, Gujarat)

In arid and semi-arid regions, ITK addresses water scarcity and soil degradation through innovative practices. In Rajasthan and Gujarat, farmers employ rainwater harvesting and moisture conservation techniques, such as using *Lawsonia inermis* (mehndi) as a cover crop and minimum tillage for crops like sugarcane (*Saccharum officinarum*), to mitigate drought impacts (Joshi 2021). In Bikaner, Rajasthan, 250 hectares of date palm (*Phoenix dactylifera*) cultivation utilizes tissue-cultured cultivars and traditional pollination with male cultivars like Ghanami and Al-in-city. The Central Arid Zone Research Institute (CAZRI) employs in situ budding for rainfed ber (*Ziziphus mauritiana*) orchards, reducing costs and enhancing adaptation. Water conservation systems, including Khadin bunds, taankas, and Johads, capture monsoon runoff to sustain orchards and recharge groundwater (Yadav 2023). Drought-tolerant crops, such as ber ("Seb," "Oola,") ker (*Capparis decidua*), pomegranate (*Punica granatum*), fig (*Ficus carica*), karonda (*Carissa carandas*), and phalsa (*Grewia asiatica*), are cultivated to align with moisture availability. Mulching with pearl millet straw, bajra haulms, and castor stalks at 5–7 cm depth reduces evaporation by 35–45% and moderates soil temperatures (Li et al. 2020). Integrated livestock systems with indigenous cattle, goats, and poultry produce organic manure, supporting a closed-loop nutrient cycle (Rajkumara et al. 2025).

7.4.3 Mountain Horticulture (Himachal Pradesh, Uttarakhand)

Himalayan states demonstrate advanced ITK practices tailored to mountainous terrains. In Himachal Pradesh, contributing 22.85% of India's fruit production area, farmers employ terracing and elevation-specific crop selection, cultivating temperate fruits like apple (*Malus domestica*), peach (*Prunus persica*), pear (*Pyrus communis*), plum (*Prunus domestica*), and apricot (*Prunus armeniaca*) at higher elevations, and subtropical fruits like mango (*Mangifera indica*), citrus (*Citrus* spp.), and litchi (*Litchi chinensis*) at lower altitudes. Mid-altitude zones support off-season vegetables, including peas (*Pisum sativum*), tomato (*Solanum lycopersicum*), cabbage (*Brassica oleracea* var. *capitata*), cauliflower (*Brassica oleracea* var. *botrytis*), capsicum (*Capsicum annuum*), beans (*Phaseolus* spp.), and garlic (*Allium sativum*) (Pathania 2025). Agroforestry systems integrate tree plantations with fruit horticulture and vegetables on terraced slopes, using contour farming and traditional soil and water conservation methods to enhance productivity and ecological balance (Rautela 2015). In Sikkim, Lepcha and Limbu communities cultivate large cardamom (*Amomum subulatum*) under alder (*Alnus nepalensis*) shade trees, optimizing shade and resource use through strict harvesting schedules (Goswami and Pariyar 2025). In Uttarakhand, terracing, gravity-driven irrigation channels, community springs, and mulching support diverse, climate-resilient crops and pseudo-cereals for food and fodder. Community seed banks preserve genetic diversity through elevation-specific seed selection and phenological forecasting, reflecting advanced ecological understanding for sustainable livelihoods (Rautela 2015; Sharma et al. 2025).

7.4.4 Integration with Modern Technologies (Sikkim)

Sikkim's transformation into the world's first fully organic state exemplifies the integration of ITK with modern technologies. Initiated in 2003 and formalized by the Sikkim Organic Mission in 2010, all 75,000 hectares of agricultural land achieved organic certification by 2015, benefiting over 66,000 farming families. The initiative phased out chemical inputs, promoting traditional practices like farmyard manure, botanical pesticides (e.g., neem-based), and crop rotation, which were validated and scaled through modern certification under the National Program for Organic Production (NPOP). Over 750 agricultural officers were trained, and village-level demonstration farms, compost, and vermicompost units were established. Market linkages via the Sikkim State Cooperative Supply and Marketing Federation Ltd. (SIMFED) enabled direct farmer purchases and premium prices for organic produce, such as maize, potatoes, and vegetables, significantly increasing farmer incomes (Bhutia 2024; Das et al. 2019; Mandal et al. 2025).

7.4.5 Farmer-Led Innovations and Community Knowledge Banks

Farmer-led innovations and community knowledge banks enhance adaptive capacity through participatory documentation and knowledge sharing. The farmer-led documentation (FLD) approach, exemplified by the Apatani tribe's Barom Nu practice in Arunachal Pradesh, involves three-year evaluation cycles to develop new rice varieties. The Baiga tribe's Bewar Kheti integrates forest regeneration with crop production (Giri et al. 2020). Community seed banks, supported by organizations like the Alliance of Bioversity International and CIAT, conserve traditional seeds and landraces, facilitating seed exchange, on-farm conservation, and value addition to enhance rural livelihoods and climate resilience (Ijatuyi et al. 2025). Additional innovations include drought-resistant crop varieties, traditional pest control paired with modern monitoring, and women-led seed banks managing regeneration cycles and knowledge exchange. Knowledge-sharing platforms, such as farmer research groups, demonstration plots, and farmer-to-farmer exchange visits, document practices in local languages and culturally relevant formats. Institutional partnerships with research institutes, policies recognizing farmers' intellectual property rights, and extension strategies amplify these innovations, enhancing their adoption and impact (Ijatuyi et al. 2025; Sharma et al. 2025; Prasad et al. 2025).

7.5 Advancing ITK in Horticulture: Institutional Support, Validation, and Dissemination

7.5.1 Role of ICAR, SAUs, and KVKs in Documentation and Validation

The Indian Council of Agricultural Research (ICAR), State Agricultural Universities (SAUs), and Krishi Vigyan Kendras (KVKs) are central to documenting, validating, and refining Indigenous technical knowledge (ITK) in horticulture. Since 2002, ICAR's Mission Mode Project under the National Agricultural Technology Project

(NATP) has cataloged 4880 ITK practices across 28 domains, including rainwater harvesting, pest management, seed storage, and post-harvest technologies. Digital inventories, seven published volumes, and e-resources ensure accessibility (Behera 2012; Das et al. 2003, 2004). SAUs and KVKs adapt ITK to regional conditions, blending traditional practices with modern technologies like drip fertigation and biopesticides to develop sustainable solutions. ICAR's Intellectual Property and Technology Management (IP&TM) unit protects farmers' intellectual contributions, ensuring equitable benefit-sharing (Suraj Bhan et al. 2016).

7.5.2 Role of PPVFRA in Promoting ITK

The Protection of Plant Varieties and Farmers' Rights Authority (PPVFRA), established under the PPV&FR Act of 2001, protects farmer-developed plant varieties and ITK related to seed conservation and traditional breeding. By registering these varieties, PPVFRA prevents biopiracy and ensures fair compensation for communities. It grants farmers and breeders intellectual property rights, allowing seed saving, use, exchange, and sale, fostering innovation and supporting biodiversity in horticulture (Mashelkar 2001; Suraj Bhan et al. 2016).

7.5.3 Participatory Research: FFS, PRA, and Innovation Platforms

Participatory approaches integrate ITK with modern science. Farmer Field Schools (FFS), developed by the FAO in the 1980s, enable experiential learning through hands-on experiments and peer exchanges, validating ITK practices like neem-based pest control (FAO 2001; Minjauw 2001). Participatory Rural Appraisal (PRA) facilitates community-driven resource mapping and problem analysis, co-creating localized solutions (Talukdar et al. 2012). Innovation platforms foster multi-stakeholder collaboration, ensuring inclusivity, gender sensitivity, and scalability of ITK-based practices, empowering farmers as co-innovators (Blackie and Gibbon 2001; Vidyawati et al. 2025; virk et al. 2005).

7.5.4 Validation and Documentation of ITK

Validation ensures ITK's integration into modern horticulture. Comparative analyses test ITK practices against conventional methods through on-station experiments for rigor and on-farm trials for applicability, assessing yield, cost-effectiveness, and socio-cultural fit (Pierotti and Wildcat 1997). Systematic documentation in digital repositories and publications preserves validated ITK, recognizing farmers' contributions to protect intellectual property and promote grassroots innovation (Gupta 2009). FFS facilitates field testing of ITK, evaluating metrics like productivity and soil health, with results shared via community and digital platforms (Ward et al. 2001; virk et al. 2005).

7.5.5 ICT Tools for ITK Dissemination

The use of Information and Communication Technology (ICT) has enhanced the documentation, preservation, and dissemination of Indigenous technical knowledge (ITK) in agriculture. ICT platforms such as mobile applications, digital databases, community radio, and online portals facilitate efficient knowledge sharing between farmers, researchers, and extension agencies. These tools improve accessibility and promote wider adoption of traditional agricultural practices while complementing modern scientific knowledge. The major ICT tools used for ITK dissemination are summarized in Table 7.2.

7.5.6 Value Addition and Product Development

ITK supports eco-friendly product development. Traditional preservation methods (sun-drying, fermentation, roasting) produce value-added goods like dried fruits, pickles, millet snacks, jaggery sweets, and herbal teas, preserving cultural identity (Shenoy 2012). Organic inputs like *panchagavya*, *jeevamrut*, and *neemastra* are standardized as alternatives to chemical inputs. Herbs such as neem (*Azadirachta indica*), tulsi (*Ocimum tenuiflorum*), and turmeric (*Curcuma longa*) are used in soaps, balms, and veterinary medicines. Aromatic crops like lemongrass (*Cymbopogon* spp.) and turmeric gain market traction through modern processing and branding, supported by scientific validation and intellectual property protection, enhancing rural livelihoods (Adefila et al. 2024; Pandey et al. 2017; Souravi and Rajasekharan 2019).

7.6 Challenges in the Indigenous Technical Knowledge (ITK)

ITK has been emphasized as having some problems because of its context of oral tradition, thus it is prone to the loss that accompanies the old age of elders or upheavals in society. The generation gap in knowledge sharing increases with westernization, the migration of the younger generation, and cultural change. Not all ITKs are scientifically validated and they are often neglected or marginalized in institutions alongside standardized technologies and they have poor intellectual property protection. Lukewarm documentation, localization, and poor digital literacy contribute to not recognizing or spreading them.

7.6.1 Challenges in ITK

Oral Tradition and Knowledge Loss: ITK's reliance on oral transmission risks loss with elders' passing, social upheavals, or endangered indigenous languages, complicating documentation (Ouma 2022). *Generation Gap and Westernization*: Youth

Table 7.2 ICT tools for ITK dissemination

Tool	Description	ITK application	Reference
NeGP-A	Unified platform for documentation and dissemination	Shares ITK on seed conservation, pest control	Punitha (2025)
Mobile apps (mKisan, farm rise)	Real-time advisories in regional languages	Promotes composting, intercropping	Priyanshu and Mrigangko (2024)
Digital platforms (digital green, e-SAP)	Participatory videos and repositories	Shares ITK via farmer-generated content	Dhulipala et al. (2021)
KVK portals and Agri-clinics	Localized advisories and validation	Documents ITK on seed drying, crop spacing	Punitha (2025)
Touchscreen kiosks and CSCs	Audio-visual dissemination in native languages	Bridges literacy gap for ITK sharing	Dlamini and Ocholla (2018)
Kisan call centers (KCC)	Expert consultations in local languages	Validates ITK on pest management	Priyanshu and Mrigangko (2024)
Kisan TV and DBT portals	Broadcasts and input-sharing platforms	Shares ITK-based success stories	Punitha (2025)
Weather apps (Meghdoot, Skymet)	Blends traditional and scientific forecasts	Integrates ITK indicators (e.g., cloud patterns)	Dhulipala et al. (2021)
Private platforms (e-Choupal)	Market and traceability platforms	Supports ITK-based marketing, rainwater harvesting	Punitha (2025)
Community radio	Folk songs, storytelling	Preserves oral ITK traditions	Hausknecht et al. (2021)
Kisaan suvidha	Links farmers to markets	Supports ITK-based organic produce marketing	Priyanshu and Mrigangko (2024)
Agrotrace	IoT-based validation tool	Validates ITK in irrigation, soil management	Punitha (2025)
myScheme	Lists government schemes	Promotes ITK-based organic farming, bio-fertilizers	Punitha (2025)
Poshan atlas	Highlights indigenous food practices	Promotes ITK-based crop diversity	Punitha (2025)
Bijak Vyapaar	Grain marketplace	Supports ITK-based millet cultivation, storage	Punitha (2025)

These tools overcome rural connectivity challenges, engage youth, and preserve oral traditions through audio-visual media (Mishra et al. 2020)

migration and modern education prioritizing Western science create gaps in ITK transfer, marginalizing traditional practices (Hausknecht et al. 2021). *Scientific Validation and Marginalization*: Unvalidated ITK faces institutional neglect, with preference for standardized technologies limiting adoption (Prinila et al. 2025).

Intellectual Property Protection: Mainstream IP systems favor individual ownership, leaving community-held ITK vulnerable to biopiracy (Newing et al. 2024). *Documentation and Digital Literacy*: Limited documentation and low digital literacy in rural areas hinder ITK recognition and dissemination (Dlamini and Ocholla 2018). *Localization and Institutional Support*: ITK's context-specific nature challenges adaptation, with insufficient institutional support for validation and integration (Imoro et al. 2021).

7.6.2 Opportunities in ITK

Sustainable and Climate-Smart Agriculture: ITK practices like mixed cropping and organic pest control align with climate-smart agriculture, enhancing resilience (Imoro et al. 2021). *Commercialization and Economic Impact*: ITK-based products (e.g., organic manure, herbal pesticides) offer premium income, empowering women and smallholders (Deißler et al. 2025). *Digital Platforms and Participatory Methods*: Platforms like Traditional Knowledge Digital Library (TKDL) and participatory validation foster ownership and knowledge transfer (Dlamini and Ocholla 2018; Ouma 2022). *Empowerment and Rural Development*: Participatory research promotes innovation and equitable benefit-sharing, boosting rural livelihoods (Newing et al. 2024). *Legal and Policy Innovation*: WIPO treaty and sui generis IP systems protect ITK, ensuring fair compensation (Newing et al. 2024).

Despite its significant role in sustainable agriculture, *Indigenous technical knowledge (ITK)* faces several challenges, including inadequate documentation, limited scientific validation, and declining intergenerational transmission. However, growing interest in sustainable, climate-resilient farming has created new opportunities to integrate ITK with modern agricultural practices. Strengthening research, policy support, and digital documentation can enhance the preservation and wider application of traditional knowledge systems. The major challenges and opportunities associated with ITK are summarized in Table 7.3.

Table 7.3 ITK challenges and opportunities

Aspect	Challenges	Opportunities
Knowledge transmission	Oral tradition, loss with elders, language extinction	Digital documentation, community archives
Generation gap	Migration, westernization, education bias	Youth engagement, school programs
Validation	Limited scientific validation, institutional bias	Participatory research, data-driven platforms
IP protection	Inadequate frameworks, biopiracy	Sui generis systems, mandatory patent disclosure
Commercialization	Poor market access, lack of branding	Premium products, women-led enterprises
Climate-smart potential	Marginalization in mainstream systems	Sustainable adaptation, ecosystem resilience

7.7 Opportunities in Indigenous Technical Knowledge (ITK)

Although these challenges are still present, ITKs hold greater potential for sustainable agriculture in the form of mixed cropping, organic fertilization, and biological controls, as well as water conservation. They are in line with climate-smart practices, and they facilitate adaptation to variability. Organic manure, herbal pesticides, and indigenous grains can be successfully commercialized to yield a positive economic impact, particularly on women farmers (Mercer et al. 2010). Digital platforms, community-based documentation, and participatory validation strengthen ownership, empowerment, and dissemination in the rural areas.

7.7.1 Climate-Smart and Adaptive Agriculture

Indigenous farmers excel in climate adaptation using time-tested methods like mixed cropping, traditional crop varieties resilient to drought or floods, organic fertilization, biological pest control, and precise water management. These practices support biocultural conservation, sustainable resource use, and adaptation to environmental variability, positioning ITK as vital for climate-smart agriculture and disaster risk reduction.

7.7.2 Value Addition and Empowerment

ITK underpins value-added enterprises, especially for women. For example, women traditionally process milk into butter, ghee, and yogurt—profitable household enterprises that contribute to income, nutrition, and empowerment but require supportive environments to sustain equitable benefit sharing as commercialization accelerates. Policies that strengthen women's land rights, access to credit, and market information, and foster cooperatives, can help maintain women's empowerment in the ITK value chain.

7.7.3 Participatory Research and Legal Innovations

Participatory action research, farmer-led innovation platforms, and direct involvement in validation and experimentation empower communities, enhance trust, and foster culturally sensitive integration of ITK. Technology—including blockchain and secure digital registries, can help document and protect ITK while allowing communities to maintain control over how their knowledge is accessed and used. Community protocols and partnerships can support equitable benefit sharing and international cooperation.

7.7.4 Education, Digital Platforms, and Youth Engagement

Innovative educational models, mentorship programs, and localized digital content reintegrate intergenerational learning and encourage youth engagement. Recognizing elders as vital knowledge holders helps restore and maintain the instruction manual for sustainable living in rapidly changing contexts.

7.8 Policy and Framework

Policy frameworks at both national and global levels underscore the critical role of ITK in advancing sustainable agriculture and biodiversity conservation. In India, the Biological Diversity Act (2002) establishes a legal basis for benefit-sharing when ITK contributes to commercial innovations, ensuring equitable returns for indigenous communities. Internationally, the Nagoya Protocol (2010) and endorsements from bodies such as the United Nations, including the UN Declaration on the Rights of Indigenous Peoples and FAO initiatives, promote the documentation, protection, and utilization of indigenous agricultural knowledge. The WIPO Treaty on intellectual property, genetic resources, and associated traditional knowledge (adopted May 2024), ratified by over 50 countries, mandates the disclosure of ITK use in patent applications to combat biopiracy (Amentae et al. 2024). These frameworks have catalyzed funding for ITK research and community-led initiatives, fostering equitable access to benefits derived from traditional knowledge.

7.8.1 Intellectual Property Rights and the Traditional Knowledge Digital Library

Legal mechanisms, particularly intellectual property rights (IPR) and India's Traditional Knowledge Digital Library (TKDL), play a pivotal role in safeguarding traditional horticultural knowledge. The TKDL, a digital repository, catalogs approximately 454,000 traditional formulations and practices, slightly lower than earlier estimates of 515,000 due to refined classification. Accessible to 17 patent offices globally, the TKDL has facilitated the withdrawal or amendment of approximately 375 patent applications, effectively curbing misappropriation. Its integration of artificial intelligence for structured classification enhances efficiency in global patent searches. Amentae et al. (2024) emphasize that IPR frameworks, while fostering innovation, risk marginalizing indigenous knowledge holders due to corporate dominance in agri-food chains, underscoring the importance of protective mechanisms like the TKDL to ensure equitable knowledge governance.

7.8.2 Role of NGOs and Grassroots Movements

Non-governmental organizations (NGOs) and grassroots movements are instrumental in preserving, disseminating, and scaling ITK for modern horticulture. Through participatory research, capacity-building workshops, and market linkage programs, NGOs bridge traditional practices with contemporary agricultural needs. For example, the Gond community in central India, supported by NGOs such as PRADAN, employs crop rotation techniques (e.g., legumes with cereals) to enhance soil fertility, yielding 10–15% income increases. In Meghalaya, the Khasi people's agroforestry model, which integrates fruit trees with staple crops, mitigates hillside erosion, with NGOs like NESFAS facilitating adoption through state organic missions. Similarly, Andhra Pradesh's community-managed natural farming (APCNF), driven by women-led groups and NGOs like Rythu Sadhikara Samstha, spans approximately 8 million hectares and achieves 10–20% higher yields than conventional systems. These findings resonate with Khatri et al. (2021), who document high ITK adoption in Nepal's Bhirkot municipality due to its cost-effectiveness and cultural compatibility, though adoption is constrained by limited market access and youth disengagement.

7.8.3 Schemes Supporting ITK: PKVY, RKVY, and State Missions

Government schemes provide critical support for ITK integration into modern farming systems. The Paramparagat Krishi Vikas Yojana (PKVY), launched in 2015, promotes organic farming clusters with financial support of ₹50,000 per hectare, covering 14.99 lakh hectares by 2025 and engaging 25.3 lakh farmers through participatory guarantee systems (PGS) and farmer producer organizations (FPOs). The Rashtriya Krishi Vikas Yojana (RKVY), relaunched as PM-RKVY in 2024 with a ₹57,074 crore allocation, funds state-driven initiatives integrating ITK, including water conservation programs like "Per Drop More Crop" and soil health management. State-level missions in regions such as Meghalaya and Sikkim have expanded organic farming and ITK documentation across over 100 districts, leveraging NGO partnerships to develop context-specific best practices. These schemes enhance the adoption of sustainable crop management, water conservation, and indigenous pest control, often integrated with technologies like bioacoustic detectors and pheromone traps in integrated pest management (IPM) (Melash et al. 2023).

7.8.4 Impact on Modern Horticulture

The integration of ITK with modern technologies has positioned traditional knowledge as a cornerstone of eco-friendly horticultural modernization. Policy interventions have driven the adoption of sustainable practices, reducing chemical pesticide use by 30–50% in South Asian organic systems (FAO 2001) and increasing rural incomes by 10–30% through organic transitions. Melash et al. (2023) advocate for the scientific validation of neglected ITK practices, such as the use of cow urine for pest control in Ethiopia, to support organic farming transitions, aligning with India's efforts to enhance ecosystem health and agricultural resilience through ITK-driven approaches.

7.8.5 Synthesis and Future Directions

The multilayered interplay of policy frameworks, legal protections, and grassroots initiatives demonstrates ITK's transformative potential in modern horticulture, improving agroecosystem stability, reducing environmental footprints, and fortifying food security. Collaborative efforts among policymakers, scientists, NGOs, and farming communities ensure the continued adaptation and relevance of ITK in a rapidly evolving agricultural landscape. Future directions include strengthening IPR-ITK frameworks to address corporate biases in agri-food chains, as highlighted by Amentae et al. (2024), and scaling successful models like APCNF globally while addressing barriers such as education and gender gaps noted in Nepal (Khatri et al. 2021). Scientific validation of neglected ITK practices, as proposed by Melash et al. (2023), should be prioritized through research partnerships to align with Sustainable Development Goals (SDGs). Additionally, digital innovations—such as AI-enhanced TKDL, blockchain for organic product traceability, and mobile applications like Krishi Kisan—alongside increased funding through PM-RKVY and global initiatives, will further institutionalize ITK, ensuring its role as a guiding force in sustainable horticulture (Fig. 7.2).

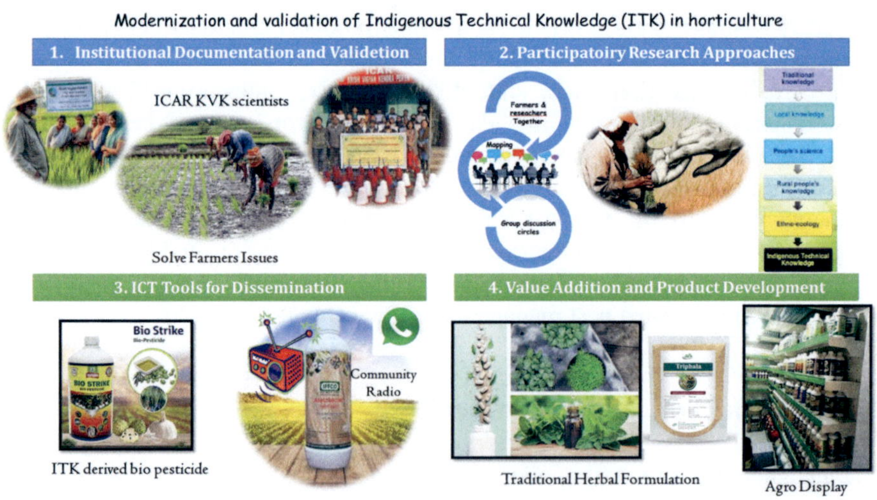

Fig. 7.2 Modernization and validation of Indigenous technical knowledge (ITK) in horticulture

7.9 Future Perspectives

The integration of Indigenous Traditional Knowledge (ITK) into modern horticulture is pivotal for fostering sustainable, resilient, and culturally enriched agricultural systems. As global challenges such as climate change, biodiversity loss, and food insecurity intensify, ITK offers a holistic framework rooted in ecological stewardship and community-driven innovation. This section outlines strategic priorities for revitalizing and mainstreaming ITK in horticulture, emphasizing decentralized knowledge hubs, educational reform, multi-stakeholder collaboration, and resilient landscape development. Drawing on recent literature (Amentae et al. 2024; Khatri et al. 2021; Melash et al. 2023) and additional sources, these perspectives provide a roadmap for ensuring ITK's enduring role in sustainable horticulture.

7.9.1 Creating Decentralized ITK Knowledge Hubs

Decentralized knowledge hubs are essential for preserving and disseminating horticultural ITK, empowering local communities, and fostering innovation. These hubs, combining physical community centers with digital platforms, serve as repositories for traditional practices, seed varieties, and agroecological knowledge. By leveraging technologies such as blockchain for intellectual property protection and AI for documentation, hubs ensure transparent benefit-sharing and safeguard indigenous wisdom (Bhattacharjee et al. 2024). Platforms facilitating farmer-to-farmer exchanges, participatory research, and videos in local dialects promote peer learning and contextual adaptation. Actively engaging women and youth, who are often primary knowledge custodians, enhances inclusivity and supports intergenerational transfer, strengthening community identity and agroecological resilience (UNDP 2025; Brijesh Kumar 2025).

7.9.2 Reviving Traditional Wisdom Through School and University Curricula

Embedding ITK in formal education systems is critical for its intergenerational transmission and cultural revitalization. Schools can integrate experiential learning modules on traditional practices such as mixed cropping, water conservation techniques (e.g., Johads), and natural pest management, fostering ecological literacy among students. Universities should develop interdisciplinary programs in ethnobotany, agronomy, and rural sociology to critically evaluate ITK alongside modern technologies (Brijesh kumar 2025). Partnerships between academic institutions and farming communities, including field internships, elder-led storytelling, and biodiversity documentation projects, cultivate respect for indigenous systems and inspire youth-driven innovation. Such initiatives bridge the gap between scientific and traditional knowledge, ensuring ITK's relevance in modern horticulture (Bhattacharjee et al. 2024).

7.9.3 Fostering Farmer–Scientist–Policy Dialogues

Continuous, participatory dialogues among farmers, scientists, and policymakers are foundational for co-creating context-specific horticultural innovations. Multi-stakeholder platforms institutionalizing farmer representation in research and policy committees ensure that ITK informs evidence-based reforms. These forums facilitate adaptive governance, addressing systemic barriers such as land tenure, gender equity, and market access, which influence ITK's viability (Khatri et al. 2021). Investments in culturally sensitive extension services, utilizing vernacular languages, enhance communication and trust, while participatory monitoring strengthens the evidence base for scaling ITK practices. Such collaboration aligns public incentives, such as subsidies for organic inputs, with farmer needs, fostering sustainable and equitable horticultural systems (Bhattacharjee et al. 2024; Brijesh Kumar 2025).

7.9.4 Building Resilient Horticultural Landscapes with ITK at the Core

ITK-based horticultural landscapes, characterized by polycultures, agroforestry, and natural pest management, exemplify ecological resilience and climate adaptability. Practices like contour bunding, mulching, and crop diversification reduce vulnerability to climate shocks and enhance ecosystem services, including soil fertility and water conservation (Melash et al. 2023; UNDP 2025). Scaling these approaches through landscape-level programs, supported by community seed banks and pollinator habitats, preserves genetic diversity and strengthens food security. Policy incentives and scientific validation, as advocated by Melash et al. (2023), can mainstream ITK-informed models, creating ecological corridors and climate-smart agricultural systems resilient to market and environmental stressors (Rainforest Alliance 2025).

7.10 Conclusion

Indigenous Traditional Knowledge holds enduring value in horticulture, integrating ecological stewardship, cultural heritage, and sustainable livelihoods. Farmers, as custodians and innovators, play a pivotal role in conserving agroecosystems and ensuring community food security. Inclusive co-creation through farmer–scientist–policy dialogues fosters dynamic adaptation, amplifying ITK's impact through mutual learning and respect. Embedding ITK in education, policy, and research frameworks positions it as an equal partner to scientific knowledge, addressing urgent challenges like climate change and resource depletion. This integrative vision ensures resilient, equitable, and culturally enriched food systems, laying a foundation for sustainable horticultural futures that honor tradition while innovating for global needs.

Declaration

Ethical Approval Not applicable.

Use of AI Tools Only minor grammar checking was performed using an automated tool. No AI tools were used for content generation or manuscript preparation.

Conflict of Interest On behalf of all the authors, the corresponding author declares that there are no conflicts of interest to disclose.

Competing Interests Not applicable.

Authors' Contributions Vasanthkumar S. S.: Conceptualization, methodology design, supervision, and preparation of the original draft.

Karthikeyan J.: Data curation, formal analysis, and review & editing of the man**uscript.**

Yazhini S. P.: Investigation, visualization, and software-relate**d tasks.**

Gowshika R.: Literature review, validation, and contribution to draft prep**aration.**

Funding This research did not receive any specific grant from funding agencies in the public, commercial, or not-for-profit sectors.

References

Absanto G, Mkunda J, Nyangarika A (2025) Modern versus traditional irrigation systems: implications for technical efficiency and input-output relationship in Northern Tanzania's horticulture sector. Glob Acad J Econ Bus 7:1. https://doi.org/10.36348/gajeb.2025.v07i01.001

Adefila AO, Ajayi OO, Toromade AS, Sam-Bulya NJ (2024) Integrating traditional knowledge with modern agricultural practices: a sociocultural framework for sustainable development. World J Biol Pharm Health Sci 20(2):125–135. https://doi.org/10.9734/WJBPHS/2024/v20i2305

Amentae TK, Song W, Wang J (2024) Intellectual property rights in the agri-food chains: a systematic review and bibliometric analysis. World Patent Inf 77:102279. https://doi.org/10.1016/j.wpi.2024.102279

Bang-Andreasen T, Nielsen JT, Voříšková J, Heise J, Rønn R, Kjøller R, Hansen HCB, Jacobsen CS (2017) Wood ash induced pH changes strongly affect soil bacterial numbers and community composition. Front Microbiol 8:1400. https://doi.org/10.3389/fmicb.2017.01400

Behera KK (2012) Indigenous technical knowledge in agriculture: a review. Indian Council of Agricultural Research

Bhan S, Bharti VK, Meetali, Deepshikha, Bharti S, Bharti S (2016) Impactful Indigenous Technical Knowledge (ITK) for sustainable agriculture. J Soil Water Conserv 15(4):349–355

Bhattacharjee P, Maity J, Biswas A (2024) Indigenous traditional knowledge (ITK) in agriculture and horticulture: a comprehensive review. World J Biol Pharm Health Sci 18(2):256–262. https://wjbphs.com/sites/default/files/WJBPHS-2024-0850.pdf

Bhutia ST (2024) Organic farming in Sikkim: impact, causes, and consequences. Libr Prog Int 44(3):29013–29020

Blackie MJ, Gibbon D (2001) Participatory research and farmer empowerment. In: Farmer innovation in Africa. Earthscan, pp 45–60

Brijesh Kumar (2025) Indigenous knowledge systems and sustainable agriculture: a review. IOSR J Biotechnol Biochem 11(1):30–38. https://www.iosrjournals.org/iosr-jbb/papers/Volume%2011,%20Issue%201/Ser-1/E1101013038.pdf

Das P, Reddy GS, Das SK, Verma LR, Mishra A (2003) Indigenous technical knowledge in agriculture: document 1. Indian Council of Agricultural Research

Das P, Reddy GS, Das SK, Verma LR, Mishra A, Rani MG, Arya HPS, Ray DP, Singh RP (2004) Inventory of indigenous technical knowledge in agriculture: document 2 (supplement 2). Indian Council of Agricultural Research

Das A, Layek J, Ramkrushna GI, Babu S, Devi M, Dey U, Suting D, Yadav G, Lyngdoh D, Prakash N (2019) Integrated organic farming system: an innovative approach for enhancing productivity and income of farmers in north eastern hill region of India. Indian J Agric Sci 89(8):1269–1274. https://doi.org/10.56093/ijas.v89i8.92843

Deißler L, Krause H, Grote U (2025) Women's empowerment and household welfare in the Kenyan African indigenous vegetable supply chain. Humanit Soc Sci Commun 12:604. https://doi.org/10.1057/s41599-025-04832-w

Dhulipala R, Gogumalla P, Rao KPC (2021) Digital innovations for scaling up agricultural knowledge. Agric Syst 190:103101. https://doi.org/10.1016/j.agsy.2021.103101

Dlamini P, Ocholla DN (2018) Information and communication technology tools for managing indigenous knowledge in KwaZulu-Natal Province, South Africa. Afr J Libr Arch Inf Sci 28(2):137–153

FAO (2001) Farmer field schools: from IPM to platforms for learning and empowerment. Food and Agriculture Organization

Ficiciyan A, Loos J, Sievers-Glotzbach S, Tscharntke T (2018) More than yield: ecosystem services of traditional versus modern crop varieties revisited. Sustainability 10(8):2834. https://doi.org/10.3390/su10082834

Giri K, Mishra G, Rawat M, Pandey S, Bhattacharyya R, Bora N, Rai JP (2020) Traditional farming systems and agro-biodiversity in eastern Himalayan region of India. In: Agro-biodiversity and agri-ecosystem management. Springer, pp 71–89. https://doi.org/10.1007/978-981-15-1902-4_5

Goswami B, Pariyar B (2025) Comprehensive analysis of large cardamom cultivation in Sikkim: challenges, opportunities, and sustainable strategies. South Asian J Agric Sci 5(1):1–10. https://doi.org/10.22271/27889289.2025.v5.i1a.172

Gupta AK (2009) From grassroots to global: the role of traditional knowledge in innovation. J Intellect Property Rights 14(5):412–420

Hausknecht S, Freeman S, Martin J, Nash C, Skinner K (2021) Sharing indigenous knowledge through intergenerational digital storytelling: design of a workshop engaging elders and youth. Educ Gerontol 47(7):285–296. https://doi.org/10.1080/03601277.2021.1923553

Ijatuyi EJ, Lamm A, Yessoufou K, Suinyuy T, Patrick HO (2025) Integration of indigenous knowledge with scientific knowledge: a systematic review. Environ Sci Pol 170:104119. https://doi.org/10.1016/j.envsci.2025.104119

Imoro ZA, Imoro AZ, Duwiejuah AB, Abukari A (2021) Harnessing Indigenous Technologies for sustainable management of land, water, and food resources amidst climate change. Front Sustain Food Syst 5:691603. https://doi.org/10.3389/fsufs.2021.691603

Joshi RB (2021) Indigenous environmental knowledge and sustainable development: a case study of traditional ecological practices in India. Ilkogretim Online—Elementary Education Online 20(6):5785–5790. https://doi.org/10.17051/ilkonline.2021.06.561

Kaur N, Kocher DK (2023) Ovicidal, larvicidal and pupicidal potential of neem seed kernel extract against *Aedes aegypti* L. Int J Trop Insect Sci 43(3):1089–1096. https://doi.org/10.1007/s42690-023-01020-8

Khatri S, Khanal S, Kafle S (2021) Perceived attributes and adoption of Indigenous Technological Knowledge on agriculture—a case study from Bhirkot municipality of Syangja District, Nepal. Cogent Food Agric 7(1):1914384. https://doi.org/10.1080/23311932.2021.1914384

Kolarkar AS, Murthy KNK, Singh N (1983) Khadin—a method of harvesting water for agriculture in the Thar Desert. J Arid Environ 6(1):59–66. https://doi.org/10.1016/S0140-1963(18)31433-2

Li R, Li Q, Pan L (2020) Review of organic mulching effects on soil and water loss. Arch Agron Soil Sci 67(1):136–151. https://doi.org/10.1080/03650340.2020.1718111

Lucy E, Wudil B, Yusuf A (2025) Insecticidal efficacy of some botanicals against *Bruchus baudni* Caill (Coleoptera: Bruchidae) on *Senegalia senegal* seeds during storage. Asian J Res Agric Forest 11(2):30–37. https://doi.org/10.9734/ajraf/2025/v11i2388

Mandal KG, Banerjee K, Samal SK, Bharti PK, Purbey SK (2025) Blending traditional knowledge of farmers in agriculture with modern scientific technologies in the state of Jharkhand. In: Acharya CL, Chaudhary R, Gurumurthy P, Rao AS (eds) Blending Indian farmers' traditional knowledge in agriculture with modern scientific technologies. Springer, pp 347–368. https://doi.org/10.1007/978-981-96-1020-4_21

Mashelkar RA (2001) Intellectual property rights and the protection of plant varieties and farmers' rights act, 2001. Curr Sci 81(5):425–430

Melash AA, Bogale AA, Migbaru AT, Chakilu GG, Percze A, Ábrahám ÉB, Mengistu DK (2023) Indigenous agricultural knowledge: a neglected human-based resource for sustainable crop protection and production. Heliyon 9(1):e12978. https://doi.org/10.1016/j.heliyon.2023.e12978

Minjauw B (2001) Farmer field schools: impact on IPM and farmer empowerment. Agric Syst 68(3):201–216

Mishra A, Singh SRK, Raut AA (2020) Traditional knowledge in agriculture. Division of Agricultural Extension, Indian Council of Agricultural Research

Mondal R (2024) Reviving Ahar-Pynes: harnessing traditional irrigation for sustainable farm production in South Bihar. SSRN Electron J. https://doi.org/10.2139/ssrn.4818281

More SS, Thokal RT, Sawant PA, Zagade PM (2025) Blending traditional knowledge of farmers in agriculture with modern scientific technologies in coastal region of Maharashtra. In: Acharya CL, Chaudhary R, Gurumurthy P, Rao AS (eds) Blending Indian farmers' traditional knowledge in agriculture with modern scientific technologies. Springer, pp 529–551. https://doi.org/10.1007/978-981-96-1020-4_32

Murry N (2021) Zabo farming system—a sustainable farming based on traditional knowledge for natural resource management practiced by tribal in Nagaland, India. Int J Agric Environ Biotechnol 14(2):163–170. https://doi.org/10.30954/0974-1712.02.2021.13

Newing H, Brittain S, Buchadas A, del Giorgio O, Fallon Grasham C, Ferritto R, Garcia Marquez JR, Khanyari M, König B, Kulkarni A, Murali R, Qin S, Rakowski J, Ghoddousi A (2024) Participatory' conservation research involving indigenous peoples and local communities: fourteen principles for good practice. Biol Conserv 296:110708. https://doi.org/10.1016/j.biocon.2024.110708

Ouma A (2022) Intergenerational learning processes of traditional medicinal knowledge and socio-spatial transformation dynamics. Front Sociol 7:661992. https://doi.org/10.3389/fsoc.2022.661992

Panda T, Mishra N, Rahimuddin S, Pradhan BK, Nayak BK, Sahu S (2024) Nurturing tradition and nature through Odisha's Shraddha rituals. J Contemporary Rituals Traditions 2(1):1–12. https://doi.org/10.15575/jcrt.467

Pandey V, Mittal R, Sharma P (2017) Documentation and application of Indigenous Traditional Knowledge (ITK) for sustainable agricultural development. Asian J Agric Ext Econ Sociol 15(3):1–9. https://doi.org/10.9734/AJAEES/2017/31481

Pathania K (2025) The role of the Himachal Pradesh horticulture development project to support small farmers and agro-entrepreneurs in Himachal Pradesh. J Agric Biotechnol Appl Sci 2(1):11–18. https://doi.org/10.63143/jabaas12202321

Pierotti R, Wildcat DR (1997) Traditional knowledge and ecological restoration. Ecol Appl 7(4):1127–1134

Pooja B, Anandaraja N, Fernandaz CC, Balasubramaniam P, Selvi RP (2025) Archiving indigenous technical knowledge (ITK) based ethnomedicinal practices among tribal communities in the Nilgiris: a perception analysis. Plant Sci Today 12(1):1–10. https://doi.org/10.14719/pst.5088

Prasad M, Kumar S, Kumar A, Khan N, Shukla SK (2025) Blending traditional knowledge of farmers in agriculture with modern scientific technologies in Central Uttar Pradesh. In: Acharya CL, Chaudhary R, Gurumurthy P, Rao AS (eds) Blending Indian farmers' traditional knowledge in agriculture with modern scientific technologies. Springer, pp 257–277. https://doi.org/10.1007/978-981-96-1020-4_16

Prinila AS, Ravi G, Preetha G, Anandham R, Manjunatha C (2025) Traditional insights into ITK and its delineation for sustainable crop protection. Plant Sci Today 12(1). https://doi.org/10.14719/pst.6110

Priyanshu, Mrigangko D (2024) Role of ICT in agricultural extension services: opportunities and challenges. J Agric Ext 28(2):45–56

Punitha P (2025) Digital agriculture and ITK dissemination: a case study of NeGP-A. Indian J Agric Sci 95(1):12–20

Rainforest Alliance (2025) The indigenous roots of regenerative agriculture. https://www.rainforest-alliance.org/insights/the-indigenous-roots-of-regenerative-agriculture/

Rajkumara S, Patil PL, Chittapur BM, Patil CR, Hegde GS, Channakeshava S (2025) Blending traditional knowledge of farmers in agriculture with modern scientific technologies in semi-arid Karnataka. In: Acharya CL, Chaudhary R, Gurumurthy P, Rao AS (eds) Blending Indian farmers' traditional knowledge in agriculture with modern scientific technologies. Springer, pp 87–108. https://doi.org/10.1007/978-981-96-1020-4_6

Rautela P (2015) Traditional practices of the people of Uttarakhand Himalaya in India and relevance of these in disaster risk reduction in present times. Int J Disaster Risk Reduct 13:281–290. https://doi.org/10.1016/j.ijdrr.2015.07.004

Risch SJ (1983) Intercropping as cultural pest control: prospects and limitations. Environ Manag 7(1):9–14. https://doi.org/10.1007/BF01867035

Salgado GP, Ambrosano EJ, Rossi F, Otsuk IP, Ambrosano GMB, Patri P, Henrique CM, Santana CA, Muraoka T, Trivelin PCO (2021) Yield and nutrient concentrations of organic cherry tomatoes and legumes grown in intercropping systems in rotation with maize. Biol Agric Horticult 38(2):94–112. https://doi.org/10.1080/01448765.2021.1992796

Sharma K, Jarial S, Patil ND (2025) Integrating indigenous technical knowledge for sustainable agriculture and horticulture: insights from India. In: Enabling indigenous knowledge systems in action research and action learning. IGI Global, pp 1–24

Shenoy S (2012) Traditional food preservation techniques in India. J Food Sci Technol 49(3):265–272

Souravi K, Rajasekharan PE (2019) Traditional knowledge in horticulture. In: Conservation and utilization of horticultural genetic resources. Springer, pp 621–636. https://doi.org/10.1007/978-981-13-3669-0_20

Talukdar RK, Barman S, Hussain A (2012) Documentation and perceived rationale of Indigenous Technical Knowledge (ITK) utilized in Boro rice cultivation by farmers of Kamrup District of Assam. J Acad Indust Res 1(7):379–382

Tynsong H, Dkhar M, Tiwari BK (2020) Traditional ecological knowledge of tribal communities of North East India. Biodivers J Biol Divers *21*(7):743. https://doi.org/10.13057/biodiv/d210743

UNDP (2025) Indigenous knowledge: crucial in the fight against climate change. https://climatepromise.undp.org/news-and-stories/indigenous-knowledge-crucial-fight-against-climate-change-heres-why

Virk DS, Chakraborty M, Ghosh J, Prasad SC, Witcombe JR (2005) Increasing the client orientation of maize breeding using farmer participation in eastern india Experimental Agriculture 41(4):413–426. https://doi.org/10.1017/S001447970500270X

Vidyawati, Singh SP, Jadoun RS (2025) Participatory agricultural extension: a catalyst for sustainability and farmer empowerment. Int J Agric Ext Soc Develop 8(4):333–341. https://doi.org/10.33545/26180723.2025.v8.i4e.1790

Vinayak M, Vani G, Subhasree N, Raju C (2022) Effect of Zai pit technology on the infiltration capacity of the soil. Int J Environ Climate Chang 12(12):1483–1489. https://doi.org/10.9734/ijecc/2022/v12i121539

Wahedi JA, Rebecca Z, Elkanah OS, Ishuwa MN, Vincent VM, Enoch TG (2017) Activities of neem and wood ash as biopesticides in the control of insect pests on vegetable crops in Mubi. GSC Biol Pharm Sci 1(1):6–10. https://doi.org/10.30574/gscbps.2017.1.1.0006

Wambugu PW, Mathenge PW, Auma EO, van Rheenen HA (2009) Efficacy of traditional maize (*Zea mays* L.) seed storage methods in western Kenya. Afr J Food Agric Nutr Dev 9(4):1110–1128. https://doi.org/10.4314/ajfand.v9i4.43882

Ward PS, Bell AR, Benton TG (2001) Farmer field schools as a tool for integrating traditional and modern agricultural practices. Agric Syst 69(3):225–240

Willey RW (1990) Resource use in intercropping systems. Agric Water Manag 17(1–3):215–231. https://doi.org/10.1016/0378-3774(90)90069-B

Yadav S (2023) Traditional rain water harvesting systems in Rajasthan: water resources conservation and its sustainable management—a review. Sustain Agri Food Environ Res 11(1):1–12. https://doi.org/10.7770/safer-v11n1-art2611

Genetic Engineering and Bio-fortification: Adding Value for Prosperity

8

Nirmal Mandal and Amit Baran Sharangi

Abstract

In the evolving landscape of horticulture, value addition through scientific innovation is redefining economic and nutritional outcomes. *Remunerative horticulture*, once centered on fruits and vegetables, now extends to a diverse array of high-value crops, including flowers, spices, plantation crops, medicinal and aromatic plants (MAPs), and emerging functional varieties. Against this backdrop, **genetic engineering** and **bio-fortification** are increasingly vital in unlocking new dimensions of prosperity, sustainability, and health security.

This chapter provides an in-depth, multidisciplinary exploration of how modern biotechnological interventions—such as transgenics, genome editing, RNAi, and omics-driven trait discovery—can reconfigure the genetic potential of horticultural crops. Traits like post-harvest longevity, disease resistance, pigment enhancement, nutraceutical enrichment, and environmental resilience are discussed through targeted case studies. These include Bt Brinjal, golden banana, and sweet potato for provitamin A, non-browning apple, anthocyanin-rich tomato, curcumin-enhanced turmeric, and engineered *Withania somnifera*, among others.

The conceptual and practical frameworks of **bio-fortification**, including classical and modern strategies, are critically analyzed for enriching essential micronutrients (Fe, Zn, vitamins A/C/E) and functional metabolites (flavonoids,

N. Mandal (✉)
Department of Agricultural Biotechnology, Faculty of Agriculture, Bidhan Chandra Krishi Viswavidyalaya, Mohanpur, West Bengal, India

A. B. Sharangi
Department of Plantation, Spices, Medicinal and Aromatic Crops, Faculty of Horticulture, Bidhan Chandra Krishi Viswavidyalaya, Mohanpur, West Bengal, India

A. Das, A. B. Sharangi (eds.), *Remunerative Horticulture*,
https://doi.org/10.1007/978-981-92-1007-7_8

antioxidants, alkaloids, terpenoids). This chapter also emphasizes the integration of **synthetic biology**, **Clustered Regularly Interspaced Short Palindromic Repeats (CRISPR)-based precision editing**, and **AI-enabled trait modeling** as transformative tools for next-generation horticulture.

Broader implications are explored through regulatory, biosafety, socioeconomic, and ethical lenses—highlighting enhanced farmer profitability, public health dividends, market readiness, and consumer acceptance. The strategic roadmap concludes with recommendations for policy, research, capacity building, and public–private–farmer partnerships to mainstream genetically engineered and bio-fortified crops within future-ready, climate-resilient horticultural systems.

Overall, this chapter positions biotechnology not just as a tool of enhancement but as a cornerstone for redefining **remunerative horticulture beyond convention**, aligning with the vision of nutrition-sensitive, high-value, and globally competitive crop systems.

Keywords

Genetic engineering · Bio-fortification · Remunerative horticulture · Genome editing · Nutritional enhancement

8.1 Introduction

8.1.1 The Expanding Scope of Horticulture in Food, Nutrition, and Livelihood Security

Horticulture has become a cornerstone of modern agriculture, transcending its traditional role of producing fruits and vegetables for subsistence to encompass **a multifaceted sector** that delivers food, nutrition, income, employment, health, and ecological services (FAO 2020). With an ever-growing global population and increasing urbanization, consumer demand for **fresh, high-quality, nutrient-rich, and safe produce** has surged dramatically. Consequently, horticulture today contributes significantly to **dietary diversity, nutritional enhancement, rural employment generation, export earnings, and agricultural sustainability** (Padulosi et al. 2013; Pingali 2015).

Globally, horticultural crops such as **fruits, vegetables, ornamentals, spices, plantation crops, and medicinal and aromatic plants** represent the **fastest-growing segment of agriculture**. Their higher economic returns per unit area, shorter gestation periods, suitability for smallholder systems, and year-round income potential make them critical to achieving **household-level food and livelihood security** (Kumar et al. 2022). In countries like India, horticulture has overtaken food grain production in terms of total output, exemplifying its **strategic role in agricultural diversification** and sustainable rural prosperity.

8.1.2 Remunerative Horticulture: A Transformative Paradigm

The term **“Remunerative Horticulture”** reflects a new vision that integrates **economic value with nutritional and qualitative excellence**. It goes beyond mere production-centric approaches to promote a market-driven, consumer-responsive, and nutrition-sensitive horticultural strategy. In this paradigm, **value creation is paramount**, and it is achieved through the following core drivers:

- Improved nutritional content (e.g., high-iron spinach, provitamin A banana).
- Extended shelf-life and post-harvest durability (e.g., delayed-ripening tomato).
- Enhanced marketability through aesthetics (color, aroma, texture, uniformity).
- Functional benefits and health attributes (antioxidants, phytonutrients).
- Resistance to pests, pathogens, and abiotic stress, ensuring consistent quality.

These traits not only increase consumer satisfaction and health benefits but also **boost profitability for producers**, enabling access to premium markets, reducing post-harvest losses, and facilitating better returns on investment. The remunerative model thus rests on the pillars of **science, sustainability, value addition, and socio-economic equity** (Kaur et al. 2020; Sharma et al. 2023).

8.1.3 Genetic Engineering and Bio-fortification: Drivers of Value Creation

Conventional breeding approaches, while valuable, often fall short when it comes to incorporating complex traits such as enhanced nutrition, abiotic stress tolerance, or shelf-life extension. This has led to the adoption of **genetic engineering** and **bio-fortification** as powerful tools to overcome genetic bottlenecks and precisely **enhance horticultural crop value**.

Genetic Engineering

Genetic engineering offers the capacity to **introduce, suppress, or modify genes** to bring about desired traits in a target crop genome. It has been instrumental in developing **Bt Brinjal for insect resistance**, **Flavr Savr tomato for delayed ripening**, and **color-modified ornamentals like blue rose and petunia**. This precision-driven technology allows scientists to create **novel phenotypes with commercial, nutritional, and ecological relevance**, often bypassing the limitations of species barriers and long breeding cycles (Nap et al. 2003; Giovannoni et al. 2017).

Bio-fortification

Bio-fortification refers to the **genetic enrichment of crops with essential nutrients**, especially micronutrients like iron, zinc, provitamin A, folate, and selenium. This approach serves as a sustainable remedy to “hidden hunger”—the widespread micronutrient deficiency affecting over two billion people globally (Bouis and Saltzman 2017). Bio-fortified horticultural crops, such as **iron-rich lettuce** or

zinc-enriched tomato, combine nutritional density with the visual and sensory appeal needed for consumer acceptance. Importantly, bio-fortification also contributes to **resilience in public health**, reducing dependence on pharmaceutical supplementation or food fortification.

When integrated, these technologies offer **multiplicative benefits**—enhancing nutrition, market value, processing quality, and ecological performance. For the horticultural sector, they open up new frontiers for developing **premium varieties**, gaining **export competitiveness**, and achieving **food-system resilience** under climate change.

8.1.4 Aligning Innovation with Sustainable Development Goals (SDGs)

The integration of biotechnological innovations into horticultural practices offers a transformative pathway toward achieving the United Nations Sustainable Development Goals (SDGs), thereby positioning scientific advancement at the core of sustainable and inclusive growth. Through the application of genetic engineering, molecular breeding, and bio-fortification strategies, biotechnology enhances the capacity of horticultural systems to address multiple global challenges in a targeted and synergistic manner.

By aligning biotechnological interventions with the SDGs, horticultural research and development gains enhanced policy relevance, and attract greater institutional and financial support. This alignment ensures that technological advancements not only drive agricultural innovation but also actively contribute to a more equitable, resilient, and sustainable global agri-food system.

8.1.5 Structure and Scope of the Chapter

This chapter provides a **comprehensive exploration** of how **genetic engineering and bio-fortification** are transforming horticultural science into a value-centric and profit-oriented discipline. It is organized into the following major thematic blocks:

- **Section 2** covers the **scientific foundations of genetic engineering**, including molecular tools, gene transfer methods, and genome editing.
- **Section 3** elaborates on the **principles and public health rationale of bio-fortification**.
- **Section 4** presents **detailed applications** of genetic engineering and bio-fortification across key horticultural groups: viz. **Fruits**, **Vegetables**, **Flowers**, **Spices, Plantation Crops**, **Medicinal and Aromatic Plants,** and **Emerging Niche Crops**.
- **Sections 5 and 6** introduce **new-generation genomic and bioengineering tools** and illustrate **trait-based case studies**.
- **Section 7** discusses the **biosafety, regulatory, and public perception aspects**, while.

- **Section 8** evaluates the **economic and market-level implications** with data-backed analysis.
- **Section 9** highlights the **existing challenges and limitations**, and.
- **Section 10** offers a **visionary roadmap** for future research, commercialization, and policy synergy.

By bridging **science, practice, policy, and market dynamics**, this chapter aims to offer a **multi-dimensional understanding** of how these advanced technologies can shape the future of **remunerative horticulture**.

8.2 Foundations of Genetic Engineering in Horticulture

8.2.1 Basic Concepts and Molecular Tools

Genetic engineering refers to the direct manipulation of an organism's genome using biotechnology, enabling the targeted introduction, deletion, or alteration of specific genes to achieve desired traits. Unlike conventional breeding, which relies on natural genetic recombination, genetic engineering permits **precise and rapid incorporation of genes**, even across taxonomic barriers (Zhang et al. 2021).

In the context of horticulture, genetic engineering is revolutionizing crop improvement by enabling:

- **Trait stacking** (e.g., combining disease resistance with extended shelf-life).
- **Novel phenotype development** (e.g., blue rose, low-ethylene vegetables).
- **Stress resilience enhancement** (e.g., drought-tolerant tomato).
- **Nutritional enhancement** (e.g., high-anthocyanin brinjal).
- **Post-harvest trait improvement** (e.g., non-browning potato).

The application of **molecular tools** underpins all genetic engineering efforts. Key molecular instruments include the following:

- **Restriction Endonucleases**: DNA-cutting enzymes used in gene cloning.
- **Ligases**: Join DNA fragments during recombinant vector construction.
- **Polymerase Chain Reaction (PCR)**: Amplifies target DNA sequences.
- **Cloning Vectors**: Such as plasmids, cosmids, and binary vectors for gene delivery.
- **Selectable Marker Genes**: Aid in the identification of transformed tissues.
- **Reporter Genes**: e.g., *GUS*, *GFP*, for visual confirmation of gene expression.
- **Ribonucleic Acid (RNA) Interference (RNAi)**: Gene silencing tool for trait suppression.
- **Genome Editing Nucleases**: Such as ZFNs, TALENs, and CRISPR/Cas systems.

Together, these tools provide the molecular backbone of gene delivery, manipulation, and validation for horticultural applications (Rao et al. 2019; Collonnier et al. 2023).

8.2.2 Milestones in Horticultural Biotechnology

The advancement of genetic engineering in horticulture is punctuated by key **scientific breakthroughs** and **regulatory developments** that have shaped its growth. The following are some pivotal milestones (Table 8.1).

These achievements illustrate the evolution from transgenic to gene-edited crops, showing a transition from whole-gene insertions to precise genomic alterations with minimal off-target effects. This progression has expanded the scope of trait engineering in a wide range of horticultural crops, including fruit trees, vegetables, ornamentals, and spices (Zaidi et al. 2020).

8.2.3 Overview of Gene Transfer Methods in Horticultural Crops

Efficient gene delivery into plant cells is central to the success of genetic engineering. Several methods have been developed and optimized for different plant types, including dicots and recalcitrant species often found in horticulture (Fig. 2.1).

8.2.3.1 *Agrobacterium*-Mediated Transformation

Agrobacterium tumefaciens is a soil bacterium that naturally transfers a segment of DNA (T-DNA) into the host genome, causing crown gall disease. This mechanism has been exploited to deliver desirable genes by modifying the **Ti plasmid** and replacing tumor-inducing genes with traits of interest.

Table 8.1 A chronological description of different achievements and their significance in the field of horticultural biotechnology

Year	Milestone	Significance
1983	First successful plant transformation (tobacco) using *Agrobacterium*	Proof of concept for genetic engineering in plants
1992	Commercial release of FlavrSavr™ tomato	First genetically modified (GM) horticultural crop for delayed ripening
1995–2000	Adoption of Bt crops begins	*Bt Brinjal* and other GM vegetables under development
2010	India's Genetic Engineering Appraisal Committee (GEAC) evaluates Bt Brinjal	Milestone in regulatory discourse on GM horticulture
2015	First gene-edited mushroom (non-browning) via CRISPR/Cas	Demonstrates genome editing in horticulture
2018–present	CRISPR-Cas technologies applied in tomato, lettuce, citrus, etc.	Precision breeding and trait-specific genome editing emerge

- **Widely used in dicots** such as tomato, brinjal, rose, and tobacco.
- **Advantages**: High transformation efficiency, single or low-copy insertions, stable expression.
- **Limitations**: Host range restriction, less effective in monocots.

In horticulture, *Agrobacterium*-mediated transformation is considered the **gold standard**, especially for **vegetables and ornamental plants** (Gelvin 2003; Hiei and Komari 2008).

8.2.3.2 Biolistics (Gene Gun)

The **particle bombardment method** uses high-velocity microprojectiles coated with DNA to physically deliver genes into plant cells. It is particularly useful for:

- **Monocots and species recalcitrant to *Agrobacterium***, such as onion, banana, and orchids.
- **Transformation of organelles** (plastids and mitochondria).
- **Gene stacking** and co-transformation with multiple genes.

However, **random insertions**, **tissue damage**, and **lower expression stability** remain key limitations.

8.2.3.3 Microinjection

In this technique, DNA is injected directly into plant protoplasts or zygotic embryos using **fine capillaries under a microscope**. Though highly precise, it is:

- **Labor-intensive and technically demanding.**
- Less popular in horticulture due to **low efficiency and high mortality.**

Still, it has been explored in elite breeding lines of fruit crops and ornamentals where **embryo-based transformation** is feasible (Datta and Potrykus 1991).

8.2.3.4 *In Planta* Transformation

This method bypasses tissue culture by directly introducing DNA into **floral meristems or embryonic tissues**. Techniques include the following:

- Vacuum infiltration of inflorescences (used in *Arabidopsis*).
- Pollen-tube pathway for in vivo fertilization-mediated transfer.
- Targeting meristematic cells in juvenile shoots.

Its use in horticulture remains limited due to plant size and developmental complexity, but it holds promise for species where regeneration is challenging.

8.2.3.5 CRISPR/Cas and Genome Editing

Genome modification started with "Transgenic approaches" in the 1980s, followed by "Zinc finger nuclease" in the 1990s, and "Genome editing" in the 2020s (Fig. 8.1).

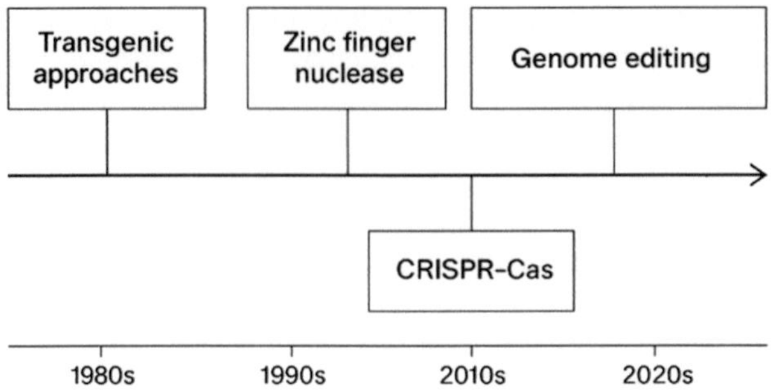

Fig. 8.1 Timeline flow chart illustrating the evolution of genetic editing technologies. An arrow indicates progression over time

CRISPR (Clustered Regularly Interspaced Short Palindromic Repeats) and associated Cas (CRISPR-associated) nucleases such as **Cas9 and Cas12a** have transformed gene editing due to their simplicity, specificity, and efficiency.

Applications in horticulture include the following:

- **Knockout of susceptibility genes** to develop disease-resistant tomatoes.
- **Editing of ripening regulators** like *SlMlo1* or *Ripening Inhibitor (RIN)* in tomato.
- **Flavor improvement in strawberries and citrus** through promoter modulation.
- **Multiplex editing** of anthocyanin biosynthesis pathways in grapes.

CRISPR/Cas has democratized genetic engineering in **diverse horticultural taxa**, eliminating the need for transgenes and accelerating regulatory clearance in certain jurisdictions (Zsögön et al. 2018).

8.3 Understanding Bio-fortification and Its Significance

This section provides a **detailed analysis of bio-fortification**, its scientific distinction from conventional approaches, its relevance to nutritional security, and highlights **major national and international initiatives**.

8.3.1 Conceptual Distinction from Conventional Fortification

Bio-fortification is defined as the **process of enhancing the nutritional quality of food crops through agronomic practices, conventional plant breeding, or modern biotechnology**, including transgenic and genome editing techniques (Bouis and Saltzman 2017).

While **conventional fortification** involves the addition of nutrients during food processing—such as adding iodine to salt or iron to flour—bio-fortification is

Table 8.2 Differences between conventional fortification and bio-fortification with respect to time of application, stability, delivery mechanism, target population, and sustainability

Aspect	Conventional fortification	Bio-fortification
Time of application	Post-harvest during food processing	During crop growth and development
Stability	Can degrade during storage or cooking	More stable as nutrients are embedded in plant tissues
Delivery mechanism	Requires continuous industrial intervention	One-time breeding or genetic engineering intervention
Target population	Urban, processed food consumers	Both rural and urban, particularly subsistence populations
Sustainability	Dependent on food industry logistics	Self-sustaining once bio-fortified variety is adopted

achieved at the **crop level**, allowing the nutrient-enriched trait to be **inherent in the harvested produce** (Table 8.2).

Thus, bio-fortification offers a **low-cost, sustainable, and rural-inclusive strategy** for addressing micronutrient deficiencies, especially in horticultural crops which are frequently consumed fresh (Nestel et al. 2006; Stein 2010).

8.3.2 Nutritional Goals of Bio-fortification in Horticulture

Horticultural crops—fruits, vegetables, spices, medicinal and aromatic plants—offer a natural platform for bio-fortification because they are rich in **health-promoting secondary metabolites**. The nutritional goals of bio-fortification in horticulture include the enhancement of:

- **Vitamin A (Provitamin A/β-carotene)**

Important for vision, immune function, and epithelial health.
Crops: Carrot, sweet potato, papaya, mango, golden banana.

- **Iron (Fe)**

Crucial for hemoglobin formation, oxygen transport, and cognitive development.
Crops: Spinach, moringa, amaranth, fenugreek.

- **Zinc (Zn)**

Supports enzymatic functions and immune defense.
Crops: Onion, pumpkin, lettuce, cucumber.

- **Folate (Vitamin B9)**

Essential for DNA synthesis and fetal development.
Crops: Broccoli, spinach, leafy greens.

- **Antioxidants (e.g., Lycopene, β-carotene, Ascorbic Acid)**

Mitigate oxidative stress and chronic disease risks.
Crops: Tomato, citrus, guava, bell pepper.

- **Flavonoids and Polyphenols**

Anti-inflammatory, anti-carcinogenic, cardioprotective roles.
Crops: Berries, grapes, tea, aromatic herbs like basil and rosemary.

The focus is not only on micronutrient density but also on enhancing **bioavailability** and **synergistic compound expression** (Saltzman et al. 2013).

8.3.3 Addressing Hidden Hunger Through Horticultural Produce

"**Hidden hunger**" refers to micronutrient malnutrition that persists even in populations consuming adequate calories. It disproportionately affects women and children in low-income, food-insecure regions, contributing to stunting, impaired cognitive development, reduced immunity, and maternal mortality (WHO 2021).

Horticultural crops provide an excellent avenue to mitigate hidden hunger due to:

- Diverse phytonutrient profiles
- Seasonal and regional adaptability
- Cultural acceptance and culinary versatility
- Fresh consumption without industrial processing

By developing **bio-fortified horticultural varieties**, nutritional interventions can be localized to suit **agro-climatic zones**, such as high-vitamin A carrots in hilly regions, iron-rich leafy vegetables in tribal belts, and zinc-enriched tomatoes in arid zones. These solutions enhance the **remunerative potential of farming**, aligning nutritional goals with **market-oriented horticulture** (Garg et al. 2018).

8.3.4 Global and Indian Initiatives on Bio-fortification

A well-structured table is included, listing key micronutrients, target crops, and associated health impacts, with all information (Table 8.3).

8.3.4.1 Global Initiatives: The Role of HarvestPlus

HarvestPlus, launched in 2003 under the Consultative Group on International Agricultural Research (CGIAR) system, is the leading global initiative in bio-fortification. It focuses on **crop breeding** for essential micronutrients through both conventional and transgenic means. While initially focused on staple cereals and legumes, the program now includes **nutrient-rich horticultural crops** such as:

- High-iron beans and leafy vegetables
- Provitamin A-rich sweet potatoes
- Bio-fortified orange maize and cassava
- Collaborative partnerships with Asian Vegetable Research and Development Center (AVRDC) for vegetables

Table 8.3 An illustration showing the target micronutrients or vitamins for bio fortification in different horticultural crops and their health benefits or relevance. The last column shows the key institutes, who have taken up the initiative

Micronutrient/ Vitamin etc.	Target crops	Health benefits	Indian/global initiative
Vitamin A (β-carotene)	Sweet potato, carrot, papaya, pumpkin	Vision, immunity, skin health	HarvestPlus, Indian Council of Agricultural Research (ICAR)-IIVR, Department of Biotechnology (DBT)
Iron (Fe)	Spinach, moringa, okra, fenugreek	Prevents anemia, supports cognitive and fetal development	ICAR-AICRP, ICMR-NIN
Zinc (Zn)	Onion, pumpkin, bitter gourd	Immunity, enzyme function, growth in children	HarvestPlus, SAUs
Folate (Vit B9)	Spinach, broccoli, leafy greens	Neural tube development, DNA synthesis	ICAR-NBPGR, DBT-KISAN
Lycopene, ascorbic acid	Tomato, bell pepper, citrus	Antioxidant, anti-inflammatory, anti-cancer effects	AICRP-vegetables
Anthocyanins, flavonoids	Grapes, berries, tea, medicinal herbs	Anti-aging, cardiovascular health, metabolic regulation	DBT, ICAR-IIHR

By 2022, over **390 bio-fortified varieties** of various crops were released globally (HarvestPlus 2023).

8.3.4.2 Indian Initiatives: ICAR and DBT Programs

India has emerged as a global leader in bio-fortification, driven by national institutions such as ICAR (Indian Council of Agricultural Research) and DBT (Department of Biotechnology).

Key initiatives include the following:

- **ICAR-AICRP on Bio-fortification** (since 2014): Coordinated evaluation and release of bio-fortified crop varieties, including horticultural components.
- **DBT's Biotech-KISAN and Nutri-Farms programs**: Support for translational research on micronutrient-rich vegetables and fruits.
- **NIN and Indian Council of Medical Research (ICMR) collaborations**: Integration of bio-fortification targets with **public health nutrition policies**.

Notable Indian bio-fortified horticultural varieties:

- **'Pusa Bhindi-5'**: Iron-rich okra
- **'KashiSharad'**: High β-carotene pumpkin
- **'Moringa PKM-1'**: Enhanced iron and calcium
- **Bio-fortified amaranth and coriander** lines under evaluation.

These efforts are strategically linked to **nutritional security, remunerative farming**, and **national dietary guidelines**.

8.4 Genetic Engineering Applications in Major Horticultural Crop Categories

Genetic engineering has emerged as a transformative tool in the development of next-generation horticultural crops, aimed not only at improving yield and resilience but also at enhancing nutritional value, aesthetic appeal, post-harvest shelf-life, and bioactive compound profiles. The strategic integration of transgenic and genome editing technologies has paved the way for crop-specific innovations across multiple horticultural sectors. This section explores the genetic engineering interventions across key horticultural categories, emphasizing trait-based advancements, critical gene targets, molecular pathways (Table 8.4), and real-world case studies.

8.4.1 Fruits

Genetically engineered fruits have demonstrated remarkable potential in improving consumer traits (flavor, appearance, nutrition), extending shelf-life, and combating post-harvest losses and diseases.

8.4.1.1 Banana (*Musa* spp.)

Provitamin A enrichment in banana is a well-documented case of bio-fortification. A notable example is the transgenic East African Highland Banana developed through the expression of a **phytoene synthase (MtPsy2a)** gene from *Musa*

Table 8.4 Summary of genetically engineered traits in key horticultural crops. The key genes associated and the approach for undertaking it has also been mentioned along with the suitable references

Crop category	Target crop	Trait engineered	Key gene(s)/ approach	Reference
Fruits	Banana	Provitamin A enrichment	*MtPsy2a*, fruit-specific promoter	Paul et al. (2017a, 2017b)
Vegetables	Tomato	Delayed ripening, anthocyanin content	*PG, Del, Ros1*	Butelli et al. (2008)
Flowers	Rose	Blue coloration	*F3'5'H* from *petunia*	Katsumoto et al. (2007)
Spices	Capsicum	Powdery mildew resistance	*CaMLO2* via CRISPR	Kim et al. (2021)
Medicinal plants	Tulsi	Ursolic acid biosynthesis	*HMGR, DXR*	Misra et al. (2014)
Niche crops	Sea buckthorn	Vitamin C enhancement	*GME, GGP* via CRISPR	Zhao et al. (2022)

troglodytarum, under the control of a fruit-specific promoter. This genetic modification significantly elevated β-carotene levels (Paul et al. 2017a, b). Other transformations include **Fusarium wilt resistance**, using *Ced9* or *Hs1pro-1* genes (Dale et al. 2017).

8.4.1.2 Papaya (*Carica papaya*)

Papaya ringspot virus (PRSV) resistance in transgenic papaya is a classic success story in plant biotechnology. The introduction of the **coat protein (CP)** gene of PRSV using *Agrobacterium tumefaciens*-mediated transformation led to durable resistance (Gonsalves 1998). This technology was first commercialized in Hawaii and subsequently adapted in other tropical countries.

8.4.1.3 Apple (*Malus domestica*)

Non-browning apples, marketed as "**Arctic™ Apples,**" were developed by silencing **polyphenol oxidase (PPO)** genes using RNA interference (RNAi) (Waltz 2015). These apples do not brown upon slicing, enhancing consumer appeal and marketability.

8.4.1.4 Citrus (*Citrus* spp.)

Citrus crops have been engineered for **Huanglongbing (HLB)** resistance and **canker tolerance**. Genes such as **pGIP (polygalacturonase-inhibiting protein)** and **SAR (systemic acquired resistance)** regulators have been introduced using citrus-specific *Agrobacterium* transformation protocols (Dutt et al. 2015).

8.4.2 Vegetables

Vegetable crops have been targeted for traits including biotic resistance, delayed ripening, and enhanced micronutrient levels.

8.4.2.1 Tomato (*Solanum lycopersicum*)

Tomato serves as a model for fruit ripening and metabolic engineering. The **FlavrSavr™** was the first commercial genetically engineered crop, wherein **polygalacturonase** gene expression was suppressed to delay softening (Martineau et al. 1994).

Recent work has focused on increasing **anthocyanin content** using genes like ***Del* and *Ros1*** from *Antirrhinum majus* (Butelli et al. 2008), yielding deep purple, antioxidant-rich tomatoes.

8.4.2.2 Brinjal (*Solanum melongena*)

Bt Brinjal expressing ***Cry1Ac*** from *Bacillus thuringiensis* was developed to combat *Leucinodes orbonalis*, a major fruit and shoot borer. Commercialized in Bangladesh in 2014, it led to reduced pesticide use and improved farmer income (Shelton et al. 2018a, b).

8.4.2.3 Potato (*Solanum tuberosum*)

Transgenic potatoes expressing ***rb* and *Rpi-blb1*** genes have shown durable resistance to late blight (*Phytophthora infestans*) (Ghislain et al. 2019). Iron and zinc-bio-fortified potatoes are also in development using iron transporters and **ferritin genes**.

8.4.2.4 Sweet Potato (*Ipomoea batatas*)

Bio-fortified sweet potatoes expressing ***AtFRO2* and *NAS1*** for enhanced iron uptake and transport, and ***PSY*** for provitamin A content, are under trial for tackling micronutrient malnutrition (Andersson et al. 2017).

8.4.2.5 Leafy Greens (*Spinach, Amaranthus*)

High-iron spinach has been engineered using **ferritin** genes from legumes and **nicotianamine synthase (NAS)** genes to improve iron bioavailability (Lucca et al. 2002).

8.4.3 Flowers and Ornamentals

In ornamental horticulture, genetic engineering targets traits that enhance visual appeal, fragrance, and vase life.

8.4.3.1 Rose (*Rosa* spp.)

Efforts to modify flower color have focused on introducing **flavonoid biosynthesis genes**, such as ***F3'5'H*** from *Petunia*, to develop blue roses (Katsumoto et al. 2007).

8.4.3.2 Chrysanthemum

Transgenic chrysanthemums have been generated to alter flower shape and shelf-life through manipulation of ethylene sensitivity and **MADS-box genes** controlling inflorescence development.

8.4.3.3 Petunia

Petunia has served as a model for **color manipulation**, especially for anthocyanin modification and fragrance biosynthesis using **PAL**, **CHS**, and **OMT** gene family members (Verdonk et al. 2005).

8.4.4 Spices and Plantation Crops

Spices and plantation crops are economically important and suitable for trait-specific bioengineering for disease tolerance and bioactive metabolite profiling.

8.4.4.1 Capsicum (Chili, Bell Pepper)

CRISPR-Cas9 has been used to knock out **CaMLO2**, conferring powdery mildew resistance. Transgenic capsicum lines have also been developed for **high capsaicinoid production** (Kim et al. 2021).

8.4.4.2 Turmeric (*Curcuma longa*)

Metabolic pathway engineering for **curcumin biosynthesis enhancement** is underway, with transgenic turmeric expressing ***Diketide-CoA Synthase (DCS)* and *Curcumin Synthase (CURS)* genes** regulated by transcription factors like **WRKY**.

8.4.4.3 Black Pepper (*Piper nigrum*)

Black pepper transformation is being attempted to combat *Phytophthora* diseases using **PR proteins**, and for improved **piperine content** via manipulation of the **shikimate pathway**.

8.4.4.4 Tea (*Camellia sinensis*) and Coffee (*Coffea* spp.)

Genetic manipulation of **caffeine synthase** and **SAMDC** (S-adenosylmethionine decarboxylase) genes is being studied to develop low-caffeine or enhanced flavor variants. Transgenics for **coffee berry disease** and **leaf rust** resistance are also being developed (López-Gómez et al. 2021).

8.4.4.5 Coconut and Oil Palm

Biotechnological efforts in coconut focus on resistance to lethal yellowing disease, whereas oil palm targets yield enhancement and **fatty acid biosynthesis** modulation through **Flavin Adenine Dinucleotide (FAD)** and **Stearoyl-Acyl Carrier Protein Desaturase (SAD)** gene editing.

8.4.5 Medicinal and Aromatic Plants

These crops are targets for metabolic engineering to enhance the production of therapeutic secondary metabolites.

8.4.5.1 Tulsi (*Ocimum sanctum*)

Overexpression of ***1-Deoxy-D-xylulose 5-phosphate Reductoisomerase (DXR)* and *3-Hydroxy-3-Methylglutaryl-CoA Reductase (HMGR)*** genes has led to increased production of **eugenol** and **ursolic acid** (Misra et al. 2014).

8.4.5.2 Ashwagandha (*Withania somnifera*)

Transgenic lines with upregulated **squalene synthase** and **HMGR** pathways have resulted in enhanced **withanolide** accumulation (Sangwan et al. 2008).

8.4.5.3 Mint (*Mentha* spp.)

Engineering of **limonene synthase** and **geranyldiphosphate synthase** genes has shown potential in boosting **essential oil content**.

8.4.5.4 *Aloe vera* and *Artemisia*

Metabolic engineering in *Aloe vera* targets **aloins** and **polysaccharides**, while in *Artemisia annua*, overexpression of ***Amorpha-4,11-diene Synthase (ADS)* and *CYP71AV1*** genes has enhanced **artemisinin** yield (Ikram and Simonsen 2017).

8.4.6 Emerging Niche Crops

Emerging horticultural crops offer high market value and nutritional significance, presenting unique opportunities for genetic interventions.

8.4.6.1 Dragon Fruit (*Hylocereus* spp.)

Efforts are being directed to enhance **betalain content**, shelf-life, and **abiotic stress tolerance** through CRISPR-based modulation of **DOPA 4,5-Dioxygenase (DODA)** and **Myeloblastosis (MYB)** transcription factor.

8.4.6.2 Quinoa (*Chenopodium quinoa*)

Although not traditionally horticultural, quinoa's introduction into high-value farming systems prompts research into genetic engineering for **saponin reduction**, **iron content enhancement**, and **salt tolerance**.

8.4.6.3 Sea Buckthorn (*Hippophae rhamnoides*)

CRISPR tools are being explored for enhancing **vitamin C biosynthesis** through manipulation of **GDP-Mannose 3′,5′-Epimerase (GME) and GDP-L-galactose Phosphorylase (GGP)** genes (Zhao et al. 2022).

8.5 Bio-fortification Approaches Across Horticultural Segments

A few Bio-fortified horticultural crops along with their traits, methods, genes and economic potential are summarized in Table 8.5.

8.5.1 Transgenic, Cisgenic, and Genome Editing Approaches

Bio-fortification in horticulture encompasses multiple biotechnological strategies tailored to enhance the nutritional profile of fruits, vegetables, and other high-value crops (Table 8.3). These include transgenic, cisgenic, and genome editing approaches, each offering varying degrees of regulatory challenges, public acceptance, and technical precision (Wang et al. 2019).

- **Transgenic approaches** involve introducing genes from unrelated species, such as bacterial genes for provitamin A synthesis in *Musa* spp. (Paul et al. 2017a, b), enabling novel biochemical pathways or enhancing existing ones.

Table 8.5 Bio-fortified horticultural crops—trait, method, genes, and economic potential

Crop	Target trait	Approach	Genes/tools used	Economic/market potential
Banana	Provitamin A content	Transgenic	*Maize psy*, *crtB*	Addressing Vitamin A Deficiency (VAD) in Africa, high health impact
Tomato	Anthocyanin, Flavonoids	Transgenic/ cisgenic	*AtMYB12*, *Delila*, *Rosea1*	Premium functional foods, export varieties
Spinach	Iron accumulation	Transgenic	*SoyFERH1*	Institutional meals, anemia mitigation
Brinjal	Anthocyanin enrichment	Transgenic	*Delila*, *Rosea1*	Dual-purpose (ornamental + nutritional)
Tulsi	Eugenol, Methyl-eugenol	Metabolic engineering	*EGS*, *CVOMT*	Ayurvedic/ phytopharma markets
Sea buckthorn	Ascorbic acid (Vit C)	Genome editing (CRISPR)	*GME*, *VTC2*	Immunity-boosting beverages, nutraceuticals
Quinoa	Iron, Zinc	Transgenic/gene stacking	*NAS*, *Zinc-Regulated Transporter/ Iron-Regulated Transporter-like Protein (ZIP)* transporters	High-end markets, international trade
Papaya	β-Carotene enhancement	RNAi + transgenic	*LCYβ*, *PSY*	Vitamin A delivery, fresh-cut produce segment
Tea	Antioxidant flavonoids	RNAi/Gene overexpression	*CHS*, *DFR*, *F3'H*	Wellness teas, specialty beverage industry

- **Cisgenic strategies** employ genes from the same or closely related species, minimizing foreign DNA content and improving consumer acceptability. For instance, *Solanum lycopersicum* modified using its own *PSY1* (phytoene synthase 1) gene for lycopene enhancement exemplifies a cisgenic system (Zhu et al. 2020).
- **Genome editing tools**, especially **CRISPR/Cas9**, have revolutionized biofortification by offering precise, DNA-sequence-specific modifications. Editing the *VTC1* gene in sea buckthorn enhanced ascorbic acid levels significantly, showing promise for editing metabolic pathways in other vitamin-rich crops (Zhao et al. 2022).

Each approach offers tailored advantages:

- **Transgenic methods** enable new pathway construction.
- **Cisgenesis** appeals to regulatory leniency in some countries (Schouten et al. 2006).
- **Genome editing** enables marker-free, DNA footprint-minimized edits.

8.5.2 Trait-Specific Case Studies by Crop Group

8.5.2.1 Fruits

- **Banana**: Provitamin A bio-fortified banana developed using maize-derived *psy* and *crtB* genes has demonstrated elevated carotenoid content (~20 μg/g dry weight), significantly addressing Vitamin A deficiency in East Africa (Paul et al. 2017a, b).
- **Papaya**: Enhanced β-carotene accumulation has been achieved through RNAi silencing of lycopene β-cyclase, redirecting flux toward provitamin A synthesis (Devani et al. 2020).

8.5.2.2 Vegetables

- **Golden Tomato**: Overexpression of *AtMYB12* and *PSY1* led to increased flavonoids and carotenoids, improving both nutritional and antioxidant quality (Butelli et al. 2008).
- **Iron-rich Spinach**: Bio-fortified lines with ferritin gene (*SoyFERH1*) showed improved non-heme iron content and absorption capacity (Lucca et al. 2002).
- **Anthocyanin-rich brinjal**: Overexpression of *Delila* and *Roseal* (from snapdragon) elevated anthocyanin levels, conferring both ornamental and nutraceutical value (Zhang et al. 2013).

8.5.2.3 Spices and Plantation Crops

- ***Capsicum annuum***: CRISPR-mediated deletion of *CaMLO2* conferred enhanced resistance and indirectly improved iron/zinc content through stress-adaptive metabolism (Kim et al. 2021).
- **Tea and Coffee**: Gene modulation of flavonoid and caffeine biosynthetic pathways is underway to enhance antioxidant load and secondary metabolite diversity (López-Gómez et al. 2021).

8.5.2.4 Medicinal and Aromatic Plants

- **Tulsi (*Ocimum sanctum*)**: Enhanced eugenol and methyl-eugenol levels achieved through plastid-targeted expression of key monoterpene biosynthesis genes (Misra et al. 2014).
- ***Artemisia annua***: Genetic engineering of the *ADS* and *CYP71AV1* genes (key in artemisinin biosynthesis) improved antimalarial compound content (Ikram and Simonsen 2017).

8.5.2.5 Emerging Crops

- **Sea buckthorn** (*Hippophae rhamnoides*): CRISPR-mediated edits in *GME* and *VTC2* genes led to vitamin C enrichment (Zhao et al. 2022).
- **Quinoa**: Engineering of iron and zinc homeostasis via overexpression of NAS (nicotianamine synthase) and ZIP transporters is under investigation to improve mineral bioavailability (Narayanan et al. 2019).

8.5.3 Bioavailability and Functional Efficacy of Engineered Nutrients

While genetic enhancement of nutrient levels is essential, it is equally vital to assess **bioavailability**—the fraction of a nutrient absorbed and utilized physiologically. Studies reveal that:

- **Ferritin-based iron** in transgenic spinach and rice maintains high bioavailability due to its encapsulated and non-reactive nature (Lucca et al. 2002).
- **Provitamin A** from golden banana and golden papaya has shown excellent uptake and conversion to retinol in human trials (Andersson et al. 2017).
- **Anthocyanins and flavonoids** from transgenic tomatoes and eggplants demonstrate antioxidative capacity and support cardiovascular and cognitive health (Butelli et al. 2008; Zhang et al. 2013).

Field validation, nutritional trials, and metabolic assimilation studies are essential to confirm the functional efficacy of bio-fortified horticultural crops under real-world consumption patterns.

8.6 New-Generation Genomic and Precision Tools

With the rapid evolution of molecular biology, next-generation precision tools have revolutionized the landscape of crop improvement, including horticultural bio-fortification. These novel genomic technologies—ranging from CRISPR/Cas systems to synthetic biology and multi-omics approaches—have accelerated the development of nutritionally enhanced horticultural crops with unprecedented specificity and efficiency.

8.6.1 CRISPR/Cas Systems and Multiplex Gene Editing

Clustered Regularly Interspaced Short Palindromic Repeats (CRISPR) and associated protein systems (Cas9, Cas12a/Cpf1, Cas13, etc.) have emerged as the most transformative tools in plant genome editing (Zhang et al. 2018). Their simplicity, precision, and multiplexing capacity enable targeted manipulation of nutritional traits.

- **Multiplexing capacity** is particularly valuable in bio-fortification, allowing the simultaneous editing of several genes involved in nutrient biosynthesis or transport. For example, simultaneous knockout of *VDE*, *LCYE*, and *ZEP* genes in tomato using CRISPR/Cas9 has enhanced provitamin A accumulation (D'Ambrosio et al. 2018).

- In banana (*Musa* spp.), CRISPR-mediated knockdown of the *MaPDS* gene confirmed system functionality and opened avenues for targeting iron and vitamin biosynthesis pathways (Tripathi et al. 2019).
- In leafy vegetables like spinach and lettuce, targeted editing of polyphenol oxidase (PPO) genes is being explored to enhance shelf-life and nutrient retention (Liu et al. 2020).

Advantages Rapid development, minimal off-target effects (with improved Cas variants), non-transgenic classification in certain regulatory environments.

8.6.2 RNA Interference (RNAi)-Based Gene Silencing

RNAi offers a post-transcriptional approach to silence undesirable genes and upregulate beneficial pathways. This method is especially effective in modulating secondary metabolism and anti-nutrient content.

- In brinjal (*Solanum melongena*), RNAi targeting the *SAD* gene reduced undesirable stearoyl-ACP desaturase activity, improving fatty acid profiles (Chakrabarti et al. 2006).
- Golden Rice is a well-documented example, where RNAi was used to suppress the *lycopene epsilon-cyclase (lcyE)* to redirect flux toward β-carotene accumulation (Paine et al. 2005).
- RNAi technology has also been used to downregulate *MATE1/2* transporters in potato to reduce glycoalkaloid content while increasing zinc levels (Bhati et al. 2021).

8.6.3 Synthetic Biology and Metabolic Pathway Redesign

Synthetic biology transcends traditional gene editing by introducing de novo pathways or reconfiguring existing metabolic circuits to enhance trait output.

- In tomato, engineering of anthocyanin pathways via transcriptional activators (*Delila* and *Rosea1*) under a fruit-specific promoter led to the development of "purple tomatoes" with significantly higher antioxidant levels (Butelli et al. 2008).
- Synthetic operons have been introduced in strawberry to coordinate flavonoid biosynthesis and increase fruit quality (Carbone et al. 2020).
- Modular pathway optimization is being explored in basil and mint to boost phenylpropanoid and terpenoid content (Luo et al. 2019).

Synthetic biology provides not only nutritional enhancement but also possibilities for designing de novo flavors, fragrances, and pharmacologically active compounds in horticultural crops.

8.6.4 "Omics" Approaches for Trait Discovery in Horticulture

The integration of high-throughput "omics" platforms—genomics, transcriptomics, proteomics, metabolomics, and ionomics—enables a comprehensive understanding of complex traits and underpins precision breeding strategies (Fig. 6.3).

- **Genomics**: High-quality genome assemblies (e.g., for pomegranate, guava, eggplant) offer valuable gene catalogs for nutrient traits.
- **Transcriptomics**: Expression profiling under mineral-deficient or stress conditions aids in identifying transcription factors (TFs) linked to nutrient homeostasis.
- **Proteomics**: Post-translational modifications and protein-protein interactions involved in vitamin and mineral transporters have been elucidated using proteomics in papaya and okra (Singh et al. 2021).
- **Metabolomics**: Targeted and untargeted metabolomics in fruits like mango and pineapple help quantify essential metabolites, enabling fine-tuning of fortification strategies (Moco et al. 2007).
- **Ionomics**: Inductively coupled plasma-mass spectrometry (ICP-MS) helps screen mineral content in large germplasm collections and mutant populations (Salt et al. 2008).

The convergence of these platforms into a systems biology framework enables predictive modeling for pathway redesign and fortification strategies (Fig. 8.2) (Rai et al. 2021).

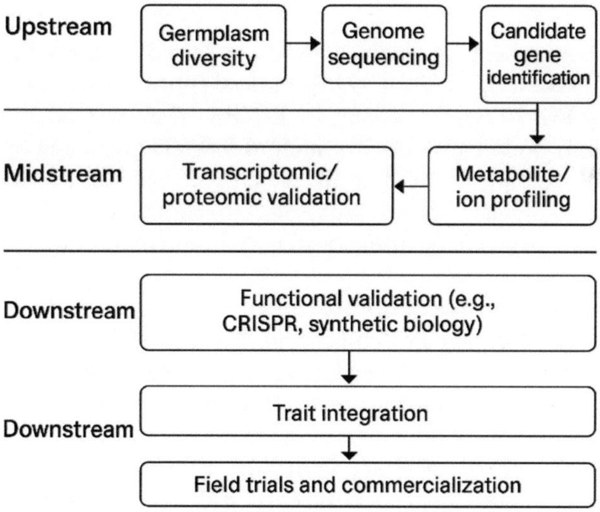

Fig. 8.2 Flow chart illustrating a scientific process divided into upstream, midstream, and downstream stages. The upstream stage includes germplasm diversity, genome sequencing, and candidate gene identification. The midstream stage involves transcriptomic/proteomic validation and metabolite/ion profiling. The downstream stage features functional validation, such as CRISPR and synthetic biology, followed by trait integration, and concludes with field trials and commercialization. Arrows indicate the flow and connection between each step

8.7 Regulatory, Biosafety, and Public Acceptance Framework

The rapid deployment of genetically engineered and bio-fortified horticultural crops calls for a carefully constructed regulatory and biosafety framework. These regulations ensure that the benefits of innovation are achieved without compromising human health, biodiversity, or societal values. This section elaborates on the structured guidelines and evaluation procedures across India and globally, coupled with a nuanced discussion on public perception and socio-ethical considerations.

8.7.1 Indian and International Regulatory Guidelines

The commercialization of genetically engineered (GE) and bio-fortified crops is governed by stringent national and international biosafety protocols that aim to ensure ecological and food safety. Both Indian and global systems mandate multi-tiered evaluations before market release.

8.7.1.1 Indian Guidelines and GEAC

In India, the **Genetic Engineering Appraisal Committee (GEAC)** under the **Ministry of Environment, Forest and Climate Change (MoEFCC)** is the apex body for approval of activities involving large-scale use and environmental release of GE organisms and products (Rules, 1989 of the Environment (Protection) Act, 1986).

- Biosafety assessment in India follows a **tiered approach**, starting with **Contained Research** (handled by Institutional Biosafety Committees or IBSCs), followed by **Field Trials** (regulated by RCGM—Review Committee on Genetic Manipulation), and finally, **Environmental Release** (approved by GEAC).
- A landmark example is **Bt Brinjal**, which received GEAC approval in 2009 based on biosafety data but was withheld from commercial release due to public opposition and demand for further testing (Ghosh 2010).

8.7.1.2 Cartagena Protocol on Biosafety

Internationally, the **Cartagena Protocol on Biosafety** (under the **Convention on Biological Diversity**, CBD) is the cornerstone of transboundary movement and risk assessment of Living Modified Organisms (LMOs).

- It emphasizes the **Precautionary Principle**, enabling countries to refuse LMO imports based on biosafety concerns (CBD 2000).
- All LMOs must undergo risk assessment covering **human health**, **biodiversity**, and **socioeconomic impacts** prior to international movement or release.

8.7.1.3 OECD Guidelines

The **Organization for Economic Co-operation and Development (OECD)** provides **science-based consensus documents** and methodologies for comparative safety assessments of novel food and feed products derived through biotechnology.

- The **concept of "substantial equivalence"** is central to OECD's food safety evaluations, requiring that a genetically modified (GM) crop must be as safe as its conventional counterpart (OECD 2001).

8.7.2 Safety Assessments for Bio-fortified and GE Horticultural Crops

Rigorous safety assessments are integral to regulatory approvals, especially in food crops where public health is directly impacted. Bio-fortified horticultural crops must undergo pre- and post-commercialization assessments to establish safety and efficacy.

8.7.2.1 Allergenicity and Toxicity

- Allergenicity testing involves **bioinformatic comparison of novel proteins** with known allergens and **in vitro assays** such as simulated gastric fluid degradation tests (Goodman et al. 2005).
- Toxicological evaluations include **acute and sub-chronic toxicity studies** in animal models, observing physiological and biochemical parameters.

8.7.2.2 Immunoreactivity and Digestibility

- Evaluations of **immune system reactivity** to introduced traits are critical. For example, immunogenic responses to lectins or novel proteins are tested using Enzyme-Linked Immunosorbent Assay (ELISA) and in vivo models.
- Digestibility studies simulate gastrointestinal environments to determine whether transgenic proteins are rapidly broken down and rendered inactive (Delaney et al. 2008).

8.7.2.3 Environmental Risk Assessment (ERA)

Environmental biosafety is evaluated by:

- **Gene flow** potential to wild relatives.
- **Impact on non-target organisms** such as pollinators and soil microbes.
- **Resistance evolution in pests** (e.g., resistance management for Bt crops).
- **Persistence and invasiveness** in non-native ecosystems.

India's ICAR and DBT collaborate to conduct **confined field trials** and monitor **unintended ecological effects** (DBT Guidelines 2021).

8.7.3 Consumer Perception, Ethical Concerns, and Labeling Regulations

Despite strong scientific consensus on the safety of approved GE crops, **public acceptance remains a critical hurdle**. Factors influencing public perception include **trust in regulatory bodies**, **transparency of information**, and **cultural or religious beliefs**.

8.7.3.1 Perception and Misinformation

- Surveys across multiple countries reveal that **consumer concern about GM foods** is often shaped by **media coverage and advocacy groups** rather than scientific data (McFadden and Lusk 2017).
- In India, public resistance to Bt Brinjal exemplified how a **lack of informed engagement** can derail policy, despite scientific approval.

8.7.3.2 Ethical and Societal Considerations

- Ethical concerns involve **"naturalness"** of food, **ownership of genetic resources**, and **control by multinational corporations**.
- For example, **cisgenesis** (gene transfer from crossable species) is often viewed more favorably than **transgenesis** (cross-kingdom gene transfer).

8.7.3.3 Labeling and Transparency

- The **Food Safety and Standards Authority of India (FSSAI)** mandates labeling for foods derived from GM crops if the GM content exceeds 1% (FSSAI Regulations 2021).
- Globally, **European Union (EU) regulations** demand strict labeling and traceability, whereas the **USA follows a voluntary disclosure model** (National Bioengineered Food Disclosure Standard 2016).

Transparent labeling is not only a regulatory requirement but also a **consumer empowerment tool**, fostering trust and informed choice.

8.8 Economic Impact and Remunerative Potential

Enhancing Farm Income Through Premium Quality and Nutritional Branding The integration of genetic engineering and bio-fortification technologies in horticulture has shown considerable potential in increasing the profitability of farming systems. Unlike traditional agricultural produce, **value-added horticultural crops**—especially those with enhanced shelf-life, nutritional density, and consumer appeal—attract **premium prices** in domestic and global markets. Genetic innovations in crops such as banana, brinjal, and tomato have made it possible to introduce **premium quality traits** like **provitamin A enrichment, anthocyanin enhancement**, and **Bt-mediated pest resistance**, resulting in superior market positioning and better income for growers (Paul et al. 2017a, b; Barry et al. 2018).

Nutritional branding of horticultural produce as a marketing strategy is gaining ground. Crops bio-fortified with micronutrients like **vitamin A, zinc, and iron** are no longer seen as just scientific innovations but as **functional foods**, demanded by increasingly health-conscious urban consumers (HarvestPlus 2020). This trend is creating new **value chains** in horticulture, rewarding producers who adopt cutting-edge biotechnology.

8.8.1 Cost–Benefit Analysis of Transgenic and Bio-fortified Varieties

Economic analyses of **transgenic horticultural crops** have shown significant improvements in both **cost-efficiency** and **gross returns**. For instance, the adoption of **Bt Brinjal in Bangladesh** yielded **up to 128% higher net income per hectare** than conventional brinjal varieties, primarily due to reduced pesticide use, lower input costs, and less crop damage (Shelton et al. 2020; Ahmed et al. 2021). Similarly, the cultivation of **Golden Banana**, bio-fortified with provitamin A in Uganda, has demonstrated **positive consumer acceptance** and potential for a **premium market segment**, leading to higher farmgate prices (Paul et al. 2017a, b).

Moreover, the **return on investment (ROI)** for bio-fortified varieties is often favorable when considering **health cost savings**, **increased productivity**, and **value retention during post-harvest handling**, which are frequently overlooked in conventional profitability calculations (Barry et al. 2018).

8.8.2 Smallholder Advantage and Market Penetration in Developing Countries

In developing countries, where **nutritional insecurity** and **smallholder dominance** are pronounced, bio-fortified and transgenic horticultural crops offer significant economic and social gains. Smallholder farmers, who typically lack access to synthetic inputs or advanced pest management technologies, benefit immensely from **trait-engineered cultivars** that reduce dependency on agrochemicals (Haque et al. 2020). This not only lowers the **cost of cultivation** but also improves **occupational health outcomes** by minimizing pesticide exposure.

The integration of these technologies into **government-supported extension networks**, as in the case of **Bt Brinjal in Bangladesh**, facilitates rapid **technology diffusion**, farmer training, and market linkages. Similar programs led by **HarvestPlus** and **ICRISAT** in Africa and Asia are demonstrating how science-backed bio-fortification can empower smallholders while contributing to public health goals (HarvestPlus 2020).

8.8.3 Success Case: Bt Brinjal in Bangladesh and Provitamin A Banana in Uganda

The **Bangladesh Bt Brinjal program**, coordinated by the Bangladesh Agricultural Research Institute (BARI) and supported by USAID and Cornell University, is widely regarded as a **landmark success** in transgenic horticulture. Farmers reported **higher yields**, **lower pest losses**, and an average **increase in net profit by 55–128%** over traditional varieties. Pesticide use was drastically cut by up to 61%, contributing to both ecological and health benefits (Shelton et al. 2020; Ahmed et al. 2021).

Similarly, the development of **provitamin A-rich banana** (commonly referred to as **golden banana**) in Uganda represents a **game-changing intervention** in combating vitamin A deficiency. Consumers showed a preference for the product not only for its **nutritional quality** but also for its **enhanced storage properties**. Pilot marketing revealed **5–15% higher prices** in retail markets, making the cultivar economically attractive to growers (Paul et al. 2017a, b).

8.8.4 Supply Chain Integration for Value Realization

The true economic potential of bio-fortified horticulture is realized when it is integrated into **value chains** supported by policy, infrastructure, and informed consumer demand. Interventions at **multiple stages of the supply chain**—including **certification**, **branding**, **storage**, **cold-chain logistics**, and **nutrition-sensitive labeling**—amplify the profitability of engineered crops. Farmers linked to cooperatives or producer companies are more likely to access these value chains and secure **remunerative prices,** as reported in different countries (Table 8.6) (HarvestPlus 2020).

Table 8.6 Comparative analysis of return from different bio fortified crops mentioning the important traits, country origin, economic benefits, and references

Crop	Trait	Country	Economic benefit	Reference
Bt Brinjal	Pest resistance	Bangladesh	55–128% higher net income; 61% lower pesticide use	Shelton et al. (2020)
Golden banana	Provitamin A enrichment	Uganda	5–15% premium pricing; enhanced shelf-life	Paul et al. (2017a, b)
Iron-rich beans	Micronutrient bio-fortification	Rwanda	Higher consumer demand; public procurement inclusion	HarvestPlus (2020)
Anthocyanin-rich tomato	Nutraceutical trait	India	Niche market pricing; increased post-harvest retention	Barry et al. (2018)
Zinc-rich sweet potato	Micronutrient density	Philippines	Improved child nutrition; integrated school meal schemes	HarvestPlus (2020)

8.9 Challenges and Bottlenecks

The deployment of genetic engineering and bio-fortification in horticultural crops, despite its promising potential, faces a host of formidable scientific, regulatory, socioeconomic, and technological challenges. These constraints can severely limit the pace and scale of adoption, particularly in the context of **remunerative horticulture**, where the goal is not only to increase yield and nutritional quality but also to ensure market profitability, export readiness, and smallholder inclusivity.

8.9.1 Transformation Inefficiencies and Genotype Dependency

One of the primary scientific hurdles in genetically engineering horticultural crops lies in the **inefficiency and genotype-specificity of transformation protocols**. Unlike model crops such as *Arabidopsis thaliana* or rice, many horticultural species (e.g., fruit crops like citrus, apple, and banana) exhibit strong **recalcitrance to in vitro regeneration**, which severely constrains stable transformation (Birch 1997; Krishna et al. 2020). Transformation success often varies dramatically across genotypes within the same species, leading to **inconsistent reproducibility and restricted germplasm utilization** (Ali et al. 2023).

Moreover, the lack of standardized **high-throughput transformation platforms** for key horticultural crops delays pipeline development for trait introgression. While techniques like Agrobacterium-mediated transformation and CRISPR/Cas-based gene editing have improved the situation, tissue culture dependence and somaclonal variations remain major constraints (Häkkinen and Palva 2019).

8.9.2 Limited Acceptance in Export Markets

Even in countries where genetically engineered or bio-fortified horticultural crops have shown success (e.g., Bt Brinjal in Bangladesh), **international trade restrictions** severely limit their market expansion. The **European Union (EU)**, in particular, maintains stringent zero-tolerance policies toward genetically modified organisms (GMOs), affecting export-oriented commodities such as **cut flowers, papaya, and mangoes** (Sharma and Ortiz 2000).

Consumer perceptions in high-value markets are largely shaped by **food safety narratives and precautionary principles**, often independent of scientific risk assessments. This leads to a fragmented global regulatory landscape that **discourages smallholders and private players** from investing in bioengineered horticultural varieties, thereby weakening the value chain from production to international marketing (Brookes and Barfoot 2020).

8.9.3 Intellectual Property Rights (IPR) Restrictions and Technology Licensing

Biotechnology in horticulture is deeply influenced by **patent landscapes and technology ownership**, often held by private multinational corporations. Access to key enabling technologies like **CRISPR-Cas9, Golden Rice genes, or vitamin A biosynthesis pathways** requires expensive licenses or contractual obligations, which become **prohibitive for public institutions and small-scale breeders** (Graff et al. 2003; Lybbert and Zilberman 2010).

Moreover, the **overlapping IPR regimes** covering gene sequences, promoters, marker systems, and transformation vectors create a legal bottleneck known as the "patent thicket," stifling innovation in public breeding systems (Fischer et al. 2015). These barriers are particularly limiting in emerging economies aiming to develop indigenous solutions for local nutritional needs.

8.9.4 Poor Extension Mechanisms and Lack of Awareness

Despite the release of bio-fortified or genetically engineered horticultural cultivars, the actual **uptake by farmers and stakeholders is minimal**, often due to weak dissemination and capacity-building systems (Fig. 8.3). In many states of India and developing countries, horticultural extension services are **severely understaffed**, and lack technological training and awareness campaigns regarding the benefits and safety of genetically engineered or fortified crops (Paroda and Kumar 2000).

Furthermore, the **limited availability of demonstration plots, training modules, and farmer field schools** on bio-fortified horticulture acts as a serious deterrent. The risk-averse nature of smallholders, combined with poor access to input credits and marketing channels, leads to **slow adoption curves** even for approved and released varieties (Bouis and Saltzman 2017).

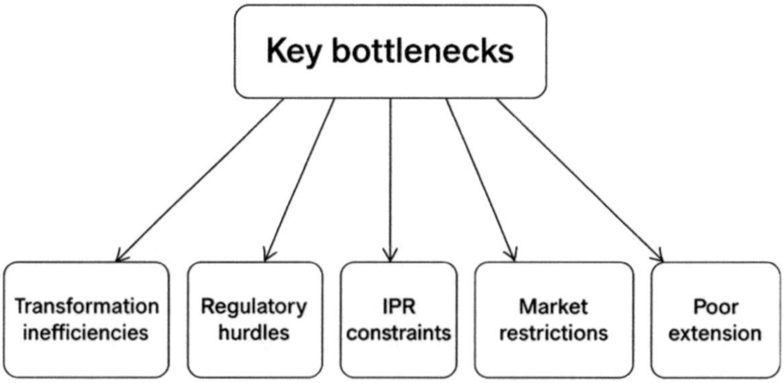

Fig. 8.3 Flow chart illustrating "Key bottlenecks" with five branches leading to: "Transformation inefficiencies," "Regulatory hurdles," "IPR constraints," "Market restrictions," and "Poor extension." Each branch represents a specific challenge

8.10 The Future of Remunerative Horticulture Through Bioengineering

The horizon of horticultural science is being reshaped by rapid advancements in biotechnology, particularly genetic engineering, genome editing, phenomics, and data-driven decision-making. As global agriculture grapples with the **dual challenges of climate change and nutritional insecurity**, bioengineering stands poised to unlock a new era of *remunerative horticulture*—one that synergizes economic viability, nutritional richness, and ecological sustainability (Fig. 8.4).

8.10.1 Climate-Resilient and Smart Bio-fortified Cultivars

Emerging climatic stressors such as **erratic rainfall, heatwaves, salinity, and drought** necessitate the development of climate-resilient cultivars that retain both yield and quality. Future bio-fortified horticultural crops will not merely aim for enhanced nutrition but will also be designed to tolerate **abiotic stress through gene stacking and multiplex gene editing** (Shinozaki and Yamaguchi-Shinozaki 2007).

For instance, **CRISPR-mediated modification of dehydration-responsive element-binding (DREB) genes**, along with introgression of **nutrient transporter genes (e.g., ZIP, IRT, NAS)**, can create dual-purpose varieties tailored for marginal environments (Langridge and Reynolds 2021). Such smart cultivars will ensure **year-round marketability**, minimize post-harvest losses under stress, and empower farmers with climate-proof solutions.

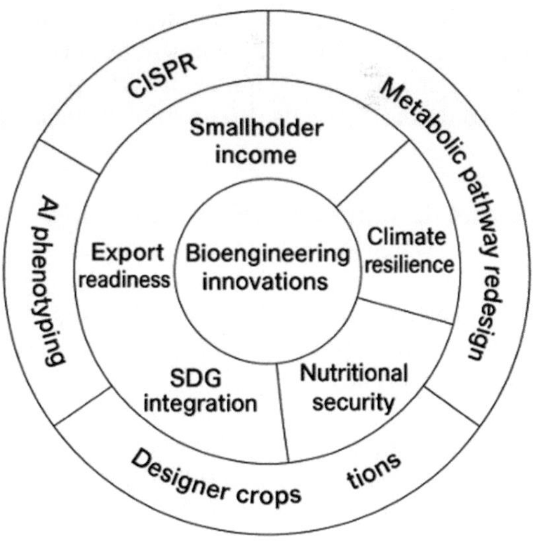

Fig. 8.4 Flow chart illustrating bioengineering innovations at the center, surrounded by six segments: smallholder income, climate resilience, nutritional security, SDG integration, export readiness, and AI phenotyping. An outer ring includes terms: Clustered Interspaced Short Palindromic Repeats (CISPR), metabolic pathway redesign, designer crops

8.10.2 AI-Driven Precision Phenotyping and Genome Design

The next frontier of genetic engineering lies in **artificial intelligence (AI)-integrated precision horticulture**. AI-driven platforms enable high-throughput phenotyping (HTP) by analyzing thousands of genotypes in varying environmental conditions through **machine vision, drones, multispectral imaging, and deep learning algorithms** (Singh et al. 2016). These platforms not only accelerate trait discovery but also inform **in silico genome design**, allowing breeders to simulate the metabolic or developmental outcomes of targeted gene edits.

Such convergence of **phenomics, genomics, and AI** supports "*design-based breeding*," enabling predictive generation of high-performing cultivars with desired nutrition, taste, and adaptability. For instance, computational modeling is being applied to optimize **vitamin biosynthesis pathways** in tomatoes and sweet potatoes, offering tremendous scope for customized horticultural solutions (Jansson et al. 2021).

8.10.3 Nutritional Phenomics and Designer Foods

An emerging concept within bio-fortified horticulture is that of **"designer foods"**—horticultural produce genetically tailored for specific **nutritional, functional, or therapeutic roles**. Through metabolic engineering, crops can be enriched with **anthocyanins, flavonoids, resveratrol, carotenoids, or anti-inflammatory compounds**, thereby moving beyond simple dietary support to disease-preventive potential (Kapoor et al. 2023).

The integration of **nutritional phenomics**—which involves high-resolution profiling of nutritional metabolites under diverse environments and genotypes—allows for the mapping of **genotype-phenotype-nutrient-health interactions**. This will ultimately lead to the development of cultivars optimized not only for high nutritional value but also for **bioavailability, palatability, and health efficacy** in different consumer segments (Bouis and Saltzman 2017).

8.10.4 Linking Genetic Engineering with Sustainable Development Goals (SDGs)

Bioengineered horticulture aligns closely with several of the **United Nations Sustainable Development Goals (SDGs)**. The contributions are multidimensional:

- **SDG 2 (Zero Hunger):** Through improved nutrient density and resilience.
- **SDG 3 (Good Health and Well-being):** By reducing micronutrient deficiencies and offering functional foods.
- **SDG 12 (Responsible Consumption and Production):** Via shelf-stable, low-waste horticultural varieties.
- **SDG 13 (Climate Action):** Through climate-resilient crops that reduce dependence on resource-intensive practices.

By aligning biotechnology interventions with sustainability targets, horticulture can move from an input-intensive model to an **impact-driven bioeconomy**—providing both prosperity to producers and health to consumers.

8.11 Conclusion

8.11.1 Summary of the Transformation Potential of Genetic Engineering and Bio-fortification

The twenty-first century marks a decisive shift in the scope and structure of horticultural science. From being a primary source of fruits, vegetables, spices, and ornamentals, horticulture is now emerging as a **strategic platform for addressing nutritional deficiencies, enhancing farmer incomes, and supporting environmental sustainability**. In this transformation, **genetic engineering and bio-fortification** have proven to be indispensable enablers.

As delineated across earlier sections, modern molecular breeding and bioengineering tools—ranging from **transgenic technologies and RNA interference to CRISPR/Cas and synthetic biology**—have enabled the targeted improvement of horticultural crops for **nutritional traits (e.g., vitamin A, iron, zinc, antioxidants)**, **shelf-life**, **resilience against biotic and abiotic stresses**, and even **consumer-centric quality attributes** such as fragrance and color.

Furthermore, these innovations are not merely laboratory constructs. Real-world success stories such as **Bt Brinjal in Bangladesh**, **provitamin A-enriched bananas in Uganda**, and **iron-rich beans and sweet potatoes under the HarvestPlus program** validate the economic and nutritional impact of these technologies on smallholder farmers and marginalized consumers alike (Bouis and Saltzman 2017; Shelton et al. 2018a, b).

The **bio-fortified horticultural segment**, in particular, has evolved from a philanthropic ideal to a **market-driven enterprise** that aligns well with the pillars of *remunerative horticulture*—namely, enhanced quality, marketability, value addition, and income upliftment.

8.11.2 Call for Integrated Strategies: Research–Policy–Market Synergy

Despite the significant strides made, the road to universal adoption of bioengineered horticulture remains punctuated with **scientific, regulatory, societal, and infrastructural challenges**. Addressing these barriers requires a **convergence of disciplines and stakeholders**, especially along three interconnected axes:

- **Research:** Continued innovation is needed in *precision gene editing*, *multi-trait stacking*, *nutrient bioavailability*, and *non-GMO molecular alternatives* to ensure both effectiveness and public acceptance.

- **Policy:** Regulatory harmonization (GEAC, OECD, Cartagena Protocol) must promote **transparent, science-based approvals**, balanced with biosafety. In parallel, **IPR frameworks must be refined** to encourage public–private partnerships without monopolizing access.
- **Market:** Building consumer confidence through **labeling**, **awareness campaigns**, and **traceability** is critical. Governments and agribusinesses must also support the development of **dedicated value chains** for bio-fortified produce—from farm gate to export markets.

A **cohesive policy–research–market framework** is vital to unlock the full potential of bioengineering technologies, ensuring they serve both social and economic objectives in horticulture.

8.11.3 Future Roadmap for Prosperity Through Innovation in Horticulture

Looking ahead, the vision of *remunerative horticulture through bioengineering* must evolve along three futuristic dimensions:

1. **Climate-smart innovation:** Bio-fortified cultivars must increasingly integrate climate-resilient traits, enabling cultivation in **stress-prone ecologies** and offering **year-round economic returns**.
2. **Digital integration:** The application of **AI, IoT, and big data** for genomic prediction, trait deployment, and market intelligence will transform traditional horticulture into a **digitally empowered value system**.
3. **Nutritional and ecological harmonization:** Through **designer crops**, **multi-nutrient stacking**, and **functional metabolite enhancement**, future horticultural innovations can simultaneously address **public health goals and environmental sustainability**, contributing directly to the **UN Sustainable Development Goals (SDGs)**.

In conclusion, the transformation of horticulture from a **subsistence-based practice to a prosperity-driven enterprise** lies in the effective deployment of **genetic engineering and bio-fortification**. These scientific tools are not mere interventions; they are strategic instruments of *economic empowerment, health assurance, and global competitiveness* for nations that invest wisely in their horticultural future.

Let us, therefore, commit to a new paradigm where **science, sustainability, and smallholders** coalesce to create the true wealth of nations—**through remunerative, resilient, and resource-smart horticulture**.

References

Ahmed AU, Malapit HJ, Laborde D, Torero M, Gelli A, de Brauw A (2021) Impacts of Bt eggplant adoption in Bangladesh: results from a randomized controlled trial. Am J Agric Econ 103(3):952–972. https://doi.org/10.1111/ajae.12184

Ali Z, Abulfaraj A, Idris A, Ali S, Tashkandi M, Mahfouz MM (2023) CRISPR/Cas-based technologies for crop improvement and future agriculture. Plant Cell Rep 42:1–14. https://doi.org/10.1007/s00299-022-02987-2

Andersson MS, Saltzman A, Virk PS, Pfeiffer WH (2017) Progress update: crop development of biofortified staple food crops under Harvest Plus. Afr J Food Agric Nutr Dev 17(2):11905–11935

Barry MA, Stagnari F, Iannetta PPM, Luti S, Truzzi F, Pisante M (2018) Biofortification of horticultural crops for alleviating micronutrient malnutrition. Agron Sustain Dev 38(2):25. https://doi.org/10.1007/s13593-018-0501-9

Bhati KK, Alok A, Kumar J, Pandey S, Sharma M (2021) Advances in biofortification of food crops via genome editing and its future prospects. Front Genet 12:637108

Birch RG (1997) Plant transformation: problems and strategies for practical application. Annu Rev Plant Biol 48(1):297–326

Bouis HE, Saltzman A (2017) Improving nutrition through biofortification: a review of evidence from Harvest Plus, 2003 through 2016. Glob Food Sec 12:49–58. https://doi.org/10.1016/j.gfs.2017.01.009

Brookes G, Barfoot P (2020) GM crops: global socio-economic and environmental impacts 1996–2018. PG Economics Ltd Report, Dorchester

Butelli E, Titta L, Giorgio M, Mock HP, Matros A, Peterek S et al (2008) Enrichment of tomato fruit with health-promoting anthocyanins by expression of select transcription factors. Nat Biotechnol 26(11):1301–1308

Carbone F, Preuss A, De Vos RC, D'Amico E, Perrotta G, Bovy AG et al (2020) Developmental, genetic and environmental control of flavonoid accumulation in strawberry fruits. Plant Cell Tissue Organ Cult 140:143–157

CBD (2000) Cartagena protocol on biosafety to the convention on biological diversity. Secretariat of the CBD, Montreal

Chakrabarti SK, Lutz KA, Larkins BA (2006) Expression of transgenes in tomato plants can be silenced by post-transcriptional gene silencing. Plant Cell Rep 25:927–935

Collonnier C et al (2023) Genome editing in horticultural crops: status and perspectives. Plant Biotechnol J 21:1–16. https://doi.org/10.1111/pbi.13997

D'Ambrosio C, Stigliani AL, Giorio G (2018) CRISPR/Cas9 editing of carotenoid genes in tomato. Transgenic Res 27:367–378

Dale J, James A, Paul JY, Khanna H, Smith M, Peraza-Echeverria S, Garcia-Bastidas F, Kema G, Waterhouse P, Mengersen K (2017) Transgenic Cavendish bananas with resistance to Fusarium wilt tropical race 4. Nat Commun 8:1496. https://doi.org/10.1038/s41467-017-01670-6

Datta SK, Potrykus I (1991) New strategies for rice transformation. Biotechnol Adv 9(4):201–217

DBT (2021) Guidelines for environmental risk assessment of genetically engineered plants. Department of Biotechnology, Government of India

Delaney B, Astwood JD, Cunny H, Conn RE, Herouet-Guicheney C, MacIntosh S et al (2008) Evaluation of protein safety in the context of agricultural biotechnology. Food Chem Toxicol 46:S71–S97

Devani RS, Pandey A, Mandal K, Nath P (2020) Enhanced provitaminA accumulation in papaya by manipulation of LCYβ gene through RNA interference. Plant Cell Rep 39:793–807

Dutt M, Barthe G, Grosser JW (2015) Transgenic citrus expressing an Arabidopsis NPR1 gene exhibit enhanced resistance against Huanglongbing (HLB; citrus greening). PLoS One 10(9):e0137134. https://doi.org/10.1371/journal.pone.0137134

FAO (2020) The state of food security and nutrition in the world 2020. FAO, Rome

Fischer K, Byerlee D, Edmeades G (2015) Crop yields and global food security: will yield increase continue to feed the world? Australian Centre for International Agricultural Research (ACIAR)

FSSAI (2021) Food safety and standards (labelling and display) regulations. Government of India
Garg M, Sharma N, Sharma S, Kapoor P, Kumar A, Chunduri V, Arora P (2018) Biofortified crops generated by breeding, agronomy, and transgenic approaches are improving lives of millions of people around the world. Front Nutr 5:12
Gelvin SB (2003) *Agrobacterium*-mediated plant transformation: the biology behind the "gene-jockeying" tool. Microbiol Mol Biol Rev 67(1):16–37
Ghislain M, Byarugaba AA, Magembe E, Njoroge A, Rivera C, Román ML, Tovar JC, Gamboa S, Forbes GA, Kreuze JF, Vishwakarma P (2019) Stacking three late blight resistance genes from wild species directly into African highland potato varieties confers complete field resistance to *Phytophthorainfestans*. Plant Biotechnol J 17(6):1119–1129. https://doi.org/10.1111/pbi.13042
Ghosh P (2010) Btbrinjal debate: India's policy paralysis on GM food. Curr Sci 99(6):717–718
Giovannoni JJ et al (2017) The epigenome and transcriptional dynamics of fruit ripening. Annu Rev Plant Biol 68:61–84
Gonsalves D (1998) Control of papaya ringspot virus in papaya: a case study. Annu Rev Phytopathol 36:415–437. https://doi.org/10.1146/annurev.phyto.36.1.415
Goodman RE, Vieths S, Sampson HA, Hill D, Ebisawa M, Taylor SL, van Ree R (2005) Allergenicity assessment of genetically modified crops—what makes sense? Nat Biotechnol 23:1233–1238
Graff GD, Cullen SE, Bradford KJ, Zilberman D, Bennett AB (2003) The public–private structure of intellectual property ownership in agricultural biotechnology. Nat Biotechnol 21(9):989–995
Häkkinen ST, Palva T (2019) Advances and challenges in plant biotechnology. Physiol Plant 166(3):450–460
Haque MS, Saha D, Talukder ZH (2020) Biosafety and socioeconomic impacts of Btbrinjal in Bangladesh. GM Crops Food 11(1):18–28. https://doi.org/10.1080/21645698.2020.1713312
HarvestPlus (2020) Biofortification progress brief: nutrient-enriched crops for global food security. IFPRI-HarvestPlus, Washington, DC. https://www.harvestplus.org/resources
HarvestPlus (2023) Biofortified crops released globally. https://www.harvestplus.org
Hiei Y, Komari T (2008) Agrobacterium-mediated transformation of rice using immature embryos or calli induced from mature seed. Nat Protoc 3(5):824–834
Ikram NKBK, Simonsen HT (2017) A review of biotechnological artemisinin production in plants. Front Plant Sci 8:1966. https://doi.org/10.3389/fpls.2017.01966
Jansson S, Douglas CJ, Myburg AA (2021) Precision breeding for sustainable horticulture: integration of systems biology and AI. Trends Plant Sci 26(5):456–470
Kapoor L, Srivastava A, Singh SK (2023) Designer horticultural crops: a vision for functional and therapeutic foods. Plant Mol Biol Report 41(1):25–41
Katsumoto Y, Fukuchi-Mizutani M, Fukui Y, Brugliera F, Holton TA, Karan M, Nakamura N, Yonekura-Sakakibara K, Togami J, Pigeaire A, Tanaka Y (2007) Engineering of the rose flavonoid biosynthetic pathway successfully generated blue-hued flowers accumulating delphinidin. Plant Cell Physiol 48(11):1589–1600. https://doi.org/10.1093/pcp/pcm131
Kaur N, Sharma V, Kumar D (2020) Remunerative horticulture: enhancing quality and profitability through genetic interventions. Hortic Int J 4(1):1–7
Kim S, Kim JH, Yoo H, Jung JH, Ahn IP (2021) CRISPR/Cas9-mediated CaMLO2 knockout confers enhanced powdery mildew resistance in pepper (*Capsicum annuum*). Plant Biotechnol Rep 15:175–183
Krishna H, Singh D, Singh SK (2020) Plant regeneration and transformation in fruit crops: advances and bottlenecks. Biotechnol Adv 44:107608
Kumar M, Singh R, Lal M (2022) Transforming horticulture for income security: an Indian perspective. Indian J Agric Sci 92(3):405–412
Langridge P, Reynolds MP (2021) Genomic tools to assist breeding for drought tolerance. Curr Opin Biotechnol 67:43–48
Liu C, Zhang L, Liu H, Zhou Q, Zhang H, Liu H (2020) CRISPR/Cas9-based targeted editing of PPO gene in spinach (*Spinaciaoleracea*) and lettuce (*Lactuca sativa*). Plant Biotechnol Rep 14:449–459

López-Gómez R, Sandoval-Mojica AF, Hernández-Soto A, Trejo-Tapia G (2021) Advances in genetic transformation of coffee (Coffea spp.) for pathogen resistance. Plants 10(3):537. https://doi.org/10.3390/plants10030537

Lucca P, Hurrell R, Potrykus I (2002) Fighting iron deficiency anemia with iron-rich rice. J Am Coll Nutr 21(3 Suppl):184S–190S

Luo X, Reiter MA, Dinh HQ, Liu Y, Phillip G (2019) Synthetic metabolic pathways for the production of plant-derived bioactive compounds in *Escherichia coli*. Nat Chem Biol 15:224–231

Lybbert TJ, Zilberman D (2010) The role of patents in the international diffusion of GM crops. AgBioforum 13(1):34–46

Martineau B, Branca F, Sayre R (1994) The FlavrSavr™ tomato: the first genetically engineered crop product. Calif Agric 48(4):13–18. https://doi.org/10.3733/ca.v048n04p13

McFadden BR, Lusk JL (2017) What consumers don't know about genetically modified food, and how that affects beliefs. FASEB J 31(1):396–403

Misra A, Sharma A, Shukla A, Tuli R (2014) Enhanced production of eugenol and methyl eugenol in transgenic lines of *Ocimum sanctum* using plastid-targeted gene expression. Plant Cell Rep 33:895–906

Moco S, Capanoglu E, Tikunov Y, Bino RJ, Boyacioglu D, Hall RD, De Vos RC (2007) Tissue specialization at the metabolite level is perceived during the development of tomato fruit. Metabolomics 3:273–288

National Bioengineered Food Disclosure Standard (2016) United States Congress. (2016), Public Law 114–216, 7 U.S.C. §1639. Washington, DC, USA: U.S. Government Publishing Office.

Nap JP, Metz PL, Escaler M, Conner AJ (2003) The release of genetically modified crops into the environment. Plant J 33(1):1–18

Narayanan N, Thiruvengadam M, Chung IM (2019) Molecular insights on biofortification in quinoa (*Chenopodium quinoa*Willd.). Biotechnol Lett 41:1393–1406

Nestel P, Bouis HE, Meenakshi JV, Pfeiffer W (2006) Biofortification of staple food crops. J Nutr 136(4):1064–1067

OECD (2001) Consensus document on compositional considerations for new varieties of tomato: key food and feed nutrients, anti-nutrients and toxicants. Organisation for Economic Co-operation and Development

Padulosi S, Thompson J, Rudebjer P (2013) Fighting poverty, hunger and malnutrition with neglected and underutilized species: needs, challenges and the way forward. Bioversity International, Rome

Paine JA, Shipton CA, Chaggar S, Howells RM, Kennedy MJ, Vernon G et al (2005) Improving the nutritional value of Golden Rice through increased pro-vitamin a content. Nat Biotechnol 23:482–487

Paroda RS, Kumar P (2000) Food production and demand in South Asia. Agric Econ Res Rev 13(2):1–24

Paul JY, Becker DK, Dickman MB, Harding RM, Khanna HK, Dale JL (2017a) Golden banana: Provitamin A biofortification through genetic engineering. Plant Biotechnol J 15(4):520–532. https://doi.org/10.1111/pbi.12650

Paul JY, Becker DK, Dickman MB, Harding RM, Khanna HK, Smith MK, Dale JL (2017b) Engineering transgenic banana resistant to Fusarium wilt: a field evaluation. BMC Plant Biol 17:1–13. https://doi.org/10.1186/s12870-017-1002-1

Pingali PL (2015) Agricultural policy and nutrition outcomes–getting beyond the preoccupation with staple grains. Food Secur 7:583–591

Rai A, Saito K, Yamazaki M (2021) Integrated omics for understanding metabolic specialization in plants. Biochem Soc Trans 49:1273–1286

Rao S et al (2019) Genetic engineering of horticultural crops: recent achievements and future perspectives. Sci Hortic 255:287–301

Salt DE, Baxter I, Lahner B (2008) Ionomics and the study of the plant ionome. Annu Rev Plant Biol 59:709–733

Saltzman A, Birol E, Bouis HE, Boy E, De Moura FF, Islam Y, Pfeiffer WH (2013) Biofortification: Progress toward a more nourishing future. Glob Food Sec 2(1):9–17

Sangwan RS, Tripathi S, Misra L, Luqman S, Sangwan NS (2008) Effect of elicitation on production of withanolide A and withaferin A in shoot cultures of Withaniasomnifera. Plant Cell Tissue Organ Cult 93:77–83. https://doi.org/10.1007/s11240-007-9254-0

Schouten HJ, Krens FA, Jacobsen E (2006) Cisgenic plants are similar to traditionally bred plants. EMBO Rep 7(8):750–753

Sharma KK, Ortiz R (2000) Transgenic crops: the Indian perspective. Electron J Biotechnol 3(2):110–119

Sharma N, Biswas S, Roy S (2023) Biotechnology for biofortification in horticultural crops: a roadmap toward nutritional security. Plant Sci Today 10(2):145–157

Shelton AM, Zhao JZ, Roush RT (2018a) Economic and ecological implications of transgenic Bt vegetables. Annu Rev Entomol 63:329–350

Shelton AM, Hossain MJ, Paranjape V, Azad AK, Rahman ML, Khan ASMMR, Prodhan MZH, Hossain MA (2018b) Bt eggplant project in Bangladesh: history, present status, and future direction. Front Bioeng Biotechnol 6:106. https://doi.org/10.3389/fbioe.2018.00106

Shelton AM, Hossain MJ, Paranjape V, Azad AK, Rahman ML, Majumder R, Hossain F, Prodhan MA (2020) Bt eggplant adoption in Bangladesh: farmer's perspectives and economic impact. Front Bioeng Biotechnol 8:498. https://doi.org/10.3389/fbioe.2020.00498

Shinozaki K, Yamaguchi-Shinozaki K (2007) Gene networks involved in drought stress response and tolerance. J Exp Bot 58(2):221–227

Singh A, Ganapathysubramanian B, Singh AK, Sarkar S (2016) Machine learning for high-throughput stress phenotyping in plants. Trends Plant Sci 21(2):110–124

Singh D, Deshmukh R, Patil C (2021) Application of proteomics in horticultural crops: current status and future perspectives. Plant Cell Rep 40:1027–1044

Stein AJ (2010) Global impacts of human nutrition and disease on economic productivity. Food Policy 35(3):179–191

Tripathi JN, Ntui VO, Shah T, Muiruri SK, Britt A, Tripathi L (2019) CRISPR/Cas9 editing of endogenous banana streak virus in the B genome of Musa spp. overcomes a major challenge in banana breeding. Commun Biol 2:46

Verdonk JC, Ric de Vos CH, Verhoeven HA, Haring MA, van Tunen AJ, Schuurink RC (2005) Regulation of floral scent production in petunia revealed by targeted metabolomics. Phytochemistry 66(18):2541–2550. https://doi.org/10.1016/j.phytochem.2005.08.014

Wang Y, Zhang Z, Zhu J (2019) CRISPR/Cas-mediated genome editing in plants: An emerging technology for crop improvement and food security. Front Plant Sci, 10:1365. https://doi.org/10.3389/fpls.2019.01365

Waltz E (2015) Nonbrowning GM apples cleared for market. Nat Biotechnol 33:326–327. https://doi.org/10.1038/nbt0315-326

World Health Organization (WHO) (2021) Micronutrient deficiencies. https://www.who.int

Zaidi SSA et al (2020) Engineering crops of the future: CRISPR-based genome editing and its applications. Front Plant Sci 11:714

Zhang Y, Butelli E, De Stefano R, Schoonbeek HJ, Magusin A, Pagliarani C, Wellner N, Hill L, Orzaez D, Granell A, Martin C (2013) Anthocyanins double the shelf life of tomatoes by delaying overripening and reducing susceptibility to gray mold. Curr Biol 23(12):1094–1100

Zhang H, Zhang J, Wei P, Zhang B, Gou F, Feng Z et al (2018) The CRISPR/Cas9 system produces specific and homozygous targeted gene editing in rice in one generation. Plant Biotechnol J 16:1122–1130

Zhang Y et al (2021) Advances in genome editing for horticultural crop improvement. Hortic Res 8:1–18. https://doi.org/10.1038/s41438-021-00534-z

Zhao M, Ren Y, Zhang X, Zhou Z, Zhang S, Zhao Y, Zhang X, Yu J (2022) CRISPR/Cas9 editing of vitamin C biosynthesis genes increases vitamin C content in *Hippophaerhamnoides*. BMC Plant Biol 22:457

Zhu C, Naqvi S, Gomez-Galera S, Pelacho AM, Capell T, Christou P (2020) Transgenic strategies for the nutritional enhancement of plants. Trends Plant Sci 25(6):500–509

Zsögön A et al (2018) De novo domestication of wild tomato using genome editing. Nat Biotechnol 36(12):1211–1216

Innovations in Harvesting: A Sustainable Approach

9

Diksha Rani, Manoj Kumar, and S. S. Vasanthkumar

Abstract

Sustainable harvesting procedures are seen to be perceived as urgent in the light of long-term growing sustainability of the horticulture production system. Horticulture production and sustainability rely on harvesting; on the other hand, conventional approaches characterized by intensive labor and inefficient resource utilization are not growing at an adequate rate to meet the world's food demands in the face of climate change and limited developing resources insufficient to provide global food requirements under climate change and resource constraints. Progress in horticulture technology and methodologies is essential for addressing global food security requirements while maintaining environmental sustainability. Recent innovations in robotics, automation, Internet of Things (IoT), precision agriculture, and biotechnology have revolutionized harvesting processes, while sustainable alternatives such as solar-powered machinery, low-till practices, and smart irrigation enhance the efficiency of resources. These additionally indicate a potential choice for automated harvesting of crops, revolutionizing current agriculture through enhanced efficiency, sustainability, and precision farming. The amalgamation of ancient practices with contemporary technologies enhances resilience and inclusion in many fields of horticulture. This chapter examines modern advances in harvesting technology and procedures that enhance environmental stewardship, economic efficiency, and social welfare in the horticulture sector. The discussion also included the importance of

D. Rani (✉) · M. Kumar
Division of Vegetable Science, Sher-E-Kashmir University of Agricultural Sciences and Technology, Jammu, Jammu and Kashmir, India

S. S. Vasanthkumar
Department of Plant and Environmental Sciences, Clemson University, Clemson, SC, USA

A. Das, A. B. Sharangi (eds.), *Remunerative Horticulture*,
https://doi.org/10.1007/978-981-92-1007-7_9

digital tools, such as IoT, data analytics, and blockchain, in improving traceability and optimizing harvest processes.

Keywords

Sustainable harvesting · IoT · Precision harvesting · Data-driven machinery · Sensor technologies · Blockchain · Robotics

9.1 Introduction

Harvesting, or the process of collecting fully mature crops, is essential to agricultural sustainability and productivity. The horticultural industry faces the dual challenges of increasing productivity and maintaining ecological balance as the population grows and resources such as water and land become more scarce. The effects of reducing the environmental impact have culminated in notable advancements in crop harvesting techniques. In addition to ensuring effective resource use, sustainable harvesting practices also safeguard ecosystems, lessen waste, and improve food security. New ways of harvesting for profitable horticulture currently focus on using modern technology and sustainable methods to make more money, cut down on waste, and have less of an influence on the environment. In light of resource depletion and climate change, traditional harvesting techniques, which are frequently time-consuming and resource-intensive, have grown more unsustainable. Innovative harvesting techniques have been brought about by recent technological advancements and process optimization, which hold promise for improving the sustainability of agricultural practices on all fronts economically, socially, and environmentally. The world's agricultural systems are under unprecedented strain due to the constantly rising demands of a rapidly expanding global population. Once the foundation of food production, traditional harvesting methods now face many difficulties because of their ineffectiveness in catering to future dietary demands, unsustainable utilization of resources, and environmental impact.

The emphasis has shifted to innovative, sustainable methods that revolutionize every component of the food supply chain, especially at the harvesting stage, as horticulture works to sustainably nourish billions of people while protecting ecosystems. Harvesting food is being redefined by the convergence of technology, data analytics, and ecological stewardship, opening the door for systems that are not only productive but also ecologically conscious and financially sustainable for future generations. In the pursuit of more sustainable alternatives, traditional harvesting techniques, which are frequently characterized by high labor intensity and resource inefficiency, have recently come under investigation. Harvesting innovations have been identified as key drivers of sustainable horticultural transformation as the industry shifts toward minimizing waste, shrinking environmental footprints, and optimizing operational efficiency.

9.1.1 Background and Rationale

Fruit crops are a central part of horticulture, providing essential vitamins and nutrients, economic security, and raw materials for numerous industries. Historically, horticulture relied heavily on manual labor and simple hand tools. These practices provided local employment but revealed limitations such as labor shortage, postharvest losses, inefficiency, and environmental harm, especially as global food demand and climate challenges increased. Unsustainable harvesting methods, such as picking too soon or too much, using machines without care, and relying too much on labor-intensive methods without enough safety measures, have often led to postharvest losses of 20–40% in many developing areas, impaired plant health from improper handling, and compromised soil integrity from using machinery without consideration. These problems highlight a broader issue: harvesting is more than merely picking fruit off of trees, but also a key critical point where human innovation, environmental stewardship, and the long-term health of the food system come together. Also, traditional mechanization made things faster and bigger, but it caused problems like soil compaction, damage to fragile crops, and a lot of fossil fuel use. Environmental factors, especially the need to protect soil health, biodiversity, and water resources, render the importance of sustainable harvesting technologies even more important.

The case for progressing horticulture toward sustainable harvesting technologies is strengthened more compellingly by changes in consumer preferences, stricter rules, and more proof of the negative effects of unsustainable practices. Human beings desire crops that are cultivated and picked with the minimum environmental damage possible. At the same time, policymakers have established ambitious targets for cutting greenhouse gas emissions, saving water, and keeping food safe. Here, innovators and stakeholders throughout the horticultural value chain have hastened to develop and adopt harvesting technologies aligned with the sustainability principles, i.e., efficiency, minimal postharvest losses, reduced energy intensity, and ecological preservation. Innovative harvesting methods directly address the important aspect of sustainability that is reducing losses after harvest. In places where 20–40% of fruit is frequently destroyed because of incorrect harvesting methods, using precision tools and mechanized systems can greatly reduce spoilage. Consequently, more food will reach consumers, and farmers will earn larger profits.

In the last 10 years, the horticultural industry has seen a technological renaissance in harvesting, due to the combination of automation, smart systems, biotechnology, and ecological design. A sustainable way to harvest uses precision agriculture, energy-efficient technologies, and eco-friendly methods that save resources and help achieve prospective environmental goals. The creation of selective mechanical harvesters with vision systems and robotics is one of the most important areas of innovation. These advanced machines use artificial intelligence and sensors to recognize the difference between ripe and unripe crops, collect them gently, and cut down on losses. This is different from traditional machines that do bulk harvesting, which often cause a lot of crop damage and waste. For example, robotic harvesters are able to locate, grab, and pick fruits like strawberries, tomatoes, and apples with amazing accuracy. This cuts down on the need for human labor and waste after harvest. Using drones and unmanned aerial vehicles (UAVs) to harvest some specialty crops is another example of this progress. These flying harvesters are still being developed, but they are meant to fly through orchard canopies and use advanced cameras and machine learning algorithms to find ripe fruit that is ready to be picked. The use of drones is especially promising for high-value crops that are hard to get to and for stepping in for workers during busy harvest times.

Smart tools that employ the Internet of Things (IoT) have rendered things even more complicated. They can now check the maturity of crops, the health of plants, and the weather in real time. Farmers can better plan the timing, intensity, and methods of harvest by linking their harvest equipment to centralized farm management systems. This cuts down on redundant passes across the field and saves fuel. These systems additionally assist with traceability and quality control by keeping track of data from each harvest event. This is beneficial for both food safety and supply chain transparency. Another important new idea is eco-friendly harvesting methods that protect soil structure, biodiversity, and resource use. A growing percentage of orchards and vegetable fields are using no-till and low-till harvest methods, along with machines that have been developed to disturb the soil as barely as possible. These methods help keep soil microbial communities safe, reduce erosion, and maintain carbon sequestration potential. Using renewable energy to power harvesting equipment, such as solar-powered harvesting carts, helps sustainability goals by cutting down on the use of fossil fuels and greenhouse gas emissions.

Innovative harvesting is very important for reducing losses after harvest. Fruits and vegetables that are fragile are known to bruise and spoil easily, which can cause substantial losses from the field and the market. Modern harvesters with soft-tissue or pneumatic grips, gentle conveyors, and microclimate monitoring make certain that the physical quality of the produce stays good from the moment it is picked. These new ideas, along with fast field-cooling systems incorporated via harvest machinery, cut down on the time that produce requires at undesirable temperatures, which keeps it fresh and nutritious. Biotechnology has additionally assisted in facilitating more sustainable. It demands fewer resources to harvest crops with better abscission characteristics, ensuring that they can easily break apart when they fully ripen. This also implies less damage. Researchers are looking into gene-editing methods to make these traits even better, which will make it easier to harvest gently and efficiently, whether by hand or with a machine.

Innovations that make harvesting safer, easier, and more accessible help social and economic sustainability. Exoskeleton suits and wearable robotics are becoming more common for workers who additionally do harvesting work. These tools lower the risk of musculoskeletal injuries while keeping productivity high. Portable digital training tools give workers the skills they need to run complex harvest systems, which contributes to maintaining occupations in rural areas even as automation takes over.

9.2 The Need for Sustainable Harvesting

Harvesting is one of the most significant components of fruit crop cultivation since it influences the quality, freshness, and market value of the fruit, as well as the balance of the ecosystem, farmers' livelihoods, and the long-term health of horticulture. Sustainable harvesting is necessary to keep ecosystems healthy, make sure resources are available for a long time, and support financial stability because of global problems like climate change, population growth, and resource scarcity. Conventional harvesting techniques are adverse to the environment and the economy, so sustainability is necessary for food security and the health of the ecosystem. Modern horticulture frequently depends on methods that deplete natural resources, resulting in enduring ecological harm. Sustainable harvesting innovations are more than just ways to improve efficiency and crop yield; they are a new approach to thinking about how to balance economics, the environment, and social justice in fruit production. Sustainable harvesting is not only needed because of its immediate advantages: higher quality produce, reduced waste, and higher farmer income, but also because of its more significant contribution to ensuring that fruit crops remain viable in a changing climate, robust in global trade networks, while remaining equitable for future generations.

9.2.1 Ecological Integrity

Sustainable harvesting assists in maintaining the population of the species and biodiversity. Sustainable harvesting implies a compromise between protecting the ecosystem and ensuring a good crop yield. New tools for precision harvesting reduce damage to fruits while ensuring the harvest is performed at the right time, which trims down on waste and keeps ecosystems healthy. Controlled atmosphere storage and smart packaging with antimicrobial coatings are examples of advanced postharvest technologies that assist in prolonging shelf life. This cuts down on spoilage and postharvest losses, which in turn mitigates resource use and environmental impact. Ecological sustainability means that the functions, relationships, and succession sequences in an ecosystem continue to remain the same, and that carbon storage, water cycling, and native species are still there. It protects the soil from degradation and keeps it fertile by rotating crops, planting cover crops, and better managing nutrients. Sustainable practices like precision agriculture contribute to making the most of resources, cut down on waste, and protect biodiversity (Kumar et al. 2024). Using controlled environments (like greenhouses) and technology-driven solutions makes horticulture even more efficient and improves the quality of the crops (Zhao et al. 2024).

9.2.2 Economic Stability

Sustainable harvesting makes sure that future generations will always have access to resources and can even increase yields over time. Modern technologies enable operations to be more efficient, cut down on waste, and raise the quality and yield of fruit crops. This enables farmers to make greater revenue and lowers the costs of inputs, which keeps productivity high and the environment in balance for a long time. The economic benefits also reach those communities that engage in wild harvesting, since ethical practices promote occupations without overexploiting resources and create markets for products that are sourced in a way that does not harm the environment (Lampert 2024). Sustainable harvesting methods focus on lowering expenditures after harvesting by using precision agriculture technologies like nanotechnology-based packaging and smart delivery systems that keep fruit fresh and in good condition (Ramya et al. 2025).

9.2.3 Community and Cultural Perspectives

Sustainable harvesting safeguards ancestral knowledge and the economic independence of people in rural and Indigenous communities. Many cultural customs include operations like replanting, scattering seeds, or rotating where crops are harvested. Climate change, unstable markets, and lower agricultural productivity make life harder for vulnerable communities, especially those in areas with few resources (Chachar et al. 2023). For instance, cooperative agricultural societies may all benefit from better tools and technologies, which lower costs and keep jobs for local workers. Communities might favor methods that keep fruit intact so that it can be used in religious, culinary, and festive traditions for a long time. By combining community involvement with scientific progress, harvesting can become more sustainable, which will lead to less waste after harvest, more biodiversity, and fairer benefits for farmers.

9.2.4 Climate Change

Climate change leads to erratic weather, higher temperatures, and changes in rainfall, which affect plant growth, flowering, and crop yield. Developments like these have an enormous impact on horticultural systems, which affects nutritional diversity and global food security. This is because conventional farming releases more greenhouse gases, thereby making climate stress worse (Chachar et al. 2023). These changes in the environment have an impact on the quality of fruit, the amount of fruit produced, and the fruit's vulnerability to pests and diseases. This means that farmers need to come up with new ways to deal with both biotic and abiotic stresses (Bhattacharjee et al. 2022). For instance, technologies that keep an eye on microclimate conditions help farmers figure out the best times to harvest to keep the fruit's taste and nutrition high. Additionally, climate-adaptive strategies such as fruit bagging, the application of anti-transpirants, and the cultivation of heat-tolerant or frost-resistant varieties enhance sustainable production by minimizing crop damage and transpiration of water (Bacelar et al. 2024).

9.2.5 Environmental Impacts of Conventional Harvesting

Traditional horticultural harvesting methods harm the environment in an enormous manner. For example, they trigger extensive forest loss and destruction of habitat to make room for more croplands, which lowers biodiversity and damages the ecological balance. Conventional farming methods frequently employ heavy machinery and chemicals that change the structure of the soil, promote erosion, and negatively impact the environment. These practices can make the soil less fertile over time and hurt the ecosystems around them. Also, taking too much water for irrigation during the harvest makes the water shortage issues worse. Sustainable methods use precision farming tools like drip irrigation, soil moisture sensors, and rainwater harvesting to make the most of water and cut down on waste. New mechanized harvesting technologies include eco-friendly machines that are meant to reduce soil compaction and energy use (Kaur et al. 2023). Nanotechnology in pest and disease management cuts down on the need for chemicals, which is better for the environment. These new ideas assist in safeguarding biodiversity, conserving soil, and utilizing water more efficiently. They also make fruit production more environmentally friendly.

9.3 Progress in the Sciences of Harvesting

Technological advances shift the way horticulture crops are selected, thereby helping farmers cope with the situation of labor shortages, work more efficiently, and ensure their produce is of better quality. Innovations in robotics, automation, data-driven machinery, and sensor-based management are disrupting the very basis of modern horticulture harvesting. These are helping to solve problems that have been around for a long time, including not having enough workers, losing crops, and wasting resources (Zhang et al. 2024a, b). Horticulture used to rely on laborious procedures that were hard work and often led to quality problems. However, in the last 10 years, there has been a big increase in advanced harvesting technologies, such as robotic automation, AI-powered systems, and the use of precision agriculture and nanotechnologies (Kaur et al. 2023; Getahun et al. 2024). Several modern innovations and smart tools utilized in horticulture for sustainable harvesting practices have been represented in flow chart representations under Figs. 9.1 and 9.2.

9.3.1 Robotics and Automated Harvesters

Robotic harvesters with smart sensors and artificial intelligence (AI) have revolutionized the way delicate fruits and vegetables are picked. These mechanisms may indicate when the best time to harvest falls, when fruit is ripe, and how to carefully take it off without hurting it. These kinds of robots, which are often used on crops

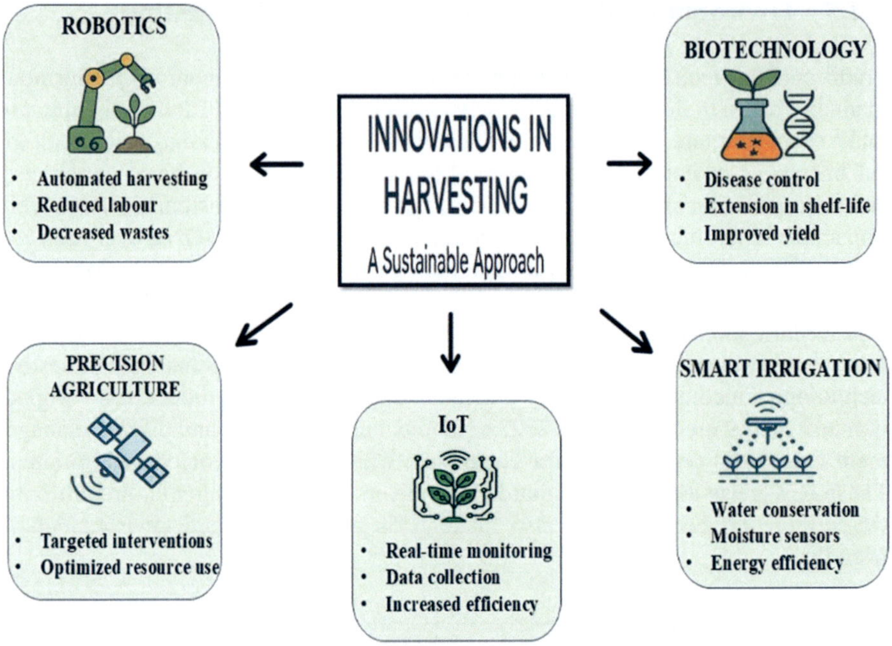

Fig. 9.1 Flow chart titled "Innovations in Harvesting: A Sustainable Approach" at the center, with arrows pointing to four surrounding concepts. Top left: "Robotics" with benefits like automated harvesting, reduced labor, and decreased waste. Top right: "Biotechnology" highlighting disease control, extended shelf-life, and improved yield. Bottom left: "Precision Agriculture" focusing on targeted interventions and optimized resource use. Bottom right: "Smart Irrigation" emphasizing water conservation, moisture sensors, and energy efficiency. Center bottom: "IoT" with real-time monitoring, data collection, and increased efficiency. Each concept is accompanied by relevant icons

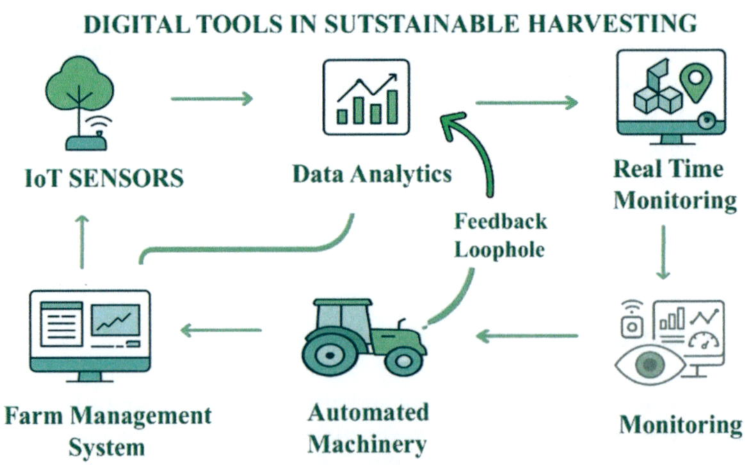

Fig. 9.2 Flow chart illustrating digital tools in sustainable harvesting. It includes interconnected elements: IoT Sensors, Data Analytics, Real-Time Monitoring, Monitoring, Automated Machinery, and Farm Management System. Arrows indicate the flow of information, with a feedback loop between Data Analytics and Real-Time Monitoring

like strawberries, peppers, and tree fruits, work all the time and cut down on the need for manual labor by 60–80%. Alaaudeen et al. (2024) reported that using robots for harvesting had a 0.75 success rate. Robotic arms and self-driving cars make harvesting more accurate, cut down on waste, and speed up the process by a lot. Modern harvesting robots now use advanced vision algorithms and specialized end effectors that can accurately identify, grab, and detach fruits or vegetables without damaging them (Gammanpila et al. 2024). These self-driving systems have sensors and AI that can tell how ripe the fruit is and when the best time to pick it is. This makes harvesting more efficient and precise than with human operators. Robots can pick fruits and vegetables with almost perfect accuracy in both protected areas and open fields. This greatly reduces losses caused by delayed or ineffective manual harvesting, while also helping to solve global labor shortages and rising operational costs (Caplan et al. 2014). Automated harvesters have computer vision and machine learning algorithms that help them find and pick ripe crops with great accuracy. This reduces damage and improves the quality of the yield.

9.3.2 Precision Agriculture

Precision agriculture represents a fundamental aspect of innovation in horticulture harvests. Farmers can employ GPS-guided machinery, remote sensing, and data analytics to assess crop welfare and maturity, hence optimizing harvest scheduling (Getahun et al. 2024; Usendi et al. 2025). Yield monitoring systems, drones, and intelligent automation enhance resource utilization, augment productivity, and reduce expense for inputs by customizing interventions to real-time field circumstances (Abobatta 2024; Usendi et al. 2025). These methods enhance environmental sustainability and profitability by minimizing waste and guaranteeing that crops are gathered at optimal quality (Abobatta 2024).

9.3.3 Smart Sensors and Data-Driven Machinery

Modern harvesting machines use smart sensors and IoT (Internet of Things) technology to keep an eye on the health of crops, their growth, moisture levels, and the state of the environment in real time. These sensors are used by precision agriculture systems to control machinery, improve timing, and cut down on losses after harvest. For example, GPS and data analytics are able to map fields, recognize ripe crops, and forecast yields. This permits farmers to harvest crops in a way that achieves the most output and quality (Getahun et al. 2024; Usendi et al. 2025). Smart machinery uses real-time data via environmental sensors, GPS, and drones to change harvesting plans based on the maturity of the crops and the conditions of the field. This makes better use of resources and lowers operating expenses (Zhang et al. 2024a, b). They help with labor shortages and keep productivity steady, and they also help with sustainable farming by making sure tasks are done at the right time and reducing waste (Gammanpila et al. 2024).

9.3.4 Advanced Machinery and Ergonomic Design

Modern harvesting tools, like telehandlers and wheel loaders, feature sophisticated hydraulic and ergonomic systems that make them comfortable and help them get work done. New design features like articulated frames and attachments that can be used in many ways make it easier to move things around and handle materials more quickly, which increases efficiency and lowers operator fatigue. Sustainable engine technologies have less of an effect on the environment, and telematics integration permits one to keep an eye on things from afar and do maintenance ahead of time to keep downtime to a minimum.

9.3.5 Vertical and Controlled-Environment Harvesting

Automated climate control, irrigation, and nutrient administration systems in smart greenhouses and vertical farming infrastructure make it possible to grow horticultural crops all year long. Sensors change the temperature, light, and moisture levels in real time to make the best conditions for plants to grow and use fewer resources. These systems work perfectly with automated harvesters, thereby making sure that the quality is always the same and that fewer people require assistance to perform the task.

9.3.6 Postharvest Innovations

In growing numbers, sorting, grading, and packing crops after they have been harvested are being done by machines. These machines use high-resolution cameras and machine vision to verify the quality and ripeness of the crops and sort them quickly. Optical sorters and automated robots make food safer by eliminating unsafe produce, lowering the cost of labor, and resulting in fewer mistakes. Digital platforms make smart supply chains possible. They follow crops from the field to the market, thereby rendering the process more transparent and reducing wastage.

9.3.7 Biotechnology and Crop Improvement

The incorporation of biotechnology to strengthen fruit crops is a major breakthrough in the harvesting of horticultural crops. It has greatly improved the efficiency, sustainability, and quality of fruit production. Biotechnology combines cutting-edge genetic, molecular, and tissue culture methods to create fruit varieties with superior characteristics that make harvesting easier or more productive, like higher yields, superior fruit quality, and longer shelf life (Rawat et al. 2024; Bhusari et al. 2025).

Genetic engineering, marker-assisted selection, and genome editing with CRISPR/Cas9 are some of the biotechnological advances that have altered the way fruit breeding programs operate. These tools permit a change in key traits in a very specific way, which helps reduce losses during harvesting and handling after harvesting (Rawat et al. 2024). For example, engineering fruit crops to be firmer and mature more slowly makes it easier to plan when to harvest them and minimizes the damage that can happen from harvesting too promptly or storing them for too long. Marker-assisted selection speeds up breeding by detecting good genes that affect fruit size, resistance to biotic and abiotic stresses, and nutritional content. This renders it easier to grow robust varieties that can be harvested by hand or by machine (Bhusari et al. 2025; Shingade and Khatri 2024). Several innovative harvesting techniques employed in horticultural crops, along with the principal technologies utilized, are compiled in Table 9.1.

9.4 Integrating Traditional and Modern Approaches

Integrating conventional and modern methods in horticulture generates cropping systems that are robust, productive, and long-lasting while respecting cultural and ecological contexts and incorporating new technology (Adefila et al. 2024). Using both traditional knowledge, like Indigenous crop rotation, agroforestry, organic pest control, and polyculture, and modern methods, like precision farming, soil analysis, and mechanized operations, leads to better yields, better use of resources, and better health for the environment (Adefila et al. 2024). This integration can be customized to accommodate the particular requirements of the area while also making the most of scientific progress. People had been picking fruit by hand, using simple tools, and relying on their deep local knowledge of ripeness and crop cycles for a long time. Modern innovations include automated machines, robots, optical sensors, and precision farming methods, all of which are meant to make up for a lack of workers and improve quality. Many experts say that instead of seeing tradition and technology as opposing forces, they should be brought together for the best results.

9.4.1 Examples of Integration

- Robotic harvesters with local knowledge: New AI-powered robots, like the Nanovel citrus harvester, may harvest fruit and move through thick foliage using real-time data from sensors. However, their use is improved by traditional field mapping and the expertise of smallholders.
- Harvest timing based on data: Farms can use optical sensors and weather forecasts to figure out the best time to harvest, but they still use traditional hand-picking for delicate fruits or special types that machines might damage.
- Mechanical and manual synergy: Tree shakers and conveyors are sometimes used to harvest large fruits like apricots or apples in bulk, and hand-picking is used for premium grades or hard-to-reach produce. This keeps both scale and quality.

Table 9.1 Compilation of horticulture harvesting techniques employed for different fruit crops

Harvesting method	Details	Fruit crop	Technology	Reference
Robotic fruit harvesting	Using AI-powered robots with cameras and sensors to find and pluck ripe fruit on their own	Apples, strawberries, kiwis	AI, computer vision, and multi-arm picking	Hua et al. (2019) and Vougioukas et al. (2025)
Nanotechnology applications	Nano-coatings, nano-fertilizers, and nano-sensors to boost production and quality after harvest	Various fruit crops	Nanoparticles for smart delivery and pest control	Ramya et al. (2025)
Mechanical shaking systems	Machines that gently shake trees to obtain the fruit detached which is subsequently picked up by machines	Tree fruits (e.g., citrus)	Shaking in a controlled way to protect the fruit	Pu et al. (2023) and Nithya et al. (2025)
Robotics	Intelligent robotic system with 3D vision and modified triangulation for apple harvesting	Apples	Robust apple recognition and localization using integrated image processing; field testing demonstrating high successful harvesting rates	Zhang et al. (2024a, b)
Robotics	Robots using computer vision and robotic arm end effectors for selective harvesting	Apples, peppers, tomatoes, yellow peaches, strawberries	Robotic arms that can select and recognize certain fruits while causing the least amount of harm, making batch harvesting easier	Huang et al. (2025)
Robotic harvesting with AI	Cuts down on labor expenses and makes harvesting more efficient and accurate	Apples	Robots that may perceive and grip goods so they can harvest crops like apples automatically	Hua et al. (2019)

The proper description of the technology and references has also been shown

Table 9.2 Case studies of the application of modern harvesting methods for horticultural crops, along with their benefits and references

Fruit crop	Modern method integrated	Benefits	Reference
Apple	Robotic harvesters using AI vision, segmentation, and smart grasping arms	More than 95% of fruits are recognized, and automated picking is quick	Alaaudeen et al. (2024)
Sweet pepper	Automated picking robots with sensors	Faster speeds and more accurate harvests, as demonstrated in the field	Hua et al. (2019)
Kiwifruit	Intelligent robotic systems	Picking fruit quickly and without damaging it	Hua et al. (2019)
Pineapple	Rotary plate robotic picker, multi-flexible finger drum	The rate of damage is 14.8%, and the rate of harvesting is 71.3%	Liu et al. (2022) and Qiu et al. 2025
Strawberry	Vision-guided picking robots	Rapid and non-damaging fruit harvesting	Hua et al. (2019)
Tomato	Automated picking robots	Better work efficiency and gentle handling	Hua et al. (2019)
Tomato	Selective harvesting robot (CV guided)	Little damage to the fruit, high commercial value maintained	Huang et al. (2025)
Strawberry	Selective robot with computer vision	Batch harvest that works well and has less bruising of fruit	Huang et al. (2025)
Cherry	Mechanical shake-and-catch, chemical loosening agent	80–90% lower labor costs and less damage to fruit	Huffman (2012)
Orange	Mobile harvesting platform	Less time than manual harvest and good selectivity	Ferreira et al. (2018)

The effectiveness of integration depends a lot on the situation. Soil testing and mapping technology can help small farms get the most out of traditional organic amendments and crop diversity. Few case studies demonstrating the successful integration of modern fruit harvesting techniques are listed below under Table 9.2. Mechanization and scientifically proven crop combinations make techniques like intercropping and cover cropping, which lower pests and improve soil health, even better. So, the two ecological knowledge and modern tools work together to increase yields and resilience.

9.4.2 Labor, Quality, and Safety Benefits

Automation immediately resolves problems with an acute shortage of workers and seasonal costs, which can be as high as 50% of annual production costs in some areas. Robots make it possible to harvest at night and in places that are hard to get to, which makes workers safer and lets them pick fruit when it is at its best. But traditional methods are still very important for specialty crops and places where technology cannot fully replace skilled manual labor. The "best" integration is always based on what works best for the area, taking into account the knowledge of farmers and new technologies for long-term productivity and sustainability.

9.5 Enhancing Resource Use Efficiency

Modern harvesting techniques, such as mechanization, automation, and digital monitoring, make fruit production more efficient by using resources more effectively. Compared with traditional methods, these techniques greatly increase productivity, lower labor costs, and cut down on waste (Nithya et al. 2025). Modern fruit harvesting combines new technologies with environmentally friendly methods to make the best use of resources. This results in higher yields, greater safety, and less waste for growers (Vougioukas et al. 2025; Nithya et al. 2025).

9.5.1 Mechanization and Automation in Harvesting

Increasing numbers of individuals are using mechanized systems and harvesting robots to deal with labor shortages, speed up the harvest, and cut costs. New tools like tree shaking machines, multilevel adjustable shelves, and ground-collecting devices make it possible to harvest fruit with little help from people. This increases the reach and decreases the amount of fruit that is not harvested (Vougioukas et al. 2025). Robots and systems that let people and robots work together can make picking more efficient by shortening the average pick cycle time and making adjustments to the configuration of the orchard (Nithya et al. 2025).

9.5.2 Labor Efficiency and Cost Reduction

Automated harvesting machines may accelerate the collection of crops by up to 40% compared to accomplishing it by hand. This means that fewer employees may encompass larger areas while becoming less exhausted. Systems with features like automatic basket unloading, the best shelf height, and stability on sloped terrain make workers even more productive and cut down on unnecessary work. This could save fruit plantations a lot of money on operating costs.

9.5.3 Digital Monitoring and Sustainability

Quality monitoring sensors and automated systems make better use of resources by ensuring the fruit is of high quality and yield, saving inputs, and lowering losses after harvest. These technologies not only help farmers cultivate additional crops but also help the environment by making farming more sustainable and less reliant on fuel-powered machines, which lowers greenhouse gas emissions.

9.5.4 Adaptation to Different Crops and Conditions

Advanced systems and robotics are made to be flexible so they can handle different types of fruit (apples, citrus, stone fruits, and nuts) and problems like tall trees or uneven ground. As navigation algorithms, exoskeleton support, and machine adaptability improve, resource use across different orchard layouts and crops is getting better and better (Nithya et al. 2025).

9.5.5 Smart Irrigation Systems and Water Management

Smart irrigation systems and water management for fruit harvesting ensure the best possible use of water through supplying it to the crops at the right time and place. These systems use sensors, AI, and real-time data to keep an eye on the weather, the soil moisture, and the needs of the crops. This cuts down on water waste and boosts the quality and quantity of fruit. Smart irrigation uses sensors, AI, machine learning, and IoT devices to keep an eye on the soil's moisture, the weather, and the crops' water needs in real time. These systems automatically set irrigation schedules so that water is only used when it is needed. This can cut water use by up to 70% compared to traditional methods while also increasing crop yields. They also help prevent over-irrigation while identifying issues early, which is important for fruit crops because it helps with long-term water management.

Features of Smart Irrigation Systems

- Sensors check the soil's moisture, temperature, and humidity in real time to figure out how much water the crops need.
- AI and machine learning use data from the environment to automatically set irrigation schedules, ensuring that water is only applied when it is needed.
- The ability to discover breaks and monitor things from afar helps keep water from getting lost and lets you fix irrigation problems quickly.
- When combined with farm management systems, it is possible to make decisions about irrigation, crop health, and sustainability as a whole.

Smart irrigation systems are extremely beneficial for fruit farming because they can save up to 70% more water than traditional methods. This is done by using real-time data about soil moisture, weather conditions, and crop needs to deliver water exactly when it is needed. This cuts down on water waste and runoff. This means that fruit trees absorb the right amount of moisture in the soil for optimal growth and fruit development, which leads to better fruit size and quality. Also, these systems cut down on the amount of labor and time needed to manage irrigation by automating and allowing remote monitoring. In addition to accomplishing this, it also helps the environment by making sure that water is used responsibly and by reducing problems like soil erosion and nutrient leaching. Smart irrigation makes fruit farms more productive and efficient, and it also saves water for future farming needs. IoT-Based Smart Water Management platforms utilize sensor data and automation for precise irrigation in agricultural trials, demonstrating practical efficacy in optimizing water usage while preserving crop health (Kamienski et al. 2019).

9.6 Energy-Efficient Harvesting Technologies

Modern fruit harvesting approaches that safeguard energy focus on automation, robotics, AI integration, sustainable power sources, and precision agriculture to boost productivity while cutting down on labor and energy use.

- Robotic harvesting systems: Dual-arm and multi-arm robots with AI-based fruit detection and picking systems make harvesting more efficient and save energy by simply grabbing ripe fruits. The goal of these robots is to speed up picking, lower fruit damage, and cut down on rest periods, which saves energy overall while they work (Vougioukas et al. 2025).
- Robots that use solar power: Some robotic harvesters have solar panels that let them use solar energy, which greatly reduces their reliance on fossil fuels and their carbon footprints. Solar energy powers different parts, which makes the harvesting process more eco-friendly (Noreen et al. 2025).
- Vibration and shake-and-catch machines: These machines gently shake fruit trees or bushes to drop fruits into collection systems. They make the process of separating fruits easier and use less energy by automating it. More advanced versions have catching systems that resemble umbrellas to protect the fruit and speed up the process.
- Integration of AI and machine learning: AI algorithms in harvesting machines let them find fruit in real time, check its ripeness, and navigate automatically, which makes the harvesting path more efficient and cuts down on unnecessary movements, saving energy and labor.
- Lightweight and Modular Robots: Portable, energy-efficient harvesting robots made for small to medium-sized orchards can be used in a variety of ways, use less power, and be more adaptable, which cuts down on the waste of resources.
- Drone-assisted harvesting: Some farms use drones to look for ripe fruit and help with precise harvesting. This can make the process more efficient and cut down on unnecessary tractor or human field movement, which saves energy in an indirect way.

9.7 Value Chain Innovations

Advanced technologies like robotics, artificial intelligence (AI), precision agriculture, and the Internet of Things (IoT) have brought significant changes to the value chain in modern fruit harvesting. These technologies render things more efficient, less reliant on workers, better at producing fruit, and better at handling fruit after it has been harvested. Value chain innovations for modern fruit harvesting include a wide range of technological, logistical, and strategic improvements that aim to make the process more efficient, lower costs, improve the quality of the fruit, and solve labor problems along the fruit supply chain. New technologies like robotics, artificial intelligence (AI), precision agriculture, advanced sensing, and data analytics are changing the way fruits are picked, processed, and shipped. These improvements not only make the physical harvesting process better, but they also combine knowledge management, sustainability practices, and real-time decision support systems to get the most value out of every step, from the orchard to the consumer. Practices that are advantageous for the environment, such as using electric and solar-powered harvesting machines, help the environment and support farming that does not add to carbon emissions.

Value chain innovation goes beyond the orchard and into the whole supply chain. AI-powered harvesting robots often have data collection and edge computing features that give farmers real-time information about the quality, quantity, and timing of their fruit harvests. This makes it easier to plan packing, marketing, and logistics more accurately, resulting in the whole supply chain working better and cutting down on losses. Farmonaut and Microsoft's Project FarmVibes are two examples of cloud-based data integration and analytics tools that help farmers, especially small and medium-sized ones, get access to cutting-edge technology. These platforms help with resource optimization, fleet and machinery management, and climate responsibility programs, which make fruit value chains more stable and long-lasting.

9.8 Postharvest Handling and Supply Chain Efficiency

Innovations advance beyond harvesting to make postharvest management, storage, and transportation better, which assists in maintaining fruit freshness and cuts down on losses. Cold chain logistics, like solar-powered refrigeration and shared cold storage spaces, are very important for keeping fruits that become rotten during transportation and distribution to markets. Grading, packaging, and mechanized sorting systems make products look better and less likely to be damaged during handling. Digital tools like inventory tracking and logistics coordination platforms make the supply chain more open and responsive. This makes it easier to match supply with demand and cuts down on waste from overstocking or spoilage (FAO 2021). These steps contribute to making fruit value chains more competitive and help growers and distributors boost further earnings.

9.8.1 Inclusive and Digital Value Chain Models

Using digital platforms and mobile services makes it easier for smallholder farmers to be an integral component of fruit value chains. Digital procurement, price alert services, and online marketplaces offer farmers access to real-time market information, which helps them negotiate better deals and find better markets. Group sales, aggregation centers, and contract farming models help small producers take advantage of new ways to harvest and process crops after they are picked. They undertake this via rendering transactions cheaper and more efficient. This openness improves the lives of people in rural areas and helps make sure that value is shared fairly throughout the supply chain.

9.8.2 Labor Optimization and Safety Enhancements

New ways of harvesting also focus on making the best use of workers and keeping them safe. Robotic harvesters cut down on the need for seasonal manual labor, which is often hard to find and expensive. Automation helps with labor shortages, especially in developed countries where the cost of harvesting is very high. Robotic systems also get rid of the need for workers to do dangerous things like climbing ladders to pick fruit, which cuts down on injuries and costs at work. These new technologies make working conditions better by automating repetitive, physically demanding harvesting tasks while keeping productivity levels high.

9.9 Barriers and Future Directions

Artificial intelligence, robotics, and big data will likely come together in the future to provide more sustainable and more precise harvesting automation. The versatility of these innovative perspectives is equally important. Even though there have been a lot of improvements, there are still numerous reasons why sustainable harvesting innovations are not commonly applied. High initial costs, cultural specificity of each market, technical complexity, and the fact that smallholder farming systems are not well integrated are all major issues. Also, due to the resulting variety of crops, field conditions, and weather patterns, not all innovations succeed in all situations or all the time. To make processes more affordable, standardized, and accessible for farmers in different parts of the world and economic levels, we need to perform research, invest, and support policies.

A sophisticated robotic harvesters might help large commercial orchards, but small and marginal fruit growers are better served by low-cost, easy-to-use tools like portable ladders, handheld electronic shears, or sharing medium-sized machinery. The combination of artificial intelligence, robotics, and big data in the future will likely yield more sustainable and accurate harvesting automation. Using renewable energy and circular waste management in harvesting operations will improve environmental results even more. To make sure that new technologies are accessible, strong, and able to adapt to real-world situations, we need collaborative innovation ecosystems that bring together farmers, tech developers, policymakers, and researchers. Accordingly, it is significant to ensure that all people have the opportunity to utilize new technologies to secure long-term utilization of new harvesting methods.

9.10 Conclusion

In conclusion, innovative methods of harvesting are also a relevant aspect of sustainable horticultural practices in that they assist in addressing the problems of insufficient workers, as opposed to having enough resources, and environmental degradation. The sector is efficient at harvesting, safely, and sustainably due to advances in robotics, smart systems, eco-friendly machinery, and biotechnology. This is for the reason they must strike a balance between productivity, efficiency, and sustainability. Traditional harvesting techniques, though proven over time, are frequently labor-intensive, inconsistent, and increasingly impractical due to diminishing agricultural labor supply, escalating costs, and the pressing need to reduce postharvest losses. This has made new ways of harvesting not only desirable but also necessary for making fruit crop production sustainable in the long term. There are still certain factors that render this challenging for all consumers to use horticultural harvests, but ongoing research, supportive policies, and working together with all stakeholders will be very important in making sure that horticultural harvests are not only plentiful but also truly sustainable in the future.

In horticulture, harvesting is especially important because it has a direct effect on the quality and marketability of the crops. New technologies like robotics, sensor integration, advanced machinery, and smart environmental controls are changing the way the industry works when it comes to harvesting horticultural crops. These technologies make it less necessary to hire people to do the work, speed up the harvesting process, and make sure that fruits are picked when they are at their best with as little damage as possible. Also, combining artificial intelligence and machine learning has made it easier for machines to tell the difference between ripe and unripe fruits, which cuts down on waste and keeps the quality of the fruit substantial. These sorts of modifications demonstrate how modern science and engineering may operate with farming to get more effective, more long-lasting results.

Policymakers, researchers, and agricultural extension services must work together to make sure that these new ideas are easy to get, cheap, and can be used in a variety of farming situations, from smallholder farms to plantations with lots of resources. In the long run, sustainable harvesting is not just about making farms more productive; it is also about making the agricultural system more resilient so that it can meet the needs of growing populations while saving natural resources and lowering environmental footprints. Fruit crops are the best example of how innovation can change horticulture because they are in high demand, have high market value, and need to be of high quality. With the right support systems, new ways of harvesting may assist make fruit farming around the world more sustainable, productive, and strong. They deal with the urgent need to reduce losses, make the best use of resources, and give consumers high-quality produce while also making sure that farming communities stay prosperous in the long term. This will render provisions that the probable future of horticulture is both efficient and resilient.

References

Abobatta WF (2024) Precision farming applications in horticulture. Hortic Int J 8(3):91–93. https://doi.org/10.15406/hij.2024.08.00309

Adefila AO, Ajayi OO, Toromade AS, Sam-Bulya NJ (2024) Integrating traditional knowledge with modern agricultural practices: a sociocultural framework for sustainable development. World J Biol Pharm Health Sci 20(02):125–135

Alaaudeen KM, Selvarajan S, Manoharan H, Jhaveri RH (2024) Intelligent robotics harvesting system process for fruits grasping prediction. Sci Rep 14(1):2820

Bacelar E, Pinto T, Anjos R, Morais MC, Oliveira I, Vilela A, Cosme F (2024) Impacts of climate change and mitigation strategies for some abiotic and biotic constraints influencing fruit growth and quality. Plants (Basel, Switz) 13(14):1942. https://doi.org/10.3390/plants13141942

Bhattacharjee P, Warang O, Das S, Das S (2022) Impact of climate change on fruit crops-a review. Curr World Environ 17(2):319

Bhusari RM, Patil SD, Dhemre JK, Shinde C (2025) Advances in biotechnological intervention in fruit crops: a review. Plant Arch 25(1):441–450

Caplan S, Tilt B, Hoheisel G, Baugher TA (2014) Specialty crop growers' perspectives on adopting new technologies. HortTechnology 24(1):81–87

Chachar M, Ahmed S, Murtazac G, Jillanib PSANS, Balocha H, Hakroa RA (2023) The impact of climate change on horticulture: a global perspective and adaptation strategies. Glob Res Environ Sustain 1(10):19–27

FAO and Ministry of Social Development and Family of Chile (2021) Promoting fruit and vegetable value chains to improve supply and consumption, Santiago de Chile. https://doi.org/10.4060/cb7954en

Ferreira MD, Sanchez AC, Braunbeck OA, Santos EA (2018) Harvesting fruits using a mobile platform: a case study applied to citrus. Eng Agric 38(2):293–299

Gammanpila HW, Sashika MN, Priyadarshani SVGN (2024) Advancing horticultural crop loss reduction through robotic and AI technologies: innovations, applications, and practical implications. Adv Agric 2024(1):2472111

Getahun S, Kefale H, Gelaye Y (2024) Application of precision agriculture technologies for sustainable crop production and environmental sustainability: a systematic review. Sci World J 2024:2126734. https://doi.org/10.1155/2024/2126734

Hua Y, Zhang N, Yuan X, Quan L, Yang J, Nagasaka K, Zhou XG (2019) Recent advances in intelligent automated fruit harvesting robots. Open Agric J 13(1):101

Huang Y, Xu S, Chen H, Li G, Dong H, Yu J, Zhang X, Chen R (2025) A review of visual perception technology for intelligent fruit harvesting robots. Front Plant Sci 16:1646871. https://doi.org/10.3389/fpls.2025.1646871

Huffman WE (2012) The status of labor-saving mechanization in U.S. fruit and vegetable harvesting. Choices. Quarter 2. Available online: http://choicesmagazine.org/choices-magazine/theme-articles/immigration-and-agriculture/the-status-of-labor-saving-mechanization-in-us-fruit-and-vegetable-harvesting

Kamienski C, Soininen JP, Taumberger M, Dantas R, Toscano A, Salmon Cinotti T, Filev Maia R, Torre Neto A (2019) Smart water management platform: IoT-based precision irrigation for agriculture. Sensors (Basel, Switzerland) 19(2):276. https://doi.org/10.3390/s19020276

Kaur B, Dimri S, Singh J, Mishra S, Chauhan N, Kukreti T, Sharma B, Singh SP, Arora S, Uniyal D, Agrawal Y (2023) Insights into the harvesting tools and equipment's for horticultural crops: from then to now. J Agric Food Res 14:100814

Kumar S, Kaushik R, Kaur H, Djajadi D, Khamidah A, Alasbali N, Alreshidi MA, Alam MW, Yadav KK, Wani AK (2024) Navigating the landscape of precision horticulture: sustainable agriculture in the digital age. Sci Hortic 338:113688

Lampert A (2024) Global non-sustainable harvest of renewable resources reduces their present price but increases their net present value. Resour Energy Econ 76:101409

Liu T, Liu W, Zeng T, Cheng Y, Zheng Y, Qiu J (2022) A Multi-Flexible-Fingered Roller Pineapple Harvesting Mechanism. Agriculture 12(8):1175. https://doi.org/10.3390/agriculture12081175

Nithya RMG, Surendrakumar A, Kavitha R, Suthakar B, Sivakumar SD (2025) Technological advances in fruit collection from the ground: a review. J Hortic Sci Biotechnol:1–14. https://doi.org/10.1080/14620316.2025.2520366

Noreen T, Ali A, Anwar S, Farooq S, Altabey WA, Kouritem SA, Noori M (2025) Implementation of fruit plucking robot in apple harvesting: a review. Results Eng 27:106575

Pu Y, Wang S, Yang F, Ehsani R, Zhao L, Li C, Xie S, Yang M (2023) Recent progress and future prospects for mechanized harvesting of fruit crops with shaking systems. Int J Agric Biol Eng 16(1):1–13

Qiu J, Duan J, Zhang Z, Xue Z (2025) Parameter design and experiment of rotary plate pineapple fruit picker. Front Plant Sci 16:1575648. https://doi.org/10.3389/fpls.2025.1575648

Ramya S, Auxcilia J, Paital B, Sharmila DJS, Vethamoni PI, Venugopal S, Indra N, Subramanian KS, Sahoo DK (2025) Application of nanotechnology in fruit crops-from synthesis to sustainable packaging. PeerJ 13:e19603. https://doi.org/10.7717/peerj.19603

Rawat M, Neha, Upasana, Bisht C, Pant S, Naithani P, Singh A (2024) Biotechnology: approaches and methods in horticultural crops. Int J Adv Biochem Res 8(8):266–274

Shingade DM, Khatri R (2024) The role of biotechnology in the future of fruit crop production: a review. J Adv Biol Biotechnol 27(9):967–987. https://doi.org/10.9734/jabb/2024/v27i91368

Usendi P, Rani AP, Mahal GS, Solanki P, Kumar S, Shukla YS, Das J, Meena KK (2025) Precision farming in horticultural crops. Int J Adv Biochem Res 9(2):565–575

Vougioukas S, Karkee M, Devadoss S, Gallardo R, Charlton D (2025) Mechanization efforts in fruit harvesting. Available Online at https://www.choicesmagazine.org/choices-magazine/theme-articles/emerging-technologies-theme/mechanization-efforts-in-fruit-harvesting

Zhang M, Han Y, Li D, Xu S, Huang Y (2024a) Smart horticulture as an emerging interdisciplinary field combining novel solutions: past development, current challenges, and future perspectives. Hortic Plant J 10(6):1257–1273

Zhang K, Lammers K, Chu P, Li Z, Lu R (2024b) An automated apple harvesting robot—from system design to field evaluation. J Field Rob 41(7):2384–2400

Zhao X, Peng J, Zhang L, Yang X, Qiu Y, Cai C, Hu J, Huang T, Liang Y, Li Z, Tian M, Liu F, Wang Z (2024) Optimizing the quality of horticultural crop: insights into pre-harvest practices in controlled environment agriculture. Front Plant Sci 15:1427471. https://doi.org/10.3389/fpls.2024.1427471

Quality Horticultural Produce: Innovations and Applications

10

Solanki Bal and Amit Baran Sharangi

Abstract

The success of horticultural ventures depends heavily on the quality of planting material used. Ensuring that this material is healthy, true to type, and free from pathogens is essential for achieving consistent productivity and superior crop performance. Time-tested propagation methods such as cutting, grafting, and budding have long supported nursery operations by enabling the localized multiplication of desirable cultivars. When applied with scientific precision, these techniques enhance plant vigor, crop uniformity, and resilience under varying agro-climatic conditions. In parallel, biotechnological advancements have introduced new possibilities in the form of plant tissue culture, with micropropagation emerging as one of the most efficient techniques. It offers rapid, large-scale, and sterile production of high-value horticultural crops, particularly those that are difficult to propagate through conventional means. Crops like banana, turmeric, strawberry, and a variety of ornamentals have shown excellent responses to tissue culture, with the added advantages of uniformity, genetic stability, and year-round availability. The growing establishment of commercial tissue culture laboratories marks a significant development in the sector, enabling not only a consistent supply of quality plantlets but also contributing to entrepreneurship and employment generation. These labs meet the increasing demand for virus-free and climate-resilient planting material, both in domestic markets and for export. When integrated thoughtfully with traditional nursery practices, such biotechnological interventions pave the way for precision horticulture. Moreover,

S. Bal
School of Smart Agriculture, Adamas University, Kolkata, West Bengal, India

A. B. Sharangi (✉)
Department of Plantation, Spices, Medicinal and Aromatic Crops, Faculty of Horticulture, Bidhan Chandra Krishi Viswavidyalaya, Mohanpur, West Bengal, India
e-mail: sharangi.amitbaran@bckv.edu.in

A. Das, A. B. Sharangi (eds.), *Remunerative Horticulture*,
https://doi.org/10.1007/978-981-92-1007-7_10

policy interventions, public–private collaborations, and financial support from government schemes are making this transition accessible to a broader range of stakeholders, including young agripreneurs and rural enterprises. The strategic combination of conventional propagation and commercial tissue culture represents a robust and future-ready approach to planting material production. It strengthens the horticultural value chain, enhances farm profitability, and supports the sector's shift toward innovation-driven, sustainable growth.

Keywords

Quality planting material · Tissue culture propagation · Nursery business · Micropropagation · Horticultural innovation · Commercial tissue culture lab

10.1 Introduction

The performance of any horticultural system is deeply tied to the quality of the planting material that enters the production chain. Healthy, disease-free, and genetically stable plant stock forms the base of all successful horticultural ventures, whether in fruit orchards, vegetable fields, floriculture hubs, or protected cultivation units. Numerous studies have shown that superior planting material can significantly enhance yield, crop uniformity, and tolerance to environmental stresses, while poor-quality plants often limit the effectiveness of even the best crop management interventions (Hartmann et al. 2018; Lee and Preece 2020). As growers have become more aware of this relationship, the demand for reliable sources of robust, true-to-type planting material has steadily increased.

Conventional nursery practices—such as cutting, grafting, budding, layering, and stooling—have served as the backbone of planting material production for decades. When implemented with scientific rigor, these techniques ensure high varietal fidelity and allow large-scale multiplication of adapted cultivars suited to specific regions (Bose et al. 2021). With rising commercial interest, traditional nurseries have gradually transitioned into organized enterprises equipped with standardized propagation protocols, certification systems, and improved business models.

Alongside these developments, plant tissue culture has emerged as one of the most transformative tools in modern horticulture. Micropropagation, in particular, enables the rapid multiplication of elite genotypes under sterile and controlled conditions, producing uniform and pathogen-free plantlets—an advantage especially important for crops vulnerable to virus transmission, such as banana, strawberry, and several ornamentals (George and Debergh 2008; Singh et al. 2019). The establishment of commercial tissue culture laboratories, ranging from small-scale modular units to advanced industrial facilities, has opened new avenues for entrepreneurship and year-round production of planting material.

The current horticultural landscape increasingly reflects a blending of traditional propagation knowledge with contemporary biotechnological innovations. This integration supports precision horticulture by enhancing propagation efficiency, ensuring genetic integrity, and building cleaner production chains (Rai et al. 2022). As consumer demand for high-quality fruits, vegetables, and ornamentals grows, and as government programs continue to promote certified planting material, the sector is expanding both technologically and economically.

This chapter examines the innovations shaping planting material production, the evolution of nursery and tissue culture enterprises, and the broader implications for productivity, sustainability, and value chain strengthening in horticulture.

10.2 Role of Quality Planting Material in Horticulture

The success of any horticultural venture fundamentally depends on the quality of the planting material used. Whether propagated through seeds or vegetatively via cuttings, grafts, or tissue culture, the initial plant stock determines not only the biological performance but also the economic viability and market potential of the crops. High-quality planting material is typically defined by several critical features: genetic purity, true-to-type characteristics, vigorous growth, freedom from pests and diseases, and optimal physiological readiness for establishment in the field (Hartmann et al. 2018). In fruit orchards, uniformity in tree growth and vigor is particularly important for synchronized flowering, simplified management, and predictable yields, whereas in vegetables and ornamentals, uniform and healthy plantlets significantly affect both yield and marketable appearance.

Maintaining genetic fidelity ensures that desirable traits—such as fruit size, color, aroma, sweetness, shelf life, and flowering patterns—are consistently expressed across all plants. This is crucial in crops and ornamental species where standardization determines marketability, including mango, banana, citrus, strawberry, and high-value ornamentals (Bose et al. 2021). Even slight deviations or off-type plants can lead to uneven grading, variable postharvest performance, and reduced market value. In systems reliant on specific rootstock–scion combinations, such as apple, citrus, and rose, the compatibility and health of both components directly impact overall performance (Lee and Preece 2020).

Physiological health is equally critical. Well-rooted, pest-free, and pathogen-free plants establish more rapidly, grow vigorously in early stages, and are less susceptible to environmental stress. Virus-free tissue-cultured bananas, for example, demonstrate superior yields and more uniform bunch development compared to field-collected suckers (Singh et al. 2019). Similar benefits have been reported for micropropagated strawberry, gerbera, chrysanthemum, and turmeric, where uniform, disease-free plantlets shorten crop cycles and enhance overall productivity.

Economically, quality planting material enables precise nutrient and water management, mechanization, and synchronized harvesting, reducing labor requirements and postharvest losses (Rai et al. 2022). Healthy seedlings also minimize transplant shock, lower mortality rates, and improve resource efficiency in both floriculture and vegetable nurseries. Furthermore, certified or accredited planting material builds confidence among farmers, processors, and exporters, often commanding higher market prices.

Given the increasing emphasis on quality in horticultural markets and the challenges posed by climate variability, the demand for scientifically produced, disease-free, and uniform planting stock is rising. Modern horticulture relies heavily on organized nurseries, tissue culture facilities, and systematic propagation strategies, which together ensure both higher productivity and long-term sustainability of the horticultural value chain.

10.3 Conventional Propagation Techniques

Vegetative propagation through traditional methods has remained central to the production of planting material in horticulture for many decades. Practices such as cuttings, grafting, and budding have allowed the reliable multiplication of superior plant types while preserving genetic uniformity and desired varietal traits (Table 10.1). Even as advanced biotechnological approaches gain prominence, these conventional techniques continue to play a vital role because of their cost-effectiveness, operational flexibility, and broad applicability across fruit, vegetable, and ornamental crops. The choice of a particular method in nursery or field conditions is largely influenced by its specific strengths, technical requirements, and inherent constraints, all of which determine its practical usefulness in commercial propagation systems.

Table 10.1 Comparison of conventional propagation methods used in horticultural crops

Propagation method	Plant part used	Major crops	Key advantages	Limitations	References
Cutting	Stem, leaf, or root segments	Grape, pomegranate, fig, chrysanthemum, hibiscus	Simple technique, low cost, true-to-type plants, minimal infrastructure	Variable rooting in woody species, disease transmission from mother plants	Hartmann et al. (2018); Puri and Verma (2019)
Grafting	Scion along with rootstock	Mango, citrus, guava, apple, tomato (vegetable grafting)	Combines desirable traits, early bearing, stress, and disease tolerance	Requires skilled labor, compatibility issues	Garner (2013); Kumar and Sharma (2021)
Budding	Single bud	Citrus, rose, peach, litchi	Efficient use of scion material, rapid multiplication	Sensitive to season and stock condition	Nanda and Mishra (2018)
Layering	Attached shoots	Litchi, guava, pomegranate, apple rootstocks	High success rate, strong root system	Labor-intensive, slow multiplication	Leakey (2014); Prasad and Soni (2020)

10.3.1 Cutting

Propagation through cuttings is based on the ability of certain plant parts—most commonly stems, but also leaves or roots—to regenerate into complete plants under suitable conditions. This method is extensively practiced in a wide range of horticultural crops, including fruits, ornamentals, medicinal plants, and plantation species. The type of cutting used—softwood, semi-hardwood, or hardwood—depends on factors such as crop species, physiological maturity, and prevailing season. Crops such as grape, pomegranate, fig, and bougainvillea show high success rates with semi-hardwood or hardwood cuttings, whereas ornamental species like coleus, chrysanthemum, and hibiscus are more commonly propagated using softwood cuttings (Puri and Verma 2019).

One of the major strengths of this method is its ability to produce uniform, true-to-type plants that faithfully express the characteristics of the parent plant. Since the new plant develops through vegetative regeneration, genetic variability is minimized, resulting in consistent growth patterns, flowering behavior, and yield performance. The use of rooting regulators such as indole-3-butyric acid (IBA) and naphthalene acetic acid (NAA) has further improved the efficiency of this technique, particularly in species that exhibit weak or slow natural root formation (Das and Rout 2020). In addition, propagation by cuttings requires relatively simple infrastructure and involves low production costs, making it suitable for both small-scale nurseries and commercial operations.

Despite these advantages, propagation through cuttings is not without constraints. Certain woody perennial species exhibit poor rooting responses, which limit the success of the method. The rooting phase is also vulnerable to fungal and bacterial infections, especially under conditions of high humidity. In sensitive crops, the requirement for mist chambers or controlled environments may increase operational costs. Furthermore, repeated harvesting of cuttings from the same mother plants without adequate rejuvenation practices can lead to reduced vigor and declining propagation efficiency over time (Ghosh et al. 2018).

10.3.2 Grafting

Grafting is a vegetative propagation technique in which two living plant parts are joined so that they grow as a single, functional unit. In this method, the upper portion (scion) represents the desired cultivar, while the lower portion (rootstock) provides the root system. Grafting has been widely adopted in perennial fruit crops such as mango, citrus, guava, aonla, ber, and apple, where seed propagation fails to ensure uniformity and predictable performance. The selection of an appropriate rootstock is a decisive factor, as it governs not only plant vigor but also tolerance to biotic and abiotic stresses, nutrient uptake efficiency, and, in some cases, fruit size and quality (Kumar and Sharma 2021). In recent years, grafting has also gained importance in vegetable cultivation, particularly in crops like tomato, cucumber, and watermelon, where resistant rootstocks help mitigate problems associated with soil-borne diseases and adverse soil conditions.

One of the key strengths of grafting lies in its ability to combine desirable characteristics from two genetically distinct plants into a single individual. A vigorous or stress-tolerant rootstock can support a high-performing scion, resulting in improved plant establishment, enhanced field uniformity, and a longer productive life for orchards. In many fruit crops, grafted plants enter the reproductive phase earlier than seedlings, thereby reducing the unproductive juvenile period. Moreover, grafting has been shown to improve plant resilience against drought, salinity, nematodes, and serious diseases such as wilt and *Phytophthora* root rot, making it an important tool for sustainable horticulture under changing climatic conditions.

At the same time, grafting is not without constraints. The technique demands skilled labor, careful handling, and close attention to environmental factors such as temperature and humidity during union formation. Successful grafting also depends on physiological compatibility between the scion and rootstock; incompatible combinations may result in weak unions, stunted growth, or delayed decline after field establishment (Rathore et al. 2020). In addition, strict nursery sanitation is essential, as grafting can inadvertently facilitate the spread of systemic pathogens if infected plant material is used. These limitations underline the need for technical expertise and well-managed nursery conditions when grafting is practiced on a commercial scale.

10.3.3 Budding

Budding is a refined vegetative propagation technique in which a single, well-developed bud is used to establish a new plant instead of a full scion shoot. Several forms of budding are practiced in nurseries, including T-budding, patch budding, and chip budding, each selected according to crop species and seasonal conditions. This method is particularly common in citrus, rose, litchi, and many temperate stone fruits such as peach, plum, and apricot. Nursery operators favor budding because it generally gives consistent success, requires very little scion material, and allows rapid multiplication of elite varieties (Nanda and Mishra 2018).

One of the major strengths of budding lies in its efficiency. Since only a single bud is used, valuable mother plants can supply a large number of propagules without being overexploited. Budded plants usually form a strong and stable union with the rootstock and tend to establish quickly once planted in the field. For commercial nurseries, this makes budding especially useful when uniform planting material is required in large quantities or when scion wood availability is limited.

Despite these advantages, budding demands precision and careful timing. Successful union formation largely depends on the physiological condition of the stock plant, particularly the ease with which the bark separates—a stage commonly referred to as "bark slipping." Adverse weather conditions, drying of budwood, or lapses in nursery sanitation can reduce success rates. In addition, certain crop species exhibit incompatibility between buds and rootstocks, which restricts the range of usable combinations and requires careful selection during nursery planning.

10.3.4 Layering

Layering is among the oldest vegetative propagation practices used in horticulture and is based on the principle of inducing roots on a shoot while it is still attached to the parent plant. During this process, the selected branch continues to receive water, carbohydrates, and nutrients from the mother plant until rooting is complete. After a functional root system develops, the newly rooted shoot is separated and planted independently. Because the propagule is never cut off during root initiation, layering generally shows a higher rate of success than many other conventional methods, especially in species that are difficult to root through cuttings.

Different forms of layering are employed depending on crop type and growth habit. Simple and serpentine layering are commonly used in climbers and trailing plants, while trench and mound (stool) layering are preferred for rootstock production. Air layering is particularly popular in woody fruit crops such as litchi, guava, citrus, and pomegranate, as it allows roots to form directly on mature branches without disturbing the plant structure. In temperate fruit crops like apple and quince, mound layering is widely used for mass multiplication of clonal rootstocks, as it enables repeated harvesting of uniform shoots from established stool beds.

One of the major strengths of layering lies in its reliability. Since rooting occurs while the shoot remains physiologically supported by the parent plant, stress during the rooting phase is minimal. As a result, layered plants often develop strong and well-distributed root systems and show rapid establishment after transplanting. The method also ensures genetic uniformity, making it suitable for maintaining elite varieties and true-to-type material. Another advantage is its low requirement for infrastructure, which allows small nurseries and farmers to adopt the technique without sophisticated facilities.

Despite these advantages, layering has certain practical constraints. The technique is labor-intensive and usually produces a limited number of propagules per mother plant, making it less efficient for crops requiring rapid, large-scale multiplication. Root initiation can also be slow in woody perennials, sometimes extending over several months. For commercial nurseries, large-scale use of layering often demands specially managed stool beds, which require space, time, and skilled maintenance. Moreover, if mother plants are infected, there is a possibility of transmitting systemic diseases to the new plants.

Even with these limitations, layering continues to play a crucial role in horticultural propagation systems. Its high success rate, simplicity, and effectiveness in difficult-to-root species ensure its continued relevance. In many fruit trees and ornamental crops, layering remains a dependable option, complementing other conventional and modern propagation approaches in nursery production systems.

10.4 Nursery Business Development

The role of nurseries in horticulture has expanded considerably over time, moving far beyond their earlier identity as small units supplying planting material. Today, nursery enterprises function as organized agri-businesses that play a decisive role in shaping the growth, diversification, and sustainability of horticulture in India and elsewhere. The quality of seedlings and grafts produced by nurseries directly affects varietal performance, orchard longevity, and overall productivity. In recent years, increasing awareness among growers about the benefits of disease-free and uniform planting material has driven nurseries toward more scientific, standardized, and commercially oriented practices.

Modern nursery development rests on a combination of physical infrastructure, skilled manpower, regulatory oversight, and effective market integration. Among these, infrastructure is fundamental. Facilities such as polyhouses, shade-net structures, mist chambers, and hardening units allow nursery operators to regulate temperature, humidity, and light—factors that are critical during the early stages of plant growth. The adoption of technologies like automated misting, controlled fertigation, and raised propagation benches has been shown to improve graft success and seedling survival, particularly in perennial fruit crops such as mango, citrus, and guava. The widespread use of root trainers, protrays, and sterile growing media has further improved root development and early plant vigor, leading to better field establishment.

Equally important is the technical capacity of those managing nursery operations. Successful nurseries rely on a sound understanding of plant propagation principles, including stock–scion relationships, seasonal suitability of propagation methods, and the judicious use of growth regulators. Strict sanitation and hygiene practices—such as maintaining healthy mother blocks, sterilizing tools and media, and monitoring insect vectors—are now recognized as essential components of nursery management. These practices are particularly important for crops vulnerable to viral and systemic diseases, where the quality of planting material determines long-term crop performance.

In response to quality concerns, regulatory mechanisms governing nursery operations have become more robust in many horticulture-producing regions. Certification and accreditation systems introduced by state departments and research organizations emphasize traceability, authentication of mother plants, and adherence to phytosanitary standards. Nurseries that comply with these norms benefit from greater credibility in the market and improved access to institutional buyers, government-supported programs, and export-linked supply chains.

From a commercial perspective, nursery profitability increasingly depends on market-oriented strategies. Many nurseries have diversified their product range to include grafted vegetable seedlings, plug-grown ornamentals, micropropagated plants, and varieties selected for climate resilience or specific market demands. Strong linkages with farmer-producer organizations, commercial orchardists, urban horticulture initiatives, and agri-entrepreneurs have expanded the reach of nursery enterprises. In addition, nurseries that offer technical advisory services, customized planting solutions, and digital sales platforms have been found to achieve better customer loyalty and financial performance.

Overall, nursery business development today represents a convergence of plant science, quality assurance, and entrepreneurship. It is no longer limited to routine multiplication of plants but involves strategic planning, adoption of technology, compliance with standards, and responsiveness to changing market needs. As horticulture faces new challenges related to climate variability, resource efficiency, and quality-driven markets, professionally managed nurseries will continue to serve as a critical foundation for sustainable growth, crop diversification, and improved farmer incomes.

10.5 Plant Tissue Culture: A Biotechnological Approach

Plant tissue culture has emerged as a cornerstone of contemporary horticultural propagation, offering possibilities that extend far beyond the limits of conventional vegetative methods. By cultivating plant cells, tissues, or organs under sterile and carefully regulated laboratory conditions, this approach allows the large-scale production of planting material that is uniform, vigorous, and free from systemic pathogens. Unlike field-based propagation, tissue culture provides a controlled environment where growth and differentiation can be precisely guided through nutrient composition and growth regulators.

Initially confined to experimental laboratories, tissue culture has gradually transitioned into a commercially viable technology over the last few decades. Its adoption has been particularly significant in crops where the maintenance of elite genetic traits is critical and where traditional multiplication is slow, inefficient, or prone to disease transmission. Through techniques such as micropropagation, horticultural enterprises are now able to preserve genetic integrity while meeting the increasing demand for high-quality planting material. As a result, tissue culture has evolved from a purely scientific tool into a key contributor to productivity, consistency, and sustainability in modern horticultural systems.

10.5.1 Micropropagation Overview

Micropropagation is a specialized in vitro technique used for clonal multiplication of plants and is generally carried out through a sequence of distinct but interconnected stages—culture initiation, shoot proliferation, root development, and hardening. The efficiency of the process is strongly influenced by the physiological condition of the source tissue, the effectiveness of sterilization procedures, and the precise balance of growth regulators supplied through the culture medium, particularly auxins and cytokinins.

In the initial stage, small explants such as shoot tips or nodal segments are collected from healthy donor plants and carefully disinfected to eliminate surface contaminants (Table 10.2). These explants are then introduced onto a nutrient medium,

Table 10.2 Stages of micropropagation and their critical requirements

Stage	Major objective	Culture conditions	Growth regulators involved	Key challenges	References
Establishment	Initiation of aseptic cultures	MS-based medium, strict sterilization	Low cytokinin	Contamination, explant necrosis	Murashige and Skoog (1962); George and Debergh (2008)
Multiplication	Rapid shoot proliferation	Controlled light and temperature	Cytokinins (BAP, kinetin)	Hyperhydricity, shoot vitrification	Bairu et al. (2011)
Rooting	Adventitious root induction	Reduced salts, auxin-rich media	IBA, NAA	Weak rooting in some species	Hazarika (2003)
Acclimatization	Adaptation to ex vitro conditions	Mist chambers, shade nets	None (physiological hardening)	High mortality due to desiccation	Hazarika (2003)

most commonly formulations derived from the Murashige and Skoog medium, which provides essential macro- and micronutrients required for early growth. Once cultures are established, multiplication is stimulated by adjusting cytokinin concentrations to promote repeated shoot formation, allowing rapid expansion of identical plantlets within a short time frame.

Rooting is achieved by transferring regenerated shoots to media enriched with auxins, which encourage adventitious root development. The final and often most critical phase is acclimatization, during which plantlets are gradually exposed to external conditions. This step is essential because in vitro-grown plants develop under high humidity and low mechanical stress, making them vulnerable when moved directly to soil or greenhouse environments.

Micropropagation is especially advantageous for crops that show poor seed viability, high genetic variability, or slow multiplication through conventional methods. It also plays a crucial role in producing uniform, disease-free planting material throughout the year, making it a reliable propagation strategy for commercial horticulture and organized nursery systems.

10.5.2 Key Crops

Micropropagation has emerged as a cornerstone of modern horticulture, particularly for crops where rapid, large-scale production of uniform and disease-free planting material is essential. Its commercial adoption spans fruit crops, vegetables, ornamentals, spices, and plantation crops, providing growers with superior planting stock that is difficult or slow to propagate through conventional means.

Banana

Banana is one of the most extensively tissue-cultured crops worldwide. Its sterile nature and susceptibility to viral and fungal diseases make conventional propagation through suckers unreliable and often risky. Tissue culture allows the production of genetically uniform, pathogen-free plantlets, ensuring consistent growth, early flowering, and uniform fruiting. Large-scale plantation programs, both for domestic cultivation and export-oriented industries, rely on micropropagated banana plantlets to maintain yield stability and minimize losses caused by disease outbreaks (Vuylsteke 1989; Hailu et al. 2019).

Ornamentals

Ornamental crops such as gerbera, orchids, anthurium, and carnation benefit significantly from micropropagation. Maintaining uniform floral traits, including flower size, color, shape, and flowering time, is critical for meeting market expectations and securing premium prices. Tissue culture enables mass production of these true-to-type plants, overcoming the limitations of vegetative propagation, which may be slow or yield variable-quality plants (Kumar and Rao 2012). In addition, micropropagation supports off-season production and year-round availability of high-demand ornamental varieties.

Potato

In potato cultivation, micropropagated plantlets form the backbone of seed-tuber production pipelines. Conventional seed systems are often affected by viruses, resulting in low yields and inconsistent crop quality. Micropropagation provides virus-free stock, enhancing tuber yield, reducing crop losses, and ensuring uniform growth across fields (Struik and Wiersema 1999). These advantages are particularly valuable for commercial seed producers and large-scale farmers aiming to maintain high productivity standards.

Spices and Plantation Crops

Tissue culture has proven highly useful in spices and plantation crops such as vanilla, cardamom, coffee, and tea. In these crops, micropropagation facilitates the rejuvenation of elite lines, rapid multiplication of high-yielding clones, and management of systemic pathogens that would otherwise reduce productivity. For instance, tissue-cultured vanilla plants provide uniform vines for commercial plantations, while disease-free cardamom and coffee plantlets ensure healthy stands for sustainable production.

High-Value Fruits

High-value fruit crops such as pineapple, grapes, and pomegranate also rely on tissue culture for rapid multiplication of superior cultivars. The method enables growers to quickly establish large orchards of uniform trees or vines, ensuring consistent fruit quality, synchronized harvests, and improved market competitiveness. Additionally, tissue-cultured plants of these crops exhibit enhanced vigor, better adaptability to local climatic conditions, and reduced susceptibility to viral infections compared to conventional propagules (Hamad and Taha 2008).

Overall, micropropagation serves as a powerful tool to meet commercial horticulture demands, offering pathogen-free, genetically uniform, and high-performing planting material. Its application across diverse crops supports large-scale cultivation, ensures consistent quality, and contributes significantly to economic returns in both domestic and international markets.

10.5.3 Quality Assurance

Ensuring quality in tissue-cultured plants is crucial for maintaining genetic uniformity, physiological vigor, and freedom from pathogens. High-quality micropropagated plants are only useful if they reliably retain the desired traits of the parent cultivar and survive well under field conditions. Quality assurance in commercial tissue culture involves multiple interrelated components:

(i) *Monitoring Genetic Stability*

Prolonged in vitro culture, especially during repeated subculturing or callus-based propagation, can lead to somaclonal variations, which may alter important traits such as growth habit, yield potential, or disease resistance. To detect such changes early, molecular markers like random amplified polymorphic DNA (RAPD), simple sequence repeats (SSR), and amplified fragment length polymorphism (AFLP) are employed. Regular genetic screening ensures that the propagated plants remain true-to-type before they are released for large-scale planting (Bairu et al. 2011; Bal et al. 2025).

(ii) *Pathogen Indexing*

Maintaining pathogen-free plants is a cornerstone of tissue culture quality. Infections by viruses, bacteria, or fungi can compromise plant growth and lead to significant economic losses. Techniques such as meristem culture combined with thermotherapy are routinely used to produce virus-free plantlets. Additionally, periodic testing through serological and molecular diagnostic tools helps detect latent infections, ensuring only clean material progresses through the propagation pipeline (Cassells and Doyle 2003).

(iii) *Media Quality and Aseptic Management*

The sterility of culture media, consistency of nutrient composition, and maintenance of aseptic conditions in growth rooms are critical to avoid contamination. Even minor lapses can result in widespread loss of cultures, as microbial contamination is one of the leading causes of failure in commercial tissue culture operations. Continuous monitoring of growth-room parameters—temperature, humidity, light intensity—and routine surface sterilization of explants help maintain high survival rates.

(iv) *Acclimatization Protocols*

The transition from in vitro to ex vitro conditions is often the most sensitive stage in tissue culture. Plants raised in highly controlled, high-humidity environments are vulnerable to desiccation and shock when transferred to greenhouses or open nurseries. Implementing structured hardening procedures—including mist chambers, shade nets, gradual humidity reduction, and careful temperature regulation—significantly improves survival and vigor during this critical phase (Hazarika 2003).

(v) *Certification and Standardization*

Certification schemes provide a formal mechanism to assure buyers of plant quality. Countries like India have introduced programs such as the National Certification System for Tissue Culture Raised Plants (NCS-TCP) under the Department of Biotechnology, which sets guidelines for clonal fidelity, pathogen indexing, and traceability. Certified plants gain market trust and often command premium prices due to verified quality standards (DBT 2020).

10.6 Commercial Tissue Culture Laboratories

Commercial tissue culture laboratories have transformed the production of high-value horticultural crops by enabling year-round propagation of uniform, disease-free plantlets on a large scale. Unlike conventional nurseries, these laboratories operate under tightly controlled conditions, regulating temperature, humidity, light intensity, and nutrient supply to optimize plant growth and rooting. The controlled environment allows precise manipulation of growth regulators, improved multiplication rates, and early detection of physiological or microbial stress.

Modern commercial units range from modular small-scale labs to fully automated industrial facilities. They are equipped with advanced infrastructure, such as autoclaves for media sterilization, laminar airflow cabinets for aseptic operations, misting systems for acclimatization, and dedicated hardening rooms to ensure high survival after transfer to soil. This level of control not only enhances productivity but also supports entrepreneurship, enabling growers and nurseries to supply consistent, high-quality planting material for fruit crops, ornamentals, vegetables, and spices throughout the year (Kumar et al. 2020a, b).

Commercial tissue culture laboratories are structured to enable the large-scale production of uniform, disease-free planting material in an efficient and controlled manner. These facilities are typically organized into distinct functional areas, including explant preparation, initiation, multiplication, rooting, hardening, and acclimatization. Each stage is closely monitored to prevent microbial contamination, preserve genetic integrity, and ensure the physiological health of the plantlets throughout the propagation cycle (Ravishankar and Bansal 2019a, b).

Modern laboratories often incorporate advanced equipment such as laminar airflow cabinets, controlled-environment growth chambers, and photoperiod-regulated incubators. These technologies allow for optimized growth conditions, enhanced multiplication rates, and reduced manual labor, making the production process both precise and cost-effective (Sharma et al. 2018a, b).

Economically, commercial tissue culture laboratories provide substantial benefits. The demand for uniform, virus-free plantlets in high-value crops—such as banana, strawberry, turmeric, and ornamental plants—remains strong in both domestic and international markets. Year-round production capacity allows entrepreneurs to supply material independent of seasonal constraints while offering opportunities to expand into export-oriented markets (Singh and Gupta 2021). Additionally, these facilities support rural livelihoods by generating employment, promoting skill development, and encouraging entrepreneurship in the horticultural sector (Patel et al. 2020).

Maintaining quality is central to the success of tissue culture operations. Regular monitoring of genetic stability, somaclonal variation, and pathogen presence ensures that only healthy and true-to-type plantlets are distributed to nurseries and farmers. Compliance with regulatory frameworks, including the Plant Quarantine Act and certification guidelines provided by agricultural authorities, further strengthens market trust and facilitates acceptance both nationally and internationally (Rani and Kumar 2019a, b).

In essence, commercial tissue culture laboratories serve as a critical link between traditional propagation methods and modern biotechnological approaches. By offering a scalable, reliable, and economically viable means of producing high-quality planting material, these laboratories not only enhance horticultural productivity but also support sustainable agricultural development and the growth of innovative agribusiness ventures.

10.7 Integration of Traditional and Modern Techniques

Combining age-old propagation methods with contemporary biotechnological approaches has proven to be an effective strategy for producing high-quality, uniform planting material in horticulture. Instead of completely replacing conventional techniques like cuttings, grafting, budding, and layering, modern interventions such as micropropagation, marker-assisted selection, and climate-controlled nursery systems enhance them by improving speed, uniformity, and disease management (Nair and Thomas 2020a, b). This hybrid strategy allows growers to select the most suitable propagation method based on the crop species, season, and market requirements.

Conventional propagation continues to hold importance, especially for crops where vegetative methods reliably maintain varietal fidelity, including guava, mango, and citrus. These methods remain cost-effective and practical for small and medium-sized nurseries, particularly in regions where access to advanced infrastructure is limited. However, traditional methods face inherent constraints, such as seasonal dependence, inconsistent rooting success, and susceptibility to soil-borne diseases (Chattopadhyay et al. 2019a, b). Introducing tissue culture-derived plantlets into these systems addresses many of these challenges, providing genetically uniform and disease-free stock that enhances the overall performance of propagated material.

For example, tissue-cultured rootstocks can serve as standardized, pathogen-free bases for grafting, effectively combining the uniformity of micropropagation with the field performance advantages of conventional grafting. Similarly, layering or budding operations can be reinforced with molecular diagnostics to verify that parent plants are genetically stable and free of latent pathogens before propagation begins (Verma and Iqbal 2021). This integration minimizes losses, increases reliability, and ensures that commercial orchards receive planting material of consistent quality.

The use of modern nursery infrastructure—such as polyhouses, mist chambers, insect-proof net houses, and fertigation systems—further strengthens traditional propagation methods. These controlled environments create optimal conditions for rooting, hardening, and seedling establishment, leading to higher survival rates and more uniform growth (Shukla et al. 2022a, b). As a result, nurseries can operate year-round and respond more efficiently to fluctuations in market demand.

Economically, a combined propagation approach reduces production risks and expands the variety of plant types that nurseries can supply. Farmers benefit from obtaining robust, uniform planting material capable of performing consistently across diverse agro-climatic regions (Pradhan and Jena 2021). By integrating conventional and modern propagation systems, horticultural enterprises can achieve sustainability, balancing technological innovation with economic viability and ecological considerations (Table 10.3).

Taken together, the synergy of traditional and biotechnological propagation techniques offers a resilient and forward-looking model for horticultural development. It preserves the proven strengths of conventional methods while leveraging the

Table 10.3 Applications and advantages of tissue culture in major horticultural crops

Crop group	Purpose of micropropagation	Benefits observed	Commercial relevance	References
Banana	Mass multiplication, virus elimination	Uniform growth, early flowering, higher yield	Large-scale plantations, export	Vuylsteke (1989); Singh et al. (2019)
Ornamentals (gerbera, orchids)	Trait uniformity	Consistent flower quality, year-round supply	Floriculture industry	Kumar and Rao (2012)
Potato	Seed-tuber production	Virus-free stock, yield stability	Seed certification programs	Struik and Wiersema (1999)
Spices (vanilla, cardamom)	Elite clone multiplication	Uniform plantations, disease reduction	High-value niche crops	Rajeevan (2016)
High-value fruits	Rapid orchard establishment	Genetic fidelity, synchronized harvest	Commercial orchards	Hamad and Taha (2008)

precision and efficiency of modern technologies to meet contemporary challenges in productivity, quality, and market competitiveness.

10.8 Government Policies and Institutional Support

Government initiatives and institutional programs have been pivotal in strengthening horticultural production systems in India, particularly in areas such as nursery development, quality planting material, and commercial-scale tissue culture. In recent decades, policy emphasis has shifted toward ensuring the availability of certified, disease-free, and genetically uniform planting stock, acknowledging that high-quality propagules form the backbone of successful horticultural ventures (Dasgupta and Roy 2018a, b).

Among the most significant programs is the Mission for Integrated Development of Horticulture (MIDH), which provides targeted financial assistance for establishing nurseries, tissue culture laboratories, mother blocks, and protected cultivation units. The scheme typically subsidizes 40–55% of infrastructure costs, enabling small and medium-scale entrepreneurs to enter the propagation sector without bearing the full burden of initial capital investments (MoAandFW 2021). Complementing national schemes, state horticulture departments provide region-specific incentives for high-value crops, minor fruits, and spices suited to local agro-climatic conditions, further stimulating nursery-based entrepreneurship (Prasad and Shankar 2020).

Research and development institutions have also played a critical role in supporting horticultural enterprises. Organizations such as ICAR, the National Horticulture Board (NHB), and the Department of Biotechnology (DBT) offer certification protocols, training programs, and regulatory guidelines to ensure the quality and reliability of propagated material. For example, the NHB's nursery accreditation system sets strict standards for mother plant management, disease testing, and propagation hygiene, enhancing farmer confidence in commercially available seedlings (NHB 2022). Similarly, DBT initiatives, particularly the National Certification System for Tissue Culture Plants (NCS-TCP), have strengthened biotech-driven propagation businesses, especially in high-demand crops like banana, pomegranate, and ornamental plants (Bhatia et al. 2019a, b).

State governments have further facilitated horticultural entrepreneurship by implementing digital tools, including online licensing platforms, traceability systems, and real-time monitoring of nurseries. Such measures prevent duplication, ensure authenticity in plant material trade, and promote transparency across the supply chain. In addition, hands-on training and incubation programs offered through agricultural universities and Krishi Vigyan Kendras (KVKs) equip aspiring entrepreneurs with skills in nursery management, plant health diagnostics, and protected cultivation techniques (Kumawat et al. 2021).

In essence, the combination of financial incentives, regulatory oversight, institutional training, and technological support has created an enabling environment for horticultural entrepreneurs. This synergy not only encourages the adoption of modern tissue culture and traditional propagation methods but also ensures sustainable, market-oriented growth and the reliable availability of high-quality planting material across India.

10.9 Future Prospects and Challenges

The production of planting material in horticulture is gradually shifting toward a more integrated and technology-driven framework, with sustainability at its core. As consumer expectations rise for high-quality fruits, vegetables, and ornamentals, nurseries and tissue culture facilities are increasingly required to supply planting stock that is uniform, genetically stable, and resilient to environmental fluctuations. The sector is also beginning to embrace advanced tools such as automated micropropagation systems, digital monitoring of growth conditions, and artificial intelligence (AI)-assisted quality evaluation, all of which promise to enhance efficiency, precision, and overall production capacity.

One of the most promising developments lies in the diversification of high-value crops propagated through a combination of traditional and biotechnological approaches. Climate-resilient varieties, stress-tolerant rootstocks, and nutritionally enriched cultivars are likely to gain importance as growers focus on crops capable of performing reliably under unpredictable environmental conditions. The expansion of protected cultivation systems, vertical farming initiatives, and peri-urban horticulture will further increase demand for compact, early-bearing, and pest-free planting material, offering nurseries new avenues for innovation and growth.

However, the sector faces several challenges that need careful management. Establishing advanced tissue culture laboratories requires substantial initial investment and skilled personnel, which can be a barrier for small and medium-scale enterprises. Market volatility, limited access to high-quality mother plants, and stringent regulatory requirements for the movement of planting material can further restrict the timely supply of newly developed varieties.

Ensuring genetic fidelity remains another key challenge, particularly for crops susceptible to somaclonal variation during in vitro propagation. Additionally, farmers often find it difficult to differentiate certified, high-quality planting material from unverified or inferior stock in informal markets. Climate change compounds these challenges, as rising temperatures, erratic rainfall patterns, and emerging pests necessitate constant adjustments in propagation strategies.

Looking forward, the long-term sustainability and growth of the planting material sector will rely on close collaboration among researchers, policymakers, nursery operators, and farmers. Strengthening certification systems, improving access to modern technologies, and enhancing skill development programs can help bridge existing gaps. By addressing these challenges proactively, the planting material industry has the potential to play a transformative role in shaping the future of horticultural production, ensuring higher productivity, reliability, and market responsiveness.

10.10 Conclusion

The production of planting material in horticulture is currently experiencing a profound transformation, driven by the integration of time-tested propagation methods with modern biotechnological innovations. Traditional techniques—such as cutting, grafting, budding, and layering—continue to form the foundation of propagation for many crops. These approaches remain widely used because they are cost-effective, adaptable to diverse conditions, and familiar to growers who have relied on them for generations.

In parallel, plant tissue culture has emerged as a powerful tool that complements conventional practices. By providing uniform, disease-free, and genetically stable plantlets, tissue culture allows for rapid multiplication at a scale that traditional nurseries alone cannot sustain. The rise of commercial tissue culture laboratories has facilitated year-round production, promoted horticultural entrepreneurship, and met the growing demand for high-quality planting material in both domestic and international markets. When combined thoughtfully with conventional propagation, micropropagation strengthens the entire horticultural value chain by ensuring reliable, diverse, and superior planting material.

Support from government programs, financial incentives, and research institutions further reinforces this evolving sector, creating opportunities for skill development, innovation, and investment. Nevertheless, the long-term success of horticultural propagation depends on how effectively these systems respond to emerging challenges such as climate variability, fluctuating market demands, and the implementation of stricter quality standards.

Overall, the sector is moving toward sustainable, technology-driven, and economically viable production systems. By integrating traditional propagation wisdom with modern biotechnological tools, growers, researchers, and entrepreneurs can ensure the consistent availability of high-quality horticultural crops. This balanced approach positions the industry to meet the demands of a rapidly changing agricultural landscape while maintaining productivity, quality, and resilience.

References

Bairu MW, Aremu AO, Van Staden J (2011) Somaclonal variation in plants: causes and detection methods. Plant Growth Regul 63(2):147–173. https://doi.org/10.1007/s10725-010-9554-x

Bal S, Sharangi AB, Saeed M, Alkhathami A, Siddiqui S, Alabdallah NM (2025) Advances in fenugreek breeding: novel genetic and omic approaches. Front Plant Sci 16:1674013. https://doi.org/10.3389/fpls.2025.1674013

Bhatia R, Singh KP, Jhang T, Sharma TR (2019a) National Certification System for tissue culture raised plants (NCS-TCP): ensuring quality planting material in India. Indian J Biotechnol 18(2):123–130

Bhatia S, Singh A, Yadav R (2019b) Regulatory advancements in micropropagated planting material in India. Biotechnol Policy Rev 7(2):112–124

Bose TK, Mitra SK, Sadhu MK (2021) Propagation of horticultural crops. Naya Udyog

Cassells AC, Doyle BM (2003) Pathogen indexing of tissue culture plants. In: Cassells DVDW (ed) Pathogen and microbial contamination management in micropropagation. Springer, pp 31–61. https://doi.org/10.1007/978-94-017-0143-6

Chattopadhyay T, Banerjee S, Roy A (2019a) Comparative evaluation of vegetative propagation techniques in tropical fruit crops. Indian J Hortic 76(4):723–731

Chattopadhyay TK, Singh S, Kumar R (2019b) Nursery management and vegetative propagation in fruit crops. Indian J Hortic 76(3):345–356

Das P, Rout S (2020) Rooting behaviour of horticultural cuttings under different hormonal treatments. Int J Plant Sci 15(2):145–152

Dasgupta M, Roy T (2018a) Policy interventions in Indian horticulture: a historical overview. Indian J Agricu Policy 23(1):41–53

Dasgupta S, Roy A (2018b) Role of quality planting material in horticultural development. J Hortic Sci 13(1):1–10

Department of Biotechnology (DBT) (2020) National Certification System for tissue culture raised plants (NCS-TCP): operational guidelines. Government of India

Garner RJ (2013) The grafter's handbook, 6th edn. Royal Horticultural Society

George EF, Debergh PC (2008) Micropropagation: uses and methods. In: Plant propagation by tissue culture. Springer

Ghosh S, Roy A, Barik D (2018) Management of mother plants for vegetative propagation in fruit crops. Indian Hortic J 75(3):230–236

Hailu G, Tadesse A, Mohammed A (2019) Banana micropropagation and its role in improving productivity. Afr J Biotechnol 18(4):77–86

Hamad AM, Taha RM (2008) Micropropagation of fruit crops: a review. Biotechnol Adv 26(6):558–569. https://doi.org/10.1016/j.biotechadv.2008.07.001

Hartmann HT, Kester DE, Davies FT, Geneve R (2018) Plant propagation: principles and practices, 9th edn. Prentice Hall

Hazarika BN (2003) Acclimatization of tissue-cultured plants. Curr Sci 85(12):1704–1712

Kumar K, Rao IU (2012) Micropropagation of ornamental plants: a review. Plant Cell Biotechnol Mol Biol 13(1–2):1–16

Kumar R, Sharma P (2021) Influence of rootstocks on growth and productivity of fruit trees. J Hortic Sci 16(1):25–39

Kumar V, Singh R (2020a) Emerging trends in nursery industry and planting material trade. Agric Rev 41(3):278–286

Kumar V, Sharma R, Meena RK (2020a) Commercial plant tissue culture: scope and constraints. Indian J Agric Sci 90(6):1057–1063
Kumar S, Mehta R, Sharma P (2020b) Advances in commercial plant tissue culture. J Hortic Sci 15(2):101–115
Kumawat V, Sharma K, Tiwari P (2021) Capacity-building programs for horticultural entrepreneurship. KVK Ext Bull 12(4):5–15
Leakey RRB (2014) Tropical trees: propagation and domestication. CABI
Lee N, Preece JE (2020) Production of horticultural crops from quality planting material. Hortic Rev 47:1–45
Ministry of Agriculture and Farmers Welfare (2021) Annual report on MIDH achievements. Government of India
Murashige T, Skoog F (1962) A revised medium for rapid growth and bioassays with tobacco tissue cultures. Physiol Plant 15(3):473–497. https://doi.org/10.1111/j.1399-3054.1962.tb08052.x
Nair P, Thomas G (2020a) Bridging traditional horticultural practices with biotechnological innovations. J Agric Biotechnol 9(2):145–158
Nair R, Thomas P (2020b) Integration of conventional and biotechnological approaches in horticultural propagation. J Appl Hortic 22(2):85–92
Nanda R, Mishra S (2018) Budding techniques in fruit crops: a practical overview. Adv Hortic Res 10(1):45–58
National Horticulture Board (2022) Accreditation guidelines for horticultural nurseries. Government of India
Patel D, Kumar N, Yadav S (2020) Entrepreneurship opportunities through commercial plant propagation. Asian J Hortic 15(1):12–22
Pradhan L, Jena D (2021) Economic impact of integrated nursery systems in horticulture. Hortic Rural Develop Rev 12(1):61–72
Prasad R, Shankar D (2020) State-level horticulture subsidies and their impact on crop diversification. J Rural Econ 45(3):77–89
Prasad B, Soni K (2020) Air layering techniques for tropical and subtropical fruit crops. J Fruit Sci 5(1):33–41
Puri S, Verma M (2019) Propagation of woody perennials through stem cuttings: a review. Asian J Hortic 14(2):89–96
Rai M, Kumar R, Singh A (2022) Integrating biotechnology with nursery management in horticulture. Adv Hortic Sci 36(2):155–168
Rani A, Kumar A (2019a) Genetic fidelity and pathogen screening in tissue culture plantlets. Plant Biotechnol J 17(2):87–98
Rani V, Kumar S (2019b) Quality control in commercial micropropagation systems. J Plant Biotechnol 46(3):213–221
Rathore DS, Meena RP, Chauhan S (2020) Graft compatibility and performance in fruit crops. Hortic Biotechnol Rep 2(1):11–21
Ravishankar GA, Bansal R (2019a) Micropropagation and mass production of horticultural crops. Plant Biotechnol Rep 13(1):45–60
Ravishankar H, Bansal YK (2019b) Infrastructure and management of commercial tissue culture laboratories. Biotechnol Today 9(1):45–52
Sharma A, Verma M, Singh P (2018a) Automation in plant tissue culture labs. Int J Plant Sci 22(4):201–212
Sharma S, Gupta N, Verma R (2018b) Automation and cost efficiency in plant tissue culture laboratories. Asian J Biotechnol 10(3):129–138
Shukla AK, Singh VK, Dwivedi BS (2022a) Climate-controlled nursery systems for sustainable horticulture. J Environ Biol 43(2):289–297
Shukla R, Mohanty A, Singh K (2022b) Protected structures for improving plant propagation efficiency. Int J Prot Agric 4(3):89–102

Singh V, Gupta R (2021) Economic feasibility of commercial tissue culture units in India. Hortic Econ Rev 8(3):55–68

Singh B, Sharma S, Thakur R (2019) Micropropagation approaches for high-value horticultural crops. J Appl Hortic 21(3):180–190

Struik PC, Wiersema SG (1999) Seed potato technology. Wageningen Pers

Verma S, Iqbal M (2021) Combining molecular diagnostics with vegetative propagation for quality assurance in horticulture. Plant Health Adv 7(1):33–47

Vuylsteke D (1989) Shoot-tip culture for the propagation, conservation and exchange of Musa germplasm. International Board for Plant Genetic Resources

Qualitative Postharvest Horticulture: Commodity Transformation for Value Addition and Quality Enhancement

11

Payal Maiti, Souvik Chakraborty, Guddu Kumar, Rajdeep Mohanta, Subhadwip Ghorai, and Shah Faisal

Abstract

Postharvest management of horticultural produce involves beyond traditional storage and transportation, involving innovative ways to enhance the quality, shelf life, and value of fruits, vegetables, and flowers. By altering the commodity's initial form, a new potential can be unlocked by extending the freshness, preserving nutrients, and producing entirely new products. Techniques such as drying and dehydration help in long-term storage while intensifying flavors. Floral potpourri offers aesthetic and aromatic appeal, turning waste into beauty. Converting vegetables into powders can simplify cooking and boost convenience for health-conscious consumers. Juices, essential oils, natural pigments, and edible dyes can preserve the essence of horticultural produce while opening a path toward health, cosmetic, and food industries. Fermentation techniques, whether alcoholic, acetic, or lactic, transform raw horticultural produce into value-added products that are not only rich in flavor but also filled with probiotics, enhancing both shelf life and nutritional benefits. Microencapsulation plays a vital role in preserving sensitive compounds such as flavors, colors, and nutrients, ensuring their stability throughout the process of postharvest handling.

P. Maiti (✉) · S. Chakraborty
Department of Post Harvest Technology, Bidhan Chandra Krishi Viswavidyalaya, Haringhata, West Bengal, India

G. Kumar
Assistant Technical Manager (ATMA), Department of Agriculture, Gopalganj, Bihar, India

R. Mohanta · S. Ghorai
Department of Agriculture, Brainware University, Kolkata, West Bengal, India

S. Faisal
Institute of Food Science and Technology, Chinese Academy of Agriculture Science, Beijing, China

A. Das, A. B. Sharangi (eds.), *Remunerative Horticulture*,
https://doi.org/10.1007/978-981-92-1007-7_11

Additionally, the use of natural fibers and applications in aromatherapy highlight the potential of nonedible parts of horticultural commodities, promoting sustainability and broadening the scope of their utilization beyond traditional consumption. This approach to postharvest handling not only reduces waste but also transforms perishable commodities into economically valuable and functionally rich products. It represents a promising avenue for farmers, processors, and entrepreneurs to tap into a growing market for health-conscious, sustainable, and innovative horticultural goods.

Keywords

Value addition · Dehydration · Fermentation · Microencapsulation · Sustainability

11.1 Introduction

A balanced diet is vital for health, with fruits and vegetables providing essential vitamins like niacin, riboflavin, thiamine, ascorbic acid, retinol, and tocopherol, as well as minerals such as calcium, phosphorus, iron, and magnesium, while also supporting gut health (Yahia et al. 2019). Postharvest management is crucial to maintain the quality, freshness, and nutritional value of horticultural crops. After harvesting, crops face both quantitative losses, which reduce the amount of food available, and qualitative losses, which impact nutritional content, safety, and marketability. These losses are particularly severe in developing countries due to inadequate infrastructure, whereas wealthier stakeholders mitigate such issues through better logistics (Yahia et al. 2019). Causes of postharvest losses include enzymatic activity leading to browning, physical damage from temperature and handling, and microbial spoilage (Yeshiwas and Tadele 2021). Because fruits and vegetables continue to respire after harvest, they lose nutrients and spoil over time, with varying shelf lives requiring tailored storage conditions. Effective postharvest practices focus on controlling humidity and temperature, using proper packaging, and applying antimicrobial treatments to slow deterioration and extend shelf life (Pokhrel 2021). Techniques include physical methods like washing, blanching, and UV irradiation, as well as chemical treatments involving preservatives, antibacterial agents, and antioxidants to inhibit spoilage. Additionally, advancements such as remote sensing for disaster mapping and crop monitoring, along with optical sensing technologies for assessing ripeness and damage, help reduce losses and improve productivity (Zhang et al. 2024).

11.1.1 Limitations of Traditional Methods

While increasing food production can address growing demand, a more sustainable solution lies in enhancing food distribution efficiency and minimizing losses throughout key stages such as harvesting, processing, storage, transportation, and retail. This approach not only improves food accessibility but also helps conserve essential natural resources like land and water. Postharvest, substantial losses can occur at multiple points, reducing the quantity of food that ultimately reaches consumers (Nanda et al. 2012).

11.1.2 Harvesting

The timing of harvest plays a crucial role in determining the quality and shelf life of fruits and vegetables (Palumbo et al. 2022). Harvesting too early, often driven by economic motives, results in immature produce with inferior taste, texture, and increased susceptibility to damage. Conversely, harvesting at full ripeness reduces shelf life and increases the risk of spoilage. Both under-ripe and overripe produce are more prone to physiological issues, which can lower their nutritional and market value, sometimes leading to total batch losses. Additionally, losses are worsened by poor agricultural practices, incorrect harvesting techniques, or failure to perform steps like pre-cooling, especially in crops such as cherries, potentially causing losses between 4% and 12% (Ozcan 2020).

11.1.3 Storage Facility

Storage technologies play a vital role in enhancing marketing, boosting product value, and meeting quarantine standards, supporting the growth and global reach of the horticulture sector. Temperature control is central to most storage systems, with additional measures like humidity regulation, better storage conditions, suitable packaging, and chemical treatments further improving effectiveness, particularly when combined with cold storage solutions (Watkins and Ekman 2004).

11.1.4 Transit

Transporting goods from producers to consumers is essential, with direct delivery often being the most efficient. However, transport costs significantly affect final prices, especially in complex supply chains serving both cities and remote areas—sometimes even exceeding production costs (Staver and Movileanu 2012). In developing countries, poor infrastructure, weak logistics, and unskilled handling during loading and unloading contribute to spoilage and physical damage of perishable food (Kaur and Watson 2024).

11.1.5 Inadequate Packaging

To minimize losses and extend the shelf life of fresh fruits and vegetables, appropriate packaging is crucial. Poor packaging methods and the use of unsuitable materials are major contributors to postharvest losses. When low-quality packaging fails to adequately protect produce, it can lead to faster spoilage and physical damage. Unfortunately, the use of cheap, low-grade packaging materials remains a widespread practice, particularly in developing and underdeveloped countries (Kitinoja and Al Hassan 2010).

11.2 Importance of Qualitative Transformation of Horticultural Produce

Horticultural crops like fruits, vegetables, flowers, and spices play a key role in food security, nutrition, and rural livelihoods (Hanif et al. 2024). However, their perishable nature leads to significant postharvest losses, especially in developing countries. To address this, qualitative transformation focuses on improving the physical, nutritional, and market value of produce.

This involves postharvest techniques such as sorting, grading, packaging, refrigeration, chemical treatments, and value-added processing to enhance shelf life, appearance, safety, and consumer appeal (Palumbo et al. 2022). These methods help preserve quality and increase product value, enabling producers to reduce losses and access premium markets.

Value addition, like making juices, dried fruits, or ready-to-eat items, also extends shelf life and offers income opportunities, especially for small-scale farmers and agro-entrepreneurs (Tanwir et al. 2025).

Furthermore, by adopting quality standards and certifications like Global GAP or HACCP, producers can meet export requirements and expand into international markets, improving competitiveness and earning potential (Lengai et al. 2022).

11.3 Concept of Altering the Initial Form

Contamination significantly reduces shelf life and crop value. According to the FAO, such losses account for about 20% of global agricultural productivity (Sellitto et al. 2021). In developing countries, 10–15% of grain legumes and cereals and 30–40% of fruits and vegetables are lost annually, with some areas in Africa and Latin America experiencing up to 50% loss. Food waste in these regions increased from 39% to 55% between 2000 and 2019. These losses are higher than in developed countries, where advanced technologies are used for storage and transport (FAO 2024). Postharvest management also aims to increase crop value from harvest to consumption. This will ultimately benefit the whole situation, such as increasing farmer profits or expanding the supply of fresh vegetables. Postharvest management occurs immediately following the harvest because fresh product begins to degrade the time it is removed from the plant, and the processing after harvest has a significant impact on the ultimate quality, whether marketed as a fresh crop or a processed product (Lalpekhlua et al. 2024).

Changing the initial form of food through postharvest methods involves applying science and technology to transform, preserve, and enhance agricultural produce after harvest for consumption. These processes turn raw agricultural products into edible, marketable, and value-added items, bridging the gap between farm output and consumer use. Techniques like freezing, canning, drying, and pasteurization help control microbial growth, enzymatic reactions, and oxidation, thereby ensuring food safety and reducing spoilage. Additionally, these methods increase the economic value of raw materials by producing high-value products such as cheese from milk or jam from fruits. Furthermore, postharvest processes can improve nutrition through food fortification, addressing deficiencies in essential nutrients. Overall, these strategies contribute to a safe, reliable, and economically sustainable food supply.

11.4 Drying and Dehydration

Drying is a traditional and highly effective physical method of food preservation, widely used for both direct consumption and further processing. It ensures year-round availability of food and pharmaceuticals by removing moisture through heat and mass transfer (Amit et al. 2017). Fresh fruits and vegetables, being highly perishable due to their high moisture content, often undergo physical or chemical pre-treatments before thermal drying to enhance efficiency, reduce energy use, and maintain quality. Drying reduces product weight and volume, cutting packaging, storage, and transport costs, while also modifying texture and flavor to create healthier snack options (Sehrawat et al. 2018). While traditionally sun-powered, drying has become more advanced, using complex equipment and optimized parameters. Innovative drying methods not only preserve food but also enhance the quality of items like herbs, spices, medicinal plants, and bioactive enzymes, contributing to value-added products.

Dehydration or drying is used to preserve food and minimize storage and transportation costs by eliminating water, thus decreasing water activity (aw). This prevents microbial and enzymatic spoilage, since most bacteria need high aw to grow normally over 0.95, with no possibility of microbial growth below 0.3. Foods are divided into high moisture (aw >0.8), intermediate moisture (0.8 > aw >0.3), and low moisture (aw <0.3) depending on aw, as presented in Fig. 11.1.

As can be seen in Fig. 11.1, enzyme-catalyzed reactions like browning, flavor loss, and breakdown of polysaccharides are reduced by lower aw. Nonenzymatic browning is also reduced. Lipid oxidation, especially in fatty foods, falls first but rises considerably below aw 0.2. Therefore, over drying may not always be desirable (Kutz 2019).

Drying methods are broadly divided into three, namely, convective, conductive, and radiative. These can either be done using natural or artificial drying methods.

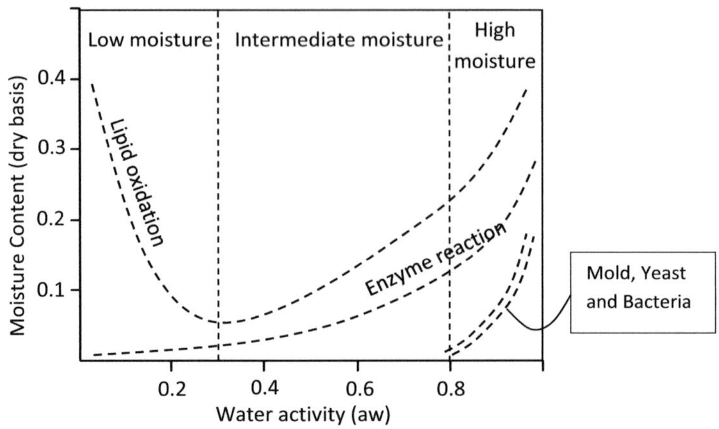

Fig. 11.1 Chart illustrating the relationship between water activity (aw) and moisture content (dry basis). The x-axis represents water activity, ranging from 0 to 1, while the y-axis shows moisture content, ranging from 0 to 0.4. Three regions are marked: low moisture, intermediate moisture, and high moisture. Two curves are depicted: one labeled "Lipid Oxidation" peaking in the low moisture region, and another labeled "Enzyme Reaction" rising in the intermediate to high moisture regions. A note indicates the presence of mold, yeast, and bacteria in the high moisture area

11.4.1 Natural Drying

Sun drying is a traditional method that removes moisture from food using solar radiation. In this technique, food products are exposed to direct sunlight for several days before packaging. It is commonly used in developing countries where fuel is scarce and costly. This method is widely adopted due to its low cost and simplicity. However, sun drying has several limitations, including long drying durations, reliance on unpredictable weather, lack of control over drying conditions, risk of insect infestation, and contamination from dust and dirt (Adeyeye et al. 2022).

11.4.2 Artificial Drying

Artificial drying is gaining popularity over natural drying as it removes moisture more quickly and efficiently, resulting in higher-quality dried products. It allows for better control over key drying parameters such as temperature, airflow, and drying time. Various artificial drying methods are available, depending on the type of food and desired outcome. The different types of artificial drying methods include:

11.4.2.1 Convectional Hot Air Drying

Convectional hot air drying (also known as convective dehydration) is a widely used and cost-effective food preservation method. It utilizes tunnel dryers or closed chambers with controlled airflow and temperature, making it more reliable than traditional sun drying. It is particularly effective for fruits like apricots and mangoes. However, this method can negatively impact the nutritional and physical properties of foods. For instance, dried raspberries showed significant reductions in anthocyanin (44–60%), flavonoids (3–25%), total phenols (32–40%), and L-ascorbic acid compared to fresh or freeze-dried samples (Stamenković et al. 2019).

11.4.2.2 Osmotic Drying

Osmotic drying is a pre-treatment method where biological tissues are immersed in a concentrated solution to reduce their moisture content. Due to the high osmotic pressure, water is drawn out from the tissue, though minimal nutrient and solute exchange may occur. This technique helps lower moisture before further drying processes like convective drying. It also enhances energy efficiency, improves texture and color, and increases the sugar-to-acid ratio of the final product (Radojčin et al. 2015).

11.4.2.3 Microwave Drying

Microwave (MW) drying uses electromagnetic waves (300 MHz–300 GHz) to heat materials by causing molecular oscillation, converting kinetic energy into heat (Laguerre and Hamoud-Agha 2020). It offers fast, energy-efficient, and volumetric heating, though achieving uniform heat distribution can be challenging due to product size, shape, and composition. MW drying has been effective for foods like potato chips and pasta, as internal pressure pushes moisture to the surface, speeding up drying (Morais et al. 2018). However, it is often combined with methods like hot air drying, as MW alone may be insufficient. This combination significantly reduces drying time; for instance, banana slices dried with both methods required 64% less time (Radojčin et al. 2021).

11.4.2.4 Freeze Drying

This method, which uses sublimation to extract frozen water, is valued for its ability to preserve significant amounts of delicate biological and nutritional substances at low temperatures (Nowak and Jakubczyk 2020). Although it is excellent at creating high-quality food, its primary disadvantages are lengthy processing periods and significant energy consumption (Lachowicz et al. 2019).

11.4.2.5 Vacuum Drying

This method of drying removes moisture by lowering the pressure within a sealed sub-atmospheric drying chamber, which reduces the boiling point of water. This allows for drying at lower temperatures, preserving the product's quality, color, and nutritional content, particularly for heat-sensitive materials (Thamkaew et al. 2021). Low pressure in vacuum drying increases the vapor pressure difference between the product and its surroundings, which speeds up the drying process. Despite its effectiveness, it is more costly than conventional methods like hot air drying due to the need for specialized equipment (Thamkaew et al. 2021). Combining vacuum and microwave drying has recently attracted interest for its synergistic benefits, especially in drying fruits and vegetables (Orphanides et al. 2016).

11.4.2.6 Other Methods

Numerous other notable drying methods have now been developed using sophisticated principles and technologies having various degrees of compatibility with different types of food products, e.g., fluidized bed drying, infrared drying, refractance window drying, low-pressure superheated steam drying, ultrasound drying, dielectric drying, and electro-hydrodynamic drying (Adeyeye et al. 2022).

11.5 Potpourri Production

11.5.1 Use of Ornamental Flowers and Aromatic Plants

The durability and freshness of flowers play a significant role in various cultural and social events, including ceremonies related to birth, marriage, and death. Potpourri represents a significant sector within the dried flower industry, offering flower growers an opportunity to make use of excess or unused floral harvests by converting them into value-added products. It typically consists of a blend of naturally fragrant dried plant materials, commonly used in bowls or sachets for decorative purposes, adding color and fragrance to homes and offices. The use of potpourri dates back to the mid-eighteenth century (Cook et al. 2015). The word "potpourri" comes from the French term "potpourri," which originally referred to a Spanish stew. In English, it describes a mixture of various items, often arranged in decorative wooden bowls or fabric sachets. Potpourri is primarily made from plant materials that contain natural aromatic oils (Cook et al. 2015). It is commonly used to freshen the air with long-lasting fragrance, scent drawers and closets, enhance room dećor, or serve as a personal gift. Typical containers include glass or wooden bowls, baskets, shells, tins, canning jars, or specially designed potpourri vases. Besides their decorative and fragrant use, potpourris also help repel moths and safeguard stored clothing. Various plant parts are used in making potpourri, including dried fruits or slices, empty or whole fruits such as legumes, pods, capsules, winged fruits, and follicles, along with fruit valves and seeds. Other components include leaves, leaflets, leafy stems, stem sections, roots, bracts, bark, wooden shavings, skeletal leaves, woven leafless stems from climbers, flower buds, flowers, inflorescences, and cones. Floral elements like lavender, jasmine, and rose promote relaxation, while camphor provides a cooling sensation. Spices such as cinnamon and vanilla offer warmth, woods and bark enhance the scent and add volume, and fruits contribute to the visual appeal.

11.5.2 Methods of Preparation and Packaging

According to Koley (2020), there are two main types of potpourris: dry and wet. Wet potpourri is prepared using fresh aromatic plant materials that are stored in a sealed container (crock) for several weeks, allowing them to ferment. This method results in a blend that holds its fragrance for a longer period. In contrast, dry potpourri is made from dried fragrant materials and is the more commonly used method due to its simplicity and faster preparation, although it tends to lose its scent more quickly.

Both methods require a fixative, which helps absorb and slowly release the essential oils, thereby extending the fragrance life. Fixatives are added to reduce the evaporation rate of volatile plant scents and ensure the aroma lingers over time. Without a fixative, the fragrance would dissipate rapidly. Common fixatives include orris root, calamus root, angelica root, frankincense, myrrh, benzoin gum, balsam of Peru, oak moss, and vanilla beans. Multiple fixatives can be used together in one potpourri blend, and they are available in various forms: liquid, dried, or powdered (Mebakerlin and Chakravorty 2015).

11.5.3 Market Potential and Aesthetic Applications

India's dried flower industry, worth Rs. 100 crores, is globally recognized for offering natural, eco-friendly, durable, and affordable decorative products. The country accounts for about 5% of global dried flower exports, with the sector growing at a rate of 15% annually. In 2020, Rimports USA LLC held the position of the second-largest potpourri and sachet manufacturer, with sales reaching 82.32 million USD, while brands under Procter & Gamble recorded sales of about 35.49 million USD in the same category. Although potpourri enjoys a strong presence in the floriculture industry, there is a lack of research on the types of plant materials used, their combinations, and the dyes applied for coloring. In response to this gap, the current study focused on developing standardized methods for producing potpourri by drying and coloring rose petals. The use of edible dyes was explored to introduce a wide range of colors, enhancing the visual appeal of potpourri and offering farmers an opportunity to increase their income through value-added floral products (Safeena and Thangam 2023).

11.6 Vegetable and Fruit Powders

The demand for fruit and vegetable powders is growing as they extend the short shelf life of fresh produce. These powders are easier to store, transport, and use as ingredients due to their low water content and activity, which enhances shelf life. Compared to liquids, fruit juice powders offer advantages like longer shelf life, easier handling, reduced volume and weight, and lower packaging needs. Their stability and ease of use make them popular in food and pharmaceutical applications for flavoring and coloring. The production process includes pre-treatment, drying, and grinding (Kucharska-Guzik et al. 2025).

Pre-treatments are essential before drying fruits and vegetables to preserve their nutritional value, color, and functional components (Ignaczak et al. 2023). These techniques improve the final powder's quality by enhancing flavor, color, and processing efficiency, while also addressing challenges like long drying times and degradation of heat-sensitive compounds (Márquez-Cardozo et al. 2021). Pulsed electric field (PEF) treatment, for example, accelerates drying and reduces processing time (Ostermeier et al. 2020). Blanching, involving brief exposure to hot water or steam, inactivates enzymes like polyphenol oxidase and peroxidase that cause browning and nutrient loss, helping preserve quality during drying and storage, particularly in enzyme-sensitive produce like apples and bananas (Ciurzyńska et al. 2021). In pomegranate arils, blanching at 90 °C for 30 s and 100 °C for 60 s reduced enzyme activity by 76% and 68%, respectively (Adetoro et al. 2021).

Various techniques such as vacuum, microwave, convection, sun, freeze, and spray drying are used based on the product's characteristics like size, structure, and nutrient content (Yao et al. 2023). Freeze drying is particularly effective in preserving nutrients, color, texture, and flavor (Guo et al. 2023). Spray drying is a key method for converting liquid fruits and vegetables into stable powders by atomizing them into fine droplets in a heated chamber. It minimizes nutrient loss due to rapid drying and often uses carriers like gum arabic or maltodextrin to improve quality and reduce stickiness (Lisiecka et al. 2024). However, it only processes liquid feeds, requiring prior juicing or extraction. Optimizing variables such as temperature, feed rate, and atomizer type is crucial for efficient spray drying.

The grinding or size-reduction stage after drying is essential for achieving high-quality fruit and vegetable powders. This process transforms brittle dried materials into fine powders, directly influencing solubility, texture, and flavor release (Ueda et al. 2023). Different milling methods, such as wet and dry milling, are used depending on the desired particle size and production scale. These techniques are illustrated in Figs. 11.2 and 11.3.

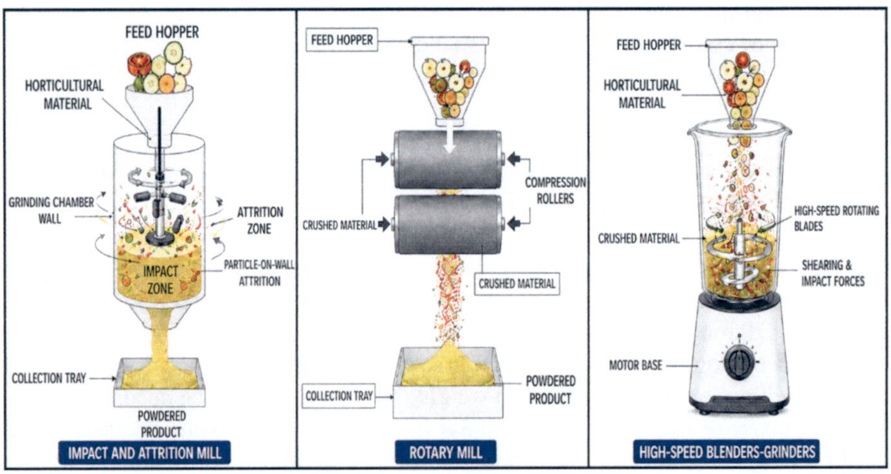

Fig. 11.2 Diagram illustrating three types of milling equipment: 1. Impact and attrition mill: features a feed hopper, rotating disks, and a collection tray. Materials are ground through impact and friction. 2. Rotary mill: includes a feed hopper, compression rollers, and a collection tray. Materials are crushed between rotating rollers. 3. High-speed blenders–grinders: shows a feed hopper, high-speed rotating blades, and a motor base. Materials are finely ground by rapid blade movement. Each subfigure is labeled and shows the flow of materials through the equipment

Fig. 11.3 Flow chart illustrating a process for converting horticultural material into fine food powder. The sequence begins with a hydration tank where material is soaked. It moves to a feed hopper, then a wet grinding mill, producing a slurry or paste. This is followed by a filtering unit separating fibrous pulp from filtrate. The filtrate is dried in a drying chamber, and finally, a sieving unit produces fine food powder. Each step is visually represented with labeled containers and arrows indicating the flow direction

(i) *Impact and Attrition Mills:* Hammer mills and pin mills are commonly utilized for large-scale production. While pin mills use high-speed revolving pins to pulverize the material, hammer mills use repetitive impact and crushing to reduce particle size (Ämmälä 2023). These techniques work especially well for turning a variety of dried materials into fine powders.

(ii) *Compression Mills:* The dried produce is crushed and ground between two or more rollers in a roller mill. In addition to being widely utilized in flour milling, this technique is also employed to produce powders with a particular distribution of particle sizes (Ueda et al. 2023).

(iii) *High-Speed Blenders and Grinders: On a lower scale, specialty spice grinders and high-capacity blenders are frequently utilized. For grinding dry goods, including those with a high oil content, into extremely fine powders without separating the oil and ruining the texture, high-speed grinders, which can achieve speeds of up to 7000 RPM, are especially useful (Jung et al.* 2018*).*

(iv) *Wet Grinding:* A common method for making food powder is wet grinding, which consists of the following five steps: hydration (sometimes called soaking), adding more water during grinding, filtering, drying, and sieving (Fig. 11.3). In order to allow the food ingredients to soften, wet grinding is often done after full hydration. In most cases, the soaking curve's moisture content finds equilibrium. Rice powder had the maximum whiteness and the lowest damaged starch content, which is beneficial for the final product's quality (Jung et al. 2018).

The choice of grinding method significantly affects the powder's physical properties and its performance in various uses. Therefore, careful control of the grinding process is vital to ensure a high-quality final product.

11.7 Juices, Essential Oils, and Extracts

A large part of postharvest processing of horticultural products for value-added foods, cosmetics, and pharmaceuticals involves producing juices, essential oils, and plant extracts, valued for their medicinal, aromatic, and nutritional benefits. Throughout the process, from raw material to final product, carefully controlled extraction and preservation methods are applied. Fruit and vegetable juices are increasingly popular as daily healthy and nutritious beverages (Ma et al. 2020). Juices are widely available and commonly consumed worldwide. Besides providing a natural blend of water, dietary fibers, sugars, and acids, they are rich sources of micronutrients such as vitamins and minerals (Lan et al. 2021). Additionally, they contain bioactive phytochemicals including polyphenols, flavonoids, carotenoids, and other organic compounds (Mapelli-Brahm et al. 2017).

11.7.1 Juice Extraction

Juice extraction begins with the proper preprocessing and handling of fruits and vegetables. Ensuring optimal ripening and storage is key to preserving flavor and aroma. After storage, the produce is cleaned, sorted, and inspected to remove foreign materials and defective fruit, preventing contamination. Effective crushing and juicing are crucial steps to maximize juice yield and minimize waste.

11.7.1.1 Cold Pressing

Cold pressing uses high pressure to extract juice from fruits and vegetables without heat or high-speed blades, unlike traditional centrifugal juicing (Khaksar et al. 2019). This method better preserves heat-sensitive nutrients like vitamins, minerals, and enzymes by minimizing heat and air exposure. Studies show cold-pressed juices retain more bioactive compounds and antioxidants than centrifugal juices. However, they have a shorter shelf life and require refrigeration to maintain quality since they are not thermally pasteurized (Khaksar et al. 2019).

11.7.1.2 Hot Pressing

This juice extraction method heats crushed fruit or vegetable pulp between 60 °C and 95 °C before pressing to deactivate enzymes like polyphenol oxidases that cause browning and reduce juice quality. Heating softens the cell walls, improving juice yield, and acts as a form of pasteurization, reducing microbial load and extending shelf life (Ahmad et al. 2021). However, high temperatures can lead to a significant loss of heat-sensitive nutrients, like vitamin C, and impart a "cooked" flavor to the juice, altering its natural taste profile.

11.7.1.3 Pulping/Beating

This method produces a thicker juice or puree rather than a clear liquid (Mboowa 2024). The fruit is typically boiled to soften it and deactivate enzymes. The pulp is then separated from seeds, stones, and peels using a pulper or finisher with revolving paddles and a fine screen, resulting in a smooth, thick puree (Reyes-De-Corcuera et al. 2014). It is commonly used for fruits like strawberries, peaches, and mangoes.

11.7.1.4 Centrifugation

The juice and pulp are separated by centrifugal force. Fruit shreds are frequently spun at a high speed in a horizontal spiral centrifuge. The lighter liquid exits down a different channel, while the heavier sediments are tossed against the drum's exterior wall and moved out by a screw. When a naturally colored, unoxidized juice is sought, this quick approach can be an excellent substitute for pressing (Ugwu et al. 2020).

11.7.1.5 Enzyme-Enhanced Extraction

This method involves adding cellulase and pectinase to the crushed fruit pulp prior to pressing. The fruits and vegetables' cell walls and other structural elements are broken down by these enzymes, which facilitates the juice's release. The end product's clarity and juice output can both be greatly enhanced by this technique (Streimikyte et al. 2022).

11.7.2 Essential Oil Extraction

Essential oils are concentrated plant extracts valued for their flavor, fragrance, and medicinal properties. These volatile aromatic liquids are obtained from various plant parts such as flowers, leaves, roots, bark, fruits, and seeds. Chemically, they are composed of different terpenoid fractions, ranging from highly volatile to heavier compounds, forming a homogeneous mixture (Machado et al. 2019).

11.7.2.1 Steam Distillation

It is the most widely used technique for producing essential oils for commercial use. Steam distillation extracts volatile compounds by passing steam through plant material, then cooling and separating the oil from water. Although simple and low cost, it involves long extraction times and risks damaging heat-sensitive compounds (Machado et al. 2024). To preserve delicate aromas and chemical profiles, alternative methods like cold pressing used for citrus peel oils are now preferred (Machado et al. 2022).

11.7.2.2 Cold Pressing

Cold pressing is commonly used to extract essential oils from citrus fruits like grapefruits, oranges, and lemons by mechanically pressing the rind to release oil from the sacs. A centrifuge is often used to separate the oil from juice and solids. Cold pressing avoids heat and chemical treatments, helping retain bioactive compounds and producing safe, high-quality edible oils (Wandhekar et al. 2023).

11.7.2.3 Solvent Extraction

Solvent extraction is used for delicate plant materials like tuberose, rose, or jasmine that are sensitive to high heat. An organic solvent, such as hexane, dissolves the plant's aromatic compounds, forming a waxy substance called "concrete." This is then washed with alcohol to separate the oils and waxes, resulting in a highly concentrated "absolute" (Souiy 2023).

11.7.2.4 Effleurage

Enfleurage is a traditional, labor-intensive method used for extracting essential oils from delicate flowers. Flower petals are placed on odorless fat spread over glass plates, allowing the fat to gradually absorb their fragrance. This process is repeated with fresh petals until the fat becomes saturated, forming a fragrant "pomade." The pomade is then treated with alcohol to extract the essential oil (Souiy et al. 2022).

11.7.2.5 Supercritical Fluid Extraction

This approach is a more costly, sophisticated, and modern method of essential oil extraction. In this procedure, carbon dioxide is used as a solvent in a "supercritical" state, which is between a liquid and a gas. The approach is mild and accurate, yielding powerful and pure oil with a greater variety of chemicals than oils made using other techniques. There is no solvent residue in the finished product since the CO_2 is readily transformed back into a gas (Souiy 2023).

Plant extracts are increasingly used in the food and cosmetics industries for their natural and functional benefits (Choi et al. 2024). In cosmetics, research is growing on their role as natural antioxidants. In the food sector, they are added to improve shelf life and enhance sensory attributes such as color and flavor, also serving as nutritional additives for humans and animals.

11.8 Natural Pigments, Edible Colors, and Dyes

11.8.1 The Shift from Synthetic to Natural Colorants

Color plays a crucial role in shaping consumer perception, especially in the food and textile sectors, as it often signals freshness, safety, and quality. The invention of the first synthetic dye, mauvine, in 1856 marked a turning point, leading to the widespread use of petroleum-based colorants due to their affordability, stability, and vibrant hues (Cortez et al. 2017). However, concerns over the health risks of synthetic dyes, such as potential toxicity and allergenic or carcinogenic effects, have sparked a return to natural colorants. Today, consumer demand for clean-label products is driving the adoption of natural pigments, which not only offer safe coloring options but also provide added health benefits like antioxidant and anti-inflammatory effects (Singh et al. 2023).

11.8.2 Sources from Fruits, Vegetables, and Flowers

Nature offers a wide range of colorful pigments, mainly produced by plants. These natural compounds give fruits, vegetables, and flowers their vivid hues and are classified into major chemical groups, each with unique characteristics and origins. Table 11.1 is a detailed list of the main plant pigment classes and their natural sources, typical colors, and defining chemical properties.

Table 11.1 Major natural pigments, their colors, sources, properties and citations

Pigment group	Colors	Sources	Properties	References
Anthocyanins (E 163)	Red, purple, and blue	Blueberries, blackberries, cherries, strawberries, grapes; vegetables like red cabbage, black carrots, radishes	Water-soluble; color changes with pH (red in acidic, blue/purple in neutral/alkaline); used as pH indicator; potent antioxidant; health benefits include vision improvement and anti-cancer activity	Shetty et al. (2017)
Carotenoids	Yellow, orange, and red	Carrots (β-carotene), tomatoes (lycopene), saffron (crocin), paprika, annatto seeds, marigolds (lutein)	Lipid-soluble; >600 types; precursors to vitamin A (β-carotene); strong antioxidants; protect against photo-oxidative damage	Koop et al. (2022)
Betalains	Red-violet and yellow	Red beet (*Beta vulgaris*), cactus fruits (prickly pear), amaranths	Water-soluble, nitrogen-containing; distinct from anthocyanins; stable at pH 4–6; heat-sensitive (limited to minimally processed foods like yogurt, ice cream, candies); antiviral & antimicrobial	Molina et al. (2023)
Chlorophylls (E 140)	Green	Spinach, nettle, alfalfa	Essential for photosynthesis; tetrapyrrole with Mg center; unstable to heat and acid → degrades to olive-brown pheophytin; limited food applications	Rymbai et al. (2011)
Curcumin	Yellow	Turmeric	Principal pigment of turmeric; vibrant yellow; widely used in foods; notable for anti-inflammatory properties	Rymbai et al. (2011)
Quinones and tannins	Brownish-red/astringent hues	Alizarin (madder root), tannins (oak galls)	Historically important as mordant dyes for textiles; not common as modern food pigments	Rymbai et al. (2011)

11.8.3 Extraction and Stabilization Methods

Turning raw plant material into usable colorants involves two key steps: extraction and stabilization, both of which influence the pigment's quality, yield, and market potential.

11.8.3.1 Extraction Techniques

The primary aim is to efficiently isolate pigments from plants using a variety of traditional and modern methods:

- *Conventional Methods:* Traditional techniques like maceration, hydro distillation, and Soxhlet extraction are widely used (Ngamwonglumlert et al. 2017). However, they are often slow, use large amounts of harmful solvents, and may degrade heat-sensitive pigments.
- *Ultrasound-Assisted Extraction (UAE):* Uses high-frequency sound waves to form cavitation bubbles that disrupt plant cells, boosting solvent penetration and pigment release. It offers faster extraction, less solvent use, and higher yields (Salauddin et al. 2021).
- *Microwave-Assisted Extraction (MAE):* Applies microwave energy to heat plant moisture, causing pressure that breaks cell walls and releases pigments efficiently and quickly (Salauddin et al. 2021).
- *Enzyme-Assisted Extraction (EAE):* Involves using enzymes like cellulases and pectinases to break down cell walls gently, allowing for selective pigment release (Ngamwonglumlert et al. 2017).

11.8.3.2 Stabilization Methods

Natural pigments are more prone to degradation than synthetic ones, being affected by light, heat, oxygen, and pH changes. Therefore, stabilizing them is essential for effective use.

- *Encapsulation:* According to Masyita et al. (2025), microencapsulation involves enclosing pigment particles with protective biopolymer coatings like starch or maltodextrin. This shields them from environmental damage, enhances stability, and allows controlled release. Spray drying and freeze drying are common methods used.
- *Co-pigmentation:* Especially useful for anthocyanin, this method uses colorless compounds such as phenolic or metal ions to form complexes with the pigment. This strengthens the color and increases its resistance to degradation (Cortez et al. 2017).
- *Chemical and Environmental Control:* Simple methods like optimizing pH and limiting oxygen exposure (e.g., vacuum packaging or nitrogen flushing) can also protect pigment stability (Molina et al. 2023).

11.8.4 Applications in Food Coloring and Textiles

Natural pigments are gaining popularity in various industries, especially food and textiles.

11.8.4.1 Direct Food Application

Pigments like beta-carotene, annatto, and betalains are used in products such as beverages, dairy items, and sweets. Anthocyanins are often found in drinks, jams, and candies (Shetty et al. 2017).

11.8.4.2 Dual Functionality

Singh et al. (2023) highlight that many natural pigments also offer health benefits like antioxidant, antimicrobial, and anti-inflammatory effects. This dual role enhances both the appearance and nutritional value of food.

11.8.4.3 Sustainable Dyeing

Pigments from madder, weld, indigo, and fruit/vegetable waste are used to dye natural fabrics like cotton and silk (Andriamanantena et al. 2019).

11.8.4.4 Mordanting

This process helps fix dyes to fabrics. Modern approaches use eco-friendly "biomordants" from sources like oak galls instead of harmful heavy metals.

11.8.4.5 Functional Textiles

As noted in Advancements in Sustainable Natural Dyes for Textile Applications, some dyes provide extra features like antimicrobial or UV protection, making them suitable for medical or outdoor clothing.

11.9 Fermentation-Based Products

Fermentation is a natural process where microorganisms break down complex organic compounds into simpler forms. This microbial and enzymatic activity alters food properties by converting proteins into peptides and amino acids, and sugars into alcohol, using enzymes like those from yeast. Key factors influencing fermentation include the type of microorganisms used and the composition of the substrates. Common fermenting microbes include Lactic Acid Bacteria (e.g., *Lactobacillus, Pediococcus*), yeasts (e.g., *Saccharomyces*), and molds (e.g., *Penicillium, Rhizopus*) (Chen 2021).

These foods enhance both nutritional and physiological functions, aiding in diabetes management, improving digestion, reducing allergies, and lowering the risk of cancer and cardiovascular diseases. Fermentation boosts nutritional value by breaking down complex compounds, increasing digestibility, and releasing bound nutrients like cellulose and hemicelluloses. It also reduces anti-nutrients such as tannins and phytic acid, enhancing the bioavailability of proteins, iron, and vitamins. For example, microbes in *tape ketan* produce thiamine (vitamin B1), and those in *tempe* fermentation increase vitamin B12, riboflavin, and niacin essential in vegetarian diets (Chadare et al. 2019). *Pulque*, a traditional Mexican drink, is rich in B vitamins and popular among low-income children for its nutritional value (Escalante et al. 2016).

11.9.1 Classification of Fermented Horticultural Produces

They may be classified by food types like dairy or beverages, by the fermenting microorganisms and final products, or by raw commodities such as cereals. Another approach groups fermentations according to the type of acid or alcohol produced.

11.9.1.1 Lactic Acid Fermentation

Lactic acid fermentation involves microbes like *Leuconostoc mesenteroides*, *Lactobacillus brevis*, and *Lactobacillus plantarum*, which convert fermentable carbohydrates into lactic acid. This process is key in preserving and processing foods such as pickles, cheeses, and sour milk. It also improves the flavor, aroma, and texture of the food (Ogrodowczyk and Drabińska 2021). For instance, a number of lactic acid bacteria in sauerkraut reduce the pH to less than 4.0, prolonging the shelf life of the cabbage.

11.9.1.2 Alcoholic Fermentations

These generate ethanol and are regarded as safe for ingestion. Wines, beers, and drinks like Mexican pulque are a few examples. Yeasts and occasionally bacteria or molds that resemble yeast are involved in these fermentations. Fermentable sugars are transformed into carbon dioxide and ethanol during the process, and the carbon dioxide creates an anaerobic environment that aids in product preservation (Zhao et al. 2025).

11.9.1.3 Acetic Acid/Vinegar Fermentation

Acetobacter bacteria turn ethanol into acetic acid when alcoholic fermentation by-products are exposed to air. The resultant acetic acid is a highly prized condiment and preservative with bacteriostatic or bactericidal properties (Román-Camacho et al. 2023).

11.9.1.4 Alkaline Fermentations

Properly conducted, these fermentations are safe and yield nutritious foods and sauces that provide low-cost protein, such as Japanese natto and Nigerian dawadawa. *Bacillus* species are the main microbes involved, breaking down proteins and releasing ammonia, which raises the pH to 8.0 or above. This high pH and elevated temperature inhibit the growth of spoilage bacteria (Owusu-Kwarteng et al. 2022).

11.9.1.5 High Salt Savory Sauces and Pastes

These are made by fermenting foods high in protein that include a lot of sodium chloride, which stops foodborne diseases and putrefaction. This method creates pastas and sauces with meaty flavors, which are staples in many diets, particularly in Asian nations (Zheng et al. 2024). Chinese soy sauce, Japanese miso, and different fish pastas and sauces are a few examples.

Fermented foods are generally safe, even when made in less sanitary conditions, because fermentation inhibits harmful microbes. For example, in tempeh, a mold lowers the pH to 5.0 or below, starting acid fermentation. *Rhizopus oligosporus* thrives in this acidic, low-oxygen environment, producing CO_2 and antimicrobial substances that prevent contamination. It grows quickly at 40–42 °C, temperatures that many other microbes cannot tolerate (Barus and Giovania 2020).

Although mycotoxins may be an issue in cereal grains and legumes, fermentation can also detoxify food substrates because it ideally does not produce any mycotoxins (Ozolina et al. 2023). It greatly lowers the remaining levels of pesticides in food products and aids in the removal of possible poisons such as phytates, hemagglutinins, and trypsin inhibitors (Armenova et al. 2023). For example, it was discovered that infants with acute diarrhea who were fed a tempeh-based formula recovered and gained weight more quickly.

11.10 Microencapsulation

11.10.1 Definition and Importance

Microencapsulation is an advanced technique where small particles or droplets are enclosed within a protective coating, forming tiny capsules. The inner content, known as the core material, can be solid, liquid, or gas, while the outer layer, or wall material, is typically made from polymers, hydrocolloids, or lipids. These capsules usually range in size from less than one micron to several hundred microns.

This method plays a vital role in postharvest handling and adding value to horticultural products. Many valuable compounds, like essential oils, natural pigments, flavors, and bioactive, are highly unstable and degrade easily when exposed to light, oxygen, heat, or moisture. Microencapsulation helps safeguard these sensitive ingredients, as noted by Calderón-Oliver and Ponce-Alquicira (2022).

11.10.2 Key Advantages of Microencapsulation

- *Protection and Stability*: It shields sensitive components such as vitamins and antioxidants from environmental damage, thereby prolonging their shelf life.
- *Physical Form Conversion*: Liquid ingredients like oils can be turned into solid powders, improving handling and ease of use in dry formulations (Desai and Jin Park 2005).

- *Taste and Odor Masking*: Strong or unpleasant flavors, such as those from fish oil or minerals, can be masked by the capsule wall to enhance product palatability (Calderón-Oliver and Ponce-Alquicira 2022).
- *Isolation of Reactive Ingredients*: In complex formulations, it prevents unwanted interactions by keeping reactive substances apart until use (Choudhury et al. 2021).

In general, microencapsulation is a useful technique that stabilizes and makes the fragile horticultural extracts usable for food, nutraceutical, and cosmetic applications. The different methods used for microencapsulation in foods and horticultural commodities are presented in Table 11.2. In addition, microencapsulation's use in preventing volatiles, flavors, and nutrients from being lost is explained in detail in Tables 11.2 and 11.3.

Table 11.2 Multiple drying techniques used for microencapsulation of food materials, with columns showing the advantages and limitations along with the references

Technique	Process steps	Key advantages	Limitations	References
Spray drying	1. Prepare emulsion/ dispersion of core (e.g., flavor oil) in wall material (maltodextrin, gum arabic) 2. Atomize liquid into fine droplets in a hot air stream 3. Drying rapid water evaporation leaves solid wall material around the core	Widely used, cost-effective Produces stable powders with good handling Suitable for heat-sensitive compounds due to short exposure time	May not form seamless coating → some core exposed on surface.	Mardani et al. (2024)
Freeze drying (lyophilization)	1. Freeze emulsion/ suspension 2. Primary drying: Ice sublimates under vacuum 3. Secondary drying: Remove unfrozen water molecules	Premium method for highly sensitive/ volatile compounds Excellent preservation of structure, aroma, and bioactivity Produces porous, stable powder	High cost. Long processing time Unsuitable for bulk, low-cost products	Gómez-Gaete et al. (2024)
Coacervation	*Simple*: Phase separation induced in a single polymer (by salt or nonsolvent) *Complex*: Two oppositely charged polymers (e.g., gelatin + gum arabic) interact; coacervate forms a uniform shell around oil droplets; hardened by cooling/pH/cross-linking	High encapsulation efficiency Provides excellent controlled release Suitable for oils, flavors	Complex and sensitive to control. Requires precise pH/ condition management.	Muhoza et al. (2023)

Table 11.3 Role of microencapsulation and its benefits in preserving volatiles, flavors, and nutrients

Preservation aspect	Mechanism	Benefits	References
Volatile compounds and flavors	Wall material reduces volatility, preventing loss Barrier prevents oxidation and off-flavors Controlled release during reconstitution (e.g., instant powders	Retains natural aroma and taste Protects delicate flavor molecules	Choudhury et al. (2021)
Nutrients (vitamins, carotenoids, omega-3 s)	Capsule wall prevents oxygen contact → oxidative stability Opaque wall protects from light Provides thermal insulation during storage/processing	Maintains nutritional value over shelf life Protects light- and heat-sensitive nutrients	Gómez-Gaete et al. (2024)
Enhanced bioavailability	Encapsulation shields nutrients from stomach acid Controlled release in the intestine improves absorption	Higher nutrient utilization by the body	Gómez-Gaete et al. (2024)

11.11 Utilization of Natural Fibers and By-Products from Waste to Wealth

The global horticulture sector, vital for food security and nutrition, produces large amounts of agricultural waste, including stems, leaves, peels, and seeds commonly known as agro-industrial by-products. Kimothi et al. (2020) note that these by-products are actually rich in valuable materials such as lignocellulosic fibers.

Utilizing natural fibers from horticultural waste plays a key role in this shift. As highlighted by Ahmed et al. (2024), this approach not only helps manage waste but also generates economic value and supports sustainable production and consumption.

11.11.1 Extraction of Fibers from Horticultural Waste

Horticultural waste serves as a significant source of lignocellulosic fibers, mainly composed of cellulose, hemicellulose, and lignin. These fibers offer strength, light weight, and biodegradability, making them ideal for various industrial uses. Common sources include banana pseudostems, pineapple leaves, citrus peels, and coconut husks. The extraction process plays a crucial role in determining fiber quality and functionality.

11.11.2 Sources of Horticultural Fibers

- *Banana Fiber:* Banana farming produces a large volume of pseudostem waste, which is often discarded postharvest. These pseudostems yield strong, lustrous, and fine-quality fibers (Alzate Acevedo et al. 2021).
- *Pineapple Leaf Fiber (PALF):* Typically discarded after fruit harvest, pineapple leaves contain durable, white, and silky fibers. Liao et al. (2025) note their long-standing use in textile production.
- *Other Fruit and Vegetable Fibers:* Siqueira et al. (2022) highlight additional sources such as peels from mango, orange, and passion fruit, as well as residues from okra and other vegetables.

11.11.3 Extraction Methods

Extraction techniques, mechanical, chemical, and biological, affect the fiber's characteristics, cost-efficiency, and environmental impact.

11.11.3.1 Mechanical Extraction (Decortication)

This physical method involves breaking down the plant material to separate fibers using decorticating machines. Iqbal et al. (2023) mention that although it is cost-effective and avoids chemicals, it can damage fibers and reduce yield.

11.11.3.2 Chemical Extraction

- *Alkali Treatment (Mercerization):* Treating plant material with alkaline solutions like sodium hydroxide removes impurities and enhances strength and dye absorption.
- *Acid Hydrolysis:* This process uses acids to break down hemicellulose and lignin. Vale et al. (2024) point out that while chemical methods produce cleaner fibers, they carry environmental concerns due to chemical use and wastewater generation.

11.11.3.3 Biological Extraction (Retting)

- *Water Retting:* Involves soaking stems in water, allowing bacteria to break down pectin.
- *Dew Retting:* Relies on fungi and natural moisture in open fields. Iqbal et al. (2023) state that although biological methods are eco-friendly and low cost, they are slow and may lead to uneven fiber quality due to inconsistent microbial activity.

11.11.4 Eco-friendly Packaging Materials

The environmental damage caused by petroleum-based plastics has led to a growing need for greener alternatives. Ncube et al. (2020) highlight how conventional plastics persist in ecosystems, causing harm. Natural fibers offer a sustainable solution.

- *Biodegradable Packaging*: These fibers can be turned into pulp and molded into items like trays, cartons, and clamshells. Oboturova et al. (2024) discuss how agricultural waste can be converted into biodegradable, compostable packaging that is safe for food use.
- *Natural Fiber Reinforced Bio Composites*: Srivastava et al. (2023) explain that natural fibers can reinforce bio plastics such as PLA or starch-based polymers, improve their strength and durability while reducing plastic use and cost. These biocomposites are biodegradable and versatile in packaging.
- *Paper and Paperboard*: Fibers from banana and pineapple leaves can replace wood pulp in making high-quality paper products, reducing the need for deforestation (Lalhruaitluangi and Mandal 2024).

11.11.5 Role in Promoting Sustainability

11.11.5.1 Using Horticultural By-Product Fibers Aligns with Circular Economy Principles and Supports Environmental, Economic, and Social Sustainability

- *Reduction of Plastic Pollution:* Fiber-based biodegradable packaging offers a solution to plastic pollution. Unlike plastics, these materials decompose naturally (Jafarzadeh et al. 2025).
- *Lower Carbon Footprint:* As renewable resources, these fibers capture CO_2 during plant growth and typically require less energy to process, lowering emissions (Boland et al. 2016).
- *Job Creation:* Fiber extraction and product manufacturing generate rural employment where waste is sourced (Kimothi et al. 2020).
- *Preservation of Traditional Skills:* Growing demand for fiber-based crafts helps sustain cultural skills and artisan livelihoods.

By integrating natural fiber use into industries, we can support a circular economy in the agri-food sector, as emphasized by Patti et al. (2020), fostering both environmental care and community resilience.

11.12 Aromatherapy and Wellness Applications

11.12.1 The Scent of Well-Being

Aromatherapy is a natural healing approach that uses plant-derived essential oils to support overall well-being. Recognized as a form of complementary medicine, it involves using aromatic oils extracted from various parts of plants to improve both mental and physical health (El 2020). Blending the art of scent with the science of chemistry, aromatherapy supports the body's self-healing abilities. Although the use of aromatic plants dates back thousands of years in spiritual, medicinal, and cosmetic practices, aromatherapy today is increasingly embraced within modern integrative healthcare (Ringu and Ranjan 2024).

11.12.2 Therapeutic Applications, Stress Relief, and Market Dynamics

Aromatherapy is valued for its diverse therapeutic benefits, delivered through inhalation, topical use, or massage. Aromatherapy is widely studied for its positive effects on mental health, especially in easing stress and anxiety.

- A meta-analysis of randomized trials showed that both inhalation and massage aromatherapy significantly reduced anxiety (Gong et al. 2020).
- Another analysis confirmed a moderate but consistent reduction in depression symptoms, with inhalation being the most effective method (Cho and Kim 2023).
- Certain essential oils are known for their calming effects. Lavender has been extensively studied and found helpful for generalized anxiety disorder (Amitani et al. 2023). Chamomile, sweet orange, and ylang-ylang are also commonly used for stress and anxiety relief.

11.12.3 A Booming Global Market

As reported, the global aromatherapy market was valued at USD 9.2 billion in 2024 and is expected to surpass USD 15.1 billion by 2030, with a CAGR of 8.9% ("Aromatherapy Market Size & Share. Industry Report-2030"). Other sources predict even higher growth, with projections reaching a CAGR of up to 12.6%.

11.12.4 Integration with Alternative and Complementary Medicine

Aromatherapy is not viewed as a standalone cure but as an effective complementary therapy that supports overall well-being (Caballero-Gallardo et al. 2025). It reflects the holistic approach of addressing the mind, body, and spirit together. Its growing inclusion in wellness practices highlights its adaptability and supportive role alongside traditional medicine.

11.13 Conclusion

Successful postharvest handling helps retain the quality, increase the shelf life, and add value to horticultural crops. Classic approaches, though fundamental, tend to lack in providing solutions to contemporary needs for sustainability, convenience, and diversification of products. Changing the primary form of horticultural products using processes such as drying, fermentation, extraction, and microencapsulation enables qualitative change that conforms to changing consumer tastes and market requirements. These techniques not only avoid postharvest loss but also create new doors to value-added food, health, wellness, and cosmetic products. Stressing sustainable methodology and creative processing, this multifaceted approach to postharvest management guarantees improved utilization of resources, producers' economic returns, and richer product variety for customers, ending up with a more dynamic and stronger horticulture industry.

References

Adetoro AO, Opara UL, Fawole OA (2021) Effect of blanching on enzyme inactivation, physicochemical attributes and antioxidant capacity of hot-air dried pomegranate (*Punica granatum* L.) arils (cv. Wonderful). Processes 9(1):25. https://doi.org/10.3390/pr9010025

Adeyeye SAO, Ashaolu TJ, Babu AS (2022) Food drying: a review. Agric Rev 1(8):1–8. https://doi.org/10.18805/ag.R-2537

Ahmad MS, Devi MP, Paul PK, Bhowmick N (2021) Study on juice extraction methods for enhancing juice yield and reducing bitterness in pummelo. Phar Innov 10(11):393–396

Ahmed B, Habib MA, Alam MS, Bhuiyan KH, Rabbani MG, Husain MT (2024) From waste to wealth: unlocking the potential of Bangladesh's okra fiber for economic prosperity. JTST 11(1):20–36

Alzate Acevedo S, Díaz Carrillo ÁJ, Flórez-López E, Grande-Tovar CD (2021) Recovery of banana waste-loss from production and processing: a contribution to a circular economy. Molecules 26(17):5282. https://doi.org/10.3390/molecules26175282

Amit SK, Uddin MM, Rahman R, Islam SR, Khan MS (2017) A review on mechanisms and commercial aspects of food preservation and processing. Agric Food Secur 6(1):51. https://doi.org/10.1186/s40066-017-0130-8

Amitani H, Nishi R, Sagiyama K, Fukumoto T, Funakoshi K, Takayanagi N, Watanabe H, Hirose M, Tagawa K, Ota K, Ito YM, Asakawa A (2023) The effect of lavender aroma for anxiety disorder: a study protocol for a multicenter, double-masked, randomized, placebo-controlled clinical trial. BMC Complement Med Ther 23(1):397. https://doi.org/10.1186/s12906-023-04231-1

Ämmälä A (2023) Comparison of pin mill and hammer mill in the fine grinding of *Sphagnum* moss. Energies 16(5):2437. https://doi.org/10.3390/en16052437

Andriamanantena M, Danth P, Cardon D, Fawbush FR, Raonizafinimanana B, Razafintsalama VE, Rakotonandrasana SR, Ethève A, Petit T, Caro Y (2019) Malagasy dye plant species: a promising source of novel natural colorants with potential applications—a review. Chem Biodivers 16(12):e1900442. https://doi.org/10.1002/cbdv.201900442

Armenova N, Tsigoriyna L, Arsov A, Petrov K, Petrova P (2023) Microbial detoxification of residual pesticides in fermented foods: current status and prospects. Foods 12(6):1163. https://doi.org/10.3390/foods12061163

Barus T, Giovania G (2020) Lactic acid bacteria from tempeh and their ability to acidify soybeans in tempeh fermentation. Microbiol Indones 14(4):4. https://doi.org/10.5454/mi.14.4.4

Boland CS, De Kleine R, Keoleian GA, Lee EC, Kim HC, Wallington TJ (2016) Life cycle impacts of natural fiber composites for automotive applications: effects of renewable energy content and lightweighting. J Ind Ecol 20(1):179–189. https://doi.org/10.1111/jiec.12286

Caballero-Gallardo K, Quintero-Rincón P, Olivero-Verbel J (2025) Aromatherapy and essential oils: holistic strategies in complementary and alternative medicine for integral wellbeing. Plants 14(3):400. https://doi.org/10.3390/plants14030400

Calderón-Oliver M, Ponce-Alquicira E (2022) The role of microencapsulation in food application. Molecules 27(5):1499. https://doi.org/10.3390/molecules27051499

Chadare FJ, Idohou R, Nago E, Affonfere M, Agossadou J, Fassinou TK, Hounhouigan DJ (2019) Conventional and food-to-food fortification: an appraisal of past practices and lessons learned. Food Sci Nutr 7(9):2781–2795. https://doi.org/10.1002/fsn3.1133

Chen P (2021) Lactic acid bacteria in fermented food. In: Dhanasekaran D, Sankaranarayanan A (eds) Advances in probiotics. Academic. https://doi.org/10.1016/B978-0-12-822909-5.00024-1

Cho K, Kim M (2023) Effects of aromatherapy on depression: a meta-analysis of randomized controlled trials. Gen Hosp Psychiatry 84:215–225. https://doi.org/10.1016/j.genhosppsych.2023.08.003
Choi HY, Lee YJ, Kim CM, Lee YM (2024) Revolutionizing cosmetic ingredients: harnessing the power of antioxidants, probiotics, plant extracts, and peptides in personal and skin care products. Cosmetics 11(5):157. https://doi.org/10.3390/cosmetics11050157
Choudhury N, Meghwal M, Das K (2021) Microencapsulation: an overview on concepts, methods, properties and applications in foods. Food Front 2(4):426–442. https://doi.org/10.1002/fft2.94
Ciurzyńska A, Falacińska J, Kowalska H, Kowalska J, Galus S, Marzec A, Domian E (2021) The effect of pre-treatment (blanching, ultrasound and freezing) on quality of freeze-dried red beets. Foods 10(1):132. https://doi.org/10.3390/foods10010132
Cook FE, Leon CJ, Nesbitt M (2015) Potpourri as a sustainable plant product: identity, origin, and conservation status. Econ Bot 69(4):330–344. https://doi.org/10.1007/s12231-015-9325-8
Cortez R, Luna-Vital DA, Margulis D, Gonzalez de Mejia E (2017) Natural pigments: stabilization methods of anthocyanins for food applications. Compr Rev Food Sci Food Saf 16(1):180–198. https://doi.org/10.1111/1541-4337.12244
Desai KGH, Jin Park H (2005) Recent developments in microencapsulation of food ingredients. Dry Technol 23(7):1361–1394. https://doi.org/10.1081/DRT-200063478
El AA (2020) Aromatherapy as complementary medicine. In: de Oliveira MS, da Costa WA, Silva SG (eds) Essential oils—bioactive compounds, new perspectives and applications. IntechOpen. https://doi.org/10.5772/intechopen.92021
Escalante A, Lopez Soto DR, Velazquez Gutierrez JE, Giles-Gómez M, Bolívar F, López-Munguía A (2016) Pulque, a traditional Mexican alcoholic fermented beverage: historical, microbiological, and technical aspects. Front Microbiol 7:1026. https://doi.org/10.3389/fmicb.2016.01026
Food and Agriculture Organization of the United Nations (2024) FAOSTAT. https://www.fao.org/faostat/en/#compare
Gómez-Gaete C, Avendaño-Godoy J, Escobar-Avello D, Campos-Requena VH, Rogel-Castillo C, Estevinho LM, Martorell M, Sharifi-Rad J, Calina D (2024) Revolutionizing fruit juice: exploring encapsulation techniques for bioactive compounds and their impact on nutrition, flavour and shelf life. Food Prod Process Nutr 6(1):8. https://doi.org/10.1186/s43014-023-00190-9
Gong M, Dong H, Tang Y, Huang W, Lu F (2020) Effects of aromatherapy on anxiety: a meta-analysis of randomized controlled trials. J Affect Disord 274:1028–1040. https://doi.org/10.1016/j.jad.2020.05.118
Guo X, Hao Q, Qiao X, Li M, Qiu Z, Zheng Z, Zhang B (2023) An evaluation of different pretreatment methods of hot-air drying of garlic: drying characteristics, energy consumption and quality properties. Lwt 180:114685. https://doi.org/10.1016/j.lwt.2023.114685
Hanif T, Murad R, Amam RS, Ishrat Fatima MZ (2024) Role of horticulture in addressing food security and global nutrition challenges. Cornous Biol 2(1):45–51. https://doi.org/10.37446/corbio/ra/2.1.2024.45-51
Ignaczak A, Salamon A, Kowalska J, Marzec A, Kowalska H (2023) Influence of pre-treatment and drying methods on the quality of dried carrot properties as snacks. Molecules 28(17):6407. https://doi.org/10.3390/molecules28176407
Iqbal SF, Kabir MH, Muhib MI, Rupok MMHK (2023) Prospect of waste banana fibre use in industry: a narrative review. Int J Sci Res Arch 10(2):828–839. https://doi.org/10.30574/ijsra.2023.10.2.1045
Jafarzadeh S, Qazanfarzadeh Z, Parandi E, Esmaeili Y, Barrow CJ, Timms W, Naebe M (2025) Eco-friendly plastics from cereal-derived by-products and waste: a circular economy approach for sustainable packaging. Mater Today Chem 47:102782. https://doi.org/10.1016/j.mtchem.2025.102782
Jung H, Lee YJ, Yoon WB (2018) Effect of moisture content on the grinding process and powder properties in food: a review. PRO 6(6):69. https://doi.org/10.3390/pr6060069
Kaur R, Watson JA (2024) A scoping review of postharvest losses, supply chain management, and technology: implications for produce quality in developing countries. J ASABE 67(5):1103–1131. https://doi.org/10.13031/ja.15660

Khaksar G, Assatarakul K, Sirikantaramas S (2019) Effect of cold-pressed and normal centrifugal juicing on quality attributes of fresh juices: do cold-pressed juices harbor a superior nutritional quality and antioxidant capacity? Heliyon 5(6):e01917. https://doi.org/10.1016/j.heliyon.2019.e01917

Kimothi SP, Panwar S, Khulbe A (2020) Creating wealth from agricultural waste. Indian Council of Agricultural Research, New Delhi, pp 0–172

Kitinoja L, Al Hassan HY (2010) Identification of appropriate postharvest technologies for small scale horticultural farmers and marketers in sub-Saharan Africa and South Asia—part 1. Postharvest losses and quality assessments. Acta Hortic 934:31–40. https://doi.org/10.17660/ActaHortic.2012.934.1

Koley P (2020) An overview of dehydration techniques. J Pharmacogn Phytochem 9(4):228–233

Koop BL, Maciel AG, Soares LS, Monteiro AR, Valencia GA (2022) Natural colorants. Nat Addit Foods:87–122

Kucharska-Guzik A, Guzik Ł, Charzyńska A, Michalska-Ciechanowska A (2025) Influence of freeze drying and spray drying on the physical and chemical properties of powders from *Cistus creticus* L. extract. Foods 14(5):849. https://doi.org/10.3390/foods14050849

Kutz M (ed) (2019) Handbook of farm, dairy and food machinery engineering. Academic

Lachowicz S, Michalska A, Lech K, Majerska J, Oszmiański J, Figiel A (2019) Comparison of the effect of four drying methods on polyphenols in Saskatoon berry. Lwt 111:727–736. https://doi.org/10.1016/j.lwt.2019.05.054

Laguerre JC, Hamoud-Agha MM (2020) Microwave heating for food. Food Preserv. Waste Exploit 35

Lalhruaitluangi N, Mandal D (2024) Utilization of pineapple leaf: an alternative for paper and textile industries. J Postharvest Technol 12(3):1–10

Lalpekhlua K, Tirkey A, Saranya S, Babu PJ (2024) Post-harvest management strategies for quality preservation in crops. Int J Veg Sci 30(5):587–635

Lan T, Bao S, Wang J, Ge Q, Zhang H, Yang W, Sun X, Ma T (2021) Shelf life evaluation of non-industrial fresh mango juice based on the microbial safety, nutritional and sensory characteristics. Food Biosci 42:101060

Lengai GM, Fulano AM, Muthomi JW (2022) Improving access to export market for fresh vegetables through reduction of phytosanitary and pesticide residue constraints. Sustainability 14(13):8183

Liao S, Chen J, Wang X (2025) An update on pineapple leaf fibers. J Nat Fibers 22(1):2509129

Lisiecka K, Dziki D, Karaś M (2024) Physical and functional properties of powders obtained during spray drying of *Cyani flos* extracts. Molecules 29(14):3400

Ma T, Wang J, Wang H, Lan T, Liu R, Gao T, Yang W, Zhou Y, Ge Q, Fang Y, Sun X (2020) Is overnight fresh juice drinkable? The shelf life prediction of non-industrial fresh watermelon juice based on the nutritional quality, microbial safety, and sensory quality. Food Nutr Res 64:10–29219. https://doi.org/10.29219/fnr.v64.4327

Machado CAT, de Andrade MAN, Lepikson HA (2019) Manufacturing 4.0: discussion on application in the extractive industry of essential oils. JBTH 2(3):86–91

Machado CA, Oliveira FO, de Andrade MA, Hodel KVS, Lepikson H, Machado BAS (2022) Steam distillation for essential oil extraction: an evaluation of technological advances based on an analysis of patent documents. Sustainability 14(12):7119. https://doi.org/10.3390/su14127119

Machado CAT, Hodel KVS, Lepikson HA, Machado BAS (2024) Distillation of essential oils: an innovative technological approach focused on productivity, quality and sustainability. PLoS One 19(2):e0299502. https://doi.org/10.1371/journal.pone.0299502

Mapelli-Brahm P, Corte-Real J, Meléndez-Martínez AJ, Bohn T (2017) Bioaccessibility of phytoene and phytofluene is superior to other carotenoids from selected fruit and vegetable juices. Food Chem 229:304–311. https://doi.org/10.1016/j.foodchem.2017.02.074

Mardani M, Siahtiri S, Besati M, Baghani M, Baniassadi M, Nejad AM (2024) Microencapsulation of natural products using spray drying: an overview. J Microencapsul 41(7):649–678. https://doi.org/10.1080/02652048.2024.2389136

Márquez-Cardozo CJ, Caballero-Gutiérrez BL, Ciro-Velázquez HJ, Restrepo-Molina DA (2021) Effect of pretreatment and temperature on the drying kinetics and physicochemical and techno-functional characteristics of pumpkin (*Cucurbita maxima*). Heliyon 7(4):e06723. https://doi.org/10.1016/j.heliyon.2021.e06802

Masyita A, Hardinasinta G, Astuti AD, Firdayani F, Mayasari D, Hori A, Nisha IN, Nainu F, Kuraishi T (2025) Natural pigments: innovative extraction technologies and their potential application in health and food industries. Front Pharmacol 15:1507108. https://doi.org/10.3389/fphar.2024.1507108

Mboowa D (2024) A review of the traditional pulping methods and the recent improvements in the pulping processes. Biomass Convers Biorefinery 14(1):1–12. https://doi.org/10.1007/s13399-020-01243-6

Mebakerlin MS, Chakravorty S (2015) Value addition in flowers. In: Sharangi A, Datta S (eds) Value addition of horticultural crops: recent trends and future directions. Springer, New Delhi. https://doi.org/10.1007/978-81-322-2262-0_5

Molina AK, Corrêa RCG, Prieto MA, Pereira C, Barros L (2023) Bioactive natural pigments' extraction, isolation, and stability in food applications. Molecules 28(3):1200. https://doi.org/10.3390/molecules28031200

Morais RM, Morais AM, Dammak I, Bonilla J, Sobral PJ, Laguerre JC, Afonso MJ, Ramalhosa EC (2018) functional dehydrated foods for health preservation. J Food Qual 2018(1):1739636. https://doi.org/10.1155/2018/1739636

Muhoza B, Yuyang H, Uriho A, Harindintwali JD, Liu Q, Li Y (2023) Spray- and freeze-drying of microcapsules prepared by complex coacervation method: a review. Food Hydrocoll 140:108650. https://doi.org/10.1016/j.foodhyd.2023.108650

Nanda SK, Vishwakarma RK, Bathla HVL, Rai A, Chandra P (2012) Harvest and post-harvest losses of major crops and livestock produce in India. AICRP, ICAR

Ncube LK, Ude AU, Ogunmuyiwa EN, Zulkifli R, Beas IN (2020) Environmental impact of food packaging materials: a review of contemporary development from conventional plastics to polylactic acid-based materials. Materials 13(21):4994. https://doi.org/10.3390/ma13214994

Ngamwonglumlert L, Devahastin S, Chiewchan N (2017) Natural colorants: pigment stability and extraction yield enhancement via utilization of appropriate pretreatment and extraction methods. Crit Rev Food Sci Nutr 57(15):3243–3259. https://doi.org/10.1080/10408398.2015.1109498

Nowak D, Jakubczyk E (2020) The freeze-drying of foods—the characteristic of the process course and the effect of its parameters on the physical properties of food materials. Foods 9(10):1488. https://doi.org/10.3390/foods9101488

Oboturova N, Povetkin S, Nikulnikova N, Lazareva N, Klopova A, Lyubchanskiy N, Sukhanova E, Lebedeva N (2024) Upcycling agricultural byproducts into eco-friendly food packaging. Potr S J F Sci 18. https://doi.org/10.5219/1949

Ogrodowczyk AM, Drabińska N (2021) Crossroad of tradition and innovation—the application of lactic acid fermentation to increase the nutritional and health-promoting potential of plant-based food products: a review. Pol J Food Nutr Sci 71(2):107–134

Orphanides A, Goulas V, Gekas V (2016) Drying technologies: vehicle to high-quality herbs. Food Eng Rev 8(2):164–180

Ostermeier R, Parniakov O, Töpfl S, Jäger H (2020) Applicability of pulsed electric field (PEF) pre-treatment for a convective two-step drying process. Foods 9(4):512. https://doi.org/10.3390/foods9040512

Owusu-Kwarteng J, Agyei D, Akabanda F, Atuna RA, Amagloh FK (2022) Plant-based alkaline fermented foods as sustainable sources of nutrients and health-promoting bioactive compounds. Front Sustain Food Syst 6:885328. https://doi.org/10.3389/fsufs.2022.885328

ÖZCAN M (2020) Turunçgil ve Çay Yetiştiriciliği Ders Notu. Ondukuzmayıs Üniversitesi sf, 105

Ozolina K, Sarenkova I, Muizniece-Brasava S (2023) The anti-nutritional factors of legumes and their treatment possibilities: a review. Res Rural Dev 29:68–76. https://doi.org/10.22616/RRD.29.2023.010

Palumbo M, Attolico G, Capozzi V, Cozzolino R, Corvino A, de Chiara MLV, Pace B, Pelosi S, Ricci I, Romaniello R, Cefola M (2022) Emerging postharvest technologies to enhance the shelf-life of fruit and vegetables: an overview. Foods 11(23):3925. https://doi.org/10.3390/foods11233925

Patti A, Cicala G, Acierno D (2020) Eco-sustainability of the textile production: waste recovery and current recycling in the composites world. Polymers 13(1):134. https://doi.org/10.3390/polym13010134

Pokhrel B (2021) Review on post-harvest handling to reduce loss of fruits and vegetables. Int J Hortic Food Sci 2(2):48–52

Radojčin M, Babić M, Babić L, Pavkov I, Bukurov M, Bikić S, Mitrevski V (2015) Effects of osmotic pretreatment on quality and physical properties of dried quinces (*Cydonia oblonga*). J Food Nutr Res 54(2):142–154

Radojčin M, Pavkov I, Bursać Kovačević D, Putnik P, Wiktor A, Stamenković Z, Kešelj K, Gere A (2021) Effect of selected drying methods and emerging drying intensification technologies on the quality of dried fruit: a review. PRO 9(1):132

Reyes-De-Corcuera JI, Goodrich-Schneider RM, Barringer SA, Landeros-Urbinan MA (2014) Processing of fruit and vegetable beverages. In: Food processing: principles and applications. Wiley-Blackwell, pp 339–362

Ringu K, Ranjan A (2024) A review on effectiveness of aromatherapy on mental health and well-being. IJSRMST 3(11):1–13

Román-Camacho JJ, García-García I, Santos-Dueñas IM, García-Martínez T, Mauricio JC (2023) Latest trends in industrial vinegar production and the role of acetic acid bacteria: classification, metabolism, and applications—a comprehensive review. Foods 12(19):3705. https://doi.org/10.3390/foods12193705

Rymbai H, Sharma RR, Srivastav M (2011) Bio-colorants and its implications in health and food industry—a review. Int J Pharm Res 3(4):2228–2244

Safeena SA, Thangam M (2023) Value addition in flower crops through production of potpourri. Bangladesh J Bot 52(1):87–96

Salauddin SM, Mia R, Haque MA, Shamim AM (2021) Review on extraction and application of natural dyes. Text Leather Rev 4(4):218–233

Sehrawat R, Nema PK, Kaur BP (2018) Quality evaluation and drying characteristics of mango cubes dried using low-pressure superheated steam, vacuum and hot air drying methods. Lwt 92:548–555

Sellitto VM, Zara S, Fracchetti F, Capozzi V, Nardi T (2021) Microbial biocontrol as an alternative to synthetic fungicides: boundaries between pre- and post-harvest applications on vegetables and fruits. Fermentation 7(2):60

Singh T, Pandey VK, Dash KK, Zanwar S, Singh R (2023) Natural bio-colorant and pigments: sources and applications in food processing. J Agric Food Res 12:100628

Siqueira MU, Contin B, Fernandes PRB, Ruschel-Soares R, Siqueira PU, Baruque-Ramos J (2022) Brazilian agro-industrial wastes as potential textile and other raw materials: a sustainable approach. Mater Circ Econ 4(1):9. https://doi.org/10.1007/s42824-021-00050-2

Shetty MJ, Geethalekshmi PR, & Mini C (2017) Natural pigments as potential food colourants: a review. Trends in Biosciences 10(21):4057–4064

Souiy Z (2023) Essential oil extraction process. In: essential oils—recent advances, new perspectives and applications. IntechOpen

Souiy Z, Zakhama N, Cheraief I, Hammami M (2022) Nutritional, physical, microbial, and sensory characteristics of gluten- and sugar-free cereal bar enriched with spirulina and flavoured with neroli essential oil. Lwt 169:113955

Srivastava V, Singh S, Das D (2023) Enhancing packaging sustainability with natural fiber reinforced biocomposites: an outlook into the future. E3S Web Conf 436:08016. EDP Sciences

Stamenković Z, Pavkov I, Radojčin M, Tepić Horecki A, Kešelj K, Kovačević DB, Putnik P (2019) Convective drying of fresh and frozen raspberries and change of their physical and nutritive properties. Foods 8(7):251

Staver CL, Movileanu P (2012) The role of transport means used in the transportation of fresh horticultural products. Sci Pap Ser Manag Econ Eng Agric Rural Dev 12(1):35–38

Streimikyte P, Viskelis P, Viskelis J (2022) Enzymes-assisted extraction of plants for sustainable and functional applications. Int J Mol Sci 23(4):2359. https://doi.org/10.3390/ijms23042359

Tanwir A, Fatima T, Noor F (2025) Quality enhancement in the food processing sector with sustainable innovations. In: Energy efficient technologies for food safety, quality, and security. IGI Global Scientific Publishing, pp 319–358

Thamkaew G, Sjöholm I, Galindo FG (2021) A review of drying methods for improving the quality of dried herbs. Crit Rev Food Sci Nutr 61(11):1763–1786. https://doi.org/10.1080/10408398.2020.1765309

Ueda JM, Morales P, Fernández-Ruiz V, Ferreira A, Barros L, Carocho M, Heleno SA (2023) Powdered foods: structure, processing, and challenges—a review. Appl Sci 13(22):12496. https://doi.org/10.3390/app132212496

Ugwu BN, Onyejiuwa CT, Ogbuke CR (2020) Review of orange juice extractor machines. Adv Sci Technol Eng Syst 5(5):485–492

Vale AC, Leite L, Pais V, Bessa J, Cunha F, Fangueiro R (2024) Extraction of natural-based raw materials towards the production of sustainable man-made organic fibres. Polymers 16(24):3602. https://doi.org/10.3390/polym16243602

Wandhekar S, Pawar V, Shinde S, Gangakhedkar P (2023) Extraction of oil from oilseeds by cold pressing: a review. Indian Food Ind Mag 4:63–69

Watkins CB, Ekman JH (2004) Storage technologies: temperature interactions and effects on quality of horticultural products. Acta Hortic 682:1527–1534. https://doi.org/10.17660/ActaHortic.2005.682.203

Yahia EM, García-Solís P, Celis MEM (2019) Contribution of fruits and vegetables to human nutrition and health. In: Postharvest physiology and biochemistry of fruits and vegetables. Woodhead Publishing, pp 19–45

Yao J, Chen W, Fan K (2023) Novel efficient physical technologies for enhancing freeze drying of fruits and vegetables: a review. Foods 12(23):4321. https://doi.org/10.3390/foods12234321

Yeshiwas Y, Tadele E (2021) an investigation into major causes for post-harvest losses of horticultural crops and their handling practice in Debre Markos, North-Western Ethiopia. Adv Agric 2021(1):1985303. https://doi.org/10.1155/2021/1985303

Zhang R, Wang M, Liu P, Zhu T, Qu X, Chen X, Xiao X (2024) Flexible Vis/NIR sensing system for banana chilling injury. Postharvest Biol Technol 207:112623. https://doi.org/10.1016/j.postharvbio.2023.112623

Zhao LZ, Chen J, Wei XY, Lin B, Zheng FJ, Verma KK, Chen GL (2025) Response of alcohol fermentation strains, mixed fermentation and extremozymes interactions on wine flavour. Front Microbiol 16:1532539. https://doi.org/10.3389/fmicb.2025.1532539

Zheng S, Zhang Z, Zhao X, Li W, Hou L (2024) A method of reducing salt content in fermented soy sauce improves its flavour and quality. Foods 13(6):971. https://doi.org/10.3390/foods13060971

Drying and Dehydration of Fruits and Vegetables

12

Gomathy Kamalakkannan

Abstract

Fruits and vegetables are highly perishable, and the need for preserving them from post-harvest spoilage using economical and efficient technology is incredible. Drying or dehydration is considered an evergreen technology as it is been practised since traditional times, as well as until recently, using modern, sophisticated equipment. Both drying and dehydration involve the process of removing moisture from horticultural commodities like fruits and vegetables. The process facilitated under controlled conditions is referred to as dehydration, and one that occurs naturally is deemed drying. The process leads to a reduction of water activity of the commodity, thereby delaying the growth of microorganisms. These drying and dehydration operations involve a complex phenomenon of heat and mass transfer, which are dependent on the heterogeneous composition and thermo-physical properties of the particular commodity. There are two stages in which moisture is removed from the biological material, where in the first stage, unbound moisture evaporates from the surface of the commodity due to the surrounding heat application, and in stage II, bound moisture is removed from the interior of the material to its surroundings. The surface of the commodity serves as the boundary for heat transfer within the commodity to the surroundings. In the case of horticultural commodities, higher dry bulb temperature, lower relative humidity, and high air velocity are ideal for better drying. Understanding the drying kinetics of each commodity is crucial for optimising the correct drying time, temperature, and airflow rate so that the product quality is maintained to the maximum. When the commodity approaches a normal equilibrium with the surrounding relative humidity and temperature, drying ceases. At this point, the

G. Kamalakkannan (✉)
Department of Fruit and Vegetable Technology, CSIR-Central Food Technological Research Institute, Mysore, Karnataka, India
e-mail: gomathyk.cftri@csir.res.in

A. Das, A. B. Sharangi (eds.), *Remunerative Horticulture*,
https://doi.org/10.1007/978-981-92-1007-7_12

microbial as well as the enzymatic activity becomes very minimal, leading to the extended shelf life of the commodity. Research advancements have credited various dryers intended for fruit and vegetable applications. However, the selection of dryers depends on the end application and the nature of the commodity chosen for drying.

Keywords

Drying · Dehydration · Drying kinetics · Heat and mass transfer · Water activity

12.1 Introduction

Vegetables and fruits are essential components of a balanced diet. Several health advantages, including lowering the risk of non-communicable diseases (NCDs), enhancing nutritional status, and encouraging sustainable development, are directly associated with their consumption. To reduce the risk of communicable diseases such as diabetes, heart disease, stroke, and some types of cancer, the Food and Agriculture Organisation (FAO) and the World Health Organisation (WHO) advise consuming 400 g of fruits and vegetables daily, while omitting starchy tubers (WHO 2003). These foods improve gut health, metabolic control, and immunological function because they are high in critical vitamins (such as A, C, and folate), minerals, dietary fibre, and a range of bioactive components such as flavonoids, antioxidants, and plant sterols. The inadequate intake is to blame for the increased rate of NCD (WHO 2023). According to recent national research, less than 10% of Indian teenagers consume the recommended daily intake of fruits and vegetables (Rathi et al. 2025). According to the Indian Journal of Medical Research, inadequate intake of fruits and vegetables adds greatly to the burden of cardiovascular disease, diabetes, and some types of cancer in India (Krishnaswamy and Gayathri 2018). Deficits in micronutrients, particularly iron, vitamin A, and folate, are prevalent in all age groups and have long-term effects on the productivity, cognitive development, and health of mothers. Barriers to adequate consumption include affordability, economic constraints, lack of awareness and nutrition education, time and convenience factors, and a shift toward unhealthy eating patterns driven by increased marketing of unhealthy options. Additionally, indirect reasons such as seasonality, poor infrastructure, inadequate storage, and inefficient supply chains also play a role.

Fruits and vegetables are highly perishable commodities that, after reaching their optimal maturity, start deteriorating. With deterioration, the original physical, biochemical, and physiological properties of fruits and vegetables are affected. Farmers and wholesalers face problems due to surplus production from competitors at the same time, and high maintenance costs associated with the storage of the commodity for a longer period of time. Not only do the sales become problematic for the seller, but also the disposal of huge quantities of unsold fruit and vegetables becomes both challenging and costly for the farmers and sellers. Fruits and vegetables have high moisture content, rich in a nutritional profile that contains simple sugars like sucrose and fructose, organic acids, vitamins, and minerals. While fruits maintain a pH (3.0–5.0), vegetables have a slightly acidic to neutral pH (5.0–7.0). All these factors serve as a favourable condition for the growth of several bacteria, moulds, and yeasts that lead to the physiological disintegration and rotting of the commodity. The high moisture content (70%–95%) combined with the high-water activity (0.80–1.00), along with the tender tissues of the fresh fruits and vegetables, increases the susceptibility to spoilage nature increases drastically. This moisture reduction can be achieved using drying and dehydration techniques. Drying or dehydration is considered an evergreen technology as it has been practised since traditional times, as well as until recently, using modern, sophisticated equipment. Both drying and dehydration involve the process of removing moisture from horticultural commodities like fruits and vegetables. The process leads to a reduction of water activity of the commodity, thereby delaying the growth of microorganisms and retarding the biochemical and enzyme activity in foods (Mercer 2024; Jafari and Malekjani 2023). According to Mujumdar (2006), drying involves heat and mass transfer, preceded by various heat and mass transfer processes, which lead to several chemical and physical transformations in the product. These changes include either shrinkage or elongation, deformation, puffing, glass or phase transition, colour changes and other physiological changes in properties of the material. The functional, nutritional, and nutritional properties of the dehydrated foods also get altered due to exposure to high temperatures or longer exposure times.

Novel drying or dehydration techniques target these issues by effectively reducing the exposure times by creating vacuum (vacuum drying), introducing reduced pressure for faster drying (low pressure superheated steam drying), electro hydrodynamic drying using high voltage electric fields, supercritical CO_2 drying utilising carbon dioxide to remove moisture in food, drying under sublimation conditions to retain physical and nutritional properties (Freeze drying) and using modified heating sources such as microwave, radiofrequency drying, refractive window drying or other dehydration methods tailored to specific products. The methods of drying vary with the mode of heat transfer, either as convective mode, conductive mode, radiation mode, or a combination of any of the other methods. These methods also provide workable ways to combat seasonal variations, enabling the preservation and year-round consumption of excess vegetables throughout harvest seasons in an economical way. Furthermore, products that have been dried and dehydrated are more portable, lighter, and easier to store and transport, particularly in areas without sufficient cold storage facilities. Consumers' preference for clean-label products, shifting lifestyles, and rising food quality perceptions are all contributing to the steady increase in the trend toward dehydrated goods (Yan et al. 2022).

12.2 Need for Drying and Dehydration

The need for drying and dehydration of fruits and vegetables remains in the context of extending shelf life and preserving commodities, reducing the post-harvest losses, thereby guaranteeing the year-round supply. When drying and dehydration are carried out in controlled parametric conditions, the commodities' nutritional and sensorial properties are well maintained. As a result of drying, the expenses associated with transportation and storage are reduced significantly. Additionally, drying is necessary to create items that are ready to eat or that are readily rehydrated, which gives customers more convenience. Additionally, it helps stabilise market supply by conserving seasonal surpluses and makes it possible to produce high-value goods like vegetable flakes, fruit powders, and instant food mixes. Drying and dehydration provide a dependable and energy-efficient substitute in areas lacking refrigeration or cold storage facilities, which makes them essential for enhancing food security, decreasing food waste, and promoting rural and industrial economic development in both developed and developing countries (Khatri et al. 2024; Martynenko and Vieira 2023; Fellows 2017). The diagrammatic representation of the advantages of drying and dehydration is provided in Fig. 12.1.

12.2.1 Prolonged Storage Life

Drying or dehydration aids in moisture removal from the commodity. This moisture becomes unavailable for the growth and multiplication of microorganisms, thereby reducing the chances of spoilage. Usually, water content in fruits and vegetables is very high, leading to an $a_w > 0.98$. Figure 12.2 represents the water content and solid content of a single orange. The single orange weighing 97 g contains approximately 72 ml of juice and the remaining weight as pomace.

The drying operation reduces the water content in foods by more than 50%. According to research studies by Scott (1957), bacterial deterioration ceases at a_w below 0.9, yeast ceases below 0.85, and moulds below 0.7. Thus, defining that food with an a_w below 0.6 as a stable dehydrated food (Labuza 1980). The other biological reactions requiring water for the enzymatic actions of polyphenol oxidase for producing brown discolouration, or the pectinases, resulting in tissue softening. Table 12.1 shows certain examples of enzymes and their effect on shelf life in the presence of high moisture. Nevertheless, techniques such as drying or dehydration help in reducing the moisture in the produce, which also reduces the rate of biochemical reactions like oxidation reactions as lipid oxidation, hydrolytic reactions, like starch, pectin, and cellulose hydrolysis. The high moisture combined with poor packaging systems could lead to anaerobic fermentations, and microbial proteolysis occurs, causing spoilage of the commodity. The high moisture in fresh produce, when reduced indirectly, reduces the water activity, thereby controlling or inhibiting insect infestation in the produce.

12.2.2 Reduction in Volume and Weight

Drying or removal of moisture in a commodity significantly reduces the water content of the produce. This implies that more storage space is available to accommodate the same produce. This attribute makes the transport and storage easier and

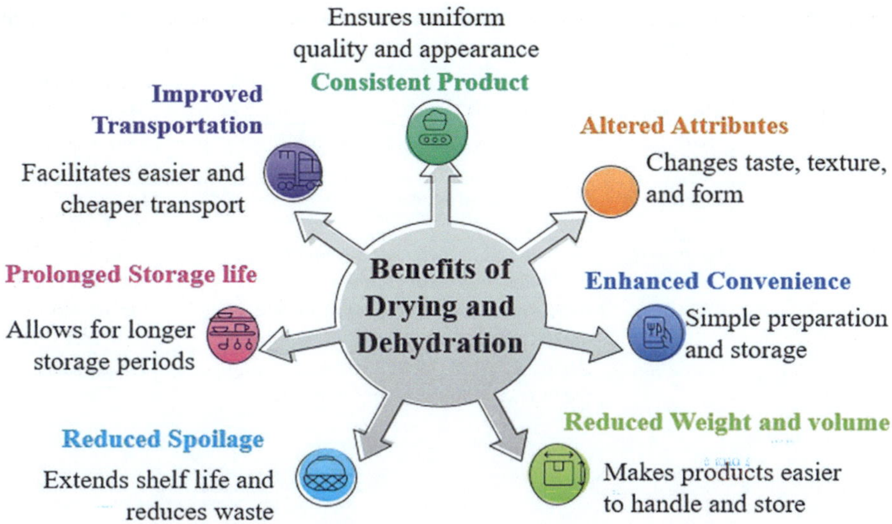

Fig. 12.1 Flow chart illustrating the benefits of drying and dehydration. Central node labelled "Benefits of Drying and Dehydration" with arrows pointing to seven benefits: "Improved Transportation" (facilitates easier and cheaper transport), "Prolonged Storage Life" (allows for longer storage periods), "Reduced Spoilage" (extends shelf life and reduces waste), "Consistent Product" (ensures uniform quality and appearance), "Altered Attributes" (changes taste, texture, and form), "Enhanced Convenience" (simple preparation and storage), and "Reduced Weight and Volume" (makes products easier to handle and store). Each benefit is accompanied by an icon and colour-coded text

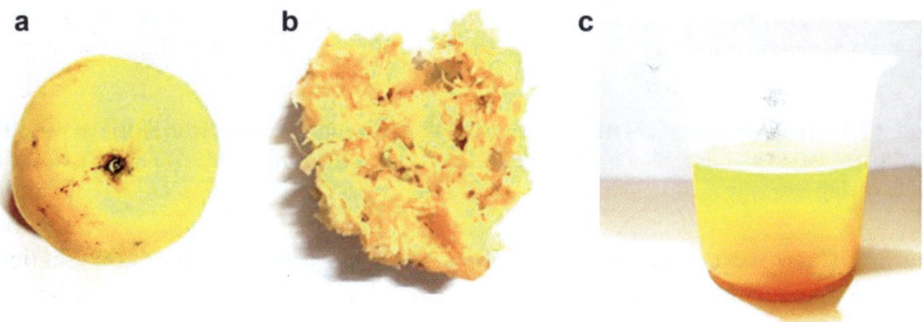

Fig. 12.2 Three items related to an orange. (**a**) A fresh orange is displayed. (**b**) There is orange pomace, which is the fibrous residue left after juicing. (**c**) A beaker contains the extracted orange juice, representing the water content

more economical. Let us discuss a true case of a carrot. 500 g of carrots were taken, peeled, and cut into small pieces. The initial weight of the pieces was 325 g. The pieces were placed in the tray drier and continuously dried at 45 °C for a period of 5 h until the consecutive weights were obtained. The final dried carrot pieces weighed 21.5 g. Figure 12.3 represents the volume reduction during drying of the

Table 12.1 The mechanism of how moisture affects different enzymes. Lastly, its effect on shelf life has also been mentioned

Enzymes	Mechanism	Moisture effect	Effect on shelf life
Pectinase	Hydrolysis	High moisture leads to faster tissue softening	Decreased shelf life
Cellulose	Hydrolysis	Accelerated structure breakdown with high moisture	Decreased shelf life
Amylase	Hydrolysis	High moisture increases sweetness and facilitates microbial growth	Decreased shelf life
Lipase	Oxidation	Increased moisture content hastens rancidity in damaged tissue	Decreased shelf life
Polyphenol oxidase	Oxidation	Higher moisture increases the browning reaction, with damage or cut tissue	Decreased shelf life

Fig. 12.3 Chopped carrots are shown in two stages. (**a**) Fresh-cut carrots with an initial weight of 325 g. (**b**) Dried carrots with a final weight of 21.5 g. The image illustrates the weight reduction after drying

carrot. Thus, we can infer that around a 93.38% reduction in weight has been achieved by means of drying.

The benefits of drying in volume reduction involve

1. Reduced storage cost: quantity of product per cubic metre increases
2. Reduced transportation cost: Lower fuel and shipment costs
3. Improved logistics: easy handling during bulk and export

12.2.3 Enhancing Convenience and Variety

The convenience of dried and dehydrated products arises from the feasibility of storing them in normal conditions without the necessity of refrigeration. By lowering weight, volume, and packing needs, dehydration not only increases shelf life but also facilitates transportation and storage of products without refrigeration (Li et al. 2022a). Most of the dried products are ready to consume, while others may require rehydration, depending on the nature of the end product. Current developments in drying technologies also contribute to the preservation of structural integrity and rehydration capacity, enabling consumers to obtain products that are ready to use or require only rehydration (Jiang et al. 2025; Santos et al. 2025). This reduces the burden of cooking and preparation time in the kitchen for consumers' use. These products can be packed according to need in small sachets or big boxes, too, as per the end use or on-the-go. Since the moisture in the products is very minimal, the handling of dried products does not cause a messy feel or cause contamination risk when packed and stored properly.

Drying or dehydration aids in the production of many value-added products such as fruit slices, fruit and vegetable bars (e.g. apple, mango, banana, guava), powders (mango powder, apple powder, beetroot powder, tomato powder), flakes (onion flakes, corn flakes), granulated bars, instant soups or mixes (using a combination of dry ingredients). These dried products can be either eaten as a whole or can be turned into a secondary product serving as an ingredient for another product to add a different taste and variety. The tailor-made recipes can be prepared using the dried products, serving the targeted client groups. The application of dry ingredients is commercially being utilised in many food industries, such as baking (fruit powder), snacks (dry fruits, powders), instant meals (dried starches), baby food (fruit powders), smoothies, and health bars (granulated dry fruits, nuts).

12.2.4 Altering Inherent Attributes

Due to modifications in cellular structure, moisture distribution, and chemical composition, drying and dehydration drastically affect the natural sensory and physical characteristics of fruits and vegetables, including their flavour, texture, and form. Certain fruits undergo such a drastic shift in qualities from their original form when they are dried that they are given a completely different name. Low-temperature techniques like freeze drying typically maintain the porosity, crispness, and structural integrity of the original product, whereas high-temperature techniques (like hot-air or microwave drying) frequently cause structural collapse, shrinkage, and case hardening, which produce firmer or tougher textures (Sagar and Kumar 2010; Petikirige et al. 2022). The removal of moisture increases the concentration of sugars and acids, making the taste sweeter and tart. However, heat-sensitive volatile flavour compounds and aroma-active substances deteriorate, particularly when exposed to high or prolonged temperatures. Bonneau et al. (2016) observed that drying of mangoes (Kent variety) resulted in a 59% reduction of volatiles. Cell wall rupture, pectin breakdown, and starch gelatinisation are the main causes of texture alterations. Gao et al. (2024) demonstrated that the pectin fractions strongly modified the textural attributes of carrot on drying. These drying processes can either improve crispness (in freeze-dried fruits) or produce unwanted leathery or rubbery textures in convective drying. Drying sometimes leads to intensified flavours and concentrates some of the nutrients. Some undesirable effects, such as the formation of wrinkles, enzymatic changes, hardening of texture, and degradation of colour, are also associated with drying. Several advanced drying technologies can be employed to prevent many of these adverse effects (Petikirige et al. 2022).

12.2.5 Improved Transportation and Transit

As drying aids in volume and weight reduction, the practicality of transportation increases. With the increase in compactness of the shipment, large quantities per consignment could be transported, boosting the space economy (Nowacka et al. 2024). This is a vital factor in streamlining logistics as well as reducing the packaging requirement. The reduced transportation cost is the main requirement in the case of bulk shipment or air transport. Also, the dehydrated products lack the requirement of refrigeration, which totally cuts the need for a refrigerated cold chain system during storage as well as transport. In general, dried fruits and vegetables are less brittle than fresh ones, which are more prone to decay or bruising. A lower moisture level makes them less prone to deterioration, physical damage, or leaking, which enhances the quality of the product when it is delivered (Fellows 2017). To prevent spoilage, fresh products often need to be transported within a few days. More flexible and cost-effective transportation routes are made possible by the ability to move dried goods over longer distances and for longer periods of time without having to meet urgent delivery deadlines (Zhu et al. 2025). Since dried products are simpler to pack, stack, and label, dried goods sometimes arrive in standardised forms (slices, powders, or flakes). They can be handled with less chance of contamination or damage, which lowers the cost of labour and materials.

12.3 Preprocessing

Drying involves the removal of moisture by natural or forced hot air. In this process, the live tissues of fresh fruit and vegetables get modified drastically, and certain undesirable quality deterioration may happen. The faster we can achieve drying with minimal heat treatment helps retain most of the desirable quality parameters of the perishable commodities. Improving the drying times and achieving the desired product quality can be achieved by means of certain pre-treatments. These pre-treatments either help in the reduction of initial moisture, modify the tissues, inhibit certain bio-enzymes or biochemical reactions leading to minimal drying times, energy consumption and enhanced quality retention of the perishables (Yu et al. 2017). In large part, the pre-treatments performed before the drying process can reduce the drying time and air-drying rates (Fernandes and Rodrigues 2008). Epidemiological studies have demonstrated that the nutritional value of fruits and vegetables has a proven inhibitory effect on a number of diseases, age-related degenerative transformations, and overall health and well-being (Schieber et al.

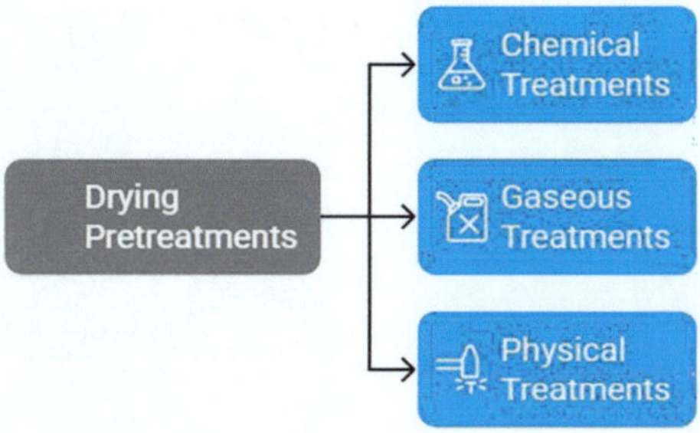

Fig. 12.4 Flow chart illustrating "Drying Pre-treatments" branching into three categories: "Chemical Treatments" with a flask icon, "Gaseous Treatments" with a gas cylinder icon, and "Physical Treatments" with a fan icon. Each category is represented in a blue box connected by arrows

2001). Retaining these nutrients even after drying or dehydration becomes mandatory for them to perform their functional properties. Pre-treatments perform various functions such as enzyme inactivation, improving rehydration property, reducing the oxygen access, fixing pigments, reducing degradation during drying, binding and removing certain moisture and hastening the drying process, making it more economical. Pre-treatments prior to drying or dehydration can be provided in three forms: physical, chemical and gaseous treatments. Figure 12.4 represents the three means of treatment with their examples.

12.3.1 Chemical Treatments

The chemical treatments include hyperosmotic solution, dip in alkaline solutions, acidic solutions and sulfite solutions. The graphical representations are provided in Fig. 12.5.

In the case of a hyperosmotic solution, the fruit tissues are submerged in a solution of a hypertonic (high solute concentration) substance, such as salt or sugar, in most cases. As water exits cells and enters the solution, some solute might get impregnated. This lessens the amount of water that needs to be eliminated through hot-air or air drying. This pre-treatment minimises drying time; lessens shrinkage, enhances texture; and influences taste through solute absorption. Under unfavourable conditions, the leaching of soluble nutrients and phenolics may happen.

12.3.1.1 Acid Dip

Utilising acids for pre-treatments aids in inhibiting microbial activity and enzymatic browning by lowering the hydrogen ion concentration of the solution (pH). These acids include citric acid, ascorbic acid, and other organic acids. It might also help in maintaining the colour of the produce upon drying or dehydration. Using an acid pre-dip before sulphuring improves a product's colour stability (Maskan 2001; Fellows 2017). Al-Jeddawi et al. (2023) inferred that the use of 2% ascorbic and 2%

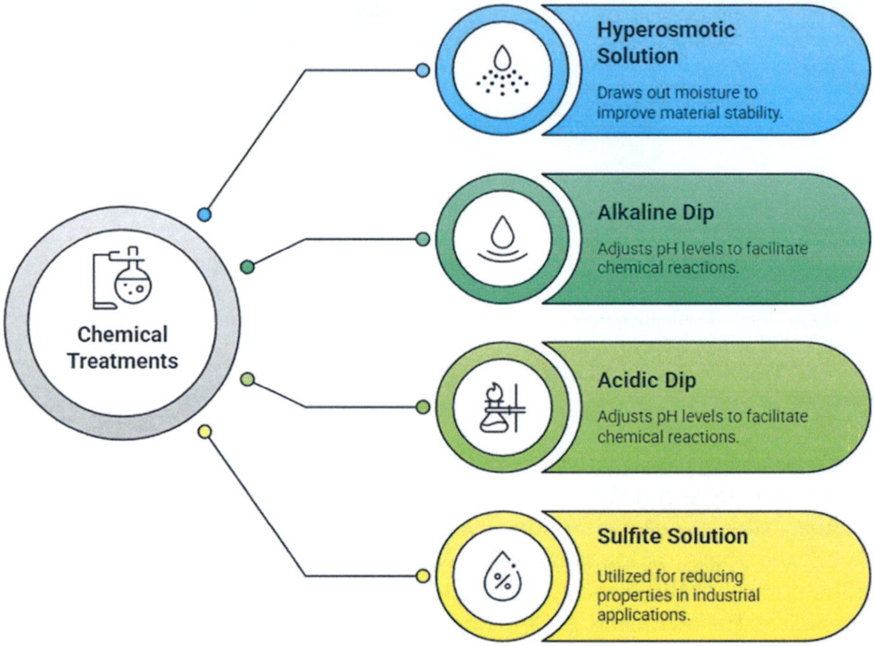

Fig. 12.5 Flow chart illustrating chemical treatments. Central circle labelled "Chemical Treatments" connects to four coloured sections: (1) Blue section: "Hyperosmotic Solution" with a droplet icon, description states it draws out moisture to improve material stability. (2) Green section: "Alkaline Dip" with a droplet icon, description mentions adjusting pH levels to facilitate chemical reactions. (3) Light green section: "Acidic Dip" with a laboratory flask icon, description also notes adjusting pH levels for chemical reactions. (4) Yellow section: "Sulfite Solution" with a droplet icon, description indicates use for reducing properties in industrial applications

citric acid reduced the browning of diced peaches, along with controlling the polyphenol oxidase activity. Better dried products in terms of colour or browning index were obtained in the case of mushrooms steeped in 0.5% citric acid (Elbah et al. 2017). Diced potatoes were pre-treated with citric acid (0.1%–0.3%) for a 10–30-min dip before hot air drying. It was observed that, in contrast to steam and water blanching, a 20-min dip at moderate concentration produced the best colour and the highest retention of vitamin C in the diced potato (Sun et al. 2020). Kuyu et al. (2018) observed that citric acid pre-treatment (3%) best retained the proximate composition, β-carotene, phenolic, and antioxidant capacity of sweet potatoes. Mango slices pre-treated with acid treatments exhibited improved colour and microstructure, increased reducing sugar, ascorbic acid, and phenols, particularly when dried using far-infrared (Udoro et al. 2020).

Constraints and Drawbacks

- Acid dips do not always have a big impact on drying rates; for example, citric acid dip did not have much of an impact on potato drying rates.

- Excessive acid flavour, changes in texture, or deterioration of delicate nutrients can result from higher concentrations or extended dip times.
- There could be a loss of solids or a leaching of soluble nutrients (like water-soluble vitamins).
- Inadequate cleaning or regulation may result in corrosion or chemical safety/residue problems.

12.3.1.2 Alkaline Dip

In case of alkali dip, the commodity is dipped in mild alkali (Potassium carbonate, sodium carbonate, wood ash, etc.) or in some cases with a combination of surfactants such as ethyl oleate, olive oil etc., the alkali treatments work by increasing the permeability of the surface or decreases the waxy surfaces, which indirectly speeds up the moisture removal (Deng et al. 2019). Deng et al. (2019) reported that alkali treatments effectively enhanced drying rates but at the same time leached some soluble nutrients. It was observed that potassium carbonate performed better than sodium carbonate as a pre-treatment agent in the case of grapes (Özcan and Gökmen 2016). Jain et al. (2018) found that alkali emulsions, such as potassium carbonate and olive oil, under warmer conditions (50°), retained the quality of raisins better with enhanced drying rates.

Drawbacks and Issues

- Excessive alkali concentrations or strength can harm fruit skin and result in unfavourable texture or flavour.
- Potential loss of some nutrients or soluble substances (should alkali lead to degradation or leaching).
- Safety and residue concerns: to guarantee there are no hazardous residues, proper washing and neutralisation are required.
- Possibility of elevated contamination risk in specific circumstances (e.g., the Ochratoxin A (OTA) study: some pre-treated samples had OTA earlier in long storage).

12.3.1.3 Sulphating

Sulphating, referred to as "sulphating," "sulfurising," or "sulphating," involves treating fruits and vegetables with sulphite, bisulphite, or metabisulphite salts before drying. These substances suppress non-enzymatic browning (Maillard reactions, etc.), act as antioxidants, inhibit enzymes (particularly oxidative enzymes), aid in microbial control, and may also help preserve colour and phenolics by halting oxidation. However, there are issues with chemical residues and regulatory limits; excessive use may impart an off-flavour, and sensitive people may experience health problems. Abbas and Soliman (2020) inferred that when dates were dipped or khalal-staged in sodium metabisulfite solution before drying, the colour greatly improved. Boudboud et al. (2023) found that $Na_2S_2O_5$-treated grapes at a 5% concentration of metabisulfite enhanced colour retention, decreased browning, and maintained the raisins' palatable appearance. Although they work well for preservation and colour retention, sulfite-based pre-treatments have a number of disadvantages.

Risks and Drawbacks

- Health and residue issues: Children and sensitive people may be at risk due to high levels of SO_2 or residual sulfite/metabisulfite, which can cause allergies requires oversight and regulation.
- Phenolic and antioxidant losses: Depending on exposure and time, some bioactive compounds may still be impacted or lost even though colour may be maintained.
- Taste/off-flavour: If excessive sulphite is not controlled, it may result in off-flavours or a "chemical smell."
- Labelling and consumer acceptance: Regulations frequently call for labelling, and some consumers steer clear of sulphites.
- Environmental and regulatory restrictions: The use of SO_2 gas or fumigation may be subject to more stringent regulations in some areas, which have limits on permissible residual levels.

12.3.2 Gaseous Pre-treatments

The gaseous pre-treatments include treatment using ozone, sulphur dioxide, or carbon dioxide gases. The graphical representation is in Fig. 12.6.

12.3.2.1 Ozone Pre-treatment

The technique of exposing the product (whole or sliced fruits and vegetables) to ozone, either gaseous or aqueous, usually for a brief period of time, prior to drying, is known as ozone pre-treatment. Ozone treatment helps in enhancing drying kinetics (removal of moisture more quickly), minimising the microbial load both before and during drying, maintaining the flavour, colour, and bioactive substances, and also minimising the degradative alterations such as enzymatic browning (Botondi et al. 2021; Xue et al. 2023). Cyclic zonation (gaseous ozone) of strawberries: 30 min every 12 h at 10 and 100 ppm for 3 days at room temperature preserved hardness, prevented enzymes that break down cell walls from deteriorating texture, etc. (Piechowiak et al. 2022). Gaseous ozone-treated fresh-cut pitaya fruit during storage (at $\approx 8 \pm 2$ °C for 4 days) was found to conserve the quality by inhibiting microbial growth, preventing weight loss, and preserving soluble solids and titratable acids. It also caused phenolic compounds to accumulate (Li et al. 2022b). Additionally, ozone-based pre-treatments have drawbacks and restrictions.

Risks and Drawbacks

- An excessive ozone concentration or exposure duration may cause tissue damage, which might affect the product's quality and structural integrity.
- Inconsistent effects on flavour and aroma compounds can occasionally result in unfavourable changes. Another issue is safety because ozone is a highly reactive

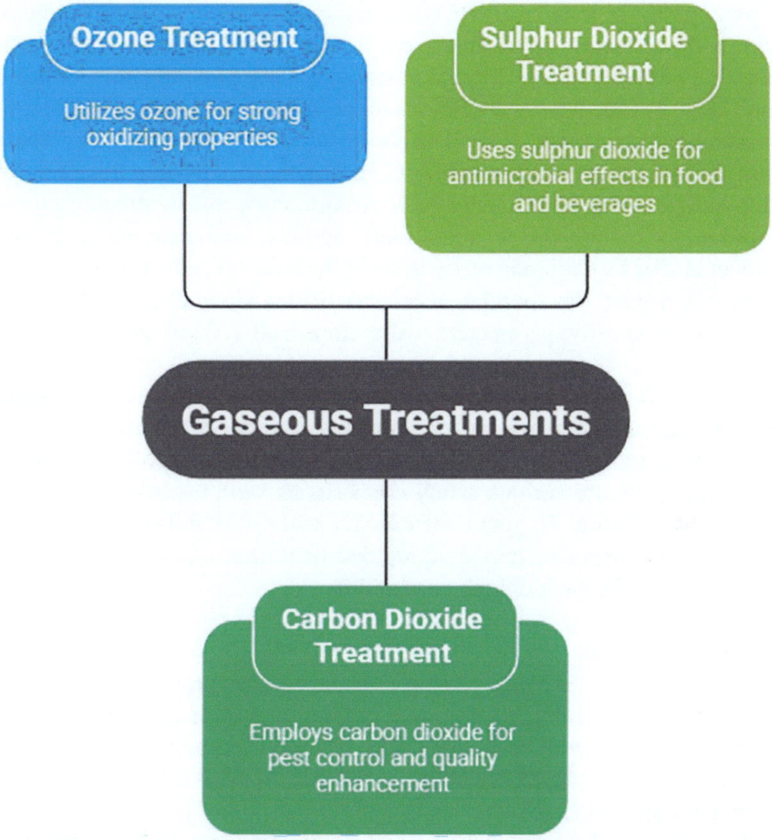

Fig. 12.6 Flow chart illustrating different gaseous treatments. The central node labelled "Gaseous Treatments" connects to three treatments: "Ozone Treatment" (utilises ozone for strong oxidising properties), "Sulphur Dioxide Treatment" (uses sulphur dioxide for antimicrobial effects in food and beverages), and "Carbon Dioxide Treatment" (employs carbon dioxide for pest control and quality enhancement). Each treatment is in a separate coloured box

gas, and handling it incorrectly can expose workers to health risks through skin exposure or inhalation.

- Ozone treatment may occasionally result in higher water loss during the drying process, which may have an impact on texture and rehydration ability.

12.3.2.2 Carbon Dioxide Pre-treatment

Prior to traditional drying, supercritical CO_2 ($scCO_2$) can remove a significant amount of moisture by penetrating the matrix and removing free or bound water under pressure (Dawson et al. 2022). By bringing the water content down to an intermediate level, this lessens the strain on the drying process that follows. Intercellular expansion, pore formation, or cell wall disruption due to CO_2 gas or CO_2 during maceration can speed up moisture diffusion (Takeya et al. 2016). CO_2

under pressure (or in a supercritical state) can inactivate enzymes or suppress metabolic activity, reducing spoilage or enzymatic browning that would otherwise degrade the product during drying or storage. This pasteurisation effect can be obtained without causing any alterations in the colour, flavour, or any other physical and mechanical properties of food (Zhou et al. 2015). Carbonic maceration at a pressure of 3.4 bar, temperature of 4.2 °C, in strawberries prior to hot air drying was found to reduce the drying time by 30%. Additionally, this treatment improved the dried product's colour (hue), antioxidant activity, and ascorbic acid retention (Ozcelik et al. 2022). Carbonic maceration in the case of seedless grapes and tomato slices prior to drying was found to accelerate drying kinetics. Also, the colour, texture, and other quality parameters were improved (Akyol et al. 2025; Turgut et al. 2017).

The use of CO_2-based pre-treatments and drying techniques has a number of drawbacks, such as expensive and complicated equipment because pressure chambers, high-pressure CO_2 supply systems, and supercritical condition maintenance are required. There are serious safety concerns as well, especially when working under extreme pressure. To guarantee efficacy and avoid unfavourable results, process variables like pressure, temperature, treatment time, and CO_2 flow must also be carefully adjusted for each unique commodity.

Risks and Drawbacks

- If these treatments are not properly managed, the product may develop gas entrapment, off flavours, texture deterioration, or structural damage.
- Another difficulty with scale-up is that, although promising in the lab or on a pilot scale, supercritical CO_2 drying techniques can be expensive to implement and run at the industrial level.
- Furthermore, careful consideration and adherence are needed to regulatory and food safety concerns, such as the possible effects of residual or decomposed CO_2 on the finished product.

12.3.2.3 Sulfur Dioxide (SO_2) Gas Fumigation or Sulphuring

One of the earliest and most popular chemical pre-treatments for drying and dehydrating fruits and vegetables is sulfur dioxide (SO_2) gas. The fruit or vegetable slices being pre-treated by sulfur dioxide (SO_2) gas are to be placed within a controlled sealed chamber for effective fumigation. The main uses of SO_2 are to stop enzymatic browning, maintain colour, stop microbes from growing, and help some vitamins—most notably vitamin C—stay in place while drying and storing (Yin et al. 2025; Hamzaoğlu et al. 2018; Sen et al. 2015; Salur-Can et al. 2017). Fruits and vegetables are usually treated by subjecting them to SO_2 gas in regulated fumigation chambers. When used as a pre-treatment for drying, sulfur dioxide (SO_2) works mainly through its antimicrobial, enzyme-inhibiting, and antioxidant qualities. It enhances the drying kinetics by modifying the tissue permeability. Fruits and vegetables retain their natural colour while drying because SO_2 prevents the activity of polyphenol oxidase (PPO), the main enzyme that causes enzymatic browning. Also, SO_2 scavenges reactive substances and oxygen radicals, which are found to interfere with pigments and nutrients. The antimicrobial effect of SO_2 is well documented with its interaction with microbial cellular metabolism. These concepts prove in the role of SO_2 in the shelf-life extension of the product. Altering the permeability of plant tissue facilitates moisture diffusion during heat transfer. Depending on the nature of the commodity and the expected duration of preservation, the SO_2 concentration may vary from 0.25% to 0.5% and the period of exposure may range from 2 to 6 h (Deng et al. 2019; Yin et al. 2025). When longan fruits were exposed to SO_2 treatment, the treatment not only prolonged the post-harvest shelf life but also delayed the deterioration of the aril and greatly enhanced the peel's appearance and colour (Mahfuzur et al. 2024).

12.3.3 Physical Treatments

The physical treatments include various simple traditional as well as advanced technologies, as listed in Fig. 12.7.

12.3.3.1 Abrasion/Peeling/Puncturing/Piercing

The mechanism of enhancing drying involves decreasing the moisture diffusion resistance by removing or weakening the outer skin or cuticle, or due to the formation of microchannels. Research has demonstrated that abrasion, such as that of grape peels, can significantly shorten drying times as much as 60%–67%. Pricking or creating holes in thick-skinned fruit and vegetables like dragon fruit, chilli, and potato promised improved drying rates (Yong et al. 2006). Physical pre-treatments like peeling, abrasion, or puncturing have a number of limitations, even if they can successfully shorten drying times and enhance rehydration properties. These techniques can be time-consuming and difficult to apply consistently on a large scale. Furthermore, the product's appearance may be compromised by the loss or disruption of the skin, which could reduce its consumer attractiveness. The quality and shelf life of the dried product may be adversely affected by exposure of the interior tissues, which can also accelerate oxidative deterioration and increase the risk of microbial contamination (Jangam et al. 2010).

12.3.3.2 Blanching and Steaming

Hot water blanching is a common, practically used thermal pre-treatment in fruits and vegetables before drying. Blanching basically softens the tissues, which facilitates the drying process. It also removes the air occupied in the intracellular spaces, thereby preventing oxidation and inactivating the enzymes (e.g. Polyphenol oxidases) involved in spoilage. An et al. (2022) found that applying a hot-air flow rolling dry-blanching method for 2–8 min significantly reduced drying time (from about 36.8% to 57.9%) as compared to untreated samples in a study on *Acanthopanax sessiliflorus* fruits. Wang et al. (2017) assessed the effects of high-humidity hot air impingement blanching (HHAIB) on red pepper. When used for brief periods of time (about 120 s), it reduced the polyphenol oxidase residual activity and also enhanced drying rates. Figure 12.8 represents the difference between the blanched

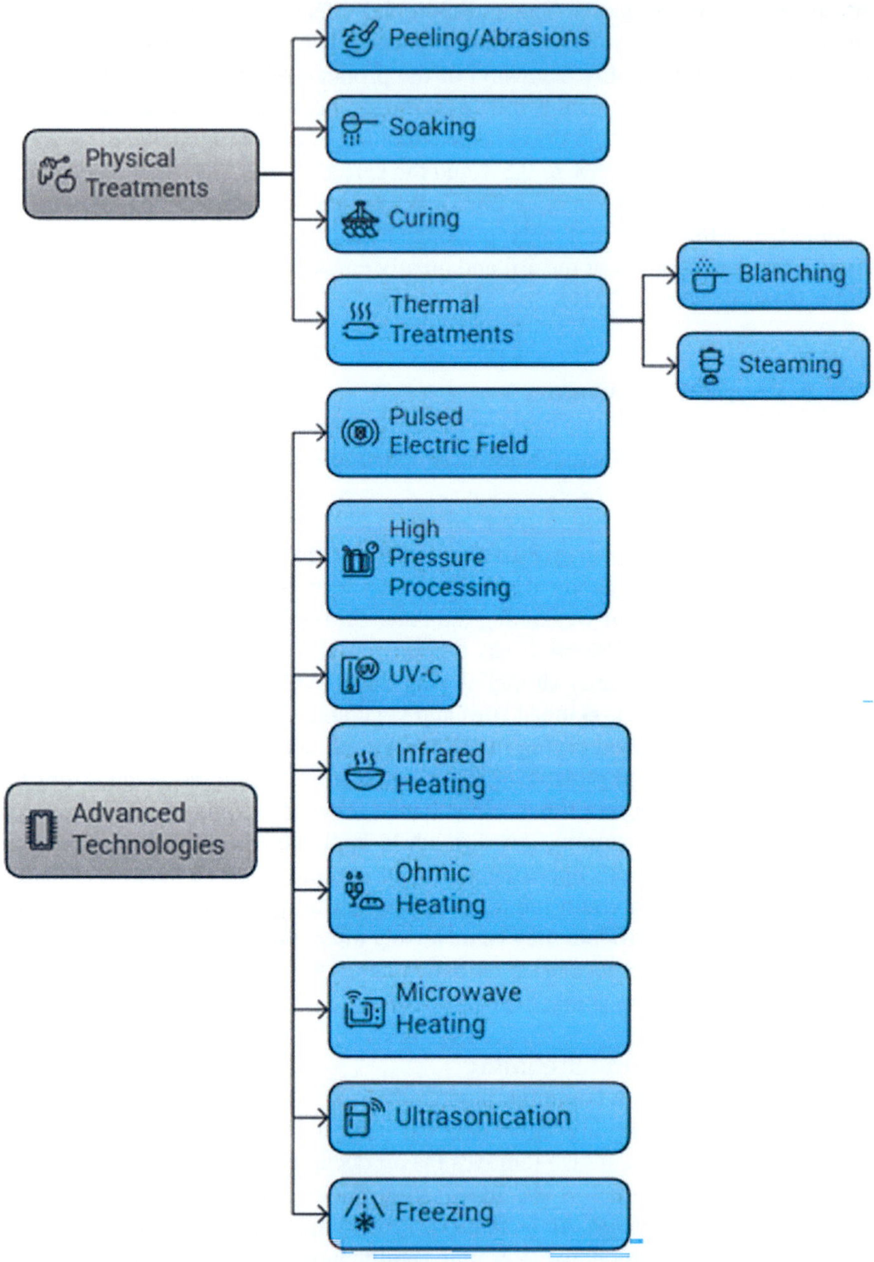

Fig. 12.7 Flow chart illustrating food processing methods. It is divided into two main categories: physical treatments and advanced technologies. Physical treatments include peeling/abrasions, soaking, curing, and thermal treatments, with subcategories of blanching and steaming. Advanced technologies encompass pulsed electric field, high-pressure processing, UV-C, infrared heating, ohmic heating, microwave heating, ultrasonication, and freezing

Fig. 12.8 Three images of broccoli are shown side by side. (**a**) Fresh broccoli with a vibrant green colour. (**b**) Blanched broccoli, appearing slightly brighter and more vivid. (**c**) Dehydrated broccoli, which is darker and more shrivelled in appearance

and unbalanced broccoli. Additionally, the texture and microstructure of the sample are negatively impacted by hot water blanching, which exacerbates the loss of soluble nutritional material (Eyarkai Nambi et al. 2016). Blanching with hot water also raises the pollutant charge and produces a lot of wastewaters. Therefore, it is quite appealing for the food business to adopt steam blanching or other alternative thermal blanching methods.

Compared to hot water blanching, steam blanching is thought to help retain the majority of minerals and water-soluble components because of its minimal leaching effect. A new thermal treatment technique termed superheated steam impingement blanching (SSIB), also known as high humidity hot air impingement blanching (HHAIB), combines the benefits of impingement with superheated steam blanching. It offers the benefit of producing a higher-quality product and blanching that is consistent, quick, and energy-efficient (Yang et al. 2017).

12.3.3.3 Ohmic Heating, Microwave Heating, and Infrared Heating

A thermal process known as ohmic heating (OH) occurs when an alternating electrical current passes through a body, such as a food system that acts as an electrical resistance, producing heat internally (Gomathy et al. 2015). For fruits and vegetables, it can serve as a substitute for a quick blanching technique (Icier et al. 2006). Regardless of the size and shape of the final product, the process can be completed in a few seconds, particularly for large, whole fruits and vegetables (Mizrahi 1996). When ohmic pre-treatment is carried out at a low frequency, the efficacy of the consequent drying is increased. Poojitha and Athmaselvi (2023) found that at an electric field strength of 26.66 V/cm, for a period of 30 s, blanching the garlic effectively enhanced the following drying process. The diallyl disulfide compounds were better retained by this ohmic pre-blanching treatment. After raisins were dried for 30 h, the samples' moisture contents decreased from an initial average of 81.2% wet basis to 23.9%, 23.8%, 38.3%, and 44.97% wet basis for the 30 Hz, 60 Hz, 7.5 kHz, and no pre-treatment, respectively (Salengke and Sastry 2005). Ohmic heating improved mass transfer during the osmotic dehydration of strawberries, according to Allali et al. (2010). As the temperature and ohmic blanching time increased, the amount of water lost and sugar gained increased proportionally.

Molecular dipole rotation and the agitation of charged ions in a high-frequency alternating electric field cause dielectric heating, which transforms the microwave energy absorbed by the materials into heat (Spigno 2016). Liu et al. (2015) found that, in comparison to steam and hot water blanching, microwave blanching shortened the drying time and improved the lightness (*L* value) of sweet potato samples by 28.6% and 24.42%–36.66%, respectively. According to Srimagal et al. (2017), samples of bitter gourds that were microwave-blanched and treated with ethyl oleate had a significantly shorter drying time overall than unblanched controls. In particular, only microwave blanching shortened the drying time from about 270 min (control) to about 240 min, and even shorter drying times were noted when ethyl oleate treatment was added. The shorter drying times associated with microwave heating might be due to the structural level alteration in tissue in the form of pore formation. This, in turn, promotes better moisture diffusion, resulting in reduced drying times. Under optimised conditions, the highest quality microwave-pre-treated dried product can be obtained. In the case of purple-fleshed sweet potatoes, which traditionally require 600 min for drying, they were dried in 6–12 min under microwave–vacuum conditions. The moisture diffusivity of the sample increased from 2.22×10^{-7} to 4.05×10^{-7} m^2/s. The final dried samples were found to retain more colour value, antioxidant and total phenolic content.

Infrared drying (IR) involves transferring radiant energy in the range of 780–1400 nm. Infrared radiation is characterised by surface-level penetration. The thickness of the commodity being dried is to be considered as an important factor to prevent uneven heating. The water activity and composition state of the food product are all factors to be considered while understanding the depth of penetration of IR radiations. The IR radiation produces heat on the product's surface by means of causing molecular vibrations of the water molecules. Foods containing higher moisture content tend to absorb more IR radiation, leading to faster heating rates, making this an efficient drying pre-treatment (Huang et al. 2021). IR radiation exposure creates micropores in the tissue cell walls, which in turn alter the microstructure, leading to improved moisture diffusion during drying. The overall product quality of infrared dried samples is superior in terms of retention of heat-sensitive materials, colour and flavour. Wu et al. (2017) observed that the oil content of IR-treated potato chips was 13.79% less compared to the other treatments performed before the frying of the chips.

The effects of ultrasound, catalytic infrared blanching, and their combination pre-treatments on the dehydration of blueberries were examined by Yu et al. in 2023. The findings demonstrated that following the osmotic dehydration of blueberries, total phenols and total anthocyanins were more easily retained following a 6-min catalytic infrared blanching. In addition to deactivating the enzyme activity in blueberries, catalytic infrared bleaching lessens the skin's hydrophobicity.

12.3.3.4 Non-thermal Physical Pre-treatments

Freezing (or freeze-thaw) as a pre-treatment alters the sample's interior microstructure (through the development of ice crystals, cell wall rupture, and microchannels), which may lessen the sample's internal resistance to moisture migration during later drying. Freezing can also weaken membranes, promote water transport, and occasionally cause partial cellular disruption. Gu et al. (2024) examined the effects of hot air (HD) and freeze-thaw pre-treatment (FT) on the drying properties, active components, and bioactivities of snake gourd. The findings demonstrated that FT pre-treatment lowered the drying time by 44% and decreased browning. The effect of freezing pre-treatment (different freezing temperatures) on vacuum microwave drying of persimmon slices was assessed in a recent paper by Tang et al. (2025). They discovered that chilling improved rehydration capacity, decreased apparent density, boosted expansion, and preserved pore structure. Better results were obtained while freezing at extremely low temperatures (such as −80 °C). Applying brief, intense electric pulses to biological materials is known as pulsed electric field treatments. Cell membranes are disrupted or permeabilised as a result of electroporation. Microchannels produced by electroporation greatly improve tissue permeability. Internal water transport is facilitated by this microstructural alteration, as seen clearly by scanning electron microscopy (SEM), due to the modifications caused at the structural level of the commodities exposed to electric fields. The resistance to transfer decreases, leading to an increase in moisture diffusivity, which leads to increased drying kinetics of the commodity (Kim et al. 2023). In the case of commodities such as parsnip and carrot, pulsed electric field pre-treatment reduced the drying times by 28% at 70 °C and 21% at 60 °C, respectively (Alam et al. 2018). Pre-treatment of plums with PEF with intensity of 1–3 kV/cm and 30 pulses at a 70 °C drying temperature resulted in shorter drying times (Rahaman et al. 2019).

High hydrostatic pressure processing is a non-thermal processing technique in which food, either liquid or solid, is exposed to high isostatic pressures (often between 100 and 800 MPa) for a brief period of time (seconds to minutes), typically at ambient temperatures. George et al. (2017) pre-treated ginger with high pressure (100–400 MPa) for 10 min before drying it. While higher drying temperatures decreased yields of both oleoresin and the bioactive molecule 6-gingerol, the moisture diffusivity increased (from ~2.03×10^{-9} to ~4.87×10^{-9} m^2/s) when dried in a cabinet dryer at 55–85 °C. Interestingly, both diffusivity (up to ~6.09×10^{-9} m^2/s) and extraction yields (6-gingerol and oleoresin) were improved when HPP pre-treatment was combined with dehydration at the lower temperature of 55 °C.

UV pre-treatments entail exposing the surfaces of food (or plant tissue) to ultraviolet radiation (usually UV-C, occasionally UV-B) prior to drying. Fernandes da Silva et al. (2025) used UV-C radiation (typically in combination with mild acids like acetic acid) as a surface pre-treatment on unripe bananas before convective hot-air dehydration. According to Fernandes da Silva et al. (2025), the UV-C pre-treatment reduced drying time by up to 12.9%, prevented darkening and browning, and showed photodegradative alterations in surface morphology associated with chemicals in banana sap. Using UV-C dosages of around 7.9–10.5 kJ/m^2 for 30–40 min before hot-air drying at 50 °C, Ramírez Fajardo et al. (2024) treated sliced vegetables (bell peppers, tomatoes, eggplants, zucchini, and melon). When compared to dried samples that were not treated, they discovered that these pre-treatments improved the preservation of colour parameters (L^*, a^*, and b^*), more successfully maintained the retention of bioactive compounds, and reduced the deterioration of colour.

12.4 Mechanisms and Principles of Drying and Dehydration

Two theories influence the drying or dehydration of any material. They are explained as follows.

12.4.1 Heat Transfer Theory

Heat is necessary to raise the temperature of food and to evaporate water at the surface. This mechanism occurs by means of molecular motion or transfer of molecules from one region to another region without any bulk material movement. This mechanism holds good for most of solid foods, whether heating or wherever the heat has to diffuse into the solid. When the heat transfer takes place by actual movement of molecules with direct contact is called conduction. The heat is transferred within the solid matrix by the actual molecular movement (Khan et al. 2022). Here, there exists direct contact between the heating surface and the food sample, and the rate of conduction is dependent on the thermal conductivity, contact resistance, and the moisture content of the sample. In the case of convective heat transfer, bulk mixing of material happens (Osama et al. 2024), which most likely appears true in the case of heat transfer from the drying medium (Hot air/gases) to the solid food surface. Another heat transfer mode is by radiation, where heat transfer happens in the space, and no medium is required (Microwave, Radiofrequency) (Fellows 2017; Jangam et al. 2010).

12.4.2 Mass Transfer Theories

The two primary phases of moisture movement during drying are as follows: first, the moisture diffuses from the food's interior to its surface, a process controlled by the effective moisture diffusivity, which is impacted by the food's structure, temperature, and any pre-treatments used (Deng et al. 2019). The mass transfer coefficient and air velocity regulate the process by which moisture evaporates into the surrounding air or vacuum after reaching the surface. Fick's second law of diffusion usually describes this moisture migration mechanism throughout the falling rate drying period. The molecular diffusion of water vapour in the air is proportional to mass transfer. The expression for molecular diffusion can be used to measure the rate of moisture diffusion. The mass diffusion of water vapour in air, the distance of water vapour movement within the product structure, temperature, and the vapour pressure gradient all affect the mass flow of moisture transportation (Deng et al. 2019; Chandramohan 2020).

The process of drying involves the combination of both heat and mass transfer, in which heat must enter the product's moist interior to evaporate moisture while the water that has evaporated simultaneously diffuses to the surface and is drawn out by the surrounding airflow. During this drying process, the moisture migration to the surface happens through a concentration gradient, whereas the moisture from the surface diffuses to the environment by the process called evaporation (Khan et al. 2018b). As drying proceeds, the product's surface gets drier, which lowers its thermal conductivity and slows down heat transfer. Consequently, the drying process gradually shifts from being heat-limited to mass-limited when internal moisture diffusion takes over as the limiting force. While the heat transfer is based on temperature gradient, the mass transfer involving moisture vapour diffusion is dependent on the vapour pressure on the liquid water surface. Initially, when the drying gets initiated, the product's surface will be saturated with moisture. When warm air is passed over the product, the dry air gains some moisture from the wet surface by means of evaporation and travels outside the product. When excess hot air is passed in the same manner against the product, the moisture from the centre of the product comes to the surface by a process called diffusion (Merger 2014). This mechanism of moisture transfer is illustrated in Fig. 12.9. The moisture evaporates from the surface at a constant rate. Initially, the surface drying takes place at a faster rate, and as soon as the moisture from the centre diffuses to the surface, the drying proceeds in a slower phase.

When we consider the plot of drying time versus the moisture content of the sample, we can obtain a typical drying process that happens in all food materials. Initially, certain cold food items require a warm-up time before getting heated up, whereas foods at room temperature do not require any warm-up time (Merger 2014). The next stage will be interrupted as a constant rate of drying, wherein the surface moisture from the material will be evaporated at a constant rate. When all the surface moisture is removed, the moisture from the centre portion of the material

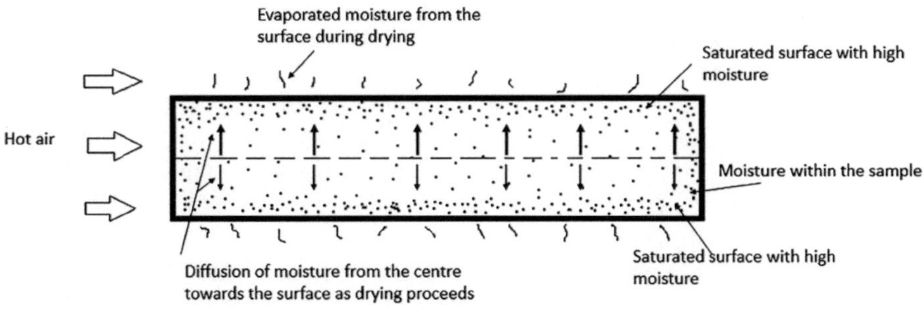

Fig. 12.9 Diagram illustrating the drying process of a sample. Hot air enters from the left, causing moisture to evaporate from the surface. The surface is labelled as saturated with high moisture. Arrows indicate the diffusion of moisture from the centre towards the surface. The process results in evaporated moisture leaving the surface

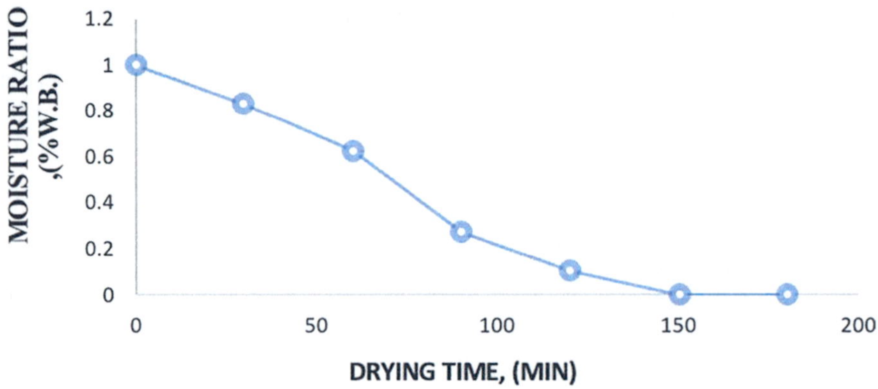

Fig. 12.10 Line graph showing the relationship between drying time in minutes (*x*-axis) and moisture ratio in percentage wet basis (*y*-axis). The graph depicts a downward trend, indicating that as drying time increases from 0 to 200 min, the moisture ratio decreases from 1.0 to near 0.0. Data points are marked along the line

diffuses to the surface and evaporates. This stage does not occur rapidly, but rather in a slow phase termed the falling rate drying period. During this phase, the heating of food material occurs as there is not much moisture left for evaporation. The transition of drying period from the constant rate towards the falling rate period is termed the critical moisture content (Fellows 2017). Figure 12.10 represents the drying rate curve generated during the drying of carrots when dried in a tray dryer.

The moisture removal from the carrot happened at a faster rate during the initial stages, and thereafter, after 120 min, the rate of removal decreased and attained a constant state after 150 min of drying.

12.5 Factors Affecting Drying and Dehydration

12.5.1 Product Properties

12.5.1.1 Shape of the Material, Size or Thickness, and Porosity of the Material

When size is considered a factor, three important dimensions are often used in the case of fruits or vegetables, which include circular, cylindrical, flat, or slab shapes (Sahay and Singh 2014). In the case of circular geometry, as in the case of drying small berries, plums, etc., the radius is the governing factor as it defines the distance required for drying the material from the centre to the surface. Even in the case of a cylinder, the radius is a vital factor, but the ratio of surface area to volume is greater in the case of cylindrical pieces. When slabs or slices are considered, the surface area to volume is still higher than a circle or a cylinder, resulting in the drying rates being much higher than the other two geometries. The flat bottom and top surfaces come in contact with the dry air inside the drier and act as a heat transfer surface area. The more the surface area, the more chance for more moisture to be evaporated at a given time. Thus, the factor controlling the drying here is one-half of the thickness. Thus, the thickness of a material also contributes to the drying rate. The thicker the material the time taken for moisture removal from centre to surface takes a longer time. Internal drying is slowed down because moisture must diffuse over a greater distance in thick materials (Fellows 2017; Mujumdar 2014). The more porous the material, the faster the drying rate. Pores allow moisture to escape more easily and can improve internal diffusion of moisture. By improving internal mass transfer, porous materials with wider pore size distributions have a major impact on drying performance (Keita 2021). According to Peters et al. (2023), the larger the pore size in the surface of the commodity better the permeability of moisture from within the commodity. The vapour and gas transport are facilitated at a faster rate with the increased pore size.

12.5.1.2 Initial Moisture Content

Moisture content is the vital factor to be considered in drying because moisture is responsible for the spoilage and reduction of shelf life of any commodity. Materials with more initial moisture content require higher drying times as well as more energy for drying. In the case of drying high moisture food, the cost drying rate period is dominated by the falling rate period as a larger amount of surface moisture removal is required than the internal moisture. The moisture diffusivity and each drying phase duration are largely dependent on the initial moisture content of any commodity, Mayor and Sereno (2004).

12.5.1.3 Surface Characteristics

The removal of moisture from a commodity and the rate of removal are also largely dependent on its surface properties, such as roughness, smoothness, texture, chemical composition, etc. A surface that is rough or porous has a greater effective surface area, which increases exposure to drying air and speeds up the removal of moisture. On the other hand, by decreasing water-air contact and obstructing heat and mass transfer, smooth or hydrophobic surfaces can prevent evaporation. Additionally, coatings or surface alterations may create resistant layers (like case hardening) that retain moisture inside and delay drying. Krokida et al. (2001) claim that changes in surface structure during drying can also affect permeability, shrinkage behaviour, and the drying phase transition. Specifically, in certain materials (like fruits or gels), surface sealing decreases moisture loss even when there is internal moisture availability. This causes the drying control to shift from surface evaporation to internal diffusion. Merger (2014) observed that drying of whole cranberries took a longer time to dry at 50 °C, as they contain a waxy cuticle that prevents the moisture removal from within the fruit. Whereas when the same fruit was sliced and dried under the same conditions, drying was completed at a faster rate.

12.5.1.4 Amount of Surface Available for Moisture Loss

A material's drying rate is largely determined by the surface area available for moisture loss. Since more moisture can always escape into the surrounding environment, a larger surface area improves evaporative mass transfer. This is particularly crucial when drying at a constant rate, when surface evaporation predominates. Increasing the exposed surface area, for example, by slicing, shredding, or spreading a material into thin layers, greatly shortens drying times and boosts energy efficiency (Mujumdar 2014). This surface area concept or influence on drying rates is considered a key phenomenon in industries to obtain a low-cost product with high quality.

12.5.1.5 Seasonal Variation of Material

With the variation in seasonal conditions, the biological and chemical character of the commodity gets altered. For instance, the size, plumpness, and juiciness of the fruit vary with the climatic conditions, rainfall, and sunshine of the respective growing seasons. Seasons accompanied by high rainfall result in a commodity with a softer texture and a high initial moisture content. This leads to increased chances of spoilage and requires more moisture removal for drying to be completed (Merger 2014). Nevertheless, products grown and harvested during the dry season will possess a denser tissue with less moisture, leading to less energy for drying and dehydration. This factor has an influential role in the drying behaviour of the material. The quality of the end product is greatly impacted by this factor. Thus, as per Simate (2001), understanding the seasonal variations serves as an important factor for obtaining a high-quality final dried product, but also for the optimal dryer selection.

12.5.2 Dryer Properties

12.5.2.1 Time, Temperature, and Relative Humidity (RH) of Air in the Drier

These properties govern the drying rates of the material within the dryer. Altering any of the above factors changes the quality of the final product. When the temperature of the dryer is increased, the moisture evaporation rate increases due to the creation of differential vapour pressure inside the drying chamber between the commodity being dried and the air being used for drying. This differential pressure serves as the driving force for the moisturisation and removal from the comfort. The heat transfer into the material increases with the increased temperature, leading to enhanced drying kinetics. However, the higher temperatures within the drying chamber due to heated air may also cause some detrimental effects on the commodity depending on the nature of the commodity, causing either shrinkage of tissue, case hardening, or tightened outer surface or even minimise the nutrition losses. The concept of drying time is considered as a function of air temperature and drying time. Higher drying times can lead to degradation in product quality or energy wastage, whereas shorter drying times lead to residual moisture in the comminuted or undried product. Thus, a careful consideration and combination of time, temperature and temperature is essential for obtaining a high-quality product with minimal energy input (Mujumdar 2014).

12.5.2.2 Volumetric Air Flow Rate and Flow Pattern

The drying efficiency and drying rate are strongly influenced by the volumetric air flow rate as well as the pattern of air flow. It is defined as the volume of air supplied into the chamber for drying, expressed usually in m^3/h. As the flow rate is increased, the mass transfer and the convective heat coefficients increase, leading to the removal of large amounts of moisture from the commodity being dried. The initial constant rate period of drying gets enhanced with increasing air flow rate. But this increased airflow can also lead to quality changes that become unacceptable, such as case hardening (Fellows 2009). The air flow inside the drying chamber can be co-current, counter-current or even cross-flow/mixed flow. Co-current air flow minimises thermal stress on the commodity, wherein counter-current flow improves drying efficiency. According to Geankoplis (2003), good-quality drying, depending on the commodity, can be guaranteed with the correct optimisation of airflow rate and direction. The velocity of air used for drying is another critical factor in drying, as the increase in air velocity absorbs and aids in moisture removal from the commodity or the surface of the food being dried.

12.6 Types of Dryers, Design, and Operation

The driers are classified into different types based on either the mode of heat transfer, as direct or indirect, based on the type of operation, as batch or continuous type, and based on the air flow pattern, as co-current or counter-current flow (Jangam et al. 2010). In direct dryers, the drying medium comes into direct contact with the material to be dried, providing the heat needed for convectional drying. Here, the mode of heat transfer is by convection. Air that has been heated before entering the drying chamber serves as the drying medium in this instance. It can be heated by passing it over heated metal surfaces, where it absorbs heat that it subsequently distributes to the object being dried, or by running it through a burner's flames. In indirect dryers, the drying material receives heat without coming into direct contact with the heat transfer medium; instead, heat is delivered by conduction from the heat transfer medium to the wet solid (Rahayu et al. 2023). One technique for indirect heating is to employ hot surfaces, like the exterior of revolving metal "drums" that have steam running through them.

In a batch dryer, a set amount of wet material is loaded, dried, and discharged before the subsequent cycle starts. The drying process takes place in distinct time intervals. The material is put into a dryer chamber and heated for a predetermined amount of time using conduction, convection, or radiation. The vapour pressure gradient, when the vapour pressure of moisture in the material surpasses the partial pressure of water vapour in the surrounding air, drives the evaporation of moisture in the substance as the temperature rises. Rahayu et al. (2023) observed that with the increase in drying temperature, the vapour pressure on the surface of chilli increased, leading to enhanced drying rates. Khan et al. (2018a) confirmed the fact of a vapour pressure gradient created due to an increase in surrounding air temperature/heating medium temperature, leading to an increase in moisture migration and evaporation.

The system is ready for the following batch when the dry material is eliminated, and the appropriate moisture content is reached. A continuous dryer is a system that operates in steady-state circumstances by continually feeding wet material into the drying unit and continuously releasing dried material. Depending on how the material is physically shaped, it usually enters the dryer via atomisers, screw feeders, or conveyors. As the material passes through the drying chamber, heat is continuously delivered, causing moisture to evaporate. The feed rate and dryer design carefully regulate the residence time, or how long the material stays within the dryer, to guarantee even and effective drying (Friso 2020; Wei et al. 2021). Continuous driers are practically applicable in the case of commercial or large-scale applications, unlike batch driers.

Hot air and wet food material flow in the same direction in co-current dryers, usually entering the drying chamber from the same end. In order to remove moisture quickly in the beginning, the air and food product enter at the same location during operation, where the warmest air meets the wettest material. Overheating is less likely as they both gradually cool down as they pass through the chamber together. From a thermodynamic standpoint, this arrangement ensures that the material is exposed to colder air as it gets drier and more heat-sensitive, preventing degradation, while the significant temperature gradient at the entry improves initial drying efficiency (Jangam et al. 2010). This setup works especially well for foods that are sensitive to heat, such as milk powder, coffee, eggs, and flavourings, which are frequently processed with spray dryers. While drum dryers may also use co-current airflow for gentle drying, fluidised bed dryers (in co-current mode) are used for drying snacks, cereals, and particle foods. In the case of counter-current drying, the hot air-drying medium as well as the food material being dried enter into the drying chamber in opposite directions. This means that the wettest material gets in contact with the coolest air at the exit of the drier, whereas the driest material gets in contact with the hottest air (Rahayu et al. 2023). This mode of heat transfer is more suitable for drying shelf-stable foods, which require a temperature differential at the drying end (Rahayu et al. 2023). This configuration's primary benefits are the utilisation of high-temperature drying air and increased energy efficiency in attaining low moisture content. It is less appropriate for food items that are sensitive to heat because of its disadvantages, which include a higher chance of overheating or quality loss at the dry end (such as case hardening or colour changes).

Though there are already different parameters on which the driers are classified, the recent terms of drier classification have been proposed based on the different generations, which depend on the type of energy used, the medium of heat transfer (Zalpouri et al. 2022; Abdul Halim et al. 2023; Kaveh et al. 2025). Thus, the classification includes the first generation, which includes the use of hot air for drying. Examples include drying of fruits and vegetables using a tunnel drier, a Kiln drier. Second-generation drying involved the utilisation of heat to dry slurry, liquid, or semi-liquids that basically resulted in powders (Langford et al. 2018). Example spray and drum driers. The third generation drying included non-thermal methods of drying that retain most of the nutrients in the foods (Freeze drying) (Kaveh et al. 2025). The latest generation of dryers, namely fourth-generation dryers, which

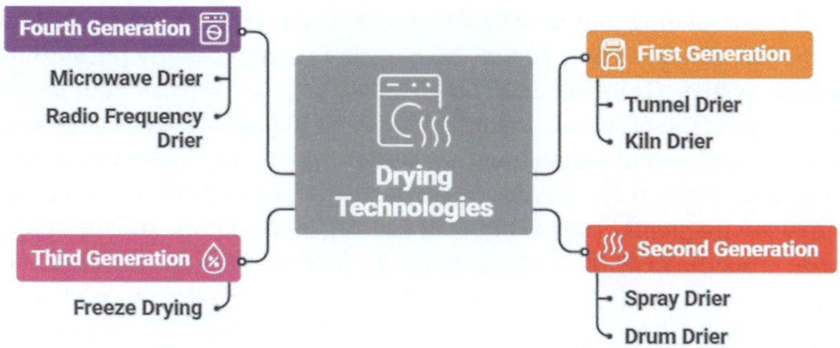

Fig. 12.11 Flow chart illustrating different generations of drying technologies. The central node labelled "Drying Technologies" branches into four categories: "First Generation" with Tunnel Drier and Kiln Drier; "Second Generation" with Spray Drier and Drum Drier; "Third Generation" with Freeze Drying; and "Fourth Generation" with Microwave Drier and Radio Frequency Drier. Each category is colour-coded and includes an icon representing the technology type

utilise less energy to produce high-quality products (Abdul Halim et al. 2023). Examples include microwave and Radio frequency driers. The schematic representation of dryers classified based on generations is provided in Fig. 12.11.

12.6.1 Solar Driers

An alternate energy source for drying is solar radiation, which is converted into solar thermal energy. The dryer cabinet consists of mirrors, a glass cover, collectors, solar trays, and black-painted trays to boost solar energy and quicken drying. A portion of the total solar radiation that strikes the glass cover of the cabinet dryer is reflected into space, while the remainder is transmitted within the cabinet. After that, some of the radiation is reflected from the crop's surface, and the remainder is absorbed by it. This raises the crop's temperature and releases long-wavelength radiation that the glass cover prevents from escaping into the atmosphere. The different types of solar dryers include direct type, indirect type and hybrid solar dryer. The selection of a solar dryer depends on the characteristics, quantity and designation of the product (Hii et al. 2012; Kilanko et al. 2019).

12.6.2 Cabinet Tray and Pan Dryers

A popular batch-type hot air dryer for dehydrating fruits and vegetables is the cabinet tray dryer. It includes a drying chamber with several racks of perforated trays, a heating unit (gas, steam, or electric), air ducts, an exhaust vent, a fan or blower for air circulation, and a thermostat to control the temperature (Jangam et al. 2010; Kilanko et al. 2019). Fruits and vegetables are sliced and cleaned before being put on trays inside the insulated cabinet. The fan draws in fresh air through the heater coils, filters it by removing dust with a screen, and then blows it across the materials on trays before sending it to an exhaust system. The produce absorbs surface moisture as the hot air moves over it, and the moisture is subsequently released through the exhaust system. This procedure of moisture removal repeats until the desired level is reached. The efficiency of the drier can be maintained by heat-proofing the doors with a gasket and maintaining an air-tight condition. The ease of operation and consistent drying make this drying method suitable for small-scale food processors. They are frequently used to dry fruit and vegetable pieces, and the drying time can range from 10 to 20 h, depending on the food and the desired final moisture content.

12.6.3 Tunnel Driers

In the case of tunnel dryers, the trays or racks with the material to be dried are placed on carts and driven through lengthy tunnels where warm air is blown over the trays or racks (Seth 2018). Convective heat and mass transport are the foundations of tunnel dryer drying. In order to facilitate the evaporation of surface moisture and subsequent interior moisture migration to the surface, heated air is blasted over or through the product trays. Regarding trolley movement, airflow can be configured as cross-flow, counter-current, or co-current.

The carts are pushed into the dryer's "front end" after being manually loaded. The wheels of the cart can be grasped in an assembly that will draw them through the tunnel, or they can be attached to a chain that will do the same. The length of time the material is in the dryer depends on how quickly they are removed. The carts are pushed out and emptied when they get to the dryer's finish. After that, the empty carts are brought back to the dryer's beginning so they can be loaded again and run through the machine with a new load of wet product. Depending on the nature of the commodity, the operating conditions can be optimised for achieving an efficient heat and mass transfer to occur for drying. The product thickness, characteristics, and the initial moisture content decide the time taken for drying the particular commodity, as in the case of longan fruit drying (Tanwanichkul and Suriyasupapong 2022) and Mango leather drying (Mohod and Khandetod 2021). However, most of the horticultural commodities, like fruits and vegetables, are suitable to be dried in a tunnel drier.

12.6.4 Fluidised Bed Drier

In a fluidised bed drier, as the name suggests, the solid material is dried by means of the fluidisation principle, namely suspending the particles in an upward direction in a stream of hot air. This is basically a convective type of drying wherein the heat and mass transfer in the material is achieved by creating a fluidisation of the sample in air. This phenomenon causes a rapid and even drying of materials (Ling et al. 2022). The fluidised bed dryer contains a perforated tray at the bottom of the dryer where the samples are initially placed. Then the hot air from the bottom of the tray is directed to lift the samples, making them suspended in the air. In this process, the density of the commodity being dried decreases and moisture is lost. The air flow rate is the critical factor controlling fluidisation of particles (Kaveh et al. 2025). Improper air flow rate may lead to blowing off the food material against the apertures of the chamber's ceiling, nor is it sufficient to lift the product from the bottom tray, but rather to maintain the samples in a suspended state. Small and intact materials are suitable for fluidised bed drying, such as curry leaves, peas, and chopped carrots. The consistent moisture removal, faster drying rate, and quality retention are the merits of fluidised bed drying. Nonetheless, the demerits include unsuitability for bulky, sticky or large irregular particles.

12.6.5 Spray Drier

Spray drying dries a liquid or mild slurry commodity into powder by means of evaporative drying. This technology is adopted in the preparation of most of the fruit and vegetable powders (Nik et al. 2024). The mechanism of drying involves feeding the material into the dryer through an atomiser in the form of fine droplets through a nozzle or rotary atomiser. The atomiser functions as a metering device as well as aids in dispersing the feed material into tiny droplets. The hot air enters the chamber either co-currently or in counter-current form, creating direct contact with the product being sprayed. This contact evaporates the moisture from the product. The fine powders after drying are collected in the cyclone separator from the air stream (More et al. 2014). Spray driers mostly operate at a drying temperature of 150–250 °C and maintain an outlet air temperature of 50–80 °C while handling food materials (Geranpour et al. 2020). Factors such as atomiser speed (Controlling droplet size), temperature of the inlet hot air, air velocity, and solid content in the initial product to be sprayed influence the quality of the final spray-dried product (Karas et al. 2023; Ur Rahman et al. 2022).

In the food business, spray drying is frequently used for items like flavourings, milk powder, coffee, egg powder, and fruit juice powders (such as orange and mango). Since the quick drying reduces thermal damage, it is especially appropriate for materials that are sensitive to heat. The creation of homogeneous, fine, free-flowing powders, continuous operation, quick drying, and compatibility with thermolabile (heat-sensitive) substances are the primary benefits of spray dryers (Assadpour and Jafari 2019). Furthermore, it is simple to regulate the process to obtain the appropriate moisture content and particle size. High energy consumption, expensive equipment and maintenance, and the requirement for exact control over feed properties (viscosity, solids content), as well as operational parameters, are drawbacks. Furthermore, spray dryers work best with liquid or slurry feeds and are not appropriate for solid or fibrous materials.

12.6.6 Drum Drier

In order to warm a set of metal drums to the appropriate temperature, drum dryers employ heat from steam or hot water. The directions in which the drums revolve are opposed. The material to be dried is placed into the "nip," or the little space, that separates the drums. Either a "mushy" solid or a sticky liquid can be this substance. For drying pulpy and extremely viscous foods, drum drying uses less energy. Conduction heat transfer is the scientific basis for drum drying. In this method, the wet feed material that forms a thin layer on the drum receives thermal energy directly from the hot, rotating drum surface (which is internally heated by steam). Drying results from the heat's quick vapourisation of the feed's moisture (Seth 2018). According to the laws of thermodynamics and heat transfer, the temperature differential between the heated drum surface and the cooler wet material causes the phase transition from liquid to vapour. Then, while the drum rotates, a stationary doctor blade removes the dried product. Advantages of drum drying include high drying rates, short processing times, relatively low energy consumption, and the ability to handle viscous or heat-sensitive materials. However, it also has disadvantages, such as limited flexibility in handling different product types, potential for uneven drying, and degradation of nutrients or flavours due to high surface temperatures.

12.6.7 Foam Mat Drier

Foam mat drying involves drying of foamed food concentrates at atmospheric pressure. Convective heat and mass transport form the basis of the drying principle in foam mat drying. When food that is liquid or semi-liquid is whipped into a froth and spread out as a thin mat, heated air can readily travel through due to the increased surface area and porous structure. The moisture within the foam evaporates when heat is transported from the heated air to the foam mat. Water vapour diffuses out via the foam's linked air cells at the same time. Compared to conventional techniques, this dual mechanism of heat conduction to the wet material and moisture diffusion through the foam structure produces faster and more consistent drying. Additionally, the foam structure lowers the drying temperature needed, preserving tastes and nutrients that are sensitive to heat. Three popular techniques are used to create food foams: (a) sparging, which involves introducing a lot of gas into the liquid through a porous sparger, turning a lot of the liquid into foam; (b) whipping (beating), which uses specialised equipment to continuously agitate the liquid to apply more mechanical stress and shear, producing a more uniform and stable foam than sparging; and (c) shaking, which is a less popular technique that involves vigorously shaking the liquid to create foam (Mounir 2018).

Foods that are high in moisture, extremely sensitive to heat, thick, sticky, and challenging to dry are all excellent candidates for this method (Prajapat et al. 2023).

By using the foam-mat drying technology, aqueous food concentrates are air-dried at a lower temperature, reducing the amount of moisture in the food and creating a stable porous sheet that resembles a honeycomb and can be ground into powders. Because of its high quality, ease of reconstitution, and comparatively quick drying time, this drying method is appropriate for heat-sensitive food ingredients.

12.6.8 Vacuum Dryers

Vacuum drying involves the drying of food materials with the aid of a vacuum created under sub-atmospheric pressure in order to remove moisture more efficiently. Due to this concept, the product gets dried at a lower temperature, resulting in a superior quality (Martínez et al. 2023). Once the product is placed within the vacuum chamber, the pressure is reduced within the chamber, and a mild heat is applied to allow for the moisture evaporation from the commodity. The vacuum system constantly extracts the vapour, ensuring effective drying while maintaining colour, flavour, aroma, and heat-sensitive nutrients. Because of the low oxygen content, vacuum drying also lessens oxidation. It is very useful for drying delicate and valuable fruits and vegetables with little loss of quality, even though it may take longer to dry and demands a larger initial expenditure.

12.6.9 Freeze-Driers

The method of sublimating water through the direct conversion of solid (ice) to vapour is known as freeze drying. There are three stages in freeze drying, which are freezing the commodity, drying by sublimation, and further desorption drying (Liu et al. 2021; Nowak and Jakubczyk 2020). This is accomplished by first freezing the product to remove all moisture, wherein ice crystals are nucleated and grow large in size, breaking the cell structure, then applying regulated heat when the surrounding pressure is lowered to below the ice's vapour pressure at that temperature (Ma et al. 2022). The food is frozen at extremely low temperatures, the majority of the ice is removed under high vacuum during primary drying, and any leftover unfrozen water is removed during secondary drying. It is frequently used to stabilise biological materials, medications, proteins, vaccines, bacteria, and animal cells, as well as high-quality food. After the water is sublimated out of the freeze-dried items, a highly porous structure is created, and lyophilisates rehydrate instantly. The quality of the freeze-dried material and the economic impact of the process should be taken into account. A product's price is mostly determined by how long it takes to freeze-dry. As a result, process parameters and other course-related factors are frequently chosen to minimise time. The qualities of the final product may deteriorate if settings are set to expedite the procedure.

12.7 Selection of Dryers

1. The production process, either batch or continuous, should be selected as the first parameter for the selection of the dryer.
2. The moisture content and water activity of the selected commodity act as the major factors in dryer selection. The suitable dryer to remove the excess moisture effectively and efficiently should be considered.
3. The thermal sensitivity of the product is a deciding factor in preserving heat-sensitive compounds from degradation at higher temperatures.
4. The final structure, shape, form and textural requirements of the final product are to be considered while selecting a dryer. A dryer that retains the cell structure, shrinkage or case hardening to achieve the target final product should be considered.
5. The stickiness of the product due to high sugars leads to a plasticisation effect in an unsuitable dryer. Dryers such as foam mat driers or vacuum driers can be considered in such cases.
6. In case of powdery or flaky products, requirements drum drying, spray drying or foam mat drying are suitable to consider, but at the same time, the most economical and efficient dryer solving.
7. The drying kinetics of the selected commodity in the particular dryer needs to be studied, as faster drying of a selected commodity may cause case hardening, whereas a very slow rate of drying also causes spoilage of the commodity.
8. The cost and energy efficiency of the dryer based on the quality attributes of the required product are to be considered.
9. A dryer that includes pre-drying stages like evaporation, preconditioning, etc., should be considered.
10. The commodity that undergoes preconditioning before drying also affects the dryer selection.
11. The selected dryer should avoid over-drying of the product.
12. The pressure inside the selected dryer suitable for the commodity of choice should be considered.
13. Safety aspects of the dryer, such as fire and toxic hazards, should be considered when operating such a dryer.
14. The need for automation under certain circumstances should be included as a potential factor in dryer selection.
15. The space occupied by the selected dryer in the plant facility acts as a practical consideration for drying installation.
16. The fuel cost and electricity cost involved in the choice of a dryer also decide the dryer selection for the selected commodity.
17. The dryer selected should comply with the environmental regulations of the society.

12.8 Commercial Application of Dried Horticultural Commodities

Dried or dehydrated products have a significant role in global food systems due to improved shelf life and availability (Shirsath 2025). The burden on transportation costs due to bulk handling also gets reduced drastically, which is the prime factor to be considered in trade and export. Drying or dehydration of the most perishable commodities, such as fruits and vegetables, turns out to be advantageous during seasonal demand and regions with inadequate cold storage facilities. Most of the dried or dehydrated products serve as a primary ingredient in many ready-to-eat, ready-to-cook, snack, bakery, and baby food products. This property facilitates their domestic as well as international demand and trade. Dried goods such as dehydrated onions, garlic, tomatoes, green peas, and fruit slices represent a significant portion of agricultural exports in export markets, particularly from the USA, China, Thailand, and India (Khante 2021). India leads in the export of dehydrated onions, and now the demand follows for spiced vegetables, mango, chips, etc., with the advancement in drying technologies and packaging advancements, the global competitiveness of the horticultural commodities has further increased (Soni and Modi 2024). These advancements have not only reduced the overall post-harvest losses but also boosted the farmers' income. New opportunities for extensive commercial uses have been created by the recent incorporation of dehydrated horticulture products into both institutional and industrial supply chains. Dried goods are used in large quantities by educational institutions, medical facilities, military installations, and disaster relief initiatives to create wholesome, long-lasting meals. Nevertheless, direct e-commerce has facilitated the growth and sales of dehydrated fruit and vegetable products, such as powders and dried snack products, which have boosted local as well as international markets.

References

Abbas M, Soliman A (2020) Impact of pretreatments and drying temperatures on quality of Siwi and Sakkoti dates. Pak J Biol Sci 23(5):658–666. https://doi.org/10.3923/pjbs.2020.658.666

Abdul Halim L, Basrawi F, Md Yudin AS, Mohd Azman NA, Oumer AN, Abd Razak SB (2023) Pulsation-assisted fluidised bed drying of heat-sensitive and sticky materials: effect of basic parameter, and pulsation-specific parameter. Part Sci Technol 41(2):163–175. https://doi.org/10.1080/02726351.2021.1977034

Akyol KN, Özçelik MM, Bardakçı MS, Küçüköner E, Karacabey E (2025) The impact of carbonic maceration pretreatment on the convective drying of seedless grapes: RSM optimization, drying characteristics and microstructure. Food Sci Nutr 13(3):e70066. https://doi.org/10.1002/fsn3.70066

Alam MR, Lyng JG, Frontuto D, Marra F, Cinquanta L (2018) Effect of pulsed electric field pretreatment on drying kinetics, color, and texture of parsnip and carrot. J Food Sci 83(9):2394–2404. https://doi.org/10.1111/1750-3841.14216

Al-Jeddawi W, Northcutt JK, Dawson P (2023) Effect of pre-treatment and freezing on the polyphenol oxidase activity and color stability of sliced peaches. J Sci Food Agric 103(7):3376–3389. https://doi.org/10.1002/jsfa.12495

Allali H, Marchal L, Vorobiev E (2010) Blanching of strawberries by ohmic heating: effects on the kinetics of mass transfer during osmotic dehydration. Food Bioprocess Technol 3(3):406–414. https://doi.org/10.1007/s11947-008-0115-5

An N-n, Zhao S-y, Li D, Wang Y, Wang L-j (2022) Hot-air flow rolling dry-blanching pretreatment improves the drying quality of Acanthopanax sessiliflorus by increasing the drying rate and inactivating enzymes. Foods 11(20):3186. https://doi.org/10.3390/foods11203186

Assadpour E, Jafari SM (2019) Advances in spray-drying encapsulation of food bioactive ingredients: from microcapsules to nanocapsules. Annu Rev Food Sci Technol 10(1):103–131. https://doi.org/10.1146/annurev-food-032818-121641

Bonneau A, Boulanger R, Lebrun M, Maraval I, Gunata Z (2016) Aroma compounds in fresh and dried mango fruit (Mangifera indica L. cv. Kent): impact of drying on volatile composition. Int J Food Sci Technol 51(3):789–800. https://doi.org/10.1111/ijfs.13038

Botondi R, Barone M, Grasso C (2021) A review into the effectiveness of ozone technology for improving the safety and preserving the quality of fresh-cut fruits and vegetables. Foods 10(4):748. https://doi.org/10.3390/foods10040748

Boudboud A, Ben Aziz M, Hajjaj H, Hajji L, de Meulenaer B, Mazouz H (2023) Impact of pretreatment and drying factors on chemical and biochemical attributes of Moroccan Thompson Seedless grapes. Int J Food Sci 2023:4438353. https://doi.org/10.1155/2023/4438353

Chandramohan VP (2020) Convective drying of food materials: an overview with fundamental aspects, recent developments, and summary. Heat Transf 49(3):1281–1313. https://doi.org/10.1002/htj.21662

Dawson BSW, Sargent R, Riley SG, Husheer S, Simpson IG (2022) Effect of supercritical CO_2 pre-treatment and kiln-drying of fresh green Pinus radiata sapwood on kiln brown stain and drying stress. Wood Sci Technol 56:1127–1148. https://doi.org/10.1007/s00226-022-01399-6

Deng LZ, Mujumdar AS, Zhang Q, Yang XH, Wang J, Zheng ZA, Gao ZJ, Xiao HW (2019) Chemical and physical pretreatments of fruits and vegetables: effects on drying characteristics and quality attributes – a comprehensive review. Crit Rev Food Sci Nutr 59(9):1408–1432. https://doi.org/10.1080/10408398.2017.1409192

Elbah AMI, Abo El Yazid A, Abdalla AE (2017) Effect of different pretreatments and drying methods on the compositional quality of oyster mushroom. J Adv Agric Res 22(1):2–23

Eyarkai Nambi V, Gupta RK, Kumar S, Sharma PC (2016) Degradation kinetics of bioactive components, antioxidant activity, colour and textural properties of selected vegetables during blanching. J Food Sci Technol 53(7):3073–3082. https://doi.org/10.1007/s13197-016-2280-2

Fellows P (2009) Food processing technology: principles and practice, 3rd edn. Woodhead Publishing

Fellows PJ (2017) Drying and dehydration of foods (Chapter 16). In: Food processing technology: principles and practice, 4th edn. Woodhead Publishing

Fernandes da Silva JH, Moura JRRO, Honorato FA, Benachour M, Azoubel PM (2025) Application of ultraviolet radiation type C and pretreatments as an emerging and innovative technology for unripe bananas drying: kinetic and diffusive study, colour, and microstructure. Food Res Int 203:115848. https://doi.org/10.1016/j.foodres.2025.115848

Fernandes FAN, Rodrigues S (2008) Application of ultrasound and ultrasound-assisted osmotic dehydration in drying of fruits. Dry Technol 26(12):1509–1516

Friso D (2020) Conveyor belt dryers with tangential flow for food drying: mathematical modeling and design guidelines for final moisture content higher than the critical value. Inventions 5(2):22. https://doi.org/10.3390/inventions5020022

Gao K, Liu B, Wu B, Guo Y, Song C, Nan S, Dai J, Shen Y, Ma H (2024) A study on the effect mechanism of pectin modification on the carrot cell wall's texture formation under ultrasonic and infrared drying. Agriculture 14(6):803. https://doi.org/10.3390/agriculture14060803

Geankoplis CJ (2003) Transport processes and separation process principles, 4th edn. Prentice Hall

George JM, Sowbhagya HB, Rastogi NK (2017) Effect of high pressure pretreatment on drying kinetics and oleoresin extraction from ginger. Dry Technol 36(9):1107–1116. https://doi.org/10.1080/07373937.2017.1382505

Geranpour M, Assadpour E, Jafari SM (2020) Recent advances in the spray drying encapsulation of essential fatty acids and functional oils. Trends Food Sci Technol 102:71–90. https://doi.org/10.1016/j.tifs.2020.05.028

Gomathy K, Thangavel K, Balakrishnan M, Kasthuri R (2015) Effect of ohmic heating on the electrical conductivity, biochemical, and rheological properties of papaya pulp. J Food Process Eng 38(4):405–413. https://doi.org/10.1111/jfpe.12172

Gu D, Li X, Dong M, Ji W, Yan Z, Zhao T, Zhang M, Liu P, Yue P, Mao G, Yang L (2024) Effects of freeze-thaw pretreatment combined with hot air on snake gourd (Trichosanthes anguina L.). Foods 13(13):1961. https://doi.org/10.3390/foods13131961

Hamzaoğlu F, Türkyılmaz M, Özkan M (2018) Effect of SO_2 on sugars, indicators of Maillard reaction, and browning in dried apricots during storage. J Sci Food Agric 98(13):4988–4999. https://doi.org/10.1002/jsfa.9033

Hii CL, Jangam SV, Ong SP, Mujumdar AS (eds) (2012) Solar drying: fundamentals, applications and innovations. TPR Group, Singapore. ISBN 978-981-07-3336-0

Huang D, Yang P, Tang X, Luo L, Sunden B (2021) Application of infrared radiation in the drying of food products. Trends Food Sci Technol 110:765–777. https://doi.org/10.1016/j.tifs.2021.02.039

Icier F, Baysal T, Icli S (2006) Effects of ohmic heating on the quality of fruit and vegetable products: a review. J Food Eng 77(4):576–585. https://doi.org/10.1016/j.jfoodeng.2005.06.020

Jafari SM, Malekjani N (2023) Drying technology in food processing. Springer

Jain P, Shrivastava SP, Pandey S (2018) Effect of pretreatment methods on fruits drying of grapes under warm air condition. Bhartiya Krishi Anusandhan Patrika 33(1 & 2):66–70

Jangam SV, Law CL, Mujumdar AS (eds) (2010) Drying of foods, vegetables and fruits, vol 1. TPR Group, Singapore. ISBN 978-981-08-6759-1

Jiang H, Zhang N, Xie L, Li G, Chen L, Liao Z (2025) A comprehensive review of the rehydration of instant powders: mechanisms, influencing factors, and improvement strategies. Foods 14(16):2883. https://doi.org/10.3390/foods14162883

Karas J, Pavloková S, Hořavová H, Gajdziok J (2023) Optimization of spray drying process parameters for the preparation of inhalable mannitol-based microparticles using a Box-Behnken experimental design. Pharmaceutics 15(2):496. https://doi.org/10.3390/pharmaceutics15020496

Kaveh M, Abbaspour-Gilandeh Y, Nadimi M, Shahbazi F, Paliwal J (2025) Review of advanced drying techniques: a path to lower greenhouse gas emissions in agriculture. Discov Sustain 6:423. https://doi.org/10.1007/s43621-025-01217-6

Keita E (2021) Particle deposition in drying porous media. Materials 14(18):5120. https://doi.org/10.3390/ma14185120

Khan MIH, Farrell T, Nagy SA, Karim MA (2018a) Fundamental understanding of cellular water transport process in bio-food material during drying. Sci Rep 8(1):15191. https://doi.org/10.1038/s41598-018-33579-6

Khan MIH, Joardder MUH, Kumar C, Karim MA (2018b) Multiphase porous media modelling: a novel approach to predicting food processing performance. Crit Rev Food Sci Nutr 58:528–546. https://doi.org/10.1080/10408398.2016.1196335

Khan MIH, Batuwatta-Gamage CP, Karim MA, Gu Y (2022) Fundamental understanding of heat and mass transfer processes for physics-informed machine learning-based drying modelling. Energies 15(24):9347. https://doi.org/10.3390/en15249347

Khante A (2021) Global market study of dried fruits. Int J Adv Eng Manag 6(7):1–10. https://ijaem.net/issue_dcp/Global%20Market%20Study%20Of%20Dried%20Fruits.pdf

Khatri B, Hamid, Shams R, Dash KK, Shaikh AM, Kovács B (2024) Sustainable drying techniques for liquid foods and foam mat drying. Discov Food 4:166. https://doi.org/10.1007/s44187-024-00223-3

Kilanko O, Ilori TA, Leramo RO, Babalola P, Eluwa SE, Onyenma FA, Ameh NI, Onwordi PN, Aworinde AK, Fajobi MA (2019) Design and performance evaluation of a solar dryer. J Phys Conf Ser 1378(3):032001. https://doi.org/10.1088/1742-6596/1378/3/032001

Kim S-Y, Lee B-M, Hong S-Y, Yeo H-H, Jeong S-H, Lee D-U (2023) A pulsed electric field accelerates the mass transfer during the convective drying of carrots: drying and rehydration kinetics, texture, and carotenoid content. Foods 12(3):589. https://doi.org/10.3390/foods12030589

Krishnaswamy K, Gayathri R (2018) Nature's bountiful gift to humankind: vegetables and fruits and their role in cardiovascular disease and diabetes. Indian J Med Res 148(5):569–595. https://doi.org/10.4103/ijmr.IJMR_1780_18

Krokida M, Maroulis Z, Saravacos G (2001) The effect of the method of drying on the color of dehydrated products. Int J Food Sci Technol 36(1):53–59. https://doi.org/10.1046/j.1365-2621.2001.00426.x

Kuyu CG, Tola YB, Mohammed A, Ramaswamy HS (2018) Determination of citric acid pretreatment effect on nutrient content, bioactive components, and total antioxidant capacity of dried sweet potato flour. Food Sci Nutr 6(6):1724–1733. https://doi.org/10.1002/fsn3.747

Labuza TP (1980) The effect of water activity on reaction kinetics of food deterioration. Food Technol 34(4):36–41, 59

Langford A, Bhatnagar B, Walters R, Tchessalov S, Ohtake S (2018) Drying technologies for biopharmaceutical applications: recent developments and future direction. Dry Technol 36:677–684. https://doi.org/10.1080/07373937.2017.1355318

Li M, Wang B, Lv W, Zhao D (2022a) Effect of ultrasound pretreatment on the drying kinetics and characteristics of pregelatinized kidney beans based on microwave-assisted drying. Food Chem 397:133806. https://doi.org/10.1016/j.foodchem.2022.133806

Li C, Wang S, Wang J, Wu Z, Xu Y, Wu Z (2022b) Ozone treatment promotes physicochemical properties and antioxidant capacity of fresh-cut red pitaya based on phenolic metabolism. Front Nutr 9:1016607. https://doi.org/10.3389/fnut.2022.1016607

Ling W, Xing Y, Hong C, Zhang B, Hu J, Zhao C, Wang Y, Feng L (2022) Methods, mechanisms, models and tail gas emissions of convective drying in sludge: a review. Sci Total Environ 845:157376. https://doi.org/10.1016/j.scitotenv.2022.157376

Liu P, Mujumdar AS, Zhang M, Jiang H (2015) Comparison of three blanching treatments on the color and anthocyanin level of the microwave-assisted spouted bed drying of purple flesh sweet potato. Dry Technol 33(1):66–71. https://doi.org/10.1080/07373937.2014.945177

Liu Y, Zhang Z, Hu L (2021) High efficient freeze-drying technology in food industry. Crit Rev Food Sci Nutr 62(12):3370–3388. https://doi.org/10.1080/10408398.2020.1865261

Ma Y, Yi J, Jin X, Li X, Feng S, Bi J (2022) Freeze drying of fruits and vegetables in food industry: effects on phytochemicals and bioactive properties attributes—a comprehensive review. Food Rev Intl 39(9):6611–6629. https://doi.org/10.1080/87559129.2022.2122992

Mahfuzur RM, Han D, Xu J, Lin Y, Guo X, Luo T, Wu Z, Huang S, Lv X, Wei J (2024) Effects of postharvest SO_2 treatment on longan aril flavor and glucosinolate metabolites. Plants 13(21):3061. https://doi.org/10.3390/plants13213061

Martínez S, Roman Chipantiza A, Boubertakh A, Carballo J (2023) Banana drying: a review on methods and advances. Food Rev Intl 40(8):2188–2226. https://doi.org/10.1080/87559129.2023.2262030

Martynenko AA, Vieira GNA (2023) Sustainability of drying technologies: system analysis. Sustain Food Technol 1:629–640. https://doi.org/10.1039/D3FB00080J

Maskan M (2001) Kinetics of colour change of kiwifruits during hot air and microwave drying. J Food Eng 48(2):169–175. https://doi.org/10.1016/S0260-8774(00)00154-0

Mayor L, Sereno AM (2004) Modelling shrinkage during convective drying of food materials: a review. J Food Eng 61(3):373–386. https://doi.org/10.1016/S0260-8774(03)00142-6

Mercer DG (2024) Chapter 1: Why food is dried. In: Concepts of dehydration and drying for small-scale food processors. Royal Society of Chemistry, pp 1–21. https://doi.org/10.1039/9781837673674-00001

Merger DG (2014) An introduction to drying and dehydration of fruits and vegetables. Donald G. Merger Publishing

Mizrahi S (1996) Osmotic dehydration of fruits and vegetables: review of concepts and applications. Food Rev Intl 12(4):423–438. https://doi.org/10.1080/87559129609541112

Mohod AG, Khandetod YP (2021) Thermal analysis of solar tunnel dryer and evaluation for commercial mango leather drying. J Agric Eng 58:90–100. https://doi.org/10.52151/jae2021581.1737

More SS et al (2014) Review on spray drying technology. Int J Pharm Chem Biol Sci 4(2):219–225

Mounir S (2018) Chapter 8: Foam mat drying (FMD). In: Drying technologies for foods: fundamentals and applications: part III. NIPA, New Delhi, pp 179–187

Mujumdar AS (2006) Fundamentals of drying. In: Mujumdar AS (ed) Handbook of industrial drying, 3rd edn. CRC Press, pp 1–20. https://doi.org/10.1201/9781420017618

Mujumdar AS (2014) Handbook of industrial drying, 4th edn. CRC Press

Nik ARNFN, Zubairi SI, Hashim H, Yaakob H (2024) Revolutionizing spray drying: an in-depth analysis of surface stickiness trends and the role of physicochemical innovations in boosting productivity. J Food Qual 2024:8929464. https://doi.org/10.1155/2024/8929464

Nowacka M, Matys A, Witrowa-Rajchert D (2024) Innovative technologies for improving the sustainability of the food drying industry. Curr Food Sci Technol Rep 2:231–239. https://doi.org/10.1007/s43555-024-00026-8

Nowak D, Jakubczyk E (2020) The freeze-drying of foods—the characteristic of the process course and the effect of its parameters on the physical properties of food materials. Foods 9(10):1488. https://doi.org/10.3390/foods9101488

Osama E, Elmardi M, Khayal O (2024) Introduction and mathematical analysis of convection heat transfer. https://doi.org/10.13140/RG.2.2.35564.37760

Özcan S, Gökmen V (2016) Alkali-based pre-treatment may prevent ochratoxin A in grapes. World Mycotoxin J 9(4):517–523. https://doi.org/10.3920/WMJ2015.2017

Ozcelik MM, Ozkan G, Karacabey E (2022) Evaluation of carbonic maceration effect as a pre-treatment on the drying process of strawberry. Agriculture 12(12):2113. https://doi.org/10.3390/agriculture12122113

Peters EAJF, Rieder DR, Kuipers JAM, van der Meer TH (2023) Modeling the drying process of porous catalysts: impact of the pore size distribution. Ind Eng Chem Res 62(38):15841–15852. https://doi.org/10.1021/acs.iecr.3c03057

Petikirige J, Karim A, Millar G (2022) Effect of drying techniques on quality and sensory properties of tropical fruits. Int J Food Sci Technol 57(11):6963–6979. https://doi.org/10.1111/ijfs.16043

Piechowiak T, Migut D, Józefczyk R, Balawejder M (2022) Ozone treatment improves the texture of strawberry fruit during storage. Antioxidants 11(5):821. https://doi.org/10.3390/antiox11050821

Poojitha P, Athmaselvi KA (2023) Effect of ohmic blanching on drying kinetics, physicochemical and functional properties of garlic powder. J Food Sci Technol 60(3):845–855. https://doi.org/10.1007/s13197-020-04676-z

Prajapat N, Patel DP, Meghwal K, Jain SK (2023) Foam mat drying of high value foods. Int J Agric Sci 15(7):12455–12457

Rahaman A, Siddeeg A, Manzoor MF, Zeng X-A, Ali S, Baloch Z, Wen Q-H (2019) Impact of pulsed electric field treatment on drying kinetics, mass transfer, colour parameters and microstructure of plum. J Food Sci Technol 56(5):2670–2678. https://doi.org/10.1007/s13197-019-03755-0

Rahayu SS, Susastriawan AAP, Surahmanto F, Firmansyah MR, Kurniawan G (2023) Energy efficiency of a biomass powered dryer: an analysis of flue gas velocity effects during chili drying. Int J Heat Technol 41(5):1365–1373

Ramírez Fajardo AF, Martín Vizcaíno C, Rodríguez García I, Guil Guerrero JL (2024) Vegetables treated before drying with natural antioxidants plus UV-C improve colour and bioactive compounds. Agriengineering 6(4):3635–3651. https://doi.org/10.3390/agriengineering6040207

Rathi N, Worsley A, Bruening M (2025) Perceived influences of fruit and vegetable consumption among Indian adolescents: a qualitative inquiry. BMC Public Health 25:271. https://doi.org/10.1186/s12889-025-21408-0

Sagar VR, Kumar PS (2010) Recent advances in drying and dehydration of fruits and vegetables: a review. J Food Sci Technol 47(1):15–26. https://doi.org/10.1007/s13197-010-0010-8

Sahay KM, Singh KK (2014) Unit operations of agricultural processing, 2nd edn. Vikas Publishing House

Salengke S, Sastry SK (2005) Effect of ohmic pretreatment on the drying rate of grapes and adsorption isotherm of raisins. Dry Technol 23(3):551–564

Salur-Can A, Türkyılmaz M, Özkan M (2017) Effects of sulfur dioxide concentration on organic acids and β-carotene in dried apricots during storage. Food Chem 221:412–421. https://doi.org/10.1016/j.foodchem.2016.10.081

Santos AADL, Leal GF, Marques MR, Cardoso Reis LC, Junqueira JRDJ, Macedo LL, Corrêa JLG (2025) Emerging drying technologies and their impact on bioactive compounds: a systematic and bibliometric review. Appl Sci 15(12):6653. https://doi.org/10.3390/app15126653

Schieber A, Stintzing FC, Carle R (2001) By-products of plant food processing as a source of functional compounds—recent developments. Trends Food Sci Technol 12(11):401–413

Scott WJ (1957) Water relations of food spoilage microorganisms. Adv Food Res 7:83–127. https://doi.org/10.1016/S0065-2628(08)60247-5

Sen F, Ozgen M, Asma BM, Aksoy U (2015) Quality and nutritional property changes in stored dried apricots fumigated by sulfur dioxide. Hortic Environ Biotechnol 56(2):200–206

Seth D (2018) Drying equipment used in the food industry: tunnel dryers. In: Seth D (ed) Unit operations in food processing (Module 13). INFLIBNET Centre. Retrieved from https://ebooks.inflibnet.ac.in/ftp02/chapter/dryers-used-in-food-industry/

Shirsath KV (2025) Dehydrated vegetables: an analysis of international trade between India and its top importing partners. Int J Res Anal Rev 12(2):1944. https://ijrar.org/papers/IJRAR25B1944.pdf

Simate IN (2001) Optimal conditions for hot air-drying of maize. J Food Eng 49(2–3):121–129. https://doi.org/10.1016/S0260-8774(00)00110-1

Soni RK, Modi G (2024) Innovations and strategies in post-harvest handling and value addition of horticultural produce. Agric Biol Res 40(4):1206–1209. https://doi.org/10.35248/0970-1907.24.40.1206-1209

Spigno G (2016) Microwave processing: impact on food product quality attributes. In: Jaiswal AK (ed) Food processing technologies: impact on product attributes. CRC Press/Taylor & Francis, pp 137–180

Srimagal A, Mishra S, Pradhan RC (2017) Effects of ethyl oleate and microwave blanching on drying kinetics of bitter gourd. J Food Sci Technol 54(5):1192–1198. https://doi.org/10.1007/s13197-017-2518-7

Sun X, Jin X, Fu N, Chen X et al (2020) Effects of different pretreatment methods on the drying characteristics and quality of potatoes. Food Sci Nutr 8(11):5767–5775. https://doi.org/10.1002/fsn3.1579

Takeya S, Nakano K, Thammawong M, Umeda H, Yoneyama A, Takeda T, Hyodo K, Matsuo S (2016) CO_2 processing and hydration of fruit and vegetable tissues by clathrate hydrate formation. Food Chem 205:122–128. https://doi.org/10.1016/j.foodchem.2016.03.010

Tang X, Zhao M, Ren A, Duan Z, Liu Y, Xian Z (2025) Mitigating browning and improving texture issues in persimmons: the impact of freezing pre-treatment on vacuum microwave drying. Front Sustain Food Syst 9:1523548. https://doi.org/10.3389/fsufs.2025.1523548

Tanwanichkul B, Suriyasupapong N (2022) Mathematical modeling of thin-layer drying of longan in hot air tunnel. KKU Res J 21:1–12. Retrieved from https://ph02.tci-thaijo.org/index.php/gskku/article/view/220243

Turgut SS, Küçüköner E, Karacabey E (2017) Improvements in drying characteristics and quality parameters of tomato by carbonic maceration pretreatment. J Food Process Preserv 41(6):e13282. https://doi.org/10.1111/jfpp.13282

Udoro EO, Anyasi TA, Jideani AIO (2020) Interactive effects of chemical pretreatment and drying on the physicochemical properties of cassava flour using response surface methodology. Int J Food Sci 2020:7234372. https://doi.org/10.1155/2020/7234372

Ur Rahman UJ, Pozarlik AK, Brem G (2022) Experimental analysis of spray drying in a process-intensified counter-flow dryer. Dry Technol 40(15):3128–3148. https://doi.org/10.1080/07373937.2021.2004160

Wang J, Fang X-M, Mujumdar AS, Qian J-Y, Zhang Q, Yang X-H, Liu Y-H, Gao Z-J, Xiao H-W (2017) Effect of high humidity hot air impingement blanching (HHAIB) on drying and quality of red pepper (Capsicum annuum L.). Food Chem 220:145–152. https://doi.org/10.1016/j.foodchem.2016.09.200

Wei X-L, Li S-P, Wu L-G, Du J-B (2021) Design of walnut belt continuous dryer based on fluent. Food Mach 37(12):80–88

World Health Organization (2003) WHO and FAO announce global initiative to promote consumption of fruit and vegetables. 11 Nov 2003. https://www.who.int/news/item/11-11-2003-who-and-fao-announce-global-initiative-to-promote-consumption-of-fruit-and-vegetables

World Health Organization (2023) Fruit and vegetables and noncommunicable diseases (NCDs). https://www.who.int/tools/elena/interventions/fruit-vegetables-ncds

Wu B, Wang J, Guo Y, Pan Z, Ma H (2017) Effects of infrared blanching and dehydrating pretreatment on oil content of fried potato chips. J Food Process Preserv 42(1):e13531. https://doi.org/10.1111/jfpp.13531

Xue W, Macleod J, Blaxland J (2023) The use of ozone technology to control microorganism growth, enhance food safety and extend shelf life: a promising food decontamination technology. Foods 12(4):814. https://doi.org/10.3390/foods12040814

Yan MR, Hsieh S, Ricacho N (2022) Innovative food packaging, food quality and safety, and consumer perspectives. Processes 10:747. https://doi.org/10.3390/pr10040747

Yang X-H, Zhang Q, Wang J, Deng L-Z, Kan Z (2017) Innovative superheated steam impingement blanching (SSIB) enhances drying rate and quality attributes of line pepper. Inf Process Agric 4(4):283–290. https://doi.org/10.1016/j.inpa.2017.07.004

Yin F, Li Z, Lai T, Long L, Liu Y, Han D, Wu Z, Shuai L, Luo T (2025) The effect of SO_2 fumigation, acid dipping, and SO_2 combined with acid dipping on metabolite profile of "Heiye" litchi (Litchi chinensis Sonn.) pericarp. Horticulturae 11(8):923. https://doi.org/10.3390/horticulturae11080923

Yong CK, Islam MR, Mujumdar AS (2006) Mechanical means of enhancing drying rates: effect on drying kinetics and quality. Dry Technol 24(3):397–404. https://doi.org/10.1080/07373930600612194

Yu Y, Jin TZ, Xiao G (2017) Effects of pulsed electric fields pretreatment and drying method on drying characteristics and nutritive quality of blueberries. J Food Process Preserv 41(6):e13303. https://doi.org/10.1111/jfpp.13303

Yu Y, Zhang L, Bei X, Wang Y, Okonkwo CE, Chen L, Zhou C (2023) Catalytic infrared blanching and ultrasound sequential synergistic pretreatment on osmotic dehydration of blueberry. Dry Technol 42(1):155–167. https://doi.org/10.1080/07373937.2023.2277774

Zalpouri R, Singh M, Kaur P, Singh S (2022) Refractance window drying–a revisit on energy consumption and quality of dried bio-origin products. Food Eng Rev 14(2):257–270. https://doi.org/10.1007/s12393-021-09295-8

Zhou L, Bi X, Xu Z, Yang Y, Liao X (2015) Effects of high-pressure CO_2 processing on flavor, texture, and color of foods. Crit Rev Food Sci Nutr 55(6):750–768. https://doi.org/10.1080/10408398.2012.677871

Zhu L, Ji X, Yang H, Cao X, Wang W, Liang M, Li J, Zhang Q, Yang X, Geng Z (2025) Heat pump technology in the field of fruit and vegetable drying: a review. Foods 14(15):2569. https://doi.org/10.3390/foods14152569

Remunerative Floriculture

13

Ashna Acharya, Dixit Chaudhary, and C. S. Chaitra

Abstract

Floriculture has emerged as a profitable option within modern agriculture, offering higher economic returns per unit area than many traditional crops. Growing demand for flowers in religious, cultural and therapeutic contexts has created vast potential for employment generation and rural income enhancement. Floriculture, a high-value enterprise, is especially relevant for small and marginal farmers aiming to move beyond subsistence farming. The chapter highlights current trends and challenges in this sector. The integration of hi-tech interventions, including protected cultivation, drip irrigation, fertigation and soilless systems, has made it possible to grow flowers round the year with improved quality. Postharvest handling and value addition techniques, including drying, oil extraction and advanced packaging, are outlined to minimize losses and thus improve marketability and returns. The chapter further explores niche segments such as edible flowers, floristry, seed and bulb production, turf grass and bonsai, which offer profitable avenues with relatively low competition. The role of supportive policy frameworks, access to institutional credit and linkage with research and development institutions is examined to understand the enabling ecosystem. Floriculture has transformed livelihoods when supported by innovation, planning and market orientation. In sum, the chapter aims to provide a comprehensive view of the floriculture sector's economic potential and practical pathways to make it a sustainable and remunerative component of diversified farming systems.

A. Acharya (✉) · D. Chaudhary
CSIR-IHBT Palampur, Palampur, Himachal Pradesh, India

C. S. Chaitra
Department of Floriculture and Landscaping, Punjab Agricultural University, Ludhiana, Punjab, India

A. Das, A. B. Sharangi (eds.), *Remunerative Horticulture*,
https://doi.org/10.1007/978-981-92-1007-7_13

Keywords

Floriculture · Employment · Hi-tech · Postharvest · Value addition

13.1 Introduction

Floriculture involves the cultivation of flowering and ornamental plants for gardens, floral arrangements, cut flowers, loose flowers, potted plants, and bedding plant markets (Ferrante and Ferrini 2023). It is a fast-growing segment of horticulture that combines cultivation, postharvest technology and value-chain marketing. Compared to many field crops, floriculture results in high per unit value, rapid turnover and diversified market channels (local retail, events, exports and value-added products), making it attractive to entrepreneurs and farmers seeking higher and more reliable returns per unit area. Improved living standards and increasing awareness to live in an eco-friendly environment have further boosted the demand for this sector (Vahoniya et al. 2018). The domestic retail market of the floriculture industry is valued at approximately USD 400 million in India and around USD 100 million in the international flower trade. Roses and dried flowers constitute the major share of flower export from India, which is estimated to be USD 30 million (Anumala and Kumar 2021). Globally, floriculture has diversified into an industry that now includes breeding, protected cultivation (greenhouses, shade houses), postharvest chain management (cold chains, controlled atmosphere storage) and direct retailing through online and institutional buyers.

13.2 Floriculture: A High-Value Crop Enterprise

13.2.1 Floriculture Versus Traditional Cropping

Floriculture crops, in comparison to traditional cropping systems, stand apart in both their economics and social relevance. Unlike crops such as rice or wheat with limited price margins, floriculture caters to an expanding, aspirational and lifestyle-driven market. As income rises, consumers increasingly associate flowers and ornamental plants with luxury, mental well-being and social status. Although the sector demands higher initial investments such as shade nets, greenhouses, irrigation systems, and technical knowledge, it offers diversified income opportunities and greater returns per unit area (floristry, landscaping, etc.) (Santosh et al. 2024). Moreover, the trend toward eco-conscious and wellness-oriented living has popularized indoor plants and home gardens, further expanding the floriculture market scope (Malviya et al. 2022).

13.2.2 Floriculture for Small and Marginal Farmers

Floriculture offers an important pathway for income diversification and rural upliftment among small and marginal farmers, who often struggle to compete with large mechanized producers in traditional field crops (Shaibur et al. 2020). Many ornamental species, such as marigold, gladiolus, tuberose, and chrysanthemum, can be cultivated profitably on small landholdings or backyards using simple and low-cost infrastructure like shade nets. Moreover, floriculture activities such as seedling production, harvesting, grading, and bouquet preparation create both on-farm and off-farm employment opportunities, particularly benefiting women and rural youth (Latulippe 2008). Recognizing this potential, various government initiatives, including those under the Indian Council of Agricultural Research (ICAR), Krishi Vigyan Kendras (KVKs), State Horticulture Missions and Floriculture Mission of the Government of India, have introduced targeted floriculture programs that promote micro-irrigation, cluster-based marketing, and training for smallholders to adopt scientific cultivation and postharvest practices (Singh et al. 2024b). These institutional efforts significantly reduce entry barriers for small farmers and promote collective marketing approaches. Ultimately, the success of smallholder floriculture depends on improving access to quality planting material, affordable protected structures, effective postharvest management, and reliable market linkages that can be bridged through coordinated extension support and farmer cooperatives (Singh et al. 2024b).

13.3 Current Trends, Challenges, and Scope

13.3.1 Present Status of Floriculture in India and the World

The floriculture market has been expanding globally, and the market research estimated global market value in the range of USD 55–64 billion in 2023–2024, with projected compound annual growth rates of 5–8% through the 2020s as urbanization and gifting culture grow. Netherlands (the trade hub), Colombia, Ecuador, Kenya (export-oriented cut flower producers), and domestic markets in the USA, Europe, Japan, and China are the major global players (Anonymous 2024).

India has seen sustained growth in both area and production of floriculture in the last decade. Government and trade statistics report that roughly 317,000 hectares were under floriculture in 2023–24, with millions of tons of production and export reaching about USD 86.6 million in 2023–24 (Anonymous 2024). Leading flower-producing states include Karnataka, Tamil Nadu, West Bengal, Maharashtra, Andhra Pradesh and parts of North-East India and Punjab for specific crops. The domestic market of loose flowers is larger in terms of value than the cut flower market in India. Despite rapid domestic demand, India's share in global floriculture export value remains limited due to quality, cold chain and phytosanitary challenges (Malviya et al. 2022).

13.3.2 Key Production Zones

Floriculture production concentrates where feasible climate, market access, and infrastructure coincide. In South India, Karnataka (Bangalore), Tamil Nadu

(Coimbatore, Erode), and Andhra Pradesh are the main players with flowers like roses, orchids, gerbera, and cut foliage; large greenhouse and nursery industries are also present in the area. West Bengal (Sunderbans belt, Nadia, Hooghly) is prominent for marigold, tuberose, and chrysanthemum; the North-East is known for high-altitude ornamentals and cut orchids. Pune and central Gujarat are the emerging pockets with expanding flower acreage in the last two decades (Chawla et al. 2016).

National Trade Hubs Karnataka Agro Industries Corporation (KAIC) and South Indian Flower Growers' Association (SIFA) collectively made efforts to establish the first flower auction center in Bangalore. Pune, New Delhi, and Hyderabad also function as distribution, auction, and retail hubs with clusters of nurseries, cold storage, and logistics (Chawla et al. 2016). Internationally, the Netherlands (Aalsmeer) remains the trade hub and auction center. Colombia, Ecuador, and Kenya are major exporters serving European and North American markets (Saha 2014).

13.3.3 Major Constraints and Opportunities

Constraints

1. *Postharvest losses*: Postharvest losses for perishable flowers are substantial where refrigerated transport, temperature-controlled packing houses, and refrigerated vans are limited. Significant losses (30–50%) have been estimated during postharvest handling of these perishable commodities (Jhanji et al. 2025).
2. *Sanitary and phyto-sanitary (SPS) measures for export*: Limited access to certified and quality planting material reduces export competitiveness. Tissue culture and biotechnological approach can eliminate the risk faced by Indian producers in terms of SPS imposed by developed countries (Anumala and Kumar 2021).
3. *Fragmented production and market access*: Lack of collective organization and strong connection with organized markets due to small and scattered producers hinders consistent supply. This also makes traceability and certification difficult (Kalmegh and Sing 2016).
4. *Knowledge gap*: Adoption of flower cultivation, especially involving hi-tech, integrated pest management and postharvest technologies, requires technical support and professional know-how often unavailable to marginal farmers (Chakraborthy 2019; Narasimham and Kishor 1997).
5. *Seasonality and market volatility*: There is a constant risk of price collapse during glut and extreme pressure on the supply chain to reach markets on time (Anumala and Kumar 2021).

Opportunities

1. *Domestic market expansion*: Domestic demand for flowers and potted plants is increasing with increasing urbanization, gifting culture and higher disposable

incomes (Chittibomma et al. 2023). This trend can further provide a steady market and encourage new retail approaches like online flower sales.

2. *Adoption of protected cultivation*: Adoption of shade nets, affordable polyhouses, and precision fertigation (micro-irrigation) can enhance crop yield quality and can extend cropping seasons. Various schemes, government initiatives, and training programs are further promoting the adoption of such technologies (De and Singh 2016).
3. *Varied agro-climatic conditions*: various agro-climatic conditions within the country can help grow a variety of ornamentals, year-round (Chawla et al. 2016).
4. *Value addition and export potential*: Value addition through drying, pigment extraction, etc., can add 30–100% value to the commodity. 70% of total floriculture export is contributed by the dry flower business, and India has a 10% share in the dry flower export industry. Other micro-enterprises involving floristry, bouquet making, vale, and the floral waste are great opportunities for women and youth (Adebayo et al. 2021).
5. *Urban greening and sustainability*: In the quest to beautify surroundings along with environmental regulations, indoor planting, vertical gardening, and landscaping in urban areas open a new high-margin market for plant nurseries (Chawla et al. 2016).

13.4 Hi-tech Interventions and Crop Regulations for Year-Round Income

The integration of advanced technological interventions and stringent crop regulations plays a pivotal role in ensuring year-round agricultural productivity and income stability for farmers. These strategies collectively aim to mitigate seasonal limitations, enhance crop resilience and promote sustainable practices. Further, recent advances in floriculture have significantly transformed the industry, incorporating high-tech interventions that enhance sustainability, productivity and quality of ornamental crops. These innovations include modified farming techniques such as greenhouse and vertical farming, soilless cultivation methods like hydroponics and aeroponics, biotechnology, nanotechnology, blockchain, precision farming and advanced postharvest technologies (Ghimire et al. 2025). Some of the major high-tech interventions in floriculture include:

(i) *Precision Agriculture*: Utilization of GPS-guided machinery, remote sensing, and drone technology enables precise input application (water, fertilizers, pesticides), optimizing resource use and increasing yields throughout the year.

(ii) *Controlled Environment Agriculture (CEA)*: Technologies such as hydroponics, aeroponics and indoor climate control systems facilitate off-season cultivation, particularly in controlled environments like greenhouses, ensuring continuous production.

(iii) *Genetic Engineering and Breeding*: Development of high-yield, disease-resistant, and low-chill or heat-tolerant crop varieties extends growing seasons and stabilizes income, especially in regions with challenging climates.
(iv) *Smart Irrigation and Soil Monitoring*: IoT-enabled sensors monitor soil moisture, weather patterns, and crop health, enabling real-time adjustments and water conservation, thus supporting year-round crop cycles.

Integration for Sustainable Income
Combining advanced technologies with supportive policies enables the creation of resilient cropping systems capable of delivering consistent income throughout the year. This approach not only enhances food security but also improves farmers' livelihoods by reducing dependence on seasonal constraints and climatic unpredictability.

13.4.1 Protected Cultivation

The liberalization of the horticultural market has paved the way for the export of high-quality crops, positioning them prominently on the global platform. The average productivity of most horticultural crops in India remains suboptimal, with a considerable disparity between actual yields and the attainable yields achievable through the adoption of improved varieties and advanced technologies. Expanding the area under cultivation to boost production is constrained by the limited availability of arable land. Therefore, enhancing productivity is imperative to satisfy future demand and to fulfill the nation's commitment toward sustainable development. This necessitates integrated strategies encompassing technological innovation, resource optimization, and eco-friendly practices to ensure increased yield, quality, and long-term sustainability of the sector while addressing environmental challenges (Reddy 2016). Protected cultivation stands out as one such technology capable of meeting these requirements. In response to the persistent challenges posed by climate change, global agriculture is increasingly transitioning toward protected cultivation systems to ensure year-round production of high-quality crops. Protected cultivation practices refer to a cropping technique where the microenvironment surrounding the plant is partially or fully controlled according to the specific requirements of the plant during its growth period to maximize yield and conserve resources. The greenhouse represents the most practical approach to protected horticulture, where the natural environment is modified using sound engineering principles to optimize plant growth and yield, resulting in higher produce per unit area and enhanced input use efficiency (Nagarajan et al. 2002).

Currently, nearly 115 countries practice protected cultivation, with China emerging as the leading contributor in terms of area and production (Singh 2014). Further, in regions of adverse climatic conditions, such as those of the hilly terrains, agricultural practices are primarily confined to the summer season, as field cultivation during winter remains largely unfeasible. Consequently, protected cultivation becomes essential to ensure continuous agricultural production throughout the year. A range of protected structures, such as greenhouses, net houses, and shade houses, play a vital role in the cultivation of ornamental crops, with each structure tailored to suit particular agro-climatic conditions. Consequently, the adoption of modern technologies, including greenhouse cultivation, micro-irrigation, fertigation, micropropagation, advanced nursery techniques, and high-tech postharvest technologies such as cold chain infrastructure, has become indispensable for improving the productivity and competitiveness of horticultural produce in India. This strategic technological integration is essential to align with global standards, reduce postharvest losses, and capitalize on emerging export opportunities under WTO frameworks. Beyond crop diversification, protected cultivation offers substantial agronomic benefits, including improved water and nutrient use efficiency, along with a potential three- to five-fold increase in productivity compared to conventional open-field cultivation systems (Sharma et al. 2024).

Further, the genetics of a plant, as well as the microclimate around it, influence its productivity. Light, temperature, humidity, air pressure, and the nature of the rooting medium are the major components of crop microclimate. Environmental control is, however, not possible in open conditions, except for some factors such as tillage, watering, and fertilization. The protected structures, like greenhouses, only allow control over any one or more of the attributes of microclimate due to their closed boundaries. The amount of sunlight entering the greenhouse depends on the transparency of the covering materials. The sunlight available inside the greenhouse is absorbed by the growing crops, the floor, and other objects. Subsequently, long-wave thermal radiation is emitted by these elements, and due to the lower transparency of the covering material to this radiation, solar energy becomes trapped in the greenhouse and increases the temperature within it. This phenomenon is commonly referred to as the greenhouse effect (Mishra et al. 2010). Greenhouse crops are successfully raised in cooler areas due to this natural increase in temperature. Numerous studies have quantified the impact of environmental conditions on crop development, individually as well as collectively.

Protected Floriculture

Trends in India indicate a steady expansion of the protected cultivation of flower crops. Globally, flower production under protected cultivation is predominantly led by cut roses, followed by carnations, chrysanthemums, gerberas, lilies, and orchids. The floriculture sector exhibits an impressive annual growth potential of 25–30%, contributing nearly 20–25 times more to the nation's foreign exchange earnings compared to conventional field crops (Sarkar 2023). Nevertheless, despite the country's abundant natural resources and favorable growing conditions, its share in the global floriculture trade remains modest (0.40–0.50%) (Ranjan et al. 2013). This limited contribution is largely due to inadequate use of protected cultivation systems required to meet the stringent quality standards for cut flowers in international markets (Nemali 2022). Recognizing the potential of this sector, the Government of India declared floriculture a "Sunrise Industry" in the early 1990s and granted it 100% export-oriented status. The growing demand for novel flower types with unique forms and colors has encouraged the introduction and cultivation of exotic species such as calla lily, bird of paradise, ornamental gingers, and heliconia under protected conditions. Protected structures play a crucial role in minimizing weather-related risks, such as pest infestations and extreme environmental conditions, which otherwise compromise flower quality under open-field cultivation (Fig. 13.1). Moreover, protected cultivation offers significant advantages, including enhanced produce quality characterized by larger flower diameter, longer stalks, and improved color stability as well as an extended postharvest shelf life.

Recent advances in protected cultivation focus on optimizing microclimate parameters. Precision light management with LED lighting delivers species-specific spectra, improving photosynthesis and growth while reducing energy consumption (Olle and Viršile 2013). Temperature regulation techniques include localized bench heating and geothermal systems (Chahidi et al. 2021), while integrated

Fig. 13.1 Protected cultivation of commercial flower crops. (**a**) Gerbera. (**b**) Anthurium. (**c**) Orchids. (**d**) Carnation

cooling methods like passive ventilation, misting, and shade structures (Tawalbeh et al. 2023) are gaining popularity, especially in warm climates. The majority of cut flower crops require carbon dioxide concentrations ranging from 1000 to 3000 ppm to enhance growth quality and nutrient enrichment (Doddrell et al. 2023). Supplemental CO_2 enrichment has also proven to boost flowering and productivity in crops like roses, *Gerbera*, and *Lilium*.

Innovations in substrate and nutrition include the shift to soilless media, such as coir, rockwool, and biochar, which offer better disease suppression, aeration, and consistent growth. Advanced fertigation techniques ensure efficient nutrient delivery, tailored to crop needs and substrate properties (Solaimalai et al. 2005). The use of plant growth regulators enables precise control over plant architecture, timing, and aesthetics, closely aligned to market and logistical requirements (Sajjad et al. 2017). Furthermore, the adoption of AI, robotics, and sensors is ushering in a new era of automation, making protected cultivation more resource-efficient, productive, and globally competitive.

13.4.2 Smart Flower Farming (Drip Irrigation, Hydroponics and Aeroponics)

Smart farming, integrating Information and Communication Technology (ICT) into agriculture, represents an advanced approach to optimize crop growth environments remotely and automatically, surpassing time and spatial constraints (Knierim et al. 2019). It utilizes a network of IoT devices such as sensors, drones, and automated systems to collect real-time data on key environmental and crop parameters including temperature, humidity, soil moisture, light intensity, and CO_2 levels (Fig. 13.2). This data is transmitted and processed via cloud-based platforms using AI and machine learning algorithms, enabling precise monitoring, predictive analytics, and decision-making for crop management (Sharma 2025). With smart farming, operations such as irrigation, fertilization, pest control, and greenhouse climate management can be automated or remotely controlled, enhancing resource use efficiency and reducing labor costs. For instance, soil moisture sensors enable automated drip irrigation systems to deliver water optimally to plant roots, minimizing waste. Drones equipped with multispectral and thermal sensors provide detailed crop health assessments for timely interventions. Smart greenhouses employ environmental sensors to maintain ideal microclimatic conditions with minimum human intervention (Table 13.1).

These technologies collectively improve crop productivity, sustainability, and resilience to climate variability. They also facilitate data-driven precision agriculture tailored for various scales from large commercial farms to smaller, high-value specialty crop systems. By integrating multiple ICT applications, smart farming transforms traditional agriculture into a tech-driven, efficient, and sustainable production system (Thongnim et al. 2023). This approach not only advances productivity but also supports environmental conservation and reduces operational risks, marking a significant paradigm shift in modern horticulture and agriculture management.

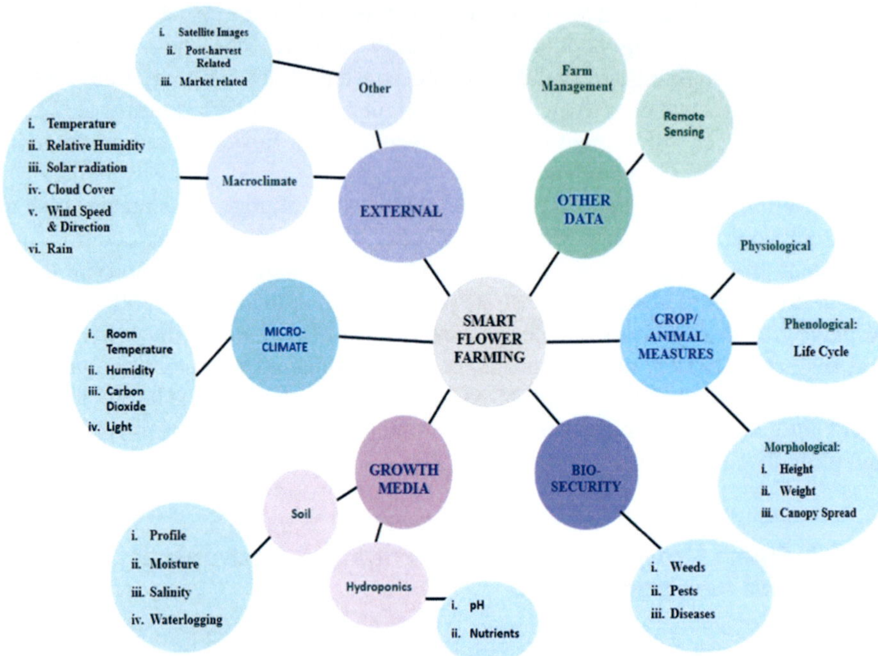

Fig. 13.2 Different components of Smart Flower Farming

13.4.3 Crop Scheduling and Off-Season Flower Production

The scheduling of plant production constitutes a pivotal component of the horticultural industry, especially within the domain of floriculture. In recent years, the limited assortment of conventional products offered to consumers has progressively been supplanted by a diverse array of commodities tailored to meet emerging consumer preferences and varied end-use requirements (Faust and Dole 2021). Crop scheduling and off-season flower production are essential practices in protected cultivation, enabling continuous and timely availability of flowers beyond their natural growing seasons. When the products comprise cut flowers or flowering plants, production scheduling primarily focuses on regulating the timing of flowering and enhancing the aesthetic quality of blooms. Among the various strategies available to achieve these objectives, the manipulation of climatic parameters within the cultivation environment represents one of the most effective and widely adopted approaches. It requires an understanding of crop-specific requirements, including light, temperature, humidity, and nutrient needs. Protected cultivation technologies such as greenhouses, net houses, and shade houses facilitate the manipulation of these environmental factors to create optimum conditions for flower growth year-round, irrespective of external climatic limitations (Erwin 2005). Overall, crop scheduling coupled with protected cultivation technologies enables floriculturists to maximize quality and quantity by producing high-value ornamental flowers

Table 13.1 Optimum light, temperature and supplemental CO2 requirements of commercial flower crops (Sharma et al. 2024)

Sr. no.	Crop	Light intensity	Temperature	Supplemental CO_2 requirement (ppm)
1.	Rose	60,000–70,000 lux	15–18 °C	1000–1500
2.	Carnation	21,500 lux (minimum)	Day temperature 20–25 °C Night temperature 10–15 °C	1000–1500
3.	Chrysanthemum	1.2–1.6 MJ/m^2/day	16–25 °C	700–900
4.	*Gerbera*	4000–6000 lux	Day temperature: 22–25 °C Night temperature 12–16 °C	800–1200
5.	*Lilium*	2000–3000 fc	Asiatic lilies require a night temperature of 13–17 °C and not higher than 21 °C. Day temperature, while oriental lilies require 17–18 °C night temperature, and day temperature should be ideally around 29 °C	1000–1500
6.	Tulip	40 Kilo lux (supplemental lighting is only required during propagation, after planting grows under natural light)	Day: 20–25 °C Night: 5–10 °C	1000–1500
7.	Anthurium	18,000–25,000 lux	Day: 24–28 °C Night: 15–22 °C	900
8.	Cymbidium	3000–6000 fc	Day: 15.5–21 °C Night: 10–12.5 °C	2000–2300

throughout the year. Production scheduling strategies in greenhouse floriculture may target one or both components of the plant-environment interaction. Accordingly, such operations can be categorized into three groups based on their mode of action: (i) those acting directly on the crop, (ii) those modifying the growth environment and (iii) those influencing the entire production process through technologies integrated within the cultivation system (e.g. hydroponics). In general, the duration of distinct developmental phases in a plant species varies according to its genotype and is modulated by environmental factors. Consequently, the selection of an appropriate variety, cultivar, or hybrid, along with the determination of optimal planting dates, significantly affects flowering earliness and winter production potential, both crucial for achieving maximum marketable value (Proietti et al. 2022).

Flower forcing, or crop regulation, is an essential horticultural strategy in commercial floriculture aimed at modifying the phenology of ornamental plants to meet specific market demands, environmental conditions, and production targets. This practice enables the timely availability of flowers, uniformity in bloom, and enhanced quality, which are critical for maintaining profitability in flower markets. Flower forcing involves manipulating environmental, cultural, and chemical factors to induce or delay flowering. These include pruning, pinching, adjusting irrigation and nutrient regimes, photoperiod management, temperature control, and the application of plant growth regulators (PGRs) (Hasna et al. 2024). The natural flowering period of many ornamental species often does not coincide with peak market demand, which is heavily influenced by festivals and special occasions such as Diwali, Christmas, Valentine's Day, and Mother's Day. Crop regulation techniques, therefore, help synchronize flowering to these high-demand windows and prevent gluts during off-peak periods, promoting a continuous and stable supply throughout the year (Song et al. 2015). This not only maximizes economic returns by allowing premium pricing but also reduces postharvest losses and pest pressures through well-timed flowering schedules (Chandel et al. 2023).

Environmental factors such as photoperiod, temperature, and light quality critically influence flowering timing. For example, short-day and long-day plants require specific day lengths to trigger blooming, which can be artificially controlled in protected environments through blackout curtains or supplemental lighting (Bhargavi et al. 2025). Vernalization (chilling treatment) and temperature manipulation are also widely used to promote or synchronize flowering (Nakano et al. 2013). Chemical treatments with fertilizers, gibberellins, ethylene, and growth retardants further assist in controlling flower induction and quality. Cultural practices like pruning and pinching enhance flowering by modifying hormonal balances within the plant. Overall, flower forcing ensures the production of uniform, high-quality flowers year-round, supporting growers in meeting quality standards for export markets and enabling continuous breeding programs (Bhargavi et al. 2025). This approach underpins the economic viability and sustainability of modern floriculture industries globally.

13.5 Postharvest Handling and Value Addition

Flower crops are perishable in nature, and the postharvest losses often range between 30% and 50% (Jhanji et al. 2025). Postharvest management serves as the bridge between flower production and profit in commercial floriculture, as it determines marketability and consumer satisfaction. Various methods targeting deterioration processes have been used to extend the postharvest life of flower crops (Kaur 2025). An efficient postharvest system is therefore crucial to convert on-farm production into profitable returns for growers.

13.5.1 Harvesting to Packaging: Minimizing Losses

Pre-harvest Considerations and Harvest Timing

Pre-harvest factors (nutrition, irrigation, disease and pest control, and varietal selection) directly influence the flowers in terms of quality, quantity, disease resistance, and postharvest life (Singh et al. 2024a). Harvesting stage and time are the primary determinants of vase life, and the flowers harvested at the proper stage tend to remain fresh for a longer duration (Mishra and Dwivedi 2015).

Harvesting and Immediate Handling

Proper harvesting techniques and careful inspection of the flowers, followed by rapid removal of field heat, are crucial. Flowers should be harvested as per the recommended stage of specific flowers, which results in optimum vase life (Jahan 2009). Using a clean, sharp tool to harvest cut stems is recommended to avoid stem crushing and microbial loading. The immediate second cut is to be given under clean water to remove field heat and air embolism. Field sorting removes damaged or diseased stems before packing, preventing cross-contamination and spoilage during transport (Mishra and Dwivedi 2015). With the advancement in technology, automated systems for stem trimming, leaf stripping, grading, and bunching of harvested flowers have improved efficiency and consistency in postharvest handling.

Conditioning Treatments

Conditioning restores turgor and prevents wilting during storage and transport. Carbohydrate in the form of sucrose, along with appropriate biocide and acidifier, reduces microbial blockage in xylem vessels, thereby sustaining water uptake (Kumar et al. 2022). Impregnation, i.e., loading of flowers with silver nitrate, nickel chloride, and cobalt chloride for a short duration to reduce microbial attack, is also recommended in flower crops like gerbera, gladiolus, chrysanthemum, and carnation (Nowak et al. 1990). Concentration of treatment must be tailored to species as well as variety and local water quality, as improper concentrations may cause phytotoxic effects.

Packaging: To Minimize Mechanical and Physiological Loss

Modern packaging reduces moisture loss, provides physical protection, and makes the commodity presentable while reducing disease attack (Rashed et al. 2024). Innovation in packaging through water tubes, insulated crates, ventilated crates, along with cushioning (newspaper/foam) minimizes mechanical damage and desiccation. Cut flowers (gladiolus, antirrhinum) prone to geotropic bending must be stored and transported in an upright position (Reid 2009). Packaging must also be cost-effective and adaptable for transport modes. Some flowers show yellowing due to a lack of light during storage and transportation; therefore, lights should be installed to avoid the same. Procona and Aquapac are some of the advanced transportation systems with flower ends dipped in water and preservatives at the bottom of the boxes (Mishra and Dwivedi 2015).

Temperature and Relative Humidity Control

Optimum temperature and relative humidity reduce respiration rate, water loss, wilting, enzymatic breakdown, pathogen growth, and ethylene production in harvested flowers (Aydin 2023). Temperature maintenance is important both before and after the packaging of flowers. The cold chain must be maintained throughout the process, as exposing flowers from cold to warm temperatures can result in condensed water on the flower surface, which may cause botrytis attack (Senapati et al. 2016). Cold storage or temperature regulation also assists in market regulation by avoiding a glut-like situation (Rudnicki et al. 1991).

Various storage methods:

Dry storage: A method used for long-term storage, where flowers are packed immediately after harvesting and pulsing treatment (Gupta and Dubey 2018).

Wet storage: A method used for short-term storage of flowers with their ends dipped in water along with a biocide. The temperature maintained during wet storage is 3–4 °C higher than dry storage (Thakur 2020).

Controlled atmospheric storage: This method is based on the regulation of temperature, oxygen, and carbon dioxide. Ethylene production and action are inhibited by increasing CO_2 concentration and reducing O_2 concentration. This method involves a high cost of application and handling (Gupta and Dubey 2018).

Modified atmospheric storage: Storage of flowers is carried out in sealed bags where O_2 is reduced, and the CO_2 level increases through respiring tissue. This variation in O_2 and CO_2 concentration reduces ethylene production. This method is cheaper than the controlled atmospheric storage (Thakur 2020).

Hypobaric storage: Storage of flowers under low pressure, low temperature, and cool, moist air. Flower life is extended by reducing ethylene biosynthesis under reduced O_2 concentration. Low-pressure storage at 180–210 mmHg is optimum for most of the cut flowers (Senapati et al. 2016).

13.5.2 Value-Added Products: Dried Flowers, Oils and Dyes

Diversification through value addition reduces risk and makes pathways for small-scale avenues.

Dried Flowers

Dried flowers and foliage are long-lasting decorative materials that can be used multiple times, retain aesthetic appeal, and have export potential (Kumar et al. 2024). Diverse flora, abundant raw material, low labor cost, and favorable climatic conditions give India an edge in the global trade of dried flowers. In India, the major exporting port of dried flowers is in Tuticorin (Sharma 2020).

Plant parts like flowers, seeds, shoots, bark, leaves, and roots are utilized extensively for aesthetic purposes (Jain et al. 2016). Durability, retention of color and form, and resistance to extreme temperature make these commodities highly suitable for long after drying. General methods followed for dry flower making are air drying, sun drying, sand/silica gel drying, press drying, oven drying, microwave drying, freeze drying, and glycerin treatment of foliage (Kumar et al. 2024). Air drying, sun drying, and sand/silica gel drying are appropriate methods for small-scale industries, while freeze drying is an advanced method that is much more efficient in retaining color and shape but costlier than the prior (Sharma et al. 2021).

Fig. 13.3 Dry flowers

Commonly used plants include Limonium, Nigella, Gypsophila, Amaranthus, Gomphrena, Helichrysum, Lotus pods, Craspedia, grasses, sedges, ferns, cones, etc. (Fig. 13.3). Various products of dry flowers in demand are potpourri, wreaths, frames, dry flower bouquets, etc.

Essential Oils, Absolutes and Floral Extracts

Many ornamental flowers are a rich source of valuable volatile compounds recoverable as essential oils, which can be used in perfumery, cosmetics, and aromatherapy. Jasmine, tuberose, rose, lavender, champaka, and magnolia are some of the common sources of essential oil extraction (Koushal et al. 2024). These oils are extracted from various parts of the plant, which gives a sweet and characteristic fragrance of the species. Flower's volatility, heat sensitivity, and required end product are some of the deciding factors for the extraction method to be used. Some of the conventional methods that are still widely adopted are steam distillation, solvent extraction, enfleurage, and hydrodistillation (Maitra et al. 2016). The supercritical CO_2 extraction method is gaining popularity due to its higher efficiency and reduced solvent residues (Cruz-Sánchez et al. 2024). The increasing demand and consumer preference for natural fragrance and sustainable cosmetic ingredients create opportunities for farmers and small enterprises.

Natural Dyes and Pigments

Flowers serve as a valuable and natural source of pigments which are used across different industries like textiles, cosmetics, food, and pharmaceuticals (Bujak et al. 2022). With increasing demand for eco-friendly, non-toxic dyes, these natural dyes are slowly replacing the carcinogenic synthetic dyes. The herbal dye market in India is expanding rapidly, with exports worth USD 47.68 million in 2016 (Waghmode et al. 2018). Waste from the temples, which includes marigolds and roses, has been successfully used for dye extraction, supporting sustainability. Despite its vast potential, the application of flower-extracted dyes remains underexploited in textile industries due to the limited knowledge on pigment stability and compatibility. Pigments can be extracted using various methods, ranging from water or solvent extraction methods to advanced ultrasound microwave-assisted techniques. Mordants like alum, iron, and tannins are used as fastening agents on fabrics (Bhosale and Gobade 2025).

13.6 Niche Markets and Export-Driven Opportunities

The global floriculture industry is projected to experience substantial expansion, with its market value anticipated to rise from USD 63.49 billion in 2025 to approximately USD 115.86 billion by 2035, reflecting a compound annual growth rate (CAGR) of 6.2% over the forecast period (Anonymous 2025b). This strong growth trajectory is primarily attributed to the increasing consumer interest in home décor and landscaping with ornamental plants and flowers, driven by rising disposable incomes and rapid urbanization. The growing preference for aesthetic enhancement of living and working spaces through diverse and vibrant floral arrangements continues to stimulate market demand. Furthermore, technological advancements in cultivation and distribution systems have significantly enhanced the availability, diversity, and quality of floricultural products across global markets. Innovations such as controlled environment agriculture, precision logistics, and optimized cold chain management have contributed to minimizing postharvest losses and improving product freshness, thereby enhancing consumer appeal.

India has matured as a key player in the global floriculture sector, leveraging biodiversity, diverse agro-climates, and skilled labor to expand exports. The Agricultural and Processed Food Products Export Development Authority (APEDA), a statutory body established by the Government of India, implements various programs to promote floriculture exports, focusing on infrastructure development like cold storage, refrigerated transport, and air cargo services alongside market expansion initiatives (Anonymous 2024). However, limitations like inadequate cold chain logistics, stringent phytosanitary regulations in importing countries, intense global competition, and dependence on traditional markets remain significant challenges to fully exploit India's export potential. Major importers include the United States, the Netherlands, the UAE, and the UK, each with distinct demand profiles ranging from high-value cut flowers for the US market to the Netherlands' role as a redistribution hub (Glover 2022). Other important importers include Germany, Japan, Australia, Singapore, and Middle Eastern countries, providing diversification opportunities. Export value growth has outpaced quantity, indicating a shift toward higher-end products and quality differentiation (Nair 2025). Export opportunities extend beyond fresh cut flowers to value-added products like dried flowers, floral extracts, essential oils, and handicrafts, diversifying market offerings and catering to evolving consumer preferences. However, challenges such as cold chain logistics, phytosanitary compliance, and heavy competition from established growers like Kenya and Ecuador persist and must be systematically addressed for continued growth.

13.6.1 Emerging Markets

Indian floriculture is poised for significant expansion into emerging niche markets such as edible flowers, floristry and religious use, driven by increasing global demand and targeted export strategies. Specialty products like traditional Indian flowers, including marigold, jasmine and tuberose, along with cut foliage, dry flowers, floral extracts and ornamental potted plants, hold premium value, especially in diaspora-centric and ethnically diverse markets. These products also encompass value-added items such as floral garlands, seeds and handicrafts, which are favored for their cultural and aesthetic significance (Meghana and Wazeed 2024).

Seasonal Demand and Market Challenges Emerging economies, particularly India, have witnessed substantial growth in floriculture demand around major cultural and social events. Flower consumption surges significantly during Hindu festivals, weddings, Valentine's Day and Christmas, creating pronounced seasonal peaks in the market. However, a recurring barrier is the underdeveloped supply chain, especially insufficient cold storage and distribution infrastructure, which impedes timely supply and results in price volatility during high-demand periods. This mirrors, to some extent, demand seasonality in mature markets: in Europe, for instance, 80% of flower imports occur in winter, with only a minor share arriving in summer, pointing to contrasting or complementary demand patterns globally.

Market Structures: From Florists to Boutiques Indian flower markets bifurcate into local florists who emphasize customer-centric services, tailored floral arrangements and competitive adaptation and flower boutiques, which leverage loyalty programs, discounts and diversified order portfolios, including large-scale decorative commissions. Modern marketing approaches have expanded to include the use of digital media, promotional events and multi-channel advertising (e.g. LCD displays, internet, print), echoing trends in established Western markets (Acharya et al. 2024).

Auction Systems: Learning from the Dutch Model Auctions remain the cornerstone of floriculture trading in advanced economies, notably the Netherlands, where this system has up-scaled efficiency and price discovery since the late nineteenth century (Feng 1999). Approximately 70% of Dutch growers' produce is auctioned, enabling transparent, competitive sales that aggregate international suppliers and buyers (Faust and Dole 2021). Major auction centers like Aalsmeer and Naaldwijk exemplify this model, though recent trends show a gradual shift toward direct sales to large retail chains. By comparison, auction-based marketing mechanisms are slowly gaining attention in India, with grower cooperatives and associations pushing for collective marketing to curtail middlemen's influence.

New Marketing Approaches for Emerging Markets For emerging floriculture markets, designing robust, export-oriented marketing systems requires:

(i) Deep market intelligence on consumer trends, cultural preferences and evolving fashion in floral consumption
(ii) Rigor in quality assurance, addressing product presentation, packaging, pre- and postharvest handling, pest and disease management, and reliable cool-chain logistics

(iii) Compliance with export logistics, documentation, phytosanitary standards, and market-specific requirements on product specifications (color, stem length, bunch weights, packaging dimensions, and consignment descriptions)
(iv) Value addition stems not only from supply reliability and product excellence but also from expertise in market access logistics, including customs procedures, non-stop international flights, proximity to ports, and regulatory clearances (Rafique et al. 2024)

Grower Associations and Collective Action Emergent grower organizations are catalyzing industry modernization by aggregating bargaining power, engaging with government and trade bodies, and collectively accessing market intelligence and export infrastructure. Notable Indian examples include the Horticulture and Floriculture Association (Belgaum), South India Floriculture Association (SIFA), and regional associations in Gujarat, Maharashtra, and Surat (Tandon and Singh 2023). These bodies are driving the adoption of best marketing practices and institutional innovations seen in leading countries.

Key Distribution Hubs and the Rise of E-Commerce India now boasts several prominent flower markets (e.g., K.R. Market Bangalore, Dadar Market, Mumbai, Gultekdi Pune, Malikghat Kolkata), each acting as regional aggregation and redistribution centers. Parallelly, online flower delivery platforms (e.g. Ferns N Petals, Flowersura, IndiaMart) are transforming urban floral consumption, opening new growth opportunities for growers and entrepreneurs.

13.6.2 Specialized Sectors

The floricultural industry comprises diverse specialized sectors integral to landscape aesthetics, urban greening, and ornamental plant cultivation. Among these, turf management, bonsai artistry, nursery production, and seed-bulb propagation represent key domains combining traditional practices with modern scientific innovation.

(a) *Turf Industry*: Turfgrass management plays a critical role in both ornamental and functional landscapes, including residential lawns, sports fields, golf courses, and environmental conservation projects (Fig. 13.4a, b). Specialized sectors within turf encompass natural turf maintenance and artificial turf development. Natural turf management involves agronomic practices like species selection, soil management, pest control, and irrigation to maintain healthy grass cover and ecosystem services such as carbon sequestration and soil erosion control (Braun et al. 2024). Artificial turf, increasingly adopted for its water-saving potential and reduced maintenance, poses environmental concerns related to microplastic pollution and end-of-life waste management (Ryan-Ndegwa et al.

Fig. 13.4 (**a**) Uniformly maintained turfgrass lawn. (**b**) Turfgrass sod rolls for lawn establishment

2024; Cheng et al. 2014). The turf industry also addresses the challenges of sustainable resource use, balancing aesthetic, functional, and ecological demands (Strandberg et al. 2012).

(b) *Bonsai Making*: Bonsai making is a specialized form of miniature tree cultivation requiring intricate horticultural skills and aesthetic judgment. Originating in East Asia, bonsai art involves pruning, wiring, root management, and soil control to simulate the appearance of mature trees at a reduced scale (Anonymous 2025a). It integrates botanical, ecological, and artistic principles, with practitioners developing techniques to control plant growth and form over long periods (Jejal and Chandr 2025). Bonsai making has been recognized not only as a cultural art but also as a therapeutic practice with benefits for mental and emotional well-being (Fig. 13.5). The industry is expanding globally, with growing markets and research emphasizing improved cultivation methods and integration of technology in bonsai care.

(c) *Nursery Production*: Nursery production is foundational to horticulture, providing planting stock ranging from seedlings and cuttings to advanced juvenile plants for food, fiber, and ornamental use (Fig. 13.6). Modern nurseries employ a range of propagation methods, including seed sowing, tissue culture, grafting, and layering, supported by environmental control technologies and substrate optimization (Yaseen et al. 2022). Challenges in nursery production include disease management, maintaining genetic purity, and ensuring quality standards for market demands. Advances in understanding soil microbial communities and their interactions with roots have informed nursery practices to enhance plant vigor and health outcomes (Baranowska et al. 2023). Training and capacity building for nursery workers are critical to improving production efficiencies and product quality.

Fig. 13.5 Container-grown bonsai

Fig. 13.6 Nursery for ornamental plant production

(d) *Seed-bulb Production*: Seed-bulb production serves as an essential segment of the floriculture sector, ensuring availability of high-quality planting material for crops such as lilies, daffodils, gladiolus, hyacinths, and hippeastrums. Bulbs, often produced from selected mother bulbs, are propagated and treated to maximize yield and minimize pathogen transmission (Benschop et al. 2010). Sustainable seed-bulb production emphasizes efficient resource utilization, postharvest handling, and storage protocols to prolong viability and vigor. In *Lilium*, propagation through scales involves separating bulb scales that develop bulblets, which grow into new bulbs (Fig. 13.7). Bulblets also form at the base of the parent bulb underground, while bulbils develop as small bulbs on the stem above ground and can be collected for planting. Tulips multiply mostly through bulb division and bulblets formed on bulbs; seeds are used, but propagation is slower. Daffodils propagate via offsets or bulblets on the scales and stems, using methods such as scooping and scaling to encourage new bulblet formation. Seed propagation in these species, while possible, is slower and less reliable in maintaining cultivar characteristics compared to vegetative methods. These methods ensure quick, true-to-type multiplication and are commonly used in the commercial production of these ornamental bulb crops. Bulbs, scales, bulblets, and bulbils all serve as vegetative organs aiding this propagation process in these crops.

The specialized sectors within turf management, bonsai making, nursery production, and seed-bulb propagation represent critical interfaces between traditional horticultural knowledge and contemporary scientific advancements. Their continuing evolution through research, technology adoption, and sustainable practices is vital to meeting the demands of urbanization, environmental stewardship, and consumer preferences.

Fig. 13.7 Seed-bulb production in *Lilium*

13.6.3 Global Demand Barriers and Breakthrough

The global floriculture industry faces several demand barriers while experiencing technological and market breakthroughs that are shaping its future. Key barriers to floriculture demand include:

(i) *Seasonal and Climatic Constraints*: Flower production is highly dependent on regional climatic conditions, which cause seasonal supply fluctuations and limit year-round availability (Chandra 2025). Extreme weather and climate change effects, such as droughts and erratic rainfall patterns, disrupt production cycles and lead to supply shortages in major producing countries like Kenya and India.
(ii) *Supply Chain and Infrastructure Limitations*: Inadequate cold chain infrastructure and postharvest handling issues result in significant losses (25–40%), reducing product quality and marketability. Fragmented marketing systems, especially in developing nations, hinder efficient trade and reduce growers' competitiveness (Saini et al. 2023).
(iii) *Regulatory and Trade Barriers*: Strict phytosanitary regulations, phytosanitary standards, and market access requirements in key export destinations pose challenges, affecting especially emerging producers. Geopolitical tensions and trade disruptions further exacerbate market access issues (Nair 2025).
(iv) *Fragmented Market and Low Integration*: Many producers struggle with a lack of organized marketing platforms, poor digital adoption, and limited integration with global value chains, restricting direct-to-consumer sales and e-commerce opportunities (Vidhate and Deshmukh 2024).
(v) *Rising Production Costs*: Inflationary pressures on inputs, energy, and logistics increase operational costs, which are not always matched by price increases, squeezing profitability and investments.

Breakthroughs Driving the Floriculture Industry

(i) *Technological Innovations*: Advances in greenhouse technologies such as LED lighting and precision drip irrigation have improved water use efficiency and enhanced crop yields by up to 30–40%. Digital tools, including AI algorithms, IoT-enabled sensors, and robotics, optimize crop management, pest control and supply chain logistics, increasing efficiency and reducing waste (Nagarsheth et al. 2025).
(ii) *Biotechnological Progress*: Gene editing techniques like Clustered Regularly Intershaped Short Palindromic Repeats (CRISPR) and molecular marker-assisted breeding are used to develop flowers with extended vase life, novel colors, better fragrance, and disease resistance. Tissue culture and micropropagation enable rapid multiplication of high-quality planting material.
(iii) *Sustainability Initiatives*: Increased consumer focus on sustainable and eco-friendly products fuels demand for organic growing methods, biodegradable packaging, and integrated pest management practices. Sustainable floricul-

ture improves environmental footprint and appeals to conscious consumers (Murthy et al. 2024).

(iv) *Digital Marketplaces and E-commerce*: The expansion of online platforms and digital marketing facilitates easier market access for producers, offering direct sales channels, better price discovery, and enhanced customer engagement worldwide.

(v) *Climate-Resilient Breeding and Protected Cultivation*: Development of climate-resilient varieties and adoption of protected cultivation systems (hydroponics, vertical gardening, etc) create control environments that optimize growth and enhance resource use efficiency and enable sustainable year-round production.

13.7 Policy Framework, Credit Access, and Institutional Support

13.7.1 Government Schemes for Floricultural Crops

The floriculture sector in India is strategically supported by a comprehensive spectrum of government schemes that are aimed at improving overall productivity, infrastructural capabilities, market access, and sustainability (Table 13.2). The following schemes/programs/missions are introduced at present:

13.7.2 Role of Banks, National Bank for Agriculture and Rural Development and Agri-fintech in Finance

The financial ecosystem supporting the floriculture sector is a combination of traditional banking, developmental institutional credit, and innovative agri-fintech platforms that collectively enable sector growth. Hi-tech floriculture operations require capital for the establishment of protected cultivation, tissue culture units, and postharvest infrastructure. Floriculture producers and entrepreneurs get credit from banks comprising commercial, regional rural, cooperative, and small finance banks, which are the frontline financiers of floriculture ventures; however, challenges include high perceived risk, collateral requirements, and seasonal production cycles.

National Bank for Agriculture and Rural Development (NABARD) significantly enhances this landscape by providing 100% refinance support to banks for loans granted to floriculture initiatives across multiple priority areas, facilitating cluster financing, and supporting the development of Farmer-Producer Organizations (FPOs). NABARD's programs encourage integrated area-based development models, capacity building, and convergence with state horticulture missions, which improve fund utilization efficiency and risk mitigation. It offers technical assistance for credit planning and promotes innovations in financing floriculture supply chains, strengthening linkages between smallholders and markets. Moreover, NABARD

Table 13.2 Schemes/programs/missions initiatives by the Government of India

Scheme/program/ mission	Objective/focus	Subsidy pattern and cost norms
Mission for Integrated Development of Horticulture (MIDH)	Integrated development of horticulture, including floriculture	Cut flowers: ₹1.00 lakh/ha cost norm; 40% subsidy for small and marginal farmers, 25% for others; 50% subsidy in NE/Himalayan/TSP areas Bulbous flowers: ₹1.50 lakh/ha; same subsidy pattern Loose flowers: ₹40,000/ha; same subsidy pattern Hi-tech nursery: ₹25 lakh/ha; 100% subsidy to public, 40% credit-linked to private (max ₹40 lakh/unit) Tissue culture: ₹250 lakh/unit for new unit; 40–50% credit-linked subsidy Protected cultivation (polyhouse): Up to ₹1650/sqm based on structure; 50% subsidy with area limits Postharvest pack houses: ₹50 lakh/unit; 35–50% credit-linked back-ended subsidy Integrated nutrient management (INM) & plant protection: 30–50% subsidy
National Horticulture Board (NHB)	Promotion of commercial horticulture and postharvest	Credit-linked back-ended subsidies for commercial horticulture of cut flowers, bulbous flowers, loose flowers, and tissue culture units. Subsidy for open-field floriculture: 40% of project cost capped at ₹30 lakh in general areas; –50% capped at ₹37.5 lakh in NE and hilly areas. Protected cultivation (polyhouse) subsidy: 50% of project cost capped at ₹56 lakh per project. Postharvest infrastructure subsidy: ranges 35–50% depending on the region. High-tech nursery and tissue culture unit subsidies at 40–50% credit-linked support. Cost norms include ₹1.00 lakh/ha for cut flowers, ₹1.50 lakh/ha for bulbous flowers, and ₹40,000/ha for loose flowers
Agricultural and Processed Food Products Export Development Authority (APEDA)	Export promotion, infrastructure, and quality improvement	40% assistance for pack houses, pre-cooling, cold storage, reefer transport, capped at ₹200 lakh per location. 90% assistance for common infrastructure by government entities, capped at ₹600 lakh. 50% assistance for quality management and traceability, capped at ₹20 lakh Up to 75% assistance for NE, Himalayan regions, SC/ST, women exporters. Financial support for market development activities, including trade exposure and capacity building

(continued)

Table 13.2 (continued)

Scheme/program/ mission	Objective/focus	Subsidy pattern and cost norms
Rashtriya Krishi Vikas Yojana (RKVY)	Infrastructure development and strengthening agri value chains	State-driven allocation supports cold storage, auction centers, and Farmer-Producer Organizations (FPO) capacity building. Funding pattern: Centre-state share 60:40 (90:10 for NE and Himalayan states) DPR (detailed project report) component funds infrastructure such as tissue culture labs, cold storages, postharvest handling, bio-control labs, vermin compost units. Assistance up to 100% for infrastructure projects executed by public sector public–private partnerships (PPP); 50% for private sector infrastructure projects. Micro-irrigation subsidy: 55% for small/marginal farmers, 45% others; limited to 5 ha per beneficiary. Unit costs and indicative bills of quantities for drip and sprinkler irrigation given (range from ₹10,000 to ₹7,00,000+ depending on area and system). Emphasis on FPO/clusters for micro-irrigation adoption. Inclusion of organic inputs and soil health components to promote sustainable floriculture practices
CSIR Floriculture Mission	Enhancing farmers' income and entrepreneurship through floriculture technology innovation and cluster development	Supports over 245 floriculture clusters covering 4121 hectares with 14,000+ farmer beneficiaries. Facilitates varietal improvement, micropropagation parks, and integrated floriculture clusters. Promotes technology dissemination, value addition, and institutional linkages. No fixed subsidy pattern; implemented as project mode funding with 100% assistance to research & development (R&D) projects and collaborative extension. Funding includes the establishment of nurseries, tissue culture units, postharvest infrastructure, and training programs
National Mission for Sustainable Agriculture (NMSA)	Promoting sustainable practices in horticulture, including efficient water use and organic farming	Subsidy support for micro-irrigation and organic inputs promotes climate resilience

Ministry of Agriculture and Farmers Welfare (2025); National Horticulture Board (2024); Agricultural and Processed Food Products Export Development Authority [APEDA] (2021); Department of Agriculture and Farmers Welfare (2024); Council of Scientific and Industrial Research [CSIR] (2020–2026); Bobde et al. (2023)

facilitates convergence between national missions like MIDH, PMKSY micro-irrigation programs, and state-level horticulture initiatives, ensuring coordinated financial and technical support for floriculture. It also champions innovative funding models such as interest subvention schemes, credit guarantee funds, and linkage with Corporate Social Responsibilit (CSR) initiatives aimed at mitigating financing risks inherent to this sector (National Bank for Agriculture and Rural Development [NABARD] 2018–2023).

Agri-fintech enterprises provide transformative solutions that streamline credit delivery and enable seamless access for small, marginal, and underserved floriculture growers. These platforms integrate alternative credit scoring methods, insurance products, secure and prompt transaction channels, and supply chain financing frameworks catered to the floriculture production and marketing seasonality. By digitizing loan disbursement, repayment, and risk mitigation, agri-fintech enhances farmer resilience and cash flow reliability, hence accelerating the adoption of modern floriculture technologies. Platforms like Jai Kisan, Samunnati, Aggois, Ayekart, Hesa, and others exemplify this accelerating trend with significant positive impact on rural financial ecosystems (Mathias 2025).

In totality, the triad of traditional banks, NABARD's institutional support, and fintech innovation forms a robust, multi-channel finance delivery system critical for expanding and upgrading India's floriculture sector. This approach ensures that floriculture producers—ranging from individual entrepreneurs to group enterprises—benefit from adequate, timely, and affordable finance, enabling sustainability, innovation diffusion, and competitiveness in domestic and global markets.

13.7.2.1 State-level Floriculture Schemes and Impact

Mizoram

Mizoram's floriculture development spans approximately 21 hectares with over 200 farmers engaged, supported by NABARD and MIDH schemes for cold storage, pack houses, and market infrastructure. A significant milestone was the first-ever export of 1024 *Anthurium* cut flowers (70 kg) to Singapore in February 2025, facilitated by APEDA in collaboration with the Mizoram Department of Horticulture and local grower cooperatives. This export success bolsters the region's floriculture profile and empowers women farmers primarily (Dayal et al. 2025).

Gujarat

Gujarat's floriculture mission presently covers around 350 hectares with a focus on protected cultivation, including polyhouse construction with subsidy rates of up to 50%. Subsidized micro-irrigation schemes enabling drip and fertigation adoption have reached 150 hectares across various floriculture clusters. The state's export-oriented floriculture benefits from APEDA-linked market development assistance, quality certification, and price support schemes. Gujarat also operates five modern nurseries with multipurpose propagation units ensuring high-quality planting material availability (MyScheme 2025b).

Meghalaya

With roughly 18 hectares under commercial floriculture, Meghalaya's approach is focused on tribal and tribal women empowerment via community-based production models. The state provides subsidies of 30–50% on nursery and tissue culture laboratory establishments. Cold storage capacity has been increased by 3000 MT through RKVY and MIDH funds to enhance postharvest preservation and export readiness. Training programs have been conducted for 400 farmers over 3 years to foster technical and business skills (MyScheme 2025a).

Karnataka

Karnataka's extensive floriculture programs incorporate subsidies for approximately 2000 hectares under traditional and hi-tech floriculture, including cut flowers, loose flowers, and bulbous varieties. The state's scheme provides a 40–50% cost subsidy for greenhouse construction and a maximum of ₹25 lakh per nursery establishment. Five Model Floriculture Centres are operational, covering 150 hectares for mother plant maintenance and mass propagation annually. The state supports the import of quality planting materials in coordination with APEDA and invests ₹100 crore annually in capacity building and market infrastructure development (Srinivasa Gowda 2009).

13.7.3 Public–Private Partnerships and R&D Support from Institutions

Public–private partnerships (PPP) and research & development (R&D) support from institutions have played a pivotal role in transforming India's floriculture sector by fostering innovation, technology transfer, and market expansion. PPPs leverage the complementary strengths of public research institutions and private enterprises, allowing for efficient commercialization of technologies and faster dissemination of innovation tailored to farmer needs.

The Indian Council of Agricultural Research (ICAR) has been instrumental in advancing PPPs by organizing industry meets and establishing guidelines to facilitate cooperative research ventures. Key areas of partnership include seed and planting material improvement, biotech product development, postharvest processing, value addition, and capacity building. Examples such as Indo-Israel and Indo-Dutch Centres of Excellence have enhanced protected cultivation, irrigation, and floriculture export readiness, especially in states like Karnataka, Haryana, and Maharashtra (Rausser and Ameden 2004).

TANFLORA in Tamil Nadu exemplifies a successful PPP venture under the state's Agri Export Zone (AEZ) strategy, jointly operated by the Tamil Nadu Industrial Investment Corporation (TIIC), government bodies, and private partners. Supported by schemes like the National Horticulture Mission (NHM) and APEDA export promotion funds, TANFLORA provides centralized facilities for grading, packing, cold storage, and market linkage for export-quality flowers, benefiting growers engaged in protected cultivation (Tanflora n.d.).

Similarly, the International Flower Auction Bangalore (IFAB), a PPP company formed by the South Indian Floriculture Association (SIFA), small growers, and the Karnataka government, offers modern auction and logistics facilities backed by APEDA and Karnataka Agro Industries Corporation. IFAB facilitates transparent pricing and direct grower-buyer interactions, enhancing competitiveness and market access for floriculture producers at both national and international levels (IFAB-International Flower Auction Bangalore 2017).

These PPP initiatives, alongside research support from institutions, are crucial drivers in India's floriculture ecosystem, enhancing infrastructure, fostering innovation, and enabling sustainable high-value agriculture through strategic collaboration between public and private sectors.

13.8 Inspirational Floriculture Ventures From India and Beyond

13.8.1 Success Stories in Indian Floriculture

The floriculture sector in India showcases exemplary models of entrepreneurial success that demonstrate the potential for high-value agriculture to generate sustainable income, empower marginalized groups, and contribute to regional economic development.

Meena Chandel, Himachal Pradesh

Meena Chandel successfully transitioned from a decade-long career in education to floriculture entrepreneurship. Operating on 22 bighas with an 11,000 square meter greenhouse, her farm produces a diverse portfolio of high-demand flowers, including roses, carnations, gerberas, and lisianthus. With an annual turnover of ₹70 lakhs, her enterprise exemplifies the integration of strategic crop scheduling aligned to peak market seasons, advanced greenhouse technology, and sustainable pest management practices. Additionally, Meena has established a cooperative society to facilitate knowledge transfer and market access for local growers, signifying her role in community empowerment and agripreneurship (Baruah n.d.).

Lohith Reddy, Bengaluru

Formerly in the technology sector, Lohith Reddy founded Lohith Flora, focusing on chrysanthemum cultivation. With a production capacity of 1500 bunches weekly and a monthly revenue of ₹7 lakhs, Lohith employs over 20 workers and has integrated cold chain logistics to maintain product freshness, demonstrating the application of modern supply chain management in floriculture (Dutta 2025).

Parvathy Mohanan, Kochi
Parvathy Mohanan transformed her passion for ornamental plants into a profitable venture during the 2020 lockdown. Starting with a modest clientele, her business scaled to generating ₹1 lakh per month across 300 cultivars through effective online marketing and home delivery, showcasing the adaptability and commercialization potential in urban floriculture (Dutta 2025).

Abhinav Singh, Uttar Pradesh
Abhinav Singh shifted from a corporate technology background to gerbera cultivation, leveraging government-supported polyhouse subsidies. Within 2 years, he achieved an annual income of ₹18 lakhs, highlighting the impact of policy incentives on promoting high-yield floriculture enterprises (Dutta 2025).

Gurwinder Singh Sohi, Punjab
Gurwinder Singh Sohi transitioned from conventional wheat farming to floriculture on 22 acres, cultivating 40 flower varieties. He innovated a semi-automated planting machine that drastically reduced labor time and has trained over 2000 farmers, underlining the role of innovation and the farmer (Dutta 2025).

13.9 Conclusion

Floriculture has become a dynamic, vibrant, and remunerative branch of modern agriculture, technology, and entrepreneurship. The integration of advanced hi-tech interventions like protected cultivation, drip irrigation, and crop scheduling ensures year-round production and income stability for growers. Small and marginal farmers benefit through crop diversification, employment, and improved market strategies supported by government schemes and cooperatives. Efficient postharvest handling and value addition, including drying, essential oil and pigment extraction, minimize losses and enhance profitability. Support from government missions, NABARD financing, and public–private partnerships has strengthened infrastructure and export potential. Growing domestic and international demand driven by urbanization, evolving lifestyle, and sustainability or eco-conscious trends continue to expand opportunities. Emerging sectors such as edible flowers, floristry, bonsai, nursery, and turf management strengthen the industry. Despite challenges in logistics, phytosanitary compliance, and cold chain, progress through innovation and collaboration is driving competitiveness. With strong policy support and technological innovation, floriculture can sustain livelihoods and enrich environmental and aesthetic well-being.

References

Acharya A, Macchindra DS, Chaudhary D (2024) Floriculture: cultivation and marketing. In: Technologies of horticulture sciences. Stella Publications, Kurukshetra, pp 179–189. [Google Scholar]

Adebayo IA, Pam VK, Arsad H, Samian MR (2021) The global floriculture industry: status and future prospects. In: Khalid Rehman H (ed) The global floriculture industry shifting directions, new trends, and future prospects. Apple Academic Press Inc., New York. https://doi.org/10.1201/9781003000723

Agricultural and Processed Food Products Export Development Authority (APEDA), Ministry of Commerce and Industry, Government of India (2021) Operational guidelines for financial assistance scheme (FAS) for the 15th Finance Commission cycle (2021–22 to 2025–26). https://apeda.gov.in/sites/default/files/Financial%20Assistance%20Schemes/FAS_Guidelines_05102021.pdf

Anonymous (2024) Floriculture. Retrieved from https://apeda.gov.in/Floriculture

Anonymous (2025a) The art of Bonsai making. Retrieved from https://www.pubs.ext.vt.edu/426/426-601/426-601.html

Anonymous (2025b) Floriculture market size and forecast outlook 2025 to 2035. Retrieved from https://www.futuremarketinsights.com/reports/floriculture-market

Anumala NV, Kumar R (2021) Floriculture sector in India: current status and export potential. J Hortic Sci Biotechnol 96(5):673–680. https://doi.org/10.1080/14620316.2021.1902863

Aydin V (2023) Factors affecting the post-harvest quality of cut flowers. In: Advances in plant research and agriculture. Iksad, p 155. [Google Scholar]

Baranowska M, Behnke-Borowczyk J, Barzdajn W, Szmyt J, Korzeniewicz R, Łukowski A, Memišević-Hodžić M, Kartawik A, Kowalkowski W (2023) Effects of nursery production methods on fungal community diversity within soil and roots of Abies alba Mill. Sci Rep 13(1):21284. https://doi.org/10.1038/s41598-023-48047-y

Baruah S (n.d.) How Meena Chandel transitioned from teaching to a thriving floriculture business – and built a Rs 70 lakh empire. Krishi Jagran. Retrieved from https://krishijagran.com/success-story/how-meena-chandel-transitioned-from-teaching-to-a-thriving-floriculture-business-and-built-a-rs-70-lakh-empire/

Benschop M, Kamenetsky R, Le Nard M, Okubo H, De Hertogh A (2010) The global flower bulb industry: production, utilization, research. Hortic Rev 36(1):1–115. [Google Scholar]

Bhargavi B, Naik MR, Kumar V, Sumathi T, Srinivas PT, Keren PJ (2025) Synchronizing blooming with market demand: crop regulation in commercial flower crops. Plant Arch 25(S2):2704–2710. https://doi.org/10.51470/PLANTARCHIVES.2025.v25.supplement-2.344

Bhosale P, Gobade N (2025) Industrial application of flowers: perfume, dye, and essential oils. In: Innovations in floriculture and landscaping: art, science and sustainability. Stella International Publication, pp 291–319. [Google Scholar]

Bobde A, Bhinde HN, Burande A, Agarwal P (2023) An analysis of government schemes promoting floriculture and horticulture: case study of India. MPJSS 28:96–107. [Google Scholar]

Braun RC, Mandal P, Nwachukwu E, Stanton A (2024) The role of turfgrasses in environmental protection and their benefits to humans: thirty years later. Crop Sci 64(6):2909–2944. https://doi.org/10.1002/csc2.21383

Bujak T, Zagórska-Dziok M, Ziemlewska A, Nizioł-Łukaszewska Z, Lal K, Wasilewski T, Hordyjewicz-Baran Z (2022) Flower extracts as multifunctional dyes in the cosmetics industry. Molecules 27(3):922. https://doi.org/10.3390/molecules27030922

Chahidi LO, Fossa M, Priarone A, Mechaqrane A (2021) Energy saving strategies in sustainable greenhouse cultivation in the Mediterranean climate – a case study. Appl Energy 282:116156. https://doi.org/10.1016/j.apenergy.2020.116156

Chakraborthy R (2019) Floriculture in northern West Bengal of India—a dream sunrise is on wait. EC Agric 5(2):1–7. [Google Scholar]

Chandel A, Thakur M, Singh G, Dogra R, Bajad A, Soni V, Bhargava B (2023) Flower regulation in floriculture: an agronomic concept and commercial use. J Plant Growth Regul 42(4):2136–2161. https://doi.org/10.1007/s00344-022-10688-0

Chandra A (2025) Floriculture: trends, challenges and opportunities. Int J Hortic Food Sci 7(8):155–159. https://doi.org/10.33545/26631067.2025.v7.i8c.379

Chawla SL, Patil S, Ahlawat TR, Agnihotri R (2016) Present status, constraints and future potential of floriculture in India. In: Commercial horticulture, 1st edn. New India Publishing Agency, New Delhi, pp 29–38. [Google Scholar]

Cheng H, Hu Y, Reinhard M (2014) Environmental and health impacts of artificial turf: a review. Environ Sci Technol 48(4):2114–2129. https://doi.org/10.1021/es4044193

Chittibomma K, Naorem C, Singh HB, Johns Tiyndel G, Patil AA, Harivignesh G, Vaishnavi CS, Kumari T (2023) Opportunities and challenges in the floriculture industry for entrepreneurship development in the north east region of India. Int J Stat Appl Math 8(6S):1340–1346. https://doi.org/10.22271/maths.2023.v8.i6Sr.1568

Council of Scientific and Industrial Research (CSIR) (2020–2026) CSIR Floriculture Mission: enhancing farmers' income and entrepreneurship development through high value floriculture utilizing CSIR technologies. India Science and Technology. https://www.indiascienceandtechnology.gov.in/technologies/csir-floriculture-mission-enhancing-farmers%E2%80%99-income-and-entrepreneurship-development-through-high

Cruz-Sánchez E, García-Vargas JM, Gracia I, Rodriguez JF, García MT (2024) Supercritical CO2 extraction of lavender flower with antioxidant activity: laboratory to a large scale optimization process. J Taiwan Inst Chem Eng 157:105404. https://doi.org/10.1016/j.jtice.2024.105404

Dayal A, Narayanan A, Biswas I (2025) First-ever export of Anthurium flowers from Mizoram to Singapore, a fillip to India's floriculture potential. Ministry of Commerce & Industry. https://apeda.gov.in/sites/default/files/press_release/PIB2113965.pdf

De LC, Singh DR (2016) Floriculture industries, opportunities and challenges in Indian hills. Int J Hortic 6(13):1–9. https://doi.org/10.5376/ijh.2016.06.0013

Department of Agriculture and Farmers Welfare, Ministry of Agriculture and Farmers Welfare, Government of India (2024) Operational guidelines for Pradhan Mantri Rashtriya Krishi Vikas Yojana (PM-RKVY). https://www.uasbangalore.edu.in/wp-content/uploads/2024/11/PM-RKVY-Guidelines-as-on-14.11.2024.pdf

Doddrell NH, Lawson T, Raines CA, Wagstaff C, Simkin AJ (2023) Feeding the world: impacts of elevated CO2 on nutrient content of greenhouse grown fruit crops and options for future yield gains. Hortic Res 10(4):uhad026. https://doi.org/10.1093/hr/uhad026

Dutta R (2025) Remarkable stories of 5 Indian flower farmers, one earning Rs 5 Cr/year. The Better India. 14 Oct 2025. https://thebetterindia.com/400395/floriculture-profitable-five-farmers-achieved-success-annual-revenue-profits/

Erwin J (2005) Factors affecting flowering in ornamental plants. In: Flower seeds: biology and technology. CABI Publishing, Wallingford, pp 87–115. https://doi.org/10.1079/9780851999067.0087

Faust JE, Dole JM (2021) The global cut flower and foliage marketplace. Cut flowers and foliages, 27:1–47.

Feng H (1999) Agricultural development in The Netherlands: an analysis of the history of Dutch agricultural development and its importance for China. Agricultural Economics Research Institute (LEI), The Hague, 104p. [Google Scholar]

Ferrante A, Ferrini F (2023) Floriculture and landscapes: perspectives and challenges. Front Hortic 2:1123298. https://doi.org/10.3389/fhort.2023.1123298

Ghimire P, Bhandari N, Lamsal G, Pun U (2025) Recent advances in floriculture industry. In: Nepalese floriculture. Floriculture Association Nepal (FAN), Lalitpur, pp 1–7. [Google Scholar]

Glover J (2022) Inserting Indian production networks into geographies of globalisation of the cut flower industry. Doctoral dissertation, The Australian National University. [Google Scholar]

Gupta J, Dubey RK (2018) Factors affecting post-harvest life of flower crops. Int J Curr Microbiol App Sci 7(1):548–557. https://doi.org/10.20546/ijcmas.2018.701.065

Hasna PM, Rafeekher M, Priyakumari I, Reshmi CR (2024) Flower forcing: a review. Int J Plant Soil Sci 36(6):592–600. https://doi.org/10.9734/ijpss/2024/v36i64662

IFAB-International Flower Auction Bangalore (2017, August 9). https://ifabindia.org/

Jahan H (2009) Production, post-harvest handling and marketing of cut-flowers in Bangladesh: an agribusiness study. SAARC J Agric 7(2):1–14. [Google Scholar]

Jain R, Janakiram T, Kumawat GL (2016) Drying techniques in ornamental plants. In: Commercial horticulture. New India Publishing Agency, New Delhi, pp 501–512. [Google Scholar]

Jejal AD, Chandr T (2025) Art and science of bonsai: horticulture in miniature. AgriFrontLine 1:10–13. [Google Scholar]

Jhanji S, Singh P, Acharya A, Dhiman A (2025) Physio-biochemical and anatomical basis of enhanced post-harvest quality of Lilium LA hybrid (Lilium longiflorum × Asiatic hybrid) cv. Parrano cut stems through pulsing with plant growth substances. N Z J Crop Hortic Sci 53:2679–2695. https://doi.org/10.1080/01140671.2025.2501282

Kalmegh S, Sing N (2016) Challenges and obstacles in Indian floriculture industry. Int J Innov Res Dev 5(7):1–20. [Google Scholar]

Kaur L (2025) Methods of enhancing post-harvest life in floriculture crops. In: Multidisciplinary research in arts, science & commerce, vol 17. The Hill Publication, p 26. [Google Scholar]

Knierim A, Kernecker M, Erdle K, Kraus T, Borges F, Wurbs A (2019) Smart farming technology innovations – insights and reflections from the German smart-AKIS hub. NJAS Wageningen J Life Sci 90:100314. https://doi.org/10.1016/j.njas.2019.100314

Koushal S, Sarvade SA, Lohar A, Kaushik K, Reddy CS, Yampalla M (2024) Value addition in floriculture: the potential of flower processing and by-products. Int J Adv Biochem Res 8(11):523–529. https://doi.org/10.33545/26174693.2024.v8.i11g.2891

Kumar PN, Raju DVS, Saha TN, Safeena SA, Gaikwad P, Mathew S (2022) Post-harvest management in commercial flower crops. Technical report. ICAR Directorate of Floricultural Research, Pune, pp 157–164. [Google Scholar]

Kumar M, Chaudhary V, Sirohi U, Srivastav AL (2024) Economically viable flower drying techniques to sustain flower industry amid COVID-19 pandemic. Environ Dev Sustain 26(9):22103–22148. https://doi.org/10.1007/s10668-023-03376-w

Latulippe N (2008) The sustainability of floriculture as a livelihood activity in West Bengal, India. Doctoral dissertation, University of Guelph

Maitra S, Swathi K, Sarkar I (2016) Essential oils: a value-addition for improvisation of commercial floriculture in India. J Agric Technol 3(1):7. [Google Scholar]

Malviya A, Vala M, Mankad A (2022) Recent floriculture in India. Int Assoc Biol Comput Dig 1(1):1–8. https://doi.org/10.56588/iabcd.v1i1.8

Mathias G (2025) 5 Agri-FinTechs powering financial growth for Indian farmers. IBS Intelligence. https://ibsintelligence.com/ibsi-news/5-agri-fintechs-powering-financial-growth-for-indian-farmers/

Meghana K, Wazeed GSA (2024) India's floral industry—present state and prospects for future. In: Collaborative novel techniques in horticulture. SR Edu Publications, Kalwakurthy, pp 161–172. https://www.researchgate.net/publication/386453622_India's_Floral_Industry-_Present_State_and_Prospects_for_Future

Ministry of Agriculture and Farmers Welfare, Government of India (2025) Mission for Integrated Development of Horticulture (MIDH): operational guidelines. https://midh.gov.in/

Mishra V, Dwivedi SK (2015) Postharvest management of fresh cut flowers. In: Postharvest biology and technology of horticultural crops. Principles and practices for quality maintenance. Apple Academic Press and CRC Press Taylor and Francis Group, Waretown, pp 347–399. [Google Scholar]

Mishra GP, Singh N, Kumar H, Singh SB (2010) Protected cultivation for food and nutritional security at Ladakh. Def Sci J 60(2):219. [Google Scholar]

Murthy P, Rani Jayadurga Nayak S, Sekhar H, Behera AG, Sherly J (2024) Sustainable floriculture: adopting eco-friendly practices in flower production. Int J Res Agron 7(10):505–511. https://doi.org/10.33545/2618060X.2024.v7.i10g.1819

MyScheme (2025a) Floriculture development scheme. Government of India. https://www.myscheme.gov.in/schemes/fcds

MyScheme (2025b) Assistance program for planting in floriculture. Government of India. https://www.myscheme.gov.in/schemes/apfpf

Nagarajan M, Senthilvel S, Palanysamy D (2002) Material substitution in greenhouse construction. Kisan World 11:57–58. [Google Scholar]

Nagarsheth S, Agbossou K, Henao N, Bendouma M (2025) The advancements in agricultural greenhouse technologies: an energy management perspective. Sustainability 17(8):3407. https://doi.org/10.3390/su17083407

Nair SP (2025) Exploring the trends in Indian floriculture exports: review of the volume and value dynamics of the key international importers in the industry. Adv Int J Res 6(5):1–8. [Google Scholar]

Nakano Y, Higuchi Y, Sumitomo K, Hisamatsu T (2013) Flowering retardation by high temperature in chrysanthemums: involvement of FLOWERING LOCUS T-like 3 gene repression. J Exp Bot 64(4):909–920. https://doi.org/10.1093/jxb/ers370
Narasimham NV, Kishor N (1997) Floriculture exports: potential and challenges. Foreign Trade Rev 32(3):42–57. https://doi.org/10.1177/001573251597030
National Bank for Agriculture and Rural Development (NABARD) (2018–2023) Area development scheme on floriculture. https://www.nabard.org
National Horticulture Board, Ministry of Agriculture and Farmers Welfare, Government of India (2024) Cost norms and pattern of assistance under MIDH for National Horticulture Board related activities. https://nhb.gov.in/pdf/Cost%20Norms.pdf
Nemali K (2022) History of controlled environment horticulture: greenhouses. HortScience 57(2):239–246. https://doi.org/10.21273/HORTSCI16160-21
Nowak J, Rudnicki RM, Duncan AA (1990) Postharvest handling and storage of cut flowers, florist greens, and potted plants, vol 1. Chapman and Hall, London. https://link.springer.com/book/9789401066761
Olle M, Viršile A (2013) The effects of light-emitting diode lighting on greenhouse plant growth and quality. Agric Food Sci 22(2):223–234. https://doi.org/10.23986/afsci.7897
Proietti S, Scariot V, De Pascale S, Paradiso R (2022) Flowering mechanisms and environmental stimuli for flower transition: bases for production scheduling in greenhouse floriculture. Plants 11(3):432. https://doi.org/10.3390/plants11030432
Rafique Z, Riaz H, Ahmad A, Iqbal T (2024) Emerging trends in the floriculture industry: adapting to market demands and consumer preferences. Biol Times 3(11):14–15. [Google Scholar]
Ranjan P, Ranjan JK, Das B, Ahmed N (2013) High value flower cultivation under low cost greenhouse in NW Himalayas. Int J ChemTech Res 5(2):789–794. [Google Scholar]
Rashed NM, Memon SA, Turki SMA, Shalaby TA, El-Mogy MM (2024) An analysis of conventional and modern packaging approaches for cut flowers: a review article. Front Plant Sci 15:1371100
Rausser G, Ameden H (2004) Public-private partnerships needed in horticultural research and development. Calif Agric 58(2):116–119. https://doi.org/10.3733/ca.v058n02p116
Reddy PP (2016) Protected cultivation. In: Sustainable crop protection under protected cultivation. Springer Singapore, Singapore, pp 1–11. https://doi.org/10.1007/978-981-287-952-3_1
Reid MS (2009) Handling of cut flowers for export. Proflora Bulletin, pp 1–26
Rudnicki RM, Nowak J, Goszczynska DM (1991) Cold storage and transportation conditions for cut flowers cuttings and potted plants. In: Hortifroid, V international symposium on postharvest physiology of ornamental plants; importance of cold in ornamental, vol 298. International Society for Horticulture Science, Wageningen, pp 225–236. https://doi.org/10.17660/ActaHortic.1991.298.27
Ryan-Ndegwa S, Zamani R, Martins T (2024) Exploring the human health impact of artificial turf worldwide: a systematic review. Environ Health Insights 18:11786302241306291. https://doi.org/10.1177/11786302241306291
Saha TN (2014) Role of all India coordinated research project in development of floriculture in India. Int J Bioresour Stress Manag 5(1):159–165. https://doi.org/10.5958/j.0976-4038.5.1.031
Saini U, Rishi M, Dwivedi P (2023) Barriers affecting cut flower transportation in India – using AHP and FAHP technique. Eur Econ Lett 13(5):806–818. https://doi.org/10.52783/eel.v13i4.663
Sajjad Y, Jaskani MJ, Asif M, Qasim M (2017) Application of plant growth regulators in ornamental plants: a review. Pak J Agric Sci 54(2):327. https://doi.org/10.21162/PAKJAS/17.3659
Santosh DT, Maitra S, Sairam M (2024) Enhancing rural economies with protected cultivation technologies. Int J Bioresour Sci 11(1):71–79. https://doi.org/10.30954/2347-9655.01.2024.8
Sarkar D (2023) Indian floriculture in global perspective. Int J Plant Soil Sci 35(14):228–238. https://doi.org/10.9734/ijpss/2023/v35i143040
Senapati AK, Raj D, Jain R, Patel NL (2016) Advances in packaging and storage of flowers. In: Commercial horticulture, vol 34. New India Publishing Agency, New Delhi, pp 473–488. [Google Scholar]
Shaibur MR, Husain H, Akter A (2020) Prospect of floriculture on socio-economic condition of farmers in Gadkhali, Jashore, Bangladesh. Bangladesh J Agric Res 45(1):77–87. [Google Scholar]

Sharma G (2020) Value addition in flowers. National Webinar on Advances in Floriculture, UHS, Bagalkot and SAAHAS

Sharma SK (2025) Smart agriculture revolution: cloud and IoT-based solutions for sustainable crop management and precision farming. In: AI-based advanced optimization techniques for edge computing. Wiley, Hoboken, pp 253–288. https://doi.org/10.1002/9781394287062.ch10

Sharma G, Singh AK, Abrol GS (2021) Avenues in value addition and processing of flowers. In: Webinar compendium on recent advances in processing of fruits and vegetables. RLBCAU, Jhansi, pp 76–87. [Google Scholar]

Sharma R, Singh S, Pant K, Mashiana HK, Dubey RK (2024) Protected cultivation of floriculture crops: innovative technologies and future challenges. In: Ornamental horticulture: latest cultivation practices and breeding technologies. Springer Nature Singapore, Singapore, pp 15–43. https://doi.org/10.1007/978-981-97-4028-4_2

Singh B (2014) Protected cultivation of horticultural crops in India: challenges and opportunities. In: 3rd International conference on agriculture & horticulture. Hyderabad International Convention Centre, Hyderabad. https://doi.org/10.4172/2168-9881.S1.010

Singh P, Singh R, Chaudhary D, Acharya A (2024a) Pre-harvest technologies: floriculture sustainability initiative. In: Ornamental horticulture: latest cultivation practices and breeding technologies. Springer Nature Singapore, Singapore, pp 211–226. https://doi.org/10.1007/978-981-97-4028-4_11

Singh S, Sharma D, Rana A, Bhargava B (2024b) Extension networks and floricultural advancement dissemination. In: Ornamental horticulture: latest cultivation practices and breeding technologies. Springer Nature Singapore, Singapore, pp 295–317. https://doi.org/10.1007/978-981-97-4028-4_16

Solaimalai A, Baskar M, Sadasakthi A, Subburamu K (2005) Fertigation in high value crops – a review. Agric Rev 26(1):1–13. [Google Scholar]

Song YH, Shim JS, Kinmonth-Schultz HA, Imaizumi T (2015) Photoperiodic flowering: time measurement mechanisms in leaves. Annu Rev Plant Biol 66(1):441–464. https://doi.org/10.1146/annurev-arplant-043014-115555

Srinivasa Gowda MV (2009) Hi-tech floriculture in Karnataka. Occasional Paper No. 49. National Bank for Agriculture and Rural Development

Strandberg M, Blombäck K, Jensen AD, Knox JW (2012) Priorities for sustainable turfgrass management: a research and industry perspective. Acta Agric Scand B Soil Plant Sci 62(sup1):3–9. https://doi.org/10.1080/09064710.2012.682163

Tandon A, Singh K (2023) Advancements, challenges, and policy interventions in agriculture-related manufacturing in India: a critical analysis. SSRN Electron J. https://doi.org/10.2139/ssrn.5070742

Tanflora (n.d.) Tanflora. Retrieved from https://tanflora.com/

Tawalbeh M, Aljaghoub H, Alami AH, Olabi AG (2023) Selection criteria of cooling technologies for sustainable greenhouses: a comprehensive review. Therm Sci Eng Prog 38:101666. https://doi.org/10.1016/j.tsep.2023.101666

Thakur N (2020) A review on the effect of storage methods and packaging material on the post-harvest longevity of cut flowers. Int J Chem Stud 8(3):2375–2379. https://doi.org/10.22271/chemi.2020.v8.i3ah.9568

Thongnim P, Yuvanatemiya V, Srinil P (2023) Smart agriculture: transforming agriculture with technology. In: Asia simulation conference. Springer Nature Singapore, Singapore, pp 362–376. https://doi.org/10.1007/978-981-99-7240-1_29

Vahoniya D, Panigrahy SR, Patel D, Patel J (2018) Status of floriculture in India: with special focus to marketing. Int J Pure Appl Biosci 6:1431–1438. https://doi.org/10.18782/2320-7051.6487

Vidhate UA, Deshmukh AA (2024) A study on marketing challenges faced by cut flower producers in Pune district. MPJSS 29(1):161–170

Waghmode MS, Gunjal AB, Nawani NN, Patil NN (2018) Management of floral waste by conversion to value-added products and their other applications. Waste Biomass Valoriz 9(1):33–43. https://doi.org/10.1007/s12649-016-9763-2

Yaseen I, Fayaz S, Khursheed A, Gupta A, Hassan GI, Rajan R, Qayoom S, Gani I, Jayarajan S (2022) Advances in propagation and nursery management: methods and techniques. In: Apples. CRC Press, pp 45–57. [Google Scholar]

Attractive Packaging for Scientific and Smart Globaltrade

14

Payel Das, Manideepa Roy, Siddhartha Sahoo, Shreya Mondol, Victry Samanta, and Ankan Das

Abstract

Packaging is integral to the global horticultural sector, enhancing product attractiveness and ensuring quality from field to home. This chapter delves into innovative and smart packaging solutions, incorporating advanced technologies like active packaging, intelligent packaging, sensors, robotics, QR codes, RFID tags, sustainable materials, real-time tracking and monitoring systems. These innovations optimize supply chain efficiency and transparency for both local and global markets. With the help of governmental policies, eco-friendly packaging that fosters sustainable trade practices minimizes environmental footprint while meeting consumer preferences. It discusses the packaging design, material, techniques, and unique functionality of each packaging that influences market expansion, offering a comparative analysis of domestic versus international horticulture market growth and trends. The chapter emphasizes the dynamic interplay between scientific progress, trade mechanisms, and visually appealing packaging, which together boost consumer interest, brand loyalty, and profitability. It also explores various packaging methods, their applications, and alignment with ecological goals. Strategies for policymakers and industry leaders to adopt inventive packaging approaches that drive economic prosperity, enhance human well-being through safer products, and prioritize environmental stewardship. These efforts will ensure sustained competitiveness and adaptability in the rapidly evolving global horticulture trade landscape.

Keywords

Innovative · Packaging · Trade · Sustainable

P. Das · M. Roy · S. Sahoo · S. Mondol · V. Samanta · A. Das (✉)
Department of Horticulture, Institute of Agricultural Science, University of Calcutta, Kolkata, West Bengal, India

A. Das, A. B. Sharangi (eds.), *Remunerative Horticulture*,
https://doi.org/10.1007/978-981-92-1007-7_14

14.1 Introduction

In today's dynamic world, advancements in science have elevated packaging from merely safeguarding products to playing a key role in their marketability (Yangang 2021). Eye-catching and innovative packaging significantly influences consumer purchasing decisions (Shukla et al. 2022). A functional magnetic resonance imaging (fMRI) study revealed that appealing packaging stimulates distinct brain activity in regions associated with visual attention and memory (Stoll et al. 2008). Research indicates that improper sealing of goods and regulatory issues during transportation pose significant challenges (Parfitt et al. 2010). Also, the accumulation of non-recyclable waste from the packaging industry is a growing environmental concern (Da Cruz et al. 2014). Additionally, insufficient understanding of packaging materials presents a critical obstacle for the food industry (Kuswandi et al. 2011). Packaging holds substantial economic value, influencing national Gross Domestic Product (GDP) by fostering world trade, enabling efficient and profitable import-export operations, and meeting consumer demands worldwide (Essuman 2018). Effective packaging is vital for maintaining strong customer relationships (Mohammadi et al. 2014). To accommodate diverse consumer preferences, export packaging must comply with international legal labeling requirements, shelf-life specifications, and cultural aesthetics of importing nations (Kader 2002). Visually appealing packaging enhances product attractiveness and influences purchasing decisions at retail (Ampuero and Vila 2006). Packaging has evolved to align with brand positioning, driving sales in competitive markets to fulfill consumer expectations (Yangang 2021). Attractive packaging serves as a critical tool in product development, supporting export promotion, attracting foreign investment, generating revenue, and contributing to balanced international trade (Ajalafshar 2017). Retailers and brand managers increasingly leverage innovative packaging as a key marketing strategy (Shukla et al. 2022). In trade, packaging enhances competitiveness in both domestic and global markets (Mohammadi et al. 2014). Advancements such as smart packaging, incorporating automation, robotics, QR codes, RFID tags, biodegradable materials, cold chain logistics, and real-time tracking and monitoring are transforming the global market (Santosh and Harshavardhan 2024). The use of IoT-based biosensors in food packaging is gaining traction, surpassing traditional methods in popularity (Sobhan et al. 2025). Since ancient times, humans have relied on natural resources like animal hides, jute, ceramics, wood, shells, bamboo, grass, and leaves for transporting and preserving food (Hussain et al. 2024). Between 1500 B.C. and 500 AD, ceramic jars played a significant role in trading wine (Twede 2002). The invention of plastics in the nineteenth century, followed by the development of polymers like polyester and polypropylene in the twentieth century, revolutionized the plastic packaging industry (Brody et al. 2008). Paper and paperboard also became widely used as packaging materials in the 1900s (Teixeira-Costa and Andrade 2021).

Eco-friendly materials, such as wooden crates and corrugated cardboard, support sustainability by being reusable and safe for both consumers and the environment, balancing functionality with ecological responsibility (Albrecht et al. 2013).

Many countries implement regulations to promote eco-friendly packaging that reflects regional cultural values (Thapliyal et al. 2024). Due to inadequate infrastructure in the processing chain, companies are shifting toward smart packaging technologies to ensure quality control from harvest to consumption (Alam et al. 2021). Emerging technologies in the packaging industry include the Industrial Internet of Things (IIoT) (Moran and Cui 2016), nanotechnology-based materials (Majid et al. 2018), and real-time monitoring systems that enhance product safety and extend shelf life (Schaefer and Cheung 2018).

This section explores trends in appealing packaging and its significance in scientific, smart domestic, and international trade within the global horticultural sector. It also addresses key issues, including packaging methods, their practical applications, and their impact on commerce, while discussing future marketing trends and regulatory considerations in the horticultural industry globally.

14.2 Foundations That Relate Science to Packaging

Highly perishable and damage-prone horticultural commodities need packaging for their safety, especially during transport (Sadhu 2022). Wood crates, corrugated shipping cartons, polymeric films, pouches, bags, baskets, carat trays are packaging materials for vegetables, and other packaging systems like film micro perforation, modification, controlled atmospheric, and active packaging systems are used to extend the shelf life of the produce (Pascall 2010). Conventional packaging materials focus on materials like paper, paperboard, plastic, glass, metals, etc., whereas passive packaging systems include polystyrene, polyurethane foams, vacuum insulated panels, and phase changing materials (Robertson 2016). Passive conventional packaging is essential for protection, preservation, distribution, and marketing by ensuring thermal stability of sensitive goods, particularly in cold chain logistics, without active power sources (Basu et al. 2018).

To illustrate, some common packaging materials are discussed below;

- *Glass*
  Glass, formed by heating a mixture of silica, sodium carbonate as a melting agent, and stabilizers like limestone, calcium carbonate or aluminum into a thick liquid poured into molds, creates airtight containers for food packaging that prevent gas exchange and enable heat sterilization of high- and moderate-acid foods while preserving their taste and quality (Marsh and Bugusu 2007).
- *Pulp, paper and paperboard*
  The pulp and paper industry, owing to its eco-friendly hallmark, made an enormous contribution to GDP (Gross Domestic Product) throughout the globe as well as to public health (Deshwal et al. 2019). Paper is a bio-based material, widely used, easily available, and has important attributes (Stark and Matuana 2021). Paper and paperboard, constituting over 30% of global packaging volume, are favored for their cost-effectiveness, printability, recyclability, and biodegradability, making them highly popular conventional packaging materials

(Kurek and Chmiel 2019). Corrugated paperboard is cost-effective and lightweight during transport, and it is efficient to carry weight during transportation (Fadiji et al. 2017). Paperboards, crafted from fibers into three-dimensional self-bonded structures, are formed as sheets or webs by layering them in a non-woven arrangement (Kirwan 2012). Various types of paper used in food packaging, including paper bags, composite cans, fiber drums, multiwall paper snacks, rigid boxes, folding cartons, corrugated fiber board (CFb), and paperboard for liquid food packaging (Deshwal et al., 2019).

- *Metal*
  Aluminum foil, the most commonly used metal foil for food packaging, offers excellent barrier properties, with thicknesses of less than 150 μm for wrapping, typically 9 μm for wraps, 50 μm for bottle caps, and 50–100 μm for ready-to-serve trays (Shin and Selke 2014).
- *Plastic*
  Polyethene (PE), polypropylene (PP), polyethene terephthalate (PET), polyvinyl chloride (PVC), and polystyrene (PS) materials contribute as 90% of the total volume of plastics used in packaging (Sadhu 2022). High-density polyethylene (HDPE) with individual seal packaging for Citrullus fruits and bell peppers effectively reduces transpiration, preserving freshness even without refrigeration and minimizing spoilage (Ben-Yehoshua 1983).
  Polypropylene (PP), a linear isotactic polymer of propylene with a low density of 0.89–0.91 g/cm^3, offers greater transparency, hardness, chemical resistance, and water vapor barrier properties compared to polyethylene, and its high melting point and biaxially oriented film (BOPP) make it ideal for thermally resistant applications, including food packaging, as noted by Mangaraj et al. (2009). Although polyvinylidene, polyester, and nylon have such low gas permeabilities that they would only be appropriate for products with extremely low respiration rates, polystyrene has also been used as a packaging material (Mangaraj et al. 2009).

The choice of key packaging materials, such as paper, paperboard, plastics, glass, and metal, is crucial for ensuring food safety and quality, with their durability being a fundamental factor in effective packaging design (Kim et al. 2014). Various packaging materials, including plastics containing toxic chemicals like bisphenol A (BPA) and bisphenol-F, metal cans with harmful monomers, and printed packaging with plasticizers and photo initiators, can contaminate food and pose health risks such as nervous, blood, and kidney issues, as well as estrogen disruption, while traditional materials like leaves and jute bags offer inadequate protection against pests, microorganisms, and environmental factors (Castle 2007; Ningwei and Mahat 2009; Mark and John 2003; Abdullahi 2014; Linda and Sarah 2010; Helen and James 2003; Bolaji 2010).

14.3 Smart Packaging

Smart packaging refers to interconnected systems that use a mix of electronics, electrochemical, and electro-optical sensors in packages to monitor the quality of fresh produce until it reaches the consumer (Alam et al. 2021). Smart packaging is an advanced way of packaging technology that comprises two broad types of techniques, such as active and intelligent packaging, and these are mostly popular as compared to the conventional method (Panou et al. 2025). Active packaging enhances the quality and shelf life of fruits and vegetables by inhibiting spoilage microorganisms while managing gas, moisture, and vapor levels within the package (Ozdemir and Floros 2004). A study revealed that a reduction in energy usage, food waste, improved food quality, and prolonged shelf life are some of the benefits that come with smart packaging (Panou et al. 2025). Active packaging includes oxygen scavengers, CO_2 emitters and absorbers, moisture regulators, ethylene scavengers, antimicrobial packaging, and flavor absorbers (Bodbodak and Rafiee 2016).

Devices of smart packaging engage, economical tiny tags onto the primary packaging commodity, such as pouches, bags, and bottles, sometimes with secondary packaging commodities such as shipping containers, to convey the inspection through the supply chain and regulate relevant action regarding food safety and quality (Yam et al. 2005). Sensor-based monitoring and smart packaging technologies significantly enhance the quality, nutritional value, and safety of fresh produce (Alam et al. 2021). Active packaging systems suppress spoilage microorganisms, thereby improving the quality and extending the storage life of fruits and vegetables by regulating gases, moisture, and vapor within the packaging (Ozdemir and Floros 2004). Active components are embedded in sachets suitable for solid foods, though not for liquids, and are also incorporated into films and coatings for food packaging (Otoni et al. 2016).

14.4 Modified Atmospheric Packaging

Modified atmospheric packaging (MAP), utilizing oxygen, nitrogen, and carbon dioxide to maintain quality and extend the shelf life of minimally processed fruits and vegetables by inhibiting spoilage microorganisms, serves as an effective alternative to traditional preservation methods, though it is best suited for short-term storage due to reduced respiration rates (Sandhya 2010). MAP is a food preservation technique in which the gas barrier inside the packaging material is modified to slow down the microbial spoilage and extend the shelf life of the food (Caleb et al. 2013). In MAP, oxygen, carbon dioxide, nitrogen, and novel gas concentrations are modified to reduce respiration rates, oxidative reactions, enzymatic browning, discoloration, and mold growth of packaged food and to maintain its shelf life in the long run (Figueroa-Enríquez et al. 2024). Modified atmosphere packaging operates through passive and active strategies, where passive MAP relies on product respiration and film permeability to achieve atmospheric balance, while active MAP employs initial gas flushing and incorporates scavengers like CO_2 emitters or oxygen and ethylene absorbers to maintain the desired environment (Charles et al. 2003). This packaging technique slows down the respiration rate of fresh fruits, berries, and leafy greens by maintaining an ideal level of carbon dioxide inside the packages (Yahia 2009). According to Caleb et al. (2013), modified atmosphere packaging significantly enhances the preservation of fresh and fresh-cut produce by slowing quality deterioration (including firmness, color, and ascorbate retention) and reducing microbial growth when temperature is controlled, and commodity-specific atmospheres are maintained. Modified atmosphere packaging (MAP) slows down the respiration rate of fresh fruits, berries, and leafy greens by maintaining an ideal level of carbon dioxide inside the packages (Yahia 2009).

14.5 Controlled Atmospheric Packaging

In controlled atmosphere storage, low oxygen and high carbon dioxide levels, combined with precisely managed temperature and relative humidity, create an environment that enhances the freshness, safety, and quality of packaged fruits and vegetables while preventing spoilage (Yahia 2009). Controlled atmospheric storage involves maintaining fruits and vegetables under precise temperature control throughout storage, utilizing CO_2 analyzers, gas injection, and removal systems within a hermetically sealed environment to preserve quality (De Siqueira Oliveira et al. 2020).

Active Packaging

In an active packaging system, various packaging methods are included, which are discussed below;

- *Controlled release packaging (CRP)*
  Controlled Release Packaging, or CRP, is well recognized for its anti-browning and antioxidant packaging component that prevents the growth of spoilage bacteria and is combined with food to extend its shelf life (Yam and Lee 2012).
- *O_2 scavengers*
  Since O_2 scavengers are a good source of ascorbic acid, ascorbate salts, or catechol—organic reducing agents applied by non-metallic scavengers, they are frequently employed in packing materials like sachets (Mishra et al. 2024).
- *Ethylene scavengers*
  The C_2H_4 scavengers that are primarily utilized in packing materials include potassium permanganate, which prevents the synthesis of ethylene and extends the shelf life of the product (Biji et al. 2015; Zagory 1995). For example, adding

1-methyl cyclopropane (1-MCP) to tomatoes can delay their ripening and increase their shelf life (Joshi et al. 2021). Using ethylene-absorbing materials like zeolite in packaging for iceberg lettuce minimizes pH changes, color degradation, and weight loss for up to 21 days (Gaikwad et al. 2019). Cherry tomatoes packed with Triticale films infused with KMnO4 can delay ripening, maintaining quality for over 20 days of storage (Aragüez et al. 2020).

- *CO_2 scavengers and emitters*
  Both calcium hydroxide, a chemical absorbent, and zeolite, a physical absorbent, are employed in packages as CO_2 absorbers to prevent the produce from swelling and bursting during storage and to preserve its freshness, while magnesium oxide, activated charcoal, and soda lime act as the most efficient forms of CO_2 scavengers (Joshi et al. 2021).
- *Moisture regulators*
  To extend the shelf life and maintain the freshness of highly perishable fruits and vegetables, moisture-absorbing pads, sheets, or sachets, such as Toppan Sheet™, Thermarite™, and Luquasorb™, are employed in packaging to control excess moisture and prevent microbial spoilage caused by fungal and bacterial growth (Bodbodak and Rafiee 2016).
- *Antimicrobial packaging*
  Amplification of active ingredients or actively functionalized polymers within the packages to control the spoilage and pathogenic fungus and bacterial growth, which leads to contaminated foods, is known as an antimicrobial packaging system (Han 2003). Microbial spoilage significantly reduces the shelf life of highly perishable fruits and vegetables, but incorporating antimicrobial agents like organic acids (sorbate, propionate, benzoate), bacteriocins (nisin), and enzymes (lysozyme) into food packaging helps mitigate this issue (Bodbodak and Rafiee 2016).

14.6 Intelligent Packaging

Intelligent packaging plays a role throughout the supply chain, from production to consumer use, incorporating data carriers, sensors, and indicators to facilitate automation, traceability, and protection of product information (Yam et al. 2005; Müller and Schmid 2019). This technology enables real-time monitoring of fruit and vegetable quality, safety, freshness, ripeness, and traceability, while also detecting potential food adulteration within packages (He et al. 2023; Dobrucka and Cierpiszewski 2014). Smart packaging employs tools like time-temperature and freshness indicators, along with QR codes, RFLP, and lot-integrated systems, to ensure real-time supply chain oversight and enhance consumer engagement (Kuswandi et al. 2011).

14.6.1 Intelligent Packaging with the Use of Targeting Metabolites

- *CO_2*
  The freshness of cut fruits and vegetables, such as peppers, is monitored by measuring CO_2 levels using natural chemical dyes or pH-sensitive indicators like methyl red and bromoethyl blue, which change color from yellow-green to orange to indicate spoilage (Chen et al. 2018). Methyl red dye serves as an indicator to monitor CO_2 levels within dragon fruit packaging, signaling changes in freshness (Warsiki and Rofifah 2018).
- *Volatile organic acid*
  Bromophenol blue, used as an indicator for volatile organic acids in guava packaging, accumulates in the upper layer of the package, reducing the pH and changing the indicator color from blue to green (Kuswandi et al. 2013). In contrast, chlorophenol red, employed as an indicator in grape packaging, detects a gradual decrease in pH and volatile organic acids in the headspace (Kuswandi and Murdyaningsih 2017).
- *Aldehyde*
  Using an aldehyde-based colorimetric sensor with methyl red dye, the ripeness and shelf life of apples can be monitored in real time (Kim et al. 2018).
- *Ethylene*
  According to a study by Pirsa and Chavoshizadeh (2018), the temperature and ethylene levels in banana packaging were monitored using an ethylene-based gas sensor integrated into bacterial cellulose-based nanofiber films enhanced with KMnO4 during the storage period.

14.6.2 Colorimetric Indicators

With growing consumer demand for fresh fruits and vegetables, advanced colorimetric intelligent packaging technologies, incorporating freshness indicators, sensors, and RFID tags, have emerged to provide detailed insights into produce quality, freshness, and traceability (Hailu et al. 2025).

- *Freshness indicator*
  Colorimetric intelligent packaging employs synthetic colorants (e.g., methyl red, phenol red, bromophenol blue) and natural colorants (e.g., anthocyanins, carotenoids, chlorophyll) within materials like biopolymers, synthetic polymers, and paper to assess the freshness of fruits and vegetables, with natural colorants being safer due to their lack of harmful effects on consumer health (Hailu et al. 2025). Bromothymol blue serves as a freshness indicator in green bell pepper packaging, monitoring freshness, pH, and CO_2 levels throughout the storage period (Chen et al. 2018).

- *pH indicator*
  Color-based pH indicators, utilized in smart intelligent packaging, detect the ripening and spoilage of fruits and vegetables by responding to the accumulation of acidic or alkaline compounds, with pH-sensitive dyes signaling deterioration due to changes in package conditions (Hailu et al. 2025). Research by Chen et al. (2018) demonstrated that pH indicators exhibit color changes triggered by CO_2 accumulation, which contributes to the spoilage of fresh produce within packages. Bromophenol blue, functioning as a pH indicator, detects spoilage in guava packaging by identifying acetic acid buildup in the headspace (Mustafa and Andreescu 2018). For instance, colorimetric pH sensors used for detecting aldehydes and formaldehyde in fruit samples reveal changes in fruit basicity through corresponding color variations (Vo et al. 2007; Feng et al. 2010).
- *Gas indicator*
  Smart packaging monitors subtle changes in O_2 and CO_2 levels within the packaging of fresh fruits and vegetables (Kuswandi et al. 2011). During storage, elevated CO_2 concentrations in packages of freshly cut or sliced produce accelerate the O_2 respiration process, leading to gradual deterioration and reduced freshness quality (He et al. 2023).

14.7 Alternate Ways for Sustainable Packaging

14.7.1 Biodegradable Packaging

With the growing global population, the demand for food and dietary supplements has surged, necessitating advanced packaging solutions to ensure product safety and market appeal (Verma et al. 2021). Bio-based polymers like cellulose, chitosan, polylactic acid (PLA), and polyhydroxyalkanoates are recognized as promising materials for sustainable packaging (Sadhu 2022). Traditional resources such as pigol leaves, clay, sal leaves, jute, and palm fronds offer innovative alternatives for eco-friendly packaging, supporting environmental sustainability while promoting India's cultural heritage and local markets (Singh et al. 2025). Modern biodegradable materials like glass, wood, and paper have gained prominence due to their significant environmental benefits compared to traditional methods (Verma et al. 2021). Currently, the packaging industry is increasingly adopting biodegradable materials for food and vegetable packaging, moving away from plastics due to their eco-friendly and recyclable nature (Du et al. 2025). Bio-based polymer matrices sourced from renewable origins, including microorganisms, animals, and plants, provide a viable alternative for sustainable packaging (Ortega et al. 2022; Wang et al. 2021).

14.7.2 Edible Film Packaging

Beyond traditional packaging methods, edible film coatings for fruits and vegetables have emerged as an innovative strategy, offering benefits such as preserving product freshness, reducing postharvest losses during storage, and minimizing waste, thereby mitigating environmental health risks (Ffadhilah et al. 2024).

- *Chitosan*
  This polysaccharide is applied as a coating for bananas, slowing the ripening process while preserving the color and thickness of the peel (Win et al. 2007).
- *Carboxymethyl cellulose*
  Used in avocados, this edible polysaccharide coating is applied by dipping the fruit, helping to reduce weight loss, respiration rate, and electrical conductivity while maintaining fruit firmness (Tesfay and Magwaza 2017).
- *Gum arabic*
  When tomatoes are coated with gum arabic, it delays ripening, reduces respiration and ethylene production, and extends shelf life while preserving the fruit's color and firmness (Ali et al. 2013). Blueberries coated with gum arabic enriched with roselle extract exhibit reduced yeast and mesophilic bacterial growth, extending shelf life to approximately 12 days (Yang et al. 2019).
- *Whey protein*
  Applied as an edible coating on apples, whey protein reduces enzymatic activity, total soluble solids, and pH levels, while also lowering water activity to enhance preservation (Azevedo et al. 2018).
- *Candelilla wax*
  In tomatoes, candelilla wax is used as a coating material to minimize weight loss, maintain firmness, and reduce titratable acidity during storage (Ruiz-Martínez et al. 2020).

14.7.3 Recyclable Waste-Based Packaging

Thivya et al. (2023) investigated the use of onion solid waste in packaging materials, highlighting its rich content of cellulose, lignocellulosic materials, polyphenols, and bioactive compounds that help maintain food freshness and quality during storage. Kumar et al. (2022) noted that incorporating onion solid waste into packaging not only adds value to the waste but also enables the creation of biodegradable active or intelligent packaging materials, potentially reducing reliance on plastic in the future. Dintcheva and Morici (2023) found that incorporating dry rose flowers and rose flower waste into polylactic acid (PLA) films enhances their resistance to UVB-induced photo-oxidation, making these materials a promising, sustainable option for biopolymer-based packaging films, despite the typical catalytic effect of additives on PLA degradation.

14.7.4 Nanotechnology-Based Packaging

Nanotechnology-based packaging utilizes sustainable, degradable, recyclable, and lightweight materials to enhance food safety, quality, and shelf life, benefiting consumer health (Aswathi et al. 2023; Hamad et al. 2018; Alfadul and Elneshwy 2010).

14.8 QR Codes and Sensors in Smart Packaging

Smart packaging significantly enhances food safety and quality by reducing waste, extending shelf life, and improving product freshness while prioritizing consumer health (Alam et al. 2021). Smart packaging integrates sensors like temperature (RTDS), humidity, and gas detectors to monitor produce freshness, displaying green for fresh, yellow for nearing expiry, and red for inedible. At the same time, QR codes provide detailed product information, tracking, achieving a 99.99% success rate compared to traditional packaging (Al and Hemalatha 2025). Incorporating freshness sensors in packaging materials allows consumers to assess fruit freshness, firmness, and ripeness, providing insights into shelf life, with the global smart packaging market reaching an estimated value of USD 26.7 billion by 2024 (Alam et al. 2021).RipeSenseTM, a New Zealand company, has developed a recyclable polyethylene terephthalate (PET) clamshell sensor that detects fruit ripeness by reacting to released aromas, changing from red for crisp fruit to orange and yellow for firm and juicy stages, while electronic nose (e-nose) sensors analyze volatile compounds to assess tomato ripeness, as studied by Gómez et al. (2006). Leak detection sensors monitoring in modified atmosphere packaging (MAP) for fruits like apples, track O_2 and CO_2 levels, help assess quality by maintaining O_2 at 0.5–5 kPa to limit ethylene sensitivity and metabolic activity, but leaks can raise O_2 and lower CO_2, fostering aerobic microbial growth (Matindoust et al. 2016; Opara et al. 2019; Beaudry 2007). Leak indicators, typically gas sensors, identify toxic substances in packaged fruits, which can jeopardize consumer health and safety (Matindoust et al. 2016). Commercial advancements in biosensors for food packaging include the Food Sentinel System by SIRA Technologies Inc. (CA, USA) and Toxin Guard by Toxin Alert (ON, Canada) (Alam et al. 2021; Mohebi and Marquez 2015). For instance, the Food Sentinel System employs a biosensor with specific antibodies integrated into a membrane, forming part of a barcode to detect food pathogens (Alam et al. 2021). Nanotechnology advancements in smart sensors for packaging enable precise detection of environmental factors like temperature, humidity, and air emissions, as well as non-invasive food quality assessment, with potential for highly accurate commercial applications, though not yet cost-effective for widespread use in mass-produced food packaging (Mahajan et al. 2014).

14.9 AI (Artificial Intelligence) and IoT (Internet of Things) Based Packaging Technology

Robotic machines should be utilized for fruit packaging to enhance product safety, such as the mGrip soft grip robotic arms developed by the US-based Soft Robotics, which enable processing plants to safely and efficiently handle, package, and process delicate items like tomatoes, persimmons, plums, strawberries, and herbs (Kondratieva et al. 2022). Advancements in smart horticulture, including data analytics, automation, robotics, IoT (Internet of things), sustainable packaging, blockchain for traceability, and circular economy principles, enable precise control, efficient handling, real-time monitoring, and enhanced quality of horticultural products while minimizing waste, conserving resources, reducing emissions, and ensuring safety through innovations like cold chain logistics, real-time tracking, and advanced postharvest processing and grading technologies (Santosh and Harshavardhan 2024). Globally, IoT-based biosensors can be a safe and effective approach, but their high cost, limited efficiency, and maintenance issues are their main concerns (Misra et al. 2020; Just and Goddard 2023).

14.9.1 Barcodes

Research is ongoing into integrating barcodes with smart food packaging, where 1D barcodes, consisting of evenly spaced parallel lines, serve as a cost-effective, widely adopted tool for tracking dispatch, reordering, and stocking across the supply chain and encoding food information for easy integration with packaging industry software, while 2D barcodes, made of dots, enhance data storage in a compact space (Palanisamy et al. 2025).

14.9.2 Digital Electronic Use in Packaging

In response to the growing demand for advanced technology, the packaging industry has embraced modern solutions like RFID tags for real-time tracking and automatic identification of fresh horticultural products, NFC labels and QR codes that allow consumers to scan product labels with smartphones for details such as ingredients, manufacturing dates, and usage instructions, and electronic shelf labels (ESL) that wirelessly update prices, promotions, and other information without manual replacement (Du et al. 2025).

14.10 Some Packaging Materials and Their Use in Trades

- In European and North American markets, packaging solutions such as plastics, corrugated fiberboard, biodegradable films, modified atmosphere packaging, and eco-friendly smart packaging are widely utilized (Thapliyal et al. 2024).
- In Ghana, robust foam boxes and cartons are used for exporting and packaging premium vegetables, while the domestic market comprises roughly 40% plastic, 25% paper, 15% metal, 10% glass, 5% wooden crates and pallets, and 5% cotton fabric for packaging materials (Kwaku and Fan 2020).
- Recent advancements enable thin natural polymer coatings to serve as primary packaging, particularly for short-distance retail markets (Bourtoom 2008).
- Apart from this, many other packaging materials are recently in use (Table 14.1).

Table 14.1 Some packaging materials and their manufacturing techniques, along with references

Materials	Manufacturing process	Reference
Glass	Melting—shaping—annealing—surface treatment	Ashby et al. (2013)
Paper and paperboard	Raw materials—pulping—sheet formation—coating and finishing	Bajpai (2016)
Plastic	Polymer synthesis—conversion (extraction, injection molding, blow molding, thermo forming)	Siracusa et al. (2008)
AgNPs-TTI (time- temperature indicator) based smart packaging	Polydiacetylene/AgNPs (PDA/AgNPs) mixed into carboxymethyl cellulose (CMC) film	Saenjaiban et al. (2020)
Oxygen scavenging sachet	Oxygen scavenging formulation coated or printed on the food-facing side of the outer top barrier film, which can be mounted inside the food package	Gohil and Wysock (2014)
Carbon dioxide scavenging package	Film bag or sachet contain with carbon dioxide absorbing component poured into flexible or rigid packages	Lee (2016)
Ethylene scavenging package	Ethylene scavengers such as potassium permanganate, silica gel, activated alumina, metal, etc., are incorporated and packed into sachets and placed in packages of fruits and vegetables	Mariah et al. (2022)
Moisture-absorbing packages	Different forms of moisture absorbers with the use of different moisture absorbents are incorporated in packages of fresh fruits, vegetables, and dried products like coffee	Gaikwad et al. (2019)
Edible film	Solvent casting—extrusion (feeding—kneading—heating)—compression molding or injection molding (filling, packing–cooling)	Ffadhilah et al. (2024)
LC sensor package	Printed interdigital capacitor (IDC) and spiral inductor attached inside the food package.	Tan et al. (2007)
Packages with an RFID tag	RFID tag comprises a chip (containing information) made of silicon and an antenna (for data transmission) made of etched copper, aluminum, or conductive ink on paper or plastic. A prepared RFID tag is incorporated with a food package	Clarke and Tanprasert (2005), Moyeenudin and Anandan (2018)
Packages with QR code	Barcode and QR codes are printed on food packaging materials	Li et al. (2024)
Nano-Biocom site: Active packaging	Preparation with the integration of nanoparticles and biodegradable polymer.	Omerović et al. (2021)
Nano composite film	Synthesis of nanoparticles such as silver zinc oxide, copper, starch nanocrystals, nanoscale cellulose, etc. (top down and bottom down method)—mixing of nanofillers with polymer matrix—Development of nanocomposite film	Akhila and Badwaik (2022)

14.11 Role of Attractive Packaging in Smart Domestic Trade

Smart domestic trade in food packaging for the horticultural industry employs cutting-edge technologies, innovative packaging, and tracking systems to improve the quality, safety, sustainability, and efficiency of products in national markets (Biji et al. 2015; Akhrib et al. 2024; Caner et al. 2024). Research on "Food Safety and Standards in India" explores AGMARK certification, safety regulations for packaging, and the role of standardized labeling and oversight in ensuring quality assurance and building consumer trust (Sharma 2022). Jairath and Purohit (2013) analyzed the Food Safety and Standards Act of 2006, emphasizing licensing requirements, enforcement challenges, and the responsibilities of food handlers in ensuring the safety of primary foods like fruits and vegetables, which enhances market competitiveness and consumer confidence. The 2019 ICRIER report, "Streamlining Food Safety Compliance Ecosystem in India," commissioned by Food Safety and Standards Authority of India (FSSAI) and supported by the UK Government's Prosperity Fund India Program, examined enforcement, licensing, packaging, and labeling requirements, addressing compliance challenges and highlighting IT-driven initiatives to streamline inspections and improve transparency.

Smart packaging and digital traceability utilize technologies such as barcodes, RFID tags, and IoT for real-time monitoring, enhancing inventory management and quality control across distribution networks (Biji et al. 2015; Rodvong and Anal 2025; Caner et al. 2024; Akhrib et al. 2024). Biji et al. (2015) investigated RFID and sensor applications for tracking crates and monitoring vegetable freshness, optimizing supply chain management in India. Rodvong and Anal (2025) showcased IoT-enabled packaging for strawberries, enabling retailers to monitor quality and prevent spoilage. Caner et al. (2024) highlighted policies offering subsidies to encourage domestic horticulture producers to adopt sustainable packaging. Akhrib et al. (2024) assessed PLA use in packaging cut flowers and fruits, with trials in local markets showing extended shelf life and reduced waste. These approaches enhance product freshness, minimize waste, improve inventory efficiency, ensure regulatory compliance, and meet consumer demand for safe and sustainable packaging solutions (Biji et al. 2015; Akhrib et al. 2024; Caner et al. 2024). Adhering to regulatory requirements and creating a strong brand identity involves complying with local food safety regulations, obtaining necessary certifications (such as India's FSSAI), and fostering consumer trust through appealing and clear packaging designs (Food Safety and Standards (Packaging and Labelling) Regulations 2011; Sharma 2022; Jairath and Purohit 2013).

14.12 Global Market Analysis of Horticultural Trade with Respect to Attractive Packaging

With rising global demand for fresh produce, the horticultural packaging industry is expanding rapidly, playing a crucial role in maintaining product quality, extending shelf life, and complying with international standards (Trade and Change 2018).

Consumers in Ghana favor products from countries such as India, China, Malaysia, Indonesia, Thailand, South Africa, the USA, and the UK, largely influenced by packaging quality (Kwaku and Fan 2020). According to Mordor Intelligence (2025), the global packaging market is valued at USD 1.18 trillion in 2025 and is expected to grow to USD 1.44 trillion by 2030, with a compound annual growth rate (CAGR) of 3.92%. The US EPA promotes sustainable packaging through its Sustainable Materials Management Program, while states like California, New York, and Oregon implement bans on single-use plastic bags and Extended Producer Responsibility laws, and the FDA requires virgin polymer or aluminum foil barriers in recycled plastic containers like PET and HDPE to ensure food safety by preventing contamination, unlike the EU's EC 10/2011 and EC 2022/1616 regulations for new and recycled plastics in food contact applications (Caner et al. 2024). Japan's Food Sanitation Law of 1947 establishes safety requirements for food packaging materials, while the 1995 Containers and Packaging Recycling Act promotes sustainable practices and recycling within the packaging industry (Thapliyal et al. 2024).

Attractive and innovative packaging plays a vital role in boosting the global horticultural trade by enhancing consumer appeal, supporting export performance, and meeting sustainability standards (Agriculture Institute 2023; Mordor Intelligence 2025). Studies show that up to 73.6% of consumers in developing nations prefer imported horticultural products due to superior packaging, highlighting its role in driving sales and fostering brand loyalty (Kwaku and Fan 2020).

14.12.1 Role of Attractive Packaging in Global Horticulture

- Effective packaging extends the shelf life of perishable horticultural goods, ensuring quality preservation for long-distance markets with minimal spoilage (Sadhu 2022; Alemu and Oanh 2025).
- Advanced packaging solutions are increasingly adopted to meet international standards and strengthen brand reputation (Data Bridge Market Research 2024).
- Packaging design serves as a key strategy for creating a compelling sales pitch and establishing brand identity in competitive markets (Kwaku and Fan 2020).
- Visually appealing packaging with clear product details significantly influences consumer purchasing decisions and elevates perceived product value, particularly in export markets (Agriculture Institute 2023; Kwaku and Fan 2020).

14.12.2 Market Trends and Regional Insights

- The global agricultural packaging market is expected to grow from approximately USD 5.48 billion in 2025 to over USD 7 billion by 2032, driven by demand for secure and sustainable packaging solutions (Mordor intelligence 2025).

- Markets in Europe and North America prioritize eco-friendly, recyclable, and biodegradable packaging, prompting exporters in countries like India, Japan, and Brazil to adopt sustainable materials to meet import requirements and enhance market appeal (Data Bridge market research 2024).
- China leads in horticultural packaging production, leveraging efficient supply chains and robust export infrastructure to supply specialized packaging globally (Accio 2023).
- The United States, Germany, and Japan are among the top importers of packaged horticultural products, fueled by demand for exotic and off-season produce (ITC Trade Map 2022).

14.12.3 Sustainability and Innovation

The shift toward sustainable, plant-based packaging is accelerating, with projected compound annual growth rates reaching up to 11.6% for plant-based packaging over the next decade (Towardspackaging 2025; Data Bridge Market Research 2024). Adopting eco-friendly and reusable materials not only enhances visual appeal but also aligns brands with environmentally conscious consumers and stringent regulations in major import markets (Future Market Insights 2025).

14.13 Regulatory Policies

14.13.1 EU Packaging and Packaging Waste Regulation (PPWR)

By 2030, all horticultural packaging must be recyclable by design, with full recycling required by 2035, and Extended Producer Responsibility (EPR) schemes holding producers accountable for recycling and disposal (Wakefield 2025).

14.13.2 USDA and FDA Standards (USA)

- Emphasize food safety, traceability, and barriers against moisture and contaminants, with QR code labeling and advanced packaging increasingly required (Mordor Intelligence 2025).
- The WTO's (World Trade Organization) sanitary and phytosanitary agreement and dispute settlement system promote fairer and more transparent global food trade, encouraging developing countries to adopt stringent safety standards to compete globally with support from international programs (Athukorala and Jayasuriya 2003).
- Developing nations can address challenges like inadequate infrastructure and food safety compliance by adopting modern safety tools, such as HACCP (haz-

ard analysis and critical control point) protocols and WTO (World Trade Organization) standards, to enhance global trade competitiveness (Van Veen 2005).
- Global trade increasingly favors biodegradable, compostable, and fiber-based packaging to meet consumer preferences and regulatory requirements (Chandra 2025; Gleebirmingham 2024).
- Policies promote reduced plastic use, molded pulp trays, edible coatings, and plant-based films to minimize postharvest losses and improve environmental outcomes (Peek 2023; Chandra 2025).

14.13.3 Barriers to Global Horticultural Trade

- Substandard or unhygienic packaging can lead to pest infestations or contamination, resulting in rejected international shipments (Henson and Humphrey 2010).
- Exporters face growing pressure to lower carbon emissions in packaging production and logistics (Siracusa et al. 2008).
- High tariffs on imported packaging materials pose significant barriers to global horticultural trade in many countries (Shepherd and Wilson 2009).
- Developed nations like the United States, Europe, and China are phasing out single-use plastics, widely used in developing countries' horticultural packaging, creating compliance challenges (Siracusa et al. 2008).

14.14 Understanding the Link between Science and Smart Trade Via Attractive Packaging

Highly perishable and damage-prone horticultural products necessitate specialized packaging to ensure their safety, particularly during transportation from farms to consumers (Sadhu 2022). Advances in packaging technologies within the horticultural industry significantly enhance smart trading, attract consumers, improve product quality and longevity, and streamline supply chain logistics while maintaining transparency and efficacy (Buragohain et al. 2025; Kwaku and Fan 2020; Du et al. 2025; Rajan and Wani 2025). Previous research highlighted that modern packaging solutions, including smart and eco-friendly methods, address both commercial objectives and public interests effectively (Ffadhilah et al. 2024; Panou et al. 2025; Ambaw et al. 2022; Sadhu 2022; Dwibedi et al. 2024; Kwaku and Fan 2020). Innovative packaging science significantly advances research by enhancing postharvest preservation and fostering sustainable trade through quality maintenance during storage and transport, with eco-friendly solutions like edible films reducing environmental impact (Shukla et al. 2022; Ambaw et al. 2022; Buragohain et al. 2025; Ffadhilah et al. 2024; Dwibedi et al. 2024). Advanced packaging with biosensors, freshness monitors, and data trackers delivers real-time quality insights, optimizes supply chain operations, reduces energy and preservative use, and enables precise interventions to minimize waste and prolong shelf life across the supply chain (Butuc et al. 2025; Panou et al. 2025; Dwibedi et al. 2024; Alemu and Oanh 2025; Biji et al. 2015; Sharma et al. 2025). By leveraging scientifically validated packaging innovations, horticultural products gain enhanced marketability and a competitive edge, influencing pricing strategies and facilitating seamless domestic and international trade. These solutions ensure product freshness, safety, traceability, and consumer appeal, thereby strengthening the global horticultural market (Drechsler and Holzapfel 2023; Kwaku and Fan 2020; Butuc et al. 2025).

14.15 In Relation to Regulation, Policy, and Future Trends

Innovative packaging science, integrated with smart trade practices in horticulture, enhances product value and marketability while aligning with modern preservation techniques, sustainability trends, and evolving global and domestic regulations to shape effective trade strategies (Dwibedi et al. 2024; Alam et al. 2021; Ffadhilah et al. 2024; Wakefield 2025; Peek 2023; Naturpac 2025). In horticulture, visually appealing packaging goes beyond aesthetics, enhancing trade by prolonging shelf life, preserving quality, minimizing waste, and enabling traceability through advanced technologies like IoT, sensors, and biodegradable materials in active, intelligent, and smart packaging, which monitor freshness, reduce spoilage, and provide supply chain transparency, boosting market trust and improving pricing (Alam et al. 2021; Ffadhilah et al. 2024; Crop Tracker 2024; Dwibedi et al. 2024).

14.16 Regulations and Policies

In 2025, global and national regulations, including India's FSSAI and MIDH policies and the EU's PPWR or The European Union's Packaging and Packaging Waste Regulation 2025, have intensified packaging standards, enforcing EPR or Extended Producer Responsibility, bans on harmful plastics, mandatory recycled content, and stringent waste management rules to drive innovation in sustainable, recyclable, biodegradable packaging solutions (Fofana 2025; Naturpac 2025; Pierce 2025; Wakefield 2025; MIDH Guidelines 2025; Food Safety and Standards Authority of India's (FSSAI) 2025). Emerging trends in horticultural packaging highlight the rise of edible films, biopolymers, and circular economy solutions that enhance resource efficiency and appeal to eco-conscious consumers, while smart packaging with real-time data and traceability capabilities meets supply chain demands for transparency and safety, with future innovations focusing on digitalization, sustainability, consumer interaction, and unified global standards (Dwibedi et al. 2024; Ffadhilah et al. 2024; Ma and Li 2024; Crop Tracker 2024; Arshad et al. 2025; Farmonaut 2025). By integrating scientific innovation with regulatory frameworks, appealing packaging science drives smarter trade practices, delivering competitive benefits, ensuring compliance, and paving the way for advanced horticultural marketing and quality assurance (Ma and Li 2024; Arshad et al. 2025; Dwibedi et al. 2024; Alam et al. 2021; Ffadhilah et al. 2024; Pierce 2025; Farmonaut 2025; Naturpac 2025; Crop Tracker 2024; Wakefield 2025; Food Safety and Standards Authority of India's (FSSAI) 2025).

14.17 Domestic and Global Packaging Market Growth in the Horticulture Sector

14.17.1 Domestic Market Growth (India)

- The Indian packaging sector was valued at INR 6656 billion in FY 2024 and is expected to grow at a 6.7% CAGR, reaching INR 8620 billion by 2028 (Brickwork Ratings 2025).
- Major growth factors include the thriving food, beverage, and horticulture industries, the rise of e-commerce, and robust government initiatives promoting sustainable packaging (Brickwork Ratings 2025; India Brand Equity Foundation 2025).
- There is a notable shift toward paper-based and recyclable materials, driven by regulatory measures and consumer demand, while plastic use, though still prevalent, faces increasing restrictions (India Brand Equity Foundation 2025; Brickwork Ratings 2025).
- India is seeing a surge in the adoption of bioplastics and plant-based packaging, supported by increased research, pilot projects, and implementation in the horticulture sector (Research Nester 2025; Grand View Research 2025).
- Prior studies on export development emphasize the importance of packaging standards, extended shelf life, and modern materials in enhancing domestic market performance and global competitiveness (Ajalafshar 2017; Buragohain et al. 2025).

14.17.2 Global Market Growth

- The global packaging industry is forecasted to increase from \$1.28 trillion in 2025 to \$1.69 trillion by 2034 (Fig. 14.1), with a CAGR exceeding 3% (Towardspackaging 2025).
- In the horticulture sector, the bioplastic packaging market is projected to grow at a 13.3% CAGR worldwide from 2025 to 2037, driven by rising sustainability demands, with the Asia-Pacific region, particularly China and India, at the forefront due to supportive policies and R&D investments (Researchnester 2025).
- The niche segment of global plantable packaging, increasingly adopted in food and horticulture, is expected to reach \$284.53 million by 2030, growing at a 12.1% CAGR from 2024, fueled by heightened eco-awareness and plastic waste

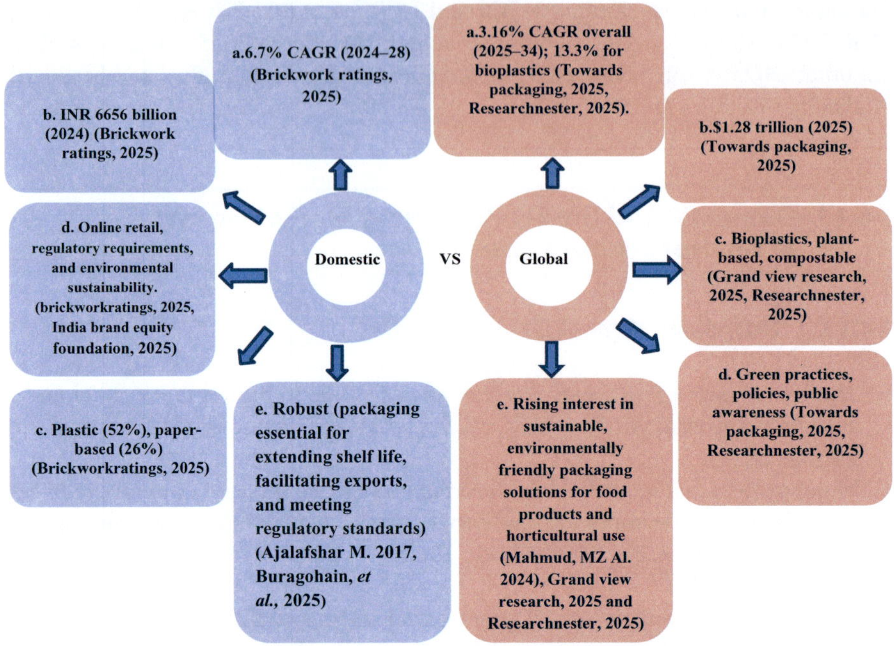

Fig. 14.1 Flow chart comparing domestic and global packaging trends. The domestic section highlights a 6.7% CAGR from 2024 to 2028, a market size of INR 6656 billion in 2024, the dominance of plastic and paper-based materials, online retail influences, and the importance of robust packaging. The global section notes a 3.16% CAGR from 2025 to 2034, a market size of $1.28 trillion in 2025, the rise of bioplastics, and a focus on sustainable practices and environmentally friendly packaging solutions. Arrows connect each section to its respective trends. (**a**) Market growth, (**b**) Size (recent), (**c**) Key materials, (**d**) Major drivers (**e**) Food/Horticultural impact. Note: The domestic market refers to the Indian market

regulations, with significant innovation occurring in Europe and North America (Grandviewresearch 2025).

- The food and beverage industry leads in adoption, closely followed by horticulture, with regulatory requirements in regions like Europe and the US accelerating the shift toward biodegradable and compostable materials (Al Mahmud et al. 2024; Grandviewresearch 2025; Researchnester 2025).

14.18 Conclusion

Attractive packaging enhances shelf life, preserves quality, serves as a shield against external environments, and protects fresh produce and horticultural products. The choice of design, materials, and preparation methods significantly influences consumer purchasing behavior. Packaging plays a crucial role in the horticultural industry, impacting both domestic and international trade and commerce, as demonstrated by various studies. Diverse packaging technologies, such as smart solutions like active and intelligent packaging, biodegradable options with waste-reduction innovations, plant-based materials, edible packaging, and nanotechnology, are now extensively applied to fruits, vegetables, ornamental plants, and more. The incorporation of smart sensors, tags, artificial intelligence, IoT, barcodes, and QR codes for tracking and monitoring is elevating standards within the horticultural sector. In trade, whether local or global, advanced packaging techniques are paving the way for a brighter future for the industry. Governments, both domestically and internationally, are promoting eco-friendly packaging through policies and regulations to support both economic and environmental sustainability. The horticultural packaging market, both in India and worldwide, is undergoing rapid expansion, driven by distinct factors, trends, and growth rates depending on the region. Research and industry reports highlight substantial advancements, particularly in sustainability, regulatory shifts, and innovative materials. These sources offer valuable insights into the rapidly transforming landscape of horticultural packaging, in both the Indian and global contexts, revealing a clear trend toward greater adoption of environmentally friendly solutions, shaped by regulatory changes and consumer preferences.

Future Considerations to Focus on

- Despite ongoing hurdles like high costs, regulatory complexities, and inadequate cold chain infrastructure, the rising demand for premium horticultural goods creates opportunities for creative and effective packaging innovations to reshape global trade. However, there are several aspects that require both governmental and industrial attention.
- Economic viability for small-scale businesses: Further research is needed to develop cost-effective strategies that account for both direct and indirect packaging expenses, enabling small horticultural enterprises to adopt advanced solutions. This includes addressing challenges related to economies of scale, bridging supply-demand mismatches, and fostering innovation in resource-constrained environments.
- Overcoming barriers to sustainable packaging adoption: Additional efforts are required to tackle financial obstacles, such as steep upfront costs and reluctance from small-scale producers, to encourage the use of reusable and environmentally friendly packaging. Initiatives like financial incentives, training programs, and collaborative research between suppliers and horticultural businesses can facilitate this transition.
- Understanding consumer attitudes toward bioplastics: Research gaps remain in exploring what motivates or discourages consumers from embracing bioplastic packaging, including their willingness to pay. Studies should focus on diverse cultural perspectives, representative sampling, and realistic scenarios to assess awareness, labeling effectiveness, and post-use behaviors like recycling or disposal.

- Establishing robust regulatory and standardization systems: There is a need for comprehensive research to create clear regulations, standards, and traceability mechanisms for packaging horticultural products, particularly in less-regulated markets. This will ensure food safety, meet export compliance needs, and promote consistency in global trade practices.

References

Abdullahi N (2014) Hazard chemicals in some food packaging materials (a review). Ann Food Sci Technol 15(1):115–120

Accio (2023). https://www.accio.com/supplier/horticultural-packaging-supplier

Agriculture Institute (2023). https://agriculture.institute/post-harvest-mgt-principles/packaging-fresh-produce-techniques-quality/

Ajalafshar M (2017) Assessment of packing as an efficacious factor in the development of horticultural product exports. QUID: Investig Cienc Tecnol 1:431–446

Akhila V, Badwaik LS (2022) Recent advancement in improvement of properties of polysaccharides and proteins-based packaging film with added nanoparticles: a review. Int J Biol Macromol 203:515–525. https://doi.org/10.1016/j.ijbiomac.2022.01.181

Akhrib S, Djellali S, Haddaoui N, Karimian D, Carraro M (2024) Biocomposites and poly (lactic acid) in active packaging: a review of current research and future directions. Polymers 17(1):3. https://doi.org/10.3390/polym17010003

Al HM, Hemalatha K (2025) Development and application of smart packaging solutions for extending the shelf life of fresh produce. SHS Web Conf EDP Sci 216:01013. https://doi.org/10.1051/shsconf/202521601013

Al Mahmud MZ, Mobarak MH, Hossain N (2024) Emerging trends in biomaterials for sustainable food packaging: a comprehensive review. Heliyon 10(1):e24122. https://doi.org/10.1016/j.heliyon.2024.e24122

Alam AU, Rathi P, Beshai H, Sarabha GK, Deen MJ (2021) Fruit quality monitoring with smart packaging. Sensors 21(4):1509. https://doi.org/10.3390/s21041509

Albrecht S, Brandstetter P, Beck T, Fullana-i-Palmer P, Grönman K, Baitz M, Fischer M (2013) An extended life cycle analysis of packaging systems for fruit and vegetable transport in Europe. Int J Life Cycle Assess 18(8):1549–1567. https://doi.org/10.1007/s11367-013-0590-4

Alemu TT, Oanh VTK (2025) Implications of developmental levels and packaging materials on the quality of iceberg lettuce for marketing system: a review. Int J Food Sci 2025(1):5548305. https://doi.org/10.1155/ijfo/5548305

Alfadul SM, Elneshwy AA (2010) Use of nanotechnology in food processing, packaging and safety–review. Afr J Food Agric Nutr Dev 10(6):2719–2739. https://doi.org/10.4314/ajfand.v10i6.58068

Ali A, Maqbool M, Alderson PG, Zahid N (2013) Effect of gum arabic as an edible coating on antioxidant capacity of tomato (*Solanum lycopersicum* L.) fruit during storage. Postharvest Biol Technol 76:119–124. https://doi.org/10.1016/j.postharvbio.2012.09.011

Ambaw A, Mukama M, Fadiji T, Opara UL (2022) Packaging of the future: innovations in packaging design and cold chain logistical management. In: V International Symposium on Pomegranate and Minor Mediterranean Fruits, vol 1349, pp 389–398. https://doi.org/10.17660/actahortic.2022.1349.53

Ampuero O, Vila N (2006) Consumer perceptions of product packaging. J Consum Mark 23(2):100–112. https://doi.org/10.1108/07363760610655032

Aragüez L, Colombo A, Borneo R, Aguirre A (2020) Active packaging from triticale flour films for prolonging storage life of cherry tomato. Food Packag Shelf Life 25:100520. https://doi.org/10.1016/j.fpsl.2020.100520

Arshad MT, Hassan S, Shehzadi R, Sani MA, Ikram A, Maqsood S, Gnedeka KT (2025) Emerging trends in sustainable packaging of food products: an updated review. J Nat Fibers 22(1):2505608. https://doi.org/10.1080/15440478.2025.2505608

Ashby MF, Shercliff H, Cebon D (2013) Materials: engineering, science, processing and design—chapters on metals, ceramics (glass) and polymers. Butterworth-Heinemann

Aswathi VP, Meera S, Maria CA, Nidhin M (2023) Green synthesis of nanoparticles from biodegradable waste extracts and their applications: a critical review. Nanotechnol Environ Eng 8(2):377–397. https://doi.org/10.1007/s41204-022-00276-8

Athukorala PC, Jayasuriya S (2003) Food safety issues, trade and WTO rules: a developing country perspective. World Econ 26(9):1395–1416. https://doi.org/10.1046/j.1467-9701.2003.00576.x

Azevedo VM, Dias MV, Elias HHDS, Fukushima KL, Silva EK, Carneiro JDDS, Soares NDFF, Borges SV (2018) Effect of whey protein isolate films incorporated with montmorillonite and citric acid on the preservation of fresh-cut apples. Food Res Int 107:306–313. https://doi.org/10.1016/j.foodres.2018.02.050

Bajpai P (2016) Pulp and paper industry: energy conservation. Elsevier. https://doi.org/10.1016/b978-0-12-803411-8.00006-8

Basu P, De S, Choudhury R (2018) Thermal management in passive cold chain packaging: A review. Renew Sust Energ Rev 82(3):1551–1565. https://doi.org/10.1016/j.rser.2017.10.037

Beaudry R (2007) MAP as a basis for active packaging. In: Intelligent and active packaging for fruits and vegetables. CRC Press, pp 31–55. https://doi.org/10.1201/9781420008678.ch3

Ben-Yehoshua S (1983) Extending the life of fruit by individual seal-packaging in plastics film; status and prospects. Plasticulture (58):45–57

Biji KB, Ravishankar CN, Mohan CO, Srinivasa Gopal TK (2015) Smart packaging systems for food applications: a review. J Food Sci Technol 52(10):6125–6135. https://doi.org/10.1007/s13197-015-1766-7

Bodbodak S, Rafiee Z (2016) Recent trends in active packaging in fruits and vegetables. In: Eco-friendly technology for postharvest produce quality. Academic Press, pp 77–125. https://doi.org/10.1016/b978-0-12-804313-4.00003-7

Bolaji PT (2010) Principles of food packaging. Becon Press Industry Limited, pp 20–22

Bourtoom T (2008) Edible films and coatings: characteristics and properties. Int Food Res J 15(3):237–248

Brickworkratings (2025). https://www.brickworkratings.com/Research/Packaging%20Industry%20Report.pdf

Brody AL, Bugusu B, Han JH, Sand CK, McHugh TH (2008) Innovative food packaging solutions. J Food Sci 73(8):107–116. https://doi.org/10.1111/j.1750-3841.2008.00933.x

Buragohain R, Gogoi A, Borah R, Goswami R, Doley D, Begum KM (2025) Effect of different packaging material on shelf life of some horticultural crops: a review. Int J Adv Biochem Res 9(5):288–292. https://doi.org/10.33545/26174693.2025.v9.i5sd.4405

Butuc O, Akkerman R, de Leeuw S (2025) Smart packaging adoption in fresh fruits and vegetables supply chains. Int J Logist Manag 36(7):284–307. https://doi.org/10.1108/ijlm-08-2024-0521

Caleb OJ, Mahajan PV, Al-Said FAJ, Opara UL (2013) Modified atmosphere packaging technology of fresh and fresh-cut produce and the microbial consequences—a review. Food Eng Rev 5(1):1–23. https://doi.org/10.1007/s12393-012-9067-2

Caner C, Yüceer M, Harte B (2024) Trends in sustainability and innovative food packaging materials: an overview. Akademik Gıda:65–77. https://doi.org/10.24323/akademik-gida.1554476

Castle L (2007) Chemical migration into food: an overview. Wood Head Publishing Limited, pp 1–13. https://doi.org/10.1533/9781845692094.1

Chandra R (2025) Sustainable post-harvest practices for horticultural produce: a review. South Asian J Agric Sci 5(1):60–64

Charles F, Sanchez J, Gontard N (2003) Active modified atmosphere packaging of fresh fruits and vegetables: modeling with tomatoes and oxygen absorber. J Food Sci 68(5):1736–1742. https://doi.org/10.1111/j.1365-2621.2003.tb12321.x

Chen HZ, Zhang M, Bhandari B, Guo Z (2018) Applicability of a colorimetric indicator label for monitoring freshness of fresh-cut green bell pepper. Postharvest Biol Technol 140:85–92. https://doi.org/10.1016/j.postharvbio.2018.02.011

Clarke RH, Tanprasert K (2005) Packaging material implications in RFID operations. J Packag Sci Technol 14:299–315

Crop Tracker (2024). https://www.croptracker.com/blog/technology-round-up-sustainable-packaging-trends-in-horticulture.html

Da Cruz NF, Ferreira S, Cabral M, Simões P, Marques RC (2014) Packaging waste recycling in Europe: is the industry paying for it? Waste Manag 34(2):298–308. https://doi.org/10.1016/j.wasman.2013.10.035

Data Bridge Market Research (2024). https://www.databridgemarketresearch.com/reports/global-agriculture-packaging-market

De Siqueira Oliveira L, Eça KS, de Aquino AC, da Silva LMR (2020) Modified and controlled atmosphere packaging. In: Fresh-cut fruits and vegetables. Academic Press, pp 151–164. https://doi.org/10.1016/b978-0-12-816184-5.00007-0

GK, Panjagari NR, & Alam T (2019) An overview of paper and paper based food packaging materials: Health safety and environmental concerns. Journal of Food Science and Technology, 56(10):4391–4403. https://doi.org/10.1007/s13197-019-03950-z

Dintcheva NT, Morici E (2023) Recovery of rose flower waste to formulate eco-friendly biopolymer packaging films. Molecules 28(7):3165. https://doi.org/10.3390/molecules28073165

Dobrucka R, Cierpiszewski R (2014) Active and intelligent packaging food-research and development–a review. Pol J Food Nutr Sci 64(1):7–15. https://doi.org/10.2478/v10222-012-0091-3

Drechsler M, Holzapfel A (2023) Horticultural supply chain network design of small and medium-sized enterprises. Sustainability Anal Model 3:100014. https://doi.org/10.1016/j.samod.2022.100014

Du L, Huang X, Li Z, Qin Z, Zhang N, Zhai X, Wang Y (2025) Application of smart packaging in fruit and vegetable preservation: a review. Foods 14(3):447. https://doi.org/10.3390/foods14030447

Dwibedi V, Kaur G, George N, Rana P, Ge Y, Sun T (2024) Research progress in the preservation and packaging of fruits and vegetables: from traditional methods to innovative technologies. Food Packag Shelf Life 46:101385. https://doi.org/10.1016/j.fpsl.2024.101385

Essuman KM (2018) Packaging and trade. Position paper. World Packaging Organization, p 11

Fadiji T, Berry T, Coetzee CJ, Opara L (2017) Investigating the mechanical properties of paperboard packaging material for handling fresh produce under different environmental conditions: experimental analysis and finite element modelling. J Appl Packag Res 9(2):3

Farmonaut (2025). https://farmonaut.com/blogs/agriculture-packaging-suppliers-7-essential-trends-for-2025

Feng L, Musto CJ, Suslick KS (2010) A simple and highly sensitive colorimetric detection method for gaseous formaldehyde. J Am Chem Soc 132(12):4046–4047. https://doi.org/10.1021/ja910366p

Ffadhilah A, Chitra R, Ganga M, Prasanthrajan M, Djanaguiraman M (2024) Edible film: a future thrust in packaging technology for horticultural crops. Plant Science Today 11:5423. https://horizonepublishing.com/index.php/PST/article/view/5423

Figueroa-Enríquez CE, Rodríguez-Félix F, Ruiz-Cruz S, Castro-Enriquez DD, Gonzalez-Rios H, Perez-Alvarez JÁ, López-Peña IY (2024) Application of active packaging films for extending the shelf life of red meats: a review. Processes 12(10) https://doi.org/10.3390/pr12102115

Fofana O (2025). https://www.packaging-gateway.com/features/new-packaging-rules-in-2025/

Food Safety and Standards Authority of India (2011) Food Safety and Standards (Packaging and Labelling) Regulations. https://fssai.gov.in/upload/uploadfiles/files/Compendium_Packaging_Labelling_Regulations_27_08_2020.pdf

Food Safety and Standards Authority of India's (FSSAI) (2025). https://agriexchange.apeda.gov.in/ImportRegulations/99999202542314275148.pdf

Future Market Insights (2025). https://www.futuremarketinsights.com/reports/plant-based-packaging-market

Gaikwad KK, Singh S, Ajji A (2019) Moisture absorbers for food packaging applications. Environ Chem Lett 17(2):609–628. https://doi.org/10.1007/s10311-018-0810-z

Gleebirmingham (2024). https://www.gleebirmingham.com/news-and-views/eco-friendly-packaging-solutions-horticulture-products

Gohil RM, Wysock WA (2014) Designing efficient oxygen scavenging coating formulations for food packaging applications. Packag Technol Sci 27(8):609–623. https://doi.org/10.1002/pts.2053

Gómez AH, Hu G, Wang J, Pereira AG (2006) Evaluation of tomato maturity by electronic nose. Comput Electron Agric 54(1):44–52. https://doi.org/10.1016/j.compag.2006.07.002

Grandviewresearch (2025). https://www.grandviewresearch.com/industry-analysis/plantable-packaging-market-report

Hailu FW, Fanta SW, Tsige AA, Delele MA (2025) Current status and challenges of colorimetric intelligent packaging for fruit and vegetables: a review. J Food Qual 2025(1):6669828. https://doi.org/10.1155/jfq/6669828

Hamad AF, Han JH, Kim BC, Rather IA (2018) The intertwine of nanotechnology with the food industry. Saudi J Biol Sci 25(1):27–30. https://doi.org/10.1016/j.sjbs.2017.09.004

Han JH (2003) Antimicrobial food packaging. Woodhead Publishing Ltd, pp 50–70. https://doi.org/10.1533/9781855737020.1.50

He X, Pu Y, Chen L, Jiang H, Xu Y, Cao J, Jiang W (2023) A comprehensive review of intelligent packaging for fruits and vegetables: target responders, classification, applications, and future challenges. Compr Rev Food Sci Food Saf 22(2):842–881. https://doi.org/10.1111/1541-4337.13093

Helen B, James W (2003) Packaged product quality and shelf life. In: Food packaging technology. Wiley-Blackwell, pp 81–89. https://doi.org/10.1002/9781444392180.ch3

Henson S, Humphrey J (2010) Understanding the complexities of private standards in global agri-food chains as they impact developing countries. J Dev Stud 46(9):1628–1646. https://doi.org/10.1080/00220381003706494

Hussain S, Akhter R, Maktedar SS (2024) Advancements in sustainable food packaging: from eco-friendly materials to innovative technologies. Sustain Food Technol 2(5):1297–1364. https://doi.org/10.1039/d4fb00084f

India Brand Equity Foundation (2025). https://ibef.org/blogs/the-future-of-packaging-india-s-role-in-global-paper-packaging-trends

ITC Trade Map (2022) International Trade statistics. International Trade Centre. https://www.trademap.org/Index.aspx

Jairath MS, Purohit P (2013) Food safety regulatory compliance in India: a challenge to enhance agri-businesses. Indian J Agric Econ 68(3):431–448. https://doi.org/10.22004/ag.econ.206346

Joshi U, Bisht TS, Mamgain LRA (2021) Smart packaging: modern way for reducing post-harvest losses of horticultural produce. Int J Agric Sci 17(AAEBSSD):331–339. https://doi.org/10.15740/has/ijas/17-aaebssd/331-339

Just DR, Goddard JM (2023) Behavioral framing and consumer acceptance of new food technologies: factors influencing consumer demand for active packaging. Agribusiness 39(1):3–27. https://doi.org/10.1002/agr.21778

Kader AA (2002) Postharvest technology of horticultural crops. University of California Agriculture and Natural Resources

Kim YT, Min B, Kim KW (2014) General characteristics of packaging materials for food system. In: Innovations in food packaging. Academic Press, pp 13–35. https://doi.org/10.1016/b978-0-12-394601-0.00002-3

Kim YH, Yang YJ, Kim JS, Choi DS, Park SH, Jin SY, Park JS (2018) Non-destructive monitoring of apple ripeness using an aldehyde sensitive colorimetric sensor. Food Chem 267:149–156. https://doi.org/10.1016/j.foodchem.2018.02.110

Kirwan MJ (ed) (2012) Handbook of paper and paperboard packaging technology. Wiley. https://doi.org/10.1002/9781118470930.ch13

Kondratieva OV, Fedorov AD, Slinko OV, Voytyuk VA, Alekseeva SA (2022) New solutions in the horticultural industry. IOP Conf Ser: Earth Environ Sci 1010(1):012103. https://doi.org/10.1088/1755-1315/1010/1/012103

Kumar M, Barbhai MD, Hasan M, Punia S, Dhumal S, Rais N, Mekhemar M (2022) Onion (*Allium cepa* L.) peels: A review on bioactive compounds and biomedical activities. Biomed Pharmacother 146:112498. https://doi.org/10.1016/j.biopha.2021.112498

Kurek M, Chmiel M (2019) An overview of paper and paper-based food packaging materials. Foods 8(9):299. https://doi.org/10.3390/foods8090299

Kuswandi B, Murdyaningsih EA (2017) Simple on package indicator label for monitoring of grape ripening process using colorimetric pH sensor. J Food Meas Charact 11(4):2180–2194. https://doi.org/10.1007/s11694-017-9603-5

Kuswandi B, Wicaksono Y, Jayus, Abdullah A, Heng LY, Ahmad M (2011) Smart packaging: sensors for monitoring of food quality and safety. Sens & Instrumen Food Qual 5(3):137–146. https://doi.org/10.1007/s11694-011-9120-x

Kuswandi B, Maryska C, Jayus, Abdullah A, Heng LY (2013) Real time on-package freshness indicator for guavas packaging. J Food Meas Charact 7(1):29–39. https://doi.org/10.1007/s11694-013-9136-5

Kwaku AR, Fan QB (2020) Effect of good product design and packaging on market value and the performance of agricultural products in the Ghanaian Market. Open Access Libr J 7:e6714. https://doi.org/10.4236/oalib.1106714

Lee DS (2016) Carbon dioxide absorbers for food packaging applications. Trends Food Sci Technol 57:146–155. https://doi.org/10.1016/j.tifs.2016.09.014

Li P, Yang J, Jiménez-Carvelo AM, Erasmus SW (2024) Applications of food packaging quick response codes in information transmission toward food supply chain integrity. Trends Food Sci Technol 146:104384. https://doi.org/10.1016/j.tifs.2024.104384

Linda S, Sarah AL (2010) Bisphenol A (BPA) in plastics and possible human health effects. Congr Res Serv 7:5700

Ma W, Li L (2024) Trends and prospects in sustainable food packaging materials. Foods 13:1744. https://doi.org/10.3390/foods13111744

Mahajan PV, Caleb OJ, Singh Z, Watkins CB, Geyer M (2014) Postharvest treatments of fresh produce. Philos Trans R Soc A Math Phys Eng Sci 372(2017):20130309. https://doi.org/10.1098/rsta.2013.0309

Majid I, Nayik GA, Dar SM, Nanda V (2018) Novel food packaging technologies: innovations and future prospective. J Saudi Soc Agric Sci 17(4):454–462. https://doi.org/10.1016/j.jssas.2016.11.003

Mangaraj S, Goswami TK, Mahajan PV (2009) Applications of plastic films for modified atmosphere packaging of fruits and vegetables: a review. Food Eng Rev 1(2):133–158. https://doi.org/10.1007/s12393-009-9007-3

Mariah MAA, Vonnie JM, Erna KH, Nur'Aqilah NM, Huda N, Abdul Wahab R, Rovina K (2022) The emergence and impact of ethylene scavengers techniques in delaying the ripening of fruits and vegetables. Membranes 12(2):117. https://doi.org/10.3390/membranes12020117

Mark JK, John WS (2003) Plastics in food packaging. In: Richard C (ed) Food packaging technology. Blackwell Publishing Ltd, London, p 214

Marsh K, Bugusu B (2007) Food packaging—roles, materials, and environmental issues. J Food Sci 72(3):R39–R55. https://doi.org/10.1111/j.1750-3841.2007.00301.x

Matindoust S, Baghaei-Nejad M, ShahrokhAbadi MH, Zou Z, Zheng LR (2016) Food quality and safety monitoring using gas sensor array in intelligent packaging. Sens Rev 36(2):169–183. https://doi.org/10.1108/SR-07-2015-0115

MIDH Guidelines (2025). https://nhb.gov.in/writereaddata/082825102800MIDH%20Guideline%202025.pdf

Mishra V, Chugh V, Gangwar S, Awasthi P, Singh AK (2024) Detection of pesticides and other contaminants in horticultural crops. In: Advances in postharvest and analytical technology of horticulture crops. Springer Nature Singapore, pp 209–242. https://doi.org/10.1007/978-981-97-7247-6_11

Misra NN, Dixit Y, Al-Mallahi A, Bhullar MS, Upadhyay R, Martynenko A (2020) IoT, big data, and artificial intelligence in agriculture and food industry. IEEE Internet Things J 9(9):6305–6324. https://doi.org/10.1109/jiot.2020.2998584

Mohammadi M, Hnzayy S, Azad N (2014) A study on packaging factors influencing on export development. Manag Sci Lett 4(10):2213–2220. https://doi.org/10.5267/j.msl.2014.9.015

Mohebi E, Marquez L (2015) Intelligent packaging in meat industry: an overview of existing solutions. J Food Sci Technol 52(7):3947–3964. https://doi.org/10.1007/s13197-014-1588-z

Moran S, Cui X (2016) The Internet of things. In: Ethical ripples of creativity and innovation. Palgrave Macmillan, pp 61–68. https://doi.org/10.1057/9781137505545_7

Mordor Intelligence (2025) Packaging market size & share analysis—growth trends and forecast (2025–2030). https://www.mordorintelligence.com/industry-reports/agricultural-packaging-market

Moyeenudin H, Anandan R (2018) Application of RFID technology for food safety. J Comput Inf Syst 14(6):153–155

Müller P, Schmid M (2019) Intelligent packaging in the food sector: a brief overview. Foods 8(1):16. https://doi.org/10.3390/foods8010016

Mustafa F, Andreescu S (2018) Chemical and biological sensors for food-quality monitoring and smart packaging. Foods 7(10):168. https://doi.org/10.3390/foods7100168

Naturpac (2025). https://naturpac.org/news/packaging-reforms-dcceew-fresh-produce/

Ningwei LD, Mahat SP (2009) Reduce, reuse and replace: a study on solutions to plastic wastes. An interactive qualifying project submitted to the faculty of Worcester Polytechnic Institute. Digital WPI

Omerović N, Djisalov M, Živojević K, Mladenović M, Vunduk J, Milenković I, Vidić J (2021) Antimicrobial nanoparticles and biodegradable polymer composites for active food packaging applications. Compr Rev Food Sci Food Saf 20(3):2428–2454. https://doi.org/10.1111/1541-4337.12727

Opara UL, Caleb OJ, Belay ZA (2019) Modified atmosphere packaging for food preservation. In: Food quality and shelf life. Academic Press, pp 235–259. https://doi.org/10.1016/b978-0-12-817190-5.00007-0

Ortega F, Versino F, López OV, García MA (2022) Biobased composites from agro-industrial wastes and by-products. Emerg Mater 5(3):873–921. https://doi.org/10.1007/s42247-021-00319-x

Otoni CG, Espitia PJ, Avena-Bustillos RJ, McHugh TH (2016) Trends in antimicrobial food packaging systems: emitting sachets and absorbent pads. Food Res Int 83:60–73. https://doi.org/10.1016/j.foodres.2016.02.018

Ozdemir M, Floros JD (2004) Active food packaging technologies. Crit Rev Food Sci Nutr 44(3):185–193. https://doi.org/10.1080/10408690490441578

Palanisamy Y, Kadirvel V, Ganesan ND (2025) Recent technological advances in food packaging: sensors, automation, and application. Sustainable Food Technol 3(1):161–180. https://doi.org/10.1039/d4fb00296b

Panou A, Lazaridis DG, Karabagias IK (2025) Application of smart packaging on the preservation of different types of perishable fruits. Foods 14(11):1878. https://doi.org/10.3390/foods14111878

Parfitt J, Barthel M, Macnaughton S (2010) Food waste within food supply chains: quantification and potential for change to 2050. Philos Trans R Soc B: Biol Sci 365(1554):3065–3081. https://doi.org/10.1098/rstb.2010.0126

Pascall MA (2010) Packaging for fresh vegetables and vegetable products. In: Handbook of vegetables and vegetable processing. Wiley-Blackwell, pp 405–422. https://doi.org/10.1002/9780470958346.ch20

Peek A (2023) Meyers. https://meyers.com/meyers-blog/sustainability-packaging-regulations-trends-across-the-world/

Pierce L (2025). https://sustainablepackaging.org/2025/08/25/packaging-policy-news/,

Pirsa S, Chavoshizadeh S (2018) Design of an optical sensor for ethylene based on nanofiber bacterial cellulose film and its application for determination of banana storage time. Polym Adv Technol 29(5):1385–1393. https://doi.org/10.1002/pat.4250

Rajan SS, Wani KM (2025) A review of smart food and packaging technologies: revolutionizing nutrition and sustainability. Food Humanity 4(22):100593. https://doi.org/10.1016/j.foohum.2025.100593

Researchnester (2025). https://www.researchnester.com/reports/horticulture-bioplastic-market/7861

Robertson GL (2016) Food packaging: principles and practice, 3rd edn. CRC Press

Rodvong T, Anal AK (2025) IoT-based traceability systems for fresh food: strategic development of policy to align with sustainable development goals. J Sci Technol Policy Manag. https://doi.org/10.1108/jstpm-01-2024-0008

Ruiz-Martínez J, Aguirre-Joya JA, Rojas R, Vicente A, Aguilar-González MA, Rodríguez-Herrera R, Aguilar CN (2020) Candelilla wax edible coating with Flourensiacernuabioactives to prolong the quality of tomato fruits. Foods 9(9):1303. https://doi.org/10.3390/foods9091303

Sadhu SD (2022) Applications of polymers in horticulture food packaging. J Hortic 9:302

Saenjaiban A, Singtisan T, Suppakul P, Jantanasakulwong K, Punyodom W, Rachtanapun P (2020) Novel color change film as a time–temperature indicator using polydiacetylene/silver nanoparticles embedded in carboxymethyl cellulose. Polymers 12(10):2306. https://doi.org/10.3390/polym12102306

Sandhya (2010) Modified atmosphere packaging of fresh produce: current status and future needs. LWT-Food Sci Technol 43(3):381–392. https://doi.org/10.1016/j.lwt.2009.05.018

Santosh DT, Harshavardhan K (2024) Smart horticulture: transformative post-harvest innovations. In: Innovations in horticulture for sustainable growth. International Books & Periodical Supply Service

Schaefer D, Cheung WM (2018) Smart packaging: opportunities and challenges. Procedia Cirp 72:1022–1027. https://doi.org/10.1016/j.procir.2018.03.240

Sharma A (2022) A study on food safety and its standards in India. Journal of Advances and Scholarly Researches in Allied Education, 19(1):632–638. https://ignited.in/index.php/jasrae/article/view/14833

Sharma K, Yadav R, Rajput R (2025) Smart packaging materials: a review of existing and emerging packaging technologies and their applications. J Sci Res Rep 31(5):53–68. https://doi.org/10.9734/jsrr/2025/v31i53004

Shepherd B, Wilson JS (2009) Trade costs and facilitation in APEC and ASEAN: delivering the goods? World Econ 32(4):535–557

Shin J, Selke SE (2014) Food packaging. In: Food processing: principles and applications. Wiley, pp 249–273. https://doi.org/10.1002/9781118846315.ch11

Shukla P, Singh J, Wang W (2022) The influence of creative packaging design on customer motivation to process and purchase decisions. J Bus Res 147:338–347. https://doi.org/10.1016/j.jbusres.2022.04.026

Singh M, Singh AK, Sambhav K, Singari RM (2025) Integrating traditional indian sustainable packaging materials, regional startups, and design thinking for a sustainable future: a review. J Philanthr Mark 5(1):254–270

Siracusa V, Rocculi P, Romani S, Rosa MD (2008) Biodegradable polymers for food packaging: a review. Trends Food Sci Technol 19(12):634–643. https://doi.org/10.1016/j.tifs.2008.07.003

Sobhan A, Hossain A, Wei L, Muthukumarappan K, Ahmed M (2025) IoT-enabled biosensors in food packaging: A breakthrough in food safety for monitoring risks in real time. Foods 14(8):1403. https://doi.org/10.3390/foods14081403

Stark NM, Matuana LM (2021) Trends in sustainable biobased packaging materials: a mini review. Mater Today Sustain 15:100084. https://doi.org/10.1016/j.mtsust.2021.100084

Stoll M, Baecke S, Kenning P (2008) What they see is what they get? An fMRI-study on neural correlates of attractive packaging. J Consum Behav 7(4–5):342–359. https://doi.org/10.1002/cb.256

Tan EL, Ng WN, Shao R, Pereles BD, Ong KG (2007) A wireless, passive sensor for quantifying packaged food quality. Sensors 7(9):1747–1756. https://doi.org/10.3390/s7091747

Teixeira-Costa BE, Andrade CT (2021) Natural polymers used in edible food packaging—history, function and application trends as a sustain-able alternative to synthetic plastic. Polysaccharides 3:32–58. https://doi.org/10.3390/polysaccharides3010002

Tesfay SZ, Magwaza LS (2017) Evaluating the efficacy of moringa leaf extract, chitosan and carboxymethyl cellulose as edible coatings for enhancing quality and extending postharvest life of avocado (*Persea americana* mill.) fruit. Food Packag Shelf Life 11:40–48. https://doi.org/10.1016/j.fpsl.2016.12.001

Thapliyal D, Karale M, Diwan V, Kumra S, Arya RK, Verros GD (2024) Current status of sustainable food packaging regulations: global perspective. Sustainability 16(13):5554. https://doi.org/10.3390/su16135554

Thivya P, Bhanu Prakash Reddy N, Bhosale Yuvraj K, Sinija VR (2023) Recent advances and perspectives for effective utilization of onion solid waste in food packaging: a critical review. Rev Environ Sci Biotechnol 22(1):29–53. https://doi.org/10.1007/s11157-022-09642-z

Towardspackaging (2025). https://www.towardspackaging.com/insights/packaging-market-sizing

Trade A, Change C (2018) Agricultural commodity markets. FAO

Twede D (2002) The packaging technology and science of ancient transport amphoras. Packag Technol Sci 15(4):181–195. https://doi.org/10.1002/pts.597

Van Veen TWS (2005) International trade and food safety in developing countries. Food Control, 16(6):491-496

Verma MK, Shakya S, Kumar P, Madhavi J, Murugaiyan J, Rao MVR (2021) Trends in packaging material for food products: historical background, current scenario, and future prospects. J Food Sci Technol 58(11):4069–4082. https://doi.org/10.1007/s13197-021-04964-2

Vo E, Murray DK, Scott TL, Attar AJ (2007) Development of a novel colorimetric indicator pad for detecting aldehydes. Talanta 73(1):87–94. https://doi.org/10.1016/j.talanta.2007.03.014

Wakefield R (2025). https://aiph.org/floraculture/news/navigating-the-eus-ppwr-new-packaging-rules-for-ornamental-horticulture/

Wang, H. M., Yuan, T. Q., Song, G. Y.& Sun, R. C. (2021). Advanced and versatile lignin-derived biodegradable composite film materials toward a sustainable world. Green Chem, 23 (11), 3790–3817.https://doi.org/10.1039/d1gc00790d

Warsiki E, Rofifah N (2018) Dragon fruit freshness detector based on methyl red colour indicator. IOP Conf Ser: Earth Environ Sci 209(1):012016. https://doi.org/10.1088/1755-1315/209/1/012016

Win NKK, Jitareerat P, Kanlayanarat S, Sangchote S (2007) Effects of cinnamon extract, chitosan coating, hot water treatment and their combinations on crown rot disease and quality of banana fruit. Postharvest Biol Technol 45(3):333–340. https://doi.org/10.1016/j.postharvbio.2007.01.020

Yahia EM (ed) (2009) Modified and controlled atmospheres for the storage, transportation, and packaging of horticultural commodities. CRC Press. https://doi.org/10.1201/9781420069587

Yam KL, Lee DS (2012) Emerging food packaging technologies: an overview. In: Emerging food packaging technologies. Woodhead Publishing, pp 1–9. https://doi.org/10.1533/9780857095664.1

Yam KL, Takhistov PT, Miltz J (2005) Intelligent packaging: concepts and applications. J Food Sci 70(1):R1–R10. https://doi.org/10.1111/j.1365-2621.2005.tb09052.x

Yang Z, Zou X, Li Z, Huang X, Zhai X, Zhang W, Tahir HE (2019) Improved postharvest quality of cold stored blueberry by edible coating based on composite gum arabic/roselle extract. Food Bioprocess Technol 12(9):1537–1547. https://doi.org/10.1007/s11947-019-02312-z

Yangang Z (2021) Research on the influence of modern consumer psychology on packaging design based on image analysis. In: Proceedings of the 2021 international conference on bioinformatics and intelligent computing. ACM, pp 327–332. https://doi.org/10.1145/3448748.3448801

Zagory D (1995) Ethylene-removing packaging. In: Active food packaging. Springer US, Boston, pp 38–54. https://doi.org/10.1007/978-1-4615-2175-4_2

Marketing and Supply Chain: Global Scenario

15

Sergio Roberto Márquez-Berber,
Natanael Magaña-Lira, Abdul Khalil Gardezi,
and Diana América Reyna-Izaguirre

Abstract

This chapter analyzes the evolving dynamics of global horticultural marketing and supply chains and their implications for achieving remunerative, sustainable, and inclusive agricultural development. It integrates theoretical perspectives from Global Value Chain (GVC) analysis, digital platform governance, climate resilience, transaction cost economics, market orientation, inclusive business models, and Environmental, Social, and Governance (ESG) frameworks to examine how value is created, governed, and distributed among actors. The chapter identifies key structural challenges, including perishability, postharvest losses, compliance costs, digital exclusion, environmental pressures, and persistent power asymmetries. It highlights the transformative role of digital technologies such as platforms, blockchain, and data-driven logistics. Comparative evidence from multiple regions, complemented by case studies of producer organizations, illustrates how institutional contexts, collective action, and market strategies shape upgrading opportunities and value capture. The chapter concludes by outlining strategic pathways to enhance competitiveness, inclusiveness, and resilience in horticultural supply chains, emphasizing coordinated public–private action, institutional strengthening, and equitable digital transformation.

S. R. Márquez-Berber (✉) · N. Magaña-Lira
Universidad Autónoma Chapingo, Departamento de Fitotecnia, Texcoco, Mexico

A. K. Gardezi
Colegio de Posgraduados, Posgrado en Hidrociencias, Montecillos, Mexico

D. A. Reyna-Izaguirre
Universidad Autónoma Chapingo, Departamento de Ingeniería Agroindustrial, Texcoco, Mexico

A. Das, A. B. Sharangi (eds.), *Remunerative Horticulture*,
https://doi.org/10.1007/978-981-92-1007-7_15

Keywords

Horticultural supply chains · Global value chains · Digital platforms · Climate resilience · ESG standards · Value capture · Collective action · Agricultural development

15.1 Introduction to Global Horticultural Markets

Horticulture has become a strategically important sector within global agri-food systems due to its high economic value, labor intensity, nutritional contribution, and growing relevance in diversified human diets. Fruits, vegetables, and other high-value crops play a central role in food security, rural development, and sustainable agriculture debates, particularly in the context of rapidly evolving global markets (Reardon et al. 2019; Barrett et al. 2022). These markets are increasingly characterized by complex governance structures, stringent quality standards, and asymmetries in power distribution among actors (Alford et al. 2024).

Over the past four decades, global horticultural trade has expanded rapidly, driven by rising incomes, urbanization, and changing consumption patterns favoring healthier and more convenient foods (HLPE 2017). This expansion has been accompanied by diversification in marketing channels and the emergence of polycentric food systems, where global, regional, and domestic value chains coexist and interact (Barrett et al. 2022). These transformations reflect shifts not only in geographic supply patterns but also in governance mechanisms, ranging from traditional spot markets to vertically coordinated supply chains and hybrid arrangements.

A defining feature of this transformation has been the consolidation of retail power through the "supermarket revolution," which has imposed increasingly strict requirements on quality, safety, and traceability (Berdegué et al. 2005). While these developments have improved efficiency and product standards, they have also created barriers to entry for small-scale producers, particularly in developing countries (Narrod et al. 2009).

At the same time, digital transformation is reshaping horticultural supply chains. Digital platforms, blockchain technologies, and Internet of Things (IoT) systems are redefining coordination mechanisms, enabling real-time information flows, and altering power relations among producers, intermediaries, and buyers (Wolfert et al. 2017; Qin et al. 2024). However, these innovations also raise concerns related to data ownership, algorithmic governance, and unequal access to digital infrastructure (McFadden et al. 2022).

Climate change further intensifies these challenges. Horticultural production is highly sensitive to environmental variability, including temperature extremes, water scarcity, and pest pressures, while supply chains are increasingly exposed to disruptions caused by extreme weather events (Thompson et al. 2023). Addressing these risks requires coordinated strategies that integrate farm-level adaptation with supply chain resilience (OECD 2025).

In this context, this chapter provides an integrated analysis of horticultural marketing and supply chains, focusing on three interconnected dimensions: (i) evolving governance structures and power relations, (ii) digital transformation and data governance, and (iii) sustainability and climate resilience. Drawing on comparative evidence and case studies, the chapter examines how value is created and captured, and identifies strategic pathways to enhance competitiveness, inclusiveness, and resilience in global horticultural supply chains.

15.2 Theoretical Frameworks for Understanding Horticultural Supply Chains

Horticultural supply chains are characterized by high perishability, strict quality requirements, and increasing integration into global markets. Understanding their organization and performance requires a multidimensional analytical framework that combines insights from global value chain (GVC) analysis, transaction cost economics, innovation and digital governance, and sustainability-oriented approaches.

15.2.1 Global Value Chain Governance

Global value chain (GVC) analysis provides a central framework for understanding the organization of horticultural markets. Gereffi et al. (2005) identify different governance structures, including market, modular, relational, captive, and hierarchical forms, which determine how value is created and distributed across actors. In horticulture, buyer-driven chains—particularly those led by supermarkets and exporters—play a dominant role in setting standards, coordinating production, and shaping upgrading opportunities. This governance structure often enables process and product upgrading but may limit functional upgrading and reinforce asymmetries in value capture.

15.2.2 Digital Platform Governance

Recent transformations in agri-food systems have introduced digital platforms as emerging governance mechanisms. Platform-based coordination relies on data aggregation and algorithmic management to facilitate transactions and market access (Kenney and Zysman 2016; Srnicek 2017). While these systems can reduce transaction costs and improve efficiency, they also shift control toward actors that manage digital infrastructure and data. The implications of this transformation for power distribution and coordination are examined further in Sect. 15.4.

15.2.3 Climate Resilience and Adaptation

Climate change introduces increasing uncertainty into horticultural production systems, affecting yields, quality, and supply stability. Resilience frameworks emphasize the capacity of systems to absorb shocks, adapt, and transform in response to environmental stress (IPCC 2023). In horticulture, this involves adjustments in production practices, infrastructure, and supply chain organization. While climate risks are increasingly recognized, their interaction with market structures and governance remains underexplored.

15.2.4 Transaction Cost Economics

Transaction cost economics (TCE) explains governance choices based on asset specificity, uncertainty, and transaction frequency (Williamson 1985). In horticulture, high perishability and quality requirements increase coordination needs, often leading to relational contracting or vertical integration. These governance arrangements can reduce uncertainty and improve efficiency but may also restrict market access for smaller producers.

15.2.5 Inclusive Business and Collective Action

Inclusive business models aim to integrate smallholders into high-value supply chains through contractual arrangements, partnerships, and producer organizations. Collective action plays a key role in improving bargaining power, facilitating access to inputs and markets, and enabling compliance with standards (Markelova et al. 2009; Fischer and Qaim 2012). However, outcomes depend on institutional capacity and governance conditions, and inclusion remains uneven across regions.

15.2.6 Market Orientation and Standards

Market access in horticulture increasingly depends on compliance with quality, safety, and sustainability standards. Certifications such as GlobalG.A.P. and organic labels function as governance tools that shape production practices and market participation. While these standards can facilitate access to high-value markets, they also impose compliance costs that may exclude smaller producers (Meemken 2020). The implications of standards for inclusion and competitiveness are further discussed in Sects. 15.3 and 15.8.

15.2.7 ESG and Sustainability Frameworks

Environmental, social, and governance (ESG) criteria are becoming central to investment decisions and market requirements in horticulture. These frameworks extend beyond traditional certifications by incorporating broader sustainability metrics and accountability mechanisms. While ESG adoption can enhance transparency and sustainability performance, it also introduces new reporting requirements and compliance costs, raising concerns about unequal participation (Singh et al. 2024).

15.2.8 Integrated Conceptual Framework

Taken together, these frameworks provide a comprehensive lens to analyze horticultural supply chains. GVC analysis explains value distribution and governance structures; TCE clarifies coordination mechanisms; digital governance highlights emerging forms of control; resilience frameworks address environmental risks; and inclusive business and ESG approaches capture social and sustainability dimensions. Table 15.1 synthesizes these perspectives, providing an integrated framework for the analysis developed in the following sections.

Table 15.1 Integrated theoretical frameworks for horticultural supply chains

Framework	Core concepts	Application to horticulture	Key limitation
GVC Governance	Buyer-driven chains, upgrading, polycentric governance	Explains power distribution and value capture across global, regional, and domestic chains	Limited attention to micro-level dynamics and farmer agency
Digital Platform Governance	Platforms, data ownership, algorithmic coordination	Analyzes how digital technologies reshape coordination and power relations	Rapid evolution limits empirical validation
Climate Resilience	Adaptation, system vulnerability, transformation	Explains responses to climate variability at farm and chain levels	Limited integration with economic frameworks
Transaction Cost Economics	Asset specificity, uncertainty, relational contracting	Explains governance choices and coordination mechanisms	Overemphasis on efficiency, less on equity
Inclusive Business Models	Smallholder integration, partnerships	Explains inclusion pathways and value distribution	Outcomes highly context-dependent
Market Orientation and Standards	Quality requirements, certifications	Explains market access and competitiveness	May reinforce exclusion
ESG and Sustainability	Environmental and social compliance	Frames sustainability as a market requirement	High data and compliance costs

15.3 Global Horticultural Trade Patterns and Market Dynamics

15.3.1 The Supermarket Revolution and Retail Concentration

The "supermarket revolution" has been one of the most significant transformations in global food systems, reshaping horticultural supply chains through retail concentration, standardization, and vertical coordination (Reardon et al. 2003, 2019). Supermarkets now dominate food retail in high-income countries and have rapidly expanded in middle-income economies, controlling a large share of fresh produce distribution (see Fig. 15.1).

This transformation has shifted bargaining power toward retailers, enabling them to impose stringent requirements on quality, safety, traceability, and delivery conditions (Dolan and Humphrey 2000). In horticultural markets, where perishability limits producers' alternatives, this buyer power is particularly pronounced. As a result, supply chains have transitioned from fragmented spot markets to more coordinated systems characterized by preferred supplier networks and centralized procurement (Reardon et al. 2019).

Fig. 15.1 The "supermarket revolution" in horticultural markets, illustrated by the retail display of fresh fruits and vegetables and its role in reshaping food distribution and supply chain coordination

Private standards play a central role in this restructuring. Retailers increasingly require compliance with certification schemes such as GlobalG.A.P. and retailer-specific protocols, which often exceed public regulatory requirements (Henson and Reardon 2005). While these standards contribute to improved product quality and safety, they also create barriers to entry, particularly for smallholders lacking the resources to meet compliance costs (Narrod et al. 2009).

The supermarket-driven restructuring has also promoted product differentiation and value addition, including pre-packaging, processing, and branding, shifting value creation upstream in the chain (Humphrey and Memedovic 2006). However, the benefits of these transformations are unevenly distributed, often favoring larger producers and well-organized supply networks, while smaller actors face exclusion or marginalization (Reardon et al. 2009).

15.3.2 Standards, Certifications, and Market Access

The proliferation of public and private standards has become a defining feature of horticultural markets, shaping access conditions and competitive dynamics (Henson and Humphrey 2010). Public standards establish minimum requirements for food safety and phytosanitary control, while private standards define more stringent criteria related to quality, sustainability, and traceability (Giovannucci and Ponte 2005).

Compliance with these standards entails significant costs, including investments in infrastructure, certification processes, and management systems (Meemken 2020). While certification can facilitate access to high-value markets and generate price premiums in some cases, evidence suggests that benefits are heterogeneous and often limited for small-scale producers (Tran and Goto 2019).

Standards can function as non-tariff barriers, contributing to the consolidation of supply chains and the exclusion of less-capitalized producers. Empirical evidence from multiple regions shows that stricter standards have led to reduced participation of smallholders and increased dominance of large-scale suppliers (Henson and Humphrey 2010).

However, collective organization and institutional support can mitigate these barriers. Producer organizations can reduce compliance costs through group certification and shared infrastructure (see Fig. 15.2), while public policies can facilitate access to training, credit, and certification systems (Markelova et al. 2009; Narrod et al. 2009). Recent efforts toward standard harmonization and technology-enabled verification aim to reduce duplication and improve efficiency, although implementation remains uneven (Loconto and Dankers 2014).

Fig. 15.2 Organized small-scale producers, in this case avocado producers from Mexico, demonstrating the capacity to meet the stringent quality, safety, and traceability requirements imposed by supermarket-led supply chains

15.3.3 Regional Value Chains and Polycentric Governance

In addition to global value chains, regional value chains (RVCs) have become increasingly important in horticultural trade, particularly in the Global South (Krishnan 2018). Rising incomes, urbanization, and the expansion of regional supermarket networks have driven demand for fresh produce within regional markets, creating new opportunities for producers.

RVCs typically exhibit less stringent quality requirements and lower compliance costs compared to global export chains, making them more accessible to smallholders (Pasquali et al. 2021). However, they also offer lower price levels and may provide fewer opportunities for upgrading. Nonetheless, RVCs can serve as entry points for producers to build capabilities before participating in more demanding global markets (Krishnan 2018).

The coexistence of global, regional, and domestic chains reflects a broader shift toward polycentric governance, where multiple coordination mechanisms operate simultaneously (Barrett et al. 2022). Producers may participate in multiple channels, allocating output across markets based on quality, price, and risk considerations. This "hybrid participation" can enhance resilience by diversifying market exposure, but it also increases complexity and transaction costs (Pasquali et al. 2021).

From a development perspective, polycentric systems offer both opportunities and challenges. While they provide flexibility and multiple pathways for market integration, they also require producers to navigate diverse standards, relationships, and logistical arrangements. Effective participation depends on institutional support, infrastructure, and organizational capacity (Barrett et al. 2022).

15.4 Digital Transformation and Platform Governance

Digital transformation is reshaping horticultural supply chains by modifying coordination mechanisms, information flows, and value capture. Digital platforms facilitate real-time coordination and reduce transaction costs, consistent with transaction cost economics (Williamson 1985), while simultaneously introducing new forms of governance based on data control and algorithmic coordination (Kenney and Zysman 2016; Srnicek 2017). These developments have important implications for power distribution within supply chains, particularly in buyer-driven systems (Gereffi et al. 2005).

15.4.1 Digital Platforms and Market Intermediation

Digital platforms increasingly act as central coordinators of market transactions, aggregating data and shaping price formation. Through algorithmic management and control over market information, platforms can influence decision-making processes and reinforce asymmetries between producers and downstream actors (Kenney and Zysman 2016). While digitalization enhances efficiency, traceability, and coordination, it also raises concerns about reduced transparency and limited access to strategic data for producers. This reflects a broader shift from relational and contract-based governance toward data-driven coordination mechanisms (Srnicek 2017).

15.4.2 Technological Innovations and Traceability

Technologies such as blockchain and digital traceability systems are increasingly promoted to improve transparency, food safety, and compliance with quality standards (Kamilaris et al. 2019; Cao et al. 2022; López-Pimentel et al. 2022). These tools can reduce information asymmetries and strengthen trust in global value chains, particularly in export-oriented horticulture. However, their adoption remains uneven due to high implementation costs, technical complexity, and limited institutional support. As a result, benefits tend to concentrate among larger actors, reinforcing existing structural inequalities (World Bank 2019).

15.4.3 Digital Inclusion and Structural Constraints

The diffusion of digital technologies in horticulture is constrained by infrastructure limitations, unequal access to digital services, and differences in technological capabilities. These barriers are particularly relevant for smallholders in developing regions, where limited connectivity and financial constraints hinder adoption (Nwangwu et al. 2024; Singh et al. 2024; FAO 2025; Abdulai et al. 2025). From a diffusion of innovation perspective (Prause et al. 2020; McFadden et al. 2022), these constraints slow adoption processes and contribute to widening gaps between technologically advanced and lagging producers. Addressing these disparities requires complementary investments in infrastructure, capacity building, and institutional support.

15.4.4 Implications for Supply Chain Governance

Digital transformation is redefining governance structures within horticultural supply chains. While it enhances efficiency and traceability, it also shifts control toward actors that manage data and digital infrastructure. This transformation reinforces buyer-driven dynamics and raises concerns about market concentration and unequal value distribution (Gereffi et al. 2005). Consequently, digitalization should be understood not only as a technological change but as a structural reconfiguration of governance, requiring regulatory and institutional responses to ensure more inclusive and equitable outcomes.

15.5 Producer Organizations and Collective Action

15.5.1 Theoretical Foundations of Collective Action

Producer organizations—such as cooperatives and farmer associations—play a central role in horticultural supply chains by enabling smallholders to overcome market failures, reduce transaction costs, and strengthen bargaining power (Hellin et al. 2009; Markelova et al. 2009). Theoretical foundations draw from collective action theory, institutional economics, and transaction cost economics.

Collective action faces classic coordination challenges, including free-rider problems and heterogeneity of member interests, which can undermine cooperation if governance mechanisms are weak (Olson 1971; Verhofstadt and Maertens 2015). However, evidence shows that well-structured organizations can effectively manage shared resources and coordinate production and marketing activities when supported by clear rules, monitoring systems, and participatory governance (Ostrom 2012).

From a transaction cost perspective, cooperatives reduce costs associated with market exchange by aggregating supply, standardizing quality, and facilitating access to inputs, credit, and technical assistance (Williamson 1985; Swinnen and Maertens 2007). These functions are particularly relevant in horticulture, where perishability and quality variability increase coordination costs.

Despite these advantages, cooperative performance varies widely. Challenges include governance inefficiencies, limited managerial capacity, and risks of elite capture (Markelova et al. 2009). Empirical studies indicate that benefits—such as improved income, productivity, and market access—are context-dependent and depend on institutional support, internal governance, and service provision (Wossen et al. 2017).

15.5.2 Case Study: La Palma–Granada Cooperative (Spain)

The La Palma–Granada cooperative illustrates how collective organization can facilitate smallholder integration into high-value European markets. With approximately 760 producers and small farm sizes, the cooperative has developed a coordinated system that integrates production, quality control, and marketing.

The cooperative provides a comprehensive service package, including input supply, technical assistance, and centralized marketing. Its packing facilities enable sorting, grading, and traceability in compliance with European standards (see Fig. 15.3), allowing members to access supermarket supply chains (Dolan and Humphrey 2000).

Collective action has enabled functional upgrading, as producers move beyond farm-gate sales to participate in higher-value stages such as packing and marketing. Process upgrading is reflected in improved production practices, including irrigation efficiency and integrated pest management, while product upgrading includes diversification and organic certification.

Governance combines democratic participation with professional management, balancing member control with operational efficiency. This structure enhances trust and coordination but requires continuous investment in communication and capacity building.

Fig. 15.3 Packing facilities of the cooperative La Palma–Granada, illustrating the role of collective organization, postharvest infrastructure, and quality control in supplying high-value European horticultural markets

Challenges include increasing compliance costs, climate risks, and dependence on concentrated retail markets. The cooperative addresses these challenges through diversification of marketing channels, investment in sustainability practices, and strengthening of technical support.

15.5.3 Case Study: TROPS Cooperative (Spain)

TROPS represents a large-scale cooperative model characterized by vertical integration, strong market positioning, and advanced technological infrastructure. With over 2800 members and significant production volumes, TROPS has developed a globally competitive system for tropical and subtropical fruits (see Fig. 15.4).

The cooperative's strategy is based on three pillars: scale, market orientation, and sustainability. Large-scale operations enable investments in advanced infrastructure, including automated sorting systems, controlled-atmosphere storage, and logistics networks that support year-round supply (Pérez-Mesa et al. 2021).

TROPS has achieved significant functional upgrading by integrating production, processing, and marketing activities, allowing greater value capture. Product upgrading is reflected in organic certification, ready-to-eat products, and branding strategies, while process upgrading includes adoption of precision irrigation and integrated pest management.

Sustainability is a core component of its competitive strategy. Initiatives include water efficiency, renewable energy use, and biodiversity conservation, aligning with ESG standards and consumer demand for sustainable products (Krishnan et al. 2023).

Market orientation is reinforced through active participation in trade fairs, continuous monitoring of consumer trends, and product innovation. This enables the cooperative to maintain strong relationships with European retailers and adapt to changing market conditions.

Fig. 15.4 Facilities of the cooperative TROPS, illustrating the role of large-scale cooperative infrastructure, vertical coordination, and market orientation in the commercialization of avocado and mango for European markets

However, TROPS faces challenges related to climate variability, market saturation, and generational renewal. Strategic responses include diversification of sourcing, investment in climate-resilient technologies, and expansion into new markets.

15.5.4 Comparative Analysis of Cooperative Models

The La Palma–Granada and TROPS cases illustrate both shared success factors and distinct strategic pathways in cooperative development.

Common elements include:

1. *Integrated service provision*, addressing production, marketing, and technical needs
2. *Investment in infrastructure and quality control*, enabling compliance with market standards
3. *Collective marketing and branding*, enhancing value capture
4. *Combination of professional management and democratic governance*, balancing efficiency and member participation
5. *Adaptive capacity*, allowing response to market, regulatory, and environmental changes

Key differences reflect scale and strategic positioning. TROPS operates at a larger scale with greater vertical integration and global sourcing, enabling stronger market power and brand development. In contrast, La Palma–Granada emphasizes diversification and local production, providing flexibility and risk management.

These differences highlight that cooperative success is context dependent. Scale can enhance competitiveness, but smaller organizations can also succeed through specialization, quality differentiation, and strong internal coordination.

15.5.5 Implications for Cooperative Development

The analysis provides several insights for cooperative development in horticultural supply chains:

1. *Collective action is a key mechanism for overcoming market barriers*, particularly for smallholders facing high transaction costs and stringent standards.
2. *Upgrading requires coordinated investments* in infrastructure, knowledge, and organizational capacity.
3. *Market orientation and innovation are critical* for maintaining competitiveness in dynamic markets.
4. *Sustainability is increasingly a competitive requirement*, not only an environmental objective.
5. *Institutional support is essential*, including policies that facilitate organization, access to finance, and technical assistance.

While cooperative models are not universally applicable, their principles—coordination, shared investment, and collective bargaining—offer valuable pathways for enhancing inclusion and value capture in horticultural supply chains.

15.6 Market Intelligence, Trade Fairs, and Value Addition

15.6.1 Market Intelligence in Horticultural Supply Chains

Market intelligence is a critical component of competitiveness in horticultural supply chains, enabling producers, cooperatives, and firms to anticipate demand trends, comply with evolving standards, and position their products in increasingly differentiated markets (Kohli and Jaworski 1990; Narver and Slater 1990).

In high-value horticulture, market requirements are dynamic and multidimensional, encompassing not only price but also quality attributes, traceability, sustainability certifications, packaging formats, and logistical reliability. Access to timely and accurate information on these dimensions is therefore essential for strategic decision-making (Pérez-Mesa et al. 2021).

Sources of market intelligence include direct buyer relationships, industry associations, trade publications, government statistics, and increasingly digital platforms and data analytics systems (Wolfert et al. 2017). However, access to and capacity to use such information remain unevenly distributed, often favoring larger firms and well-organized producer groups (Ponte and Gibbon 2005).

In this context, intermediated spaces for information exchange—particularly international trade fairs—play a central role in reducing information asymmetries and facilitating strategic coordination across value chain actors.

15.6.2 Trade Fairs as Strategic Coordination Platforms

International trade fairs have become key institutional mechanisms for coordinating horticultural supply chains, functioning as platforms for business-to-business (B2B) interaction, market intelligence generation, and strategic positioning (Gereffi and Lee 2014). Events such as Fruit Attraction (Madrid, see Fig. 15.5) and Fruit Logistica (Berlin) bring together producers, exporters, retailers, input suppliers, logistics firms, and policymakers from around the world.

These events perform multiple functions. First, they enable direct negotiation and relationship-building between suppliers and buyers, reducing search and transaction costs. Second, they facilitate market intelligence gathering, allowing participants to observe trends in product differentiation, packaging, branding, and sustainability requirements. Third, they serve as arenas for technological diffusion, showcasing innovations in postharvest handling, cold chain logistics, digital platforms, and sustainability practices.

Participation in trade fairs enhances firms' capacity for market orientation, enabling them to align production and marketing strategies with evolving consumer

Fig. 15.5 View of the Americas section at Fruit Attraction, highlighting the diversification of supply origins and the growing role of international trade in global horticultural markets

preferences and regulatory requirements (Pérez-Mesa et al. 2021). For exporters and cooperatives, these events are also critical for maintaining existing commercial relationships and accessing new market segments.

15.6.3 Value Addition and Competitive Differentiation

Trade fairs and market intelligence processes are closely linked to value addition strategies in horticultural supply chains. As competition intensifies, firms increasingly rely on differentiation mechanisms to capture value beyond commodity production (Humphrey and Memedovic 2006).

Key forms of value addition include:

- *Product differentiation*: organic certification, specialty varieties, geographic indications
- *Process upgrading*: improved postharvest handling, cold chain management, and quality control
- *Functional upgrading*: integration into packing, processing, branding, and direct marketing
- *Marketing innovation*: packaging, labeling, and branding strategies aligned with consumer preferences

These strategies allow producers and cooperatives to access higher-value market segments and reduce dependence on price-based competition. However, they require investments in infrastructure, knowledge, and organizational capacity, which may limit participation by smallholders (Ponte et al. 2023).

15.6.4 Constraints and Uneven Access to Market Intelligence

Despite their importance, access to trade fairs and market intelligence remains uneven. Participation costs—including travel, exhibition fees, and logistical expenses—can be prohibitive for small-scale producers and emerging cooperatives. Moreover, effective participation requires not only presence but also absorptive capacity, including language skills, technical knowledge, and negotiation capabilities.

As a result, benefits from trade fairs tend to be concentrated among larger firms and well-supported organizations, reinforcing existing inequalities in value chains (Ponte and Gibbon 2005). Public support programs, industry associations, and collective participation mechanisms can partially address these barriers by facilitating access and reducing costs.

15.6.5 Strategic Implications

From a strategic perspective, trade fairs and market intelligence platforms play a critical role in shaping competitiveness in horticultural supply chains. They

1. Reduce information asymmetries and transaction costs
2. Facilitate upgrading and innovation
3. Enable relationship-building and trust formation
4. Support market diversification and risk management

However, maximizing their benefits requires complementary investments in organizational capacity, collective action, and institutional support. Without such investments, market intelligence systems may reinforce existing power asymmetries rather than promote inclusive development.

15.7 Comparative Regional Dynamics in Horticultural Supply Chains

Comparative regional analysis reveals how global trends in horticultural supply chains—such as supermarket expansion, standardization, digitalization, and sustainability pressures—are shaped by distinct institutional environments, market structures, and policy frameworks. While these trends are widely observed, their impacts vary significantly across regions, influencing upgrading opportunities, value capture, and inclusion patterns (Reardon et al. 2019; Barrett et al. 2022).

15.7.1 Asia: Export Expansion and Contract Farming

Asia has emerged as a dynamic region in horticultural markets, driven by economic growth, urbanization, and integration into global value chains (OECD 2021). Countries such as China, Thailand, and Vietnam have expanded exports through investments in cold chains, logistics, and quality upgrading, while contract farming has played a central role in linking smallholders to export markets (Barrett et al. 2012; Meemken and Bellemare 2020).

These arrangements can improve access to inputs, credit, and technical assistance, but their inclusiveness depends on contract design and institutional conditions. Evidence shows that benefits are uneven, with risks of exclusion and unequal bargaining power. Despite rapid modernization, fragmentation at the farm level continues to constrain coordination, reinforcing the importance of producer organizations and state support.

15.7.2 Africa: Export Specialization and Structural Dependence

African horticultural supply chains are characterized by specialized export-oriented clusters integrated into European and Middle Eastern markets (FAO 2025). Countries such as Kenya, Ethiopia, Morocco, and Egypt have developed competitive export sectors based on contract farming, compliance with private standards, and coordination with international buyers.

From a GVC perspective, these systems are strongly buyer-driven, with market access conditioned by compliance with stringent quality, traceability, and sustainability requirements (Gereffi et al. 2005; Henson and Humphrey 2010). While exports generate income and employment, they also reproduce structural dependence and vulnerability due to logistics constraints, climate risks, and asymmetries in power and value distribution (Barrett et al. 2022).

Digitalization and ESG standards are reshaping these systems, but uneven access to infrastructure and capabilities limits inclusive upgrading. Institutional capacity remains a key determinant of long-term development outcomes.

15.7.3 Latin America: Diversification and Regional Integration

Latin America combines export-oriented production with growing regional markets and diversified value chain structures (OECD-FAO 2022). Countries such as Chile, Peru, Brazil, and Mexico have developed competitive advantages through investments in irrigation, postharvest infrastructure, and certification systems.

Regional trade agreements and logistics integration have strengthened participation in global and regional value chains (Brenton et al. 2022). However, structural heterogeneity persists, with advanced export firms coexisting alongside smallholders facing constraints in access to finance, technology, and market information.

Digitalization and sustainability pressures are increasing, requiring continuous upgrading and institutional support to ensure inclusive participation.

15.7.3.1 Mexico: Export Competitiveness and Structural Duality

Mexico has emerged as one of the world's leading horticultural exporters, ranking among the top global suppliers of avocados, berries, tomatoes, peppers, and other high-value crops. Mexican horticultural exports reached around $15 billion in 2025, with the United States absorbing over 80% of exports. Its competitiveness is based on agroecological diversity, year-round production, geographic proximity to North American markets, and strong integration under NAFTA and USMCA frameworks (Zahniser et al. 2023; Servicio de Información Agroalimentaria y Pesquera (SIAP) 2025).

Mexican supply chains exhibit a dual structure. Advanced agribusiness firms and organized producer groups operate highly coordinated export systems (see Fig. 15.6), while many small- and medium-scale producers remain in traditional channels with limited market access and bargaining power.

Compliance with international standards has driven modernization. Investments in cold chains, traceability systems, and phytosanitary infrastructure have enhanced

Fig. 15.6 Large-scale Mexican vegetable producers primarily supplying the United States market, illustrating the importance of export-oriented production systems, advanced technology, and vertical coordination in modern horticultural supply chains

export reliability. At the same time, environmental sustainability, labor conditions, and climate risks are becoming increasingly central to competitiveness.

Digitalization is expanding through traceability systems and logistics platforms, although adoption remains uneven. Persistent challenges include market concentration, infrastructure gaps, and limited access to finance. Addressing these constraints requires integrated strategies combining institutional strengthening, digital inclusion, and sustainability governance.

15.7.4 Europe: Technological Leadership and Regulatory Power

Europe plays a dual role as a major producer and a global hub for horticultural trade, logistics, and governance (OECD-FAO 2022). Countries such as Spain and the Netherlands combine high-technology production systems with advanced logistics and strong integration with retail chains (see Fig. 15.7).

European supply chains are characterized by coordinated governance structures, stringent standards, and advanced digital systems for traceability and logistics (Trienekens 2011). The region also leads in sustainability regulation, with ESG standards shaping production practices both within Europe and in exporting countries.

Fig. 15.7 High-technology greenhouse production of chrysanthemum in the Kingdom of the Netherlands

Table 15.2 Comparative regional dynamics in horticultural supply chains

Region	Key features	Governance	Upgrading	Main constraint
Asia	Rapid growth, smallholder base	Mixed (traditional + modern retail)	Process and product upgrading	Fragmentation, weak logistics
Africa	Export-oriented, smallholder dominance	Polycentric, buyer-driven	Limited upgrading	Infrastructure and compliance costs
Latin America	Strong exports, institutional coordination	Buyer-driven (exports), relational (domestic)	Process and functional upgrading	Inequality, environmental pressures
Europe	High standards, strong cooperatives	Retail-led, cooperative governance	Advanced upgrading	Retail concentration, high costs

Cluster dynamics and innovation systems support competitiveness, but challenges such as climate variability, labor shortages, and rising costs are increasing. In response, firms are intensifying automation, diversification, and sustainability investments.

15.7.5 Synthesis: Divergent Pathways and Common Challenges

Across regions, horticultural supply chains exhibit diverse trajectories shaped by institutional contexts, market structures, and technological capabilities. Despite these differences, common challenges emerge, including power asymmetries, compliance costs, environmental pressures, and uneven access to digital technologies.

Comparative analysis highlights that successful upgrading depends not only on market integration but also on organizational capacity, institutional support, and governance arrangements that enable equitable value distribution. These findings reinforce the need for integrated strategies that combine competitiveness, sustainability, and inclusion in global horticultural supply chains. Table 15.2 provides a comparative overview of regional dynamics.

15.8 Climate Risks, Sustainability, and Inclusive Transformation

Horticultural supply chains face increasing exposure to climate variability, environmental constraints, and evolving sustainability requirements. Rather than reiterating conceptual frameworks, this section focuses on mechanisms through which climate and sustainability pressures affect production systems, market participation, and value distribution.

15.8.1 Climate Impacts on Production Systems

Climate variability affects horticulture through changes in temperature patterns, rainfall variability, and the increased frequency of extreme events. These factors influence crop yields, product quality, and production stability, particularly in systems dependent on narrow climatic windows. Empirical evidence indicates that climate stress can disrupt production cycles and increase variability in supply, with implications for both domestic and export markets (IPCC 2023; FAO 2025).

15.8.2 Water Constraints and Resource Efficiency

Water availability has emerged as a critical constraint for horticultural production. In many regions, increasing competition for water resources and declining availability are affecting irrigation-dependent systems. Inefficient irrigation practices further exacerbate these challenges, reducing resilience to climate variability. Improving water-use efficiency through technological and management innovations is therefore central to adaptation strategies (Frija et al. 2009; Mancosu et al. 2015; Dinar et al. 2019).

15.8.3 Sustainability Standards and Market Access

Sustainability requirements increasingly shape access to high-value markets. Compliance with environmental, social, and food safety standards has become a prerequisite for participation in global horticultural chains. While these standards can enhance product differentiation and market access, they also impose compliance costs and institutional requirements that may limit participation by smaller producers (van Wijk 2014; Akrong et al. 2023; Boonaert and Maertens 2023). As a result, sustainability-driven market integration remains uneven across regions and production systems.

15.8.4 ESG and Evolving Market Requirements

Environmental, social, and governance (ESG) criteria are expanding beyond traditional certification schemes to include broader metrics of sustainability performance. Buyers, retailers, and investors increasingly require transparency and accountability in production practices. Although ESG frameworks can promote improved environmental and social outcomes, they also introduce additional reporting burdens and reinforce barriers to entry for less-resourced actors (Singh et al. 2024). Table 15.3 summarizes the main sustainability standards and their implications.

Table 15.3 Sustainability standards and market implications in horticulture

Standard	Focus	Market effect	Inclusion challenge
GlobalG.A.P.	Food safety, traceability	Access to high-value export markets	High compliance costs for smallholders
Organic	Soil health, no synthetic inputs	Price premiums, niche markets	Transition costs and yield variability
Fair Trade	Social standards, price floor	Income stability potential	Requires collective organization
ESG Compliance	Environmental and governance metrics	Increasing buyer and investor requirements	High data and reporting barriers

15.8.5 Digitalization for Sustainability

Digital technologies support sustainability objectives by enabling traceability, monitoring, and more efficient resource use. However, as discussed in Sect. 15.4, access to these technologies is uneven, and their benefits are not uniformly distributed. Consequently, digitalization may simultaneously enhance sustainability outcomes and deepen existing inequalities if complementary support mechanisms are not in place (McFadden et al. 2022; Chowdhury et al. 2023; Nwangwu et al. 2024; Abdulai et al. 2025).

15.8.6 Toward Inclusive and Resilient Supply Chains

Addressing climate and sustainability challenges requires integrated approaches that combine technological innovation, institutional support, and collective action. Investments in infrastructure, capacity building, and governance mechanisms are essential to enhance resilience and ensure more inclusive participation (OECD 2021; Akrong et al. 2023; Abdulai et al. 2025). Strengthening producer organizations and facilitating access to markets can help mitigate asymmetries and improve value distribution across supply chains (Hellin et al. 2009; Verhofstadt and Maertens 2015; Wossen et al. 2017).

Overall, climate and sustainability pressures are reshaping horticultural supply chains by influencing production conditions, market requirements, and governance structures. Effective responses depend on aligning environmental objectives with economic viability and social inclusion, ensuring that transitions toward sustainability do not exacerbate existing inequalities.

15.9 Conclusions

Horticultural supply chains are undergoing profound transformations driven by the interaction of global market integration, technological change, and increasing environmental constraints. This chapter has shown that these transformations cannot be understood through a single analytical lens but rather require an integrated perspective that combines governance structures, coordination mechanisms, and sustainability considerations.

A central finding is that power asymmetries remain a defining feature of horticultural markets, particularly within buyer-driven chains where standards, market access, and value capture are largely controlled by downstream actors. While these structures have enabled significant process and product upgrading, they have also limited the capacity of producers—especially smallholders—to capture higher value and achieve more equitable participation.

The emergence of digital technologies and platform-based coordination is further reshaping these dynamics. Although digitalization enhances efficiency, traceability, and market connectivity, it also shifts control toward actors that manage data and digital infrastructure. This reconfiguration of governance introduces new forms of asymmetry that extend beyond traditional market relationships.

At the same time, climate variability and sustainability requirements are redefining the conditions under which horticultural production and commercialization take place. Increasing pressures related to water availability, environmental performance, and compliance with evolving standards and ESG criteria are raising both the complexity and the cost of participation in high-value markets.

These transformations highlight a critical tension between competitiveness and inclusion. While global integration and technological innovation create opportunities for upgrading, they also risk reinforcing structural inequalities if access to resources, information, and markets remains uneven. Addressing this tension requires coordinated interventions that go beyond firm-level strategies.

Policy and institutional responses should therefore focus on improving enabling conditions for inclusive upgrading. This includes investments in infrastructure and water management, strengthening producer organizations, facilitating access to digital tools and market information, and designing regulatory frameworks that ensure more balanced governance structures.

Ultimately, the transition toward more sustainable and resilient horticultural supply chains will depend on the ability to align economic efficiency with environmental sustainability and social inclusion. Future research should further explore how different governance configurations and policy interventions can support this alignment across diverse regional contexts.

Acknowledgments The authors would like to express their sincere appreciation to **Fatima Paola Amaro-Rodríguez** for her valuable support in the preparation and refinement of the case study on the cooperative La Palma–Granada. We also gratefully acknowledge **María Guadalupe Meras-Rodríguez** and **Thelma Guadalupe Valdez-Reyes** for their contributions to the development of the case study focused on the TROPS cooperative.

We thank **Universidad Autónoma Chapingo** for its institutional support, which made possible academic visits to Fruit Attraction and to key horticultural production and postharvest facilities in Mexico, Spain, and the Kingdom of the Netherlands. These field experiences provided essential empirical insights that enriched the comparative and applied perspective of this chapter.

The authors acknowledge the use of ChatGPT (OpenAI) as a supporting tool for language refinement, structural organization, and editorial consistency during the preparation of this chapter. The authors also acknowledge the use of SciSpace for facilitating the identification and retrieval of relevant academic literature. All referenced sources were critically assessed and verified by the authors. Full responsibility for the content, interpretations, and conclusions presented in this chapter rests solely with the authors.

Note to the Editor All photographs included in this chapter (Figs. 15.1–15.7) are original and were taken by the first author. The author holds full copyright of all images and grants permission for their reproduction and publication in this volume.

References

Abdulai AR, Gibson R, Fraser EDG (2025) Digitalizing traditional agri-food supply chains in developing economies: a pathway to regulated sustainable governance. Sustain Futur 10:101356. https://doi.org/10.1016/j.jrurstud.2023.103019

Akrong R, Akorsu AD, Jha P, Agyenim JB (2023) Towards environmental sustainability: the role of certification in the adoption of climate-smart agricultural practices among Ghanaian mango farmers. Cogent Food Agric 9:2174482. https://doi.org/10.1080/23311932.2023.2174482

Alford M, Pasquali GP, Barrientos S, Opondo M (2024) Governance of intersecting value chains in an era of polycentric trade: the case of Kenyan fresh fruit and vegetables. J Econ Geogr 24:663–678. https://doi.org/10.1093/jeg/lbae018

Barrett CB, Bachke ME, Bellemare MF et al (2012) Smallholder participation in contract farming: comparative evidence from five countries. World Dev 40:715–730. https://doi.org/10.1016/j.worlddev.2011.09.006

Barrett CB, Reardon T, Swinnen J, Zilberman D (2022) Agri-food value chain revolutions in low- and middle-income countries. J Econ Lit 60:1316–1377. https://doi.org/10.1257/jel.20201539

Berdegué JA, Balsevich F, Flores L, Reardon T (2005) Central American supermarkets' private standards of quality and safety in procurement of fresh fruits and vegetables. Food Policy 30:254–269. https://doi.org/10.1016/j.foodpol.2005.05.003

Boonaert E, Maertens M (2023) Voluntary sustainability standards and farmer welfare: the pathways to success? Food Policy 121:102543. https://doi.org/10.1016/J.FOODPOL.2023.102543

Brenton P, Ferrantino MJ, Maliszewska M (2022) Reshaping global value chains in light of COVID-19. Implications for trade and poverty reduction in developing countries. The World Bank Group, Washington, DC

Cao Y, Yi C, Wan G et al (2022) An analysis on the role of blockchain-based platforms in agricultural supply chains. Transp Res E Logist Transp Rev 163. https://doi.org/10.1016/j.tre.2022.102731

Chowdhury MR, Sourav MSU, Bin Sulaiman R (2023) The role of digital agriculture in transforming rural areas into smart villages. In: Technology and talent strategies for sustainable smart cities: digital futures, vol 7, pp 57–78. https://doi.org/10.1108/978-1-83753-022-920231003

Dinar A, Tieu A, Huynh H (2019) Water scarcity impacts on global food production. Glob Food Sec 23:212–226. https://doi.org/10.1016/J.GFS.2019.07.007

Dolan C, Humphrey J (2000) Governance and trade in fresh vegetables: the impact of UK supermarkets on the African horticulture industry. J Dev Stud 37:147–176. https://doi.org/10.1080/713600072

FAO IUW and WHO (2025) The state of food security and nutrition in the world 2025. FAO; IFAD; UNICEF; WFP; WHO, Rome

Fischer E, Qaim M (2012) Linking smallholders to markets: determinants and impacts of farmer collective action in Kenya. World Dev 40:1255–1268. https://doi.org/10.1016/j.worlddev.2011.11.018

Frija A, Chebil A, Speelman S et al (2009) Water use and technical efficiencies in horticultural greenhouses in Tunisia. Agric Water Manag 96:1509–1516. https://doi.org/10.1016/J.AGWAT.2009.05.006
Gereffi G, Lee J (2014) Economic and social upgrading in global value chains and industrial clusters: why governance matters. J Bus Ethics 133:25–38. https://doi.org/10.1007/S10551-014-2373-7
Gereffi G, Humphrey J, Sturgeon T (2005) The governance of global value chains. Rev Int Polit Econ 12:78–104. https://doi.org/10.1080/09692290500049805
Giovannucci D, Ponte S (2005) Standards as a new form of social contract? Sustainability initiatives in the coffee industry. Food Policy 30:284–301. https://doi.org/10.1016/J.FOODPOL.2005.05.007
Hellin J, Lundy M, Meijer M (2009) Farmer organization, collective action and market access in Meso-America. Food Policy 34:16–22. https://doi.org/10.1016/j.foodpol.2008.10.003
Henson S, Humphrey J (2010) Understanding the complexities of private standards in global agri-food chains as they impact developing countries. J Dev Stud 46:1628–1646. https://doi.org/10.1080/00220381003706494
Henson S, Reardon T (2005) Private agri-food standards: implications for food policy and the agri-food system. Food Policy 30:241–253. https://doi.org/10.1016/J.FOODPOL.2005.05.002
HLPE (2017) Nutrition and food systems. A report by the high level panel of experts the high level panel of experts on food security and nutrition of the Committee on World Food Security, Rome
Humphrey J, Memedovic O (2006) Global value chains in the Agrifood Sector, Vienna
IPCC (2023) AR6 Synthesis Report: Climate Change 2023 — IPCC. https://www.ipcc.ch/report/sixth-assessment-report-cycle/. Accessed 13 Feb 2026
Kamilaris A, Fonts A, Prenafeta-Boldύ FX (2019) The rise of blockchain technology in agriculture and food supply chains. Trends Food Sci Technol 91:640–652. https://doi.org/10.1016/j.tifs.2019.07.034
Kenney M, Zysman J (2016) The rise of the platform economy. Issues Sci Technol XXXII:61–69
Kohli AK, Jaworski BJ (1990) Market orientation: the construct, research propositions, and managerial implications. J Mark 54:1–18. https://doi.org/10.1177/002224299005400201
Krishnan A (2018) The origin and expansion of regional value chains: the case of Kenyan horticulture. Global Netw 18:238–263. https://doi.org/10.1111/glob.12162
Krishnan A, De Marchi V, Ponte S (2023) Environmental upgrading and downgrading in global value chains: a framework for analysis. Econ Geogr 99:25–50. https://doi.org/10.1080/00130095.2022.2100340
Loconto A, Dankers C (2014) Impact of international voluntary standards on smallholder market participation in developing countries
López-Pimentel JC, Alcaraz-Rivera M, Granillo-Macías R, Olivares-Benitez E (2022) Traceability of Mexican Avocado Supply Chain: a microservice and blockchain technological solution. Sustainability 14:14633. https://doi.org/10.3390/su142114633
Mancosu N, Snyder RL, Kyriakakis G, Spano D (2015) Water scarcity and future challenges for food production. Water 7:975–992. https://doi.org/10.3390/W7030975
Markelova H, Meinzen-Dick R, Hellin J, Dohrn S (2009) Collective action for smallholder market access. Food Policy 34:1–7. https://doi.org/10.1016/j.foodpol.2008.10.001
McFadden J, Casalini F, Griffin T, Antón J (2022) The digitalisation of agriculture: a literature review and emerging policy issues. OECD food, agriculture and fisheries paper. https://doi.org/10.1787/285cc27d-en
Meemken EM (2020) Do smallholder farmers benefit from sustainability standards? A systematic review and meta-analysis. Glob Food Sec 26:100373. https://doi.org/10.1016/J.GFS.2020.100373
Meemken EM, Bellemare MF (2020) Smallholder farmers and contract farming in developing countries. Proc Natl Acad Sci USA 117:259–264. https://doi.org/10.1073/pnas.1909501116
Narrod C, Roy D, Okello J et al (2009) Public–private partnerships and collective action in high value fruit and vegetable supply chains. Food Policy 34:8–15. https://doi.org/10.1016/j.foodpol.2008.10.005

Narver JC, Slater SF (1990) The effect of a market orientation on business profitability. J Mark 54:20–35. https://doi.org/10.1177/002224299005400403
Nwangwu KN, Onyenekwe CS, Opata PI et al (2024) Can digital technology promote market participation among smallholder farmers? Int Food Agribusin Manag Rev 27:706–728. https://doi.org/10.22434/IFAMR2023.0065
OECD (2021) Building more resilient and sustainable global value chains through responsible business conduct. OECD Publishing
OECD (2025) Agricultural policy monitoring and evaluation 2025: making the most of the trade and environment Nexus in agriculture. Agricultural Policy Monitoring and Evaluation 2025. https://doi.org/10.1787/A80AC398-EN
OECD-FAO (2022) OECD-FAO Agricultural Outlook 2022–2031. OECD Publishing, Paris
Olson Mancur (1971) The logic of collective action : public goods and the theory of groups. Harvard University Press
Ostrom Elinor (2012) Governing the commons: the evolution of institutions for collective action. Cambridge University Press
Pasquali G, Barrientos S, Maggie O (2021) Polycentric governance of global and domestic horticulture value chains: shifting standards in Kenyan fresh fruit and vegetables. The University of Manchester, Manchester
Pérez-Mesa JC, Piedra-Muñoz L, Galdeano-Gómez E, Giagnocavo C (2021) Management strategies and collaborative relationships for sustainability in the agrifood supply chain. Sustainability 13:749. https://doi.org/10.3390/SU13020749
Ponte S, Gibbon P (2005) Quality standards, conventions and the governance of global value chains. Econ Soc 34:1–31. https://doi.org/10.1080/0308514042000329315
Ponte S, das Nair R, Chisoro S (2023) Is sustainability governance abetting inequality? Reflections from the South African wine value chain. Geoforum 147. https://doi.org/10.1016/j.geoforum.2023.103877
Prause L, Hackfort S, Lindgren M (2020) Digitalization and the third food regime. Agric Hum Values 38:641–655. https://doi.org/10.1007/S10460-020-10161-2
Qin L, Xie W, Jia P (2024) Value chain digitalization, global value chain embeddedness, and distributed innovation in value chains. Sustainability (Switzerland) 16. https://doi.org/10.3390/su16072845
Reardon T, Timmer CP, Barrett CB, Berdegué J (2003) The rise of supermarkets in Africa, Asia, and Latin America. Am J Agric Econ 85:1140–1146. https://doi.org/10.1111/J.0092-5853.2003.00520.x
Reardon T, Barrett CB, Berdegué JA, Swinnen JFM (2009) Agrifood industry transformation and small farmers in developing countries. World Dev 37:1717–1727. https://doi.org/10.1016/J.WORLDDEV.2008.08.023
Reardon T, Echeverria R, Berdegué J et al (2019) Rapid transformation of food systems in developing regions: highlighting the role of agricultural research & innovations. Agric Syst 172:47–59. https://doi.org/10.1016/j.agsy.2018.01.022
Servicio de Información Agroalimentaria y Pesquera (SIAP) (2025) Panorama Agroalimentario 2025. Ciudad de Mexico
Singh A, Dwivedi A, Agrawal D et al (2024) Can sustainable and digital objectives synchronize? A study of ESG activities for digital supply chains using multi-methods. Bus Strateg Environ 33:8413–8435. https://doi.org/10.1002/BSE.3925
Srnicek N (2017) Platform Capitalism. Polity Press, Cambridge
Swinnen JFM, Maertens M (2007) Globalization, privatization, and vertical coordination in food value chains in developing and transition countries. Agric Econ 37:89–102. https://doi.org/10.1111/J.1574-0862.2007.00237.X
Thompson WJ, Varma V, Joerin J et al (2023) Smallholder farmer resilience to extreme weather events in a global food value chain. Clim Chang 176:152. https://doi.org/10.1007/s10584-023-03586-1

Tran D, Goto D (2019) Impacts of sustainability certification on farm income: evidence from small-scale specialty green tea farmers in Vietnam. Food Policy 83:70–82. https://doi.org/10.1016/J.FOODPOL.2018.11.006

Trienekens JH (2011) Agricultural value chains in developing countries a framework for analysis. Int Food Agribusin Manag Assoc 14(2):51–82

van Wijk MT (2014) From global economic modelling to household level analyses of food security and sustainability: how big is the gap and can we bridge it? Food Policy 49:378–388. https://doi.org/10.1016/J.FOODPOL.2014.10.003

Verhofstadt E, Maertens M (2015) Can agricultural cooperatives reduce poverty? Heterogeneous impact of cooperative membership on farmers' welfare in Rwanda. Appl Econ Perspect Policy 37:86–106. https://doi.org/10.1093/aepp/ppu021

Williamson OE (1985) The economic institutions of capitalism. Firms, market, relational contracting. The Free Press, Macmillan Inc

Wolfert S, Ge L, Verdouw C, Bogaardt MJ (2017) Big data in smart farming—a review. Agric Syst 153:69–80. https://doi.org/10.1016/j.agsy.2017.01.023

World Bank (2019) Enabling the Business of Agriculture. https://eba.worldbank.org/en/eba. Accessed 7 Feb 2026

Wossen T, Abdoulaye T, Alene A et al (2017) Impacts of extension access and cooperative membership on technology adoption and household welfare. J Rural Stud 54:223–233. https://doi.org/10.1016/J.JRURSTUD.2017.06.022

Zahniser S, Avendaño Ruiz BD, Astill G (2023) How Mexico's horticultural export sector responded to the Food Safety Modernization Act. USDA. https://doi.org/10.32747/2023.8134140.ERS

Climate Resilience in Horticulture: Future-Ready Horti-farming

16

Debapriya Sarkar, Ritoban Pandit, Raj Kumar, Debarpita Datta Ray, Arju Ali Khan, Anwesha Das, Arindam Das, and Manas Kumar Pandit

Abstract

Climate change is a concerning issue that has manifested as a significant threat to horticulture, disrupting productivity and resilience through rising temperatures, weather extremes, erratic rainfall, and shifting pest and disease patterns. With increasing climate variability across regions such as South and Southeast Asia, including India, traditional open-field farming is becoming increasingly vulnerable and particularly concerning, given rapid population growth and declining cultivable land. As a result, ensuring food security demands a strategic shift toward more climate-resilient and sustainable farming practices. Crucial adaptive strategies like the development of stress-tolerant crop varieties, improved irrigation efficiency, integrated nutrient and pest management, and the adoption of smart farming tools like remote sensing, AI in decision-making, and

D. Sarkar · R. Kumar · A. Das (✉)
Department of Vegetable and Spice Crops, Uttar Banga Krishi Viswavidyalaya, Pundibari, West Bengal, India
e-mail: arindamdas@ubkv.ac.in

R. Pandit
Department of Agricultural Meteorology and Physics, Bidhan Chandra Krishi Viswavidyalaya, Mohanpur, Haringhata, West Bengal, India

D. D. Ray
Department of Genetics and Plant Breeding, Bidhan Chandra Krishi Viswavidyalaya, Mohanpur, Haringhata, West Bengal, India

A. A. Khan · A. Das
Department of Fruit Science, Bidhan Chandra Krishi Viswavidyalaya, Mohanpur, Haringhata, West Bengal, India

M. K. Pandit
Department of Vegetable Science, Bidhan Chandra Krishi Viswavidyalaya, Mohanpur, Haringhata, West Bengal, India

A. Das, A. B. Sharangi (eds.), *Remunerative Horticulture*,
https://doi.org/10.1007/978-981-92-1007-7_16

Internet of Things (IoT) have to be tried in a nation like India, emerging toward a better economy. The integration of traditional knowledge with modern innovations, including genomics, phenotyping, and decision-support systems, is a necessary pathway to enhance adaptability and productivity under changing climatic conditions. This chapter presents a detailed evaluation of climate resilience in horticulture, highlighting the integration of indigenous knowledge with advanced technologies such as protected cultivation, digital farming, and the use of renewable resources, while also drawing attention to the socioeconomic and policy dimensions essential for widespread adoption. The chapter also addresses the role of agro-biodiversity, traditional knowledge, and policy interventions in building robust and sustainable horticultural systems. Climate-resilient horticulture practices offer a viable solution to ensure food security for the growing population, economic stability, reduce drudgery, enhance farmers' livelihoods, and support environmental sustainability and ecological well-being in a changing climate.

Keywords

Climate resilience · Sustainable farming · Resource efficiency · Adaptive practices · Technology enabled Horticulture

16.1 Introduction

Climate resilience refers to the capacity of social, ecological, and production systems to anticipate, absorb, adapt to, and recover from climate-induced shocks while maintaining essential functions and structures. The Intergovernmental Panel on Climate Change defines resilience as the ability of interconnected systems to cope with hazardous events, trends, and disturbances while reorganizing to sustain productivity, biodiversity, and ecosystem services (Summary for Policymakers 2023). In agricultural contexts, climate resilience extends beyond short-term coping mechanisms to include long-term adaptive and transformative responses that enhance system stability under persistent climatic stress.

Horticultural produce of fruits, vegetables, ornamentals, spices, medicinal, and plantation crops plays a critical role in global food and nutritional security, income generation, and rural employment. Horticultural crops are among the most climate-sensitive agricultural commodities due to their narrow temperature and moisture optima, shallow root systems, and high dependence on precise management practices (Zhao et al. 2023). Observations by the IPCC (2021) indicate rising global temperatures, altered precipitation patterns, increased frequency of extreme weather events, and shortened cold seasons, all of which disproportionately affect horticultural productivity and quality.

Climate-induced stresses directly influence physiological processes such as photosynthesis, respiration, transpiration, and source-sink relationships, thereby affecting yield, quality, and postharvest performance (Mohanty et al. 2015). Extreme events like heat waves, droughts, floods, and unseasonal rainfall disrupt flowering, fruit set, pigment synthesis, and nutritional composition while accelerating pest and disease outbreaks (Sharma et al. 2022). While short-term impacts are often visible at the seasonal scale, the long-term consequences of climate change on yield stability, cropping patterns, and regional suitability pose far greater risks to horticultural sustainability.

Given the background, climate-resilient horticulture has emerged as a central pillar of climate-smart agriculture, promoted globally through frameworks such as the Paris Agreement and the Sustainable Development Goals (Grasham et al. 2021). Climate-resilient horti-farming integrates improved crop genetics, resource-efficient technologies, diversified production systems, and supportive policy environments to buffer climate risks while enhancing productivity. This chapter focuses on understanding climate impacts on horticulture and outlines pathways toward future-ready horticultural systems capable of sustaining production under increasing climatic uncertainty.

16.2 Climate Change and Its Impacts on Horticulture

Climate change affects horticultural crops through complex interactions among temperature, rainfall, atmospheric composition, and biotic stress factors. Both short-term weather variability and long-term climatic trends shape crop performance, nutritional quality, and economic returns. Elevated temperatures, erratic rainfall, increased frequency of droughts and floods, and rising atmospheric CO_2 concentrations collectively influence growth, phenology, and biotic interactions across horticultural systems (Ray et al. 2015; Ainsworth and Long 2005).

Long-term climate change is associated with declining nutrient availability, altered mineral uptake, and reduced produce quality, while short-term stress events disrupt stomatal conductance, transpiration, and assimilate partitioning (Sharma et al. 2022). High temperatures adversely affect net photosynthesis and fruit biochemistry, leading to disorders such as sunburn and cracking in fruits (Manzoor et al. 2024). In vegetables, climate variability has pronounced effects on color development, vitamin content, and antioxidant properties, thereby influencing nutritional quality (Gruda et al. 2025). Increased pest and disease pressure under warmer and more humid conditions has resulted in excessive pesticide use, raising concerns related to environmental contamination and human health (Pugliese et al. 2024).

16.2.1 Temperature Variability and Crop Responses

Temperature is a dominant climatic factor governing growth, flowering, and fruit development in horticultural crops. Heat stress reduces photosynthetic efficiency, alters metabolic pathways, and disrupts reproductive processes (Hatfield et al. 2011). Elevated temperatures increase sunburn and cracking in temperate fruits such as apple, apricot, and cherry, while high temperatures during fruit maturity lead to cracking and scorching in litchi (Kumar and Kumar 2007). Citrus fruit set is particularly sensitive to heat stress, with extreme temperatures causing floral damage and reduced fertilization (Rao and Laxman 2013).

Temperate fruit and nut crops require adequate winter chilling to break dormancy and ensure synchronized flowering. Global warming threatens winter chill accumulation, especially in marginal growing regions, leading to irregular flowering and yield reductions (Luedeling et al. 2001). Rising night-time temperatures increase respiration rates, reducing assimilate availability for reproductive sinks and resulting in smaller fruits and inferior quality in crops such as tomato and cucurbits (Wahid et al. 2007).

16.2.2 Elevated CO_2 and Crop Responses

Atmospheric CO_2 concentrations have increased markedly over recent decades, influencing crop growth through enhanced photosynthetic rates and biomass accumulation (Le Quere et al. 2018). While CO_2 fertilization can increase canopy size and yield potential, its benefits are often constrained by interactions with temperature stress, water availability, nutrient limitations, and pest pressure (McGrath and Lobell 2013). Denser canopies under elevated CO_2 create humid microclimates conducive to diseases such as powdery mildew, rusts, and blights (Pangga et al. 2013).

A critical concern associated with elevated CO_2 is the "carbon dilution effect," whereby increased carbohydrate accumulation reduces mineral nutrient concentrations in edible plant parts. Studies have reported significant declines in micronutrients such as zinc, iron, calcium, and magnesium in vegetables and fruits under elevated CO_2 and ozone conditions (Scheelbeek et al. 2018), posing risks to nutritional security.

16.2.3 Reduced Rainfall and Water Stress

Altered rainfall patterns and prolonged dry spells significantly affect horticultural productivity, particularly in rainfed systems. In apple-growing regions of Himachal Pradesh, reduced monsoon rainfall and hotter summers have increased canker incidence, while decreased winter rainfall has reduced apple scab severity by limiting pathogen sporulation (Sharma 2012). Drought stress also influences viral disease dynamics, as observed with beet yellows virus under moisture-limited conditions (Clover et al. 1999). Water stress weakens plant defense mechanisms, increasing susceptibility to pests and diseases while reducing yield stability (Farooq et al. 2009).

16.2.4 Insect–Pests and Disease Outbreaks

Climate change strongly influences pest and disease dynamics by accelerating insect development, increasing the number of generations and expanding geographical ranges. Many pests are now able to complete additional life cycles due to extended periods of favorable temperatures, intensifying crop damage. The diamondback moth, a major pest of Brassica vegetables, has shown increased severity across subtropical and temperate regions under warming climates (Deuter 2008).

Changes in temperature and rainfall patterns also disrupt pollinator activity, leading to poor fruit set in fruits, vegetables, and nuts. Moreover, climate-induced stress expands host ranges for several pests, further complicating management strategies (Singh et al. 2010). There is an urgent need for climate-resilient pest management approaches integrated with broader adaptive strategies.

16.3 Climate-Resilient Crop Management Practices and Good Agricultural Practices (GAP)

Climate-resilient crop management integrates stress-tolerant cultivars, efficient resource utilization, and adaptive production systems to sustain horticultural productivity under rising temperatures, erratic rainfall, and increasing soil degradation. Within this framework, Good Agricultural Practices (GAP) provides standardized guidelines to ensure food safety, environmental sustainability, and economic viability through certification systems such as GLOBAL GAP and India GAP. The convergence of climate-resilient agronomy with GAP principles forms the foundation of climate-smart, market-oriented, and future-ready horticultural systems (Kumar et al. 2013).

Core resilience strategies include improved water management, soil and nutrient conservation, crop diversification, and precision input application. Micro-irrigation technologies, particularly drip irrigation, improve water-use efficiency by 30–50% in vegetable crops (Table 16.1) such as tomato and bell pepper while stabilizing yields under moisture stress (Patel et al. 2023). Rainwater harvesting and on-farm storage structures further enhance water availability in rainfed regions. Conservation agriculture practices such as reduced tillage, residue retention, and mulching improve soil organic carbon, moisture retention, and erosion control. Integrated nutrient management and site-specific fertilization reduce nutrient losses and sustain long-term soil health. Diverse horticultural cropping systems involving short-duration cultivars and agroforestry act as a buffer to climate risks, and do so by regulating microclimates and stabilizing farm income. National initiatives such as NICRA and extension through Krishi Vigyan Kendras have demonstrated income gains of 20–30% through the adoption of climate-resilient practices, underscoring their role in transforming climate vulnerability into adaptive capacity (Steduto et al. 2012).

16.3.1 Crop Diversification and Mixed Cropping

Crop diversification and mixed cropping enhance climate resilience by distributing production risk across crops with contrasting growth habits, rooting depths, and maturity periods. Diversified horticultural systems improve yield stability under

Table 16.1 Different climate resilient crop management practices

Practice category	Technique	Climate risk mitigated/benefit	Crop examples	References
Water management	Drip irrigation + mulch	40% water saving, reduced heat stress	Tomato, bell pepper	Kirnak et al. (2003) and Adombilla (2023)
Soil health	Zero tillage + residue retention	Moisture conservation, erosion control	Okra-based rotations	Durga and Anith (2021)
Nutrient management	Site-specific N management	25% fertilizer reduction, less leaching	Brinjal, chili	Deshmukh et al. (2009) and Nisar et al. (2025)
Cropping systems	Diversification/ intercropping	Yield stability under variability	Cowpea + baby corn	Bijarnia et al. (2024)
Varietal choice	Stress-tolerant cultivars	Drought, heat, disease tolerance	Tomato (Arka Rakshak)	Mooventhan et al. (2019)

Table 16.2 Benefits of mixed cropping systems in horticulture

Cropping system	LER	Key climate resilience benefit	References
Okra + cowpea	1.4	Drought buffering, nitrogen fixation	Maduwanthi and Karunarathna (2019)
Tomato + marigold	1.3	Pest suppression, microclimate moderation	Chaturvedi et al. (2024)
Dragon fruit + legumes	1.5	Improved water-use efficiency, soil cover	Yusoff et al. (2008)

erratic rainfall, temperature extremes, and pest pressure while enhancing soil health and economic returns through staggered harvests. Legume inclusion improves biological nitrogen fixation, soil structure, and moisture retention, often reducing erosion by 30–50% (Mihrete and Mihretu 2025).

Horticulture-based diversification models such as vegetable–legume intercrops, fruit-based agri-horti systems, and plantation crop intercropping are particularly effective in semi-arid and rainfed regions. Land equivalent ratio (LER) analyses consistently demonstrate superior land-use efficiency and climatic buffering in diversified systems (Table 16.2). Successful implementation requires careful crop compatibility, optimized row ratios, and integration with integrated nutrient management practices. NICRA-validated diversification models have reported significant gains in yield stability and resilience across climate-vulnerable districts (Bhandari et al. 2014).

16.3.2 Improved Irrigation and Water Management

Improved irrigation and water management constitute a cornerstone of climate-resilient horticulture under increasing water scarcity driven by erratic rainfall, rising evapotranspiration, and competition for freshwater resources (Jain et al. 2024).

Table 16.3 Advanced irrigation strategies for climate smart crop production

Technique	Water saving (%)	Crop examples	Recent advancements	Reference
Drip irrigation with fertigation	40–60	Tomato, okra, dragon fruit	Solar-powered systems, PMKSY expansion	Sharma and Kaushal (2015)
Sprinkler irrigation	30–40	Fenugreek, garlic, potato	Orchard-scale pivot systems	Patel and Parmar (2020)
Subsurface drip irrigation	50	Grapes, pomegranate	Reduced evaporation, IoT integration	Bhavitha et al. (2024)
Sensor- and AI-based irrigation	25–35	Bell pepper, ginger	Real-time decision-support tools	Stojanova et al. (2025)
Mulching with basin irrigation	50–70	Aonla, ber	Biodegradable mulches	Krishna et al. (2018)

Precision irrigation technologies save 30–60% water while increasing yields by 20–40% in fruits and vegetables (Lakhiar et al. 2024). Modern systems deliver water close to crop evapotranspiration requirements, minimizing conveyance and application losses common in surface irrigation (Hiekal 2020).

Advances from conventional drip and sprinkler systems to subsurface drip, deficit irrigation, and sensor-based automated scheduling have strengthened adaptive capacity (Table 16.3). Integration of soil-moisture sensors, IoT platforms, and AI-driven decision tools enables real-time irrigation and fertigation, reducing water and nutrient losses. Mulching combined with drip irrigation significantly reduces evaporative losses, while policy-driven expansion under national missions has accelerated adoption. These innovative irrigation practices are efficient and are the central pillar of GAP-compliant, climate-resilient horticulture (Jaiswal et al. 2025).

16.3.3 Soil Health and Fertility Management

Soil health governs water retention, nutrient availability, and biological activity under climate stress, making it a foundational pillar of climate-resilient horticulture. Rising temperatures, erratic rainfall, salinity, and intensive cultivation have accelerated soil degradation, reducing the resilience of horticultural systems. Healthy soils buffer climatic extremes by improving infiltration, stabilizing nutrient cycling, and supporting diverse microbial communities. A 1% increase in soil organic matter can sequester approximately 20 t CO_2 ha^{-1}, contributing to climate mitigation alongside adaptation (Lal 2004).

Integrated soil fertility strategies, including organic amendments, green manuring, biofertilizers, and conservation tillage, enhance water-use efficiency by 20–30% and improve yield stability under stress (Akchaya et al. 2025). Precision nutrient management through soil testing, site-specific fertilization, and emerging inputs such as nano-fertilizers increases nutrient-use efficiency while minimizing losses

Table 16.4 Sustainable soil management approaches in horticulture

Practice	Key input/approach	Climate stress mitigated	Horticultural crop example	References
Integrated nutrient management	FYM (10 t ha^{-1}) + biofertilizers	Drought, nutrient stress	Bell pepper, okra	Sharma et al. (2023) and Kumar et al. (2013)
Residue mulching	Crop residues (5–8 t ha^{-1})	Heat stress, moisture loss	Baby corn, ginger	Mishra et al. (2025) and Chandra and Govind (2001)
Green manuring	Dhaincha (20–25 days)	Soil erosion, low fertility	Cowpea-based systems	Singh et al. (2017)
Biofertilizers + AMF	PSB + Azotobacter + *Glomus*	Nutrient uptake, drought	Ginger, vegetables	Virginie et al. (2024) and Osillos and Nagpala (2014)
Precision nutrition	SSNM/nano-urea (foliar)	Leaching, inefficiency	Tomato, dragon fruit	Khamarunneesa et al. (2023) and Sethunath et al. (2025)

and pollution (Table 16.4). National initiatives such as NICRA and the Soil Health Card Scheme promote these practices at scale (Yadav and Mandaliya 2025).

16.3.4 Integrated Pest and Disease Management

Integrated Pest and Disease Management (IPM/IDM) is central to climate-resilient horticulture, combining cultural, biological, varietal, physical, and need-based chemical interventions. Climate change accelerates pest development and increases generation cycles by 10–25% per 1 °C rise in temperature (Skendzic et al. 2021), making exclusive reliance on chemical control unsustainable.

Climate-resilient IPM emphasizes prevention through resistant varieties, timely planting, sanitation, and crop rotation, supported by surveillance and threshold-based interventions. Pheromone traps, sticky traps, and digital advisory platforms enable early detection and climate-informed forecasting. Preference for biopesticides and biological control agents reduces chemical load and spray frequency by up to 50%. India's ICAR–National Centre for Integrated Pest Management has operationalized crop- and region-specific IPM modules and weather-based forewarning systems, resulting in reduced pesticide use and improved market acceptance of residue-safe produce (Sarangi et al. 2025).

16.3.5 Precision Horticulture and Smart Farming

Precision horticulture employs GPS-enabled machinery, remote sensing, drones, IoT sensors, and AI/ML platforms to enable site-specific management of high-value horticultural crops. By integrating weather forecast, crop model and pest surveillance, AI-driven decision-support platforms generate predictive advisories for improved crop management (Enyiukwu et al. 2025). In India, satellite-based advisories, drone-assisted spraying, and solar-powered automated irrigation systems are accelerating adoption. Inputs with spatial and temporal crop demand and precision tools improve yield stability by 20–30% while reducing water, nutrient, and pesticide losses (Table 16.5). Remote sensing facilitates early detection of heat, water, and nutrient stress, while IoT-based sensors enable automated fertigation and irrigation scheduling (Table 16.6) (Mishra et al. 2025; Jena et al. 2025).

Emerging technologies such as AI-based yield forecasting and blockchain-enabled traceability align precision horticulture with GAP compliance and export standards.

16.4 Protected Cultivation and Climate-Smart Horticultural Technologies

Protected cultivation has emerged as one of the most effective adaptation strategies for minimizing climate-induced risks in horticultural production. By modifying the crop microclimate, protected structures buffer extreme temperatures, heavy rainfall,

Table 16.5 Crop protection and management strategies in climate-resilient horticulture

Component	Key techniques	Horticultural crop examples	Climate resilience benefit	References
Cultural	Crop rotation, mulching, sanitation, trap crops	Okra against soil-borne wilt	Breaks pest cycles, reduces vector build-up	Chandra and Govind (2001) and Singh et al. (2024)
Biological	*Trichoderma*, NPV, predators (ladybird beetles)	Bell pepper against thrips	Heat-tolerant, eco-safe suppression	Messelink et al. (2013) and Beltrà et al. (2018)
Varietal	Resistant/tolerant hybrids (e.g., Pusa Sadabahar, Arka Rakshak)	Tomato against *Tuta absoluta*	Stable resistance under heat stress	Balyan et al. (2020) and Mishra et al. (2019)
Physical	Sticky traps, light traps, netting	Chili against fruit borer	Non-chemical, low residue	Hossain et al. (2020) and Sunitha (2018).
Chemical (need-based)	Botanicals (neem), Bt, biopesticides	Ginger against rhizome rot	Resistance management, reduced residues	Marwan et al. (2023) and Sudha (2024).

Table 16.6 Examples of different technologies and their applications in agriculture, their benefits in climate resilience, and horticultural crop example

Technology	Application	Climate resilience gain	Horticultural crop example	References
Multispectral drones	Early detection of water and heat stress	Improved irrigation timing, 20% water saving	Bell pepper	Al-Najadi et al. (2025)
Soil sensors + VRT	Precision fertigation	30% reduction in nutrient losses	Okra	Raj et al. (2024)
GPS-enabled automation	Uniform planting and orchard management	Reduced fuel use and operational overlap	Dragon fruit	Mondol et al. (2025)

Table 16.7 Different protected structures and their benefits in climate mitigation

Structure type	Climate stress mitigated	Crop examples	Yield advantage	Reference
Polyhouse	Heat, rainfall, wind	Capsicum, tomato	3–5 times	Ali (2024)
Shade-net house	High radiation	Leafy vegetables	2–3 times	Barche et al. (2015)
Low tunnel	Cold and frost	Cucurbits	Early harvest	Barche et al. (2015)
Naturally ventilated greenhouse	Heat stress	Cucumber	Energy-efficient	Kumar et al. (2024)

The crop suitable for each structure and its advantages in terms of yield have also been shown

hailstorms, and high wind speeds, thereby ensuring stable yields and superior produce quality. In climate-vulnerable regions, protected cultivation enables off-season production, efficient resource utilization, and enhanced economic returns from high-value crops (Singh and Peter 2014).

Greenhouses, polyhouses, shade-net houses, and low tunnels regulate temperature, humidity, radiation, and pest entry, offering yield advantages of 2–5 times over open-field cultivation in vegetables (Table 16.7) such as capsicum, tomato, and cucumber (Barche et al. 2015). Recent innovations include naturally ventilated structures, climate-adaptive coverings, and energy-efficient designs that reduce input costs and carbon footprints. Protected cultivation also enables precise fertigation, integrated pest management, and uniform crop growth, strengthening climate resilience at the farm level (Gowdhaman 2024).

16.4.1 Climate-Smart Technologies for Horticulture

Climate-smart horticultural technologies integrate productivity enhancement, climate adaptation, and mitigation within a single framework (Table 16.8). These technologies aim to optimize input efficiency while reducing greenhouse gas emissions

Table 16.8 Climate-smart technologies, their applications, and benefits in horticulture

Technology	Application	Climate resilience benefit	References
Solar-powered drip irrigation	Precision water delivery	Reduced energy and water use	Jain et al. (2023)
Renewable cold storage	Postharvest management	Lower losses, income stability	Sadiq et al. (2025)
Weather-based advisories	Crop planning	Reduced climate risk	Gangopadhyay et al. (2019)
Low-energy protected structures	Microclimate regulation	Lower carbon footprint	Li et al. (2025)

Table 16.9 Potential of soilless and urban horticulture systems depicting the water-saving percentage, key advantages, and suitable crops

System	Water saving (%)	Suitable crops	Key advantage	References
Hydroponics	80–90	Leafy greens	Precise nutrition	Al-Kodmany (2018)
Aeroponics	95	Seed potato	Rapid multiplication	Buckseth et al. (2016)
Vertical farming	>90	Herbs, greens	High land-use efficiency	Jadhav et al. (2025)
Rooftop gardening	50–60	Vegetables	Urban food security	Lal (2020)

and environmental footprints. Solar-powered irrigation systems, renewable energy-based cold storage units, and climate-informed decision-support tools are increasingly adopted across horticultural value chains (Ouda and Zohry 2022).

Solar-powered drip irrigation reduces dependence on fossil fuels while enabling timely irrigation under water-scarce conditions. Renewable energy-based cold storage facilities reduce postharvest losses estimated at 20–30% in fruits and vegetables while stabilizing farmer income (Jain et al. 2023). Climate advisory platforms combining weather forecasts with crop models help farmers adjust sowing dates, irrigation schedules, and pest management strategies. Collectively, these technologies improve system efficiency and resilience while aligning horticulture with low-carbon development goals (Advisory C I 2024).

16.4.2 Urban Horticulture, Hydroponics, and Vertical Farming

Urban horticulture, hydroponics, and vertical farming represent transformative pathways toward future-ready horticultural systems under climate change, land scarcity, and rapid urbanization (Table 16.9). These systems decouple crop production from climatic variability by providing controlled environments with precise nutrient and water management. Hydroponic systems require up to 90% less water than conventional soil-based cultivation and enable year-round production with minimal land use (Al-Kodmany 2018).

Vertical farming further enhances productivity per unit area by stacking production layers, making it suitable for peri-urban and urban landscapes. These systems are particularly effective for leafy vegetables, herbs, and microgreens, offering high returns and proximity to consumers. Although initial investment costs remain high, declining technology costs, renewable energy integration, and policy support are improving feasibility. Urban horticulture contributes to nutritional security, employment generation, and reduced food miles, strengthening the resilience of urban food systems (Nutakor et al. 2020).

16.4.3 Sustainable Resource Management and Energy Efficiency

Sustainable management of water, soil, and energy resources is integral to climate-resilient horticulture (Table 16.10). Rising input costs and climate uncertainty necessitate efficient resource use across production systems. Renewable energy integration through solar, biogas, and biomass reduces dependence on fossil fuels and lowers greenhouse gas emissions (Roy and Rani 2018).

Energy-efficient irrigation pumps, solar dryers, and low-energy cold storage systems enhance postharvest value while reducing carbon footprints. Conservation practices such as mulching, rainwater harvesting, and precision nutrient management reduce resource losses and enhance resilience. Integration of renewable energy within protected cultivation and precision irrigation systems exemplifies the synergy between adaptation and mitigation in horticulture (Kassam et al. 2015).

16.4.4 Agroforestry and Conservation-Based Horticulture

Agroforestry-based horticultural systems enhance climate resilience by integrating trees, crops, and sometimes livestock within a single land-use system. Tree components moderate microclimates, reduce wind speed, enhance soil carbon sequestration, and diversify income sources. Fruit-based agroforestry systems such as mango, guava, or litchi intercropped with vegetables or legumes are particularly effective in buffering climatic variability in tropical and subtropical regions (Gupta et al. 2023).

Table 16.10 An outlook of different resources, technology to manage them, and their benefits

Resource	Technology/practice	Climate benefit	Reference
Water	Rainwater harvesting	Drought buffering	Kumar et al. (2006)
Energy	Solar pumps, dryers	Reduced emissions	Terang and Baruah (2023)
Soil	Organic amendments	Improved moisture retention	Bhattacharyya et al. (2016)
Postharvest	Low-energy cold storage	Reduced losses	Alonzo and Buban (2025)

Table 16.11 Key components of agroforestry and conservation horticulture benefits

System	Key component	Climate resilience benefit	Reference
Fruit-based agroforestry	Tree-crop integration	Microclimate moderation	Nair (2012)
Alley cropping	Trees + vegetables	Reduced heat stress	Montes et al. (2019)
Conservation tillage	Residue retention	Moisture conservation	Lal (2015)
Living mulches	Cover crops	Soil protection	Pfeiffer et al. (2016)

Conservation horticulture emphasizes minimal soil disturbance, continuous soil cover, and crop diversification (Table 16.11). These practices improve soil structure, enhance water infiltration, and stabilize yields under climate stress. Agroforestry and conservation-based approaches also contribute to biodiversity conservation and ecosystem services, aligning horticulture with sustainable landscape management (FAO 2013).

16.5 Climate-Resilient Breeding, Biotechnology, and Genetic Resources

Genetic improvement is a cornerstone of climate-resilient horticulture, offering durable solutions to abiotic and biotic stresses through varietal adaptation. Climate-resilient breeding focuses on traits such as heat and drought tolerance, salinity resistance, early maturity, yield stability, and resistance to emerging pests and diseases. Given the perennial nature and long breeding cycles of many horticultural crops, integrating conventional breeding with modern molecular tools is essential to accelerate genetic gains (Jain et al. 2024).

Crop wild relatives (CWRs) and landraces represent invaluable reservoirs of adaptive traits evolved under diverse and often harsh environments. Systematic collection, characterization, and utilization of CWRs have contributed to stress tolerance in tomato, chili, cucurbits, and fruit crops. Pre-breeding programs and participatory varietal selection strengthen local adaptation while enhancing adoption among farmers. Advances in high-throughput phenotyping, marker-assisted selection, and genomic selection (Fig. 16.1) have reduced breeding timelines and improved precision under climate variability (Bhatta and Malla 2020; Mishra et al. 2024; Kaur et al. 2021; Karkute et al. 2017).

Biotechnological approaches, including tissue culture, doubled haploids, transgenics, and genome editing, complement conventional breeding by enabling targeted trait improvement. Genome-editing tools such as CRISPR/Cas can be exploited for precise modification of stress-responsive genes without introducing foreign DNA, thereby addressing regulatory and consumer concerns (Fig. 16.2). Climate-resilient rootstocks, somatic hybrids, and grafting technologies have further expanded adaptive options in vegetables and fruit crops. Together, genetic

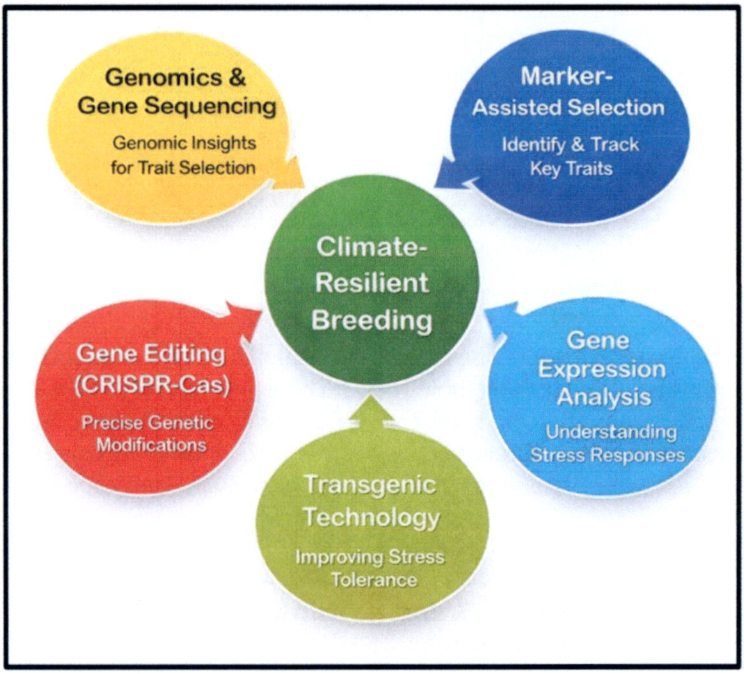

Fig. 16.1 Flow chart illustrating "Climate-Resilient Breeding" at the center, surrounded by five interconnected concepts: "Genomics & Gene Sequencing" for trait selection, "Marker-Assisted Selection" for identifying key traits, "Gene Expression Analysis" for understanding stress responses, "Transgenic Technology" for improving stress tolerance, and "Gene Editing (CRISPR-Cas)" for precise genetic modifications. Each concept is in a colored bubble with arrows indicating its relationship to the central theme (Collard and Mackill 2008)

resources and modern breeding technologies underpin long-term resilience and productivity in horticulture (Table 16.12) (Varshney et al. 2021).

16.6 Socioeconomic, Institutional, and Policy Dimensions

Climate resilience in horticulture extends beyond technologies to encompass socioeconomic, institutional, and policy frameworks that enable adoption at scale. Small and marginal farmers, who dominate horticultural production in many developing countries, face constraints related to capital access, market volatility, and climate risk exposure. Climate change poses multifaceted threats to horticulture, including more extreme temperatures, altered precipitation, and increased pest pressure. A climate-smart farming system is a multidimensional and complex process. The concept of Climate-Smart Agriculture (CSA) has been developed to address three pillars: food security, adaptation, and mitigation (FAO 2013). Building resilience requires integrating social dimensions like farmer identity, knowledge exchange, gender budgeting, and community governance alongside technical innovations that

Fig. 16.2 Flow chart illustrating "Advanced Breeding Strategies for Climate-Resilient Crops" at the center, surrounded by five strategies: "Genomic-Based Strategies" (genomic selection, marker-assisted selection, QTL mapping), "Harnessing Wild Crop Relatives" (introgression of beneficial traits, expanding genetic diversity), "Precision Phenotyping & Data-Driven Analysis" (high-throughput digital phenotyping, AI, big data, predictive breeding), "Speed Breeding Approaches" (accelerated breeding cycles, optimized growth conditions), and "Genome Editing Technologies" (CRISPR/Cas systems, base and prime editing). Each strategy is represented in a distinct color segment (Razzaq et al. 2021)

Table 16.12 Genetic and biotechnological approaches showing the target trait, crop example, and benefit toward climate resilience

Approach	Target trait	Crop example	Climate resilience benefit	References
Conventional breeding	Heat and drought tolerance	Tomato	Yield stability	Ayenan et al. (2019)
Marker-assisted selection	Disease resistance	Chili	Reduced losses	Suwor et al. (2017)
Grafting	Salinity tolerance	Tomato, brinjal	Stress buffering	Singh et al. (2020)
Tissue culture	Rapid multiplication	Banana	Uniformity	Saraswathi et al. (2016)

shape adaptation in horti-farming systems for equitable, future-ready practices. Innovation diffusion and learning systems that connect researchers, extension agents, and farmers accelerate the diffusion of resilient practices. Agricultural advisory services must be context-specific, culturally sensitive, and economically accessible, ensuring access to climate information like weather forecasts, risk maps, and early warning systems. Participatory bottom-up approach models, where farmers co-design adaptation strategies by climate analysis, resource mapping, and better cropping calendars, can improve relevance and context-specific innovative climate-smart agricultural technologies (Mwongera et al. 2017). Social networks and farmer field schools facilitate trust, reduce risk, and enable collective action in resource management and also redesign their farming systems (Bakker et al. 2022). Strengthening extension systems, farmer organizations, and value-chain linkages is critical for translating climate-resilient innovations into field-level impact.

Institutional support through climate-smart policies, risk-sharing mechanisms, and incentive-based schemes accelerates adoption of resilient practices. Crop insurance, weather-indexed advisories, and credit-linked subsidies reduce farmers' vulnerability to climate shocks. Public-private partnerships and farmer-producer organizations facilitate access to quality inputs, technologies, and markets. Gender-inclusive approaches and youth engagement enhance social resilience while fostering innovation and entrepreneurship in horticulture (FAO 2013).

Policy frameworks aligned with national climate action plans and sustainable development goals play a pivotal role in mainstreaming climate-resilient horticulture. Investments in research, infrastructure, digital platforms, and capacity building create enabling environments for future-ready horti-farming systems.

16.7 Role of Women and Marginalized Communities in Resilient Horticulture

Women farmers are the "the Backbone of Resilience"; they perform approximately 70% of agricultural labor globally and are the primary agents in transitioning toward climate-smart practices 2016. (FAO, 2020). Women and marginalized communities are central to building resilient horticultural systems for ensuring food security, climate adaptation, value addition, and sustainable livelihoods. Women contribute significantly to horticulture through activities such as seed selection, planting, harvesting, processing, and marketing of fruits, vegetables, and flowers. Women are often the primary keepers of traditional seed varieties. They diversify crops from season to season, which acts as a natural insurance policy against total crop failure during droughts or floods. Their traditional knowledge of local crops, soil health, and water conservation helps in adopting climate-resilient practices. Women are often the custodians of traditional landraces and community gardens. Empowering women through access to land rights, training, credit, and technology improves productivity and strengthens resilience against climate change. Women-led horticulture often prioritizes "kitchen gardens" by micro-level adoptions of vertical gardening, value-added products, hanging pot cultivation in saline areas, and vermicomposting, ensuring local households have access to nutrient-dense food and provide income during climate stress or economic shocks (FAO 2011, 2014).

Marginalized communities, including small and marginal farmers, tribal groups, and landless laborers, also play a crucial role in resilient horticulture. These communities often cultivate diverse crops using indigenous seeds and low-input farming methods that are naturally adaptable to changing climatic conditions. Their reliance on traditional varieties of fruits and vegetables helps preserve biodiversity and reduces dependence on chemical inputs. Gendered vulnerabilities are not the simple outcome of a social categorization but are created and recreated through social practices that operate at scales as small as the household. Gender norms influence access to resources, information, and decision-making, affecting resilience outcomes; targeted inclusion of women can broaden adaptive capacity (Ravera et al. 2016). The local constitution of gender and the intersection of gender roles, in all their diversity, become somewhat obscure by the mainstream approaches to gender equality. Community-based farming, collective marketing, and cooperative models enable marginalized groups to share resources, reduce risks, and improve market access (Ravera et al. 2016, 2019).

16.8 Capacity Building, Extension Services, and Knowledge Dissemination of Resilient Practices

Extension services play a pivotal role in scaling climate-resilient practices in horticulture by bridging knowledge gaps between research and farmers. Training programs equip farmers with skills for adaptive strategies, including soil conservation and resilient grafting in fruit crops. Participatory approaches ensure practices fit local contexts, enhancing uptake in diverse horticultural systems.Agents facilitate access to seeds, credit, and insurance, addressing barriers like low awareness and financial constraints. They deliver tailored training on techniques like drought-tolerant varieties, efficient irrigation, and integrated pest management, boosting adoption rates among smallholders, fostering behavioral change, and collective action for long-term resilience (Anteneh and Melak, 2024) Extension agents conduct field demonstrations, workshops, and farmer field schools to teach climate-smart horticultural practices such as mulching, shade netting, and identifying drought-tolerant varieties. Through hands-on training programs, farmers gain practical skills for things like soil conservation, smarter irrigation methods, integrated pest management, and grafting resilient fruit crops that can better withstand tough conditions. What works best is when these approaches involve farmers directly, making sure the techniques actually suit local soils, weather, and needs by means of personalized advice that scales up better and higher adoption rates, especially among smallholder farmers, for better resource management and risk preparedness. Extension services and farmer training programs disseminate knowledge on precision horticulture and controlled-environment systems (Table 16.13). Governments support community-led initiatives like seed banks for climate-tolerant varieties. Insurance schemes cover losses from extreme weather, boosting confidence in adaptive strategies (Mwongera et al. 2018)

Table 16.13 Institutional and policy interventions, roles, and outcomes for climate-resilient horticulture

Intervention	Role	Outcome	Reference
Climate-smart subsidies	Technology adoption	Reduced vulnerability	Ravera et al. (2016)
Crop insurance	Risk mitigation	Income stability	
Extension and KVKs	Knowledge dissemination	Higher adoption	
FPOs and PPPs	Market access	Value-chain resilience	

16.9 Global Collaboration and Knowledge Sharing

Global collaboration and knowledge sharing are powerful levers to prevent climate-driven damage in horticultural enterprises by accelerating learning and spreading resilient practices. Community knowledge and social learning are central to the rapid adoption of climate-smart practices. In 2014, the Global Alliance for Climate-Smart Agriculture (GACSA) was launched with the goal of helping 500 million farmers practice climate-smart agriculture (CSA), defined as "agriculture that sustainably increases productivity, enhances resilience (adaptation), reduces/removes GHGs (mitigation) where possible, and enhances achievement of national food security and development goals" (FAO 2013). Resilience in horticulture requires supportive policy frameworks that incentivize climate-adaptive practices. Government schemes, self-help groups, and nongovernmental organizations further support capacity building and social inclusion. Such integration enhances farmer agency by providing tools for risk management and innovation. However, barriers such as digital literacy gaps, leveraging AI for predictive analytics, public-private partnerships, unreliable internet, and institutional silos impede progress; digital tools like mobile advisories and apps provide real-time climate forecasts and crop advice, reaching remote areas effectively (Vincent and Balasubramani 2021). Payment for Ecosystem Services (PES) schemes reward farmers for carbon sequestration and biodiversity efforts in orchards and vegetable fields.

16.10 Future-Ready Horti-Farming: An Integrated Framework

Future-ready horti-farming represents an integrated, adaptive, and technology-enabled approach that aligns productivity, resilience, and sustainability under changing climatic conditions. It combines climate-resilient genetics, precision management, protected cultivation, digital tools, and supportive institutional frameworks into a cohesive system. Emphasis is placed on diversification, resource-use efficiency, low-carbon technologies, and data-driven decision-making (FAO 2013).

Key elements of future-ready horticulture include climate-informed planning, adoption of resilient varieties, precision irrigation and fertigation, integrated pest management, renewable energy use, and value addition through postharvest management. Digital agriculture platforms, artificial intelligence, and remote sensing enhance predictive capacity and responsiveness to climate risks (FAO 2017). At the landscape level, agroforestry and conservation-based horticulture strengthen ecosystem services and biodiversity. The transition toward future-ready horti-farming requires coordinated efforts among researchers, policymakers, extension agencies, and farmers. Continuous learning, innovation, and adaptive governance are essential to respond to evolving climate challenges (FAO 2020).

16.11 Conclusions

Climate resilience in horticulture is not a singular intervention but a suite of integrated strategies encompassing crop choice, water management, technology adoption, pest management, and institutional support. Strengthen farmer-to-farmer networks and participatory learning to speed adaptive uptake. Gender-embedded social equity goals ensuring inclusive access to resources, local data collection, and transparent benefit-sharing mechanisms must be the policymakers' priority. Environmental regulations enforce water-use efficiency and limit pesticide overuse, mandating adoption of integrated pest management. As climate variability intensifies, future-ready horti-farming must emphasize sustainability, innovation, and adaptability. Resilient horticultural systems can secure food and nutritional security, enhance livelihoods, and contribute to ecological sustainability in an uncertain climatic future. Climate change poses unprecedented challenges to horticultural systems due to their high sensitivity to temperature, water availability, and biotic stresses. However, horticulture also offers significant opportunities for climate adaptation and mitigation through diversified cropping systems, efficient resource use, and high-value production. Climate-resilient horticulture integrates adaptive agronomic practices, improved genetics, protected cultivation, precision technologies, and supportive policy environments to enhance productivity and stability under climatic uncertainty.

Future-ready horti-farming emphasizes system-level resilience, innovation, and sustainability, moving beyond incremental adaptation toward transformative change. Community knowledge and social learning are central to rapid adoption of climate-smart practices. Effective climate resilience requires multilevel governance that aligns agricultural policy with local needs. Local co-management, extension service support, and micro-finance mechanisms enable smallholders to invest in climate-resilient infrastructures. Climate-smart strategic investments in research, infrastructure, digital platforms, and capacity building are critical to scaling climate-resilient practices. By adopting integrated and inclusive approaches, horticulture can play a pivotal role in ensuring food and nutritional security, improving livelihoods, and promoting environmental sustainability in a changing climate.

References

Adombilla R (2023) Effect of drip irrigation application regimes and mulching on soil properties, growth and yield traits of tomato (Solanum lycopersicum L.) varieties. Doctoral dissertation, University for Development Studies, West African Centre for Water, Irrigation and Sustainable Agriculture

Advisory CI (2024) Quality of science: evaluative evidence from CGIAR portfolio summary

Agriculture Organization (Fao) (Ed.). (2014) Food and nutrition in numbers: 2014. Food & Agriculture Organization of the UN (FAO)

Ainsworth EA, Long SP (2005) What have we learned from 15 years of free-air CO2 enrichment (FACE)? A meta-analytic review of the responses of photosynthesis, canopy properties and plant production to rising CO2. New Phytol 165(2):351–372

Akchaya K, Parasuraman P, Pandian K, Vijayakumar S, Thirukumaran K, Mustaffa MRAF et al (2025) Boosting resource use efficiency, soil fertility, food security, ecosystem services, and climate resilience with legume intercropping: a review. Front Sustain Food Syst 9:1527256

Al-Kodmany K (2018) The vertical farm: A review of developments and implications for the vertical city. Buildings 8(2):24

Ali UM (2024) Climate-smart horticultural practices: building resilience in a changing environment: a scoping review. Agric For Fish 13(5):202–209. https://doi.org/10.11648/j.aff.20241305.19

Al-Najadi R, Al-Mulla Y, Al-Abri I, Al-Sadi AM (2025) Effectiveness of drone-based thermal sensors in optimizing controlled environment agriculture performance under arid conditions. Sci Rep 15:9042. https://doi.org/10.1038/s41598-025-94432-0

Alonzo JGM, Buban DJ (2025) Design and development of solar powered low-cost cold storage system for agricultural products. Int J Comput Exp Sci Eng 11(2):2713–2720

Anteneh A and Melak A (2024) ICT-based agricultural extension and advisory service in Ethiopia: A review. Cogent Food & Agriculture, 10(1):2391121

Ayenan MAT, Danquah A, Hanson P, Ampomah-Dwamena C, Sodedji FAK, Asante IK, Danquah EY (2019) Accelerating breeding for heat tolerance in tomato (*Solanum lycopersicum* L.): an integrated approach. Agronomy 9(11):720

Bakker T, Dugué P, de Tourdonnet S (2022) How do farmers change their practices at the farm level after co-design processes in Farmer Field Schools? Agric Syst 201:103457. ISSN 0308-521X. https://doi.org/10.1016/j.agsy.2022.103457

Balyan S, Rao S, Jha S, Bansal C, Das JR, Mathur S (2020) Characterization of novel regulators for heat stress tolerance in tomato from Indian subcontinent. Plant Biotechnol J 18(10):2118–2132. https://doi.org/10.1111/pbi.13371

Barche S, Nair R, and Jain PK (2015) A review of mulching on vegetable crops production. Ecology. Environ Conserv 21(2):859–866

Beltrà A, Wäckers FL, Nedvěd O, Pekas A (2018) Predation rate and performance of three ladybirds against the green peach aphid *Myzuspersicae* in sweet pepper. Entomol Exp Appl 166(6):491–499

Bhandari DC, Meghwal PR, Lodha S (2014) Horticulture based production systems in Indian arid regions. In: Sustainable horticultural systems: issues, technology and innovation. Springer International Publishing, Cham, pp 19–49

Bhatta BP, Malla S (2020) Improving horticultural crops via CRISPR/Cas9: current successes and prospects. Plants 9(10):1360

Bhattacharyya R, Ghosh BN, Dogra P, Mishra PK, Santra P, Kumar S, Fullen MA, Mandal UK, Anil KS, Lalitha M et al (2016) Soil conservation issues in India. Sustainability 8(6):565. https://doi.org/10.3390/su8060565

Bhavitha K, Pasha ML, Ramulu V, Prakash TR, Rajaiah P, Revathi P (2024) Impact of AI based irrigation scheduling approaches and drip irrigation methods on yield of Chilli (Capsicum annum L.) and chemical properties of soil. Int J Environ Clim Chang 14(7):540–547

Bijarnia A, Tetarwal JP, Yadav RK, Bijarnia AL, Singh D, Saini Y (2024) Effect of fertility levels and stress mitigating chemicals on nutrient content, uptake, intercropping advantage and competition effect in cowpea–baby corn intercropping. Heliyon 10:e38194. https://doi.org/10.1016/j.heliyon.2024.e38194

Buckseth TS, Sharma AK, Pandey KK, Singh BP, Muthuraj R (2016) Methods of pre-basic seed potato production with special reference to aeroponics—a review. Sci Hortic 204:79–87. https://doi.org/10.1016/j.scienta.2016.03.041

Chandra R, Govind S (2001) Effect of mulching on yield of ginger (*Zingiber officinale*Rosc.). J Spices Aromat Crops 10(1):13–16

Chaturvedi A, Tripathi AN, Singh AP (2024) Intercropping of African marigold with tomato for better productivity, pest control and economic returns in Arunachal Pradesh, India. J Biochem Int 11(1):7–12. https://doi.org/10.56557/jobi/2024/v11i18819

Clover GRG, Smith HG, Azam-Ali SN, Jaggard KW (1999) The effects of drought on sugar beet growth in isolation and in combination with beet yellows virus infection. J Agric Sci 133(3):251–261

Collard BC, Mackill DJ (2008) Marker-assisted selection: an approach for precision plant breeding in the twenty-first century. Philos Trans R Soc Lond B Biol Sci 363(1491):557–572

Deshmukh AK, Shashidhara GB, Bidari BI (2009) Effect of site-specific nutrient management (SSNM) through targeted yield approach on yield and nutrient uptake of chilli (*Capsicum annuum* L.). Asian J Soil Sci 4(2):287–289

Deuter P (2008) Defining the impact of climate change on horticulture in Australia. Garnaut Climate Change Review, Department of Primary Industries and Fisheries, Queensland, pp 1–23

Durga C, Anith S (2021) Effect of conservation agriculture practices on weed management in okra under rice–okra–green manure cropping system. Indian J Weed Sci 53(2):164–168. https://doi.org/10.5958/0974-8164.2021.00030.7

Enyiukwu DN, Bassey IN, Nwaogu GA, Chukwu LA (2025) Intelligent machines and plant disease forecasting systems: digital solution for managing crop-stressors toward food security in Africa. J Agric Res Dev Ext Technol 7(1):11–38

FAO (2013) Climate-smart agriculture sourcebook. CCAFS: CGIAR Research Program on Climate Change, Agriculture and Food Security. Available online at: https:// ccafs.cgiar. org/ publications/ climate-smart-agriculture-sourcebook #. Wa QfGT5JbIU

FAO (2020) The state of food and agriculture 2020. Overcoming water challenges in agriculture. Food and Agriculture Organization of the United Nations

Farooq M, Wahid A, Kobayashi NSMA, Fujita DBSMA, Basra SM (2009) Plant drought stress: effects, mechanisms and management. In: Sustainable agriculture. Springer Netherlands, Dordrecht, pp 153–188

Food O (2016) The state of food and agriculture. Climate Change Agri 78

Food and Agriculture Organization (2011) The state of food and agriculture: women in agriculture

Food and Agriculture Organization (2013) Climate-smart agriculture sourcebook. FAO

Food and Agriculture Organization (2017) The future of food and agriculture

Food and Agriculture Organization, F. A. A. O. (2017) The state of food and agriculture 2017: leveraging food systems for inclusive rural transformation

Gangopadhyay PK, Khatri-Chhetri A, Shirsath PB, Aggarwal PK (2019) Spatial targeting of ICT-based weather and agro-advisory services for climate risk management in agriculture. Clim Chang 154(1):241–256

Gowdhaman P (2024) Adaptation to progressive climate change, technology development, agricultural practices. Dr. KL Chaudhary, 207

Grasham CF, Calow R, Casey V, Charles KJ, De Wit S, Dyer E, Fullwood-Thomas J, Hirons M, Hope R, Hoque SF, Jepson W, Korzenevica M, Murphy R, Plastow J, Ross I, Ruiz-Apilánez I, Schipper ELF, Trevor J, Walmsley N, Zaidi H (2021) Engaging with the politics of climate resilience towards clean water and sanitation for all. Npj Clean Water 4(1):42. https://doi.org/10.1038/s41545-021-00133-2

Gruda NS, Samuolienė G, Dong J, Li X (2025) Environmental conditions and nutritional quality of vegetables in protected cultivation. Compr Rev Food Sci Food Saf 24(2). https://doi.org/10.1111/1541-4337.70139

Gupta SR, Dagar JC, Sileshi GW, Chaturvedi RK (2023) Agroforestry for climate change resilience in degraded landscapes. In: Agroforestry for sustainable intensification of agriculture in Asia and Africa. Springer, pp 121–174

Hatfield JL, Boote KJ, Kimball BA, Ziska LH, Izaurralde RC, Ort D, Antle JM (2011) Climate impacts on agriculture: implications for crop production. Agron J 103(2):351–370

Hiekal HADM (2020) Improving performance of surface irrigation system by designing pipes for water conveyance and on-farm distribution. In: Technological and modern irrigation environment in Egypt: best management practices & evaluation. Springer International Publishing, Cham, pp 223–256

Hossain MM, Khalequzzaman KM, Alam MS, Mondal MTR, Islam MM (2020) Efficacy of different coloured sticky traps against thrips of chilli. Int J Appl Sci Biotechnol 8(2):187–190

IPCC (2021) Climate change 2021-the physical science basis. Interaction 49(4):44–45

Jadhav V, Grondona T, Pistillo A, Pennisi G, Ghio M, Gianquinto G, Orsini F (2025) Optimizing planting density for increased resource use efficiency in baby-leaf production of lettuce (*Lactuca sativa* L.) and basil (*Ocimumbasilicum* L.) in vertical farms. Horticulturae 11(4):343

Jain S, Singh H, Kore DS, Mishra S, Sharma R (2023) A comprehensive review on post harvest physiological disorders in citrus fruit crops (Citrus spp.). Int J Environ Clim Chang 13(10):2863–2873

Jain S, Lamo K, Walling S, Imchen A, Tirkey JF, Singh A (2024) Climate-resilient horticulture: adapting to climate change through innovative practices and technologies. Int J Environ Climate Chang 14(11):219–233. https://doi.org/10.9734/ijecc/2024/v14i114541

Jaiswal N, Kumar TV, Shukla C (2025) Smart drip irrigation systems using IoT: a review of architectures, machine learning models, and emerging trends. Discov Agric 3(1):253

Jena SR, Saha S, Agarwal S (2025) Smart innovations for climate resilience, forest sustainability, and green economy. Srjx Research and Innovation Lab Llp

Karkute SG, Singh AK, Gupta OP, Singh PM, Singh B (2017) CRISPR/Cas9 mediated genome engineering for improvement of horticultural crops. Front Plant Sci 8:1635

Kassam A, Friedrich T, Derpsch R, Kienzle J (2015) Overview of the worldwide spread of conservation agriculture. Field Actions Sci Rep 8. http://journals.openedition.org/factsreports/3966

Kaur H, Pandey DK, Goutam U, Kumar V (2021) CRISPR/Cas9-mediated genome editing is revolutionizing the improvement of horticultural crops: recent advances and future prospects. Sci Hortic 289:110476

Khamarunneesa M, Sajeena S, Hakkim VM, Prashanth K (2023) Field evaluation of site specific drip fertigation in tomato. Int J Environ Climate Chang 13(7):351–362

Kirnak H, Tas I, Kaya C, Higgs D (2003) Responses of drip irrigated bell pepper to water stress and different nitrogen levels with or without mulch cover. J Plant Nutr 26(2):263–277. https://doi.org/10.1081/PLN-120017135

Krishna H, Saroj PL, Jatav MK, Awasthi OP (2018) Fruit based cropping models to increase farmer's income-an experience of arid region. CIAH/Tech 70:1–61

Kumar R, Kumar KK (2007) Managing physiological disorders in litchi. Indian Hortic 52(1):22–24

Kumar MD, Patel A, Singh OP (2006) Rainwater harvesting in India: some critical issues for basin planning and research. Land Use Water Resour Res 6:1–17

Kumar T, Kumar M, Singh MK, Kumar V, Kumar A, Kumar S, Singh B (2013) Impact of integrated nutrient management (INM) on growth and economic yield of okra. Ann Hortic 6(1):107–114

Kumar P, Khapte PS, Singh A, Saxena A (2024) Optimization of low-tech protected structure and irrigation regime for cucumber production under hot arid regions of India. Plants 13(1):146. https://doi.org/10.3390/plants13010146

Lakhiar IA, Yan H, Zhang C, Wang G, He B, Hao B, Rakibuzzaman M (2024) A review of precision irrigation water-saving technology under changing climate for enhancing water use efficiency, crop yield, and environmental footprints. Agriculture 14(7):1141

Lal R (2004) Soil carbon sequestration to mitigate climate change. Geoderma 123(1–2):1–22

Lal R (2015) A system approach to conservation agriculture. J Soil Water Conserv 70(4):82A–88A. https://doi.org/10.2489/jswc.70.4.82A

Lal R (2020) Home gardening and urban agriculture for advancing food and nutritional security in response to the COVID-19 pandemic. Food Secur 12:871–876. https://doi.org/10.1007/s12571-020-01058-3

Le Quéré C, Andrew RM, Friedlingstein P, Sitch S, Pongratz J, Manning AC, Korsbakken JI, Peters GP, Canadell JG, Jackson RB, Boden TA (2018) Global carbon budget 2017. Earth Syst Sci Data 10(1):405–448

Li N, Xu W, Zhang Y, Ma W, Ren Y (2025) Energy saving technologies and practices in facility agriculture in cold regions. Agronomy 15(1):204

Luedeling E, Girvetz EH, Semenov MA, Brown PH (2001) Climate change affects winter chill for temperate fruit and nut trees. PloS One 6:e20155

Maduwanthi AKMRB, Karunarathna B (2019) Biological and economic benefit of okra (*Abelmoschus esculentus* L.)–cowpea (*Vigna unguiculata* L. Walp) intercropping in sandy regosol. Middle East J Agric Res 8(1):28–34

Manzoor MA, Xu Y, Lv Z, Xu J, Shah IH, Sabir IA, Wang Y, Sun W, Liu X, Wang L, Liu R, Jiu S, Zhang C (2024) Horticulture crop under pressure: unraveling the impact of climate change on nutrition and fruit cracking. J Environ Manag 357:120759. https://doi.org/10.1016/j.jenvman.2024.120759

Marwan H, Hayati I, Mulyati S (2023) Effectiveness of biofungicide formula on rhizome rot disease of red ginger and its plant growth. Biodiversitas 24(4):2143–2148. https://doi.org/10.13057/biodiv/d240425

Mcgrath JM, Lobell DB (2013) Reduction of transpiration and altered nutrient allocation contribute to nutrient decline of crops grown in elevated CO_2 concentrations. Plant Cell Environ 36(3):697–705

Messelink GJ, Bloemhard CM, Sabelis MW, Janssen A (2013) Biological control of aphids in the presence of thrips and their enemies. BioControl 58(1):45–55

Mihrete TB, Mihretu FB (2025) Crop diversification for ensuring sustainable agriculture, risk management and food security. Global Chall 9(2):2400267

Mishra D, Mohanty AK, Mukhi SK, Singh DV (2019) Assessing the performance and adoption rate of tomato hybrid "ArkaRakshak" having multiple disease resistance in Jagatsinghpur district of Odisha, India. Int J Curr Microbiol App Sci 8(9):2458–2464. https://doi.org/10.20546/ijcmas.2019.809.284

Mishra S, Nayak S, Tuteja N, Poosapati S, Swain DM, Sahoo RK (2024) CRISPR/Cas-mediated genome engineering in plants: application and prospectives. Plants 13(14):1884

Mishra A, Mohapatra A, Bairagya MD, Nayak RD, Nandi R, Gulati J, Behera B (2025) Effect of irrigation, tillage, and mulch management on the growth, yield, and quality of baby corn (*Zea mays* L.). Int J Plant Soil Sci 37(8):671–679

Mohanty NM, Sinha NNK, Hati NK, Reddy NKS, Chaudhary NR (2015) Elevated temperature and carbon dioxide concentration effects on wheat productivity in Madhya Pradesh: a simulation study. J Agrometeorol 17(2):185–189. https://doi.org/10.54386/jam.v17i2.1001

Mondol MSA, Akbar U, Mandal O, Rani A, Dar AH, Chatterjee A, Abdi G (2025) Advances in agronomic practices, postharvest technologies, and medicinal potential of dragon fruit (*Hylocereus* spp.): a comprehensive updated review. J Agric Food Res 22:102157

Montes O, Castro R, Villanueva C, Pérez M, Uribe M (2019) Tree used in horticulture based alley cropping. J Appl Hortic 21(2):87–100

Mooventhan P, Kumar J, Dixit A, Sharma KC, Sivalingam PN, Gupta AK, Singh U, Singh SRK, Venkatesan P, Kaushal P (2019) ArkaRakshak tomato to solve farmer woes. Indian Hortic 64(3). https://epubs.icar.org.in/index.php/IndHort/article/view/93683

Mwongera C, Shikuku KM, Twyman J, Läderach P, Ampaire E, Van Asten P, Twomlow S, Winowiecki LA (2017) Climate smart agriculture rapid appraisal (CSA-RA): a tool for prioritizing context-specific climate smart agriculture technologies. Agric Syst 151:192–203

Mwongera C, Nowak A, Notenbaert AM, Grey S, Osiemo J, Kinyua I, Lizarazo M and Girvetz E (2018) Climate-smart agricultural value chains: Risks and perspectives. In The climate-smart agriculture papers: investigating the business of a productive, resilient and low emission future (pp. 235-245). Cham: Springer International Publishing

Nair R, Barche S (2011) Protected cultivation of vegetables – Present status and future prospects in India. Indian J Appl Res 4(6):245–247

Nair PKR (2012) Climate change mitigation: a low-hanging fruit of agroforestry. In: Nair P, Garrity D (eds) Agroforestry—the future of global land use, Advances in agroforestry, vol 9. Springer. https://doi.org/10.1007/978-94-007-4676-3_7

Nair PKR, Kumar BM, Nair VD (2009) Agroforestry as a strategy for carbon sequestration. J Plant Nutr Soil Sci 172(1):10–23. https://doi.org/10.1002/jpln.200800030

Nisar S, Mavi MS, Singh J, Srivastava S, Dey P (2025) Optimizing nutrient management strategies to achieve higher productivity, greater nutrient use efficiency in eggplant and maintenance of soil health. J Plant Nutr 48(11):1817–1831

Nutakor F, Bizumuremyi S, Li J, Liu W (2020) Does the environmental kuznets curve for CO2 emissions exist for Rwanda? Evidence from bootstrapped rolling-window granger causality test. Sustainability 12(20):8636

Osillos PL, Nagpala AL (2014) The effects of arbuscular mycorrhizal fungi (AMF) as biofertilizer on the growth, yield and nutrient uptake of tomato (*Lycopersicon esculentum* Mill.). Int J Sci Eng Res 3(11):49–65

Ouda S, Zohry AEH (2022) Climate-smart agriculture. Springer International Publishing

Pangga IB, Hanan J, Chakraborty S (2013) Climate change impacts on plant canopy architecture: implications for pest and pathogen management. Eur J Plant Pathol 135(3):595–610

Patel PK, Parmar BJ (2020) Sprinkler irrigation system and its adoption by potato growers: a case of Banaskantha District. Ind J Adult Edu 81(4):161–168

Patel A, Kushwaha NL, Rajput J, Gautam PV (2023) Advances in micro-irrigation practices for improving water use efficiency in dryland agriculture. In: Enhancing resilience of dryland agriculture under changing climate: interdisciplinary and convergence approaches. Springer Nature Singapore, Singapore, pp 157–176

Pfeiffer A, Silva E, Colquhoun J (2016) Living mulch cover crops for weed control in small-scale applications. Renew Agric Food Syst 31(4):309–317

Pugliese M, Gilardi G, Garibaldi A, Gullino ML (2024) The impact of climate change on vegetable crop diseases and their management: the value of phytotron studies for the agricultural industry and associated stakeholders. Phytopathology 114(5):843–854. https://doi.org/10.1094/phyto-08-23-0284-kc

Raj J, Khalkho D, Tripathi MP, Bisen Y (2024) Optimizing water use for okra cultivation: a soil moisture sensor-based approach. J Sci Res Rep 30(4):153–161. https://doi.org/10.9734/jsrr/2024/v30i41900

Rao NKS, Laxman RH (2013) Phenotyping horticultural crops for abiotic stress tolerance. In: Climate-resilient horticulture: adaptation and mitigation strategies. Springer India, pp 147–157

Ravera F, Iniesta-Arandia I, Martín-López B, Pascual U, Bose P (2016) Gender perspectives in resilience, vulnerability and adaptation to global environmental change. Ambio 45(Suppl 3):235–247. https://doi.org/10.1007/s13280-016-0842-1

Ravera F, Reyes-García V, Pascual U, Drucker AG, Tarrasón D, Bellon MR (2019) Gendered agrobiodiversity management and adaptation to climate change: differentiated strategies in two marginal rural areas of India. Agric Hum Values 36(3):455–474

Ray DK, Gerber JS, MacDonald GK, West PC (2015) Climate variation explains a third of global crop yield variability. Nat Commun 6(1):5989

Razzaq A, Kaur P, Akhter N, Wani SH, Saleem F (2021) Next-generation breeding strategies for climate-ready crops. Front Plant Sci 12:620420. https://doi.org/10.3389/fpls.2021.620420

Roy P, Rani A (2018) Agromet advisory for empowering farmers to mitigate climate change. BIotechArticles:1-6. https://www.biotecharticles.com/

Sadiq SM, Singh IP, Ahmad MM, Sani BS (2025) Cold storage solutions to reduce post-harvest loss: start-ups for youth in the agricultural supply chain. New Countryside 4(2):37–44

Sarangi S, Bhavana P and Padhan EK (2025) Insecticide resistance and management strategies. Advances in Applied Entomology; Dash SS, Pradhan PP, Sahoo S, Nag YK, Sorahia D Eds: 184-208

Saraswathi MS, Uma S, Kannan G, Selvasumathi M, Mustaffa MM, Backiyarani S (2016) Cost-effective tissue culture media for large-scale propagation of three commercial banana (*Musa* spp.) varieties. J Hortic Sci Biotechnol 91(1):23–29

Scheelbeek PF, Bird FA, Tuomisto HL, Green R, Harris FB, Joy EJ, Chalabi Z, Allen E, Haines A, Dangour AD (2018) Effect of environmental changes on vegetable and legume yields and nutritional quality. Proc Natl Acad Sci 115(26):6804–6809

Sethunath K, Bhaskar J, Jayasree P, Gomez S, Sreekumar PM, Paul R (2025) Effect of NPK fertilizer application on growth and yield of dragon fruit in Kerala. Int J Res Agron 8(6):877–886. https://doi.org/10.33545/2618060X.2025.v8.i6k.3121

Sharma IM (2012) Changing disease scenario in apple orchards: perspective, challenges and management strategies. In: Proceedings of the national symposium on blending conventional and modern plant pathology for sustainable agriculture, Indian Institute of Horticultural Research, Bengaluru, vol 123

Sharma P, Kaushal A (2015) Growing okra with drip fertigation-a review. Inter J Eng Sci Invent 4(9):2319–6734

Sharma G, Sharma A, Sinha NK, Sharma OP, Singh A, Pandey AK, Kumar A, Trivedi SK, Sao B, Sahu MK (2022) Assessment of long-term climate variability and its impact on the decadal growth of horticultural crops in Central India. Ecol Process 11(1):61. https://doi.org/10.1186/s13717-022-00405-w

Sharma N, Kumar S, Yadav D, Kumar R, Kumar A (2023) Effect of integrated nutrient management (INM) practices on growth, yield, economics and soil properties of sweet pepper under open condition at mid hill of Himachal Pradesh, India. Int J Plant Soil Sci 35(19):773–782. https://doi.org/10.9734/ijpss/2023/v35i193610

Singh DK, Peter KV (2014) Protected cultivation of horticultural crops. New India Publishing Agency

Singh HP, Singh JP, Lal SS (2010) Challenges on climate change-Indian horticulture. Westville Publishing House, New Delhi, p 224

Singh G, Ram H, Aggarwal N (2017) Agronomic management of cowpea for high grain yield and sustainable agriculture in Western Indo-Gangetic Plain of India. Int J Curr Microbiol App Sci 6(6):2633–2647. https://doi.org/10.20546/ijcmas.2017.606.314

Singh H, Kumar P, Kumar A, Kyriacou MC, Colla G, Rouphael Y (2020) Grafting tomato as a tool to improve salt tolerance. Agronomy 10(2):263

Singh RP, Tathode VR, Purushotham P, Sahu N, Karthik S (2024) Plant protection insights: diverse perspective on plant pathology and entomology. 1st edn. Stella International Publication

Skendžić S, Zovko M, Živković IP, Lešić V, Lemić D (2021) The impact of climate change on agricultural insect pests. Insects 12(5):440

Steduto P, Hsiao TC, Fereres E, Raes D (2012) Crop yield response to water. FAO

Stojanova S, Volk M, Balkovec G, Kos A, Stojmenova Duh E (2025) The future of vineyard irrigation: AI-driven insights from IoT data. Sensors 25(12):3658

Sudha PG (2024) Management of rhizome rot of ginger by Trichoderma asperellum Kjr and its effect on yield. Master's thesis, Odisha University of Agriculture and Technology, Bhubaneswar

Summary for policymakers (2023) Cambridge University Press eBooks (pp. 3–34). https://doi.org/10.1017/9781009325844.001

Sunitha ND (2018) Ecofriendly management of thrips in capsicum under protected condition. J Entomol Zool Stud 6(1):617–621

Suwor P, Sanitchon J, Thummabenjapone P, Kumar S, Techawongstien S (2017) Inheritance analysis of anthracnose resistance and marker-assisted selection in introgression populations of chili (*Capsicum annuum* L.). Sci Hortic 220:20–26

Terang B, Baruah DC (2023) Techno-economic and environmental assessment of solar photovoltaic water pumps for irrigation in Assam, India. Energy Policy. https://doi.org/10.1016/j.enpol.2023.113807

Varshney RK, Bohra A, Yu J, Graner A, Zhang Q, Sorrells ME (2021) Designing future crops: genomics-assisted breeding comes of age. Trends Plant Sci 26(6):631–649

Vincent A, Balasubramani N (2021) Climate-smart agriculture (CSA) and extension advisory service (EAS) stakeholders' prioritisation: a case study of Anantapur district, Andhra Pradesh, India. J Water Clim Change 12(8):3915–3931

Virginie TIA, Rosalie MM, Fabien DDF, Armand ND, Duthie TV, Inocent G (2024) Biofertilizers effects on some agromorphological parameters and harvest stage on nutrients composition of ginger (*Zingiber officinale* Rosc.). Int J Agric Nutr 6(1):121–128. https://doi.org/10.33545/26646064.2024.v6.i1b.151

Wahid A, Gelani S, Ashraf M, Foolad MR (2007) Heat tolerance in plants: an overview. Environ Exp Bot 61(3):199–223

Yadav KSR, Mandaliya VB (2025) Nano-fertilizer for climate-smart agriculture: a comprehensive patent analysis for sustainable soil management practices. In: Contemporary suitability of nanobionics in agriculture: nanotechnology in horticultural crops. Springer Nature Switzerland, Cham, pp 195–213

Yusoff MM, Halim M, Mohamed M, Rastan S, Meon Z (2008) Growth, yield and fruit quality of red dragon (*Hylocereus polyrhizus*) as affected by plant support system and intercropping with long bean (*Vigna sinensis*). J Food Agric Environ 6:305–311

Zhao J, Liu D, Huang R (2023) A review of climate-smart agriculture: recent advancements, challenges, and future directions. Sustainability 15(4):3404

Are We Future-Ready for Extreme Weather Calamities?

17

Sagar Maitra, Sumit Ray, Anjana J. Atapattu, Bisruti Maity, Masina Sairam, Lalichetti Sagar, Satya Prasad Nanda, D. T. Santosh, Dinkar Jagannath Gaikwad, Harun I. Gitari, Shaik Rishitha, Luciana L. Chappa, Nithar Ranjan Madhu, and Akbar Hossain

Abstract

Horticulture, being deeply reliant on ecology, is under shock due to the current climatic anomalies; hence, climate resilience is more imperative than an option, which demands a strategic reconfiguration of production systems for long-term sustainability. The increasing population and land shrinkage resulting from industrialization and urbanization created challenges to food and nutritional security, underscoring the importance of climate-resilient horticulture (CRH) as a vital option. CRH encompasses key interventions, including crop improvement targeting resistance to both biotic and abiotic stress; controlled environment cultivation; smart irrigation and automation; drone use for input delivery; the application of decision support tools, satellite imagery, weather analytics, and warning systems; as well as precision management of inputs. Furthermore, to reduce postharvest

S. Maitra · S. Ray · B. Maity · M. Sairam (✉) · L. Sagar · S. P. Nanda · S. Rishitha
M. S. Swaminathan School of Agriculture, Centurion University of Technology and Management, Paralakhemundi, Odisha, India
e-mail: sagar.maitra@cutm.ac.in; sumit.ray@cutm.ac.in; 240805190007@centurionuniv.edu.in; sairam.masina@cutm.ac.in; lalichetti.sagar@cutm.ac.in; spnanda@cutm.ac.in; 250506192001@centurionuniv.edu.in

A. J. Atapattu
Agronomy Division, Coconut Research Institute of Sri Lanka, Lunuwila, Sri Lanka

D. T. Santosh · D. J. Gaikwad
Research Centre of Excellence for Smart Agriculture, Centurion University of Technology and Management, Paralakhemundi, Odisha, India
e-mail: santosh.dt@cutm.ac.in; gaikwad@cutm.ac.in

H. I. Gitari
Department of Agricultural Science and Technology, School of Agriculture and Environmental Sciences, Kenyatta University, Nairobi, Kenya
e-mail: harun.gitari@ku.ac.ke

A. Das, A. B. Sharangi (eds.), *Remunerative Horticulture*,
https://doi.org/10.1007/978-981-92-1007-7_17

damages, proper steps must be adopted, including efficient packaging and IoT (Internet of Things)-enabled cold storage automation, to ensure a steady income for farmers. Several agroecological steps are to be considered such as crop diversification, agroforestry, specialty crop cultivation, urban horticulture like rooftop, vertical, and terrace gardening, soilless cultivation, community or institutional gardens, edible landscapes, green walls, etc., which can contribute to landscape-level adaptations and carbon sequestration while aligning with various United Nations Sustainable Development Goals (SDGs). Policy support for climate-smart horticultural crop cultivation, along with its exposure to farmers, can ensure the sustainable growth of horticulture in the current era of climate change (CC). Now, the present situation warrants a paradigm shift, integrating innovative technologies, ecological stewardship, and earning consumer currencies for a future-ready horticulture that can flourish under the current scenario of adverse climatic conditions. The chapter focuses on climate-smart horticultural crop production, integrating all possible innovative technologies with future directions.

Keywords

Automation · Climate resilience · Future-ready horticulture · Protected cultivation · Remunerative horticulture · Sustainable development goals · Yields

Abbreviations

ABA	Abscisic Acid
ACC	1-Aminocyclopropane-1-Carboxylic Acid
AI	Artificial Intelligence
APSIM	Agricultural Production Systems sIMulator
APX	Ascorbate Peroxidase
ATP	Adenosine Triphosphate
BR	Brassinosteroid
BS	Biostimulant

L. L. Chappa
Department of Agricultural Sciences, College of Agriculture, Mwalimu Nyerere University of Agriculture and Technology, Butiama, Tanzania
e-mail: luciana.chappa@mjnuat.ac.tz

N. R. Madhu
Department of Zoology, Acharya Prafulla Chandra College, Kolkata, West Bengal, India
e-mail: nithar@apccollege.ac.in

A. Hossain
Soil Science Division, Bangladesh Wheat and Maize Research Institute, Dinajpur, Bangladesh

CBF/DREB	C-repeat Binding Factor/Dehydration-Responsive Element-Binding
CBF	C-repeat Binding Factor
CC	Climate change
CEC	Controlled Environment Cultivation
Cl	Chloride ion
CO_2	Carbon Dioxide
CRH	Climate-resilient Horticulture
CRISPR-Cas9	Clustered Regularly Interspaced Short Palindromic Repeats-CRISPR Associated Protein 9
CSH	Climate-Smart Horticulture
DREB	Dehydration-Responsive Element-Binding protein
DSSAT	Decision Support System for Agro-technology Transfer
DST	Decision Support Tools
GHG	Greenhouse Gas
GWAS	Genome-Wide Association Studies
HSP	Heat Shock Protein
IAA	Indole-3-Acetic Acid
IBM	International Business Machines
IMD	India Meteorological Department
IoT	Internet of Things
JA	Jasmonic Acid
LED	Light Emitting Diode
MAP	Modified-Atmosphere Packaging
MAS	Marker-Assisted Selection
ML	Machine Learning
NHX1	Vacuolar Na^+/H^+Antiporter 1
NUE	Nutrient Use Efficiency
O_2	Oxygen
PDR	Pathogen-Derived Resistance
PGR	Plant Growth Regulators
PHM	Postharvest Management
ppm	parts per million
PPO	Polyphenol Oxidase
PSII	Photosystem II
QTLs	Quantitative Trait Loci
RNA	Ribonucleic Acid
RNAi	RNA interference
ROS	Reactive Oxygen Species
RZO	Root Zone Oxygen
SDG	Sustainable Development Goals
SLC	Soilless Cultivation
SMS	Short Message Service
SNP	Single-Nucleotide Polymorphisms
SOD	Superoxide Dismutase
SOS1	Salt Overly Sensitive 1
UV-C	Ultraviolet-C

17.1 Introduction

The recent agriculture is experiencing unprecedented climatic worries that include temperature fluctuations, inconsistent rainfall issues, extreme weather conditions, and enhanced threats from biotic agents harming crop yields. The climatic vagaries are triggered by increasing greenhouse gas (GHG) emissions, deforestation, growing global population, and shrinkage and misuse of natural resources, which raise the question of achieving food and nutritional security while focusing on agricultural sustainability by 2030 (Ray et al. 2025a; Maity et al. 2025; Santosh et al. 2023a; Ghosh et al. 2021; Hemasree et al. 2025; Kabir et al. 2023; Maitra et al. 2024a). Among various crop cultivations, horticulture is the cornerstone for diversity and nutrient-enriched food production. However, horticulture is the most affected sector in the changing climatic scenario (Chandio et al. 2025). The reliance on ecological balance and a congenial environment favors the growth and yield of horticultural crops (Kumar et al. 2023). The adverse effect of CC is pronounced in vegetables, fruits, spices, and plantation crops. An earlier study documented that the temperature variation in fruit crops, drought or waterlogging in vegetables, and disturbances in pollination in nuts and berries are common due to CC (Liu et al. 2023; Tchonkouang et al. 2024). In tropical regions, pest and disease problems in vegetables occur with an enhancement in humidity and temperature (Bhandari et al. 2021). Hence, amid such tumult, an adaptive ingenuity strategy is climate-resilient horticulture (CRH). CRH addresses these challenges by integrating modern technologies and traditional knowledge (Sharma 2013). It stands on multiple key pillars, each of which indicates specific vulnerabilities and aims to develop production systems that can withstand plants in biotic (such as pests and diseases) and abiotic (including drought, elevated CO_2 levels, temperature stresses) stresses (Fig. 17.1).

Agroecological principles, genetic innovation, and precision management are adopted for adaptability against climatic volatility (Sairam et al. 2023a, 2025a; Sagar et al. 2024). Cultivation of abiotic stress-tolerant varieties such as heat-adaptive tomato hybrids and flood-resistant leafy greens, adoption of multifunctional agroforestry models for microclimate modification, development and integration of data-driven tools such as remote sensing and AI-optimized irrigation, enhancing soil organic carbon through cover cropping and biochar amendments, and adoption of protected cultivation and postharvest handling are commonly practiced (Singh et al. 2021a, b; Datta et al. 2023; Sheeba et al. 2025). CRH is an integrated approach that combines innovative technologies, adaptive strategies, and ecological stewardship, as well as socioeconomic adoption of technologies such as remote sensing, IoT-based monitoring and management, and AI-based resource use, not only to buffer against CC but also to promote long-term sustainability (Ray et al. 2024; Roy et al. 2025). Climate-resilient horticulture is not for the enhancement of production, but it also extends beyond the productivity of horticultural crops. Due to a lack of proper handling, inadequate storage, transportation, and processing infrastructure globally, almost 20–40% of horticultural produce is lost (Jarman et al. 2023). Postharvest management is meaningful in reducing postharvest losses and increasing crop profitability. Moreover, CRH is aligned with the United Nations

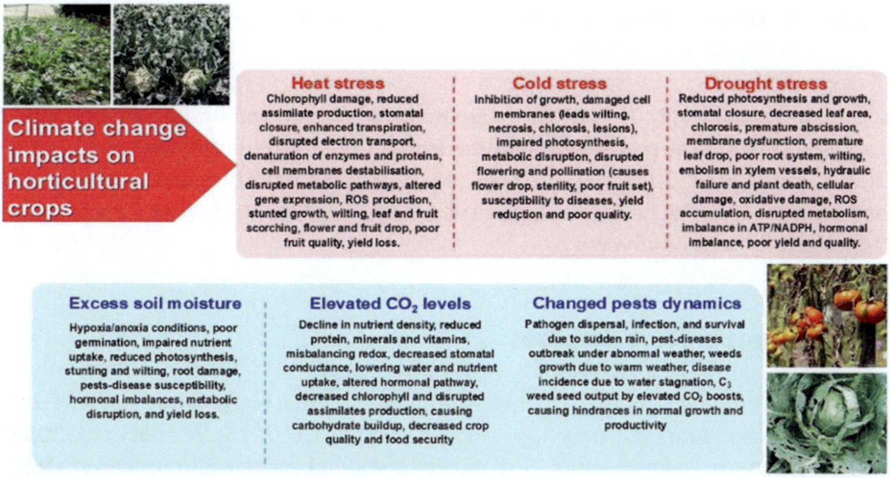

Fig. 17.1 Flow chart illustrating the impacts of climate change on horticultural crops. The chart is divided into sections detailing various stress factors: heat stress, cold stress, drought stress, excess soil moisture, elevated CO_2 levels, and changed pest dynamics. Each section lists specific effects on crops, such as chlorophyll damage, reduced photosynthesis, and increased susceptibility to diseases. The chart highlights the complex interactions between environmental stressors and crop health, emphasizing yield loss and quality reduction. Images of affected crops are included for visual context

Sustainable Development Goals (SDG), serving as a linchpin for interconnecting it with sustainability. CRH aligns with several SDGs such as "Zero Hunger (SDG 2) by safeguarding food and nutritional security, Good Health and Well-Being (SDG 3) by enhancing nutritional content and delivery, Sustainable Cities and Communities (SDG 11) through vertical and soilless farming, Responsible Consumption and Production (SDG 12) by ensuring precise input use and good quality produce, Climate Action (SDG 13) through low GHG emission practices and carbon sequestration and Life on Land (SDG 15) via rejuvenating various biodiversity corridors and shelters for pollinators." The interconnection of horticulture with several SDGs creates synergies for sustainable development. Furthermore, such efforts support ecosystem services, namely, "provisioning (abundance and quality of produce), regulating (climatic and microclimatic moderations), supporting (nutrient cycling and shelters for various birds and pollinators) and cultural (recreational greenspace for nurturing cultural heritage)." Additionally, such a strategy provides a year-round income to the farming community. The adaptation and success of CSH focus on technological awareness and adoption among farmers, knowledge sharing among themselves, research and development, blending technology with local validation, and strong policy support. In this chapter, CRH evolved as a proactive blueprint for horticultural farming while being synonymous with planetary stewardship, equitable prosperity, and enduring vitality.

17.2 Climate Change Impacts on Horticultural Crop Production

CC manifests through distinct yet interconnected abiotic and biotic stresses, which decline horticultural production and the growth of several crops (Manepalli et al. 2022; Bhutia et al. 2023). The impact of CC disrupts physiological, phenological, and ecological dimensions, resulting in declining yields of horticultural crops, nutritional quality, and biodiversity.

17.2.1 Temperature Stresses

Since the pre-industrial era, global warming has caused a temperature rise of 1.1 °C and is predicted to increase by 1.8–4 °C by 2100 (IPCC 2018; Warren et al. 2024). Such an increase in temperature will negatively affect the horticultural crop yields. High temperature causes flower drop in some fruits like apples, cherries, and almonds, resulting in an inferior fruit set (Singh et al. 2016; Sønsteby and Heide 2019; González-Martínez et al. 2026). An atmospheric temperature of more than 30 °C results in embryo abortion in peppers and tomatoes, which lowers yield by 20–50% (Tandel et al. 2025). Heat stress reduces sugar content in strawberries and grapes (Lv et al. 2022; Hewitt et al. 2023). Moreover, high temperatures exceeding 30 °C in the reproductive stage decline yield by 20–50% in several horticultural crops (Beltrán et al. 2019). For flowers such as petunia and rose, reduced dry matter accumulation and flower quality are observed under heat stress conditions (Kim 2018; Borghi et al. 2019). High temperatures malform the tuber and lower the nutrient content in potatoes (Mukherjee et al. 2024).

Cold stress occurs when temperature reaches below the threshold levels and it affects the normal growth of horticultural crops. In general, low-temperature stress is of two types, namely chilling (0–15 °C) and freezing (less than 0 °C). Under cold stress, cell membranes in plants are damaged, which reduces photosynthesis and assimilate production. Cold temperature also alters metabolism and plants may wilt under severe conditions. Tomatoes and potatoes exhibit sensitivity to chilling temperatures (Roychowdhury et al. 2025). Tomatoes are sensitive when exposed to temperatures below 10 °C (Atayee and Noori 2020). Under cold stress, *Brassica* vegetables are adversely affected by a decline in chlorophyll and photosynthesis (Raza et al. 2024). Further, *Brassica* vegetables enhance lipid peroxidation, which checks the growth. Seed germination is another physiological process that is hampered in *Brassica rapa* when the atmospheric temperature comes down to below 12 °C. Cabbage shows a delayed head formation due to cold stress (Ahmad et al. 2024). Olive trees grown in the Mediterranean region are affected by cold stress when the temperature ranges between 0 °C and − 3 °C, with mild damage, resulting in lowered growth; however, when the temperature falls below −18 °C, severe damage is noted (Petruccelli et al. 2022). Chilling injury in *citrus* reduces leaf growth and chlorophyll content and creates tissue necrosis. Freezing injury is known to damage the cell membrane due to cold-induced ROS production (Kim 2018).

17.2.2 Drought and Flood Stresses

Unstable hydrological regimes and high evaporation due to enhanced temperature and solar radiation, and irregular rain lead to drought and flood, which adversely impact the normal growth process of crops (Nabinejad and Schüttrumpf 2023). Most vegetables, which are shallow-rooted crops, require frequent irrigation and suffer severely due to soil moisture shortage (Fanourakis et al. 2025). Drought is a shortage of soil moisture, which reduces photosynthate production, restricts CO_2 uptake and ROS production, alters growth hormone production, and initiates stomatal closure (Li and Liu 2022). Severe drought causes wilting, and in extreme cases, plant death occurs. Beyond the general symptoms and physiological incidences, different crop species show specific symptoms. In Brassica crops, poor germination, seedling health, photosynthesis, stomatal conductance and transpiration, elevated ROS production and lipid peroxidation, changes in hormonal regulation, stunted growth, and bolting are very common effects (Mohan et al. 2025). In almonds, yield reduction is recorded due to deficient rain (Osorio-Marín et al. 2024). In ornamental crops such as *Begonia semperflorens,* reduced chlorophyll content and, for *Chrysanthemum morifolium,* ROS-induced petal wilting were observed (Fanourakis et al. 2025; Karagüzel 2025). Flooding and water stagnation inhibit root respiration, reduce nutrient uptake under hypoxic conditions, and cause root decay with increased ethylene concentration in crops (Manghwar et al. 2024). Crops such as papaya, banana, onion, potato, and garlic are sensitive to waterlogging, triggering root rot and reduced yields (Ghodke et al. 2018). In nuts, waterlogging can reduce yield by 25% because of hypoxia-caused root dieback (Ananth et al. 2025). Furthermore, drought and flood conditions directly affect microbial symbioses and nutrient availability, where drought leads to a decrease in phosphorus uptake, while flood triggers the leaching of nitrogen.

17.2.3 Shifts in Pest, Disease, and Weed Population Dynamics

A warm climate and an increase in relative humidity promote weeds, insects, and diseases and accelerate life cycles (Subedi et al. 2023). CC reshaped the dynamics of pest, disease, and weed in horticultural systems. Irregular rainfall patterns favor pathogen dispersal, infection, and survival (Lahlali et al. 2024). Elevated moisture levels promote spore germination in fungi such as powdery mildew on cucurbits (Rivera-Romero et al. 2024). Soil-borne diseases also increase, such as *Phytophthora* root rot in citrus, which weakens the root system (Chaudhary et al. 2016). *Fusarium species* in banana cause a severe problem in warm climates, and an increased temperature favors weed infestation (Mohite et al. 2024; Ramesh et al. 2017). Furthermore, temperature increases also lead to early emergence of *Cydia pomonella*, *Anarsia lineatella*, and *Grapholita molesta* in almond and walnut orchards (Osorio-Marín et al. 2024). In addition, higher CO_2 boosts the C_3 weed seed output by 44%, while warming triggers the population of C_4 weed species, such as *Digitaria sanguinalis*, in Christmas tree production (Mou et al. 2025).

Moreover, a synergistic effect of an increase in weed and pest populations was observed in the North Carolina Fraser fir plantation (Jha et al. 2023). Consequently, this shift in biotic dynamics not only disrupts the food supply chain but also declines beneficial insects and increases the pest population.

17.2.4 Elevated Carbon Dioxide Levels

The atmospheric CO_2 has risen from 280 ppm to 420 ppm in 2025 and is projected to reach 800 ppm by the end of the century (Statista 2025). The increase in CO_2 enhances biomass production but decreases the nutritional quality and yield, hence acting as a double-edged sword in horticulture (Dong et al. 2018). Increased atmospheric CO_2 produces more biomass in C_3 horticultural crops (Rangaswamy et al. 2021; Namizaki et al. 2022). In strawberries, high CO_2 concentration degrades nutritional qualities by reducing vitamin C and phenols (Balasooriya et al. 2017). Leafy greens under 800 ppm of CO_2 level increase biomass by 17–19%; however, nutrient qualities were dropped, probably due to the dilution effect (Doddrell et al. 2023). Additionally, in mango, an elevated CO_2 level creates spongy tissue and malformation (Ullah et al. 2010). Changing the C:N ratio increases carbohydrate storage and sugar synthesis over protein synthesis. Strawberries produce reduced levels of antioxidants and vitamin C under elevated CO_2 levels, whereas ornamental crops produce inferior foliage and inflorescence (Croonenborghs et al. 2009). Rising CO_2 levels dilute nutrient concentration in several fruits and vegetables and decline their nutritional quality.

17.2.5 Extreme Weather Calamities

CC induces storms, heatwaves, frost, hailstorms, and cyclones, which directly affect horticultural crops (Ray et al. 2025b). Hailstorms reduce the yields of grape, mango, and litchi; however, high wind speed causes fruit damage, yield reduction, and economic loss (Ruland et al. 2023; Müller et al. 2023). Due to heatwaves, tomatoes, capsicum, and mangoes experience fruit drop (Raju et al. 2024). Frost causes delayed germination, flower dropping, and dieback to several horticultural crops. Cyclones uproot perennial fruits, and the extreme climatic events disturb the supply chain, causing the vulnerability of smallholders involved in horticulture, hampering their livelihood.

17.3 Crop Improvement for Climate-Resilient Horticulture

CC causes physiological changes in horticultural crops. Hence, the crop improvement strategies can enhance resilience to the adverse effects of CC. The following parts are mimicked to create crop improvement for climate-resilient horticulture.

17.3.1 Conventional Breeding and Germplasm Enhancement

Conventional breeding techniques, namely, hybridization and selection, are important for developing climate-resilient horticultural cultivars (Acquaah 2015). Over the years, newly developed cultivars have become popular among farmers (Table 17.1). In tomato, breeding for drought stress tolerance incorporates genes from *Solanum pennellii,* which enhance osmotic regulation and maintain turgor pressure (Pessoa et al. 2023). For heat-stress-tolerant varieties selected for heat shock protein (HSP) expression, which improve pollen germination and fruit set under increasing temperatures (Qian et al. 2025). Hybridization among varieties and mutation breeding can create new traits for stress tolerance. Marker-assisted selection (MAS) can support to develop new cultivars by linking with stress tolerance genes (Kumari et al. 2024). This technique facilitates early and precise selection of genes. Breeding and germplasm enhancement develop new varieties that align with local climate conditions while mitigating stresses. Presently, several cultivars of horticultural crops have been developed with varied levels of stress tolerance, some of which are presented below (Table 17.1).

17.3.2 Molecular and Genomic Tools

Genomic advancement is a technique of gene manipulation for plant breeding. Genome-wide association studies (GWAS) and next-generation sequencing facilitate the identification of single-nucleotide polymorphisms (SNPs) and quantitative trait loci (QTLs) targeting stress tolerance (Harris et al. 2024). Genomic selection models also develop multi-stress-tolerant varieties. Transcriptomics and proteomics reveal stress-response pathways. In brassicas, the implementation of aquaporin genes enhances water transport ability under salt stress, and proteomics identifies ROS enzymes such as superoxide dismutase (SOD) for the correction of heat stress (Yang et al. 2022). These tools integrate with breeding techniques to enhance the breeding efficiency of crops toward climate resilience.

Genetic engineering has revolutionized breeding techniques by modifying genetic material to develop crops with desirable traits (Hamdan and Tan 2024). Such improved breeding eradicated the limitations of conventional breeding techniques by altering stress-responsive genes. Integrating modern tools with conventional breeding approaches enhanced yield stability against climatic stresses (Table 17.2).

17.4 Controlled Environment Cultivation

CC reshapes agriculture as well as horticulture. The cultivation of horticultural crops under a controlled environment, such as a greenhouse, where crops can adapt regardless of natural climatic variables, is also known as controlled environment cultivation (CEC) (Espeland and Kettenring 2018). CEC encompasses greenhouses,

Table 17.1 Table for stress-tolerant cultivars of horticultural crops developed through the conventional breeding technologies, showing the crop, variety, breeding specification and parental lines, year of release, abiotic stress tolerance, and references

Sl. No.	Crop	Variety	Breeding specification and parental lines	Year of release	Abiotic stress tolerance	References
1	Coriander	RD 385 (RPCAU DHANIA-1)	Selection from local germplasm	2019	Resistance to lodging	Aravind et al. (2020)
2	Fenugreek	Narendra Richa	Selection from local landrace	2018	Alkaline tolerance	
3	Black pepper	Panniyur 9	Open-pollinated progeny of Panniyur 3	2018	Drought-tolerant	
4	Tomato	Pusa Shakti	Open-pollinated variety	2024	Tolerant to high temperature	PIB (2024)
5	Pomegranate	Solapur Anardana	Three-way cross of Bhagawa × [(Ganesh × Nana) × Daru].	2024	Drought-tolerant	
6	Velvet bean	Arka Dhanvantari	Open Pollinated	2024	Highly drought-tolerant	
7	Cocoa	Vittal cocoa Hybrid-2	IC 565551 × IC 565556	2024	Moisture-stress tolerant	
8	Potato	Kufri Bhaskar	Kufri Khyati × Kufri Surya	2024	Tolerant to high temperature	
9	Muskmelon	Thar Mahima	Selection process from local landraces	2024	Tolerant to high temperature	
10	Brinjal	VRM1	Selection from local landrace	2010	Drought-tolerant	TNAU portal
11	Bottle gourd hybrid	CO1	F_1 hybrid crossing NDBG 121 x Arka Bahar	2012	Drought- and heat-tolerant	
12	French beans	Ooty 3	Pure line selection	2016	Salinity-tolerant	
13	Onion	K41	Mass selection	2019	Salt tolerance	
14	Tapioca	YTP 2	Selection from germplasm collection	2019	Drought-tolerant	
15	Coconut	VPM 5	Hybrid between LOT X CCHT	2014	Drought-tolerant	

Table 17.2 Table mentioning different crops, improved cultivars developed by modern breeding techniques, genetic engineering techniques, resistance developed, and references

Crop	Improved cultivars developed by modern breeding techniques	Genetic engineering technique	Resistance/tolerant	Reference
Apple	Arctic® Granny, Arctic® Golden, Arctic® Fuji, Arctic® Gala	RNA interference (RNAi) silencing of PPO genes	Polyphenol oxidase (PPO) enzymatic browning	Lobato-Gómez et al. (2021); Zhang (2023)
Banana	Rasthali	Agrobacterium-mediated transformation	Fusarium wilt resistance	Chakrabarti et al. (2002); Tripathi et al. (2019)
	Williams	Agrobacterium-mediated transformation	Nematode resistance	Atkinson et al. (2004)
Potato	GM potato, Victoria, Shangi, russet Burbank, ranger russet, Atlantic, Katahdin, Desiree,	RNA interference (RNAi), also gene silencing of PPO and acrylamide-forming enzymes, and transgenic R-gene transfer (*RB*, *Rpi-blb2*)	Late blight, polyphenol oxidase (PPO) enzymatic browning	Van Der Vossen et al. (2005); Haverkort et al. (2008); Haesaert et al. (2015); Orbegozo et al. (2016); Paluchowska et al. (2022); Moyo et al. (2024)
Tomato	GABA tomato, virus-resistant GM tomato, Moneymaker, UC82, Roma, Pusa Ruby,	Conventional transgenic approach	Tomato yellow leaf curl virus, γ-aminobutyric acid	Kunik et al. (1994); Ammara et al. (2015); Wang et al. (2019); Pramanik et al. (2021); Ezura (2022)

(continued)

Table 17.2 (continued)

Crop	Improved cultivars developed by modern breeding techniques	Genetic engineering technique	Resistance/tolerant	Reference
Brinjal	Bt brinjal	*Cry1Ac* insertion (Bt expression)	Brinjal fruit borer	Shelton et al. (2019); Kumar et al. (2025)
Papaya	SunUp, rainbow, Khak Dum, Khak Nual, Huanong –1	Agrobacterium-mediated transformation + pathogen-derived resistance (PDR) via coat protein/RNA silencing, coat protein-mediated RNA-silencing mechanism, and replicase gene-mediated	Papaya ringspot virus	Gonsalves et al. (2004); Sakuanrungsirikul et al. (2005); Azad et al. (2014); Hamim et al. (2018)
Carnation	Florigene moon	Conventional transgenic metabolic engineering (*Agrobacterium*)	Redirection of anthocyanin biosynthesis to enhance flower color	Anderson and Walker (2013)
Squash	ZW-20, CZW-3	Conventional transgenic pathogen-derived resistance	Potyviruses such as zucchini yellow mosaic virus, watermelon mosaic virus, cucumber mosaic virus	Prendeville and Pilson (2009)
Grape	Thompson seedless, VvMLO6, VvMLO7	CRISPR-Cas9	Powdery mildew	Scintilla et al. (2021); Wan et al. (2020); Moffa et al. (2024)
Cucumber	Chinese long 9930, Xintai Mici, R1461	CRISPR-CsMLO8	Powdery mildew	Zhang et al. (2020); Shnaider et al. (2022); Tek et al. (2022); Dong et al. (2023)
Strawberry	Chandler	Conventional transgenesis through *Agrobacterium tumefaciens*-mediated transformation	Salt stress tolerance	Husaini and Abdin (2008)
	Woodland	CRISPR/Cas9	Cold and drought tolerance	Han et al. (2023a)
	Benihope	*Agrobacterium*-mediated CRISPR/Cas9	Powdery mildew resistance	Akter et al. (2024)
	Honeoye	*Agrobacterium*-mediated transformation	Frost tolerance	Owens et al. (2002)

hydroponics, aquaponics, and vertical farming, all of which support ecological balance and alignment with the SDGs. CEC enhances land, water, and input use efficiency, stabilizing yields and quality as an assured income.

17.4.1 Greenhouses

Greenhouse restricts the high intensity of solar radiation and excess rainfall, as crops grown inside enjoy a soothing environment and microclimate. In this way, it protects crops grown in the CEC against these two climatic extremes (Maitra et al. 2024b). In recent times, automation has added further value to CEC. Integrating climate sensors, ventilation, and IoT-enabled irrigation for microclimate modification in an inner environment creates favorable conditions. It ensures congenial conditions for crops such as cucumbers, strawberries, capsicum, roses, and gerbera in the tropical and subtropical conditions (Bisht et al. 2025). In gerbera, crops grown under protected cultivation recorded the highest plant growth parameters and flower yield (Sairam et al. 2022, 2025b; Maitra et al. 2024b). Capsicum grown under protected cultivation also enhanced the yield and quality (Maitra et al. 2024c; Santosh et al. 2024). Greenhouses protect crops from pests. Adopting the greenhouse technique creates a favorable microclimate for crops irrespective of seasonal limitations. The systems regulate temperature, humidity, and light, while also protecting plants from climatic abnormalities. Such structures are suitable for the cultivation of off-season and high-value horticultural crops (Fig. 17.2).

17.4.2 Urban Horticulture and Soilless Cultivation

Urban horticulture is the concept of crop cultivation in city spaces, including smart farming techniques, rooftop gardening, vertical farming, hydroponics, aeroponics, etc., which maximize the utilization of limited land and resources (Sashika et al. 2024). Rooftop gardens and vertical farms optimize space utility in cities. Vertical farms, using stacked hydroponics or aeroponics systems, can produce leafy vegetables with less water than soil-based farming, countering the water scarcity problem in urban areas (Garzón et al. 2023; Roy et al. 2024). Due to the lack of space and water for crop production in urban areas, specialty crops such as microgreens, herbs, and medicinal plants (e.g., tulsi) are highly valued. Tulsi can tolerate urban salinity, stabilize soil, and fetch premium prices (Jagdish 2025). Urban horticulture also synergizes with climate-smart technologies. IoT sensors monitoring temperature, humidity, and nutrient levels, along with optimized irrigation systems, improve water-use efficiency as well as boost NUE and support crop quality under stresses (Kumar et al. 2025; Khambalkar et al. 2025). Economically, urban horticulture boosts local economies while enhancing urban biodiversity (Fig. 17.3).

Soilless cultivation (SLC) emerged at a high pace in urban areas, aligning with the principles of climate-smart horticulture (Lakhiar et al. 2025). SLC is the cultivation of crops using nutrient-enriched water or inert media rather than using natural

Fig. 17.2 A collage of four photographs showing different plants in a greenhouse setting. Top left: A colored bell pepper plant with a ripe red pepper. Top right: A bed of vibrant red and pink gerbera flowers. Bottom left: Dendrobium orchids with purple blooms in pots under a green shade net. Bottom right: Parthenocarpic cucumber plants with lush green leaves growing in a covered area. Each section is labeled with a yellow tag indicating the plant type

Fig. 17.3 Hydroponic systems in a greenhouse setting. The left image shows a blue and white pipe structure supporting lush green leafy plants. The right image features a similar setup with a different type of leafy greens, both thriving in a controlled environment. The greenhouse background includes other plants and equipment

soil (Tuxun et al. 2025). SLC provides optimal root-zone conditions through precise management of nutrients, water, oxygen, and pH (Rajaseger et al. 2023). The root system acts as both an anchor and a nutrient extraction organ, supported by physical media (such as perlite, cocopeat, vermiculite, rock wool, and so on) and nutrient-enriched solutions (Baiyin et al. 2021; Sanchez et al. 2021). SLC includes hydroponics, aeroponics, aquaponics, and vertical farming. The hydroponic technique eliminates soil-related constraints and grows plants in nutrient-rich solutions (Rajendran et al. 2024). In natural soil, during waterlogging, the root zone oxygen (RZO) drops below 5 mg L^{-1}, leading to fermentative metabolism, ethylene accumulation, and root decay (Zhang et al. 2025); whereas in aeroponics, RZO 18 mg L^{-1} withstands cytochrome oxidase activity, which increases the ATP yield by 25–40% (Vanlerberghe 2013). In the present day, commercially managed SLC farms have been automated with sensors, IoT-enabled management, and AI-driven input supply (Mondal et al. 2024; Rajendran et al. 2024).

Crops like leafy vegetables and herbs grow faster in SLC. Lettuce recorded an enhanced number of leaves, height, and leaf fresh weight in hydroponic conditions (Munda et al. 2025; Sahu et al. 2025). Hydroponics ensures consistent production in regions where drought or soil degradation is more prevalent (Rajaseger et al. 2023). On the other hand, an aquaponics system creates a symbiotic relationship between plants and fish by integrating fish farming in agriculture (Ibrahim et al. 2023). Horticultural crops, namely, basil and tomatoes, are paired with tilapia and thrive in fish aquaponic systems. Aeroponic systems mist nutrient-rich solution, keeping the plant roots moist and supplying sufficient nutrients (Mahmood and Al-Ansari 2023).

17.4.3 Vertical Farming

Vertical farming is a multilayered indoor system that uses artificial lights and automated techniques (Roy et al. 2024). This method is suitable for urban areas by optimizing light spectra and growth cycles of the crop. Microgreens, herbs, saffron, and strawberries can be grown with low-carbon agriculture systems (Erekath et al. 2024). It may replace sunlight with LED lights and soil with nutrient-rich mist, which supports crop growth in extreme weather conditions (Gunapala et al. 2025). Solar energy can be considered a source of energy for LED lights, which are ecologically friendly. Vertical farming maximizes productivity per unit area while reducing or minimizing resource use (Fig. 17.4).

Fig. 17.4 Rows of potted plants with lush green foliage are arranged on a metal rack in a greenhouse setting. On the right, a vertical garden setup displays multiple tiers of leafy greens growing in a controlled environment. The scene suggests a focus on efficient space utilization for plant cultivation

17.5 Smart Irrigation Systems

Herbaceous horticultural crops are sensitive to water-deficient conditions. CC exaggerates extreme weather conditions, such as uneven distribution of rain and intermittent dry spells (Nhamo et al. 2025). Traditional irrigation methods are associated with low water-use efficiency when losses are featured due to conveyance losses of water in irrigation channels through seepage, percolation, and evaporation and energy waste and GHG emissions. Conventional irrigation systems cannot provide site- and plant-specific water as per crop needs. In this regard, the precision irrigation system plays a vital role, as IoT remains at the cornerstone (Gupta et al. 2025). Soil moisture sensors, microcomputers, and actuator-equipped IoT systems provide the precise quantity of water for crop needs, or based on an AI-enabled irrigation schedule and regime (Jeanguenin et al. 2016; Sagar et al. 2025). It has been proven that sensor-based instruments enable the indication of the water needs of the crop. Moreover, to enhance irrigation efficiency, a micro-irrigation system is to be adopted, which is an appropriate method of irrigating the rhizosphere with less water (Obaideen et al. 2022). An intervention like a smart irrigation system can ensure customized irrigation targeting climate-resilient horticulture.

17.5.1 Micro-irrigation Systems

Drip and sprinklers are micro-irrigation systems that ensure efficient water use. Drip irrigation system pours water dropwise to the rhizosphere directly and hydrates the active root zone (Santosh et al. 2021). Micro-irrigation contains a network of tubes and emitters, saving about 30–60% water loss. Drip irrigation can be integrated with soil-moisture sensors and timers for enhancing resilience against various stresses (Santosh et al. 2023b). However, a sprinkler system mimics rainfall by evenly distributing water (Santosh et al. 2021). In specialty crops, drip and spray stakes in soilless media cut off 20–50% water by reducing over-irrigation, and sensor-based scheduling can be installed (Majsztrik et al. 2017). In crops such as mango, yield increases by 40% under drip irrigation while decreasing labor needs for surface irrigation (Sowjanya 2012). Adding IoT and sensors into a modern sprinkler system for better water management based on soil moisture levels. These systems prevent the possibility of waterlogging and prevent nutrient leaching (Oğuztürk 2025). Integrating weather data with sprinkler systems improves resource use efficiency while reducing energy costs. It ensures the uniform distribution of water and increases crop yield. Moreover, integration of mulching and drip irrigation facilitates an efficient water and nutrient management as drip irrigation creates the scope of fertigation, which enhances nutrient use efficiency. Earlier studies proved that drip irrigation and mulching were beneficial for fruits, vegetables, and flowers, even in controlled environment cultivations (Santosh et al. 2021, 2022a, b, 2023a, b, 2024) (Fig. 17.5).

Fig. 17.5 A series of eight labeled photographs (A to H) depicting different agricultural scenes. (**a**) A grove of trees with visible trunks and sparse ground cover. (**b**) A small orchard with young trees planted in rows on dry soil. (**c**) A banana plantation with trees bearing fruit, surrounded by mulch. (**d**) Rows of plants covered with white plastic mulch, showing young seedlings. (**e**) A field of leafy green vegetables, possibly cabbage, in neat rows. (**f**) A field with young plants in evenly spaced rows on bare soil. (**g**) A greenhouse interior with rows of plants growing in raised beds. (**h**) A small circular garden bed with dense green foliage, highlighted by a red circle. Each image highlights different agricultural practices and plant growth stages

17.5.2 Subsurface Irrigation

Subsurface irrigation delivers irrigation directly to the rhizosphere, checking over-watering and water losses (Santosh et al. 2022a). Adding IoT and multiple sensor-enabled irrigation systems ensures producing high-quality production with judicious water and energy investment.

17.6 Climate-Smart Crop Management for Remunerative Horticulture

Climatic variability registers an impact on the horticulture landscape. Horticultural crops are essential for global food security and economic stability. CC disrupts physiological processes, resulting in yield loss and qualitative degradation of horticultural crops (Shah et al. 2024). In this regard, climate-smart crop management is essential for sustainable horticultural crop production, where decision support tools (DSTs) are integral components.

DSTs gather data from soil sensors, nearby or connected weather stations, satellites, and crop models to provide real-time recommendations for crop management (Sagar et al. 2024). DSTs fit in information from several sources to assist all stakeholders in making informed decisions based on models simulated (Mabhaudhi et al. 2023). Crop responses to climatic conditions, nutrient levels, and pest incidence possibilities can be predicted by DSTs using AI and ML (Kinsky 2025). Currently, technological developments have supported using proximal and remote sensing, including drones equipped with multispectral and hyperspectral cameras, which can monitor crop health, pests, and disease attacks (Alsadik et al. 2024). Several mobile apps are available that alert farmers to weather and pest attacks. Crop models like APSIM and DSSAT can support farmers in scheduling crop management practices and selecting varieties with predicted output for horticultural crops (Wajid et al. 2021; Sairam et al. 2023b; Banerjee et al. 2025; Sagar et al. 2025).

17.7 Weather Analytics and Warning Systems

CC imposes significant physiological changes on horticultural crops. Extreme climatic factors cause yield loss, increasing pathogen susceptibility, and reducing plant biomass (Mukesh et al. 2024). Flooding increases anaerobic conditions, inducing root rot and elevating reactive oxygen species (ROS) (Sasidharan et al. 2017). Weather analytics and warning systems alleviate the effects of CC to some extent by enabling preemptive strategies to maintain yield stability (Reichstein et al. 2025). Weather analytics systems refer to the integration of data driven from meteorological stations, satellites, drones, and real-time weather monitoring (Wu and Xue 2024). By processing, modeling, and interpretation techniques, it monitors temperature, humidity, rainfall, and solar radiation. AI and machine learning are also used for a weather analytics system to predict climate risks and weather patterns (Waqas et al. 2024). Weather stations and sensors continuously collect weather data for interpretation and to mitigate climate risks. Platforms such as IBM's Watson Decision Platform and India's Agromet Services (IMD) integrate local data with crop models to predict climate risks (Das and Nayak 2024). The India Meteorological Department, in collaboration with State Agricultural Universities/Indian Council of Agricultural Research, implemented a scheme in Maharashtra named IMD's Gramin Krishi Mausam Sewa, which delivers 5-day district-level forecasts and helps in scheduling sowing and irrigation (Sharma 2024).

Weather warning systems provide alerts via SMS for extreme climate events. Kisan Suvidha is one app (quite popular in India) that delivers alerts via SMS by using automated weather stations and satellite (Kisan Suvidha App Kisan Suvidha). Weather analytics improve climate resilience for managing crops under adverse conditions (Tamru et al. 2025). In nuts and grapes, farmers avail AI advisories for frost and heatwaves, which help them cover the crop canopy in a timely manner and decrease yield loss by 20% (Farmonaut 2026). The crop model APSIM predicts erratic rainfall and optimizes irrigation for yield stability (Mamassi et al. 2025). Scaling these technologies through policy will enhance resilient horticulture farming and food security (Harsanto et al. 2025). Integrating meteorological forecasts with offers timely alerts to stakeholders.

17.8 Crop Diversification

Horticultural production systems experience hazards under CC. Crop diversification is a keystone for mitigating the negative impacts of CC (Santosh et al. 2023a). Crop diversification is a system of cultivating multiple species or varieties of crops cultivated in a farming system (Maitra and Gitari 2020). It is crucial in climate-resilient horticulture by involving intercropping, crop rotation, and agroforestry (Mihrete and Mihretu 2025). Integrating legumes in cropping systems enhances soil health as legume crops fix atmospheric nitrogen into the soil. Intercropping with drought-tolerant and water-sensitive millets and vegetables stabilizes productivity during dry spells (Maitra et al. 2024b). Diversification balances the nutrient demands of the crops and reduces pest and disease attacks. It also increases the economic incomes of the farmers and reduces the possibility of crop failure. Different root architectures improve water and nutrient uptake and maintain photosynthesis under stress conditions (Shoaib et al. 2022).

Integrating different species that have synergistic effects enhances crop resilience (Maitra et al. 2021a, b; Manasa et al. 2021). Intercropping of pineapple, ber, and dragon fruit in mango-based systems achieved a land equivalent ratio of 1.68, which was more efficient than monocultures (Kishore et al. 2025). In drought-prone areas, the combination of high-value pomegranate with pulse crops optimizes water-use efficiency and increases pomegranate yield (Bhatia et al. 2024). Cultivating the same crop with different varieties to minimize different stress risks. In the case of grape cultivation, a combination of early- and late-maturing varieties helps mitigate heat stress and provides a better fruit quality (FarmAtma 2025). Varietal diversity increases physiological stability by maintaining turgor pressure and reducing ROS-induced damage during stress. Crop rotation improves soil fertility and breaks the disease cycle. Cultivating cover crops such as mustard before vegetable crops enhances water retention, maintains nutrient balance, and suppresses weeds (Brennan and Smith 2018). Enhancing system efficiency through better use of space, time, and inputs can also be done through intercropping systems (Maitra et al. 2019; Nori et al. 2025). The capability of intercropping systems by creating microclimate, nutrient cycling, enhancing soil fertility, supporting ecosystem services, etc., can be integrated in a horticultural system (Ray et al. 2025b; Santosh et al. 2022b; Maitra et al. (2021c). The spacing between trees in an orchard provides sufficient space to integrate other crops and generate income during the initial period, which helps in decreasing the economic pressure on the farmer (Burgess et al. 2022). When the orchard canopy increases, intercropping shade-loving crops, namely, turmeric, ginger, spinach, and other leafy vegetables, can be done. Crop diversification enhances climate resilience by increasing yields and soil health while reducing input costs. It increases cropping intensity and boosts food security. Diversification increases farmers' income while minimizing market risks. It is pivotal for climate-resilient horticulture to mitigate different climate risks.

17.9 Biostimulants

Aside from CC, another reason for horticultural yield losses is disruption in plant physiology. Horticultural crops are more vulnerable to physiological changes due to high water and nutrient demand and are sensitive to abiotic stress. The smallest deviation in climate can lead to a larger physiological disorder. In such a context, biostimulants (BS) and plant growth regulators (PGRs) emerge as pivotal tools to foster climate resilience. Biostimulants are natural derivatives or synthetic materials that stimulate plant growth and enhance resilience, crop productivity, and product quality. BS includes humic acid, fulvic acid, seaweed extracts, microbial inoculants, amino acids, and PGRs, which modulate physiological processes at cellular and molecular levels, fine-tune the development processes, and stress-response pathways.

Seaweed extracts are rich in auxin, and cytokinin upregulates antioxidant enzyme cascade and scavenges reactive oxygen species (ROS), whereas microbial consortia enhance rhizosphere competence by solubilizing phosphates and producing siderophores. BS, such as *Kappaphycus alvarezii,* is rich in polysaccharides and phytohormones, which induce the expression of superoxide dismutase (SOD) and ascorbate peroxidase (APX). In tomatoes, the application of *Ascophyllum nodosum* stabilized PSII efficiency, enhancing proline by 45%, fruit set by 28%, and yield by 17% under combined heat and drought stress. Humic acid and fulvic acid directly modify the rhizosphere by improving soil structure and stimulating microbial activity. Fulvic acid activates the plasma membrane to hyperpolarize cells, which facilitates $K+$ influx under salinity through the penetrating root membrane. In an apple orchard grown in saline soil, application of humic acid mitigated $Cl-$ toxicity by downregulating SOS1 and upregulating NHX1 and preserving photosynthetic rates, which in turn decreased yield loss by 25% and enhanced fruit size (Eissa et al. 2007). In strawberries, pre-chilling spray of 1% proline BS-activated CBF/DREB regulons enhances cryoprotectant synthesis, i.e., raffinose and unsaturated fatty acids in phospholipids, which delays supercooling and ice nucleation (Buler and Nowak 2014). In the case of PGRs, the phytohormones reprogram the phytohormone signaling, enhance stress priming and tolerance through stomatal closure and osmolyte accumulation, and mediate the ROS burst. Under flooding or heat, PGRs such as ethylene inhibit ACC synthase, curbing climacteric bursts, which enhance abscission, delay ripening, and yield stability at 90%, which is observed in avocado. Under salinity, PGRs such as auxin stimulate lateral root emergence and enhance nutrient foraging, while cytokinin preserves shoot biomass and maintains meristematic activity. Under salinity conditions, application of IAA and benzyladenine enhanced yield by 80% in cucumber (Chen et al. 2019). The emerging PGRs such as jasmonic acid, strigolactones, brassinosteroids, and salicylic acid enhance climatic resilience through ROS homeostasis, root-microbe symbiosis, and osmotic defenses against drought, salinity, and temperature extremes (Hossain et al. 2021a, b, 2022a, b; Maitra et al. 2021a, b). ABA and jasmonic acid (JA) enhanced the yield by 15% while saving 18% of water in a citrus orchard, underscoring the ABA and JA impacts on water-stress conditions (Garner and Ingrao 2015). PGRs such as brassinosteroid (BRs) lead to upregulation of proline and antioxidants and reduced pitting and decay (Bisht et al. 2025).

17.10 Agroforestry Systems and Specialty Crops

Agroforestry is a multilayered system that integrates trees with horticultural crops, such as alley cropping or silvihorticulture. Integrating agroforestry and specialty crops, their synergistic effects are the heart of climate-resilient horticulture (Maitra et al. 2025). The designed system supports ecological balance, better resource use efficiency, and socioeconomic stability. The mimicking of natural forest through agroforestry not only enhances soil health but also supports carbon sequestration and maintains a microclimate (Maitra et al. 2025). On the other hand, inclusion of speciality crops in agroforestry systems can fetch a better economic return. Specialty crops can fetch a premium price in the market but are highly susceptible to minor changes in the climatic conditions (Rathore et al. 2022). Agroforestry operates through various diversities as per the local agroecology (Datta et al. 2023). Pairing annual or perennial crops along with woody plants helps in nutrient cycling through trees such as *Faidherbia albida* and *Gliricidia sepium* (Bhatia et al. 2024; Alamu et al. 2023). A yield enhancement of 15–20% was observed in Chinese chestnut when grown as an alley crop compared to monoculture, signifying yield enhancement while aligning with economic and environmental sustainability (Afsah-Hejri et al. 2025). Coffee in an agroforestry system, which includes trees such as *Inga* sp., provides shade, reducing heat stress on berries by 5–10 °C and extending the harvest window by 20–30 days in Latin America (Nasiro 2024). Additionally, biodiversity enrichment is observed in agroforestry systems. Population buildup of pollinators, which helps increase fruit set in orchards by 15–25% and natural pest suppression by 30%, is observed in polycultures rather than monocultures (Taillandier et al. 2023; Abebaw et al. 2025). In agroforestry, under mango or guava fruit trees, cultivation of shade-loving crops like ginger and turmeric improves soil moisture content and soil organic carbon content and reduces soil erosion (Murmu et al. 2022).

17.11 Postharvest Management

Postharvest losses are common in the era of CC. Reducing postharvest losses is a key strategy for climate-resilient horticulture. It ensures productivity and farmers' income (Dhingra et al. 2021). Mitigation of postharvest losses begins after harvest. Harvesting at the right temperature is important. Cooler periods, such as early morning, decrease respiration rates and extend shelf life (Gidado et al. 2024). Controlling temperature and humidity in storage is necessary. Adjusting O_2 and CO_2 levels slows down respiration (Liu et al. 2025). Leafy vegetables have a short shelf life, with about 48 hours (Kakade et al. 2015). Modified-atmosphere packaging (MAP) helps with these vegetables. It extends shelf life, stabilizes metabolic rates, and maintains output quality (Kargwal et al. 2020). Most postharvest losses result from physiological and biochemical changes in plant tissues of fruits and vegetables after harvest (Gidado et al. 2024). New technologies help overcome these losses, such as chitosan-based films and semi-permeable barriers (Flórez et al. 2021). For example, strawberry packaging uses silver nanoparticles. These prevent fungal growth and reduce losses (Zhang 2023). Another example is the use of ozone in figs, which enhances shelf life. Application of an aqueous solution in blueberries and strawberries preserves phenolics and texture. Cold plasma and UV-C can inactivate molds in blueberries by 70% (Ji et al. 2020). IoT-based sensors monitor the storage atmosphere and give alerts to prevent spoilage (Tailor et al. 2025). Cold storage and MAP can increase market value and stabilize crop production. Transportation is also important for reducing postharvest losses. Delayed transport induces microbial growth and rotting (Al-Dairi et al. 2022). AI-based transport systems guide for the best routes, reducing spoilage and delays.

Value addition and processing also minimize postharvest losses. It also adds nutritional value at market price. These techniques stabilize production, utilize excess produce, enhance the economic conditions of the farmers, and ensure sustainable farming (Urugo et al. 2024). Cold storage and advanced infrastructure are high costs that farmers cannot accommodate, which need government and policy support. These techniques are used to prevent postharvest losses, preserving nutrient content, and ensuring food security.

CC drastically increases postharvest losses due to a 1 °C temperature increase in the respiration rate, facilitating faster ripening and a reduced shelf life. Heat stress induces ethylene synthesis in fruits and triggers cuticular microcracking (Srivastava 2019). Hence, Postharvest Management (PHM) becomes a keystone for climate-resilient horticulture (Singh et al. 2021a, b). Understanding plant physiology is essential for designing effective postharvest management techniques. The rate of various physiological processes, such as respiration, ethylene production, and transpiration, determines the product's shelf life (Valenzuela 2023). Therefore, in tropical countries, where temperature and humidity levels fluctuate, postharvest losses are more due to increased respiration and transpiration rates, enhanced pest and disease pressure, reduced quality parameters, and supply chain disruptions (Gidado et al. 2024). Considering such climatic variability, postharvest management must be energy-efficient, decentralized, and adaptable to varied conditions. In this scenario, solar-powered cold storage, evaporative cool chambers, controlled-atmosphere storage, and modified-atmosphere packaging are gaining popularity. Along with storage, packaging is another important postharvest management technique that acts as a buffer between produce and the environment (El-Ramady et al. 2014). Climate-smart packaging materials, which are biodegradable, breathable, and moisture-retentive, can prevent the dehydration and bruising of produce (Han et al. 2023b). Additionally, technologies such as nanocoatings, antimicrobial films, and biodegradable trays can help maintain freshness while reducing plastic waste pollution (Olawore et al. 2024). It has been observed by several researchers that tomato, lettuce, and strawberry exhibit enhanced dry matter, cuticle thickness, and stress-primed metabolism, which reduces downstream postharvest losses (Nie et al. 2024). Postharvest management exploits this priming through rapid field heat removal (vacuum cooling), respiration suppression, humidity control (via foggers and alginate coatings), and chilling injury mitigation, extending the shelf life of crops 2–3 times greater at ambient temperature (Lalpekhlua et al. 2024). Due to CC, the intensification of pest and disease incidence can trigger postharvest problems, especially under high-humidity conditions (Lahlali et al. 2024). Integrated management, including biological control agents, natural antimicrobials, nonchemical treatments, and climate-controlled storage, plays a significant role in climate-resilient horticulture.

17.12 Future Perspective

To combat such a climate-change situation, innovative strategies need to be adopted. The involvement of the latest crop improvement approaches such as CRISPR-Cas9 and molecular breeding for climate-resilient cultivars and the use of beneficial microbes, microbial consortia, growth promoters, and biostimulants can create a healthy synchrony in the soil–plant–atmospheric continuum. On the other hand, controlled environment cultivation and soilless cultivation can flourish in urban clusters. The application of AI and ML will evolve predictive analysis and automate farm management. AI-driven platforms will also help in pest and disease incursion, detection, prediction, and management. Imagery-based and real-time crop health monitoring and decision support systems can flourish for the enhancement of input use efficiency. Regenerative practices such as cover cropping, agroforestry hybrids, and the application of biochar amendments will not only restore degraded soils but also improve water retention and biodiversity. The rise of "Cyborg Farm" will integrate digital systems with robotics, which will advance from single-task machines to multifunctional and capable of selective harvesting and will also reduce labor dependency and postharvest losses. The tech-driven future and resilience of a farm will be linked directly with the resilience of its community and landscape. Formalization of "Carbon farming" and ecosystem service markets can easily provide farmers with extra income while practicing a climate-resilient horticulture system. A policy shift can also be constructed from subsidizing input policies to incentivizing ecosystem stewardship. Hence, the future of climate-resilient horticulture will shift from food production to a regenerative, equitable, sustainable, and intelligent food production system, which can thrive in climatic anomalies.

17.13 Conclusion

CC impacts horticulture with several adverse effects and it causes havoc by lowering the productivity of horticultural crops, quality, and economic output. Horticultural crops face abiotic and biotic stresses under CC, namely, temperature variations, fluctuation in rainfall and humidity, soil moisture stresses, elevated CO_2 levels and enhanced solar radiation, changes in biotic stress dynamics, etc., and to cope with the situation, corrective measures, such as mitigation and adaptation options, are to be considered. Among adaptation strategies, CRH has gained attention, where the focus is centered on the development of stress-tolerant cultivars and the adoption of climate-resilient technologies such as CEA, including SLC, micro-irrigation, and automation in farm management, including support of IoT, AI, ML, DSTs, crop models, and robotics. Time has changed the pace of technological interventions, where smart farming of horticultural crops will emerge as a prime factor, which facilitates horticultural crop management. Moreover, horticultural farming is not merely a passion; rather, it is remunerative farming. Future directions of the sustainable horticultural crop cultivation are aligned with SDG, suggesting smart and climate-resilient horticulture.

Author's Contribution Sagar Maitra[1], Sumit Ray[1], Anjana J. Atapattu[2], Bisruti Maity[1], Masina Sairam[1*], Lalichetti Sagar[1], Satya Prasad Nanda[1], Santosh D.T.[3], Dinkar Jagannath Gaikwad[3], Harun I. Gitari[4], Shaik Rishitha[1], Luciana L. Chappa[5], Nithar Ranjan Madhu[6], Akbar Hossain[7]

Review of Literature: Sagar Maitra, Santosh D.T., Dinkar Jagannath Gaikwad, Masina Sairam.

Initial Draft: Sumit Ray, Anjana J. Atapattu, Sarthak Pattanayak, Nithar Ranjan Madhu, Lalichetti Sagar, Shaik Rishitha, Harun I. Gitari.

Figures/Visualization and Tables: Sumit Ray, Bisruti Maity, Satya Prasad Nanda, Luciana L. Chappa.

Final Draft and Editing: Sagar Maitra, Santosh D.T., Dinkar Jagannath Gaikwad, Akbar Hossain.

References

Abebaw SE, Yeshiwas EM, Feleke TG (2025) A systematic review on the role of agroforestry practices in climate change mitigation and adaptation. Climate Res and Sust 4(2):e70018. https://doi.org/10.1002/cli2.70018

Acquaah G (2015) Conventional plant breeding principles and techniques. In: Advances in plant breeding strategies: breeding, biotechnology and molecular tools. Springer International Publishing, Cham, pp 115–158. https://doi.org/10.1007/978-3-319-22521-0_5

Afsah-Hejri L, Maharlooei M, Ehsani R, Maboudian R (2025) Promoting sustainable food production through waste reduction and valorization: a California case study. npj Mater Sustain 3(1):1–21. https://doi.org/10.1038/s44296-025-00060-0

Ahmad S, Maryam M, Mansoor U, Sumbal K, Ihtisham W, Nida S, Zabeehullah B, Muhammad U, Muhammad MZ, Amrat E (2024) Crop responses and strategies for mitigating cold, salt, and drought stress in vegetables: a review. Asian J Res Crop Sci 9(2):168–180. https://doi.org/10.9734/ajrcs/2024/v9i2277

Akter F, Wu S, Islam MS, Kyaw H, Yang J, Li M, Fu Y, Wu J (2024) An efficient Agrobacterium-mediated genetic transformation system for gene editing in strawberry (Fragaria × ananassa). Plants 13(5):563. https://doi.org/10.3390/plants13050563

Alamu EO, Adesokan M, Fawole S, Maziya-Dixon B, Mehreteab T, Chikoye D (2023) Gliricidia sepium applications for enhancing soil fertility and crop nutritional qualities: a review. Forests 14(3):635. https://doi.org/10.3390/f14030635

Al-Dairi M, Pathare PB, Al-Yahyai R, Opara UL (2022) Mechanical damage of fresh produce in postharvest transportation: current status and future prospects. Trends Food Sci Technol 124:195–207. https://doi.org/10.1016/j.tifs.2022.04.018

Alsadik B, Ellsäßer FJ, Awawdeh M, Al-Rawabdeh A, Almahasneh L, Elberink SO, Abuhamoor D, Asmar YA (2024) Remote sensing technologies using UAVs for pest and disease monitoring: a review centered on date palm trees. Remote Sens 16(23):4371. https://doi.org/10.3390/rs16234371

Ammara UE, Mansoor S, Saeed M, Amin I, Briddon RW, Al-Sadi AM (2015) RNA interference-based resistance in transgenic tomato plants against tomato yellow leaf curl virus-Oman (TYLCV-OM) and its associated betasatellite. Virol J 12(1):38. https://doi.org/10.1186/s12985-015-0263-y

Ananth SRJ, Kurien S, Suresh V, Kumar PR, Petchiammal KI (2025) Flood stress and its effects on fruit crops: a critical review. J App Biol Biotechnol 13(5):22–34. https://doi.org/10.7324/jabb.2025.243581

Anderson NO, Walker NJ (2013) Marketing genetically modified organism carnations by future floral designers: student-designed policy formulation. HortTechnol 23(5):683–688. https://doi.org/10.21273/HORTTECH.23.5.683

Aravind S, Akshitha HJ, Sarathambal C, Krishnamurthy KS, Radha E, Alagupalamuthirsolai M, Thankamani CK, Santhosh JE, Nirmal Babu K, Johny AK, Ravindran PN (Eds.) (2020) ICAR-AICRPS varieties. ICAR- All India Coordinated Research Project on Spices, ICAR-IISR, Kozhikode, Kerala, India 53

Atayee AR, Noori MS (2020) Alleviation of cold stress in vegetable crops. J Sci Agric 4:38–44. https://doi.org/10.25081/jsa.2020.v4.6110

Atkinson HJ, Grimwood S, Johnston K, Green J (2004) Prototype demonstration of transgenic resistance to the nematode Radopholus similis conferred on banana by a cystatin. Transgenic Res 13(2):135–142. https://doi.org/10.1023/b:trag.0000026070.15253.88

Azad MAK, Amin L, Sidik NM (2014) Gene technology for papaya ringspot virus disease management. Sci World J 2014:1–11. https://doi.org/10.1155/2014/768038

Baiyin B, Tagawa K, Yamada M, Wang X, Yamada S, Shao Y, An P, Yamamoto S, Ibaraki Y (2021) Effect of nutrient solution flow rate on hydroponic plant growth and root morphology. Plants 10(9):1840. https://doi.org/10.3390/plants10091840

Balasooriya HN, Dassanayake KB, Bruce-Tomkins S, Seneweera S, Ajlouni S (2017) Impacts of elevated carbon dioxide and temperature on physicochemical and nutrient properties in strawberries. J Hort Sci Res 1:004. https://doi.org/10.36959/745/394

Banerjee K, Dutta S, Das S, Sadhukhan R (2025) Crop simulation models as decision tools to enhance agricultural system productivity and sustainability: a critical review. Technol Agron 5(1):e002. https://doi.org/10.48130/tia-0024-0032

Beltrán R, Valls A, Cebrián N, Zornoza C, García Breijo F, Reig Armiñana J, Garmendia A, Merle H (2019) Effect of temperature on pollen germination for several Rosaceae species: influence of freezing conservation time on germination patterns. PeerJ 7:e8195. https://doi.org/10.7717/peerj.8195

Bhandari R, Neupane N, Adhikari DP (2021) Climatic change and its impact on tomato (*Lycopersicum esculentum* L.) production in plain area of Nepal. Environ Chall 4:100129. https://doi.org/10.1016/j.envc.2021.100129

Bhatia, AK, Pant, KS, Prakash, P, Kumar, P, Sharma, H, Chauhan, S, Bishnoi, P, Kumari, B. (2024). Fruit and Pulse Synergy: Evaluating Vigna mungo performance in Himalayan wild pomegranate-based agroforestry systems. Research Square, 1–19. https://doi.org/10.21203/rs.3.rs-4887733/v1

Bhutia KL, Pramanick B, Maitra S, Ghosh S, Hossain A (2023) Molecular mechanisms and strategies in response to abiotic stresses for the sustainability of crop production under changing climate. In: Plant small RNA in food crops. Academic Press, pp 45–67. https://doi.org/10.1016/B978-0-323-91722-3.00004-X

Bisht A, Thakur N, Kumar A, Singh S, Anmol, Singh SK, Bakshi M (2025) Application of brassinosteroid to enhance abiotic stress tolerance in fruit crops: a review. Int J Adv Biochem Res 9(6):1016–1022. https://doi.org/10.33545/26174693.2025.v9.i6l.4652

Borghi M, Perez de Souza L, Yoshida T, Fernie AR (2019) Flowers and climate change: a metabolic perspective. New Phytol 224(4):1425–1441. https://doi.org/10.1111/nph.16031

Brennan EB, Smith RF (2018) Mustard cover crop growth and weed suppression in organic strawberry furrows in California. HortSci 53(4):432–440. https://doi.org/10.21273/HORTSCI12576-17

Buler Z, Nowak J (2014) Biostimulant usage for preserving strawberries to climate damages. Hort Sci (Prague) 41(4):151–156. https://doi.org/10.17221/161/2014-HORTSCI

Burgess AJ, Cano MEC, Parkes B (2022) The deployment of intercropping and agroforestry as adaptation to climate change. Crop Environ 1(2):145–160. https://doi.org/10.1016/j.crope.2022.05.001

Chakrabarti A, Ganapathi TR, Mukherjee PK, Bapat VA (2002) MSI-99, a magainin analogue, imparts enhanced disease resistance in transgenic tobacco and banana. Planta 216(4):587–596. https://doi.org/10.1007/s00425-002-0918-y

Chandio AA, Ozdemir D, Tang X (2025) Modelling the impacts of climate change on horticultural crop production: evidence from Turkiye. Food Energy Secur 14(1). https://doi.org/10.1002/fes3.70040

Chaudhary S, Kusakabe A, Melgar JC (2016) Phytophthora infection in flooded citrus trees reduces root hydraulic conductance more than under non-flooded condition. Sci Hortic 202:107–110. https://doi.org/10.1016/j.scienta.2016.02.031

Chen L, Liu Y, Ma Q, Wu H, Mao X (2019) Mechanisms of the IAA and ACC-deaminase producing strain of endophytic bacterium on enhancing the salt tolerance of cucumber. BMC Plant Biol 19(1):21. https://doi.org/10.1186/s12870-018-1618-5

Croonenborghs S, Ceusters J, Londers E, De Proft MP (2009) Effects of elevated CO_2 on growth and morphological characteristics of ornamental bromeliads. Sci Hortic 121(2):192–198. https://doi.org/10.1016/j.scienta.2009.01.018

Das SK, Nayak P (2024) Integration of IoT-AI powered local weather forecasting: a game-changer for agriculture. arXiv:1–14. https://doi.org/10.48550/arXiv.2501.14754

Datta P, Behera B, Rahut DB (2023) India's approach to agroforestry as an effective strategy in the context of climate change: an evaluation of 28 state climate change action plans. Agric Syst 214:103840. https://doi.org/10.1016/j.agsy.2023.103840

Dhingra ND, Alagusundaram NK, Jha NSN (2021) Post-harvest management: a strategy for climate resilient agriculture. Agric Eng Today 45(4):16–22. https://doi.org/10.52151/aet2021454.1543

Doddrell NH, Lawson T, Raines CA, Wagstaff C, Simkin AJ (2023) Feeding the world: impacts of elevated CO_2 on nutrient content of greenhouse grown fruit crops and options for future yield gains. Hortic Res 10(4):uhad026. https://doi.org/10.1093/hr/uhad026

Dong J, Gruda N, Lam SK, Li X, Duan Z (2018) Effects of elevated CO_2 on nutritional quality of vegetables: a review. Front Plant Sc 9:924. https://doi.org/10.3389/fpls.2018.00924

Dong S, Liu X, Han J, Miao H, Beckles DM, Bai Y, Zhang S (2023) CsMLO8/11 are required for full susceptibility of cucumber stem to powdery mildew and interact with CsCRK2 and CsRbohD. Hortic Res 11(2):uhad295. https://doi.org/10.1093/hr/uhad295

Eissa FM, Fathi MA, El-Shall SA (2007) The humic acid and rootstock in enhancing salt tolerance of 'Anna' apple seedlings. J Plant Prod 32(5):3667–3682. https://doi.org/10.21608/jpp.2007.78759

El-Ramady HR, Domokos-Szabolcsy É, Abdalla NA, Taha HS, Fári M (2014) Postharvest management of fruits and vegetables storage. In: Sustainable agriculture reviews, vol 15. Springer, pp 65–152. https://doi.org/10.1007/978-3-319-09132-7_2

Erekath S, Seidlitz H, Schreiner M, Dreyer C (2024) Food for future: exploring cutting-edge technology and practices in vertical farm. Sustain Cities Soc 106:105357. https://doi.org/10.1016/j.scs.2024.105357

Espeland EK, Kettenring KM (2018) Strategic plant choices can alleviate climate change impacts: a review. J Environ Manag 222:316–324. https://doi.org/10.1016/j.jenvman.2018.05.042

Ezura H (2022) The world's first CRISPR tomato launched to a Japanese market: the socio-economic impact of its implementation on crop genome editing. Plant Cell Physiol 63(6):731–733. https://doi.org/10.1093/pcp/pcac048

Fanourakis D, Tsaniklidis G, Makraki T, Nikoloudakis N, Bartzanas T, Sabatino L, Ntatsi G (2025) Climate change impacts on greenhouse horticulture in the Mediterranean Basin: challenges and adaptation strategies. Plants 14(21):3390. https://doi.org/10.3390/plants14213390

FarmAtma (2025) Grape cultivation in India: complete guidance. FarmAtma, Available at: https://www.google.com/search?q=FarmAtma+%282025%29+Grape+cultivation+in+India%3A+complete+guidance

Farmonaut (2026) Agricultural weather news: top ag weather stations 2026. Farmonaut Available at: https://farmonaut.com/news/agricultural-weather-news-top-ag-weather-stations-2026

Flórez M, Guerra-Rodríguez E, Cazón P, Vázquez M (2021) Chitosan for food packaging: recent advances in active and intelligent films. Food Hydrocoll 124:107328. https://doi.org/10.1016/j.foodhyd.2021.107328

Garner LC, Ingrao A (2015) The effects of foliar synthetic abscisic acid (s-ABA) on citrus nursery trees and mature orchard trees. Proc Plant Growth Regul Soc Am 42:150–155

Garzón J, Montes L, Garzón J, Lampropoulos G (2023) Systematic review of technology in aeroponics: introducing the technology adoption and integration in sustainable agriculture model. Agron 13(10):2517. https://doi.org/10.3390/agronomy13102517

Ghodke P, Shirsat D, Thangasamy A, Mahajan V, Salunkhe V, Khade Y, Singh M (2018) Effect of water logging stress at specific growth stages in onion crop. Int J Curr Microbiol App Sci 7(1):3438–3448. https://doi.org/10.20546/ijcmas.2018.701.405

Ghosh D, Brahmachari K, Das A, Hassan MM, Mukherjee PK, Sarkar S, Hossain A (2021) Assessment of energy budgeting and its indicator for sustainable nutrient and weed management in a rice–maize–green gram cropping system. Agron 11(1):166. https://doi.org/10.3390/agronomy11010166

Gidado M, Gunny AAN, Gopinath SC, Ali A, Wongs-Aree C, Salleh NHM (2024) Challenges of postharvest water loss in fruits: mechanisms, influencing factors, and effective control strategies. J Agricul Food Res 17:101249. https://doi.org/10.1016/j.jafr.2024.101249

Gonsalves C, Lee DR, Gonsalves D (2004) Transgenic virus-resistant papaya: the Hawaiian 'rainbow' was rapidly adopted by farmers and is of major importance in Hawaii today. APSnet Features 10. https://doi.org/10.1094/APSnetFeature-2004-0804

González-Martínez E, Lordan J, Munné-Bosch S, Miarnau X (2026) Impact of rising temperatures during dormancy on key yield components in almond. Sci Hort 355:114513. https://doi.org/10.1016/j.scienta.2025.114513

Gunapala R, Gangahagedara R, Wanasinghe W, Samaraweera AU, Gamage A, Rathnayaka C, Hameed Z, Baki ZA, Madhujith T, Merah O (2025) Urban agriculture: a strategic pathway to building resilience and ensuring sustainable food security in cities. Farm Sys 3(3):100150. https://doi.org/10.1016/j.farsys.2025.100150

Gupta S, Chowdhury S, Govindaraj R, Amesho KT, Shangdiar S, Kadhila T, Iikela S (2025) Smart agriculture using IoT for automated irrigation, water and energy efficiency. Smart Agric Technol 12:101081. https://doi.org/10.1016/j.atech.2025.101081

Haesaert G, Haverkort A, Gheysen G, Vossen J (2015) Transformation of the potato variety Desiree with single or multiple resistance genes increases resistance to late blight under field conditions. Crop Prot 77:163–175. https://doi.org/10.1016/J.CROPRO.2015.07.018

Hamdan MF, Tan BC (2024) Genetic modification techniques in plant breeding: a comparative review of CRISPR/Cas and GM technologies. Horti Plant J 11(5):1807–1829. https://doi.org/10.1016/j.hpj.2024.02.012

Hamim I, Borth WB, Marquez J, Green JC, Melzer MJ, Hu JS (2018) Transgene-mediated resistance to Papaya ringspot virus: challenges and solutions. Phytoparasitica 46(1):1–18. https://doi.org/10.1007/s12600-017-0636-4

Han J, Li X, Li W, Yang Q, Li Z, Cheng Z, Lv L, Zhang L, Han D (2023a) Isolation and preliminary functional analysis of FvICE1, involved in cold and drought tolerance in Fragaria vesca through overexpression and CRISPR/Cas9 technologies. Plant Physiol Biochem 196:270–280. https://doi.org/10.1016/j.plaphy.2023.01.048

Han Y, Wang S, Cao Y, Singh GP, Loh SI, Cheerlavancha R, Marelli B (2023b) Design of biodegradable, climate-specific packaging materials that sense food spoilage and extend shelf life. ACS Nano 17(9):8333–8344. https://doi.org/10.1021/acsnano.2c12747

Harris L, McDonagh EM, Zhang X, Fawcett K, Foreman A, Daneck P, Sergouniotis PI, Parkinson H, Mazzarotto F, Inouye M, Hollox EJ, Birney E, Fitzgerald T (2024) Genome-wide association testing beyond SNPs. Nat Rev Genet 26(3):156–170. https://doi.org/10.1038/s41576-024-00778-y

Harsanto B, Kasumaningrum Y, Arviansyah MR, Siregar AY, Purnomo D, Freddy N, Iskandar Y, Utama ID, Sari DI (2025) Leveraging disruptive technologies for food security: a systematic review on agricultural supply chain resilience to climate change. Curr Res Food Sci 10:101079. https://doi.org/10.1016/j.crfs.2025.101079

Haverkort AJ, Boonekamp PM, Hutten R, Jacobsen E, Lotz LAP, Kessel GJT, Visser RGF, Van Der Vossen EAG (2008) Societal costs of late blight in potato and prospects of durable resistance through cisgenic modification. Potato Res 51(1):47–57. https://doi.org/10.1007/s11540-008-9089-y

Hemasree KR, Sairam M, Maitra S, Ray S, Maheswari N, Gaikwad DJ, Santosh DT (2025) Effects of Sulphur and zinc on growth and productivity of summer cowpea under conditions of southern Odisha. Crop Res 60. https://doi.org/10.31830/2454-1761.2025.CR-1012

Hewitt S, Hernández-Montes E, Dhingra A, Keller M (2023) Impact of heat stress, water stress, and their combined effects on the metabolism and transcriptome of grape berries. Sci Rep 13:9907. https://doi.org/10.1038/s41598-023-36160-x

Hossain A, Ahmad Z, Moulik D, Maitra S, Bhadra P, Ahmad A, Garai S, Mondal M, Roy A, Sabagh AE, Aftab T (2021a) Jasmonates and salicylates: mechanisms, transport and signalling during abiotic stress in plants. In: Aftab T, Yusuf M (eds) Jasmonates and salicylates signaling in plants. Springer, pp 1–29. https://doi.org/10.1007/978-3-030-75805-9_1

Hossain A, Raza A, Maitra S, Asaduzzaman M, Islam MR, Hossain MJ, Sabagh AE, Garai S, Mondal M, Latef AAHA, Aftab T (2021b) Strigolactones: a novel carotenoid-derived phytohormone—biosynthesis, transporters, signalling, and mechanisms in abiotic stress. In:

Aftab T, Hakeem KR (eds) Plant growth regulators. Springer Nature, pp 275–303. https://doi.org/10.1007/978-3-030-61153-8_13

Hossain A, Pamanick B, Venugopalan VK, Ibrahimova U, Rahman MA, Siyal AL, Maitra S, Chatterjee S, Aftab T (2022a) Emerging roles of plant growth regulators for plants adaptation to abiotic stress-induced oxidative stress. In: Emerging roles of plant growth regulators. Academic Press, pp 1–72. https://doi.org/10.1016/B978-0-323-91005-7.00010-2

Hossain A, Venugopalan VK, Rahman MA, Alam MJ, Al-Mahmud A, Islam MA, Siyal LA, Maitra S, Aftab T (2022b) Physiological, biochemical, and molecular mechanisms of plant steroid hormones brassinosteroids under drought-induced oxidative stress in plants. In: Aftab T, Naeem M (eds) Emerging plant growth regulators in agriculture. Elsevier, pp 99–130. https://doi.org/10.1016/b978-0-323-91005-7.00011-4

Husaini AM, Abdin MZ (2008) Development of transgenic strawberry (Fragaria × ananassa Duch.) plants tolerant to salt stress. Plant Sci 174(4):446–455. https://doi.org/10.1016/j.plantsci.2008.01.007

Ibrahim LA, Shaghaleh H, El-Kassar GM, Abu-Hashim M, Elsadek EA, Hamoud YA (2023) Aquaponics: a sustainable path to food sovereignty and enhanced water use efficiency. Water 15(24):4310. https://doi.org/10.3390/w15244310

IPCC (2018) Global warming of 1.5°C: An IPCC Special Report on the impacts of global warming of 1.5°C above pre-industrial levels and related global greenhouse gas emission pathways, in the context of strengthening the global response to the threat of climate change, sustainable development, and efforts to eradicate poverty. In: Masson-Delmotte V, Zhai P, Pörtner H-O, Roberts D, Skea J, Shukla PR, Pirani A, Moufouma-Okia W, Péan C, Pidcock R, Connors S, Matthews JBR, Chen Y, Zhou X, Gomis MI, Lonnoy E, Maycock T, Tignor M, Waterfield T (eds). Available at: https://www.ipcc.ch/site/assets/uploads/sites/2/2019/06/SR15_Full_Report_High_Res.pdf

Jagdish (2025) Tulsi plant care at home in India: complete guide for healthy growth. Agri Farm https://www.agrifarming.in/tulsi-plant-care-at-home-in-india

Jarman A, Thompson J, McGuire E, Reid M, Rubsam S, Becker K, Mitcham E (2023) Postharvest technologies for small-scale farmers in low- and middle-income countries: a call to action. Postharvest Biol Technol 206:112491. https://doi.org/10.1016/j.postharvbio.2023.112491

Jeanguenin L, Mir A, Chaumont F (2016) Uptake, loss and control. In: Encycl Appl Plant Sci, pp 135–140. https://doi.org/10.1016/B978-0-12-394807-6.00087-3

Jha PK, Zhang N, Rijal JP, Parker LE, Ostoja S, Pathak TB (2023) Climate change impacts on insect pests for high value specialty crops in California. Sci Total Environ 906:167605. https://doi.org/10.1016/j.scitotenv.2023.167605

Ji Y, Hu W, Liao J, Jiang A, Xiu Z, Gaowa S, Liu C (2020) Effect of atmospheric cold plasma treatment on antioxidant activities and reactive oxygen species production in postharvest blueberries during storage. J Scie Food Agric 100(15):5586–5595. https://doi.org/10.1002/jsfa.10611

Kabir M, Habiba UE, Khan W, Shah A, Rahim S, De Los Rios-Escalante PR, Farooqi Z, Ali L, Shafiq M (2023) Climate change due to increasing concentration of carbon dioxide and its impacts on environment in 21st century; a mini review. J King Saud Univ Sci 35(5):102693. https://doi.org/10.1016/j.jksus.2023.102693

Kakade A, More P, Jadhav S, Bhosle V (2015) Shelf life extension of fresh-cut spinach. Int J Agric Environ Biotechnol 8:609. https://doi.org/10.5958/2230-732X.2015.00067.4

Karagüzel ÜÖ (2025) Blossoms amid drought: a bibliometric mapping of research on drought stress in ornamental plants. Front Plant Sci 16:1644092. https://doi.org/10.3389/fpls.2025.1644092

Kargwal R, Garg MK, Singh V, Garg R, Kumar N (2020) Principles of modified atmosphere packaging for shelf life extension of fruits and vegetables: an overview of storage conditions. Int J Chem Stud 8:2245–2252. https://doi.org/10.22271/chemi.2020.v8.i3af.9545

Khambalkar PA, Agrawal S, Dhaliwal S, Yadav SS, Sadawarti MJ, Singh A, Yadav IR, Yadav K, Shivansh N, Prasad D, Singh A, Afreen N (2025) Sustainable nutrient management balancing soil health and food security for future generations. Appl Food Res 5(2):101087. https://doi.org/10.1016/j.afres.2025.101087

Kim WS (2018) Air temperature and humidity affect petunia ornamental value. Hortic Sci Techno 36(1):10–19. https://doi.org/10.12972/kjhst.20180002

Kinsky G (2025) Balancing water supply and demand with soil and ET data. CropX Digital Agronomy Platform and Farm Management System. Available at: https://cropx.com/2025/11/10/balancing-water-supply-and-demand-with-soil-and-et-data/

Kisan Suvidha App Kisan Suvidha. Mobile application overview. IndiaFilings. Available at: https://kisansuvidha.gov.in/.

Kishore K, Sangeetha G, Rupa TR, Ganeshmurthy AN, Samant D, Acharya GC, Srinivas P (2025) Assessment of mango-based intercropping systems for productivity, resource use efficiency and environmental sustainability in tropical region of India. Next Sustain 5:100121. https://doi.org/10.1016/j.nxsust.2025.100121

Kumar M, Kaushik K, Singh S, Kumar S, Rai A, Kumar R, Kumar A (2023) Sustainable horticulture practices: an environmental-friendly approaches. Pharma Innov J 12:4777–4782

Kumar R, Prabu J, Kumar A, Yadav S, Khanikar S, Singh A, Hansda S, Verma S (2025) Advancing horticulture through IoT and sensor technologies: trends, challenges and future directions. J Exp Agric Int 47(2):389–416. https://doi.org/10.9734/jeai/2025/v47i23299

Kumari M, Dubey AK, Kumar R, Kumar A (2024) Marker-assisted selection in plant breeding for stress tolerance. In: Elsevier eBooks, pp 371–387. https://doi.org/10.1016/b978-0-443-18927-2.00005-4

Kunik T, Salomon R, Zamir D, Navot N, Zeidan M, Michelson I, Gafani Y, Czosnek H (1994) Transgenic tomato plants expressing the tomato yellow leaf curl virus capsid protein are resistant to the virus. Nat Biotechnol 12(5):500–504. https://doi.org/10.1038/nbt0594-500

Lahlali R, Taoussi M, Laasli S, Gachara G, Ezzouggari R, Belabess Z, Aberkani K, Assouguem A, Meddich A, Jarroudi ME, Barka EA (2024) Effects of climate change on plant pathogens and host–pathogen interactions. Crop Environ 3(3):159–170. https://doi.org/10.1016/j.crope.2024.05.003

Lakhiar IA, Yan H, Syed TN, Zhang C, Shaikh SA, Rakibuzzaman M, Vistro RB (2025) Soilless agricultural systems: opportunities, challenges, and applications for enhancing horticultural resilience to climate change and urbanization. Hortic 11(6):568. https://doi.org/10.3390/horticulturae11060568

Lalpekhlua K, Tirkey A, Saranya S, Babu PJ (2024) Post-harvest management strategies for quality preservation in crops. Int J Veg Sci 30(5):587–635. https://doi.org/10.1080/19315260.2024.2392551

Li S, Liu F (2022) Crop exposure to drought stress under elevated CO_2: responses at physiological, biochemical, and molecular levels. In: Elsevier eBooks, pp 21–42. https://doi.org/10.1016/b978-0-323-85449-8.00012-9

Liu X, Gao T, Liu C, Mao K, Gong X, Li C, Ma F (2023) Fruit crops combating drought: physiological responses and regulatory pathways. Plant Physiol 192(3):1768–1784. https://doi.org/10.1093/plphys/kiad202

Liu L, Lin H, Zhou X, Zhang Z, Zhang Y, Deng S, Peng S, Gong S, Guo S, Fan W (2025) Application of modified atmosphere preservation technology in cherry storage: a review. Agri 15(5):462. https://doi.org/10.3390/agriculture15050462

Lobato-Gómez M, Hewitt S, Capell T, Christou P, Dhingra A, Girón-Calva PS (2021) Transgenic and genome-edited fruits: background, constraints, benefits and commercial opportunities. Hortic Res 8:166. https://doi.org/10.1038/s41438-021-00601-3

Lv J, Zheng T, Song Z, Pervaiz T, Dong T, Zhang Y, Jia H, Fang J (2022) Strawberry proteome responses to controlled hot and cold stress partly mimic post-harvest storage temperature effects on fruit quality. Front Nutr 8:812666. https://doi.org/10.3389/fnut.2021.812666

Mabhaudhi T, Dirwai TL, Taguta C, Sikka A, Lautze J (2023) Mapping decision support tools on agricultural water productivity: a global systematic scoping review. Agric Water Manag 290:108590. https://doi.org/10.1016/j.agwat.2023.108590

Mahmood F, Al-Ansari T (2023) Renewable energy-driven sustainable greenhouse systems: an overview. In: Elsevier eBooks, pp 652–665. https://doi.org/10.1016/b978-0-323-90386-8.00073-5

Maitra S, Gitari H (2020) Scope for adoption of intercropping in organic agriculture. Indian J Nat Sci 11(36):28624–28631

Maitra S, Palai JB, Manasa P, Kumar DP (2019) Potential of intercropping system in sustaining crop productivity. Int J Agric Environ Biotechnol 12(1):39–45. https://doi.org/10.30954/0974-1712.03.2019.7

Maitra S, Hossain A, Sahu C, Mishra UN, Banarjee P, Bhadra P, Praharaj S, Shankar T, Bhattachrya U (2021a) The role of gibberellin against abiotic stress tolerance in plants. In: Fahad S, Sönmez O, Shah S, Wang D, Wu C, Adnan M, Turan V (eds) Plant growth regulators for climate-smart agriculture. CRC Press, pp 1–18. https://doi.org/10.1201/9781003109013-5

Maitra S, Hossain A, Banerjee P, Bhadra P (2021b) The role of phytohormones in heat stress tolerance in plants. In: Fahad S, Sönmez O, Shah S, Wang D, Wu C, Adnan M, Turan V (eds) Plant growth regulators for climate-smart agriculture. CRC Press, pp. 1–? https://doi.org/10.1201/9781003109013-10

Maitra S, Hossain A, Brestic M, Skalicky M, Ondrisik P, Gitari H, Brahmachari K, Shankar T, Bhadra P, Palai JB, Jena J, Bhattacharya U, Duvvada SK, Lalichetti S, Sairam M (2021c) Intercropping—a low-input agricultural strategy for food and environmental security. Agron 11(2):343. https://doi.org/10.3390/agronomy11020343

Maitra S, Sain S, Sairam M, Shankar T, Ray S, Banerjee M, Malik GC (2024a) Impact of rabi maize–legume intercropping system on growth, yield, and crop competition in maize. Res Crop 25(3):409–415. https://doi.org/10.31830/2348-7542.2024.roc-1093

Maitra S, Sairam M, Santosh DT (2024b) Evaluation of the growth, flower quality, yield and vase life of gerbera (*Gerbera jamesonii* L.) cultivars under open-ventilated saw-tooth type polyhouse. Res Crop 25:104–109. https://doi.org/10.31830/2348-7542.2024.roc-1055

Maitra S, Sairam M, Santosh DT, Gaikwad DJ, Sahoo U (2024c) Growth, productivity and quality of colored capsicum (*Capsicum annuum* L.) as influenced by hybrids and plant growth regulators under controlled environment conditions. Res Crop 25(2):328–335. https://doi.org/10.31830/2348-7542.2024.roc-1079

Maitra S, Sairam M, Sagar L, Ray S, Santosh DT, Gaikwad DJ, Gitari HI (2025) Alley cropping: a sustainable agroforestry system for enhancing productivity, soil health, and biodiversity. In: Plant biotechnology and sustainable agriculture. CABI, pp 235–248. https://doi.org/10.1079/9781800627659.0016

Maity B, Sairam M, Ray S, Maitra S, Dash B, Guchhait D (2025) Evaluation of polyhalite as a potassium and Sulphur source for improving growth and productivity of summer groundnut (*Arachis hypogaea* L.). Crop Res 60:357–362. https://doi.org/10.31830/2454-1761.2025.CR-1069

Majsztrik JC, Fernandez RT, Fisher PR, Hitchcock DR, Leach Cox J, Owen JS Jr, White SA (2017) Water use and treatment in container-grown specialty crop production: a review. Water Air Soil Pol 228(4):151. https://doi.org/10.1007/s11270-017-3272-1

Mamassi A, Nhamo N, Gummadi S, Ammar K, Dhanhani MAHA, Hashmi HAA, Bouras H, Accatino F, Zaaboul R (2025) Climate change and irrigation management shape crop resilience in UAE arid agriculture: An APSIM model-driven assessment. Agric Water Manag 319:109819. https://doi.org/10.1016/j.agwat.2025.109819

Manasa P, Sairam M, Maitra S (2021) Influence of maize–legume intercropping systems on crop growth and productivity. Int J Biores Sci 8(1). https://doi.org/10.30954/2347-9655.01.2021

Manepalli SB, Tomar S, Gaikwad DJ, Maitra S (2022) Abiotic stress signaling in plants and transgenic technology as a triumph: a review. J App Biol Biotechnol 10(5):5–13. https://doi.org/10.7324/jabb.2022.100501

Manghwar H, Hussain A, Alam I, Khoso MA, Ali Q, Liu F (2024) Waterlogging stress in plants: unraveling the mechanisms and impacts on growth, development, and productivity. Environ Exp Bot 224:105824. https://doi.org/10.1016/j.envexpbot.2024.105824

Mihrete TB, Mihretu FB (2025) Crop diversification for ensuring sustainable agriculture, risk management, and food security. Global Chall 9(2):2400267. https://doi.org/10.1002/gch2.202400267

Moffa L, Mannino G, Bevilacqua I, Gambino G, Perrone I, Pagliarani C, Bertea CM, Spada A, Narduzzo A, Zizzamia E, Velasco R, Chitarra W, Nerva L (2024) CRISPR/Cas9-driven double modification of grapevine MLO6-7 imparts powdery mildew resistance, while editing of NPR3 augments powdery and downy mildew tolerance. Plant J 122(2):e17204. https://doi.org/10.1111/tpj.17204

Mohan VR, MacDonald MT, Abbey L (2025) Impact of water deficit stress on brassica crops: growth and yield, physiological and biochemical responses. Plants 14:1942. https://doi.org/10.3390/plants14131942

Mohite DP, Kavino M, Nakkeeran S, Raveendran M, Raghu R, Vethamoni PI, Saranya N (2024) Biohardening with endomicrobiome: a novel approach to develop Fusarium wilt resistance in banana (Musa spp.). Microbe 4:100109. https://doi.org/10.1016/j.microb.2024.100109

Mondal S, Mondal A, Rathore J, Koundinya KK, Sharma AK (2024) India's shift toward sustainable energy: a comprehensive approach to renewable energy integration and environmental sustainability. J Renew Sustain Energy 16:042701. https://doi.org/10.1063/5.0220054

Mou SS, Haus MJ, Hayden ZD, Patterson EL, Saha D (2025) Climate-driven challenges in weed management for ornamental crop production in the United States: a review. Front Agron 7:1556418. https://doi.org/10.3389/fagro.2025.1556418

Moyo M, Magembe E, Mwaura L, Byarugaba AA, Barekye A, Nyongesa M, Ghislain M (2024) Nutritional and anti-nutritional compositional analysis of transgenic potatoes with late blight resistance. Front Bioeng Biotechnol 12:1432079. https://doi.org/10.3389/fbioe.2024.1432079

Mukesh G, Sairam M, Maitra S, Gaikwad DJ, Sagar L, Ray S (2024) Agrometeorological indices, physiological growth parameters, and performance of finger millet as influenced by different cultivars under hot and sub-humid regions of Odisha. Int J Exp Res Rev 42:148–161. https://doi.org/10.52756/ijerr.2024.v42.013

Mukherjee A, Sreekumar J, Sheela MN, Immanuel S, Sahoo MR, Hegde V (2024) Enhancing adaptability, nutritional quality in tropical tuber crops: source for adaptive food and nutrition. Food Nut Sci 15:1141–1169. https://doi.org/10.4236/fns.2024.1511074

Müller K, Keller M, Stoll M, Friedel M (2023) Wind speed, sun exposure, and water status alter sunburn susceptibility of grape berries. Front Plant Sci 14:1145274. https://doi.org/10.3389/fpls.2023.1145274

Munda SC, Maitra S, Gaikwad DJ, Duvvada SK, Sairam M, Santosh DT, Ray S (2025) Influence of growing media and nutrient levels on growth and yield of lettuce (Lactuca sativa L.) in hydroponics. Crop Res 60:185–190. https://doi.org/10.31830/2454-1761.2025.cr-1033

Murmu S, Padhi GK, Murmu P, Roy D, Dhara PK (2022) Potential of different fruit-based agroforestry models for improving livelihood in humid and subtropical regions of India. Indian J Ecol 49:1861–1869. https://doi.org/10.55362/IJE/2022/3748

Nabinejad S, Schüttrumpf H (2023) Flood risk management in arid and semi-arid areas: a comprehensive review of challenges, needs, and opportunities. Water 15:3113. https://doi.org/10.3390/w15173113

Namizaki H, Iwasaki Y, Wang R (2022) Effects of elevated CO_2 levels on growth and yield of summer-grown cucumbers cultivated under different day and night temperatures. Agron 12:1872. https://doi.org/10.3390/agronomy12081872

Nasiro K (2024) The benefits of agroforestry coffee production systems: a review. World J Food Sci Technol 8:86–105. https://doi.org/10.11648/j.wjfst.20240804.13

Nhamo G, Chapungu L, Mutanda GW (2025) Trends and impacts of climate-induced extreme weather events in South Africa (1920–2023). Environ Dev 55:101183. https://doi.org/10.1016/j.envdev.2025.101183

Nie H, Yang X, Zheng S, Hou L (2024) Gene-based developments in improving tomato quality: focus on firmness, shelf life, and pre- and post-harvest stress adaptations. Horticulturae 10:641. https://doi.org/10.3390/horticulturae10060641

Nori M, Masina S, Sagar M, Sumit R, Hemasree KR, Kumar RM, Dinkar JG, Shriram KR (2025) Influence of precision nitrogen management in rabi maize and cowpea intercropping system under varied planting proportions in South Odisha. Plant Sci Today 12(3):1–9. https://doi.org/10.14719/pst.8194

Obaideen K, Yousef BA, AlMallahi MN, Tan YC, Mahmoud M, Jaber H, Ramadan M (2022) An overview of smart irrigation systems using IoT. Energy Nexus 7:100124. https://doi.org/10.1016/j.nexus.2022.100124

Oğuztürk GE (2025) AI-driven irrigation systems for sustainable water management: a systematic review and meta-analytical insights. Smart Agric Technol 11:100982. https://doi.org/10.1016/j.atech.2025.100982

Olawore O, Ogunmola M, Desai S (2024) Engineered nanomaterial coatings for food packaging: design, manufacturing, regulatory, and sustainability implications. Micromachines 15:245. https://doi.org/10.3390/mi15020245

Orbegozo J, Roman ML, Rivera C, Gamboa S, Tovar JC, Forbes GA, Lindqvist-Kreuze H, Kreuze JF, Ghislain M (2016) Rpi-blb2 gene from Solanum bulbocastanum confers extreme resistance to late blight disease in potato. Plant Cell Tissue Organ Cult 125:269–281. https://doi.org/10.1007/s11240-016-0947-z

Osorio-Marín J, Fernandez E, Vieli L, Ribera A, Luedeling E, Cobo N (2024) Climate change impacts on temperate fruit and nut production: a systematic review. Front Plant Sci 15:1352169. https://doi.org/10.3389/fpls.2024.1352169

Owens CL, Thomashow MF, Hancock JF, Iezzoni AF (2002) CBF1 orthologs in sour cherry and strawberry and the heterologous expression of CBF1 in strawberry. J American Soc Hortic Sci 127:489–494. https://doi.org/10.21273/jashs.127.4.489

Paluchowska P, Śliwka J, Yin Z (2022) Late blight resistance genes in potato breeding. Planta 255:127. https://doi.org/10.1007/s00425-022-03910-6

Pessoa HP, Dariva FD, Copati MGF, De Paula RG, De Oliveira Dias F, Gomes CN (2023) Uncovering tomato candidate genes associated with drought tolerance using Solanum pennellii introgression lines. PLoS One 18:e0287178. https://doi.org/10.1371/journal.pone.0287178

Petruccelli, R, Bartolini, G, Ganino, T, Zelasco, S, Lombardo, L, Perri, E, Durante, M, Bernardi, R. (2022). Cold Stress, Freezing Adaptation, Varietal Susceptibility of Olea europaea L.: A Review. Plants, 11(10):1367. https://doi.org/10.3390/plants11101367

Pramanik D, Shelake RM, Park J, Kim MJ, Hwang I, Park Y, Kim J (2021) CRISPR/Cas9-mediated generation of pathogen-resistant tomato against tomato yellow leaf curl virus and powdery mildew. Int J Mol Sci 22:1878. https://doi.org/10.3390/ijms22041878

Prendeville HR, Pilson D (2009) Transgenic virus resistance in cultivated squash affects pollinator behaviour. J Appl Ecol 46:1088–1096. https://doi.org/10.1111/j.1365-2664.2009.01698.x

Press Information Bureau, Government of India (2024) Details of 109 varieties of field and horticultural crops released by Prime Minister Shri Narendra Modi on 11th August 2024. Ministry of Agriculture & Farmers Welfare. Available at: https://www.pib.gov.in/PressReleasePage.aspx?PRID=2044754&reg=3&lang=2

Qian D, Li T, Zheng C, Wang M, Chen S, Li C, An J, Yang Y, Niu Y, An L, Xiang Y (2025) Heat-stable protein PGSL1 enhances pollen germination and tube growth at high temperature. Nat Commun 16:3642. https://doi.org/10.1038/s41467-025-58869-1

Rajaseger G, Chan KL, Tan KY, Ramasamy S, Khin MC, Amaladoss A, Haribhai PK (2023) Hydroponics: current trends in sustainable crop production. Bioinfo 19:925–938. https://doi.org/10.6026/97320630019925

Rajendran S, Domalachenpa T, Arora H, Li P, Sharma A, Rajauria G (2024) Hydroponics: exploring innovative sustainable technologies and applications across crop production, with emphasis on potato mini-tuber cultivation. Heliyon 10:e26823. https://doi.org/10.1016/j.heliyon.2024.e26823

Raju C, Pazhanivelan S, Perianadar IV, Kaliaperumal R, Sathyamoorthy NK, Sendhilvel V (2024) Climate change as an existential threat to tropical fruit crop production: a review. Agriculture 14:2018. https://doi.org/10.3390/agriculture14112018

Ramesh K, Matloob A, Aslam F, Florentine SK, Chauhan BS (2017) Weeds in a changing climate: vulnerabilities, consequences, and implications for future weed management. Front Plant Sci 8:95. https://doi.org/10.3389/fpls.2017.00095

Rangaswamy TC, Sridhara S, Manoj KN, Gopakkali P, Ramesh N, Shokralla S, El-Abedin TKZ, Almutairi KF, Elansary HO (2021) Impact of elevated CO2 and temperature on growth, development and nutrient uptake of tomato. Horticulturae 7:509. https://doi.org/10.3390/horticulturae7110509

Rathore SS, Babu S, El-Sappah AH, Shekhawat K, Singh VK, Singh RK, Upadhyay P, Singh R (2022) Integrated agroforestry systems improve soil carbon storage, water productivity, and economic returns in the marginal land of the semi-arid region. Saudi J Biol Sci 29:103427. https://doi.org/10.1016/j.sjbs.2022.103427

Ray S, Maitra S, Sairam M, Sravya M, Priyadarshini A, Shubhadarshi S, Padhi DP (2024) An unravelled potential of foliar application of micro and beneficial nutrients in cereals for ensuring food and nutritional security. Int J Exp Res Rev 41:19–42. https://doi.org/10.52756/ijerr.2024.v41spl.003

Ray S, Hemasree KR, Maheswari N, Maity B, Sairam M, Santosh DT, Maitra S (2025a) Optical sensing and site-specific nutrient management for sustainable precision agriculture. Int J Biores Sci 12:135–146. https://doi.org/10.30954/2347-9655.01.2025.15

Ray S, Maitra S, Sairam M, Sameer S, Sagar L, Divya B, Gitari HI (2025b) The nexus between intercropping systems, ecosystem services and sustainable agriculture: a review. Res Crop 26. https://doi.org/10.31830/2348-7542.2025.ROC-1166

Raza, MA, Sohail, H, Hassan, MA, Sajad S, Xing Y, Song J (2024). Cold stress in Brassica vegetables: Morpho-physiological and molecular responses underlying adaptive mechanism. Sci Hortic, 329, 113002. https://doi.org/10.1016/j.scienta.2024.113002

Reichstein M, Benson V, Blunk J, Camps-Valls G, Creutzig F, Fearnley CJ, Han B, Kornhuber K, Rahaman N, Schölkopf B, Tárraga JM, Vinuesa R, Dall K, Denzler J, Frank D, Martini G, Nganga N, Maddix DC, Weldemariam K (2025) Early warning of complex climate risk with integrated artificial intelligence. Nat Commun 16:2564. https://doi.org/10.1038/s41467-025-57640-w

Rivera-Romero CA, Palacios-Hernández ER, Vite-Chávez O, Reyes-Portillo IA (2024) Early-stage identification of powdery mildew levels for cucurbit plants in open-field conditions based on texture descriptors. Inventions 9:8. https://doi.org/10.3390/inventions9010008

Roy AS, Das S, Saha D, Barat S (2024) Vertical farming: a sustainable agriculture format of the future. Int J Res Agron 7:308–314. https://doi.org/10.33545/2618060X.2024.v7.i4Sd.641

Roy S, Hoque A, Saikia P, Padhiary M (2025) Climate-smart agriculture: AI-based solutions for enhancing crop resilience and reducing environmental impact. Asian Res J Agric 18:291–310. https://doi.org/10.9734/arja/2025/v18i1665

Roychowdhury R, Das SP, Sarkar P, Khan Z, Kumar A, Sarker U, Sajeevan, RS. (2025). Physiological, biochemical and molecular signaling basis of cold stress tolerance in plants. Front Plant Sci, 16, 1707204. https://doi.org/10.3389/fpls.2025.1707204

Ruland K, Montague T, Helwi P (2023) Impact of hail-netting on Vitis vinifera L. canopy microclimate, leaf gas exchange, fruit quality, and yield in a semi-arid environment. Vitic Data J 4:e108805. https://doi.org/10.3897/vdj.4.e108805

Sagar L, Sairam M, Reddy MD, Sindhu L (2024) Calibration and validation of CERES-rice model using varied transplanting dates and seedling ages of RNR 15048 and assessing high temperature sensitivity in the North Eastern Ghat region of Odisha. Mausam 75:869–876. https://doi.org/10.54302/mausam.v75i3.6042

Sagar L, Sairam M, Reddy MD (2025) Estimation of irrigation requirement and schedule in southern Odisha for major rabi cereal crops using FAO CROPWAT 8.0 model. Indian J Agric Res 59. https://doi.org/10.18805/IJARe.A-5941

Sahu C, Gaikwad DJ, Nanda S, Dash GK, Maitra S, Munda SC, Mullapudi S (2025) Comparative performance of lettuce cultivars Batavia and Romain in nutrient film technique under hydroponics system. Crop Res 60:61–67. https://doi.org/10.31830/2454-1761.2025.cr-1009

Sairam M, Maitra S, Shankar T, Sagar L, Santosh DT (2022) Performance of different gerbera (Gerbera jamesonii L.) cultivars under protected cultivation in South Odisha conditions. Crop Res 57:211–215. https://doi.org/10.31830/2454-1761.2022.032

Sairam M, Maitra S, Sahoo U, Sagar L, Krishna TG (2023a) Evaluation of precision nutrient tools and nutrient optimization in maize (Zea mays L.) for enhancement of growth, productivity and nutrient use efficiency. Res Crop 24:666–677. https://doi.org/10.31830/2348-7542.2023.ROC-1016

Sairam M, Sagar L, Devender Reddy M, Neelima TL, Ramulu V (2023b) Yield prediction, sensitivity analysis and calibration of DSSAT crop model for different cultivars, establishment methods and nitrogen levels in rice (Oryza sativa L.) under normal and varied temperatures. Int J Biores Scie 10:149–157. https://doi.org/10.30954/2347-9655.02.2023.1

Sairam M, Maitra S, Sagar L, Biswas T, Bárek V, Brestic M, Hossain A (2025a) Application of precision nutrient tools for the optimization of fertilizer requirements and assessment of the growth and productivity of maize (Zea mays L.) in the northeastern Ghat of India. J Agric Food Res:101958. https://doi.org/10.1016/j.jafr.2025.101958

Sairam M, Ray S, Padhi DP, Dayal PAS, Santosh DT, Maitra S (2025b) Growth, productivity and quality of gerbera cultivars under forced- and open-ventilated polyhouse conditions. Res Crop 26:675–681. https://doi.org/10.31830/2348-7542.2025.roc-1256

Sakuanrungsirikul S, Sarindu N, Prasartsee V, Chaikiatiyos S, Siriyan R, Sriwatanakul M, Lekananon P, Kitprasert C, Boonsong P, Kosiyachinda P, Fermin G, Gonsalves D (2005) Update on the development of virus-resistant papaya: virus-resistant transgenic papaya for people in rural communities of Thailand. Food Nutr Bull 26:422–426. https://doi.org/10.1177/15648265050264S310

Sanchez E, Berghage R, Ford T, Gioia FD (2021) Hydroponics systems and principles of plant nutrition: essential nutrients, function, deficiency, and excess. Penn State Extension:1–4. Available at: https://extension.psu.edu/hydroponics-systems-and-principles-of-plant-nutrition-essential-nutrients-function-deficiency-and-excess

Santosh DT, Tiwari KN, Maitra S (2021) Influence of drip irrigation and plastic mulch on young cashew (Anacardium occidentale L.). Int J Biores Sci 8:87–92. https://doi.org/10.30954/2347-9655.02.2021.4

Santosh DT, Tiwari KN, Maitra S (2022a) Influence of different levels of irrigation and black plastic mulch on the performance of banana under drip irrigation. Crop Res 57(3):202–210. https://doi.org/10.31830/2454-1761.2022.031

Santosh DT, Souvik S, Maitra S (2022b) Estimation of crop water requirement in maize and chickpea for different patterns of intercropping. Int J Biores Sci 13:41392–41398

Santosh DT, Debnath S, Maitra S, Sairam M, Sagar LL, Hossain A, Moulick D (2023a) Alleviation of climate catastrophe in agriculture through adoption of climate-smart technologies. In: Sustainable development goals series, pp 307–332. https://doi.org/10.1007/978-3-031-44397-8_17

Santosh DT, Maitra S, Tiwari KN (2023b) Optimizing nutrients through drip fertigation and plastic mulch on the growth and yield in banana (Musa acuminata L.). Res Crop 24:330–336. https://doi.org/10.31830/2348-7542.2023.roc-11163

Santosh DT, Hadole M, Maitra S (2024) Enhancing capsicum cultivation under polyhouse, shade net house and open field conditions with drip fertigation. Int J Exp Res Rev 43:190–200. https://doi.org/10.52756/ijerr.2024.v43spl.014

Sashika MN, Gammanpila H, Priyadarshani S (2024) Exploring the evolving landscape: urban horticulture cropping systems–trends and challenges. Sci Horti 327:112870. https://doi.org/10.1016/j.scienta.2024.112870

Sasidharan R, Hartman S, Liu Z, Martopawiro S, Sajeev N, Van Veen H, Yeung E, Voesenek LACJ (2017) Signal dynamics and interactions during flooding stress. Plant Physiol 176:1106–1117. https://doi.org/10.1104/pp.17.01232

Scintilla S, Salvagnin U, Giacomelli L, Zeilmaker T, Malnoy MA, Van Der Voort JR, Moser C (2021) Regeneration of plants from DNA-free edited grapevine protoplasts. bioRxiv:1–5. https://doi.org/10.1101/2021.07.16.452503

Shah IH, Manzoor MA, Jinhui W, Li X, Hameed MK, Rehaman A, Li P, Zhang Y, Niu Q, Chang L (2024) Comprehensive review: effects of climate change and greenhouse gases emission relevance to environmental stress on horticultural crops and management. J Environ Manag 351:119978. https://doi.org/10.1016/j.jenvman.2023.119978

Sharma J (2013) Climate-resilient horticulture: adaptation and mitigation strategies. In: Singh HCP, Rao NKS, Shivashankar KS (eds) climate-resilient horticulture. springer, India, pp 81–88. https://doi.org/10.1007/978-81-322-0974-4
Sharma S (2024) Gramin Krishi Mausam Sewa. Launchpad IAS. Available at: https://launchpadeducation.in/gramin-krishi-mausam-sewa/
Sheeba SM, Nirmala MD, Premavathi R, Shanmugasundaram KA, Dheebakaran G, Vanitha G (2025) Opportunities and challenges in the adoption of protected technologies in horticultural crops. Plant Sci Today 12(2):1–12. https://doi.org/10.14719/pst.6254
Shelton AM, Hossain MJ, Paranjape V, Prodhan MZ, Azad AK, Majumder R, Sarwer SH, Hossain MA (2019) Bt brinjal in Bangladesh: the first genetically engineered food crop in a developing country. Cold Spring Harbor Perspect Biol 11(10):a034678. https://doi.org/10.1101/cshperspect.a034678
Shnaider Y, Elad Y, Rav-David D, Pashkovsky E, Leibman D, Kravchik M, Shtarkman-Cohen M, Gal-On A, Spiegelman Z (2022) Development of powdery mildew resistance in cucumber using CRISPR/CAS9-mediated mutagenesis of CSAMLO8. Phytopath 113:786–790. https://doi.org/10.1094/phyto-06-22-0193-fi
Shoaib M, Banerjee BP, Hayden M, Kant S (2022) Roots' drought adaptive traits in crop improvement. Plants 11(17):2256. https://doi.org/10.3390/plants11172256
Singh N, Sharma DP, Chand H (2016) Impact of climate change on apple production in India: a review. Curr World Environ 11:251–259. https://doi.org/10.12944/CWE.11.1.31
Singh AK, Shikha K, Shahi JP (2021a) Hybrids and abiotic stress tolerance in horticultural crops. In: Elsevier eBooks, pp 33–50. https://doi.org/10.1016/b978-0-12-822849-4.00015-2
Singh D, Sharma RR, Devappa V, Kamil D (eds) (2021b) Postharvest handling and diseases of horticultural produce. CRC Press, pp 43–56. https://doi.org/10.1201/9781003045502
Sønsteby A, Heide OM (2019) Temperature effects on growth and floral initiation in sweet cherry (Prunus avium L.). Sci Hortic 257:108762. https://doi.org/10.1016/j.scienta.2019.108762
Sowjanya B (2012) Economics of drip irrigation in mango cultivation -a study in Krishna district of Andhra Pradesh. Andhra Agric J 59(3):479–483
Srivastava Y (2019) Climate change: a challenge for postharvest management, food loss, food quality, and food security. In: Elsevier eBooks, pp 355–377. https://doi.org/10.1016/b978-0-12-816483-9.00019-0
Statista (2025) Global atmospheric carbon dioxide concentration 1959–2024. Available at: https://www.statista.com/statistics/1091926/atmospheric-concentration-of-co2-historic/
Subedi B, Poudel A, Aryal S (2023) The impact of climate change on insect pest biology and ecology: implications for pest management strategies, crop production, and food security. J Agric Food Res 14:100733. https://doi.org/10.1016/j.jafr.2023.100733
Taillandier C, Cörvers R, Stringer LC (2023) Growing resilient futures: agroforestry as a pathway towards climate resilient development for smallholder farmers. Front Sustain Food Sys 7:1260291. https://doi.org/10.3389/fsufs.2023.1260291
Tailor R, Parikh S, Kumar K, Collins T, El-Ocla H (2025) IoT-based food quality monitoring system. IET Wireless Sensor Syst 15(1):e70008. https://doi.org/10.1049/wss2.70008
TNAU portal (Tamil Nadu Agricultural University Portal). Horticultural College and Research Institute, Coimbatore Technologies developed/varieties released Available at: https://tnau.ac.in/site/hcri-coimbatore/technologies-developed/
Tamru S, Hansen J, Zebiak S, Tesfaye A, Minten B, Demissie T, Radeny M, Tesfaye K, Solomon D (2025) Climate information services enhance farmers' resilience to climate change: impacts on agricultural productivity. Clim Risk Manag 49:100724. https://doi.org/10.1016/j.crm.2025.100724
Tandel YN, Zala VR, Koti S (2025) Impact of high-temperature on growth and development of fruit crops. In: Soto-Hernández M, Palma-Tenango M, Aguirre-Hernández E (eds) Growth and development in plants and their medicinal and environmental impact physiology series 1009830. https://doi.org/10.5772/intechopen.1009830
Tchonkouang RD, Onyeaka H, Nkoutchou H (2024) Assessing the vulnerability of food supply chains to climate change-induced disruptions. Sci Total Environ 920:171047. https://doi.org/10.1016/j.scitotenv.2024.171047

Tek MI, Calis O, Fidan H, Shah MD, Celik S, Wani SH (2022) CRISPR/Cas9-based mlo-mediated resistance against Podosphaera xanthii in cucumber (Cucumis sativus L.). Front. Plant Sci 13:1081506. https://doi.org/10.3389/fpls.2022.1081506

Tripathi L, Ntui VO, Tripathi JN (2019) Application of genetic modification and genome editing for developing climate-smart banana. Food Energy Sec 8:e00168. https://doi.org/10.1002/fes3.168

Tuxun A, Xiang Y, Shao Y, Son JE, Yamada M, Yamada S, Tagawa K, Baiyin B, Yang Q (2025) Soilless cultivation: precise nutrient provision and growth environment regulation under different substrates. Plants 14:2203. https://doi.org/10.3390/plants14142203

Ullah H, Ahmad S, Anwar R, Noor-Un-Nisa N, Nafees M, Thompson A (2010) Effect of "oxygen and carbon-dioxide" on the post-harvest management in tree-ripe mango storage. J Chem Soc Pakistan 32(4):485–491. https://inis.iaea.org/records/qzgbk-cjg36

Urugo MM, Yohannis E, Teka TA, Gemede HF, Tola YB, Forsido SF, Tessema A, Suraj M, Abdu J (2024) Addressing post-harvest losses through agro-processing for sustainable development in Ethiopia. J Agric Food Res 18:101316. https://doi.org/10.1016/j.jafr.2024.101316

Valenzuela JL (2023) Advances in postharvest preservation and quality of fruits and vegetables. Foods 12:1830. https://doi.org/10.3390/foods12091830

Van der Vossen EA, Gros J, Sikkema A, Muskens M, Wouters D, Wolters P, Pereira A, Allefs S (2005) The Rpi-blb2 gene from Solanum bulbocastanum is an Mi-1 gene homolog conferring broad-spectrum late blight resistance in potato. Plant J 44:208–222. https://doi.org/10.1111/j.1365-313x.2005.02527.x

Vanlerberghe G (2013) Alternative oxidase: a mitochondrial respiratory pathway to maintain metabolic and signaling homeostasis during abiotic and biotic stress in plants. Int J Mole Sci 14:6805–6847. https://doi.org/10.3390/ijms14046805

Wajid A, Hussain K, Ilyas A, Habib-Ur-Rahman M, Shakil Q, Hoogenboom G (2021) Crop models: important tools in decision support system to manage wheat production under vulnerable environments. Agri 11:1166. https://doi.org/10.3390/agriculture11111166

Wan D, Guo Y, Cheng Y, Hu Y, Xiao S, Wang Y, Wen Y (2020) CRISPR/Cas9-mediated mutagenesis of VvMLO3 results in enhanced resistance to powdery mildew in grapevine (Vitis vinifera). Hortic Res 7:116. https://doi.org/10.1038/s41438-020-0339-8

Wang T, Zhang H, Zhu H (2019) CRISPR technology is revolutionizing the improvement of tomato and other fruit crops. Hortic Res 6:77. https://doi.org/10.1038/s41438-019-0159-x

Waqas M, Humphries UW, Chueasa B, Wangwongchai A (2024) Artificial intelligence and numerical weather prediction models: a technical survey. Nat Hazards Res 5:306–320. https://doi.org/10.1016/j.nhres.2024.11.004

Warren R, Price J, Forstenhäusler N, Andrews O, Brown S, Ebi K, Gernaat D, Goodwin P, Guan D, He Y, Manful D, Yin Z, Hu Y, Jenkins K, Jenkins R, Kennedy-Asser A, Osborn TJ, Van Vuuren D, Wallace C, Wang D, Wright R (2024) Risks associated with global warming of 1.5 to 4 °C above pre-industrial levels in human and natural systems in six countries. Clim Chang 177:48. https://doi.org/10.1007/s10584-023-03646-6

Wu Y, Xue W (2024) Data-driven weather forecasting and climate modeling from the perspective of development. Atmos 15:689. https://doi.org/10.3390/atmos15060689

Yang X, Li J, Ji C, Wei Z, Zhao T, Pang Q (2022) Overexpression of an aquaporin gene EsPIP1;4 enhances abiotic stress tolerance and promotes flowering in Arabidopsis thaliana. Plant Physiol Biochem 193:25–35. https://doi.org/10.1016/j.plaphy.2022.10.019

Zhang S (2023) Recent advances of polyphenol oxidases in plants. Molecules 28:2158. https://doi.org/10.3390/molecules28052158

Zhang X, Qu J, Li H, La S, Tian Y, Gao L (2020) Biochar addition combined with daily fertigation improves overall soil quality and enhances water-fertilizer productivity of cucumber in alkaline soils of a semi-arid region. Geoderma 363:114170. https://doi.org/10.1016/j.geoderma.2019.114170

Zhang Y, Chen X, Geng S, Zhang X (2025) A review of soil waterlogging impacts, mechanisms, and adaptive strategies. Front Plant Sci 16:1545912. https://doi.org/10.3389/fpls.2025.1545912

Reshaping Global Horticulture: Climate Resilience Through Technology and Sustainable Practices

18

Arijit Ghosh, Rohit Kumar Choudhury, Somnath Pati, Suchandra Dutta, Amlanjyoti Sar, Bibekananda Maiti, Souptik Sarkar, Sumana Balo, and Prakash Kumar Jha

Abstract

Climate resilience in horticulture requires transformational strategies in light of the increasing impacts of climate change on production in India, which accounts for 30.4% of agricultural GDP from just 11% of land. Increased temperature (1.5–5.4 °C by 2080), erratic rainfall, and extreme events have reduced yield by 10–40%, caused intensification of pest-disease outbreaks, such as Phytophthora epidemics, and increased carbon footprint through intensive inputs like fertilizers, which are also a major source of N_2O, irrigation, and postharvest cold chains. Key mitigation depends on carbon sequestration in perennials, from 2.5 to 12 t C/ha/yr via biomass and soil organic carbon (SOC), from agroforestry, and zero-emission practices that include conservation tillage, mulching, precision fertiga-

Arijit Ghosh and Rohit Kumar Choudhury are co-first authors.

A. Ghosh · R. K. Choudhury (✉)
Department of Agricultural Chemistry and Soil Science, Faculty of Agriculture, Bidhan Chandra Krishi Viswavidyalaya, Nadia, West Bengal, India
e-mail: choudhurykr.rohit@bckv.edu.in

S. Pati
Department of Soil and Water Conservation, Faculty of Agriculture, Bidhan Chandra Krishi Viswavidyalaya, Nadia, West Bengal, India

S. Dutta
Ramakrishna Mission Vivekananda Educational and Research Institute (RKMVERI), Narendrapur, Kolkata, West Bengal, India

A. Sar
Department of Fruit Science, Faculty of Horticulture, Bidhan Chandra Krishi Viswavidyalaya, Nadia, West Bengal, India

B. Maiti
Department of Plant Physiology, Faculty of Agriculture, Bidhan Chandra Krishi Viswavidyalaya, Nadia, West Bengal, India

A. Das, A. B. Sharangi (eds.), *Remunerative Horticulture*,
https://doi.org/10.1007/978-981-92-1007-7_18

tion, and renewables. Adaptive measures involve drought/salt-tolerant cultivars, such as Solar Fire tomato and Dog Ridge grape rootstock, protected cultivation, and climate-resilient Integrated Pest Management (IPM) using predictive modeling, biological controls, and digital tools for real-time monitoring. While protected systems with high embodied energy boost yields, critically, Life Cycle Assessment (LCA) must guide shifts to low-carbon alternatives. Policy gaps in missions like NHM limit the scaling of Measurement, Reporting, Verification (MRV) protocols for carbon credits. Integration through reformed subsidies like PM-KUSUM, FPOs, and horticulture-specific standards will help reposition the sector as a net sink, aligning with India's Net Zero 2070 goals.

Keywords

Climate change adaptation · Carbon footprint · Carbon capturing and sequestration · Zero emission agriculture · Precision horticulture · Carbon and climate change policies

18.1 Introduction

Indian agriculture has grown fast over the years and has become largely self-sufficient in food production. In this progress, horticulture has played a major role. It is today one of the best ways to diversify farming because besides giving food and nutrition, it helps in health, gives more income from the same land, and provides more employment. Even a small investment in horticulture yielded very good returns. Production and productivity increased, exports went up, and India emerged as the second-largest producer of fruits and vegetables in the world. It was no accident. New technology combined with relentless effort pushed horticulture from old-fashioned backyard gardens to a fast-paced, commercial business. If we take a look backward, there are five large chapters in this story. Horticulture, before independence, was largely a matter of home gardens and growing things for consumption and just out of love. Things changed when, after 1948, there was a demand from the

S. Sarkar
Department of Agricultural Chemistry and Soil Science, Faculty of Agriculture, Bidhan Chandra Krishi Viswavidyalaya, Nadia, West Bengal, India

School of Agriculture, Swami Vivekananda University, Barrackpore, West Bengal, India

S. Balo
Department of Soil Science and Agricultural Chemistry, Institute of Agricultural Sciences (IAS), Siksha O Anusandhan (Deemed to be University), Bhubaneswar, Khorda, Odisha, India

P. K. Jha
Department of Plant and Soil Sciences, Mississippi State University, Starkville, MS, USA
e-mail: pj442@msstate.edu

farm community that agricultural produce such as potatoes, coconuts, areca nuts, and spices be grown for the market, and the first support structures came into place.

However, it was only in the 1980s that they began to actually appreciate just how important the field of horticulture was to their diet, employment, and agricultural incomes. It is here that research received its major fillip and the field began to become more organized. The '90s were when increased budget allocations and missions began by the government to promote better technology dissemination and research. Finally, the 2000s were when the innovation explosion began: micro-irrigation, protected farming, micro-propagation, and the use of high-technology planting material.

Horticulture production has expanded by almost eight times since 1950. Today, India is estimated to be cultivating 234.5 million tons of horticultural commodities on merely 11% of agricultural land, and this is accounting for a massive 30.4% of agricultural GDP.

"Golden Revolution" is a name some people have assigned to it, and it is definitely justified. But this is not where it ends. The demand for fruits, vegetables, and other horticultural crops keeps increasing both in the domestic and foreign markets. Despite all this progress, there remains a mismatch between what consumers aspire for and what farmers produce. And land and water are becoming increasingly scarce, while climate change is making things more difficult. If Indian horticulture is to continue its winning streak, smart, resource-efficient, and innovative solution(s) are required.

18.2 Climate Change and Indian Horticulture

Since the industrial era began, huge amounts of fossil fuel have been burned. The result is that greenhouse gases such as carbon dioxide, methane, nitrous oxide, and sulfur dioxide have built up in the atmosphere. That buildup is the primary reason why, at present, climate change is occurring at such a rapid rate. According to Article 1 of the United Nations Framework Convention on Climate Change (UNFCCC), climate change refers to any change in the global climate system brought about by human activities that alter the composition of the atmosphere, in addition to natural changes over time. At present, the concentration of carbon dioxide is well above its historical record. Before factories and cars, CO_2 concentration was around 280 parts per million. In 2012, this had risen to about 393 ppm. Methane, nitrous oxide, and other greenhouse gases have been following a similar upward trend. According to the IPCC (2007), this increase in CO_2 increased global temperatures by 0.74 °C from 1906 to 2005. During the last three decades, frequency and severity of droughts, not only globally but also in India, have increased. The sea levels are rising, and frost, heat waves, and cold snaps continue to rise—an added stress to farmers and rural communities. The prospects, according to the IPCC (2007), indicate that South Asia will continue to warm up: the temperatures could increase by 0.5–1.2 °C by 2020, 0.88–3.16 °C by 2050, and 1.56–5.44 °C by 2080.

Rainfall will probably become more variable, and extreme weather events are likely to increase.

Global warming is not only a rise in temperature or a change in rainfall patterns. Its signature is seen in the wild swing and increased frequency of extreme weather events. Such changes in climate have taken a heavy toll on horticultural crops in India (Samra and Singh 2004). During March 2004, northern India experienced unusually high temperatures and it was enough to destroy tomatoes, onions, garlic, fruits, and vegetables across this belt. Cold waves in December 2002 and January 2003 destroyed mango, guava, papaya, brinjal, tomato, and potato crops in wide areas. Again, cold waves in 2006 damaged tomato, potato, mango, guava, banana, papaya, tamarind, wood apple, and brinjal. There is similar loss and damage associated with rain. Excessive rain in 1998, 2005, and 2010 damaged kharif and late-kharif onion crops. Unseasonal heavy rains along the West Coast during March 2008 resulted in low yield and poor quality of cashew nuts. Similarly, the apple orchards in Himachal Pradesh have moved further up into the hills because the old locations are not getting requisite chilling hours. This adjustment indicates how changing climate conditions are altering what farmers can grow and where they can grow it.

Overall, projected temperature increases by 2080–2100 are estimated to lead to yield losses of 10–40% in India (Rosenzweig and Parry 1994; Fischer et al. 2002; IPCC 2007). Most Indian studies also confirm these trends, reporting that future climates could result in reduced productivity (Aggarwal and Sinha 1993; Rao and Sinha 1994; Lal et al. 2008; Saseendran et al. 2000; Mall and Aggarwal 2002). However, for specific crops like sorghum (Srivastava et al. 2010), maize (Byjesh et al. 2010), and a wide array of others (Naresh Kumar et al. 2011), substantial spatial variability has been reported. Acceleration of the already strong growth rate of horticulture is required to meet the national aspiration of 4% growth in agriculture. Add to it the increasing demand, both domestic and international, particularly for quality produce. Climate change further complicates these challenges by increasing the variability in quality and productivity of horticulture crops. Thus, a systematic assessment of regional climate impacts on horticultural crops backed by developing robust and location-specific adaptation strategies is critical (Fig. 18.1). Available information on the response of horticultural crops to elevated CO_2 and temperature from Indian studies and also some of the observed impacts of climate-related events on horticultural crops in India has been documented earlier (Aggarwal 2009; Naresh Kumar and Aggarwal 2009; Chadha and Naresh Kumar 2010; Naresh Kumar et al. 2010, 2011, 2012).

18.3 Impact of Climate Change on Pest and Disease Dynamics

Research evidence indicates that climate change, through increased atmospheric CO_2, changes in temperature regimes, and altered precipitation, can have a significant impact on the developmental biology and pathogenicity of plant pathogens.

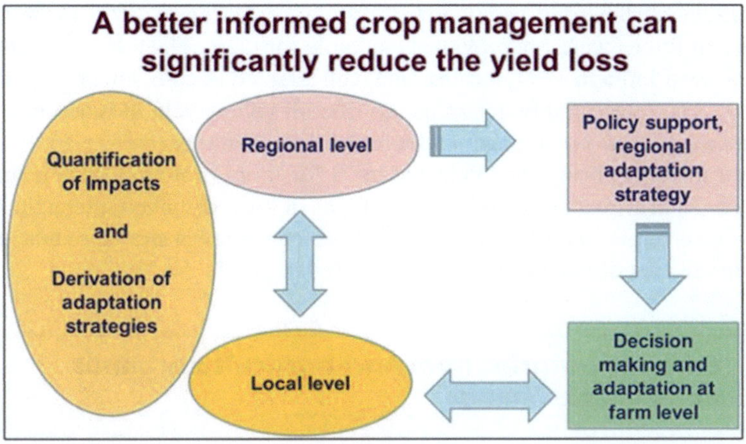

Fig. 18.1 Flow chart illustrating crop management strategies to reduce yield loss. The process begins with "Quantification of Impacts and Derivation of Adaptation Strategies" in an oval shape. This leads to two pathways: "Regional level" and "Local level," each represented by arrows. The "Regional level" connects to "Policy support, regional adaptation strategy," while the "Local level" connects to "Decision making and adaptation at farm level." The chart emphasizes informed crop management's role in reducing yield loss

These changes also influence the physiology and resistance of host plants. Studies carried out on important diseases of horticultural crops in India have shown that climate change has been a factor in the epidemiology and outbreak of several pathogens. These include fruit rot (*Phytophthora meadii*) and leaf spot (*Phyllosticta arecae*, *Colletotrichum gloeosporioides*) in arecanut, bud rot (*Phytophthora palmivora*) in coconut, and a range of diseases in grapes such as yellow rust, *Alternaria* bunch rot, *Botryodiplodia* dieback, and *Grenaria* leaf and fruit rot in grapes. Others include *Macrophoma* infection in banana, *Cercospora*, and wilt in chili; blossom blight in mango; and wilt and nodal blight in pomegranate. However, since 2008, serious outbreaks of different *Phytophthora* species have been reported on a wide range of crops. These include late blight on potato and tomato, *P. infestans*; fruit rot in brinjal, *P. parasitica*; wilts in chili and capsicum, *P. capsici*; blights in cucurbits, *P. capsici*; fruit rot in okra, *P. parasitica*; and root rot in cabbage and cauliflower (*P. megasperma*, *P. drechsleri*) (Chowdappa 2010).

Detailed analysis of the *P. infestans* populations associated with late blight in tomato revealed that there was a distinct lineage from the earlier global US-1 lineage and the IN-1 and IN-2 genotypes predominant in Indian potatoes and also established that this newly evolved population was responsible for the severe epidemics in tomato since 2009. It is projected that each 1 °C increase in temperature advances late blight onset by 4–7 days with lengthening of the susceptible period up to 10–20 days. Similarly, during 1997–1998, the AICRP (Tropical Fruits), Jalgaon Centre, reported Sigatoka leaf spot in banana for the first time. Analysis of data for the years 2000–2007 revealed that disease incidence was positively correlated with higher afternoon relative humidity and increased rainfall. Climate change may also favor the establishment of newly introduced viruses or their vectors in areas where climate conditions were not previously suitable for them. The tobacco streak virus, which entered India in 1997, has become well adapted to chili, cucurbits, ornamentals, and okra. Interestingly, resistance to Bhendi yellow vein mosaic virus in okra has been reported to break down under higher temperatures.

These incidents bring into light the need for more profound knowledge about how changes in climate influence various types of host–pathogen interactions. Such knowledge is required for the design of disease management strategies that are resilient in the face of climate change (Krishna Reddy 2010).

18.4 Carbon Footprint from Agri-horticultural Lands and Climate Change

18.4.1 Concept of Carbon Footprint

The concept of "carbon footprint" originally evolved from the term "ecological footprint" proposed by Rees (1992) and Wackernagel and Rees (1996). Ecological footprint can be defined as the area of biologically productive land and water required to support a particular human population, usually expressed in global hectares (Pandey and Agrawal 2014). Similarly, few studies also reported that the biologically productive global hectares that are required to assimilate the carbon dioxide (CO_2) produced during the lifespan of a human or human population (Pandey and Agrawal 2014; Browne et al. 2009). As global warming has emerged as a serious global concern, attention has been directed toward understanding its relevance to carbon footprint and the definition has been developed independently in a modified way (East 2008). A carbon footprint can now be understood as a hybrid concept, which takes its terminology from the "ecological footprint" but acts as a measure of the global warming potential (Pandey and Agrawal 2014). Carbon footprint is a concept that has been around for a few decades; however, it is sometimes referred to as global warming potential or life cycle impact category indicator (Finkbeiner 2009). A survey was carried out by Wiedmann and Minx (2007), and based on this survey, different definitions of carbon footprint have been recognized from different points of view. According to them, the term "carbon footprint" may be defined as the total amount of CO_2 released into the atmosphere exclusively caused by both direct and indirect emissions stemming from an activity or those accumulated along the whole life cycle of a product. Due to the complexity of accounting for all indirect emissions, many carbon footprint studies reported only direct emissions or first-order indirect emissions (Carbon Trust 2007; Wiedmann and Minx 2007; Matthews et al. 2008). Still, indirect emissions can represent a significant portion of the total carbon footprint in many cases (Matthews et al. 2008). Initially, the CO_2 emission was the only gas considered when assessing carbon footprints. Later, besides CO_2 emissions, several key GHGs have been taken into consideration, which include CH_4, N_2O, etc. That is why Wiedmann and Minx (2008a, b) slightly modified the classical definition and stated carbon footprint as the emission of a certain quantity of gases produced from several human activities and their impact on climate change. A carbon footprint measures the GHG emissions within a specific boundary related to any individual, product, organization, or any system during its entire life cycle—from production to end-of-life disposal in terms of CO_2 equivalent (CO_2e) (Pandey et al. 2011). According to the IPCC (2014), CO_2-e is the concentration of CO_2 that would have the same radiative potential as a given mixture of CO_2 and other greenhouse gases (Choudhury et al., 2026).

There are formulas to quantify this, one such being by Pandey and Agrawal (2014), which calculates global warming potential (GWP) in kg CO_2-e per hectare by summing emissions/removals of CH_4 multiplied by 25, N_2O multiplied by 298, and CO_2:

$$\text{Spatial Carbon Footprint} / CO_2\text{e}\left(\text{kgha}^{-1}\right) = \sum_{i=1}^{n}[GWP]_i$$
$$= \sum_{i=1}^{n}\left(\text{emission of } CH_4 \times 25 + \text{emission of N}_2\text{O} \times 298 + \text{emission of } CO_2\right)_i \quad (18.1)$$

$$\text{Yield Scaled Carbon Footprint}\left(\text{kg}\, CO_2\ \text{ha}^{-1}\text{kg}^{-1}\right) = \frac{\sum_{i=1}^{n}[GWP]_i}{\text{Grain Yield}} \quad (18.2)$$

Agricultural carbon footprint can be computed by multiplying agricultural inputs by corresponding greenhouse gas emission coefficients divided by the grain yield, as described by Lal (2004a, b, c, d):

$$\text{Carbon Footprint from Agriculture} = \frac{\sum_{i=1}^{n}\left(\text{Agricultural Inputs} \times \text{GHGs Emission Coefficients}\right)}{\text{Grain Yield}} \quad (18.3)$$

18.4.2 Contribution of Agri-horticultural Practices to Carbon Footprint

Most horticultural systems are input-intensive and energy-dependent, so footprints reflect both on-farm biophysical processes (soil and plant–microbe interactions) and off-farm processes such as the manufacture of fertilizers, agrochemicals, plastics, and cold-chain logistics. Global estimates of Agriculture, Forestry, and Other Land Use (AFOLU) indicate that anthropogenic GHG emissions from managed lands amount to nearly a quarter, with a share directly linked to fertilizer-related N_2O, soil carbon loss, and rice-like CH_4 emissions under waterlogged conditions represented by croplands and plantations. In this aggregate, the area covered by horticultural crops is small compared to cereals (Fig. 18.2a), but very often, the emissions intensities per unit product are higher due to protected cultivation, irrigation, cold storage, and packaging needs. Agriculture covers approximately 35% of

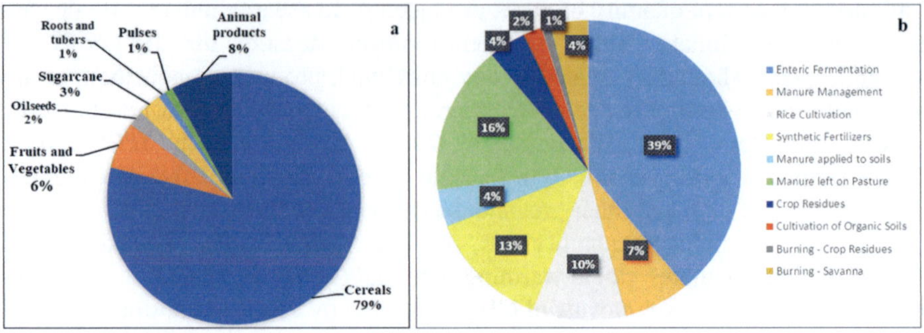

Fig. 18.2 Two pie charts labeled (**a**) and (**b**). Chart (**a**) shows the distribution of agricultural products: cereals (79%), animal products (8%), fruits and vegetables (6%), sugarcane (3%), oilseeds (2%), roots and tubers (1%), and pulses (1%). Chart (**b**) illustrates sources of emissions: enteric fermentation (39%), manure left on pasture (16%), rice cultivation (10%), synthetic fertilizers (13%), manure applied to soils (4%), crop residues (4%), cultivation of organic soils (2%), manure management (7%), burning of crop residues (1%), and burning of savanna (4%). ((**a**) *Source*: Adopted from Kashyap and Agarwal (2020) with permission from the publisher. *Licence no*: 6170610242866. ©Springer Nature, 2025. (**b**) *Source*: FAOSTAT (2014))

land globally and contributes around 13.5% of the total anthropogenic greenhouse gas (GHG) emissions, responsible for about 25%, 50%, and 70% of anthropogenic CO_2, CH_4, and N_2O, respectively (Montzka et al. 2011). Agriculture is dominated by CH_4 emissions from rice cultivation and livestock, while N_2O production originates principally from fertilizer use. An increase in atmospheric CO_2 is also likely to enhance N_2O emissions from upland cropland soils (Van Groeningen et al. 2011). For CO_2, it is the soil respiration that plays the dominant role, though most cultivation activities, as well as inputs such as fertilizers, pesticides, and fuel, have significant embedded CO_2. Of the practices, mulching and organic manure are more likely to increase the CH_4 emission, but midseason drainage significantly reduces emission (Zou et al. 2005; Ma et al. 2007). Contribution of different farming operations to the carbon footprint has been depicted via Figs. 18.2b and 18.3.

18.4.2.1 Soil Management and Tillage

Soil disturbance through conventional tillage accelerates decomposition of soil organic matter, increasing CO_2 emissions from microbial respiration and weakening the soil's role as a long-term carbon sink. Repeated deep tillage in orchards and plantations, especially on sloping lands, also enhances erosion and depletes soil organic carbon, indirectly raising the carbon footprint by lowering inherent fertility and increasing dependence on external inputs.

18.4.2.2 Fertilizer and Pesticide Use

Nitrous oxide is the dominant non-CO_2 gas from cropped soils, arising from nitrification and denitrification when mineral nitrogen is present in excess of plant demand or under conditions of alternating wetting and drying. High-value horticultural crops frequently receive heavy nitrogen dressings and fertigation, so their N_2O

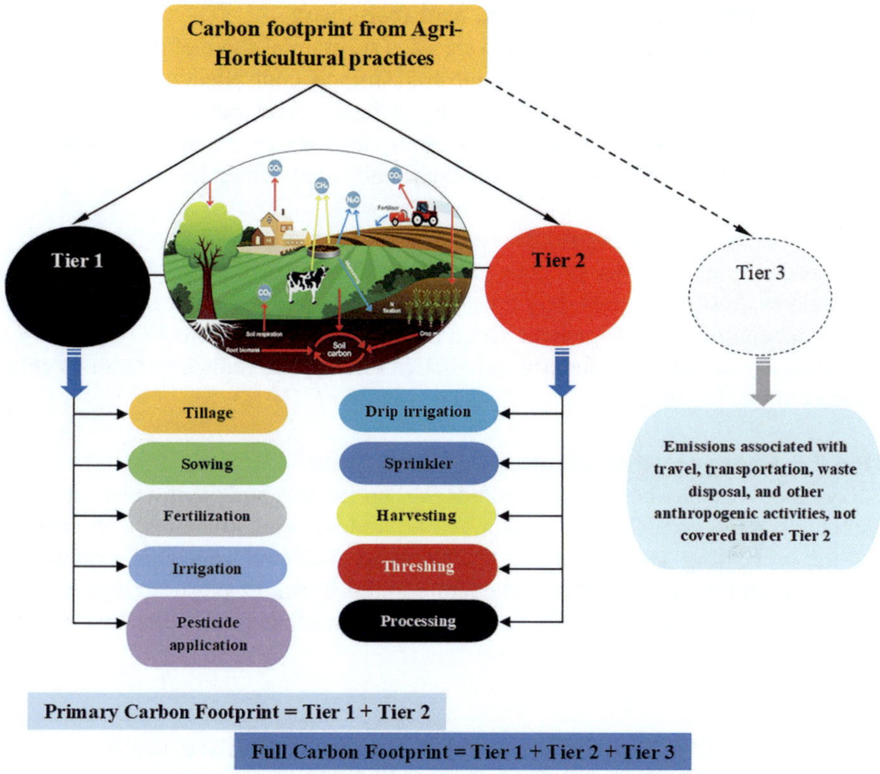

Fig. 18.3 Flow chart illustrating the carbon footprint from agri-horticultural practices. The chart is divided into three tiers. Tier 1 includes activities like tillage, sowing, fertilization, irrigation, and pesticide application. Tier 2 covers drip irrigation, sprinkler use, harvesting, threshing, and processing. Tier 3 involves emissions from travel, transportation, waste disposal, and other activities not covered in Tier 2. The primary carbon footprint is the sum of Tier 1 and Tier 2, while the full carbon footprint includes all three tiers. An illustration in the center depicts a farm with various agricultural activities contributing to soil carbon. (*Source*: Modified from Pandey and Agrawal (2014); *Concept*: Carbon Trust (2007), BSI (2008), and Lynas (2007))

footprint per hectare and per tonne can exceed that of many field crops if nutrient management is inefficient.

18.4.2.3 Irrigation Practices and Water Management

Irrigation contributes to the carbon footprint both directly through energy used for pumping, pressurization, filtration, and water conveyance and indirectly by altering soil aeration and microbial activity. Surface or flood irrigation can enhance N_2O and, in ponded systems or poorly drained soils, CH_4 emissions by creating anaerobic microsites; it also tends to be less water-efficient, increasing energy use per unit yield.

18.4.2.4 Postharvest, Processing, and Transport

For fruits, vegetables, cut flowers, and spices, a substantial share of the carbon footprint arises beyond the farm gate through cold storage, controlled-atmosphere facilities, processing, packaging, and often long-distance refrigerated transport. Energy demand along these stages results in CO_2 emissions whose magnitude depends on grid carbon intensity, refrigeration efficiency, and supply chain length.

18.4.2.5 Crop Residues and Biomass Management

Agri-horticultural systems produce diverse residues—prunings, leaves, rejected fruits, stalks, husks, and pack-house waste—which can either act as a carbon source or sink depending on management. Open burning of residues emits CO_2, CH_4, N_2O, and aerosols, contributing directly to radiative forcing and indirectly to air-quality-related climate feedbacks.

18.4.2.6 Energy Use and Embedded Emissions

Energy inputs in agri-horticulture include fuel for tractors and pumps, electricity for greenhouses and protected cultivation, plastic mulches and structures, and machinery for processing and packaging. In intensive fruit and vegetable production, embedded emissions from the manufacture and disposal of plastics (mulch films, trays, greenhouse covers) and from frequent mechanized operations can be substantial.

18.5 Adaptive Strategies in Horticulture and Agronomic Practices Toward Climate Resilience

18.5.1 Drought and Heat-Tolerant Varieties

Various crops have also been developed with the potential to withstand the harsh weather conditions. These crops include tomato “Solar Fire,” which is resistant to high temperatures with the ability to yield even when the temperature goes beyond 90 °F or 32 °C. The variety also exhibits some level of resistance to drought conditions, which makes it more ideal for cultivation in regions with limited or drought-ridden weather conditions (Foolad 2007). The Vaishali hybrid variety of tomato gives medium-sized (100 g) premium quality with good yield potential and adaptability to hot, humid conditions. The variety is resistant to Fusarium and Verticillium wilt. The Kufri Bhaskar potato variety performs well under heat stress. The variety is also resistant to mites and hopper-burn insects. Kufri Pukhraj and Kufri Khyati varieties showed maximum resistance to drought, with a drought tolerance index of 0.98, which can be grown under drought conditions of Rajasthan, Haryana, Punjab, Uttar Pradesh, Bihar, West Bengal, Gujarat, Madhya Pradesh, Odisha, and Chhattisgarh (Kumar et al. 2013). The “Tasty Green” cucumber variety is also resistant to heat with the potential to yield even in hot weather conditions (Collado-González 2021). Sierra lettuce varieties are heat-resistant varieties developed to withstand hot climatic conditions, which makes them ideal to grow in areas that experience excessive heat during planting seasons and cannot develop heads of lettuce due to the heat from the climate (Jenni et al. 2013). Jammu Mango "Selection-05," developed by the Sher-e-Kashmir University of Agricultural Sciences and Technology, Jammu, is drought-resistant and ideal for planting in a high-density manner in semiarid, semi-tropical climatic zones with irrigation (Sharma et al. 2020). The drought indices use genotypes of Vitis parviflora, Pusa Navrang, and Dogridge (Vitis × champini) as drought-resistant varieties of grapes, which are sensitive to drought conditions, while Pusa Trishar and Flame Seedless are sensitive to drought conditions of grape varieties (Prakash et al. 2025). Grenache and Cabernet Sauvignon are grape species, and they are known for their high resistance to drought and their suitability for growing in hot and dry climates (Hochberg et al. 2015). Ganesh and Wonderful are varieties of pomegranates that grow well in hot and arid areas and are also known to have high resistance levels to drought (Levin 2006). *Portulaca grandiflora* (Moss Rose) is an ornamental crop and is suitable to grow in hot and dry climates, and it also requires very little water (Huang et al. 2005).

18.5.2 Salt-Tolerant Varieties

Salt tolerance is an important factor in maintaining the robust growth of horticultural crops, especially under circumstances when saline soil is becoming an increasing concern. Nowadays, it is seen that many crops are being grown using varieties and rootstocks that are not affected by rising salt concentrations in soil. To cite an example, fruit crops are grown using Rangpur lime, Cleopatra mandarin, rough lemon, tangelo, sour orange, sweet orange, and citranges. These rootstocks are not reluctant to grow in tough or saline soil and hence help in maintaining growth and fruiting. Mango rootstocks also have their varieties, including Gomera, 13/1, Nekkare, Movandan, and Kurukkan varieties, which are tolerant of high salt concentrations compared to other varieties. Stone fruits, which are peaches and plums, are grown by grafting them onto Marianna, Lovell, Shalil, and Yunnan rootstocks, mainly because of their potential in maintaining growth and making them salt-tolerant. Avocado rootstocks are West Indian and Mexican varieties because they are relatively salt-tolerant compared to other varieties. Grape rootstocks are those of Salt Creek, 1613-3, and Dog Ridge because of their high salt-tolerance levels. Varieties like Thompson Seedless, Perlette, Cardinal, and Black Rose are also unyielding when it comes to saline soil. Berries, including boysenberry and Indian Summer raspberry, also adapt quickly when exposed to saline soil. Lassen and Shasta strawberry varieties are relatively more tolerant (Singh and Rani 2023). Vegetables also benefit when varieties and hybrids are tolerant of saline soil. In the case of tomatoes, 'Pusa Rohini' varieties are quite effective. 'Pusa Sadabahar' and 'Arka Neelkanth' are suitable for growing chili and brinjal varieties. If it is grown in an environment that is shielded from stress and other elements, it too would protect them from other stresses and hence their productivity levels would increase (Meena et al. 2025). Okras also have some varieties that protect them from being affected by saline soil when watered using saline solutions. Arka Anamika and Parbhani Kranti are some of them and are suitable because when watered using saline water, they also cope well with other stresses, and hence, their productivity levels increase. Cabbages also have some varieties, which are the Golden Acre and Pusa Mukta varieties that are able to grow in saline soil because other varieties are shriveled and hence their growth is impeded. Cauliflower varieties belonging to Pusa Snowball can adapt when grown under saline soil and, hence, are able to grow without much stress. Beetroot naturally adapts itself and hence continues growing whether it is saline or not because its growth continues without being affected by saline soil.

As regards carrots, it is decently tolerant, hence a good choice where salinity is an issue but not unmanageable, as per (Yadav et al. 1998).

18.5.3 Water Management Strategies

Vegetable growing is mainly affected by the quality and use made of water, more particularly because most varieties are eaten raw and contain a high amount of water. Most vegetable farms in the world still make use of surface irrigation methods, although this method is very inefficient and provides the farm with only half the water it may use ideally (Von and Chieng 2004). The water situation in India is already under stress due to increasing temperatures that affect the amount and rate of evaporation and the level of groundwater. Some regions in India experience patchy water shortages because most of the country's yearly rainfall is received for an average short period during the monsoon seasons. The effect on communities due to changes in the level of the water balance caused by climate change is mainly land use-specific.

Using emitters and pipes, drip irrigation, also known as micro/trickling irrigation, delivers water slowly and directly to the root zone. In such irrigation, more control of fertilizer and water is accomplished. There is also less runoff and percolation of water when compared to the surface irrigation technique, with drip irrigation resulting in large reductions of 50–80% of water loss. Research studies have indicated that drip irrigation requires about 50% less water compared to furrow irrigation in the case of chilies, especially when higher rates of discharge are used by emitters (AVRDC 2005). Reduction of 70% in nitrous oxide emissions has also been linked to drip irrigation in relation to melons (Sanchez-Guerrero et al. 2009). An increase in efficiency is also realized when drip irrigation is installed below the ground (Sharma et al. 2005). In the case of tomatoes, efficient drip irrigation not only increases the efficiency of irrigation but also improves fruit quality, and irrigation scheduling needs consideration of large fluctuations in demand based on the different growth stages of the crop (Rudich et al. 1977).

18.5.4 Soil Management Strategies

The basic goals of soil management are to (i) preserve or improve productivity per unit input, area, and time; and (ii) to support environmental regulatory processes, such as reduction of GHG emissions, enhanced air quality, and filtering out pollutants from air and water (Lal and Kimble 1997). Vegetables can withstand heat, heavy rain, and moisture loss with the aid of mulching, raised beds, protective shelters, and erosion-control techniques. Because they lower soil temperature, minimize evaporation, prevent erosion, and keep fruits clean while suppressing weeds, both organic and synthetic mulches are frequently utilized in high-value vegetable systems. Additionally, these advantages might reduce irrigation requirements by almost 25%. Rice straw is a popular option for summer tomatoes in tropical locations, and organic mulches enhance the soil's fertility and structure. When compared to unmulched fields, mulching has increased the development of crops like eggplant, okra, gourds, and melon in India (Pandita and Singh 1992). Tomato yields can also be greatly increased by combining drip watering with black plastic mulch (Singh et al. 2009). It is advised to use rice straw in conjunction with dark plastic mulch in hot lowland regions because the straw keeps the soil from scorching while the plastic stops light from penetrating the soil. Simple plastic rain shelters have also been demonstrated to increase yields for tomatoes cultivated during the rainy season, when waterlogging and fruit loss are frequent (Midmore et al. 1992). Leaving agricultural leftovers or cultivating cover crops aids in the replenishment of soil organic matter, but heavy tillage rapidly mineralizes it (Rahn et al. 2003). It has long been demonstrated that conservation tillage lowers carbon emissions and increases soil carbon storage while also strengthening farm resilience and mitigating the adverse effects of climate change. Early planting (80.2%), mixed farming (77.2%), appropriate seed storage (75.8%), and the combined use of organic and inorganic manures are some of the adaptation techniques that farmers in the Niger Delta are currently implementing, according to studies. In order to adapt to the changing climate, many additionally use mulching, low or zero tillage, cover crops, regular weeding, watershed protection, and even afforestation (Nzeadibe et al. 2011).

18.5.5 Efficient Use of Fertilizer

In most crop production systems, N is supplied in larger quantities through inorganic sources (chemical fertilizers) due to its faster availability to plants, while small quantities are added through organic sources (crop residues, organic manure, biological N fixation, and organic manures). The first and most important step in the N cycle is the addition of N to the soil–plant system through either organic sources or inorganic sources. While crop growth and development rely on the uptake and utilization of soil-applied nitrogen, a considerable portion of this nitrogen can be lost from the soil-plant system due to a variety of processes, including soil erosion, surface runoff, leaching, de-nitrogenization, and ammonia volatilization. N adsorbed on soil particles can be lost by wind erosion in addition to water erosion, soil erosion and surface runoff. While water erosion is the most often documented method of N loss in humid and sub-humid locations, wind erosion is more common in dry and semiarid climatic zones. Nitrate that has been put to the surface can dissolve in the water after a significant downpour and be lost by drainage (Wei et al. 2024). Fertilizers containing urea and urea derivatives that are surface-applied and do not immediately seep into the soil are more vulnerable to losses due to volatilization. In warm, sunny conditions and alkaline soil, this process of N loss is more severe; in only 1 week, up to 20% of N may volatilize and be lost to the sky. Denitrification is the anaerobic, microbially driven conversion of nitrate nitrogen to various gaseous forms of nitrogen (NO, N_2O, and N_2). Deep placement, the use of super granules, and foliar spraying of N fertilizer are some of the ways to apply N fertilizer that can improve its recovery (Wen et al. 2024).

18.5.6 Reducing Nitrous Oxide Emissions

Possible methods of N_2O emission from the soil include a biological reaction (chemodenitrification) and biological processes (denitrification, nitrification, dissimilatory nitrate reduction, and assimilatory nitrate reduction). Nonetheless, it has been determined that nitrification and denitrification are the most significant processes, with the other mechanisms making up only 1% of the overall output. When nitrate is available in anaerobic microsites formed wherever microbial demand for O_2 surpasses diffusion-mediated supply, denitrification takes place. This might happen in situations where water is impeding the diffusion of oxygen, such as in water-saturated areas or the cores of soil aggregates, or in situations where there is an abnormally high local demand for oxygen. N_2O is also used up by denitrification in soils when N_2O is reduced to N_2. Therefore, this bacterial process might be used as a supply or as a sink of N_2O.

18.5.7 Conservation Tillage

Conservation tillage (CT) is a term used to describe an assortment of tillage techniques that improve soil aggradation and shift soil-degrading processes, thereby increasing soil organic carbon (SOC). These techniques reduce soil and water losses when compared to conventional or plow-based tillage methods and use crop residue mulch to protect against raindrop impact. It is important for sequestering carbon in the soil (Lal 2004a, b, c, d). Earthworms and termites are two examples of the species richness and activity of soil fauna that are typically beneficial to conservation tillage. It significantly increases earthworm activity, which leads to the enhancement of soil structure (Katyal et al. 2001). Conservation tillage also enhances soil hydrologic characteristics and soil tilth because of the advantages of soil fauna; it typically has a positive impact on SOC due to deep placement and mixing. The transport of SOC from the surface to the subsoil is facilitated by the burrowing activities of soil fauna. Additionally, conservation tillage enhances the stability and agglomeration of aggregates, which leads to the enhancement of water-holding capacity (Katyal et al. 2001).

The FAO stated that the conservation agriculture system facilitates environmentally friendly land management, boosts crop productivity without harming the environment, and lessens the negative consequences of climate change. Due to its many advantages, which include higher soil fertility and water quality, improved soil physiochemical and biological quality, long-term sustainable productivity, and climate change mitigation. Conservation agriculture approaches, in contrast to traditional systems, increase soil water, especially soil infiltration rate, nutrient status, residual water content, and organic carbon content in upland crop production systems. The three main tenets of CA are: minimizing soil disturbance through no-till practices, maintaining soil cover through mulching, and adjusting crop rotation and intercropping techniques (Hussain et al. 2021). Certain structural traits in woody crops give them the ability to potentially retain a large amount of atmospheric carbon (Carbonell-Bojollo et al. 2010; Gonzalez-Sanchez et al. 2019). They can store carbon in the soil (by rhizodeposition, for example) and in permanent organs like trunks, branches, and roots due to their prolonged life cycle. These perennial woody crops also permit direct SOC transfer into the subsurface, reducing their susceptibility to mineralization, thanks to their vast and deeply ingrained root systems. Enhancing the rate of SOC sequestration is advised by the implementation of cover crops in the cropping system as a management approach (Ruibérriz et al. 2012; Mazzoncini et al. 2011). Some climate-resilient adaptive strategies of conservation agriculture under potato and mango cultivation have been explored in Table 18.1.

18.5.8 Crop Management Practices and Carbon Emissions

Growing a variety of plant types on the same piece of land is known as crop rotation. This is in contrast to continuous monoculture, the practice of continuously cultivating a single species on the same land, or intercropping, which is the practice of growing two or more crops simultaneously on the same land. Models utilize the idea of crop rotation to depict the time aspect of cropping plan decision-making (Yang et al. 2024). Agronomists have traditionally established a method to determine cropping plans based on crop proportions in crop rotation since crop succession in a particular region affects productivity and, in turn, cropping plan decisions. Certain writers have contended that crop selections obtained from crop rotation are the only way to guarantee the repeatability of a cropping system across time. Establishing conclusions on cropping plans so necessitates glancing back and forth in time. Hence, building cropping systems that are stable throughout time naturally starts with crop rotation as a specific crop sequence. Crop rotation is seen as a necessary component of integrated farming, which runs counter to the idea that monocropping is a sustainable farm practice. Crop rotation is a notion that offers an intriguing way to have a different crop on a certain plot of land every year. It can lessen agriculture's negative effects on the environment while preserving output and advancements throughout time. Crop rotations are also used to disrupt the cycles of weeds

Table 18.1 Climate-resilient adaptive strategies in potato and mango cultivation under diverse agroecological conditions of West Bengal, along with benefits, key findings, and references

Crop/region/ condition	Adaptive strategy	Benefit	Key findings	Reference
Potato—West Bengal (alluvial and general)	Selection of suitable cultivar + shifting planting date under climate change	To mitigate yield loss from rising temperature and CO_2 stress	Potential yield reduction under stress can be limited to 1.5% (2020) or 7.1% (2055), instead of 23–29%	Dua and Sharma (2017a, b)
Potato—Coastal Sundarbans, saline soils	Zero-tillage + Paddy-straw mulching (ZTSM) after rice	Save tillage, conserve moisture, allow rice-potato rotation, suitable for marginal saline soils	Good tuber yield under mulch; adoption increased from 23 to >1100 farmers over 7–8 years	Dey et al. (2022)
Potato—Coastal Sundarbans	Resource recycling (straw, manure, pond water) + farm-type-based land management	To ensure resource sustainability, labor balance, facilitate smallholder adoption	Patterns show viability of scaling ZT-potato, altering seasonal land use, enhancing livelihood resilience	Goswami et al. (2024)
Mango—Lower Indo-Gangetic plains (WB agroecological zones)	Use of deep, well-drained, coarse-loamy soils + integrated nutrient/soil management	Soil suitability is critical for stable mango yield; INM + cover crops to sustain SOC and fertility	Soil pH 5.6–7.9; OC 4.2–9.0 g/kg; high base status supports sustainable mango cultivation	Gangopadhyay and Bhattacharyya (2021)
Mango (general)—growers facing climate stress	Adoption of climate-resilient orchard practices (summer irrigation, KNO_3 spray, IPM)	To reduce fruit drop and mitigate heat stress and pest/disease outbreaks under climate variability	85% of growers used summer irrigation; 74% used KNO_3 for fruit retention; >50% used IPM—but overall adoption is only "moderate" (52.5%)	Ahsan and Begum (2023)

and diseases and to lessen reliance on outside resources. However, the idea of crop rotation offers very little guidance on how to arrange crops over various, uneven areas of land (Pi et al. 2024).

18.5.9 Protected Cultivation and Climate-Controlled Systems

One of the most effective methods available to safeguard horticultural crops against unforeseen changes in climate is through protected farming. Control buildings, such as greenhouses, net houses, and low tunnels, provide insulation against unforeseen climatic changes. These protected methods decrease unforeseen risks caused by climatic changes. These protected methods provide an environment ideal for crop growth. It is accomplished through the regulation of light, temperature, humidity, and CO_2 concentration. Thus, controlled environment agriculture (CEA) is an essential mechanism in climate-resilient horticulture (Sharath Kumar et al. 2020). Farmers are in a position to decrease heat stress in crops, stop soil leaching due to heavy rainfall, and protect crops from wind by covering crops with transparent materials. Cracking and dropping, common in open fields, are also avoided in greenhouses. To provide ideal temperatures during nights, they use passive heating methods with sun-trapping materials, insulated walls, and thermal blankets. These varieties significantly improved winter yields and were successful in terms of efficiency in cold climatic regions (Chen et al. 2000; Gao et al. 2010). One of the significant advantages offered by protected farming is its ability to provide regulation of crop temperatures. Techniques such as forced ventilation, shade, and evaporative cooling pads provide ideal tropical temperate climatic conditions in fields with high limitations. Evaporative cooling paddings are still useful in reducing heat stress. Initially, they were incorporated in early greenhouse inventions. Later on, they were introduced in Indian farming (Cunningham 1985).

Soilless culturing methods such as hydroponics, aeroponic systems, and aquaculture, which enable accurate control of light, humidity, and nutrient levels, are also promoted with the aid of protected farming. Hydroponics could be easily adopted due to the higher use efficiency and lower incidence of diseases; now, it is extensively used in the cultivation of leafy vegetables (Singh 2017; Tiwari and Bisht 2015). The conservation of water and improved use of nutrients in aeroponic and aquaculture systems offer other advantages and make them suitable for climate-resilient farming (Sulaiman and Thomas 2020). Precision farming methods, which involve the use of sensors and data analysis applications for monitoring microclimate, soil moisture, and nutrient status, are also aided by protected farming practices and environmental modification. As per present climate-smart approaches, this enhanced resistance to the changes in climate patterns also helps to improve the efficiency of use of all resources (Smith et al. 2021; FAO 2020).

18.6 Reduced Pesticide Use and Integrated Disease and Pest Management (IDPM)

There is a greater need to reduce the carbon footprint of agricultural practices than ever before, given the increasing environmental issues. Carbon dioxide and pollution are increased by agricultural practices, especially those requiring large amounts of pesticides. Artificial pesticides damage the environment, soils, and aquatic environments in addition to carbon dioxide release throughout their production, packaging, and use. Modern agricultural practices are essential while the world is in search of a long-term remedy to slow down climatic issues. Minimizing the need for chemical pesticide control through the application of Integrated Pest Management (IPM) is also another effective way of addressing such problems. To address the problem of minimizing the need for chemical pesticide control, Integrated Pest Management (IPM) applies a range of methods to manage pests. Through Integrated Pest Management (IPM), a balance is made between ecological and health concerns to minimize negative impacts on the environment resulting from agricultural practices (Peshin and Zhang 2014). Unlike conventional ways of managing pests, which entirely rely on chemical pesticide control, a range of methods and processes is used by Integrated Pest Management (IPM) to manage pests to avoid destroying useful microbes.

The use of Integrated Pest Management strategies has very beneficial effects on the environment as well as health. This is because Integrated Pest Management makes the carbon footprint of agricultural systems lower due to the reduction in the emissions that are generated through the use, production, and application of synthetic pesticides (Pretty et al. 2018). Apart from that, Integrated Pest Management techniques such as crop rotation and the use of organic matter help improve soil health and its potential for carbon sequestration through the promotion of more ecologically friendly agricultural systems. Furthermore, the reduction in the use of pesticides through the application of Integrated Pest Management strategies is beneficial in helping beneficial species such as pollinators and natural pest controls, including insects as well as other species that are crucial in the reduction of carbon through the promotion of ecological processes (Bianchi et al. 2006).

Climate change is, therefore, introducing a new pests and diseases paradigm to the horticultural crops, which have, prior to this, never been encountered. With the rise in temperatures, precipitation variability, heat waves, and humidity, there is, therefore, over-multiplication of pests and diseases, which are finding their way into new locations, also eventually gaining new life spans as a consequence of variations in climatic seasons. For the farmer, as well as the researcher, it is, therefore, insufficient if traditional approaches are employed in efforts to control pests. In their place, there is, therefore, a need *for* a new strategy, termed Integrated Disease Pest Management (IDPM), which is able to adapt itself according to the changing nature of environmental conditions within a short span of time. IDPM, as such, is a new paradigm toward controlling pests, whereby ecologists, effective crop management, and new surveillance systems, as well as actions, are, therefore, appropriately incorporated (Chakraborty and Newton 2011).

18.6.1 Climate-Resilient IPM Approaches

18.6.1.1 Redesigning Crop Protection for a Changing Climate

Climate-resilient IPM focuses on helping crops withstand new patterns of pest and disease pressure. Probably the most important single step is the development and deployment of varieties and rootstocks that perform well under heat, drought, or excess moisture. These enhanced genotypes often display more stable resistance against pathogens even when plants are challenged by environmental stress. Heat-tolerant grape rootstocks and vegetable hybrids bred with an emphasis on stress adaptation are good examples of how plant breeding supports climate-ready management of pests and diseases.

18.6.1.2 Cultural Practices Tailored to New Pest Cycles

Changing life cycles due to fluctuating climatic conditions are a characteristic of pests. Planting date adjustments, field hygiene, pruning to reduce humidity in orchards, and the use of mulches or drip irrigation may disrupt the pest buildup. Such simple, though strategic, changes create less favorable conditions for diseases like powdery mildew, blights, and wilt complexes, reducing the chances of explosive outbreaks.

18.6.1.3 Stronger Surveillance and Early Warning Systems

Climate variability has made real-time monitoring very critical. Current tools include automated insect traps, portable diagnostic kits, remote sensing platforms, and weather-based disease forecasts that provide much earlier warning to farmers well in advance of any threat becoming serious. Degree-day models, for example, are commonly applied to predict the arrival of fruit flies, whiteflies, and thrips, since their populations build up more rapidly under warmer conditions (Deutsch et al. 2018). Such forecasting systems have helped farmers to apply control measures only when necessary, thus avoiding excessive use of chemicals and saving money.

18.6.1.4 Moving Toward Low-Carbon Pest Management

The rising frequency of pest outbreaks often leads farmers to repeated pesticide spraying. However, these inputs have a high environmental cost. Climate-resilient IDPM promotes precision spraying, selective pesticides, botanicals, and better calibration of equipment. These measures will protect the environment, reduce carbon emissions, and preserve helpful organisms that form the backbone of natural pest control.

18.6.2 Enhancing Biological Controls Under Changing Climates

Another technique that is considered quite environmentally friendly is biological control, involving natural predators, parasitoids, and microbial biocontrol agents. Yet its effectiveness is greatly dependent on the climate.

18.6.2.1 Natural Enemies Under Climatic Stress

Temperature, humidity, and rainfall can affect the movement, feeding habits, and reproduction of beneficial insects. In most instances, pests get an advantage because their population increase is faster than that of their natural enemies. In a similar process, entomopathogenic fungi such as Beauveria bassiana and Metarhizium anisopliae can be less successful during times of heat stress or drought (Roy et al. 2023).

18.6.2.2 Developing Climate-Compatible Biocontrol Agents

To make biological control methods more effective, scientists have been working to isolate and multiply heat and moisture-tolerant strains of Trichoderma, Pseudomonas, Bacillus, and other beneficial microorganisms. These robust microorganisms will continue to safeguard agri-crops even in this ever-changing weather pattern.

18.6.2.3 Ecological Engineering: Working with Nature

While conventional agri-environmental strategies focus only on inputs, biological engineering enhances conditions in the agricultural field for natural enemies. Flower stripes, refuges, crop association patterns, and hedgerows increase the number of pollinators and parasitoids, thus maintaining a balanced biological control community. These methods reduce the severity of climate change and make environments more resilient.

18.6.2.4 Safeguarding Beneficial Fauna

Insecticides need to be used properly so as not to affect natural enemies. Therefore, Climate-Resilient IDPM promotes a change toward safe molecules, rejecting broad-spectrum insecticides and botanical insecticides. Thus, natural enemies will be in abundance and working properly, particularly when pest infestations are high due to warmer and humid seasons.

18.6.3 Host–Pathogen Interactions Under Climate Change

To design efficient IDPM strategies, it is important to comprehend the behavior of plants, pathogens, and the environment.

18.6.3.1 Plant Defenses Under Stress

Heat, dryness, and high levels of CO_2 can impact the plant's immune response. Plants can allocate less attention to the immune response and more attention to survival, making it easier for the disease to attack the tissues. This is the reason the disease conditions of heat, dryness, and wilt increase after an extreme weather event (Velásquez et al. 2018).

18.6.3.2 New Strains and Widening Distribution

Climate change favors the evolution of new, more virulent strains of pathogens. Moreover, warmer winters and longer warm seasons enable viral insect vectors, such as thrips and whiteflies, to overwinter in areas in which they were previously absent. This is associated with a rise in begomoviral, tospovirus, and other rapidly spreading disease complexes in vegetables.

18.6.3.3 Shifting Plant Microbiomes

Beneficial microorganisms inhabiting roots, leaves, and fruits protect plants against diseases. But harsh changes in climatic conditions affect microorganisms and, at times, impede their ability to protect plants. Climate shift analysis is helpful in building microorganisms capable of surviving in stressed conditions.

18.6.4 Predictive Modeling for Future-Ready Pest Management

Predictive models are becoming an essential tool because pest behavior is becoming more unpredictable due to climate change.

18.6.4.1 Mapping Future Pest Distributions

Ecological niche modeling techniques like MaxEnt and CLIMEX forecast potential patterns of temperature and rainfall distributions and their impacts on the spread of pests. These models enable the identification of susceptible regions for incursions by fruit flies, mealybugs, or fungal pathogens and enable the formulation of preventive strategies by policymakers.

18.6.4.2 Phenology and Degree-Day Tools

Temperature prediction models assist in estimating the stage when important developmental milestones will be achieved by the pests. These models assist farmers by helping them target the pests more accurately.

18.6.4.3 Machine Learning and Digital Decision Systems

The new data-driven systems have the capacity to analyze weather data, soil variables, and pest prevalence data, making accurate predictions of the outbreak of diseases possible. The predictive systems are now being incorporated into mobile-based advising systems, whereby real-time guidelines are provided to the farmers.

A pictorial representation of carbon-neutral agriculture has been depicted in Fig. 18.4.

18.7 Carbon Capture and Sequestration—Toward Zero Emission Agriculture

18.7.1 Role of Perennial Horticultural Systems in Carbon Sequestration

Climate change in recent years has become prominent and is now among the most pressing global challenges, and agriculture is both the source of and the solution for the reduction of greenhouse gases (Lal 2004a, b, c, d). Agricultural systems are major sources of carbon dioxide (CO_2), methane (CH_4), and nitrous oxide (N_2O), and they also offer the prospect of becoming carbon sinks (Powlson et al. 2011a, b; Nair et al. 2010a, b, c). The perennial horticultural systems comprising orchards,

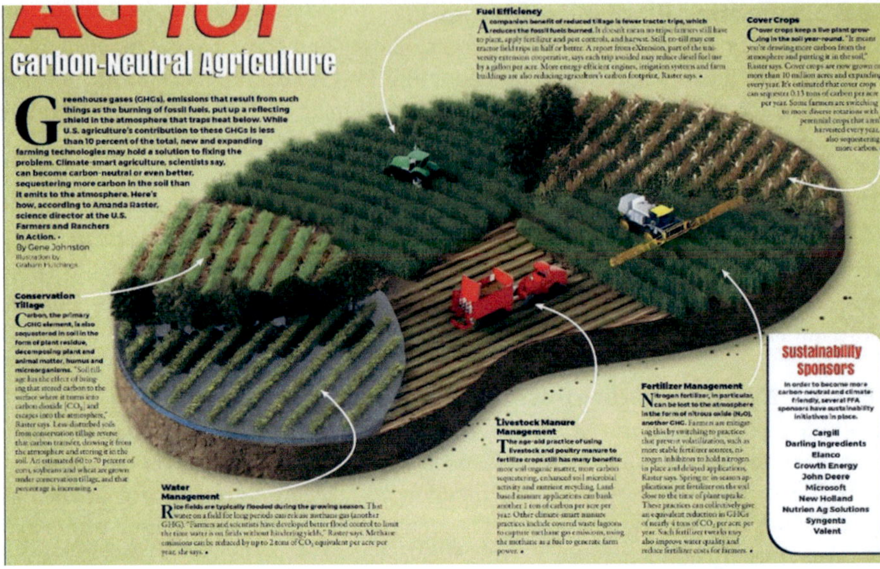

Fig. 18.4 Illustration titled "AG 101: Carbon-Neutral Agriculture" depicting a cross-section of farmland with various sustainable farming practices. The image includes tractors and fields with different crops, highlighting methods like conservation tillage, cover crops, and livestock manure management. Text boxes provide information on fuel efficiency, water management, and fertilizer management. A list of sustainability sponsors is shown on the right. The visual emphasizes reducing greenhouse gas emissions and improving agricultural sustainability

fruit plantations, and vegetables could be an effective carbon sink because of the continuous growth of vegetation with long-term persistence of plant residues (Bhattacharyya et al. 2015; Chaturvedi et al. 2017a, b). Carbon sequestration refers to the capacity of trees, plants, and crops that absorb carbon dioxide (CO_2) from the atmosphere through the photosynthesis process and store carbon in tree trunks, branches, and roots (Lal 2004a, b, c, d). Horticultural crops have immense opportunities for the sequestration of carbon in the terrestrial ecosystem, more than the agricultural and agroforestry crops. Several studies have shown that CO_2 sequestration in perennial horticultural crops is significantly higher compared to the annual ones (Fig. 18.5). The carbon sequestration capacity in different horticultural crops and the addition of more residues in the perennial horticultural crops and the addition of more residues of plant materials in the soil level leads to less emission of carbon dioxide gas from the soil of the annual horticultural crops (Ganeshamurthy et al. 2020). The carbon sequestration of fruit trees is dependent on the age of the trees, the size of the trees, and the land area. The rate of carbon is high in juveniles. However, the value of carbon sequestration per unit area is reduced after reaching maturity (Jana et al. 2009). The carbon sequestration capacity of the tree is increasing continuously and depends on the time and age of the tree until it reaches maturity. However, the photosynthesis process of carbon in plants is dependent on different factors, including age, leaf area, and photosynthesis. Unlike crops that are

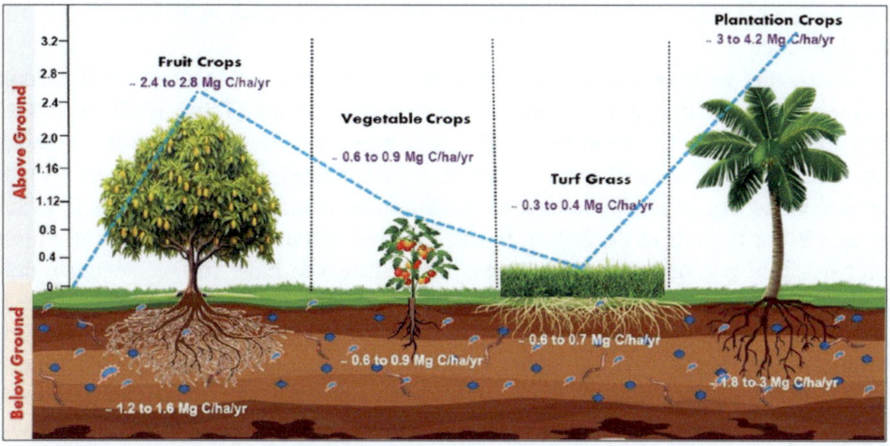

Fig. 18.5 Diagram illustrating carbon sequestration in different crops. The image shows four types of crops: fruit crops, vegetable crops, turfgrass, and plantation crops. Each crop is depicted with its root system and associated carbon sequestration rates, measured in megagrams of carbon per hectare per year (Mg C/ha/yr). Fruit crops sequester 2.4–2.8 Mg C/ha/yr above ground and 1.2–1.6 Mg C/ha/yr below ground. Vegetable crops sequester 0.6–0.9 Mg C/ha/yr above ground and below ground. Turf grass sequesters 0.3–0.4 Mg C/ha/yr above ground and 0.6–0.7 Mg C/ha/yr below ground. Plantation crops sequester 3–4.2 Mg C/ha/yr above ground and 1.8–3 Mg C/ha/yr below ground. The diagram highlights the differences in carbon sequestration potential among these crop types. (*Source*: Adopted from Ilakiya et al. (2024). License under CC-BY. ©MDPI, 2025)

harvested and re-established every year, perennials conserve above and below-ground tissues for several years, thus providing the potential to accumulate carbon in vegetation as well as the soil (Montagnini and Nair 2004a, b; Singh et al. 2015a, b, c). Perennial farming entails the presence of deeper root vegetation than that found in other farming systems, thus accounting for increased soil carbon stock at deeper soil layers, hence increased soil organic carbon stability (Lal 2005; Jose 2009). The regular deposition of leaf litter, as well as fruit and pruning, results in increased organic matter inputs to the soil, hence increased soil fertility as well as soil functionality and carbon sequestration (Nair et al. 2010a, b, c; Bhattacharyya et al. 2015). Furthermore, perennial horticultural farming yields multiple products, including vegetables, timber, and forage, among others, thus forming an efficient and sustainable economic and ecological project (Chaturvedi et al. 2017a, b; Powlson et al. 2011a, b).

The realization about the potential of perennial horticulture in acting as a tool to mitigate climate change has been recognized globally and in the context of the Indian scenario, especially in those parts where the soils are degraded or of low quality (Montagnini and Nair 2004a, b; Singh et al. 2015a, b, c). These practices have been found to be useful in addition to carbon sequestration; they are also useful in maintaining soils and water resources, biodiversity, and sustainability of livelihoods (Jose 2009; Bhattacharyya et al. 2015). This chapter intends to highlight the mechanisms, extent, and implications of carbon sequestration in perennial horticultural practices.

There are various studies that demonstrate the general capacity for carbon sequestration existing within perennial horticultural crops (Table 18.2). In Indian agroforestry ecosystems where horticultural plant species have been used, it has been shown to increase soil carbon content between 30 and 38 Mg C per hectare annually due to a 25–35% increase above levels found in agricultural lands (Dhyani et al. 2015a, b; Nair et al. 2010a, b, c). The aboveground biomass content within orchards and perennial plantations has been shown to have further sequestration potential between 4 and 7 t CO_2 per hectare per year, based on plant species and ages and accordingly, on management techniques (Montagnini and Nair 2004a, b; Powlson et al. 2011a, b). Globally, perennial crop integration within agricultural activity has been reported to have a mitigation potential of 1–2 Gt CO_2 per year, especially if such agricultural techniques and approaches are applied on already degraded lands (Jose 2009; Chaturvedi et al. 2017a, b). Perennials with deep-rooting systems can obtain soil water from lower soil profiles, and thus, they increase overall drought resistance (Bhattacharyya et al. 2015; Singh et al. 2015a, b, c). Deep roots enable perennials to capture water from subsoil layers, thereby increasing their drought resilience (Montagnini and Nair 2004a, b; Singh et al. 2015a, b, c). At the same time, their canopies may provide essential support by ensuring reduced evaporation and runoff losses so as to maintain overall soil levels (Powlson et al. 2011a, b; Bhattacharyya et al. 2015).

18.7.2 Carbon Sequestration and Climate Action in Agriculture

Climate change, which is influenced mostly by the higher emission of greenhouse gases into the atmosphere, has become a threat to the existence and survival of both natural and human resources. Among the highest greenhouse gases emitted into the atmosphere, carbon dioxide is the most prominent one, which mostly results from the burning of fossil fuels, deforestation, and ill management of natural resources. However, agricultural systems, which emit high levels of carbon dioxide, also possess potential capabilities that can be utilized as carbon sinks by the process known as carbon sequestration, the long-term capture and storage of carbon from the atmosphere into organic forms such as plants and soils. At the agricultural level, perennial horticultural crops can be a prospective means to counter climate change through various carbon sequestration processes. The perennial crops can be ideally cultivated in semiarid climates, which fail most because most agricultural crops can be easily uprooted due to high evaporation levels in such climates. These crops not only survive but also increase soil stability and fix carbon both in their biomass and in soil. It is rather imperative to know the accurate carbon sequestration process in perennial horticultural crops in order to develop climate change mitigation plans on agricultural lands.

Table 18.2 Carbon sequestration potential and ecosystem benefits of perennial horticulture and agroforestry systems with key carbon pools, mechanisms of carbon sequestration, approx. C-sequestration rate, additional ecosystem benefits, and references

Perennial system	Key carbon pools	Mechanisms of carbon sequestration	Approx. C-sequestration rate	Additional ecosystem benefits	References
Fruit Orchards (Mango, Apple, Citrus, Litchi)	Aboveground biomass, root biomass, soil organic carbon (SOC)	Continuous accumulation of woody biomass; deposition of leaf litter; turnover of deep roots	2.5–6.0 t C ha^{-1} yr^{-1}	Improvement of soil fertility, provision of habitat for biodiversity, moderation of microclimate	Lal (2004a, b, c, d), Sharma et al. (2019), and Chavan et al. (2021)
Agroforestry (Silvopomology, Alley orchards)	Tree biomass, understory vegetation, SOC	Synergistic carbon storage from trees and crops; reduced disturbances; enhanced humification	3.0–10.0 t C ha^{-1} yr^{-1}	Reduction of erosion, enhancement of infiltration, improvement of nutrient cycling	Nair et al. (2009), Kumar and Nair (2011a, b), and Pandey et al. (2020a, b)
Perennial Plantation Crops (Tea, Coffee, Rubber, Cocoa)	Woody stems, shading trees, soil carbon	High litterfall; biomass from roots; permanent canopy cover	1.8–7.2 t C ha^{-1} yr^{-1}	Regulation of shade, refuges for biodiversity, stabilization of slopes	Ahmed et al. (2014), Nath et al. (2015), and Pandey and Singh (2021)
Vineyards (Grapes)	Vine wood biomass, permanent trellis biomass, SOC	Accumulation of pruned wood, carbon-rich leaf litter, minimal tillage practices	1.2–3.5 t C ha^{-1} yr^{-1}	Retention of soil moisture, reduction of erosion, improvement of soil aggregates	Novara et al. (2012), Brilli et al. (2019), and Gatti et al. (2021)

(continued)

Table 18.2 (continued)

Perennial system	Key carbon pools	Mechanisms of carbon sequestration	Approx. C-sequestration rate	Additional ecosystem benefits	References
Nut Orchards (Cashew, Walnut, Almond)	Long-lived wood biomass, SOC	Large and dense biomass; slow decomposition rates; extensive root systems	4.0–12.0 t C ha^{-1} yr^{-1}	Long-term carbon storage; nutrient recycling; windbreak effects	Jose and Bardhan (2012), Mohan et al. (2020)
Coconut and Arecanut Plantations	Vast stem biomass, fronds, SOC	Exceptionally long-lived biomass; continuous frond loss; minimal soil disturbance	3.5–8.0 t C ha^{-1} yr^{-1}	Soil conservation, coastal protection, enhanced water-use efficiency	Nair et al. (2010a, b, c), George et al. (2015), and Upadhyay et al. (2021)
Perennial Medicinal and Aromatic Plant Systems (Vetiver, Lemongrass, Patchouli)	Root biomass, SOC	Dense fibrous roots support SOC stabilization; high turnover rate	1.0–2.8 t C ha^{-1} yr^{-1}	Strong soil binding, slope stabilization, water retention	Lavania (2003), Singh et al. (2011), and Bhat et al. (2019)
Mixed Perennial Home Gardens (Tropical Multistrata Systems)	Multi-tier biomass, SOC, detritus layers	Ongoing stratified carbon input; minimal soil disturbance; forest-like structure	7.0–15.0 t C ha^{-1} yr^{-1}	High biodiversity, year-round production, climate resilience	Kumar et al. (1994a, b), Kehlenbeck et al. (2015), and Saha et al. (2020a, b)

18.7.3 Strategies for Climate Change Mitigation Through Agroforestry

Protecting and maintaining the existing carbon sinks within forests, grasslands, and agroforestry systems through the prevention of deforestation, soil erosion, and biomass destruction (Khan et al. 2024), enhanced carbon sequestration through the promotion of additional tree cover, increasing soil organic matter, and inducing deep-rooted perennials to sequester carbon in biomass and soils. Replacing fossil fuel–based energy sources with renewable energy derived from biomass and agroforestry systems, as well as promoting the use of wood as a sustainable alternative to energy-intensive construction materials, can reduce greenhouse gas emissions, enhance carbon sequestration, improve energy sustainability, and support climate change mitigation (Irache Cabello II 2023). Methods one and two entail an accumulation of the total carbon sinks, but they are liable to cause an accumulation of released carbon emissions in the future when they are not adequately handled through changes in land use or the biodegradation process of the biomass. These may entail an accumulation of released carbon emissions in the future when they are not adequately handled through changes in land use or the biodegradation process of the biomass. However, method three presents a long-lasting, sustainable option to the problem, in which the carbon sequestration process continues indefinitely. Carbon sinks are diversified through agroforestry systems in the following ways: Tree Biomass: Trees employed in agroforestry systems absorb carbon dioxide from the atmosphere through photosynthesis and fix the gases within the leaves, stems, and roots of the trees. Aboveground tree biomass is an efficient carbon sink that reduces carbon concentrations in the atmosphere (Atangana et al. 2014). Beneath-Ground Carbon Sequestration: The agroforestry system enhances the fixation of carbon within the soil through the development of tree roots to considerable depth, leaving organic materials in the soil and increasing its carbon stock (Dissanayaka et al. 2024). This process also enhances the fertility and health of the agroecosystem. Litter and Mulch Contribution: The leaves, twigs, and organic matter from the plants fall to the ground and contribute to build up the soil organic matter content within the agroforestry system.

18.7.4 Agroforestry and Mixed Cropping for Enhanced Ecosystem Services

Agroforestry is not only an economically viable but also a highly effective and sustainable approach to mitigate climate change. The advantage here comes from carbon sequestration. Essentially, plants remove a chemical called carbon dioxide (CO_2) from the atmosphere and store it away, mostly in their leaves, trunks, roots, and soil. Of course, not all agroforestry methods sequester the same amount of carbon. Monocultures, mixtures of crop agroforestry, pure agroforestry, and integrated agroforestry practices sequester varied levels of carbon within the soil and vegetation (Fahad et al. 2022). Another factor adding to the advantage of agroforestry comes from the amount of carbon sequestered within soil and vegetation. Essentially, what makes agroforestry an advantage comes from sequestering carbon above the ground through leaves and trunks of crops and plants and below the ground within roots and soil organic matter. Moreover, agroforestry helps mitigate the rising levels of GHG emissions by enhancing the seepage of carbon into the soil, reducing disturbance to the soil, and reducing the emissions of methane (CH_4) and nitrous oxide (N_2O) gases from agricultural soils (Kim et al. 2016). This process relies on the type of agricultural trees to cultivate and how to develop the agroforestry technology, soil type, and climatic factors (Newaj and Dhyani 2008). Agroforestry not only helps to mitigate environmental challenges—it also offers several other economic and social advantages to agricultural farmers while mitigating climate change through several approaches (Jhariya et al. 2019). Agroforestry helps to reduce the amount of atmospheric carbon dioxide and enhance the sequestration process by plants and soils (IPCC 2021).

18.7.5 Mechanisms of Carbon Sequestration

Perennial horticultural systems sequester carbon through a variety of pathways (Figs. 18.5 and 18.6):

18.7.5.1 Aboveground Biomass

These perennial plants sequester carbon through their aboveground parts, such as stems, branches, leaves, and fruits, and can do so for many years (Montagnini and Nair 2004a, b; Powlson et al. 2011a, b). Their woody components are a good long-term carbon pool (Lal 2005; Chaturvedi et al. 2017a, b). However, the rate of carbon accumulation varies according to what kind of plants were converted, their age, density of growth, and what can be done with them (Singh et al. 2015a, b, c; Bhattacharyya et al. 2015). To determine an approximate value of the aboveground carbon sequestration of such carbon sinks, we can multiply the volume of woody parts by density (Ravindranath and Ostwald 2008). The volume of the aboveground parts of a plant was estimated from the following equation:

$$AGB(\text{g}) = \text{Volume of biomass}\left(\text{cm}^3\right) \times \text{wood density}\left(\text{g}/\text{cm}^3\right).$$

18.7.5.2 Below-Ground Biomass and Rhizodeposition

However, one cannot ignore the role of the roots. Perennials have deep roots that penetrate the soil. These roots help release nutrients into the soil, and in return, the soil sustains them (Jose 2009; Lal 2005). This carbon remains in the soil for a significantly longer time than carbon present on the surface because it decomposes slowly and is entrapped in soil aggregates (Montagnini and Nair 2004a, b; Singh et al. 2015a, b, c). Tree species like "amla" and "tamarind" keep extending their roots up to two or three meters deep into the earth, and thus carbon is also stored deep, where it remains safe for a longer period. Roots, when they decompose and

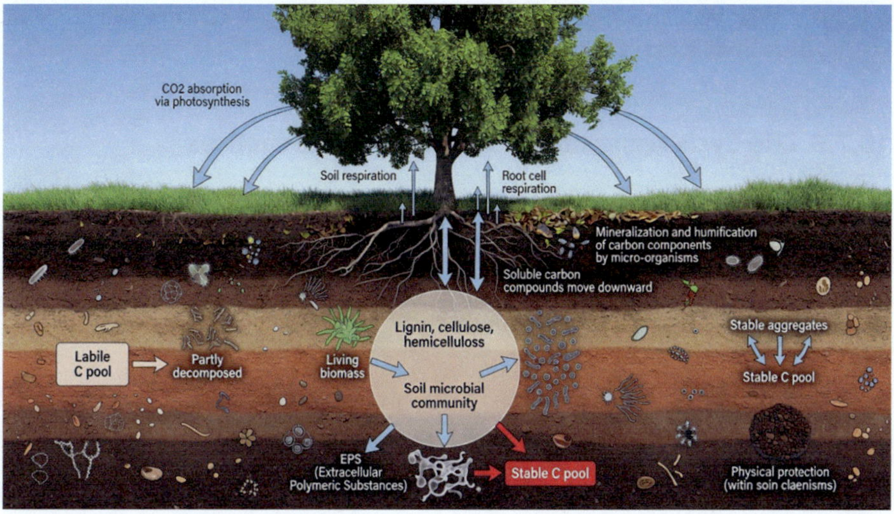

Fig. 18.6 Illustration of biological carbon sequestration. The top section shows a tree and grass with arrows indicating the flow of CO_2 from the atmosphere into the soil and back. Labels describe processes like soil respiration and root cell respiration. The middle section depicts soil layers with terms like "Soil Organic Matter" and "Soluble carbon compounds." The bottom section illustrates the soil microbial community, highlighting components like lignin, cellulose, hemicellulose, and stable carbon pools. Arrows indicate processes such as mineralization, immobilization, and microbial growth. (Modified from Fiorentino et al. (2025). License under CC-BY. ©MDPI, 2025)

release exudates, help microorganisms along with them to create soil structures that resist the decomposer action of microorganisms and prevent them from decomposing the carbon present in the soil (Choudhury et al., 2026).

18.7.5.3 Soil Organic Carbon Accumulation

There is more. When leaves fall, branches are trimmed, or fruits rot, all decompose and add to the soil, making it more fertile, better at holding water, and home to more microbial activity (Bhattacharyya et al. 2015; Powlson et al. 2011a, b). Because perennial farming patterns do not significantly disturb soil, soil carbon is retained in its position (Jose 2009; Chaturvedi et al. 2017a, b). Soil organic carbon is essential because it supports soil in both holding more water and decomposing nutrients and maintaining an active microbial community. Higher SOC content not only encourages more productivity in poor soil but is also an effective long-term storage of carbon. It is important to build high amounts of stable organic matter, or humus, in this farming method, especially in SSA (Mishra and Sharma 2020).

18.7.5.4 Permanence and Stability

This is because perennials remain on the land for many decades. Hence, the carbon sequestration process is relatively more permanent than annual crops (Montagnini and Nair 2004a, b; Lal 2005). Also, with reduced tillage and an open canopy, loose soil loses relatively less carbon due to soil respiration and erosion (Singh et al. 2015a, b, c; Bhattacharyya et al. 2015).

18.7.6 Life Cycle Assessment of Horticulture Practices

Life Cycle Assessment, or LCA, has definitely gained traction in measuring horticultural systems' effects on the environment, such as greenhouse gas emissions and carbon storage, not only in food production but in the whole industry. These days, many nations apply LCA results to inform large-scale decisions in agricultural planning (Sanderson et al. 2019). See Table 18.3 for better clarification. In every nation, specialists apply exergy analysis in tracing how these energies are spent. The more we know about energy efficiency, the better we become at using it. Yildizhan and Taki (2018) emphasized that Cumulative Energy Demand is a crucial measure in determining if agricultural production is sustainable enough. In Cumulative Exergy Demand (CExD), all forms of energy (renewable and nonrenewable) are accounted for from application start (Ghannadzadeh 2018). LCA lets us focus on any item, process, or production system so we can calculate its various impacts on the environment at varying lengths of a particular item's life cycle, depending on whatever form of data is being gathered. An LCA traces pollutants such as carbon dioxide, methane, ammonia, nitrogen oxide, and sulfur dioxide emissions from entering their pathways through air, water, and soil, providing us a better perspective on their influence on the surroundings.

18.8 Future-Ready Technologies and Innovations

18.8.1 Digital Transformation and Smart Horticulture

Right now, it is the technological aspect, data, and automation that are changing the horticulture industry. This is where it is all heading—technology adapted for the climate, protecting the environment, and catering to growing demands for food. This is characterized by a feature of increased precision, sustainability, and flexibility. A number of technologies are enabling this, including Artificial Intelligence (AI), Internet of Things (IOT), and Remote Sensing (RS), among a myriad of others, allowing for proper apportioning of resources on a real-time basis within different agri-production systems in the horticulture sector (Kumar et al. 2024a, b, c). With horticulture system transformations to digital technology, agricultural operations on farms become well-planned, reduced, and less laborious. Farmers will enjoy increased outputs with reduced negative impacts on the environment. Technologies applied will enable efficiency in input resources, reduced pesticide waste, and improved crop quality (Bernhardt et al. 2021).

Table 18.3 Life Cycle Assessment (LCA) components and environmental impacts of horticultural Production Systems, as per previous findings and literature

LCA component	Environmental indicators assessed	Major findings from literature	References
Open-field horticulture	GHG emissions, energy use, land occupation	High emissions from fertilizer manufacture and field N_2O; major energy demand from irrigation	Clune et al. (2017), Ingram and Fernandez (2012a, b), and Fan et al. (2024a, b)
Protected cultivation (polyhouse/ greenhouse)	Carbon footprint, cumulative energy demand, plastic waste impacts	Very high embodied energy and plastic load; emission per kg yield may decrease due to higher productivity	Müller et al. (2023a, b), Bonaguro et al. (2016a, b), and van Rijn et al. (2021a, b)
Organic vs. conventional horticulture	GWP, eutrophication, acidification, ecotoxicity	Organic reduces ecotoxicity and fossil energy use; may have higher N_2O emissions from slow-release nutrients	Ali et al. (2024a, b), Alhashim et al. (2021a, b), and Clune et al. (2017)
Agroforestry-based horticulture	Carbon sequestration credit, land-use change impacts	Strong negative emissions due to SOC buildup and biomass C storage; major long-term mitigation potential	Lahtinen et al. (2025), Nair (2012), and Sharma et al. (2025a, b)
Peri-urban horticulture/ Urban farming	Water use, waste recycling, nutrient circularity	Enhances resource recovery and reduces food miles; infrastructure increases energy load	Ali et al. (2024a, b), Urban Farms LCA (2023), and van Rijn et al. (2021a, b)
Drip- and micro-irrigation systems	Energy consumption, water use, indirect fertilizer impacts	Reduces water use and N_2O intensity; manufacturing of drip lines adds upfront carbon cost	Fan et al. (2024a, b), Alhashim et al. (2021a, b), and Ingram and Fernandez (2012a, b)
Mulching and soil-cover systems	Plastic footprint, soil GHG emissions, residue breakdown	Plastic mulches dominate LCA impacts; organic mulching increases SOC but adds transport-related emissions	Bonaguro et al. (2016a, b), Clune et al. (2017), and Fan et al. (2024a, b)
Postharvest handling and cold storage	Electricity use, refrigerant leakage, transport emissions	Cold chains raise total GWP significantly; logistical efficiency lowers energy burden	Johnson et al. (2025a, b), Alhashim et al. (2021a, b), and Clune et al. (2017)
Composting/ waste-to-energy systems	Methane emissions, nutrient recovery potential	Reduces landfill emissions; composting may emit CH_4/ NH_3; anaerobic digestion provides renewable energy credits	Johnson et al. (2025a, b), van Rijn et al. (2021a, b), and Fan et al. (2024a, b)
Fertigation and precision nutrient management	N fertilizer-related emissions, nutrient leaching	Strong reduction in N_2O and embodied fertilizer emissions; highly efficient mitigation per unit of nutrient input	Kumar et al. (2024a, b, c), Fan et al. (2024a, b), and Alhashim et al. (2021a, b)

18.8.2 Precision Agriculture

The key components of precision farming were information and tech, as well as management. A lot of data is required to achieve this. Precision farming involves the extraction of key information from all corners using tech tools and effective decision-making. Farmers use tech tools such as Global Positioning System (GPS), Geographic Information System (GIS), yield monitors, remote sensors, and guidance systems to modify the way they deal with fields under certain conditions.

18.8.3 Artificial Intelligence and Machine Learning Applications

AI is now a prominent factor in horticulture because of its ability to process large amounts of data, identify patterns, and make independent decisions. AI-based technology is now manifesting in every nook and cranny, from harvesting using robots and crop modeling to crop estimation and identification of pests and diseases (Choudhury and Dasgupta, 2026; Kumar et al. 2024a, b, c). "Machine learning, via neural networks and support vector machines, assists with early detection of plant distress through image and sensor processing." AI-assisted pest management utilizes optical and acoustic sensors that detect the presence of pests, meaning that fewer pesticides are now employed. However, with AI vision, harvesting robots can distinguish between fruit and vegetables and trees, meaning that fewer commodities are damaged, along with fewer workers, because of AI-assisted robots.

18.8.4 Internet of Things (IoT) and Precision Automation

Smart agriculture is being powered by the Internet of Things technology. IoT technology integrates data from sensors, drones, robots, and weather stations on a real-time basis (Choudhury and Dasgupta, 2026). This data enables the farm to take care of soil moisture, nutrient levels, microclimate, and crop health in an effective and more focused manner (Bhujel et al. 2020; Khan et al. 2020).

Integrating IoT and AI goes even further to improve control of greenhouse environments, management of fertilizer, and irrigation and monitoring of crops by means of drones and self-driving robots. It also makes way for lower labor costs, reduction of inefficiencies, and improvements in control of pests, yield, and crop monitoring (El Hoummaidi et al. 2021).

18.8.5 Remote Sensing and Geospatial Technologies

In large-scale and orchard-scale horticulture monitoring, remote sensing has grown into an important tool. Mapping, canopy analysis, stress analysis, and yield prediction are all done accurately by multispectral and thermal remote sensing using satellites and UAVs (Usha and Singh 2013).

However, block-level mapping of apple orchards with more than 90% accuracy has been achieved by high-resolution satellites, thus improving market planning and fertilizer estimates (Sharma and Panigrahy 2007). It is crucial in horticultural applications focusing on resource use and flexibility.

18.8.6 Global Positioning System (GPS)

GPS consists of satellites orbiting the earth, which flash back precise location information to earth. It can provide a bird's-eye view from space. GPS helps in receiving these satellite signals to determine their precise position and time (Singh 2024). The GPS provides accuracy from 100 m to 0.01 m. Depending on a particular site, it assists farmers in fetching information about fields like soil type, insect density, weed distribution, water resources, and so on. This becomes particularly important when farmers are required to choose the type of seeds to be planted, insecticides, herbicides, fertilizers, and watering necessary for the fields.

18.8.7 Geographic Information System (GIS)

It is a system developed to collect, organize, deal with, process, appraise, and display geographic information. A Geographic Information System, or GIS, combines the use of hardware, software, and techniques to allow us to gather, store, sort, and analyze any kind of location-related data for mapping (Choudhury et al., 2025; Singh 2024). The beauty of GIS lies in how it allows the use of the gathered information to be used by the user any time they want to use it. Just imagine computerized GIS maps as layer cakes—with each layer, the data is different: it could be crop yield, soil survey, crop type, nutrient levels, or even the susceptibility of the area to pests. If we explore a farming GIS database, much more interesting data will be found: topographical details of the area, soil condition, drainage systems, rainfall, irrigation systems, amount, and type of pesticides used, and data from crop production.

Just by combining all this data and analyzing it, Benjamin starts to realize the relationship of all these crops in one area.

18.8.8 Blockchain and Smart Supply Chains

Blockchain technology provides horticultural value chains with some level of clarity, trackability, and trust. Blockchain technology improves food safety, quality, and fair remuneration for farmers by offering safe and immutable storage solutions for data in decentralized systems (Zhang et al. 2022). Blockchain technology applications create an opportunity for tracking products from farms to final consumer consumption, reducing both farm-level and postharvest losses and increasing smart contracts and index insurance in the horticultural sector (Kumar et al. 2024a, b, c;

Abbas et al. 2017). Future-readiness in local and international horticultural supply chains requires such technological innovations.

18.8.9 Vertical Farming and Soilless Cultivation Systems

Vertical farming is an innovative game-changer for both urban and peri-urban horticulture. Vertical farms, by the use of hydroponics, aeroponics, and aquaponics, have allowed high-density crop production regardless of soil and climate limitations (Lakhiar et al. 2018). Nutrient Film Technique (NFT), Deep Water Culture (DWC), and drip systems are hydroponic systems that maximize water use and nutrients for improved productivity and quality of produce (Niu and Masabni 2022). Vertical farming has been included in future food systems since the integration of IoT and automation will enable precision control of growth parameters (Palande et al. 2018).

18.8.10 Grid Soil Sampling and Variable-Rate Fertilizer (VRT) Application

Grid soil sampling is another approach used in the toolbox. This is a fast and easy approach used for managing the soil according to the distinct nature of each spot in the farm. Some farm functions require the implementation of variable-rate technology. They rely on the type of soil used by the farm, hence determining the speed at which farm inputs are applied. The processes involved, such as crop choice, planting density, or the variable rate application of fertilizer, insecticide, or herbicide at the right spot at the right time, can all be controlled using data extrapolated by GIS. The approach is mostly used by developed countries. Through implementing similar ideas, grid soil sampling increases the intensity level of sampling (Pawase et al. 2024).

18.9 Policy and Institutional Framework for Climate-Resilient Horticulture

18.9.1 Existing National and State-Level Horticulture Missions

Horticulture has now emerged as the backbone of Indian agriculture, sharing 30.4% of agricultural GDP from just 11% of the cultivated area (Saxena et al. 2018). It is this vital sector that faces unprecedented challenges emanating from a changing climate, indicated by rising temperatures, variable rainfall patterns, and increased frequency of extreme weather events. Projections are indeed indicative of a 3–6 °C rise in temperature and a 15–40% increase in rainfall over India by the end of the twenty-first century, with notable regional variations (Prasad and Rana 2006).

The Mission for Integrated Development of Horticulture (MIDH) was an umbrella scheme launched in 2014 for holistic horticultural growth in the areas of expansion, production enhancement, postharvest management, and market linkages (Government of India 2014). Integrating six schemes, namely, NHM, HMNEH, NBM, and three Central Sector Schemes, namely, NHB, CDB, and the CIH, Nagaland, MIDH caters to cluster-based interventions, protected cultivation, nursery accreditation, and value-chain infrastructure. It promotes diversified horticulture crops with enhanced productivity through quality planting material, micro-irrigation, and integrated nutrient and pest management. Likewise, the recently initiated HCDP under MIDH enhances competitiveness by integrating production, processing, and marketing of selected clusters. It has contributed significantly to the growth of horticulture in India, and national production increased from about 280 million tons in 2013–2014 to nearly 368 million tons in 2024–2025 (National Horticulture Board 2024). In this respect, MIDH plays a very critical role in promoting sustainable, resilient, and market-oriented horticulture development in India by facilitating technology adoption, the creation of infrastructure, and farmer-centric interventions.

State-specific horticulture missions, concurrently with the central MIDH, have been formulated by state governments situated in different agro-climatic zones of India. These are envisaged for regional strengthening of production systems in different states, promotion of climate-resilient practices, and value-chain competitiveness (Shah 2008). State missions supplement national goals with local planning, crop-oriented initiative programs, and human resource development activities that address agro-climatic variations (Table 18.4).

However, reviews show that while these missions have historically targeted productivity and infrastructural expansion, their guidelines often lack explicit integration of climate risk management such as region-specific vulnerability assessments, risk-informed cropping calendars, or incentives for low-carbon/resource-efficient practices (Sharma 2013). Thus, there is a clear need for reorienting existing horticulture missions from a productivity-only mandate to a resilience-oriented framework that embeds climate risk, sustainability, and mitigation (low-carbon) goals.

18.9.2 Integrating Climate Resilience into Horticulture Policy

To effectively integrate climate resilience into horticulture, policy must shift from a traditional "disaster management" approach toward a proactive "risk mitigation" framework. This transition aligns with the pillars proposed under the National Mission for Sustainable Agriculture (NMSA) and strategic guidance from NITI Aayog. Current cultivars often lack tolerance to heat, floods, salinity, or drought, making them unsuitable for future climate scenarios (Kumar et al. 2013). Therefore, State Agricultural Universities must be incentivized to prioritize breeding for abiotic stress resilience rather than solely targeting yield maximization. Evidence from the National Innovations in Climate Resilient Agriculture (NICRA) program, which has successfully developed heat-tolerant tomato and frost-resistant mango varieties, highlights the need for policy mechanisms that scale these innovations from "lab to land" through subsidized seed systems (Bal et al. 2024).

Table 18.4 State-wise horticulture missions and key development priorities of India

State/UT	Name of state horticulture mission	Key focus areas
Andhra Pradesh	Andhra Pradesh State Horticulture Mission	Mango, citrus, banana clusters; protected cultivation; micro-irrigation; postharvest infra
Arunachal Pradesh	State Horticulture Mission Arunachal Pradesh	Kiwi, orange, large cardamom; tribal horticulture; nursery development
Assam	Assam State Horticulture Mission	Citrus revival, banana, arecanut; supply chain strengthening
Bihar	Bihar State Horticulture Mission	Litchi, mango, makhana; precision farming; processing units
Chhattisgarh	Chhattisgarh State Horticulture Mission	Vegetables, tuber crops, mango, cashew; cold storage infra
Goa	Goa State Horticulture Mission	Cashew, coconut, floriculture; homestead horticulture; processing
Gujarat	Gujarat Horticulture Development Mission	Date palm, banana, onion; PPP-based cold chain and packhouses
Haryana	Haryana State Horticulture Mission	Orchard development; polyhouses; micro-irrigation; horticulture clusters
Himachal Pradesh	Himachal Pradesh Horticulture Mission	Apple, stone fruits, off-season vegetables; high-density orchards
Jharkhand	Jharkhand State Horticulture Mission	Mango, litchi, vegetables, spices; tribal horticulture; protected structures
Karnataka	Karnataka State Horticulture Mission	Floriculture, polyhouses, precision farming, protected cultivation
Kerala	Kerala State Horticulture Mission	Spices (pepper, cardamom), banana, tuber crops, homestead gardens
Madhya Pradesh	Madhya Pradesh State Horticulture Mission	Citrus, vegetable clusters, potato; micro-irrigation
Maharashtra	Maharashtra State Horticulture Mission	Mango, pomegranate, banana, vegetables; export-oriented infra
Manipur	Manipur State Horticulture Mission	Pineapple, passionfruit, cabbage, ginger; protected farming
Meghalaya	Meghalaya State Horticulture Mission	Ginger, turmeric, citrus; high-value horticulture
Mizoram	Mizoram State Horticulture Mission	Passion fruit, bird's eye chili, vegetables; tribal cluster farming
Nagaland	Nagaland State Horticulture Mission	Naga chili, pineapple, large cardamom, vegetables; community nurseries
Odisha	Odisha State Horticulture Mission	Mango, cashew, banana, vegetables, spices; floriculture
Punjab	Punjab State Horticulture Mission	Kinnow, vegetables, polyhouses; postharvest infra
Rajasthan	Rajasthan State Horticulture Mission	Pomegranate, date palm, citrus, spices; arid zone horticulture

(continued)

Table 18.4 (continued)

State/UT	Name of state horticulture mission	Key focus areas
Sikkim	Sikkim State Horticulture Mission	Organic horticulture: cardamom, ginger, vegetables, floriculture
Tamil Nadu	Tamil Nadu State Horticulture Mission	Precision horticulture, banana, vegetables; greenhouse farming
Telangana	Telangana State Horticulture Mission	Vegetables, mango, citrus; greenhouse farming; Rythu Vedika integration
Tripura	Tripura State Horticulture Mission	Pineapple, rubber-based horticulture; value-added units
Uttar Pradesh	Uttar Pradesh State Horticulture Mission	Malihabad mango belt, guava, vegetables, floriculture
Uttarakhand	Uttarakhand State Horticulture Mission	Apple, stone fruits, vegetables, herbs; postharvest facilities
West Bengal	State Horticulture Mission West Bengal (SHM-WB)/Banglar Bagan	Mango, litchi, guava

The second pillar is water-use efficiency and protected cultivation, requiring a policy shift from broad "area coverage" subsidies to targeted "precision efficiency" interventions. Given the water-intensive nature of horticultural crops, the Pradhan Mantri Krishi Sinchayee Yojana (PMKSY) should be more tightly integrated with the Mission for Integrated Development of Horticulture (MIDH), making micro-irrigation systems such as drip and sprinkler irrigation mandatory for water-demanding crops in vulnerable agro-climatic zones. Protected cultivation structures, including polyhouses and shade nets, should be promoted as essential tools to shield high-value crops from extreme weather variability.

A third priority area concerns postharvest and cold-chain resilience, where policy must encourage decentralization of cold storage infrastructure. Rising temperatures under climate change increase spoilage rates and expose the vulnerability of India's predominantly centralized cold storage system, which is often inaccessible to smallholders. Policies should therefore promote low-cost, farm-gate solutions such as the Pusa Zero Energy Cool Chamber and solar-powered cold rooms, recognizing them as vital climate resilience assets rather than solely as marketing infrastructure (Sadiq et al. 2025). This decentralized, climate-sensitive postharvest ecosystem is essential for reducing losses, strengthening value chains, and ensuring that horticulture systems remain viable under future climate uncertainties.

Simultaneously, there is a strong push for Agroforestry Integration, specifically the promotion of "Horti-Silvi-Pastoral" systems that create protective microclimates against heat waves, a strategy highlighted in the Greening and Restoration of Wasteland with Agroforestry (GROW) Report, Niti Aayog (2024). To financially de-risk these transitions, Weather-Index Insurance Reform is essential, as the current PMFBY structure often overlooks specific horticultural risks; new policies must introduce "add-on" covers that trigger during critical phenology stages, such as hailstorm damage during flowering. Finally, bridging the implementation gap requires Extension Services 2.0, which envisions a dedicated cadre of "Climate-Horticulture Sakhis" to provide on-ground, weather-based disease forecasting, ensuring farmers can preemptively manage risks like blight during unseasonal rains.

18.9.3 Policy Reforms Toward Low-Carbon Emission Horticulture

Agriculture is often seen as a major source of greenhouse gas (GHG) emissions; however, horticulture has the unique potential to shift from a net emitter to a significant carbon sink. With India's horticulture output now exceeding food grains, integrating low-carbon practices is essential for meeting the nation's Nationally Determined Contributions (NDCs) and its broader goal of Net Zero emissions by 2070 (Youdon and Bajaj 2022). This requires moving from energy-intensive systems to resource-efficient and renewable-energy-based horticulture.

The first reform dimension involves incentivizing renewable energy infrastructure, with policy reforms aimed at redirecting subsidy structures from diesel or grid-powered systems to renewable energy alternatives. This includes expanding the use of solar irrigation pumps through the PM-KUSUM scheme (Rathi and Soni 2024), particularly in high-water-use horticultural regions; promoting solar-powered cold storages and packhouses to reduce spoilage-related emissions; and mandating energy-efficient motors for postharvest processing. These interventions directly reduce GHG emissions linked to fossil fuel dependence.

The second reform dimension focuses on mandating resource efficiency, where principles such as "More Crop Per Drop" and soil carbon enhancement must shift from advisory guidelines to enforceable standards (Pound et al. 2018). This involves compulsory micro-irrigation for water-intensive crops in water-scarce regions, broad adoption of conservation tillage, mulching, and biochar to reduce reliance on synthetic fertilizers, and the promotion of integrated nutrient management (INM) to enhance soil organic carbon and reduce nitrous oxide emissions.

A third reform dimension emphasizes supporting perennial systems and agroforestry, recognizing the strong mitigation potential of perennial fruit orchards and horticulture-based agroforestry models (Raj and Jhariya 2023). Expanding orchards helps sequester carbon in long-lived woody biomass, while integrating hedgerows, boundary plantations, and agroforestry elements offers additional climate buffering and carbon storage benefits.

The fourth reform dimension calls for the integration of horticulture into national climate policy, including the development of a "Horticulture Carbon Standard" to enable farmers to earn credits for maintaining orchards and adopting low-carbon vegetable production systems. Together, these reforms chart a pathway to reposition horticulture as a key driver of India's low-carbon future through better convergence between MIDH and renewable energy programs. Reimagining horticulture as both a livelihood source and a carbon sink can help advance a truly "green" Green Revolution.

18.9.4 Carbon Credit and Carbon Market Policies in Horticulture

The horticulture sector offers considerable potential for generating carbon credits, owing to its ability to enhance soil carbon stocks, reduce emissions, and expand perennial biomass (Addai and Agrib 2013). Practices such as the application of organic amendments, compost, and biochar; conservation tillage; micro-irrigation; and the establishment of orchards and agroforestry systems significantly contribute to carbon sequestration and reduced greenhouse gas emissions.

Carbon markets provide an important opportunity for the horticulture sector to earn additional income by adopting sustainable practices that reduce greenhouse gas emissions or enhance carbon sequestration. Policies now support both compliance and voluntary carbon markets, allowing growers to generate and sell carbon credits to entities seeking to meet regulatory requirements or voluntarily offset their emissions (Reddy et al. 2025). India's Carbon Credit Trading Scheme (CCTS) establishes a formal compliance market by regulating emissions from energy-intensive sectors (Sarkar and Dash 2011), while the Energy Conservation (Amendment) Act, 2022, provides the legal foundation for issuing carbon credit certificates.

Horticulture growers can also participate in the Voluntary Carbon Market (VCM) by implementing climate-smart practices such as improving energy efficiency in greenhouses through renewable energy systems, LED lighting, heat pumps, and efficient refrigeration units. Soil and land management strategies such as cover cropping, reduced or zero tillage, hedgerow planting, agroforestry, and avoidance of postharvest land clearing help build soil carbon and reduce emissions. Fertilizer-related interventions, including the use of organic or bio-based fertilizers, precision nutrient application, and diversified crop rotations, similarly contribute to carbon savings. In carbon markets, farmers first adopt eligible low-carbon practices, after which the resulting emission reductions or sequestration gains undergo third-party verification to generate certified carbon credits (Gorain et al. 2021). These credits are then traded either on government-regulated platforms or within voluntary exchanges, where typical buyers include companies seeking to comply with emission norms or offset their carbon footprint. Through this mechanism, carbon markets create a financial incentive for horticulture growers to transition toward sustainable, low-carbon production systems while contributing to broader climate mitigation goals.

18.10 Conclusions and Recommendations

18.10.1 Synthesis of Key Findings

Evidence shows that India's rapidly growing horticulture sector is highly vulnerable to climatic variability. Heat stress, erratic rainfall, dry spells, and shifting phenology reduce yields, increase pests and diseases, and affect quality. Studies and modeling indicate that without effective adaptation, major fruit and vegetable systems may face significant yield losses (Challinor et al. 2014). Despite the scientific maturity of various climate-resilient horticultural interventions, their adoption at the field level has been slow and uneven. A spectrum of adaptation options ranging from climate-resilient and stress-tolerant cultivars, refined planting and harvesting calendars, protected cultivation structures, micro-irrigation and precision irrigation systems, to energy-efficient and resource-efficient postharvest technologies has been validated in diverse agroecological contexts.

18.10.2 Strategic Research and Development Needs

A critical examination of the prevailing policy frameworks reveals several research priorities that must guide future work in climate-resilient horticulture. Future research (Table 18.2) in climate-resilient horticulture should focus on developing crop- and region-specific vulnerability and risk assessment models to improve policy targeting and advisory services. Research must also quantify soil carbon enhancement and carbon sequestration across horticultural systems while establishing robust MRV (Measurement, Reporting, Verification) protocols to support carbon market participation. Digital decision-support tools integrating climate data, remote sensing, crop modeling, and pest forecasting will be crucial for real-time adaptive management. Finally, socioeconomic and institutional studies are needed to understand adoption barriers and to design effective farmer-centered models such as FPOs, cooperatives, and strengthened extension systems to scale climate-smart horticulture practices. Collectively, these research priorities underscore the need for interdisciplinary, multi-scalar, and policy-aligned initiatives to strengthen the climate resilience and sustainability of the horticulture sector.

18.10.3 Policy, Institutional, and Technological Roadmaps for Resilience

To transform horticulture into a climate-resilient, low-emission, and economically viable sector, the following roadmap is proposed:

(a) Policy Roadmap

Climate-risk assessment and mitigation objectives should be integrated into all national and state-level horticulture mission guidelines, supported by incentives for climate-resilient cultivars, protected cultivation, micro-irrigation, renewable-energy infrastructure, and other low-carbon practices. Horticulture must also be formally recognized in climate policies and carbon-credit mechanisms, with clear regulatory frameworks for horticulture-based carbon projects (Fig. 18.7). Additionally, strong institutional backbones such as extension systems and farmer organizations are essential to effectively deliver and scale climate-smart horticulture.

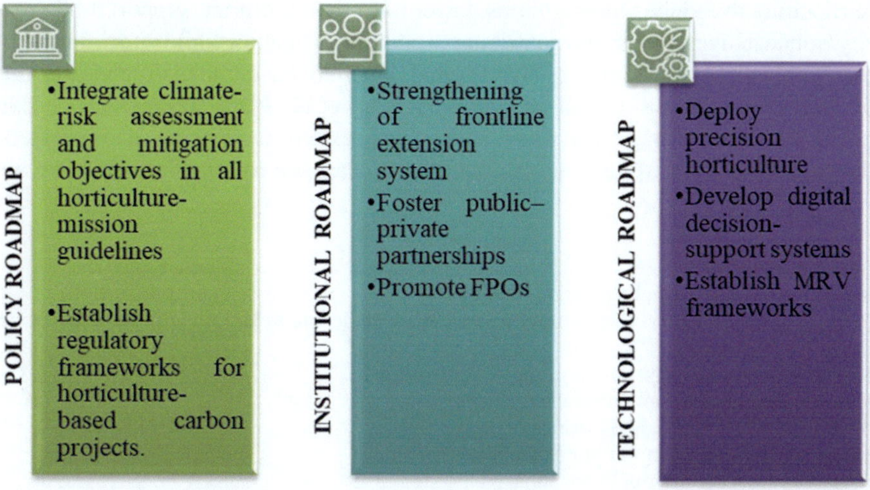

Fig. 18.7 Flow chart depicting three roadmaps for horticulture. The "Policy Roadmap" includes integrating climate-risk assessment and establishing regulatory frameworks for carbon projects. The "Institutional Roadmap" focuses on strengthening extension systems, fostering public–private partnerships, and promoting FPOs. The "Technological Roadmap" involves deploying precision horticulture, developing digital decision-support systems, and establishing MRV frameworks. Each roadmap is represented in a distinct colored box with relevant icons

(b) Institutional Roadmap

Frontline extension system, such as Krishi Vigyan Kendras and state horticulture departments, should be strengthened as climate-advisory hubs. Collective farmer institutions, including FPOs and cooperatives, must be promoted to support the adoption of climate-smart technologies, resource pooling, and carbon-credit initiatives. Public–private partnerships are essential for advancing precision horticulture, protected cultivation, renewable-energy-based cold chains, and carbon-farming interventions (Fig. 18.7).

(c) Technological Roadmap

Precision horticulture should be expanded through drip and micro-irrigation, fertigation, protected cultivation, and climate-adaptive structures such as polyhouses and shade nets. Digital decision-support systems integrating weather forecasts, remote-sensing-based risk assessment, crop modeling, and advisories must be developed. Research and dissemination of stress-tolerant cultivars, perennial or agroforestry-based horticulture systems, and soil carbon sequestration practices, including biochar, compost, and conservation tillage, should be prioritized (Fig. 18.7). Additionally, MRV frameworks and carbon accounting protocols tailored to horticultural systems are essential for credible climate action.

A sustainable, future-ready horticulture sector requires the convergence of scientific innovation, policy reforms, institutional strengthening, and technological adoption. Aligning climate-resilient strategies with low-carbon pathways enables horticulture to reduce vulnerabilities, lower its carbon footprint, support food security, enhance livelihoods, and contribute to climate mitigation while remaining commercially viable. In India's diverse and smallholder-dominated context, this transformation is both urgent and essential (Pati et al. 2025). It demands coordinated, sustained engagement from researchers, extension systems, policymakers, the private sector, and farming communities to translate reforms into effective on-ground action.

Author Contribution
Arijit Ghosh: Manuscript Preparation, Editing
Rohit Kumar Choudhury: Manuscript Preparation, Compilation, Revision, Editing, Visualization, Copyright borrowings
Suchandra Dutta: Manuscript Preparation, Visualization
Somnath Pati: Manuscript Preparation, Editing
Amlanjyoti Sar: Manuscript Preparation
Bibekananda Maiti: Manuscript Preparation
Souptik Sarkar: Manuscript Preparation
Sumana Balo: Manuscript Preparation
Prakash Kumar Jha: Revision, Validation

References

Abbas N, Zhang Y, Taherkordi A, Skeie T (2017) Mobile edge computing: a survey. IEEE Internet Things J 5(1):450–465

Addai D, Agrib BA (2013) The economics of adaptation to climate change by broadacre farmers in western Australia. University of Western Australia

Aggarwal PK (2009) Global climate change and Indian agriculture-case studies from ICAR network project. ICAR Pub, New Delhi

Aggarwal PK, Sinha SK (1993) Effect of probable increase in carbon dioxide and temperature on productivity of wheat in India. J Agric Meteorol 48:811–814

Ahsan TMS, Begum RU (2023) The role of agroforestry approach as a potential tool for attaining climate smart agriculture framework: Bangladesh perspectives. Russ J Agric Soc Econ Sci 135(3):88–103

Ahmed N, Borah P, Gogoi R (2014) Climate variability and its influence on tea production in Assam. Indian J Plant Physiol 19(3):223–230

Alhashim M, Chen Q, Li Y (2021a) Environmental impacts of mixed and legume-based cropping systems: a life-cycle perspective. Agric Syst 191:103–121

Alhashim R, Deepa R, Anandhi A (2021b) Environmental impact assessment of agricultural production using LCA: a review. Climate 9(11):164

Ali M, Koh L, Acquaye A et al (2024a) Sustainability assessment of peri-urban organic horticulture: a case study in the United Kingdom. Int J Life Cycle Assess 29(3):456–468

Ali M, Rahman S, Gupta R (2024b) Urban and peri-urban horticulture: resource use efficiency and circularity benefits. J Sustain Agric 46(2):155–172

Atangana A, Khasa D, Chang S, Degrande A (2014) Carbon sequestration in agroforestry systems. In: Tropical agroforestry. Springer Netherlands, Dordrecht, pp 217–225

AVRDC (Asian Vegetable Research and Development Centre) (2005) Annual report. Shanhua, Taiwan

Bal SK, Subba Rao AVM, Sarath Chandran MA, Timmanna (2024) National Innovations in Climate Resilient Agriculture - AICRPAM Component, Annual Report 2023. ICAR – Central Research Institute for Dryland Agriculture, Santoshnagar, Hyderabad, Telangana, India. Pg.88.

Bernhardt H et al (2021) Challenges for agriculture through Industry 4.0. Agronomy 11(10):1935
Bhat R, Harish D, Srinivasan V (2019) Soil carbon dynamics under aromatic crop systems. J Environ Biol 40(6):1203–1210
Bhattacharyya R, Singh A, Pal D (2015) Carbon sequestration potential of perennial horticultural systems in India. J Environ Manag 154:25–35
Bhujel A et al (2020) Sensor systems for greenhouse microclimate monitoring and control. J Biosyst Eng 45:341–361
Bianchi FJJA, Booij CJH, Tscharntke T (2006) Sustainable pest regulation in agricultural landscapes: a review on landscape composition, biodiversity, and natural pest control. Proc R Soc B Biol Sci 273(1595):1715–1727. https://doi.org/10.1098/rspb.2006.3530
Bonaguro JE, Coletto B, Sambo P, Zanin G et al (2016a) Environmental impact in floriculture: LCA approach at farm level. Acta Hortic 1112:1–10
Bonaguro P, Conti F, Marini L (2016b) Environmental footprint of mulching systems in horticulture: a comparative LCA. J Clean Prod 112:3652–3663
Brilli L, Llopart MF, Moriondo M (2019) Carbon sequestration and greenhouse gas emissions in Mediterranean vineyards. Agric For Meteorol 276:107–118
Browne D, O'Regan B, Moles R (2009) Use of carbon footprinting to explore alternative household waste policy scenarios in an Irish city-region. Resour Conserv Recyl 54:113–122
BSI (2008) Publicly available specification 2050. Specification for the assessment of the life cycle greenhouse gas emissions of goods and services. British Standards Institute
Byjesh K, Naresh Kumar S, Aggarwal PK (2010) Simulating impacts, potential adaptation and vulnerability of maize to climate change in India. Mitig Adapt Strateg Glob Change 15:413–431
Carbon Trust (2007) Carbon footprint measurement methodology, version 1.1. The Carbon Trust, London
Carbonell-Bojollo R, Ordóñez-Fernández R, Rodríguez-Lizana A (2010) Influence of olive mill waste application on the role of soil as a carbon source or sink. Clim Chang 102:625–640. https://doi.org/10.1007/s10584-009-9703-2
Chadha KL, Naresh Kumar S (2010) Climate change impacts on production of horticultural crops. Policy paper. In: Souvenir and abstracts, national seminar on impact of climate change on fruit crops at Pau, Ludhiana from 6–8 Oct 2010. Sponsored by National Horticultural Mission, pp 10–15
Challinor AJ, Watson J, Lobell DB, Howden SM, Smith DR, Chhetri N (2014) A meta-analysis of crop yield under climate change and adaptation. Nature Clim Change 4(4):287–291
Chakraborty S, Newton AC (2011) Climate change, plant diseases and food security: an overview. Plant Pathol 60(1):2–14
Chaturvedi OP, Prasad R, Handa AK (2017a) Tree-based horticultural systems and their carbon storage potential. Indian J Agrofor 19(2):12–20
Chaturvedi RK, Lal R, Singh B (2017b) Horticultural crops and carbon sequestration: a review. Agric Ecosyst Environ 237:55–65
Chavan SB, Dhyani SK, Handa AK (2021) Carbon sequestration in fruit orchards: a synthesis. Indian J Agrofor 23(1):1–12
Chen TY, Yamaguchi T, Kuroyanagi T (2000) Energy saving sun light greenhouse in China. J Soc Agr Struc 31:113–118
Choudhury RK, Dasgupta S (2026) Role of AI, remote sensing, and digital tools in regenerative farming: With a special focus on West Bengal region of Eastern India. SATSA Mukh Ann Tech Iss 30:359–374
Choudhury RK, Balo S, Sarkar A, Mukhopadhyay D, Bhattacharyya K, Dasgupta S, Biswas A, Biswas A (2025) Concept of soil quality, it's assessment and mapping through geoinformatics and GIS based platforms. In Soil degradation, desertification and restoration: application of geospatial technology and modeling. Springer Nature Switzerland AG
Choudhury RK, Sarkar D, Chowdhuri A, Balo S, Ghosh S, Singh S, Mukhopadhyay D, Bhattacharyya K, Mondal S, Dasgupta S, Dey A, Bashir O, Sanyal SK, Biswas A (2026) Soil Organic matter: dynamics and management for sustainable agriculture. In: Bashir O, Bangroo,

SA, Senesi N, Kader S, Hakeem KR Sustainable soil management: bridging the sustainable development goals. Elsevier, In Press
Chowdappa P (2010) Impact of climate change on fungal diseases of horticultural crops. In: Singh HP, Singh JP, Lal SS (eds) Challenges of climate change- Indian horticulture. Westville Publishing house, New Delhi
Clune S, Crossin E, Verghese K (2017) Systematic review of greenhouse gas emissions for different fresh food categories. J Clean Prod 140:766–783
Collado-González J (2021) Drought-tolerant cucumber (Cucumis sativus L.) genotypes perform better under long-term salt stress conditions. Horticulturae 7(4):74
Cunningham B (1985) The box that broke the barrier: the swamp cooler comes to southern Arizona. J Ariz Hist 26:163–174
Dalwai Committee (2018) Report of the Committee on Doubling Farmers' Income: Volume VIII – Risk Management. Ministry of Agriculture & Farmers' Welfare
Deutsch CA, Tewksbury JJ, Tigchelaar M et al (2018) Increase in crop losses to insect pests in a warming climate. Science 361(6405):916–919
Dey S, Sreenivasulu A, Veerendra GTN, Rao KV, Babu PSSA (2022) Renewable energy present status and future potentials in India: an overview. Innov Green Dev 1(1):100006. https://doi.org/10.1016/j.igd.2022.100006
Dhyani SK, Ram A, Dev I (2015a) Potential of agroforestry systems in carbon sequestration in India. Indian J Agric Sci 86(9):1102–1109
Dhyani SK, Handa AK, Uma A (2015b) Silvopastoral systems for ecological restoration and livelihood security in India. Range Manag Agrofor 36(2):230–240
Dissanayaka DMNS, Udumann SS, Atapattu AJ (2024) Synergies between tree crops and ecosystems in tropical agroforestry. Agroforestry:49–87
Dua VK, Sharma J (2017a) Forecasting impact of climate change on potato productivity in West Bengal and adaptation strategies. Indian J Hortic 74(4):533–540. https://doi.org/10.5958/0974-0112.2017.00103.7
Dua VK, Sharma J (2017b) Performance of improved potato cultivars under changing winter temperatures in eastern India. Indian J Hortic 74(4):560–567
East AJ (2008) What is a carbon footprint? An overview of definitions and methodologies. Vegetable industry carbon footprint scoping study–discussion papers and workshop, 26 September 2008. Horticulture Australia Limited, Sydney
El Hoummaidi L et al (2021) Using unmanned aerial systems and deep learning for agriculture mapping. Heliyon 7(10):e08154
Fahad S, Chavan SB, Chichaghare AR, Uthappa AR, Kumar M, Kakade V et al (2022) Agroforestry systems for soil health improvement and maintenance. Sustainability 14(22):14877
Fan Y, Li J, Wu Y (2024a) Carbon and energy footprints of irrigation, fertigation, and mulching technologies in horticulture. Agric Water Manag 296:108–212
Fan Z, Li B et al (2024b) Life cycle assessment of agricultural production: methodology, application, and challenges. Agric Ecosyst Environ 347:108–119
FAO (2020) Overcoming water challenges in Agriculture. The State of Food and Agriculture
FAOSTAT (2014) Food and agriculture organization of the United Nations. http://www.fao.org/3/a-i3671e.pdf
Finkbeiner M (2009) Carbon footprinting – opportunities and threats. Int J Life Cycle Assess 14(2):91–94. https://doi.org/10.1007/s11367-009-0064-x
Fiorentino A, Rajput FZ, Di Serio A, Baldi V, Guarino F, Baldantoni D et al (2025) Role of plants and urban soils in carbon stock: status, modulators, and sustainable management practices. Plants 14(4):546
Fischer G, Shah M, van Velthuizen HV (2002) Climate change and agricultural vulnerability. International Institute for Applied Systems Analysis, Laxenburg
Foolad MR (2007) Genome mapping and molecular breeding of tomato for abiotic stress tolerance. Plant Mol Biol Report 25(1–2):25–49
Ganeshamurthy AN, Kalaivanan D, Rajendiran S (2020) Carbon management in tropical and subtropical terrestrial systems. Springer Singapore, pp 333–348

Gangopadhyay SK, Bhattacharyya T (2021) Mangogrowing soils in Eastern India, West Bengal: A case study. Agropedology 30:168–186

Gao LH, Qu M, Ren HZ, Sui XL, Chen QY, Zhang Z (2010) Structure, function, application and ecological benefit of single slope energy-efficient solar greenhouse in China. Hort Technol 20:626–631. https://doi.org/10.21273/HORTTECH.20.3.626

Gatti M, Pirez F, Poni S (2021) Long-term carbon storage in perennial grapevine systems. Sci Hortic 280:109–131

George SJ, Thomas R, Mathew T (2015) Biomass and carbon storage in coconut-based agroecosystems. Indian J Agrofor 17(2):54–61

Ghannadzadeh A (2018) Exergy-aided environmental sustainability assessment of an ethylene dichloride–vinyl chloride production process. Chem Eng Res Des 130:109–128

Gorain S, Malakar A, Chanda S (2021) An analysis of carbon market and carbon credits in India. Asian J Agric Ext Econ Sociol 39(2):40–49

Gonzalez-Sanchez EJ, Veroz-Gonzalez O, Conway G, Moreno-Garcia M, Kassam A, Mkomwa S, Ordoñez-Fernandez R, Triviño-Tarradas P, Carbonell-Bojollo R (2019) Meta-analysis on carbon sequestration through Conservation Agriculture in Africa. Soil Tillage Res 190:22–30

Goswami R, Roy R, Gangopadhyay D, Sen P, Roy K, Sarkar S, Mainuddin M (2024) Understanding resource recycling and land management to upscale zero-tillage potato cultivation in the coastal Indian Sundarbans. Land 13(1):108

Hochberg U et al (2015) Grapevine Xerophysiology: current knowledge and future challenges. OENO One 49(3):1–20

Huang B et al (2005) Physiology, breeding, and genetic engineering of drought resistance in turfgrass. Plant Sci 168(4):887–896

Hussain S, Hussain S, Guo R, Sarwar M, Ren X, Krstic D, Aslam Z, Zulifqar U, Rauf A, Hano C et al (2021) Carbon sequestration to avoid soil degradation: a review on the role of conservation tillage. Plants 10:2001. https://doi.org/10.3390/plants10102001

Ilakiya T, Parameswari E, Swarnapriya R, Yazhini G, Kalaiselvi P, Davamani V, Singh S, Vinothini N, Dharani C, Garnepudi SL et al (2024) Unlocking the carbon sequestration potential of horticultural crops. C 10(3):65

Ingram DL, Fernandez RT (2012a) Life cycle assessment: a tool for determining the environmental impact of horticultural crop production. HortTechnology 22(3):414–420

Ingram J, Fernandez DS (2012b) Environmental outcomes of vegetable–legume intercropping: a systems analysis. Eur J Agron 40:117–125

IPCC report (2007) Climate change 2007: Synthesis report. Contribution of Working Groups I, II and III to the Fourth Assessment Report on Climate Change (Core Writing Team, RK Pachauri, A Reisinger Eds.). IPCC. IPCC AR4 Synthesis Report PDF. https://www.ipcc.ch/site/assets/uploads/2018/02/ar4_syr_spm.pdf

IPCC (2014) Fifth assessment report of the intergovernmental panel on climate change. http://www.ipcc.ch/report/ar5/

IPCC (2021) Climate change 2021: the physical science basis. Cambridge University Press

Irache Cabello II C (2023) From resource exploitation to nature restoration: unlocking the potential of agroforestry systems as feedstock provisioners for sustainable composite manufacturing

Jain S, Lamo K, Walling S, Imchen A, Tirkey JF, Singh A (2024) Climate-resilient horticulture: Adapting to climate change through innovative practices and technologies. International Int J Environ Clim Chang 14(11):219–233

Jana BK, Biswas S, Majumder M, Roy PK, Mazumdar A (2009) Carbon sequestration rate and aboveground biomass carbon potential of four young species. J Ecol Nat Environ 1(2):15–24

Jenni S et al (2013) Heat-tolerant lettuce cultivars subjected to high day temperatures. Hort Sci 48(9):1197–1202

Jhariya MK, Banerjee A, Yadav DK, Raj A (2019) Agroforestry and climate change: issues, challenges, and the way forward. In: Agroforestry and climate change. Apple Academic Press, pp 1–34

Johnson P, Mukherjee S, Wang L (2025a) Cold-chain development and its climate impacts in horticultural value chains. Food Syst Environ 18(1):45–61

Johnson T, Mensah J et al (2025b) Comparative and prospective life cycle assessment of agricultural fruit waste disposal via anaerobic digestion vs landfill: a case study from Ghana. Int J Life Cycle Assess (forthcoming)
Jose S (2009) Agroforestry for ecosystem services and environmental benefits: an overview. Agrofor Syst 76(1):1–10
Jose S, Bardhan S (2012) Agroforestry for biomass production and carbon sequestration. J Environ Qual 41(5):1445–1454
Kakumanu KR, Gavali RS, Babu VS, Rao VS, Ghate U (2020) Capacity building impacts on adaptation of climate-resilient agriculture interventions in India. climate change adaptation and sustainable livelihoods, 1.
Kashyap D, Agarwal T (2020) Food loss in India: water footprint, land footprint and GHG emissions. Environ Dev Sustain 22(4):2905–2918
Katyal JC, Rao NH, Reddy MN (2001) Critical aspects of organic matter management in the tropics: the example of India. Nutr Cycl Agroecosyst 61:77–88. https://doi.org/10.1023/A:1013320502810
Kehlenbeck K, Kindt R, Sinclair F (2015) Biodiversity and multifunctionality of tropical homegardens. Int J Agric Sustain 13(1):1–15
Khan FA et al (2020) Environmental monitoring and disease detection in smart greenhouses. J Phys Commun 4(5):055008
Khan IA, Khan SM, Jahangir S, Ali S, Tulindinova GK (2024) Carbon storage and dynamics in different agroforestry systems. Agroforestry:345–374
Kim DG, Kirschbaum MU, Beedy TL (2016) Carbon sequestration and net emissions of CH4 and N2O under agroforestry: synthesizing available data and suggestions for future studies. Agric Ecosyst Environ 226:65–78
Krishna Reddy M (2010) Climate change and virus diseases of horticultural crops. In: Singh HP, Singh JP, Lal SS (eds) Challenges of climate change-Indian horticulture. Westville Publishing, New Delhi
Kumar BM, Nair PKR (2011a) Carbon sequestration potential of agroforestry systems in dry regions. J Arid Environ 75:134–142
Kumar S, Nair PKR (2011b) Carbon sequestration potential of agroforestry systems in India. Agrofor Syst 84(3):323–335
Kumar NK (2013) Climate-Resilient horticulture: adaptation and mitigation strategies. Int J Innov Hortic 2(1)
Kumar BM, Nair PKR, Sreedharan C (1994a) Homegardens in tropical South India: characteristics and carbon storage. Agrofor Syst 26(1):107–125
Kumar S, Yadav RK, Singh R (1994b) Productivity and sustainability of multistrata homegardens in humid tropics. Agrofor Today 6(3):10–13
Kumar R, Kang GS, Pandey SK, Gopal J (2013) Kufri khyati: a new early maturing potato variety for indian plains. Potato J 36(1–2). https://epubs.icar.org.in/index.php/PotatoJ/article/view/32689
Kumar M et al (2024a) Role of cutting-edge technology in horticulture crops for shaping the future smart farming ecosystem. Environ Ecol 42(4B):1933–1942
Kumar R, Bhardwaj A, Singh LP, Singh G, Kumar A, Pattnayak KC (2024b) Comparative life cycle assessment of environmental impacts and economic feasibility of tomato cultivation systems in northern plains of India. Scientific Reports
Kumar R, Singh V, Pal D (2024c) Precision fertigation for reducing nitrogen losses and greenhouse gas emissions. Nutr Cycl Agroecosyst 129:155–170
Lahtinen T, Virkajärvi P, Niskanen M (2025) Carbon balance of agroforestry-based horticultural systems. Agrofor Syst 99(2):215–229
Lakhiar IA et al (2018) Modern plant cultivation technologies under controlled environments. J Plant Interact 13(1):338–352
Lal R (2004a) Carbon emission from farm operations. Environ Int 30(7):981–990
Lal R (2004b) Soil carbon sequestration and ecosystem services. Geoderma 123:1–22

Lal R (2004c) Soil carbon sequestration impacts on global climate change and food security. Science 304(5677):1623–1627
Lal R (2004d) Soil carbon sequestration to mitigate climate change. Geoderma 123(1–2):1–22
Lal R (2005) Forest soils and carbon sequestration. For Ecol Manag 220(1–3):242–258
Lal R, Kimble J (1997) Conservation tillage for carbon sequestration. Nutr Cycl Agroecosyst 49:243–253. https://doi.org/10.1023/A:1009794514742
Lal SS, Govindakrishnan PM, Dua VK, Singh JP, Pandey SK (2008) Impact assessment of climate change for research priority planning in horticultural crops. CPRI Pub, Shimla
Lavania S (2003) Vetiver root systems for soil carbon storage. Curr Sci 84(12):1523–1525
Levin GM (2006) Pomegranate roads: a Soviet botanist's exile from Eden. Pulses 1:89–104
Lynas M (2007) Carbon counter. Harper Collins, Glasgow
Ma J, Li XL, Xu H, Han Y, Cai ZC, Yagi K (2007) Effects of nitrogen fertilizer and wheat straw application on CH4 and N2O emissions from a paddy rice field. Aust J Soil Res 45:359–367
Malhotra SK (2017) Horticultural crops and climate change: A review. Indian J Agric Sci 87(1):12–22
Mall RK, Aggarwal PK (2002) Climate change and rice yields in diverse agro-environments of India. I. Evaluation of impact assessment models. Clim Chang 52:315–330
Matthews SC, Hendrickson C, Weber CL (2008) The importance of carbon footprint estimation boundaries. Environ Sci Technol 42:5839–5842
Mazzoncini M, Sapkota TB, Barberi P, Antichi D, Risaliti R (2011) Long-term effect of tillage, nitrogen fertilization and cover crops on soil organic carbon and total nitrogen content. Soil Tillage Res 114:165–174
Meena RL, Meena BL, Chandra P, Sanyal SK, Narjary B (2025) Commercial vegetable production under protected structures in saline environments: a promising strategy for livelihood security. Indian Farm 75(11):52–55
Midmore DJ, Roan YC, Wu MH (1992) Management of moisture and heat stress for tomato and hot pepper production in the tropics. In: Kuo CG (ed) Adaptation of food crops to temperature and water stress. AVRDC, Shanhua, pp 453–460
Mishra VK, Sharma NK (2020) Role of agroforestry in mitigating climate change in semi-arid regions. J Soil Water Conserv 19(4):275–282
Mohan J, Sheoran V, Singh R (2020) Carbon stock estimation in cashew orchards of coastal regions. Indian J Soil Conserv 48(3):215–221
Montagnini F, Nair PKR (2004a) Carbon sequestration in agroforestry systems: opportunities and challenges. Mitig Adapt Strateg Glob Chang 9:441–455
Montagnini F, Nair PKR (2004b) Carbon sequestration: an under-exploited environmental benefit of agroforestry systems. Agrofor Syst 61:281–295
Montzka SA, Dlugokencky EJ, Butler JH (2011) Non-CO2 greenhouse gases and climate change. Nature 476:46–50
Müller A, Patel V, Schneider U (2023a) Environmental burdens of protected cultivation structures: a global LCA synthesis. J Environ Manag 330:117–128
Müller F, Schmidt S et al (2023b) Life cycle assessment of different strawberry production methods in Germany with a particular focus on plastic emissions. Int J Life Cycle Assess 28:611–625
National Horticulture Board (2024) Annual report 2023–24. National Horticulture Board, Ministry of Agriculture and Farmers Welfare
Nair PKR (2012) Carbon sequestration studies in agroforestry systems: a reality check. Agrofor Syst 86:243–253
Nair PKR, Mohan Kumar B, Nair VD (2009) Agroforestry as a strategy for carbon sequestration. J Plant Nutr Soil Sci 172(1):10–23
Nair PKR, Kumar BM, Nair VD (2010a) Carbon sequestration in agroforestry systems. Adv Agron 108:237–307
Nair PKR, Nair VD, Kumar BM, Showalter JM (2010b) Carbon sequestration in tropical agroforestry systems. Plant Soil 328:1–19
Nair PKR, Kumar BM, Nair VD (2010c) Agroforestry as an approach to ecosystem management. J Soil Water Conserv 65(5):127A–136A

Naresh Kumar S, Aggarwal PK (2009) Impact of climate change on coconut plantations. In: Aggarwal PK (ed) Global climate change and Indian agriculture-case studies from ICAR network project. ICAR pub, New Delhi
Naresh Kumar S, Kasturi Bai KV, Thomas GV (2010) Climatic change and plantation crops: impact, adaptation and mitigation with special reference to coconut. In: Singh HP, Singh JP, Lal SS (eds) Challenges of climate change- India horticulture. Westville Publishing, New Delhi
Naresh Kumar S, Aggarwal PK, Swaroopa Rani DN, Jain S, Saxena R, Chauhan N (2011) Impact of climate change on crop productivity in Western Ghats, coastal and northeastern regions of India. Curr Sci 101(3):33–42
Naresh Kumar S, Singh AK, Aggarwal PK, Rao VUM, Venkateswarlu B (2012) Climate change and Indian agriculture: salient achievements from ICAR network project. IARI Pub, New Delhi
Nath AJ, Lal R, Das AK (2015) Carbon sequestration in tea-dominated agroforestry systems. Agric Ecosyst Environ 199:123–131
Newaj R, Dhyani SK (2008) Agroforestry systems for carbon sequestration: present status and scope. Indian J Agrofor 10(1):1–9
NITI Aayog (2024) Greening and Restoration of Wasteland with Agroforestry (GROW): Report and portal. Government of India. NITI Aayog
Niu G, Masabni J (2022) Hydroponics. In: Plant factory basics, applications and advances. Elsevier Science, pp 153–166
Novara A, Gristina L, Saladino SS (2012) Soil organic carbon in Mediterranean vineyards. Soil Res 50(7):634–642
Nzeadibe TC, Egbule CL, Chukwuone N, Agu V (2011) Farmers' perceptions of climate change governance and adaptation constraints in Niger Delta Region of Nigeria. African Technology Policy Network, p 7
Palande V et al (2018) Fully automated hydroponic system for indoor plant growth. Procedia Comput Sci 129:482–488
Pandey D, Agrawal M (2014) Carbon footprint estimation in the agriculture sector. In: Assessment of carbon footprint in different industrial sectors, vol 1. Singapore, Springer Singapore, pp 25–47
Pandey R, Singh H (2021) Carbon dynamics and sequestration potential of coffee-based agroecosystems. Agrofor Syst 95(4):789–804
Pandey D, Agrawal M, Pandey JS (2011) Carbon footprint: current methods of estimation. Environ Monit Assess 178(1–4):135–160. https://doi.org/10.1007/s10661-010-1678-y
Pandey R, Bohra B, Jat ML (2020a) Cereal–legume intercropping for sustainable intensification. Field Crop Res 258:107–954
Pandey R, Verma P, Nair PKR (2020b) Agroforestry systems and global carbon stocks. Clim Dev 12(9):833–845
Pandita ML, Singh N (1992) Vegetable production under water stress conditions in rainfed areas. In: Kuo CG (ed) Adaptation of food crops to temperature and water stress. AVRDC, Shanhua, pp 467–472
Pati A, Das S, Zaidi KH, Tahir M (2025) A bibliometric review on the inter-connection between climate change and horticulture sector. Discov Agric 3(1):12
Pawase PP, Nalawade SM, Walunj AA, Bhanage GB, Kadam PB, Durgude AG, Patil MR (2024) Comprehensive study of on-the-go sensing and variable rate application of liquid nitrogenous fertilizer. Comput Electron Agric 216:108482
Peshin R, Zhang W (2014) Integrated pest management and pesticide use. In: Integrated pest management: dissemination and impact. Springer, pp 1–23. https://doi.org/10.1007/978-94-007-7796-5_1
Pi H, Zhang X, Li S, Webb NP (2024) Influence of crop rotation, irrigation, fertilization, and tillage on the aggregate property and soil wind erosion potential in the floodplain of the Yellow River. Aeolian Res 67:100925. https://doi.org/10.1016/j.aeolia.2024.100925
Pound B, Lamboll R, Croxton S, Gupta N, Bahadur AV (2018) Climate-Resilient Agriculture in South Asia: An analytical framework and insights from practice. Action on Climate Today

Powlson DS, Stirling CM, Jat ML, Gerard BG, Palm C, Sanchez P, Cassman KG (2011a) Limited potential of no-till agriculture for climate change mitigation. Nat Clim Chang 4:678–683
Powlson DS, Addiscott TM, Whitmore AP (2011b) Nitrogen fixation and soil fertility in cereal–legume systems. Plant Soil 340:1–3
Prabhakar M (2019) Climate smart agriculture: policies and practices for adaptation and mitigation to climate change in india. climate smart agriculture: strategies to respond climate change in south asia. SAARC Agriculture Centre, Dhaka, Bangladesh, 62.
Prakash J, Amulya S, Verma MK, Singh SK, Bharadwaj C, Thakre M, Kumar C (2025) Assessing effectiveness of drought tolerance indices in selecting grape (Vitis vinifera) under induced drought stress conditions. Indian J Agric Sci 95(4):425–431
Prasad R, Rana R (2006) A study on maximum temperature during March 2004 and its impact on rabi crops in Himachal Pradesh. J Agrometeorol 8(1):91–99
Pretty J, Benton T, Bharucha Z, Dicks L, Flora C, Godfray C, Goulson D, Hartley S, Lampkin N, Morris C, Pierzynski GPV, Prasad PV, Reganold J, Rockström J, Smith P, Thorne P, Wratten S (2018) Global assessment of agricultural system redesign for sustainable intensification. Nat Sustain 1:441–450. https://doi.org/10.1038/s41893-018-0114-0
Rahn C, Bending GD, Turner MK, Lillywhite R (2003) Management of N mineralization from crop residues of high N content using amendment materials of varying quality. Soil Use Manag 19:193–200
Raj A, Jhariya MK (2023) Carbon Sequestration in agroforestry and horticulture-based farming systems: mitigating climate change and advancing food and nutrition security. In Emerging solutions in sustainable food and nutrition security (pp. 143-182). Cham: Springer International Publishing
Rao GD, Sinha SK (1994) Impact of climatic change on simulated wheat production in India. In: Rosenzweig C, Iglesias I (eds) Implications of climate change for international agriculture: crop modelling study. EPA, USA, pp 1–10
Rathi G, Soni MK (2024) Impact Analysis of PM KUSUM Scheme Individual Pump Solar Component C. In 2024 International Conference on Sustainable Energy: Energy Transition and Net-Zero Climate Future (ICUE) (pp. 1–7). IEEE
Ravindranath NH, Ostwald M (2008) Carbon inventory methods handbook for greenhouse gas inventory. In: Carbon mitigation and round wood production projects, vol 29. Springer
Reddy K, Srikanth G, Sreeja Reddy CM, Parihar, Panneerselvam Peramaiyan, K. Sreenivasa Reddy, Kiranmoy Patra, Ayan Sarkar, Alok Sinha, B. Mahesh Naik, and M. Karthika (2025) "Policies and technological intervention for regenerative agriculture and carbon credits to achieve carbon-negative agriculture." In Carbon-Negative Agriculture, pp. 59-96. Singapore: Springer Nature Singapore
Rees WE (1992) Ecological footprints and appropriated carrying capacity: what urban economics leaves out. Environ Urban 4:121–130
Rosenzweig C, Parry ML (1994) Potential impact of climate change on world food supply. Nature 367:133–138
Roy HE et al (2023) Biological control and climate change: implications for natural enemies and pest suppression. Biol Control 182:105123
Rudich J, Zamski E, Regev Y (1977) Genotype variation for sensitivity to high temperature in the tomato: pollination and fruit set. Bot Gaz 138:448–452
Ruibérriz MAR, Bojollo RMC, Braña CA, Lizana AR, Fernández RMO (2012) Carbon sequestration potential of residues of different types of cover crops in olive groves under Mediterranean climate. Span J Agric Res 10:649–661
Sadiq SM, Singh IP, Ahmad MM, Sani BS (2025) Cold Storage Solutions to Reduce Post-Harvest Loss: Start-ups for Youth in the Agricultural Supply Chain. New Countryside 4(2):37–44. https://doi.org/10.55121/nc.v4i2.453
Saha S, Ghosh S, Chattopadhyay S (2020a) Carbon storage and ecosystem services in multistrata homegardens of eastern India. Environ Dev Sustain 22(4):3711–3728
Saha S, Nair PKR, Kumar BM (2020b) Carbon storage and multifunctionality of tropical homegardens. Environ Conserv 47(2):85–94

Samra JS, Singh G (2004) Heat wave of March 2004: impact on agriculture. ICAR Pub, New Delhi
Sanchez-Guerrero MC, Lorenzo P, Medrano E, Baille A, Castilla N (2009) Effects of EC-based irrigation scheduling and CO2 enrichment on WUE of a greenhouse cucumber crop. Agric Water Manag 96:429–436
Sanderson V, Bamber N, Pelletier DN (2019) Cradle-to-market life cycle assessment of Okanagan (Canada) cherries: helicopters, seasonal migrant labour and flying fruit. J Clean Prod 229:1283–1293
Sarkar AN, Dash S (2011) Emissions trading and carbon credit accounting for sustainable energy development with focus on India. Asia Pac Bus Rev 7(1):50–80
Saxena M, Bhattacharya S, Malhotra SK (2018) Horticultural statistics at a glance. Horticulture Statistics Division, Department of Agriculture, Cooperation and. Farmers' Welfare.
Shah D (2008) Synthesising investment initiatives for horticulture sector in india: a crucial component for agricultural growth. Indian J Agric Econ 63(3):354
Sharath Kumar M, Heuvelink E, Marcelis LF (2020) Vertical farming: moving from genetic to environmental modification. Trends Plant Sci 25(8):724–727. https://doi.org/10.1016/j.tplants.2020.05.012
Sharma A, Panigrahy S (2007) Apple orchard characterization using remote sensing and GIS
Sharma J, Upadhyay AK, Adsule PG (2005) Effect of drip water application at subsurface on grapevine performance- a case study. J Appl Hortic 7:137–131
Sharma SD, Singh R, Saini P (2019) Biomass accumulation and carbon stock in apple orchards. Sci Hortic 246:789–797
Sharma J (2013) Climate-resilient horticulture: adaptation and mitigation strategies (pp. 81–88). In: Singh HCP, Rao NKS, Shivashankar KS (eds) Bangalore, India: Springer India.
Sharma A, Wali VK, Singh VB, Salgotra R (2020) Jammu mango: new mango variety for subtropical region of Jammu and Kashmir. Indian Hortic 65(2)
Sharma A, Tewari R, Singh S (2025a) Agri-horticultural systems under climate stress: ecosystem multifunctionality insights. Curr Agric Rep 11:45–62
Sharma R et al (2025b) Long-term agroforestry enhances soil organic carbon pools and deep soil carbon sequestration in the Indian Himalayas. Front Environ Sci 13:1568564
Singh HCP, Rao NKS, Shivashankar KS (eds) (2013) Climate-resilient horticulture: adaptation and mitigation strategies (pp 81–88). Bangalore, India, Springer India
Singh HP (2017) Hydroponics – a technique of soilless cultivation. National Horticulture Board (NHB)
Singh S (2024) Remote sensing GPS and GIS in agriculture. Int J Res Publ Semin 15(2):102–106
Singh K, Rani D (2023) Climate change impacts on orchard management: a review. Planta 7:1428–1446
Singh R, Kumar S, Nangare DD, Meena MS (2009) Drip irrigation and black polyethylene mulch influence on growth, yield and water-use efficiency of tomato. Afr J Agric Res 4(12):1427–1430
Singh B, Ramesh K, Verma V (2011) Soil organic carbon under perennial aromatic crops. J Soil Sci Plant Nutr 11(4):53–63
Singh A et al (2015a) Assessment of soil and biomass carbon stocks under perennial and agroforestry systems. J Environ Qual 44(5):1537–1546
Singh A, Rani S, Bhattacharyya R (2015b) Biomass and soil carbon stock assessment in perennial horticultural systems. J Environ Qual 44(5):1537–1546
Singh AK, Patra S, Sen S (2015c) Horti–silvicultural systems for enhanced carbon stocks. Indian J Hortic 72(3):487–492
Smith HE, Sallu SM, Whitfield S, Gaworek Michalczenia MF, Recha JW, Sayula GJ, Mziray S (2021) Innovation systems and affordances in climate smart agriculture. J Rural Stud 87:199–212
Srinidhi A, Meuwissen MP, Jadhav S, Ludwig F, Dadas D, Nikam N, Werners SE (2025) Climate resilient development pathways for farmer producer organizations in semi-arid India. Clim Dev 1–16
Srivastava A, Naresh Kumar S, Aggarwal PK (2010) Assessment on vulnerability of sorghum to climate change in India. Agric Ecosyst Environ 138:160–169

Sulaiman M, Thomas GV (2020) Aquaponics: principles and practices. CMFRI
Tiwari KN, Bisht TS (2015) Hydroponic techniques for horticultural crops. Indian Council of Agriculture & Research (ICAR)
Upadhyay R, Thomas L, George J (2021) Carbon sequestration potential of arecanut plantations in humid tropics. J Plant Crop 49(1):9–15
Urban Farms LCA (2023) Life cycle assessment of rooftop and peri-urban farming systems. Int J Life Cycle Assess 28(4):355–372
Usha K, Singh B (2013) Potential applications of remote sensing in horticulture. Sci Hortic 153:71–83
Van Groeningen KJ, Osenberg CW, Hungate BA (2011) Increased soil emissions of potent greenhouse gases under increased atmospheric CO2. Nature 475:214–216
van Rijn F, Schenck R, de Boer I (2021a) Environmental hotspots in protected horticulture: a meta-LCA analysis. Resour Conserv Recycl 170:105–563
van Rijn L, Tuomisto H et al (2021b) Environmental and social life cycle assessment of growing media for urban rooftop farming. Int J Life Cycle Assess 26:2085–2102
Velásquez AC, Castroverde CDM, He SY (2018) Plant–pathogen warfare under changing climate conditions. Curr Biol 28(10):R619–R634
Von WS, Chieng SS (2004) A comparison between low-cost drip irrigation, conventional drip irrigation, and hand watering in Nepal. Agric Water Manag 64:143–160
Wackernagel M, Rees WE (1996) Our ecological footprint: reducing human impact on the earth. New Society Publishers
Wei Q, Xu J, Liu Y, Wang D, Chen S, Qian W et al (2024) Nitrogen losses from soil as affected by water and fertilizer management under drip irrigation: development, hotspots and future perspectives. Agric Water Manag 296:108791
Wen S, Cui N, Gong D, Xing L, Wu Z, Zhang Y et al (2024) Effect of nitrogen fertilizer management on N_2O emission and NH_3 volatilization from orchards. Soil Tillage Res 242:106165
Wiedmann T, Minx JA (2007) Definition of carbon footprint. ISAUK Research Report 07-01. ISAUK Research & Consulting, Durham
Wiedmann T, Minx J (2008a) A definition of 'carbon footprint'. Ecol Econ Res Trends 1:1–11
Wiedmann T, Minx J (2008b) A definition of 'carbon footprint'. In: Pertsova CC (ed) Ecological economics research trends. Nova Science Publishers, pp 1–11. https://www.novapublishers.com/catalog/product_info.php?products_id=5999
Yadav V, Singhal SL, Grover A (1998) Salt tolerant vegetables. Ann Agric Zool 4(1–2):206–214
Yang X, Xiong J, Du T, Ju X, Gan Y, Li S et al (2024) Diversifying crop rotation increases food production, reduces net greenhouse gas emissions and improves soil health. Nat Commun 15(1):198
Yildizhan H, Taki M (2018) Assessment of tomato production process by cumulative exergy consumption approach in greenhouse and open field conditions: case study of Turkey. Energy 156:401–408
Youdon C, Bajaj P (2022) India's approach and position on climate change governance. National Maritime Foundation
Zhang Y et al (2022) Blockchain: an emerging technology to upgrade fresh fruit supply chains. Trends Food Sci Technol 124:1–12
Zou JW, Huang Y, Jiang JY, Zheng XH, Sass RL (2005) A 3-year field measurement of methane and nitrous oxide emissions from rice paddies in China: effects of water regime, crop residue, and fertilizer application. Glob Biogeochem Cycles 19:GB2021

Natural Pigments from Edible Flowers: Extraction and Applications

19

Anshul Kumar, Bharati Kashyap, S. R. Dhiman, and Y. C. Gupta

Abstract

Chemical compounds known as pigments have the ability to absorb light in the visible spectrum. Pigmentation is the fundamental process for producing flower colour. Flavonoids, betalains, chlorophylls, and carotenoids are the four main pigment classes that make up the wide variety of floral colours. Each pigment class has a unique chemical composition that ultimately influences the particular wavelength it absorbs and, consequently, the colour it produces. The most common and varied class of pigments in angiosperms is flavonoids. Anthocyanins, aurones, chalcones, flavanols, flavones, and flavanones are a few notable categories of flavonoids. However, anthocyanins are the flavonoids that give flowers their widest range of colour. Considering the stability and appropriate extraction technique to be used is among the most important aspects of exploring the extraction of pigments from plant sources. Both conventional and unconventional methods of extraction are employed. The advantages and disadvantages of each of these approaches vary. After decades of study and development, these natural pigments have been identified for their health benefits, and prospective use in food, medicine, cosmetics, and other segments has been emphasised. The demand for natural pigments has grown as a result of public awareness of the negative consequences of synthetic colourants.

Keywords

Commercial applications · Edible flowers · Flower pigments · Pigment extraction

A. Kumar (✉) · B. Kashyap · S. R. Dhiman · Y. C. Gupta
Department of Floriculture and Landscaping, Dr. YS Parmar University of Horticulture and Forestry, Solan, Himachal Pradesh, India

A. Das, A. B. Sharangi (eds.), *Remunerative Horticulture*,
https://doi.org/10.1007/978-981-92-1007-7_19

19.1 Introduction

Chemical compounds known as pigments are capable of absorbing light in the visible spectrum (Kumari et al. 2017). According to Van-der-Kooi et al. (2019), pigmentation mostly produces flower colour. Flavonoids, betalains, chlorophylls, and carotenoids are four key pigment groups that contribute to the wide variety of flower colours (Narbona et al. 2021). Each pigment class has a unique chemical composition, which ultimately influences the particular wavelengths it absorbs and, consequently, the colour it produces. In angiosperms, flavonoids are the most prevalent and varied class of pigments. Important subgroups of flavonoids include anthocyanins, aurones, chalcones, flavanols, flavones, and flavanones. However, anthocyanins are the flavonoids that give flowers their most diverse coloration (Narbona et al. 2021). The main floral pigments, their production, and regulation have been documented (Forkmann 1991; Tanaka et al. 2008; Tripathy and Pattanayak 2012). Studies in model systems like maize, Arabidopsis, petunia, and snapdragon are where most of the molecular data on the control and production of flower colours come from. The eco-physiological regulation and purposes of floral coloration, however, continue to be better understood through nonclassical plant models (Grotewold 2006). Pigments are broadly divided into water-soluble pigments that are located in cell sap and fat-soluble pigments that are found in protoplasts of plant protoplasm (Humphrey 2004; Kong et al. 2003; Zhoh et al. 2010). Chlorophylls and carotenoids are fat-soluble pigments, and flavonoids (anthocyanins) and betalains are water-soluble pigments (Zhao et al. 2022).

Considering the stability and appropriate extraction process to be used is one of the most important aspects to consider when investigating the extraction of pigments from plants (Saini and Keum 2018). Both conventional and unconventional methods of extraction are available. Each of these approaches has unique advantages and disadvantages (Ngamwonglumlert et al. 2017). Soxhlet extraction and maceration are examples of conventional techniques that are easy, affordable, and useful (Veggi et al. 2013). There are a number of unconventional extraction techniques, viz., enzyme-assisted extraction method, microwave-assisted extraction method, pulsed electric field extraction method, ultrasound-assisted extraction method, supercritical fluid extraction method, and pressurised liquid extraction method (Usman et al. 2022). Raw pigments and other substances are frequently extracted using conventional techniques. However, non-traditional extraction methods—also referred to as green extraction techniques—have supplanted conventional extraction in recent years due to their lower solvent use, shorter extraction times, and lower environmental impact (Cheok et al. 2014). Carotenoids, chlorophylls, and anthocyanins appear to be properly extracted with MAE; however, heat-sensitive pigments like betalains are not. It is possible to extract betalains using the non-thermal extraction procedures UAE, EAE, and PEF (Ngamwonglumlert et al. 2017).

Natural pigments are an appealing alternative and are gaining popularity as a result of modern customers' increased concern for their health. The FDA and Codex have approved the use of extracts from four important families of naturally occurring pigments found in nature: anthocyanins, betalains, carotenoids, and chlorophylls (Francis 1996). Natural colourants continue to have issues with their lack of stability and limited colour shade variation (Delgado-Vargas and Paredes-Lopez 2003).

Demand for natural pigments has increased due to growing awareness among the public of the negative consequences of synthetic dyes. Due to the growing consumer desire for more natural products and natural pigments during the past 35 years, a trend towards replacing synthetic dyes with natural pigments has been observed (Cai et al. 2005). Since most natural colours already have antibacterial and antioxidant qualities, they might also have a significant therapeutic potential.

19.2 Pigments

They are molecules that only reflect certain wavelengths of light while absorbing others. A pigment is any chemical molecule that absorbs visible light between 380 and 760 nm. They are sometimes referred to as biochrome or biological pigment. The pigment theory explains how pigments give plants their colour. According to this idea, a coloured pigment, like red, absorbs the majority of the frequencies of light that are not red and only reflects the red light frequency. This is also known as the subtractive colour theory since all hues other than the pigment colours are absorbed. According to Narbona et al. (2021), there are four major pigment classes widely present in nature, viz., flavonoids, betalains, chlorophylls, and carotenoids.

19.2.1 Classification of Pigments

Pigments can be classified into five different categories (López-Cruz et al. 2023). On the basis of origin, pigments are natural or synthetic (Bauernfeind 1981). Natural ones are produced by plants, e.g. flavonoids, betalains, chlorophyll, and carotenoids. Synthetic ones are made in the laboratories, e.g. titanium dioxide, lead carbonate, etc.

The chromophore's chemical structure can be used to categorise pigments into metal-coordinated porphyrins and chromophores with conjugated systems (Wong 1989). Caramel carotenoids, betalains, anthocyanins, and synthetic colours are examples of chromophores having conjugated systems. Chlorophyll and its derivatives are metal-coordinated porphyrins.

Furthermore, based on their structural characteristics, pigments can be classified as benzopyran derivatives (oxygenated heterocyclic compounds), n-heterocyclic compounds different from tetrapyrroles, isoprenoid derivatives, and tetrapyrrole derivatives (Bauernfeind 1981; Hari et al. 1994). Different from tetrapyrroles, N-heterocyclic chemicals include betalains and benzopyran derivatives (oxygenated heterocyclic compounds), which consist of anthocyanins and other flavonoids. Tetrapyrrole derivatives include chlorophylls, whereas isoprenoid derivatives include carotenoids.

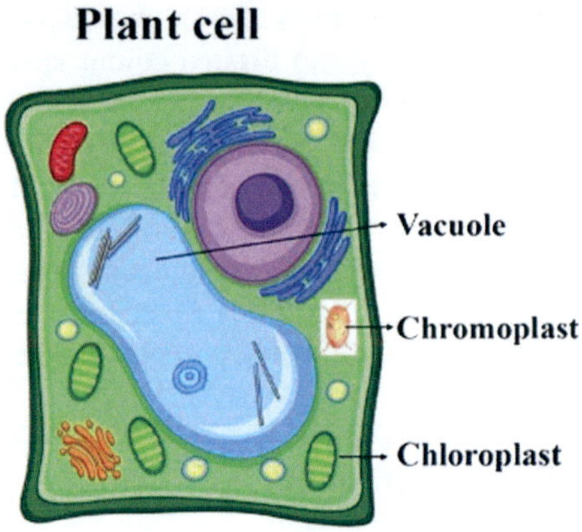

Fig. 19.1 Diagram of a plant cell showing labelled components: a large central vacuole, a chromoplast, and a chloroplast. The cell is depicted with various organelles, including a nucleus, mitochondria, and endoplasmic reticulum, within a rectangular cell wall. The illustration highlights the structure and organisation of a typical plant cell. (Source: brgfx/"Freepik")

The FDA has classified the pigments as food additives, making them certifiable and exempt from certification (Frick and Meggos 1988; Wong 1989). Certifiable are artificial pigments, which are further divided into synthetic pigments and lakes. Exempt from certification are both synthetic versions of natural colours and pigments originating from natural sources like minerals, plants, or animals.

Plant pigments can be divided into water-soluble and lipid-soluble categories based on the ease with which they dissolve in water (López-Cruz et al. 2023). Water-soluble pigments are flavonoids and betalains. Lipid-soluble pigments are chlorophyll and carotenoids.

19.2.2 Site of Pigment Accumulation

Chlorophylls as well as carotenoids dissolve in fat and further accumulate in plastids (chromoplast/chloroplast). These pigments, anthocyanins and betalains, can be found in the cell's vacuole and are soluble in water (Fig. 19.1) (Kumari et al. 2017).

19.3 Major Flower Pigment Classes

19.3.1 Flavonoids

The class of non-nitrogenous biological pigments known as flavonoids is one that is widely prevalent in plants (Dias et al. 2021). The vacuoles of plant cells contain phenolic chemicals called flavonoids, which are water-soluble and have an aromatic ring with an attached –OH group (Panche et al. 2016). There have been described over 8000 distinct types of flavonoids (Wang et al. 2022). According to Martín et al. (2017), anthocyanins, a significant category, are mostly responsible for the red colour of flowers as well as the purple and purple-red hues of fall leaves. Other flavonoids, including chalcones, aurones, and flavones (Fig. 19.2), are either invisible

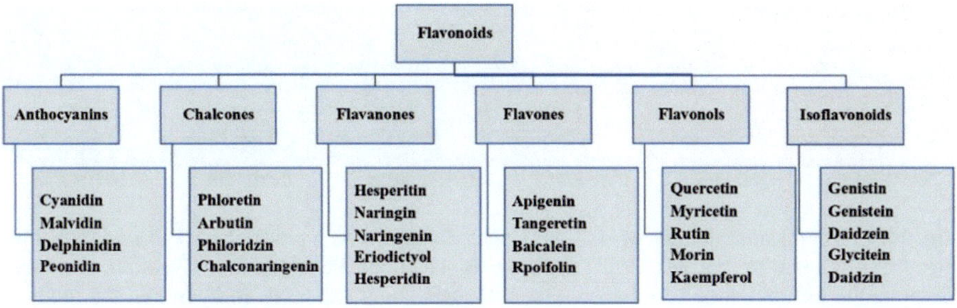

Fig. 19.2 Diagram illustrating the classification of flavonoids. The main category, "Flavonoids," branches into six subcategories: Anthocyanins, Chalcones, Flavanones, Flavones, Flavonols, and Isoflavonoids. Each subcategory lists specific compounds. Anthocyanins include Cyanidin, Malvidin, Delphinidin, and Peonidin. Chalcones include Phloretin, Arbutin, Phloridzin, and Chalconaringenin. Flavanones include Hesperitin, Naringin, Naringenin, Eriodictyol, and Hesperidin. Flavones include Apigenin, Tangeretin, Baicalein, and Rpoifolin. Flavonols include Quercetin, Myricetin, Rutin, Morin, and Kaempferol. Isoflavonoids include Genistin, Genistein, Daidzein, Glycitein, and Daidzin. (Redrawn from Panche et al. (2016))

Pigment	R1 position	R2 position	Flavylium ion
Pelargonidin	-H	-H	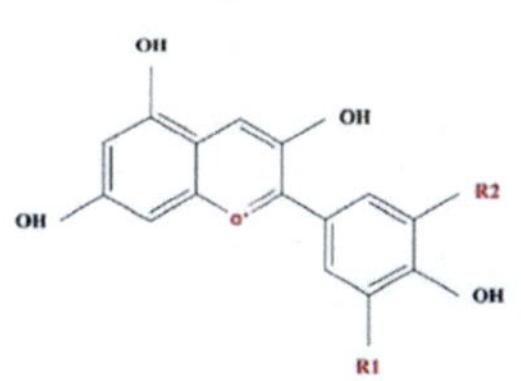
Cyanidin	-OH	-H	
Peonidin	$-OCH_3$	-H	
Delphinidin	-OH	-OH	
Petunidin	$-OCH_3$	-OH	
Malvidin	$-OCH_3$	$-OCH_3$	

Fig. 19.3 Table showing different pigments and their chemical structures. The table lists pigments such as Pelargonidin, Cyanidin, Peonidin, Delphinidin, Petunidin, and Malvidin, with their corresponding R1 and R2 positions. The R1 and R2 positions include groups like –H, –OH, and –OCH. On the right, there is a diagram of a flavylium ion structure, indicating where R1 and R2 are located

to humans or pale yellow to off-white (Santos et al. 2017). Anthocyanins will be briefly discussed in this section.

19.3.1.1 Anthocyanins

A subclass of phenolic compounds is called anthocyanins (Greek: anthos, which means flower, and kyaneos, which means dark blue) (Nassour et al. 2020). In epidermal cells, anthocyanins are produced in the cytoplasm and stored in the vacuolar lumen (Passeri et al. 2016). Flavylium cation (C6–C3–C6) is the basic building block of anthocyanins, which is linked to sugars, methoxyl, or hydroxyl groups to produce more than 635 anthocyanins that have been discovered so far (Mattioli et al. 2020). According to Khoo et al. (2017), there are six important anthocyanins (Figs. 19.3 and 19.4): delphinidin, peonidin, pelargonidin, cyanidin, malvidin, and petunidin.

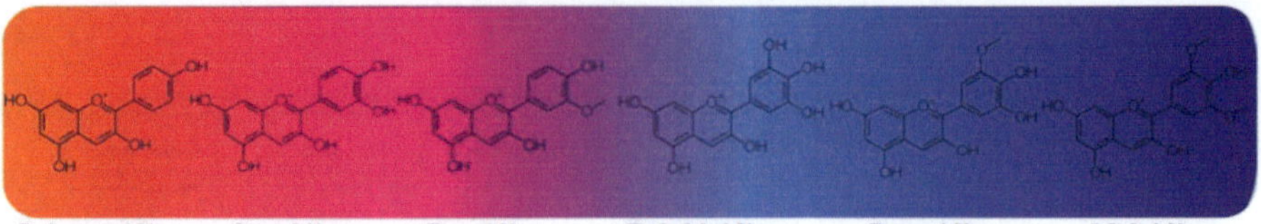

Fig. 19.4 Chemical structures of six compounds displayed on a gradient background transitioning from orange to purple. Each structure is labelled: Pelargonidin, Cyanidin, Peonidin, Delphinidin, Petunidin, and Malvidin. The structures show variations in hydroxyl and methoxy groups attached to a common core (Content adaptation & reuse: © 2013, Author's: Anthony Ananga, Vasil Georgiev, Joel Ochieng, Bobby Phills and Violeta Tsolova. Originally published in book: The Mediterranean Genetic Code - Grapevine and Olive, Chapter: 11, Publisher: InTech, under CC BY 4.0 l license. Available from: doi: 10.5772/3442)

Table 19.1 The edible flower's colour, anthocyanin concentration, and antioxidant activity

Species	Colour	Anthocyanin content (mg/100 g FW)	Antioxidant activity (mmol/100 g FW)
Ageratum houstonianum	Blue	2.99 ± 0.2	27.85 ± 3.3
Antirrhinum majus	Red White	7.37 ± 0.5 0.70 ± 0.1	21.18 ± 2.6 4.71 ± 0.6
Begonia semperflorens	Red	5.09 ± 0.4	13.24 ± 1.7
Borago officinalis	Blue	1.43 ± 0.1	0.55 ± 0.1
Calendula officinalis	Orange	0.47 ± 0.1	3.68 ± 0.3
Dianthus barbatus	Red White	13.35 ± 1.2 0.73 ± 0.1	38.67 ± 3.0 4.36 ± 0.8
Fuchsia hybrida	Red	7.58 ± 0.6	47.52 ± 3.2
Pelargonium peltatum	Red	12.52 ± 1.1	34.78 ± 2.9
Petunia hybrida	Red White	14.44 ± 1.2 2.5 ± 1.1	10.22 ± 0.8 5.40 ± 1.3
Tagetes erecta	Orange	0.75 ± 0.1	70.42 ± 4.1
Tropaeolum majus	Orange	8.27 ± 0.7	10.05 ± 0.8
Viola wittrockiana	Red Yellow	12.4 ± 1.1 2.93 ± 0.2	36.55 ± 3.0 3.31 ± 0.3

Adapted from Benvenuti et al. (2016)

Anthocyanins are major bioactive pigments present in edible flowers and contribute significantly to their antioxidant potential. The concentration of these compounds and the associated antioxidant activity vary widely among species and flower colours. Table 19.1 presents a comparative overview of anthocyanin content and antioxidant activity in selected edible flowers.

19.3.1.2 Biosynthetic Pathway

The endoplasmic reticulum produces anthocyanins, which are then transported to collect in the vacuoles of various cells as well as tissues in each of the plant's photosynthetic and reproductive organs (Poustka et al. 2007; Landi et al. 2015). Four steps can be identified in the production of anthocyanins in plants, which are regulated by structural genes (Fig. 19.5) (Kitamura et al. 2004; Routaboul et al. 2012). Early stages are successively catalysed by phenylalanine ammonia-lyase, cinnamate 4-hydroxylase, and 4-coumarate-CoA ligase from phenylalanine to 4-coumaryl CoA. Then, chalcone synthase, chalcone isomerase, and flavanone 3-hydroxylase control the conversion of 4-coumaroyl CoA to dihydrokaempferol. DHK can also be transformed into dihydroquercetin by the enzyme flavonoid 3′-hydroxylase or into dihydromyricetin by the enzyme flavonoid 3′5′-hydroxylase. Then, dihydroflavonol 4-reductase converts DHK, DHQ, and dihydromyricetin into leucopelargonidin, leucocyanidin, and leucodelphinidin, respectively. Anthocyanidin synthase/leucoanthocyanidin dioxygenase then catalyses the formation of these leucoanthocyanidins into coloured anthocyanidins (pelargonidin, cyanidin, and delphinidin) (Chaves-Silva et al. 2018). To enhance pigment stability and water solubility,

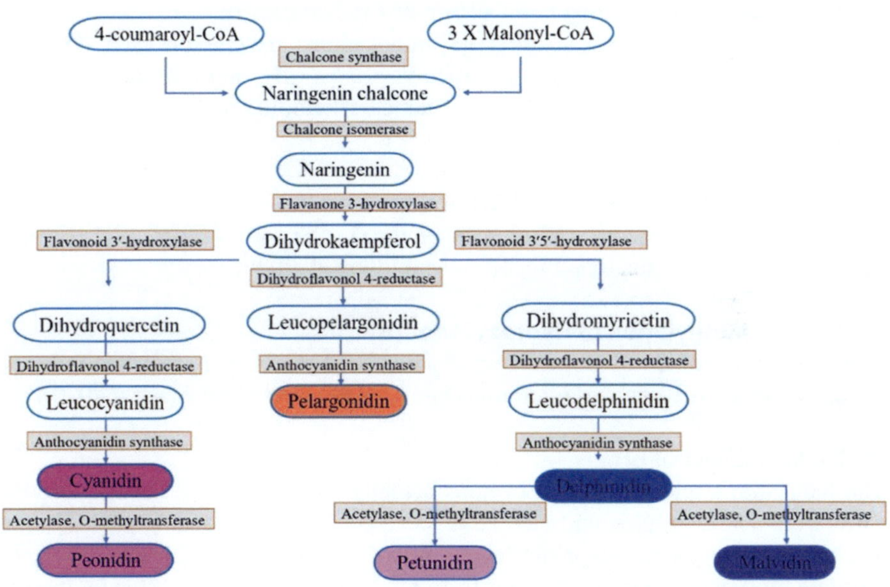

Fig. 19.5 Flow chart illustrating the biosynthesis pathway of flavonoids. It begins with 4-coumaroyl-CoA and 3 X Malonyl-CoA, leading to the formation of naringenin chalcone via chalcone synthase. Naringenin chalcone is converted to naringenin through chalcone isomerase. Naringenin is then transformed into dihydrokaempferol by flavanone 3-hydroxylase. Dihydrokaempferol can be further processed into dihydroquercetin, leucopelargonidin, or dihydromyricetin through different hydroxylases and reductases. These compounds are then converted into anthocyanidins such as cyanidin, pelargonidin, and delphinidin, which are further modified by acetylase and O-methyltransferase to produce peonidin, petunidin, and malvidin. Each step is labelled with the corresponding enzyme. (Redrawn from Ma et al. (2021))

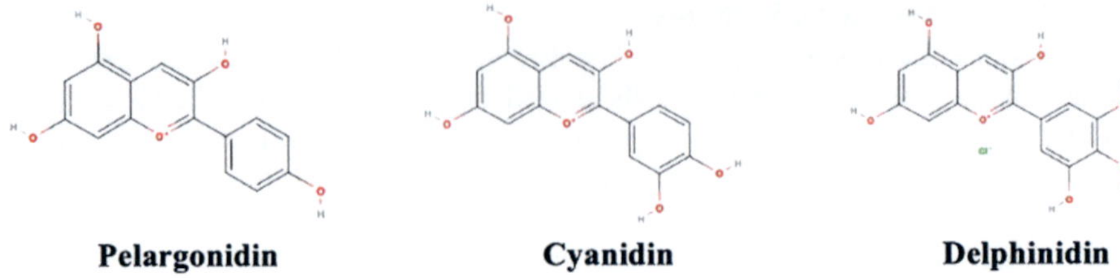

Fig. 19.6 Three chemical structures are shown, each labelled with a name: Pelargonidin, Cyanidin, and Delphinidin. Each structure consists of aromatic rings with hydroxyl groups and other functional groups attached. The structures are depicted in a linear format, highlighting the molecular connectivity and arrangement of atoms

anthocyanidins undergo further enzymatic modifications. These reactions are mediated by enzymes such as UDP-glucose flavonoid glucosyltransferase, acetylases, O-methyltransferases, and anthocyanin transferases, which introduce sugar, acyl, and methyl groups, resulting in more stable and highly soluble anthocyanin pigments (Winkel-Shirley 2001).

19.3.1.3 Factors Affecting the Colour of Anthocyanins

The colour as well as the stability of anthocyanins depend on their structure (Fig. 19.5), quantity of hydroxyl and methyl groups, pH, and intramolecular and intermolecular interactions with other substances, including metal ions and climate change (Enaru et al. 2021).

19.3.1.3.1 Number of Hydroxyl Groups

With increase in the number of oxygen atoms (in the form of hydroxyl groups), there is increase the blue tone Fig. 19.6 (Mattioli et al. 2020).

19.3.1.3.2 Methylation of Hydroxyl Groups

The replacement of the hydrogen atom of a hydroxyl group by a $-CH_3$/methyl radical results in an increase in the redness (Fig. 19.7) (Alappat and Alappat 2020).

19.3.1.3.3 Effect of pH

The molecular structure of anthocyanins actually varies as the pH of the solution shifts from acidic to basic and vice versa (Khoo et al. 2017), causing anthocyanins to change colour at different pH levels. According to Yoshida et al. (2009), the morning glory's petal colour changes from red (at the bud stage) to blue (at the full bloom stage). Vacuolar pH in surface-pigmented cells was estimated using a microelectrode (pH-sensitive). The results showed a clear developmental shift in vacuolar pH. At the bud stage, cells were red in colour, exhibiting a pH of approximately 6.6, whereas the fully expanded flower stage had blue-coloured cells and displayed a more alkaline pH of around 7.7. The pH of the vacuole (bud stage) was near 6.6 through the activity of proton pumps. As flowering progresses, expression of the NHX1 gene increases, promoting an exchange of H^+ with K^+ ions across the vacuolar membrane, which leads to vacuolar pH elevation to about 7.7.

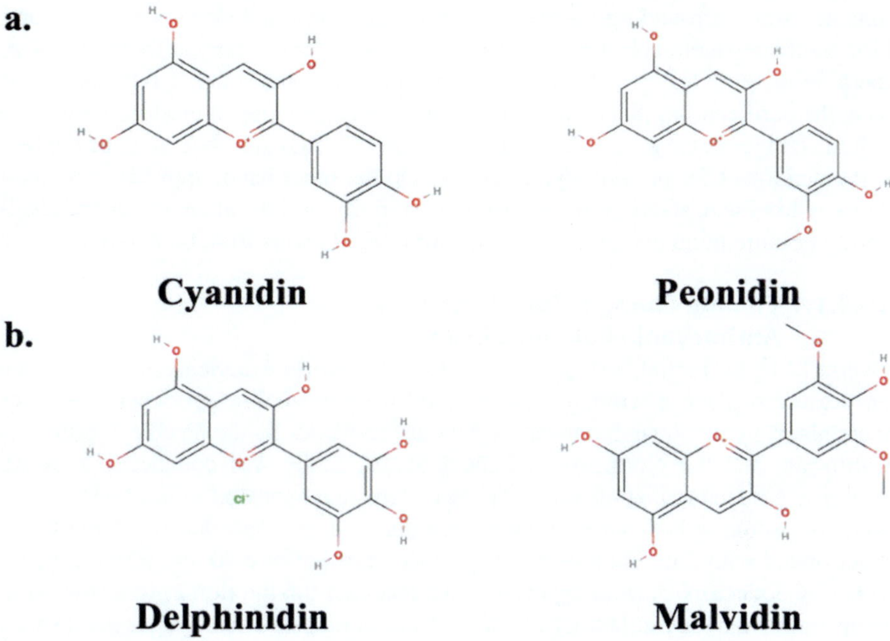

Fig. 19.7 The image is a panel figure showing chemical structures of four compounds. (**a**) Two structures labelled "Cyanidin" and "Peonidin," each with a three-ring backbone and various hydroxyl groups attached. (**b**) Two structures labelled "Delphinidin" and "Malvidin," each with a similar three-ring backbone, with Delphinidin having an additional chlorine atom and Malvidin having methoxy groups

19.3.1.3.4 Co-pigmentation Basics of Blue Flower Colouration in Nature

The mechanism of blue colour generation is typically quite complex; the presence of delphinidin-type pigments alone is insufficient; instead, pH, the anthocyanin complex, and metal ions all play significant roles (Enaru et al. 2021; Iwashina 2013). The various types of co-pigmentation include self-association, intermolecular co-pigmentation, and intramolecular co-pigmentation.

19.3.1.3.5 Metalloanthocyanins

They are stoichiometric mixtures of anthocyanins, flavones, and metal ions that self-assemble into supramolecular metal complex pigments (Fedenko et al. 2022). Examples of flowers that contain metalloanthocyanins include *Commelina communis*, *Salvia patens*, *Centaurea cyanus*, and *Salvia uliginosa*. The hydrangea is the most amazing plant that contains metalloanthocyanins. One pigment, delphinidin-3-glucoside, is responsible for both colours in hydrangeas (blue flower colour in acidic soil and red colour in basic soil). The underlying chemical mechanism causing the colour change is not the acidity of the soil. The secret is a metal ion. The roots of the hydrangea release citric acid in response to the stimulus of Al^{3+}, which is mobile in acidic soil conditions. As a result, a mixture of citric acid and citrate ions develops surrounding the roots. Citrate ions and aluminium ions (Al^{3+}) combine to create stable complexes that make it easier for hydrangea roots to absorb them. While both leaves and sepals contain similar amounts of Al^{3+}, only the sepals have the particular pigments that can interact with these ions to produce noticeable colour changes in the flower. Aluminium sulfate [$Al_2(SO_4)_3$] is thought to be the best soil amendment for promoting a blue hue. On the other hand, applying lime (calcium hydroxide, $Ca(OH)_2$) to increase the alkalinity of the soil decreases the availability of aluminium and changes the colour of the flowers from blue to red.

19.3.1.4 Climate Change's Impact on Flower Anthocyanin Polymorphisms

Over a 124-year period, a study was conducted in North America, and information on preserved plant specimens was obtained from the online specimen databases maintained by regional botanical collection networks in the Pacific Northwest, California, and the Northeastern United States. Study was conducted in North America. Collectors identified certain specimens that appeared to have white blossoms but actually had purple flowers, for instance, proving that pigments likely deteriorated with time. As a result, they resorted to flower colour descriptions provided by collectors. According to previous research linking floral pigmentation to temperature and the availability of water, it was anticipated that species that experience greater temperature increases over time would show a decrease in pigmentation, while those that experience greater precipitation or humidity decreases over time would show an increase in the frequency of pigmentation (Sullivan and Koski 2021). Three target species under study—*Allium cernuum*, *Raphanus sativus*, and *Hydrophyllum virginianum*—have examples of floral anthocyanin polymorphisms.

19.3.1.5 Anthocyanin Degradation

The popular ornamental plant *Brunfelsia acuminata*, which has an array of colours due to a quick shift in colour after blooming, was the subject of a study by Li et al. (2019b). Through the use of quantitative real-time PCR, gas chromatography combined with a time-of-flight mass spectrometer, and comparative proteomics, the reason behind the colour changes on the first (purple), third (white and purple), and fifth (white) days of the flower's life cycle was ascertained. Fifty-two proteins recorded could be divided into eight categories, with 6% being associated with the metabolic pathway of anthocyanin. The abundance of anthocyanin-related proteins declined markedly between day 1 and day 5, paralleling the changing transcript levels of the major structural genes involved in the pathway, including CHI, CHS, and F3′5′H, in petal tissues. The main volatile compounds at day 5 grew significantly. The results indicated that shared metabolic intermediates compete with one another, thereby redirecting metabolite flux towards either the floral volatile biosynthesis pathway or lignin formation or leading them to degrade into other metabolic pathways. As a result, the anthocyanin level markedly dropped, and the deep purple hue of the petals gave way to white.

19.3.2 Betalains

One of the main pigment families that give plant organs their striking colours is betalains, which are nitrogen-containing, water-soluble pigments responsible for red-violet and yellow coloration in plants (Tanaka et al. 2008). Besides imparting vivid coloration, betalains have outstanding antioxidant activity, per studies by Kanner et al. (2001). Thus, their potential health-promoting qualities, viz., hepatoprotective, anti-inflammatory, chemopreventive, hypolipidemic, and antidiabetic nature, have been extensively researched (Stintzing and Carle 2004; Esatbeyoglu et al. 2015; Khan 2016). The food industry makes extensive use of betalain-based products as natural food colouring because of their stability across a broad pH range, and the dietary supplement industry has commercialised a number of betalain-based products due to their purported nutritional benefits (Stintzing and Carle 2004). Betalains are relatively uncommon in nature and are confined exclusively to plants belonging to the order Caryophyllales. In members of the order Caryophyllales, the presence of betalains excludes the occurrence of anthocyanins, as these two pigment types are not produced together in the same species. According to Polturak and Aharoni (2018), the Molluginaceae and Caryophyllaceae families are the exceptions, whose plants only produce anthocyanins and betalains. *Portulaca oleracea* L. flowers, *Gomphrena globosa* L. flowers, *Echinopsis tubiflora* flowers, *Dianthus caryophyllus* L. flowers and stems, *Bougainvillea* flowers, *Mirabilis jalapa* L. flowers, etc., are some examples of flowers that contain betalains (Li et al. 2019a, b).

19.3.2.1 Classification of Betalains

Betalains are often divided into two classes (betacyanins and betaxanthins) based on their structural features and light-absorption qualities (Khan and Giridhar 2015; Polturak and Aharoni 2018). Betacyanins are red to violet in colour. The common examples of betacyanins are *Amaranthus caudatus* (amaranthin and iso-amaranthin), *Gomphrena globosa* (gomphrenin-V and gomphrenin-I), *Bougainvillea globosa* (bougainvillein-r-1 and bougainvillein-V), etc. Betaxanthins are yellow in colour. The common examples are *Mirabilis jalapa* (miraxanthin-V and miraxanthin-III), *Portulaca grandiflora* (portulacaxanthin-II and portulacaxanthin-III), etc.

19.3.2.2 Biosynthetic Pathway

The shikimate pathway produces betalains, which are then produced in the cytoplasm and endoplasmic reticulum. Betalains eventually become glycosides and are stored in the vacuole (Chen et al. 2017; Grotewold 2006). The mechanism of betalain production in plants is summarised in Fig. 19.8. Because just three essential enzymatic catalysts are needed for the production of betalains—tyrosine hydroxylation, DOPA oxidase, and DOPA-4,5-dioxygenase—it is less complicated than the synthesis of anthocyanins and carotenoids (Timoneda et al. 2019). Cytochrome P450 enzymes first convert tyrosine to L-DOPA, which is further changed by DOPA-OX into dopaquinone and finally cyclo-DOPA. Furthermore, L-DOPA is converted to 4,5-seco-DOPA by 4,5-dioxygenase, which then cyclises to yield betalamic acid. Betalamic acid serves as a central intermediate in the betalain

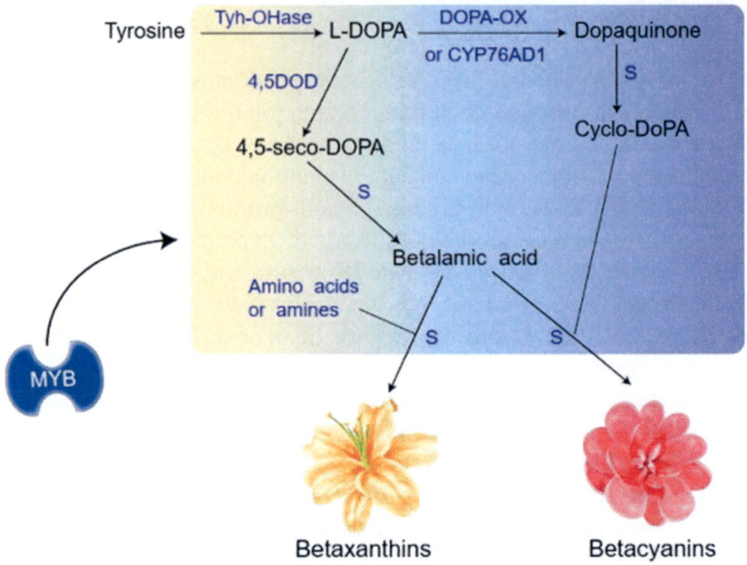

Fig. 19.8 Flow chart illustrating the biosynthesis pathway of betalains from tyrosine. Tyrosine is converted to L-DOPA by Tyh-OHase, then to dopaquinone by DOPA-OX or CYP76AD1. L-DOPA also converts to 4,5-seco-DOPA via 4,5DOD, leading to betalamic acid. Betalamic acid combines with amino acids or amines to form betaxanthins, depicted with a yellow flower, and with cyclo-DOPA to form betacyanins, shown with a pink flower. The MYB factor influences the pathway. (Figure reproduced from Zhao et al. (2022), Metabolites (MDPI), CC BY 4.0)

biosynthetic pathway. It reacts non-enzymatically with cyclo-DOPA to produce red-violet betacyanins or combines with amino acids and amines to generate yellow betaxanthins (Polturak and Aharoni 2019).

19.3.3 Chlorophyll

Phyllon, which means "leaf," and chloros, which means "green," are the Greek words from which the word chlorophyll is derived (Pareek et al. 2017). In 1817, French chemists Pierre-Joseph Pelletier and Joseph Bienaime Caventou first isolated and identified chlorophyll (Gopi et al. 2014). In addition to a magnesium ion in the middle, chlorophyll is composed of carbon and nitrogen atoms. The chlorophyll pigment reflects the green spectrum at wavelengths between 430 and 660 nm and absorbs blue and red solar radiation at the same wavelengths (İnanc 2011).

19.3.3.1 Classification of Chlorophyll

Chlorophyll can be classified into mainly two categories (Fig. 19.9). Chlorophyll a has the corresponding molecular formula $C_{55}H_{72}MgN_4O_5$. It is the primary pigment, blue-green in colour and absorbs light in the range of 430–660 nm. Chlorophyll b has the corresponding molecular formula $C_{55}H_{70}MgN_4O_6$. It is an accessory pigment, yellow-green in colour and absorbs light in the range of 450–650 nm.

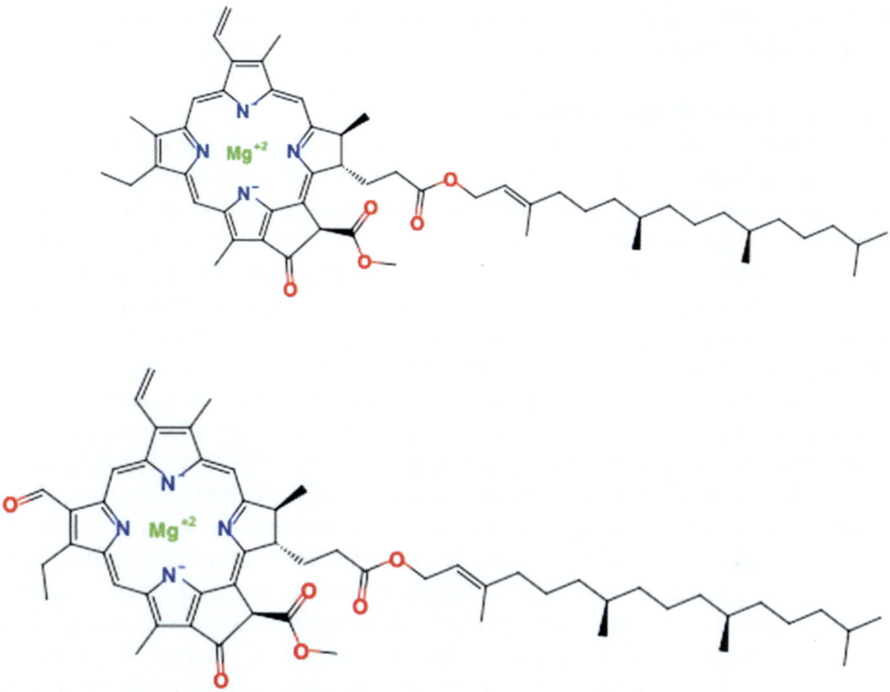

Fig. 19.9 Chemical structure diagram showing two similar complex organic molecules. Each molecule features a central ring with alternating single and double bonds, containing nitrogen atoms and a magnesium ion (Mg^{2+}). Attached to the ring are long hydrocarbon chains with various functional groups, including oxygen atoms. The structures are depicted with lines and letters representing bonds and elements

Numerous studies have investigated the genetic regulation of chlorophyll accumulation in floral tissues. In Rosa chinensis cv. "Viridiflora," the phyllody phenotype has been attributed to altered expression of floral organ identity genes (Yan et al. 2016). Abnormal transformation of floral organs into leaf-like structures is associated with the down-regulation of RcPISTILLATA, RcSEPALLATA3, and RcAPETALA3, which likely leads to the loss of petal identity. Moreover, ectopic expression of RcAPETALA2 has been linked with the loss of stamen and pistil identity, further contributing to leaf-like organ development (Yan et al. 2016).

To elucidate the factors influencing chlorophyll content in petals of Dianthus caryophyllus, Ohmiya et al. (2014) compared the expression of chlorophyll metabolism-related genes in leaves (high chlorophyll content), pale-green petals (low chlorophyll content), and white petals (very low chlorophyll content). Their findings indicated that non-green petals exhibited reduced expression of chlorophyll biosynthetic genes, particularly those encoding Mg-chelatase, compared with leaves. Conversely, genes involved in chlorophyll degradation, such as STAY-GREEN and pheophytinase, were more highly expressed in non-green petals. The authors concluded that the lack of chlorophyll in carnation petals is likely due to both diminished biosynthesis and enhanced degradation (Ohmiya et al. 2014).

19.3.3.2 Biosynthetic Pathway

The biosynthesis of chlorophyll occurs in several stages, involving various enzymes and intermediates (Fig. 19.10). Biosynthesis occurs in multiple stages. The italicised genes encode enzymes involved in chlorophyll biosynthesis and degradation, including glutamyl-tRNA reductase (1), glutamate-1-semialdehyde 2,1-aminotransferase (2), 5-aminolevulinate dehydrogenase (3), porphobilinogen deaminase (4), uroporphyrinogen III synthase (5), uroporphyrinogen III decarboxylase (6), coproporphyrinogen III oxidase (7), protoporphyrinogen oxidase (8), Mg-chelatase (9), Mg-protoporphyrin IX methyltransferase (10), Mg-protoporphyrin IX monomethyl ester cyclase (11), protochlorophyllide oxidoreductase (12), divinyl chlorophyllide an 8-vinyl reductase (13), chlorophyll synthase (14), geranylgeranyl diphosphate reductase (15), chlorophyllide an oxygenase (16), chlorophyll b reductase (17), hydroxymethyl chlorophyll a reductase (18), chlorophyllase (19); pheophytinase (20), pheophorbide a oxygenase (21), and red chlorophyll catabolite reductase (22). STAY-GREEN (SGR) is linked to the breakdown of chlorophyll, whereas MCS indicates a material that chelates metals. The production of chlorophyll starts with the formation of 5-aminolevulinic acid (ALA), an amino acid, from glutamate. The ALA synthesis route is another name for this phase. According to Wu et al. (2019), the enzyme glutamyl-tRNA synthetase is involved in the process of turning glutamate into ALA. The second stage involves the transformation of ALA into protoporphyrin IX, a cyclic tetrapyrrole molecule created through a sequence of enzyme processes. The synthesis of porphyrins is a multi-enzyme and multi-intermediate process. Porphobilinogen deaminase, which catalyses the production of uroporphyrinogen III, is one important enzyme implicated (Beale 1999). Magnesium protoporphyrin IX (MgP) is created by adding magnesium to protoporphyrin IX in the following procedure. The magnesium chelatase enzyme facilitates this process. The subsequent transformation of MgP into chlorophyllide is followed by the addition of a phytyl tail to produce chlorophyll a. Chlorophyll b and chlorophyll c are two more types of chlorophyll that can be created by additional changes (Tanaka and Tanaka 2011). When chlorophyll molecules are created, they bind to particular proteins known as apoproteins or apoprotein-binding proteins. In the photosynthetic apparatus, such as the reaction centre complexes in thylakoid membranes, these proteins aid in the integration of chlorophyll (Kouril et al. 2005).

19.3.4 Carotenoids

The name carrot (*Daucus carota*) served as the basis for the term “carotenoids” (Wackenroder 2009). In nature, carotenoids are widely dispersed. According to Johnson et al. (2018), they are found in lipid membranes and are entirely hydrophobic (fat-soluble). In addition to giving flowers colour, carotenoids safeguard the photosynthesis process from damage and provide the necessary substrates required

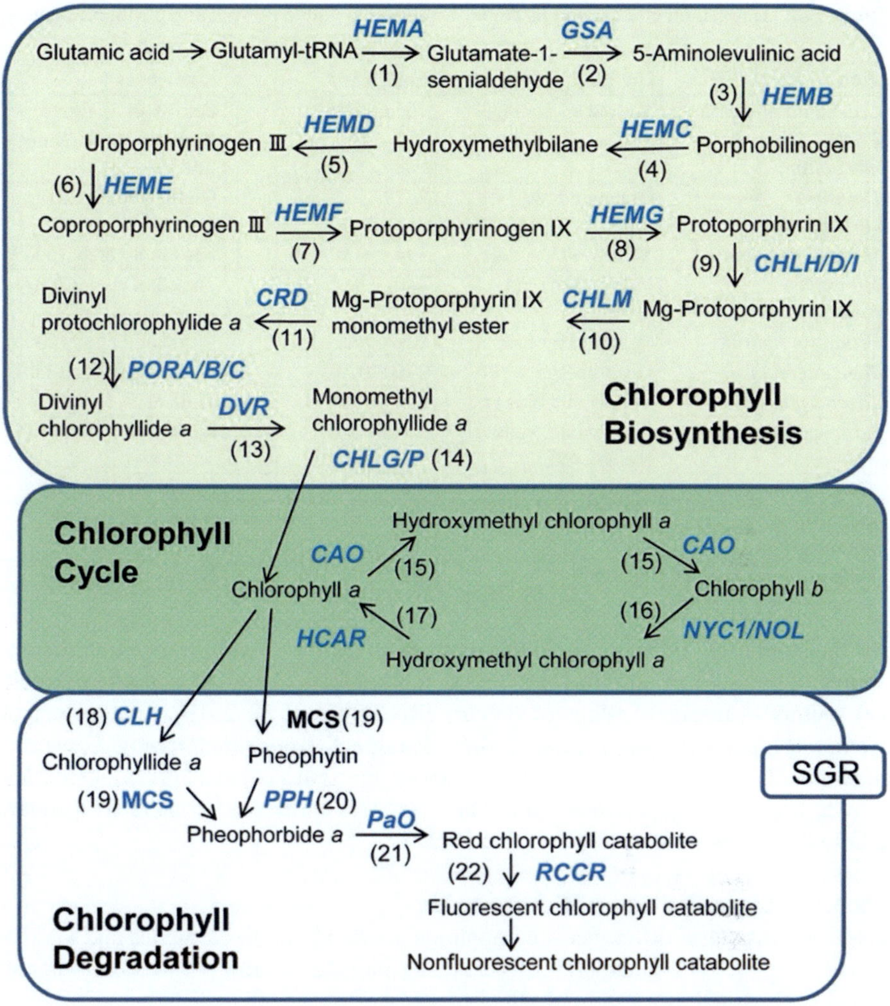

Fig. 19.10 Flowchart illustrating the process of chlorophyll metabolism, divided into three sections: Chlorophyll Biosynthesis, Chlorophyll Cycle, and Chlorophyll Degradation. The biosynthesis section shows the conversion of glutamic acid to divinyl chlorophyllide a through a series of steps involving various enzymes like HEMA, HEMD, and CHLG/P. The cycle section details the conversion between chlorophyll a and b, involving enzymes such as CAO and HCAR. The degradation section outlines the breakdown of chlorophyll a to nonfluorescent chlorophyll catabolite, with steps involving enzymes like CLH and RCCR. Each step is numbered and involves specific chemical transformations. (Figure reproduced from Ohmiya et al. (2014). Licensed under CC BY 4.0)

Table 19.2 Carotenoid content and its colour for different edible flowers along with references

Flower	Colour	Total carotenoids (μg/g FW)	References
Calendula officinalis	Yellow and orange	1073–1696	Petrova et al. (2016)
Chrysanthemum morifolium	Yellow and orange	122–343	Kishimoto and Ohmiya (2006)
Dianthus caryophyllus	Green and red	2–12	Ohmiya et al. (2013)
Helianthus annuus	Yellow and orange	144–1600	Kishimoto et al. (2007)
Hibiscus rosa-sinensis	Yellow, red, and orange	140–270	Trivellini et al. (2007)
Petunia × *hybrida*	Solid or bicolour	0.32–96.8	Murakami et al. (2003)
Tagetes erecta	Deep orange	270–2020	Kishimoto et al. (2007)
Tagetes patula	Yellow and orange	6.3–1304	Tinoi et al. (2006)
Rosa spp.	Pink, white, yellow, orange, red	0.1–61.7	Prata et al. (2017)
Antirrhinum majus	Orange, yellow	29	Barrio et al. (2018)
Tropaeolum pentaphyllum	Orange	396	Bona et al. (2017)
Viola × *wittrockiana*	Yellow	146	Barrio et al. (2018)

for the synthesis of abscisic acid (Swapnil et al. 2021). A well-known commercial source of natural carotenoids for the food, animal feed, chemical, and pharmaceutical sectors is marigold (*Tagetes erecta*) (Rodrigues et al. 2019). Others include *Gentiana lutea*, *Chrysanthemum* "Yellow Paragon," *Tropaeolum majus*, *Lotus corniculatus*, *Lilium bulbiferum*, *Narcissus pseudonarcissus*, *Crocus biflorus*, *Oncidium* Gower Ramsey hybrid, *Rosa* "Sun city," etc. The carotenoid content of different edible flowers is summarised in Table 19.2.

19.3.4.1 Classification of Carotenoids

Based on structural differences, carotenoids are divided into carotenes and xanthophylls (Fig. 19.11). Examples of carotenes include α- and β-carotene, whereas lutein is classified as a xanthophyll.

19.3.4.2 Biosynthetic Pathway

The carotenoid biosynthetic pathway (Fig. 19.12) involves several key enzymes and intermediates, including methylerythritol 4-phosphate (MEP) and geranylgeranyl pyrophosphate (GGPP). Major enzymes participating in this pathway are phytoene synthase (PSY), phytoene desaturase (PDS), ζ-carotene desaturase (ZDS), carotenoid isomerase (CRTISO), lycopene β-cyclase (LYCB), and lycopene ε-cyclase (LYCE). Further hydroxylation steps are catalysed by β-carotene hydroxylase (HYDB, also referred to as BCH) and by heme-containing cytochrome P450 enzymes, CYP97A and CYP97C, which function as carotene β- and ε-ring hydroxylases.

Carotenoid biosynthesis is initiated by the condensation of geranylgeranyl pyrophosphate (GGPP), which is derived from the upstream methylerythritol 4-phosphate

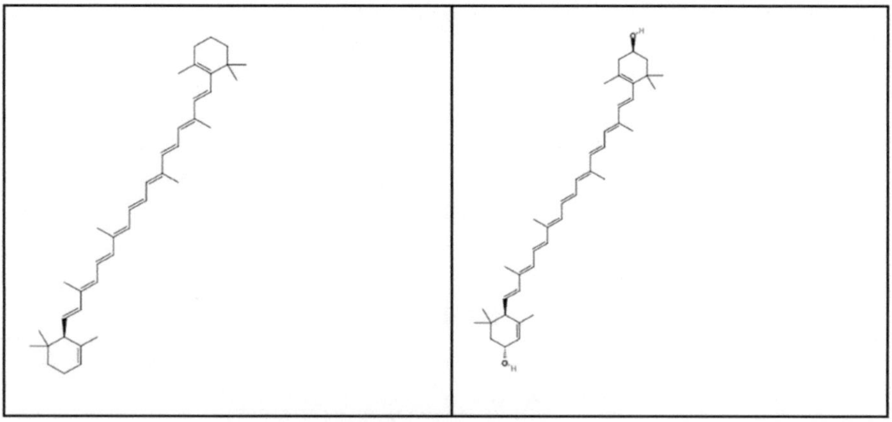

Fig. 19.11 Two-panel image showing chemical structures. The left panel displays a long hydrocarbon chain with alternating single and double bonds, ending in a cyclohexene ring with methyl groups. The right panel shows a similar structure with an additional hydroxyl group attached to the cyclohexene ring

(MEP) pathway. The overall metabolic flux of this pathway is strongly influenced by the availability of precursors supplied by the MEP route, making it a critical regulatory point (Rodriguez-Concepcion 2010). Supporting this, Suwarno et al. (2015) reported a positive association between carotenoid accumulation and the expression levels of key MEP pathway genes, including 1-deoxy-D-xylulose-5-phosphate synthase (DXS), 1-deoxy-D-xylulose-5-phosphate reductoisomerase (DXR), and 1-hydroxy-2-methyl-2(E)-butenyl 4-diphosphate reductase (HDR). The condensation of GGPP leads to the formation of the first committed carotenoid intermediate, C_{40} 15-cis-phytoene (Fig. 19.1).

This step represents the primary rate-limiting reaction in carotenoid biosynthesis within the endosperm and is catalysed by phytoene synthase (PSY) (Zhai et al. 2016). Subsequently, phytoene undergoes a series of four desaturation reactions mediated by phytoene desaturase (PDS) and ζ-carotene desaturase (ZDS), resulting in the formation of lycopene. During this process, carotenoid isomerase (CRTISO) converts cis-configured intermediates into their all-trans forms. In the final cyclisation steps, lycopene is transformed into ε-carotene and β-carotene through the action of lycopene ε-cyclase (LCYE) and lycopene β-cyclase (LCYB), respectively.

The double hydroxylation of α-carotene and β-carotene is catalysed by carotenoid hydroxylase enzymes specific to the β-ring (HYDB or CYP97A) and the ε-ring (CYP97C), which results in the synthesis of the xanthophylls lutein and zeaxanthin, respectively. The β-β/β-ε branch ratio is significantly modulated by the cyclisation of lycopene. Enhanced expression of the LCYB gene redirects carotenoid biosynthesis towards the β–β branch, whereas increased expression of LCYE shifts metabolic flux towards the alternative pathway (Cazzonelli 2011).

Floral coloration results from the interaction of three principal pigment groups—flavonoids, carotenoids, and betalains. Anthocyanins such as pelargonidin,

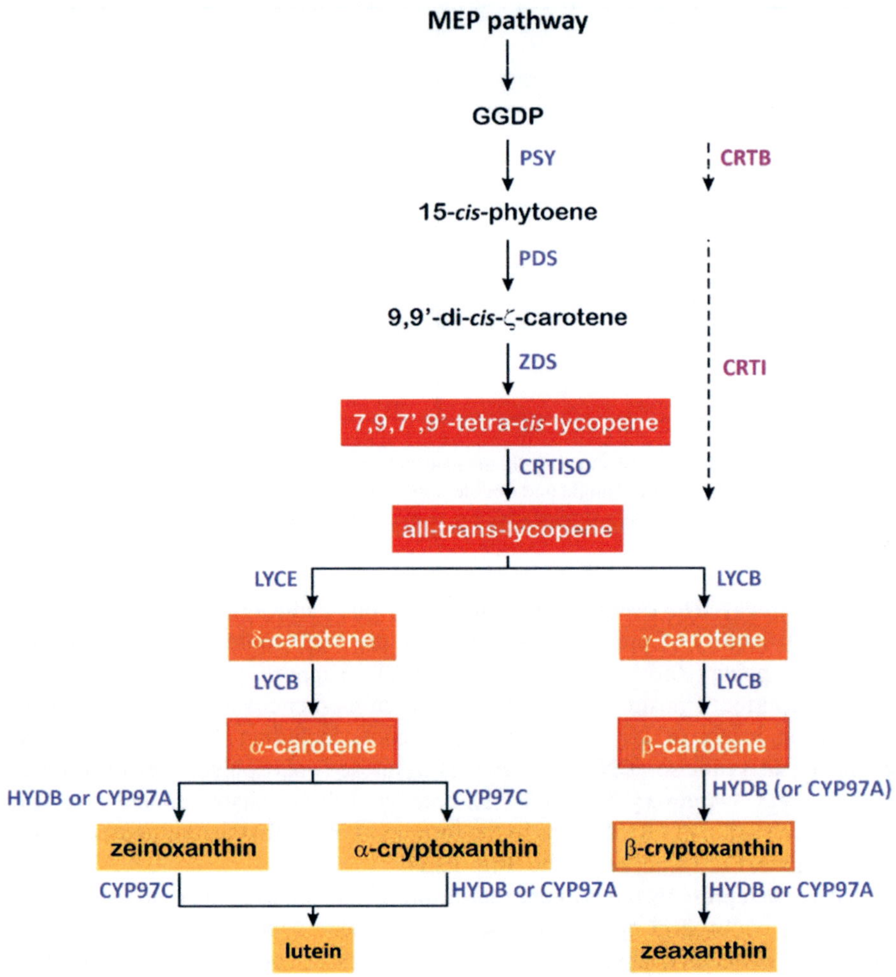

Fig. 19.12 Flow chart illustrating the MEP pathway for carotenoid biosynthesis. It begins with GGDP, leading to 15-cis-phytoene via PSY. Subsequent steps involve phytoene desaturase (PDS) converting it to 9,9′-di-cis-ζ-carotene, then ZDS to 7,9,7′,9′-tetra-cis-lycopene, and CRTISO to all-trans-lycopene. The pathway branches into δ-carotene and γ-carotene, further converting to α-carotene and β-carotene, respectively. Enzymes like LYCE, LYCB, HYDB, CYP97A, and CYP97C facilitate conversions to zeinoxanthin, α-cryptoxanthin, lutein, and zeaxanthin CRTB and CRTI are alternative pathways. (Adapted from Trono (2019), Plants, MDPI, licensed under CC BY 4.0)

cyanidin, and delphinidin predominantly govern red, pink, blue, and purple hues. Carotenoids are mainly responsible for yellow to orange shades, while betalains contribute to red-violet (betacyanins) and yellow (betaxanthins) colours in specific plant taxa. In several species, colour expression arises from co-pigmentation or the combined accumulation of multiple pigment classes (Samanta et al. 2011).

19.4 Extraction of Flower Pigments

Numerous techniques can be used to extract pigments from plant materials. To remove raw pigments, conventional techniques are frequently utilised. Since they use less solvent, take less time to extract, and are more environmentally friendly, non-conventional extraction methods or green extraction techniques have recently replaced conventional extraction methods (Mohammad-Azmin et al. 2022). The selection of an appropriate method is determined by the heat sensitivity of the pigment to be extracted (anthocyanins have high to moderate heat stability, betalains have moderate to low heat stability, and carotenoids have moderate heat stability).

19.4.1 Conventional Methods

19.4.1.1 Maceration

It is a widely used method that can be performed either at ambient temperature or with the application of mild external heat. This technique is particularly suitable for isolating heat-sensitive pigments. In this process, plant material is finely ground and mixed with an appropriate solvent to facilitate the release of crude pigments. The mixture is then kept in an extraction vessel and intermittently agitated through shaking or stirring to enhance solvent penetration. After completion of the extraction, the liquid extract is separated by filtration, while the remaining solid residue may be pressed mechanically or centrifuged to recover additional extract (Ngamwonglumlert et al. 2017).

19.4.1.2 Soxhlet Extraction

In Soxhlet extraction, target compounds are isolated by continuously washing the sample with condensed solvent vapours. The powdered plant material, placed in a thimble, comes into direct contact with the condensed solvent, allowing soluble constituents to dissolve into the extracting medium. This process involves repeated cycles in which the solvent vapour condenses and percolates through the sample until extraction is complete. Completion is typically indicated when the solvent siphoning back into the flask becomes colourless, signifying that no further compounds are being solubilised (Gopalasatheeskumar 2018).

19.4.2 Unconventional Methods

19.4.2.1 Supercritical Fluid Extraction

This method exploits the unique physicochemical characteristics of supercritical fluids, which combine gas-like diffusivity with liquid-like solvating ability, thereby enhancing extraction efficiency. Carbon dioxide (CO_2) is commonly regarded as an effective supercritical solvent because its transition point to the supercritical phase occurs at approximately 31 °C and 74 bar. CO_2 exhibits high diffusivity and low viscosity in its supercritical state, comparable to gases, while maintaining strong solvation capacity similar to liquids. Typically, SFE using CO_2 is performed within pressure ranges of 8–40 MPa and temperatures between 30 and 60 °C. These mild operating conditions make the technique particularly suitable for isolating heat-sensitive pigments. Moreover, as supercritical CO_2 is inherently non-polar, SFE is especially effective for the extraction of low-polarity pigments such as carotenoids as well as chlorophylls (Khaw et al. 2017).

19.4.2.2 Microwave-Assisted Extraction

The electromagnetic radiation is used in this process. The solvent's polar molecules absorb this energy, leading to the reorientation of polar molecules and the movement of charged species. By vapourising the water in the matrix and creating high pressure in the cell wall, microwaves specifically affect plant cells. This effect generates heat, alters the cell wall's physical characteristics, and ultimately results in cell wall rupture. As a result, the solvent enters the plant cell more readily, favouring the mass transfer of the solvent from the cell (Akhtar et al. 2019).

19.4.2.3 Pulsed Electric Field Extraction

Most extraction media used for betalains and anthocyanins are polar solvents that possess electrical conductivity, enabling the passage of electric current through the sample matrix. In contrast, non-polar solvents exhibit high electrical resistance and negligible conductivity, which prevents the transmission of an electric field through such media. Therefore, compared to chlorophylls and carotenoids, betalains and anthocyanins are better extracted using PEF (Azmir et al. 2013).

19.4.2.4 Enzyme-Assisted Extraction

Enzymes are used in enzyme-assisted extraction to speed up the removal of pigments from plant material. In order to hydrolyse plant cellulose cell walls and assist extraction, a variety of enzymes are utilised, including pectinase, cellulase, and hemicellulase (Marathe et al. 2017).

19.5 Commercial Uses of Flower Pigments

19.5.1 Food Industry

Colour is the single most important indicator of food quality because it can either increase or decrease appetite (Schlintl and Schienle 2020). Food colouring increases the visual appeal, flavour appeal, and recognizability of food. Increasing concern for the harmful effects associated with unnatural colourants has enhanced the global preference for natural alternatives (Juric et al. 2022; Lu et al. 2021). This shift is largely attributed to the recognised therapeutic and medicinal properties of natural pigments. Regular intake of these colourants has been associated with a reduced incidence of several human diseases (Ghosh et al. 2021). Researchers and consumers worldwide are increasingly paying close attention to pigments, which are intriguing functional components.

Pigments are sensitive to environmental and processing obstacles, have low bioavailability, and have low water solubility. These are only a few of the characteristics that prevent pigments from being used in food. In order to completely replace synthetic colourants, natural pigments will eventually be developed in response to the growing customer desire for healthier, more natural foods. The commercial development of natural pigments is hampered by several factors, viz., the narrow spectrum of natural colours permitted for use in food, the higher price, the lengthy regulatory approval process for new colourants, and the significant amounts of raw material needed for extraction (Ghosh et al. 2022). Rising consumer preference for naturally derived pigments over synthetic colourants is fuelling rapid expansion in the food industry focused on the development and production of natural food colourants.

The global market for natural pigments is supported by numerous companies operating across different regions of the world, particularly in Asia, Europe, and North America. In the Asian region, China hosts several key manufacturers such as Anhui Uherbsol Biotechnology Co., Ltd., Shanghai Zhiqi Biotechnology Co., Ltd., and Chenguang Biotech Group Co., Ltd., while India contributes through companies like Aarkay Food Products and ITC Colours. Europe is represented by firms such as Allied Biotech Corp., located in Denmark. The United States accounts for a large share of industrial activity in this sector, with several well-established companies including Chr. Hansen Inc., DDW "The Color House," GNT Group USA Inc., Sensient Flavours & Technologies, Kalsec Inc., LycoRed Corp., ROHA Food Colors USA, DSM, FMC Corporation, and Naturex Inc. Overall, the presence of these enterprises highlights the broad international engagement and growing commercial significance of natural pigment production and trade (Cortez et al. 2017).

19.5.2 Cosmetic Industry

The market for natural colours in the form of pigments used in cosmetics has become more popular due to the negative effects of continuous use of synthetic colouring compounds (Brudzynska et al. 2021). To add colour to the face, lips, hair, etc., dermatological products are formulated using plant-derived pigments. Plant pigments have been usefully used in cosmetics due to the antibacterial characteristics and radical scavenging activities (Boo et al. 2012). In order to create an insoluble colouring compound and obtain a new cosmetic alternative, Chanel Cosmetics collaborated with CRITT Horticole of Rochefort. The majority of plant-based extracts that contain varied colouring agents are generally soluble dyes; however, Chanel and CRITT produced insoluble pigments required for cosmetic formulations by binding naturally derived pigments from plants onto a mineral-based carrier material. Using kaolin minerals, sappan wood was selected among various plant samples that give it its red hue. The decision was affected by a number of variables, including colour intensity, spectrocolorimetric readings, ease of distribution within a powdered base, safety profile based on toxicity assessments, consistency in coverage during use, and compatibility with oil-based systems. Intended for use in products such as lip colour formulations, shine-enhancing lip preparations, eye makeup powders, and cheek-enhancing cosmetics (Brudzynska et al. 2021).

19.5.3 Animal Feed

Numerous impacts of plant pigments in animal nutrition are already well-known, such as the antioxidant capabilities of polyphenolic flavones and flavonols, the appetite-stimulating colours of anthocyanins, and the stable colours of animal products provided by carotenoids due to their fat solubility. When given the option to select their own fodder plants, ruminants were found to balance the selection of plants based on secondary plant metabolites and their own personal health requirements. "Self-medication" is the term used to characterise this behaviour. The potential antioxidative capabilities of betalains and carotenoids, for example, are among the intended benefits of feeding plant pigments that have not yet been fully realised (Faehnrich et al. 2016).

Pigments added to feed are typically used to improve the aesthetics and nutritional value of animal feed. Carotenoids play the most significant role among all feed pigments and control the feed pigments market (Langi et al. 2018). Animals must obtain their carotenoids from their diet because they are unable to synthesise them. The carotenoid pigment is frequently used in the poultry and fish industries to improve the colour of egg yolks, chicken skin, and fish skin (Grashorn 2016; Oehlenschlager and Ostermeyer 2016).

19.5.4 Pharmaceutical Industry

In the pharmaceutical industry, the chemical composition of tablets, tablet coatings, capsules, liquid orals, toothpaste, and ointments is usually coloured to enhance visual appeal, extend durability, offer standard preparations, or serve as a means of identification (Salawi 2022). The active ingredients in natural pigments are used to meet medical demands. For example, carotenoids are used as vitamin supplements (Elvira-Torales et al. 2019), bixin is used as one of the natural colouring agents for ointments (Vilar et al. 2014), and chlorophyll, carotenoids, and anthocyanins are the main ingredients in plasters and film coating for tablets (Mohd-Nasir et al. 2018).

19.5.5 Green Energy

Dye-sensitised solar cell (DSSC), commonly referred to as Gratzel cells, represents an innovative class of photovoltaic devices. DSSCs offer several advantages, including low manufacturing costs, rapid energy payback, structural flexibility, effective operation under diffuse light conditions, and the ability to produce multicoloured devices. These features make them particularly attractive, especially because a wide variety of sensitisers—both synthetic and natural—can be used as light-harvesting agents responsible for generating charge carriers (Jamalullail et al. 2017). In DSSCs, charge separation and transport occur through a wide band-gap semiconductor, most commonly titanium dioxide (TiO_2). Functionally, DSSCs resemble natural photosynthetic systems. For natural dyes to be effective in DSSCs, they must exhibit strong light absorption, environmental sustainability, low cost, and ease of disposal. Among the natural pigments explored, carotenoids, betalains, flavonoids, and chlorophylls have shown considerable potential (Hug et al. 2014).

19.5.6 Fabric Dyeing

Natural dyes are widely used to colour most natural textiles, as well as some synthetic fibres (Bechtold et al. 2003). However, the main problems with naturally dyed textiles are shade repeatability and the shade's transient performance in the presence of light and water, as well as the absence of a clear standard application technique (Affat 2021). Achieving good colour fastness against light and washing is a difficulty for the dyer.

Even if a number of researchers had suggested different dyeing techniques and process parameters, this information was still insufficient, which is why a study is needed to produce a standard dye extraction methodology and standardise the complete process of natural dyeing on textiles (Samanta et al. 2019). 50 of the almost 450 species that are known to produce dyes in India alone are thought to be the most significant (Siva 2007).

19.5.7 Holi Colours

Following the introduction of artificial colours in the mid-1800s, natural colours were forgotten and rejected as part of history (Krizova 2015). Because of their lower costs, simple manufacturing processes, wide variety of vibrant and attractive colours, and long shelf lives, synthetic colours have gained popularity in the market. However, the majority of the synthetic Holi colours currently on the market are made of dangerous chemicals that can lead to skin allergies and are also carcinogenic. In addition to damaging skin and hair, synthetic colours pollute the environment (Ardila-Leal et al. 2021).

According to a recent study by the Lucknow-based Industrial Toxicology Research Centre, the popular red pigment used for Holi contains Rhodamine B, a dye whose usage in cosmetics and food colouring has been outlawed in various countries due to its carcinogenic qualities. Marigolds, roses, carnations, garden poppies, and flame of the forest (Butea monosperma) are examples of flowering plants that are used to prepare herbal Holi colours (Birsanta 2022).

19.6 Conclusion

Flowers' brilliant colours and ability to draw pollinators are both greatly influenced by their pigment content. Carotenoids, flavonoids, and anthocyanins are a few of the major pigments in charge of the vast spectrum of shades seen in flowers. Due to their vivid hues and varied chemical compositions, floral pigments have proven to be useful resources for a variety of enterprises. These pigments can be obtained and used in several commercial applications through conventional and non-conventional extraction procedures. Flower pigments provide an environmentally beneficial alternative to synthetic compounds for anything from natural cosmetic and textile colours to food colourants and pharmaceutical components. In addition to helping to preserve floral biodiversity, their commercialisation and extraction open up new potential for the creation of environmentally friendly and cutting-edge goods.

As we continue to explore the potential in order to maximise the commercial feasibility of flower pigments and build a more vivid and sustainable future, additional study and improvements in extraction methods are required as we continue to explore their possibilities of flower pigments, further research and advancements in extraction techniques are necessary to maximise their commercial viability and create a more vibrant and sustainable future.

References

Affat SS (2021) Classifications, advantages, disadvantages, toxicity effects of natural and synthetic dyes: a review. Univ Thi-Qar J Sci 8(1):130–135

Akhtar I, Javad S, Yousaf Z, Iqbal S, Jabeen K (2019) Microwave assisted extraction of phytochemicals an efficient and modern approach for botanicals and pharmaceuticals. Pak J Pharma Sci 32(1):223–230

Alappat B, Alappat J (2020) Anthocyanin pigments: beyond aesthetics. Molecules 25(23):1–17

Ardila-Leal LD, Poutou-Piñales RA, Pedroza-Rodríguez AM, Quevedo-Hidalgo BE (2021) A brief history of colour, the environmental impact of synthetic dyes and removal by using laccases. Molecules 26(13):3813

Azmir J, Zaidul ISM, Rahman MM, Sharif KM, Mohamed A, Sahena F, Jahurul MHA, Ghafoor K, Norulaini NAN, Omar AKM (2013) Techniques for extraction of bioactive compounds from plant materials: a review. J Food Eng 117:426–436

Barrio GR, Periago MJ, Luna-Recio C, Javier GAF, Navarro-González I (2018) Chemical composition of the edible flowers, pansy (*Viola × wittrockiana*) and snapdragon (*Antirrhinum majus*) as new sources of bioactive compounds. Food Chem 252:373–380

Bauernfeind JC (1981) Natural food colors. In: Bauernfeind JC (ed) Carotenoids as colorants and vitamin A precursors. Academic Press, New York

Beale SI (1999) Enzymes of chlorophyll biosynthesis. Photosynth Res 60:43–73

Bechtold T, Turcanu A, Ganglberger E, Geissler S (2003) Natural dyes in modern textile dyehouses-how to combine experiences of two centuries to meet the demands of the future? J Clean Prod 11(5):499–509

Benvenuti S, Bortolotti E, Maggini R (2016) Antioxidant power, anthocyanin content and organoleptic performance of edible flowers. Sci Hortic 199:170–177

Birsanta G (2022) Studies on the preparation of herbal gulal. MSc thesis, Dr. YS Parmar University of Horticulture and Forestry Nauni, Solan, 65p

Bona GS, Boschetti W, Bortolin RC, Vale MGR, Moreira JCF, Rios OA, Flôres SH (2017) Characterization of dietary constituents and antioxidant capacity of *Tropaeolum pentaphyllum* Lam. J Food Sci Technol 54:3587–3597

Boo HO, Shin JS, Hwang SJ, Bae CS, Park SH (2012) Antimicrobial effects and antioxidative activities of the cosmetic composition having natural plant pigments. Korean J Plant Res 25(1):80–88

Brudzynska P, Sionkowska A, Grisel M (2021) Plant-derived colorants for food, cosmetic and textile industries: a review. Materials 14(13):1–18

Cai YZ, Sun M, Corke H (2005) Characterization and application of betalain pigments from plants of the Amaranthaceae. Trends Food Sci Technol 16:370–376

Cazzonelli CI (2011) Carotenoids in nature: insights from plants and beyond. Funct Plant Biol 38:833–847

Chaves-Silva S, dos Santos AL, Chalfun A, Zhao J, Peres LEP, Benedito VA (2018) Understanding the genetic regulation of anthocyanin biosynthesis in plants-tools for breeding purple varieties of fruits and vegetables. Phytochemistry 153:11–27. https://doi.org/10.1016/j.phytochem.2018.05.013

Chen N, Yu ZH, Xiao XG (2017) Cytosolic and nuclear co-localization of betalain biosynthetic enzymes in tobacco suggests that betalains are synthesized in the cytoplasm and/or nucleus of betalainic plant cells. Front Plant Sci 8:831

Cheok CY, Salman HA, Sulaiman R (2014) Extraction and quantification of saponins: a review. Food Res Int 59:16–40

Cortez R, Luna-Vital DA, Margulis D, Gonzalez de Mejia E (2017) Natural pigments: stabilization methods of anthocyanins for food applications. Compr Rev Food Sci Food Saf 16(1):180–198

Delgado-Vargas F, Paredes-Lopez O (2003) Natural colorants for food and nutraceutical uses. CRC Press, Boca Raton

Dias MC, Pinto DC, Silva A (2021) Plant flavonoids: chemical characteristics and biological activity. Molecules 26(17):5377

Elvira-Torales LI, García-Alonso J, Periago-Castón MJ (2019) Nutritional importance of carotenoids and their effect on liver health: a review. Antioxidants 8(7):229

Enaru B, Dretcanu G, Pop TD, Stanila A, Diaconeasa Z (2021) Anthocyanins: factors affecting their stability and degradation. Antioxidants 10(12):1–24

Esatbeyoglu T, Wagner AE, Schini-Kerth VB, Rimbach G (2015) Betanin--a food colorant with biological activity. Mol Nutr Food Res 59(1):36–47

Faehnrich B, Lukas B, Humer E, Zebeli Q (2016) Phytogenic pigments in animal nutrition: potentials and risks. J Sci Food Agric 96(5):1420–1430

Fedenko VS, Landi M, Shemet SA (2022) Metallophenolomics: a novel integrated approach to study complexation of plant phenolics with metal/metalloid ions. Int J Mol Sci 23(19):1–40

Forkmann G (1991) Flavonoids as flower pigments: the formation of the natural spectrum and its extension by genetic engineering. Plant Breed 106:1–26

Francis FJ (1996) Safety of food colorants. In: Hendry GAF, Houghton JD (eds) Natural food colorants, 2nd edn. Chapman & Hall, New York

Frick D, Meggos H (1988) FD&C colors. Food Technol 7:49–56

Ghosh S, Sarkar T, Das A, Chakraborty R (2021) Micro and nanoencapsulation of natural colors: a holistic view. Appl Biochem Biotechnol 193:3787–3811

Ghosh S, Sarkar T, Das A, Chakraborty R (2022) Natural colorants from plant pigments and their encapsulation: an emerging window for the food industry. LWT 153:112527

Gopalasatheeskumar K (2018) Significant role of soxhlet extraction process in phytochemical research. MJPMS 7(1):43–47

Gopi S, Varma K, George R (2014) A short review on the medicinal properties of chlorophyll juice. AJPTI 2(9):89–94

Grashorn M (2016) Feed additives for influencing chicken meat and egg yolk color. In: Handbook on natural pigments in food and beverages. Woodhead Publishing

Grotewold E (2006) The genetics and biochemistry of floral pigments. Annu Rev Plant Biol 57:761–780

Hari RK, Patel TR, Martin AM (1994) An overview of pigment production in biological systems: functions, biosynthesis and applications in food industry. Food Rev Int 10(1):49–70
Hug H, Bader M, Mair P, Glatzel T (2014) Bio-photovoltaics: natural pigments in dye-sensitized solar cells. Appl Energy 115:216–225
Humphrey AM (2004) Chlorophyll as a color and functional ingredient. J Food Sci 69(5):422–425
İnanc AL (2011) Chlorophyll: structural properties, health benefits and its occurrence in virgin olive oils. Acad Food J 9(2):6–32
Iwashina T (2013) Flavonoid properties of five families newly incorporated into the order Caryophyllales. Bull Natl Mus Nat Sci 39:25–51
Jamalullail N, Mohamad IS, Norizan MN, Baharum NA, Mahmed N (2017) Short review: natural pigments photosensitizer for dye-sensitized solar cell (DSSC). In: 2017 IEEE 15th student conference on research and development (SCOReD). IEEE, pp 344–349
Johnson QR, Mostofian B, Gomez GF, Smith JC, Cheng X (2018) Effects of carotenoids on lipid bilayers. Phys Chem Chem Phys 20(5):3795–3804
Juric S, Juric M, Krol-Kilinska Z, Vlahovicek-Kahlina K, Vincekovic M, Dragovic-Uzelac V, Donsi F (2022) Sources, stability, encapsulation and application of natural pigments in foods. Food Rev Int 38(8):1735–1790
Kanner J, Harel S, Granit R (2001) Betalains a new class of dietary cationized antioxidants. J Agric Food Chem 49(11):5178–5185
Khan MI (2016) Plant betalains: safety, antioxidant activity, clinical efficacy and bioavailability. Compr Rev Food Sci Food Saf 15(2):316–330
Khan MI, Giridhar P (2015) Plant betalains: chemistry and biochemistry. Phytochemistry 117:267–295
Khaw KY, Parat MO, Shaw PN, Falconer JR (2017) Solvent supercritical fluid technologies to extract bioactive compounds from natural sources: a review. Molecules 22(7):1–22
Khoo HE, Azlan A, Tang ST, Lim SM (2017) Anthocyanidins and anthocyanins: colored pigments as food, pharmaceutical ingredients and the potential health benefits. Food Nutr Res 61(1):1–22
Kishimoto S, Ohmiya A (2006) Regulation of carotenoid biosynthesis in petals and leaves of chrysanthemum (*Chrysanthemum morifolium*). Physiol Plant 128:436–447
Kishimoto S, Sumitomo K, Yagi M, Nakayama M, Ohmiya A (2007) Three routes to orange petal color via carotenoid components in 9 compositae species. J Jpn Soc Hortic Sci 76:250–257
Kitamura S, Shikazono N, Tanaka A (2004) TRANSPARENT TESTA 19 is involved in the accumulation of both anthocyanins and proanthocyanidins in Arabidopsis. Plant J 37:104–114
Kong JM, Chia LS, Koh NK, Chia TF, Brouillard R (2003) Analysis and biological activities of anthocyanins. Phytochemistry 64:923–933
Kouril R, Zygadlo A, Arteni AA, de Wit CD, Dekker JP, Jensen PE, Scheller HV, Boekema EJ (2005) Structural characterization of a complex of photosystem I and light-harvesting complex II of Arabidopsis thaliana. Biochemistry 44(33):10935–10940
Krizova H (2015) Natural dyes: their past, present, future and sustainability. In: Recent developments in fibrous material science. Kosmas Publishing, Prague
Kumari P, Panwar S, Namita T, Kaushal S, Ullas PS (2017) Pigment profiling of flower crops: a review. Ecol Environ Conserv 23(2):1000–1008
Landi M, Tattini M, Gould KS (2015) Multiple functional roles of anthocyanins in plant-environment interactions. Environ Exp Bot 119:4–17
Langi P, Kiokias S, Varzakas T, Proestos C (2018) Carotenoids: from plants to food and feed industries. In: Barreiro C, Barredo JL (eds) Microbial carotenoids, Methods in molecular biology. Humana Press, New York
Li G, Meng X, Zhu M, Li Z (2019a) Research progress of betalain in response to adverse stresses and evolutionary relationship compared with anthocyanin. Molecules 24(17):3078
Li M, Sun Y, Lu X, Debnath B, Mitra S, Qiu D (2019b) Proteomics reveal the profiles of color change in Brunfelsia acuminata flowers. Int J Mol Sci 20(8):1–14
López-Cruz R, Sandoval-Contreras T, Iñiguez-Moreno M (2023) Plant pigments: classification, extraction and challenge of their application in the food industry. Food Bioprocess Technol 15:1–77

Lu W, Shi Y, Wang R, Su D, Tang M, Liu Y, Li Z (2021) Antioxidant activity and healthy benefits of natural pigments: a review. Int J Mol Sci 22(9):1011–1028

Ma Y, Ma X, Gao X, Wu W, Zhou B (2021) Light induced regulation pathway of anthocyanin biosynthesis in plants. Int J Mol Sci 22(20):11116

Marathe SJ, Jadhav SB, Bankar SB, Singhal RS (2017) Enzyme-assisted extraction of bioactives. In: Puri M (ed) Food bioactives: extraction and biotechnology applications. Springer, Cham

Martín J, Navas MJ, Jiménez-Moreno AM, Asuero AG (2017) Anthocyanin pigments: importance, sample preparation and extraction. In: Marcos Soto-Hernandez M, Palma-Tenango M, Garcia-Mateos MDR (eds) Phenolic compounds-natural sources, importance and applications. IntechOpen, Croatia

Mattioli R, Francioso A, Mosca L, Silva P (2020) Anthocyanins: a comprehensive review of their chemical properties and health effects on cardiovascular and neurodegenerative diseases. Molecules 25(17):1–42

Mohammad-Azmin SNH, Sulaiman NS, Mat Nor MS, Abdullah PS, Abdul Kari Z, Pati SA (2022) Review on recent advances on natural plant pigments in foods: functions, extraction, importance and challenges. Appl Biochem Biotechnol 194(10):4655–4672

Mohd-Nasir H, Abd-Talib N, Mohd-Setapar SH, Wong LP, Idham Z, Chavez Casillas A, Ahmad A (2018) Natural colorants from plants for wellness industry. Int J Pharm Sci Res 9(3):836–843

Murakami Y, Fukui Y, Watanabe H, Kokubun H, Toya Y, Ando T (2003) Distribution of carotenoids in the flower of non-yellow commercial petunia. J Hortic Sci Biotechnol 78:127–130

Narbona E, Del-Valle JC, Arista M, Buide ML, Ortiz PL (2021) Major flower pigments originate different colours. Front Ecol Evol 9:1–14

Nassour R, Ayash A, Al-Tameemi K (2020) Anthocyanin pigments: structure and biological importance. J Chem Pharm Sci 13:45–57

Ngamwonglumlert L, Devahastin S, Chiewchan N (2017) Natural colorants: pigment stability and extraction yield enhancement via utilization of appropriate extraction methods. Crit Rev Food Sci Nutr 57(15):3243–3259

Oehlenschlager J, Ostermeyer U (2016) Feed additives for influencing the color of fish and crustaceans. In: Handbook on natural pigments in food and beverages. Woodhead Publishing

Ohmiya A, Tanase K, Hirashima M, Yamamizo C, Yagi M (2013) Analysis of carotenogenic gene expression in petals and leaves of carnation (*Dianthus caryophyllus* L.). Plant Breed 132:423–429

Ohmiya A, Hirashima M, Yagi M, Tanase K, Yamamizo C (2014) Identification of genes associated with chlorophyll accumulation in flower petals. PLoS One 9(12):1–16

Panche AN, Diwan AD, Chandra SR (2016) Flavonoids: an overview. J Nutr Sci 5:47–62

Pareek S, Sagar NA, Sharma S, Kumar V, Agarwal T, González-Aguilar GA, Yahia EM (2017) Chlorophylls: chemistry and biological functions. In: Yahia EM (ed) Fruit and vegetable phytochemicals: chemistry and human health, 2nd edn. John Wiley & Sons, Ltd., Hoboken

Passeri V, Koes R, Quattrocchio FM (2016) New challenges for the design of high value plant products: stabilization of anthocyanins in plant vacuoles. Front Plant Sci 7:153

Petrova I, Petkova N, Ivanov I (2016) Five edible flowers-valuable source of antioxidants in human nutrition. Int J Pharmacogn Phytochem Res 8:604–610

Polturak G, Aharoni A (2018) "La Vie en Rose": biosynthesis, sources and applications of betalain pigments. Mol Plant 11(1):7–22

Polturak G, Aharoni A (2019) Advances and future directions in betalain metabolic engineering. New Phytol 224(4):1472–1478

Poustka F, Irani NG, Feller A, Lu Y, Pourcel L, Frame K, Grotewold E (2007) A trafficking pathway for anthocyanins overlaps with the endoplasmic reticulum-to-vacuole protein-sorting route in Arabidopsis and contributes to the formation of vacuolar inclusions. Plant Physiol 145(4):1323–1335

Prata GGB, Souza KO, Lopes MMA, Oliveira LS, Aragao FAS, Alves RE, Silva SM (2017) Nutritional characterization, bioactive compounds and antioxidant activity of Brazilian roses (*Rosa* spp.). J Agric Sci Technol 19:929–941

Rodrigues DB, Mercadante AZ, Mariutti LR (2019) Marigold carotenoids: much more than lutein esters. Food Res Int 119:653–664
Rodriguez-Concepcion M (2010) Supply of precursors for carotenoid biosynthesis in plants. Arch Biochem Biophys 504:118–122
Routaboul JM, Dubos C, Beck G, Marquis C, Bidzinski P, Loudet O, Lepiniec L (2012) Metabolite profiling and quantitative genetics of natural variation for flavonoids in Arabidopsis. J Exp Bot 63:3749–3764
Saini RK, Keum YS (2018) Carotenoid extraction methods: a review of recent developments. Food Chem 240:90–103
Salawi A (2022) Pharmaceutical coating and its different approaches, a review. Polymers 14(16):3318
Samanta A, Das G, Das SK (2011) Roles of flavonoids in plants. Carbon 100(6):12–35
Samanta AK, Awwad N, Algarni HM (2019) Chemistry and technology of natural and synthetic dyes and pigments. IntechOpen, London
Santos EL, Maia BH, Ferriani AP, Teixeira SD (2017) Flavonoids: classification, biosynthesis and chemical ecology. In: Goncalo CJ (ed) Flavonoids-from biosynthesis to human health. IntechOpen, Croatia
Schlintl C, Schienle A (2020) Effects of coloring food images on the propensity to eat: a placebo approach with color suggestions. Front Psychol 11:2966
Siva R (2007) Status of natural dyes and dye-yielding plants in India. Curr Sci 10:916–925
Stintzing FC, Carle R (2004) Functional properties of anthocyanins and betalains in plants, food, and in human nutrition. Trends Food Sci Technol 15(1):19–38
Sullivan CN, Koski MH (2021) The effects of climate change on floral anthocyanin polymorphisms. Proc Biol Sci 288(1946):20202693
Suwarno WB, Pixley KV, Palacios-Rojas N, Kaeppler SM, Babu R (2015) Genome-wide association analysis reveals new targets for carotenoid biofortification in maize. Appl Genet 128:851–864
Swapnil P, Meena M, Singh SK, Dhuldhaj UP, Marwal A (2021) Vital roles of carotenoids in plants and humans to deteriorate stress with its structure, biosynthesis, metabolic engineering and functional aspects. Curr Plant Biol 26:1–11
Tanaka R, Tanaka A (2011) Chlorophyll cycle regulates the construction and destruction of the light-harvesting complexes. Biochim Biophys Acta Bioenerg 1807(8):968–976
Tanaka Y, Sasaki N, Ohmiya A (2008) Biosynthesis of plant pigments: anthocyanins, betalains and carotenoids. Plant J 54(4):733–749
Timoneda A, Feng T, Sheehan H, Walker-Hale N, Pucker B, Lopez-Nieves S, Guo R, Brockington S (2019) The evolution of betalain biosynthesis in Caryophyllales. New Phytol 224(1):71–85
Tinoi J, Rakariyatham N, Deming RL (2006) Determination of major carotenoid constituents in petal extracts of eight selected flowering plants in the north of Thailand. Chiang Mai J Sci 33:327–334
Tripathy BC, Pattanayak GK (2012) Chlorophyll biosynthesis in higher plants. In: Eaton-Rye J, Tripathy B, Sharkey T (eds) Photosynthesis: advances in photosynthesis and respiration. Springer International Publishing, Cham
Trivellini DA, Vernieri P, Ferrante A, Serra G (2007) Physiological characterization of flower senescence in long life and ephemeral Hibiscus (*Hibiscus rosa-sinensis* L.). Acta Hortic 755:457–464
Trono D (2019) Carotenoids in cereal food crops: composition and retention throughout grain storage and food processing. Plants 8(12):551
Usman I, Hussain M, Imran A, Afzaal M, Saeed F, Javed M, Afzal A, Ashfaq I, Al Jbawi E, Saewan S (2022) Traditional and innovative approaches for the extraction of bioactive compounds. Int J Food 25(1):1215–1233
Van-Der-Kooi CJ, Dyer AG, Kevan PG, Lunau K (2019) Functional significance of the optical properties of flowers for visual signalling. Ann Bot 123(2):263–276

Veggi PC, Martinez J, Meireles MAA (2013) Fundamentals of microwave extraction. In: Chemat F, Cravotto G (eds) Microwave-assisted extraction for bioactive compounds: theory and practice. Springer, New York

Vilar DD, Vilar MS, Raffin FN, Oliveira MR, Franco CF, de Athayde-Filho PF, Diniz MD, Barbosa-Filho JM (2014) Traditional uses, chemical constituents and biological activities of *Bixa orellana* L.: a review. ScientificWorldJournal 2014:857292

Wackenroder HW (2009) The discovery and early history of carotene. Bull Hist Chem 34(1):32–38

Wang L, Chen M, Lam PY, Dini-Andreote F, Dai L, Wei Z (2022) Multifaceted roles of flavonoids mediating plant-microbe interactions. Microbiome 10(1):1–3

Winkel-Shirley B (2001) Flavonoid biosynthesis. A colorful model for genetics, biochemistry, cell biology, and biotechnology. Plant Physiol 126:485–493

Wong DWS (1989) Colorants. In: Mechanism and theory in food chemistry. AVI, Westport

Wu Y, Liao W, Dawuda MM, Hu L, Yu J (2019) 5-Aminolevulinic acid (ALA) biosynthetic and metabolic pathways and its role in higher plants: a review. Plant Growth Regul 87:357–374

Yan H, Zhang H, Wang Q, Jian H, Qiu X, Baudino S, Just J, Raymond O, Gu L, Wang J, Bendahmane M (2016) The *Rosa chinensis* cv. Viridiflora phyllody phenotype is associated with misexpression genes. Front Plant Sci 7:1–14

Yoshida K, Miki N, Momonoi K, Kawachi M, Katou K, Okazaki Y, Uozumi N, Maeshima M, Kondo T (2009) Synchrony between flower opening and petal-color change from red to blue in morning glory, *Ipomoea tricolor* cv. "Heavenly Blue". Proc Jpn Acad Ser B Phys Biol Sci 85(6):187–197

Zhai S, Xia X, He Z (2016) Carotenoids in staple cereals: metabolism, regulation and genetic manipulation. Front Plant Sci 7:1197

Zhao X, Zhang Y, Long T, Wang S, Yang J (2022) Regulation mechanism of plant pigments biosynthesis: anthocyanins, carotenoids and betalains. Metabolites 12(9):871

Zhoh CK, Kwon HJ, Ahn SR (2010) Antioxidative and antimicrobial effects to skin flora of extracts from peel of *Allium cepa* L. Korean J Aest Cosm Soc 8:49–58

Waste Management in Horticulture: A Remunerative Approach

20

Ankan Das, Payel Das, and Amit Baran Sharangi

Abstract

Since a growing number of people are worried about their health and realize that horticultural products are a potent source of useful elements, the horticultural industry is expanding quickly. Nonetheless, a different group of people suffers from dietary deficiencies and is totally ignorant of nutritional security. Horticultural items have a higher moisture content than agricultural products, which speeds up spoiling and loss, hence resulting in a significant loss in this sector. Additionally, a good bulk of waste is produced in this field by the loss resulting from the disposal of several anatomical components of them, both in direct use and in industrial processing and academic study. However, because they include a variety of bioactive components, the undesirable materials—such as peel, seed, pomace, cores, etc.—have important health benefits. Examples of these components include colors, carbohydrates, dietary fibers, and polyphenols. Therefore, the issue of loss and nutritional instability can be somewhat reduced if the discarded item can be employed to create edible and nonedible items. In this context, the chapter concentrates on the several horticultural waste management components. There has been much discussion on the bioactive substances found in the waste parts as well as contemporary methods for extracting them. The wastes' health-promoting qualities have been explained. The research being conducted on the products made from leftover fruits and vegetables has received attention. The chapter also emphasizes the scopes or future requirements needed for this subject.

Keywords

Wastes · Bioactive compounds · Extraction · Properties · Value-addition

A. Das (✉) · P. Das
Department of Horticulture, Institute of Agricultural Science, University of Calcutta, Kolkata, West Bengal, India

A. B. Sharangi
Department of Plantation, Spices, Medicinal and Aromatic Crops, Faculty of Horticulture, Bidhan Chandra Krishi Viswavidyalaya, Mohanpur, West Bengal, India

A. Das, A. B. Sharangi (eds.), *Remunerative Horticulture*,
https://doi.org/10.1007/978-981-92-1007-7_20

20.1 Introduction

The yield of a number of horticultural crops has increased significantly as a result of farmers' acceptance of technology and scientists' ongoing research. The bulk of crop residue and postharvest waste has also increased correspondingly to increases in acreage and productivity (Garg 2023). Over a third of the 1.3 billion tons of food that is created for people to eat globally is wasted or lost, with the highest rates of waste occurring in fruits, vegetables, roots, and tubers (Lobo and Dorta 2019). Food waste from different industrial, agricultural, and domestic sources has significantly increased because of the world's population growth and lifestyle changes (Krishnamoorthi et al. 2023). The amount of consumer agricultural trash produced annually per person in North America and Europe is from 95 to 115 kg, whereas in sub-Saharan Africa, the south, and Southeast Asia, it is just 6–11 kg. Forty percent of losses worldwide happen at the postharvest and processing stages in underdeveloped nations, whereas the same losses happen at the consumer and retail levels in developed nations (Lobo and Dorta 2019).

According to estimates from the Food and Agriculture Organization (FAO), over 40% of India's food production is wasted (Plazzotta et al. 2017; NAAS 2019). Furthermore, according to the Food Corporation of India, this loss ranged from 10% to 15% of the entire output (Kumar et al. 2020). Sadly, the findings of some other reputable institutions' analyses of loss in our nation are concerning. For instance, the Ministry of Food Processing Industries (MFPI) in India calculated that the losses of horticultural commodities were 12 and 21 million tons, respectively, with cumulative food value loss and waste output of 10.6 billion USD, or almost 4.4 billion USD (NAAS 2019; Kumar et al. 2020). In horticulture, loss and waste can happen at any point in the chain of activities, which is increasingly a serious worry. Indigestible portions that are discarded at various phases, including collecting, handling, shipping, and processing, are referred to as "fruit and vegetable waste" (FVW), a more general term (Chang et al. 2006; Kumar et al. 2020).

Food processing industry creates multiple food products, employs a lot of people, and improves their economic standing. Food production, processing, and preparation all produce massive amounts of leftovers, which pollute the environment and pose health risks (Egyankosh 2022). Resources and invested capital are being depleted because of the exponential growth of waste. Leaf/straw, harvesting waste, food preparation waste, post-processing trash, leftover material as garbage, food processing operation waste, etc., are some examples of the waste material (Egyankosh 2022). Furthermore, once the materials undergo color changes, biochemical reactions, microbial infestation, breakage or frostbite, heat treatment, or reach ripeness levels that consumers find unacceptable, they are frequently deemed waste (Liu et al. 2023). In the handling of food processing waste, it is particularly crucial to use trash to create goods with additional value. Garbage from every business is produced in different forms, and each type of garbage needs a distinct treatment before it can be used or disposed of (Egyankosh 2022). For both financial and environmental reasons, horticultural wastes, including peels, seeds, and other parts of fruits and vegetables with a high phytochemical content and essential nutrients, must be disposed of appropriately (Raj et al. 2016; Sinha et al. 2024).

The waste derived from vegetables and fruits has a broad spectrum of characteristics, which are dependent on the species of plant and tissues (Ganesh et al. 2022). Also, many other wastes have been discussed in Table 20.1. Important bioactive compounds like polyphenols, dietary fiber, proteins, lipids, vitamins, organic acids, and minerals are present in different food waste types, including waste from fruit, vegetables, grains, and other production and processing. Some of these substances can be found in higher concentrations in the parts that are thrown away than in the parts that are sold (Liu et al. 2023). For example, leftover peels and seeds contain a lot of phytochemical substances that can be employed as food flavorings and preservation agents (Sridhar et al. 2021). Similar to this, the antioxidant and antidiabetic qualities of vegetable-thrown-away tissues are high in carotenoids, vitamins, and fibers and are intended to prevent human illnesses and disorders (Gowe 2015; Yusuf 2017; Ganesh et al. 2022). However, the lack of knowledge about the nutritional and economic worth of byproducts limits their use (Schroeder 1999). But fruit and vegetable solid waste has good economic value and can be turned into a variety of byproducts (Gray 2006; Joshi 2020). Things are slowly changing now, and people are becoming more conscious of their health and eating to satiate not only their appetite but also their nutritional needs. Because of the high value of these products and the potential for economic recovery, there has been a growing interest in the novel aspects of using these leftovers as byproducts for additional plundering during the manufacturing of food additives or dietary supplements with significant nutritional value (Joshi 2020). Thus, the potential of the most significant byproducts of processing fruits and vegetables as a source of valuable compounds is covered in this chapter.

In horticulture, waste output fluctuates based on crop variety, production intensity, and on-farm activities such as pruning or selective harvesting. Under high-input systems, residues, including trimmings, branches, foliage, and peels, commonly reach 20–50 tons per hectare per year (Table 20.2). Insights are drawn from worldwide and India-specific agricultural assessments that highlight these organic byproducts (Dharmaraj 2022; Slatter 2024; Moreno et al. 2021).

20.2 Bioactive Compounds Present in Fruit and Vegetable Wastes

A diverse range of substances derived from plants, polyphenols, have drawn significant attention due to their potential to improve health (Fig. 20.1). These substances are present in many different diets and are widely dispersed across the plant kingdom (Liu et al. 2023). Since polyphenols are a group of naturally occurring micronutrients found in plants, over five hundred phytochemicals have been identified. These substances are responsible for a plant's color and can help protect it from a variety of dangers. Additionally, eating foods that contain polyphenols has several health benefits for us (Hasan et al. 2017; Krishnamoorthi et al. 2023). One of the

Table 20.1 Wastes across greenhouses and open-field orchard systems, their composition, general management practices, and references

Type of waste	Sources	Composition	General management practices	References
Plant trimmings and residues (leaves, stems, branches, prunings)	Greenhouses, orchards, nurseries	Primarily lignocellulosic: cellulose (glucan ~13–22%), hemicellulose (~5–6%), lignin (~9–26%), soluble extracts (up to ~40%), ash (~15–17%)	Composting or vermicomposting, mulching, anaerobic digestion	Moreno et al. (2021), Federal TVET Agency (2020), Indian Institute of Horticulture Technology (2025)
Discarded or substandard produce (e.g., damaged, overripe, or culled fruits/ vegetables like tomato plant waste)	Harvesting, sorting, packing areas	High in fermentable sugars, carbohydrates, proteins, ash	Juice/sugar extraction via centrifugation or pressing, enzymatic processing of remaining pulp for further value, safe use as livestock feed after appropriate treatment, or composting where suitable	Moreno et al. (2021), Federal TVET Agency (2020), Indian Institute of Horticulture Technology (2025)
Packaging and containment materials (plastic trays, pots, nets, films)	Nursery propagation, greenhouse growing, postharvest packing	Primarily synthetic/ inorganic: various plastics and polymers (nonbiodegradable types common)	Reuse after cleaning/ sterilization when feasible, participate in dedicated recycling schemes, shift toward biodegradable or reusable alternatives to reduce long-term waste	Indian Institute of Horticulture Technology (2025)
Chemical-related wastes	Application in pest and nutrient management	Hazardous: mixtures with active agrochemical ingredients (liquids, solids, or contaminated materials)	Prioritize integrated pest management (IPM) to reduce chemical reliance, use secure collection/ return-to-supplier programs for containers, follow specialized hazardous waste protocols; strictly avoid inclusion in composting or organic recycling streams	Indian Institute of Horticulture Technology (2025)

Table 20.2 Type of waste and waste levels across different horticultural systems in different regions

Systems	Main components	Waste levels	Region
Tomato greenhouse production	Stems, leaves, pruning debris from plants; substandard or damaged fruits	Plant biomass residues commonly around 49–73 t/ha/year; fruit loss 25%	Almería region, Spain; total regional plant waste from tomatoes estimated at roughly 470,000 t/year based on cultivated area and per-hectare generation rates (Moreno et al. 2021)
Fruit tree orchards (e.g., mango, citrus)	Pruning branches, dropped or culled fruits due to quality issues, pests, or weather	On-farm losses frequently in the 20–37% range of potential crop	Australian commercial orchard (Slatter 2024)
Open-field vegetable crops (e.g., onion, potato)	Peel/skin layers, leafy tops, stems, rejected tubers/bulbs	Organics often 65–75% of total discards; field-level generation roughly 20–30 t/ha/year	Typical in Indian vegetable farming areas, significant byproduct from peeling contributes to high organic fractions (Dharmaraj 2022)
Flower cultivation and nurseries	Cut stems/foliage trimmings, unsellable wilted flowers, leftover potting substrates	Organics typically 25–80% of waste, around 10–20 t/ha/year	Indian floriculture operations (Dharmaraj 2022)
Spice crop production (e.g., chili/pepper)	Stems, leaves, outer husks, postharvest processing remnants	Part of India's broader ~500 million t annual agricultural residues, horticultural crops (including spices) represent a meaningful share	Widespread in Indian spice-growing regions (Dharmaraj 2022)
Vineyard systems (grapes for table, wine, or drying)	Winter/annual cane prunings, diseased or undersized rejected bunches/berries	Farm-level losses ranging 12–46%	Australia (Slatter 2024)

Note: Estimates derived from audits revealed that horticulture makes up about 29% of total agricultural waste in India, where annual agricultural residue generation reaches approximately 500 million tons. In comparison, Australian horticulture farms report an average loss rate of 20%, based on data from over 10,645 operations (Slatter 2024; Reddy et al. 2025; Dharmaraj 2022)

most well-known and prevalent groups of polyphenols, flavonoids are a crucial component of fruit and vegetable waste (Krishnamoorthi et al. 2023). Up to 12 million tons of fruit waste are produced each year globally by the fruit industry, of which apple pomace is a major component (Martău et al. 2021). Apple peel has three to six times as many flavonoids as apple flesh, including special flavonoids like quercetin glycosides that are absent from apple flesh (Khalid et al. 2021; Liu et al. 2023).

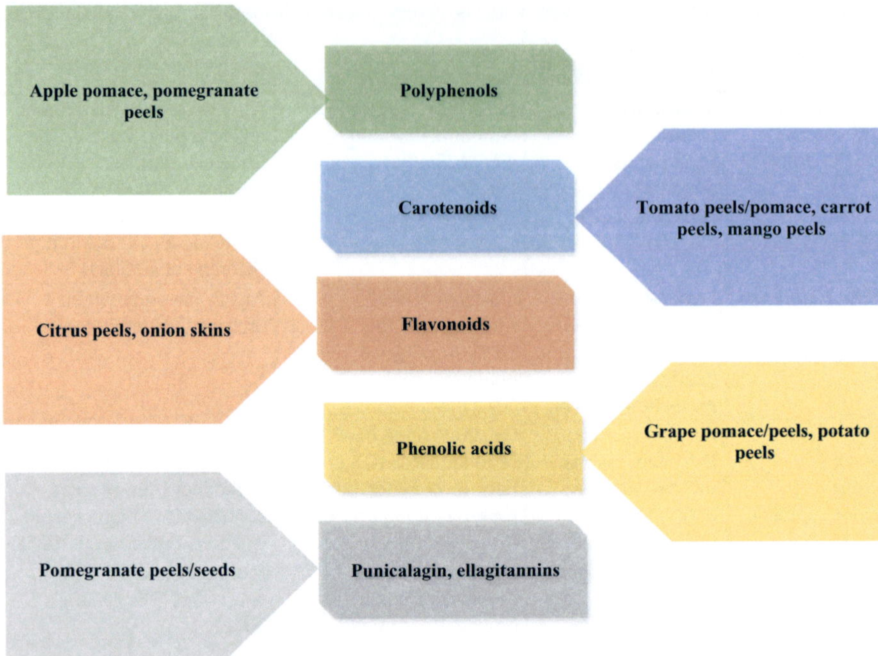

Fig. 20.1 Flow chart illustrating the sources and types of bioactive compounds. "Apple pomace, pomegranate peels" lead to "Polyphenols." "Citrus peels, onion skins" lead to "Flavonoids." "Pomegranate peels/seeds" lead to "Punicalagin, ellagitannins." "Tomato peels/pomace, carrot peels, mango peels" lead to "Carotenoids." "Grape pomace/peels, potato peels" lead to "Phenolic acids." Each source is connected to its corresponding compound type with arrows. *Note*: Source references include Carvalho et al. (2025), Patel et al. (2021), Yadav et al. (2024), Hussain et al. (2022), Parra-Pacheco et al. (2024), Das et al. (2024), Papoutsis et al. (2025), Castillejo and Martínez-Zamora (2024)

Numerous horticultural products are enhanced using the same active component; some of the additional studies are included below. A byproduct of processing cranberries, cranberry pomace is made up of the fruit's seeds, skins, and stems. It is usually thrown out because of its unpalatability, even though it contains healthy polyphenols such as proanthocyanidins and anthocyanins (Liu et al. 2023). Citrus fruits frequently contain flavanones, which include substances like naringenin and hesperidin (Livadariu et al. 2023; Krishnamoorthi et al. 2023). Because citrus peel has more polyphenols than the fruit's edible portion, citrus trash is a good source of phenolic compounds (Balasundram et al. 2006; Sagar et al. 2018). Numerous further studies have been carried out to ascertain whether flavanones are present in other horticultural wastes. When comparing the peels of apples, peaches, and pears to the peeled fruits, Gorinstein et al. (2001) discovered twice as many total phenolics. Only around 25% of the phenolic compounds found in the peel of bananas (*Musa cavendish*) are present in the pulp, which has been reported to have 232 mg/100 g dry matter (DM) of these compounds (Someya et al. 2002; Sagar et al. 2018). The peels, pulp remnants, seeds, and stems that make up 15–55% of the berry's initial mass are byproducts of processing blueberries, and they have higher concentrations of natural anthocyanins as well as phenolic compounds than the pulp (Bener et al. 2013; Liu et al. 2023). Given that it contains 50% phenolics out of the total concentration of bioactive components, potato peel has been identified as a good source of phenolic compounds in the context of vegetable waste (Friedman 1997; Sagar et al. 2018). Up to 175,000 tons of carrot trash could be produced yearly by a large commercial carrot processing facility. A very sound concentration of total phenolic content was discovered in the unpressed carrot trash (Liu et al. 2023). Waste from fruits and vegetables was found to contain a lot of phenolic compounds in the order of increasing sequence, as per a study by Zeyada et al. (2008): Olive leaves > tomato peel > cucumber peel > watermelon peel > potato peel (Sagar et al. 2018). About 6% and 4% of all polyphenols and tannins, respectively, are found in coffee waste produced during the manufacturing of soluble coffee. Theaflavins and catechins, two types of polyphenols, are also abundant in discarded tea leaves (Liu et al. 2023).

20.2.1 Proteins

The most important macromolecules for the creation of body muscles are proteins, which are also required building blocks for many other bodily components. Numerous harmful disorders and diseases can result from a protein deficit. A number of fruits and vegetables have been identified as good suppliers of proteins, despite their inedible sections and waste (Sagar et al. 2018). Although spent tea leaves contain a variety of amino acids and good levels of protein (18–35% of crude protein), their potential for application in human food supplements has not yet been realized (Negi et al. 2022; Liu et al. 2023). Using the Lowry method for protein measurement, Chitturi et al. (2013) examined the peels of a variety of fruits and discovered that the peels of papaya, avocado, and kiwifruit had high protein contents (1.79%, 1.57%, and 1.55%, respectively). According to Dugo and Di Giacomo (2002), Pfaltzgraff et al. (2013), Mamma and Christakopoulos (2014), and Sagar et al. (2018), the protein content of citrus peels ranged from 2.5% to 9.0%. In addition, studies have been conducted to determine the protein content of additional fruit and vegetable waste samples. According to Sharma et al. (2016) and Sagar et al. (2018), for example, the protein content is 10.06 g in carrot pomace, 4.45 g in apple pomace, 5.4 g in mosambi peel, 13.27 g in green pea peels, 9.5 g in mango peel, 8.7 g in pineapple peel, 6.02 g in banana peel, 5.97 g in orange peel, 3–5 g in potato solid waste, 17–22 g in tomato solid waste, 20.4 g in cabbage leaves, 16.1 g in cauliflower leaves, and 20.2 g in pea pods, per 100 g.

20.2.2 Dietary Fiber

Dietary fiber is made up of cellulose, noncellulosic polysaccharides including hemicellulose, pectic compounds, gums, mucilages, and the noncarbohydrate component lignin. It is the indigestible content of plant material (Dhingra et al. 2012; Liu et al. 2023). The aforementioned component can also be found in large quantities in the trash or abandoned products produced from horticulture products. For example, the peel of bananas has a lot of dietary fiber, which is not soluble, with beneficial qualities like the capability to store water, oil, and glucose (Wang et al. 2022; Liu et al 2023). Jaime et al. (2002) inspected the said substance content of three distinct types of onions, examining the entire vegetable from the skin to the interior layers. The "Grano de Oro" onion's skin had the maximum number of total dietary fibers (TDFs) (68.3% dry matter [DM]), while the inner portion had the lowest amount (11.6% DM) (Sagar et al. 2018). The remnants from Jerusalem artichokes are usually thrown away as waste, which pollutes the environment and wastes resources. Nonetheless, there is a chance that these residues could be used to extract high-performance, high-purity soluble dietary fiber, which could have exciting uses in the food packaging sector (Zhou et al. 2022; Liu et al 2023). According to Ncobela et al. (2017), the dry matter of potato peels includes 61.0–125 g/kg of crude fiber. Additionally, it was noted that potato solid waste was a good source of fiber (27–35 g/100 g) (Afifi 2011; Sharoba et al. 2013; Sagar et al. 2018). On the other hand, garlic peel is frequently thrown away as industrial waste even though it has a large amount of nutritional fiber (about 70%) (Liu et al. 2023). When nonstarch polysaccharides (NSPs) were examined in cauliflower byproducts (stems and florets), stems had higher concentrations (3.11% fresh weight, FW) than florets (2.31% FW) (Sagar et al. 2018). A different study looked at the amount of dietary fiber in waste materials such as carrot and grape pomace. It was discovered to be 63.6% DM in carrot pomace. Of this, 13.50% of the fibers were soluble and 50.10% were insoluble (Chau et al. 2004). It was discovered that grape pomace was a rich source of dietary fibers, specifically cellulose, hemicelluloses, and trace amounts of pectins (Kammerer et al. 2005; Sagar et al. 2018). Other experts have offered further information about the amount of dietary fiber in fruit and vegetable waste. Tomato pomace has up to 50% fiber on a dry weight basis (Del Valle et al. 2006; Sagar et al. 2018). Ajila and Rao (2013) examined the dietary fiber of mango peels and discovered that the main neutral sugars in both soluble and insoluble dietary fibers were galactose, glucose, and arabinose. The TDF concentration ranged from 40.6% to 72.5% (Sagar et al. 2018).

20.2.3 Vitamins and Minerals

Vitamins are vital, natural components that are present in extremely small quantities yet are vital for life and for the body's numerous chemical processes. Vitamins are essential for cell growth, maintenance, and process regulation in the body

(Krishnamoorthi et al. 2023). Vitamins that are water-soluble (B-complex vitamins and vitamin C) and vitamins that are fat-soluble (A, D, E, and K) are among the various kinds of vitamins (Liu et al. 2023). Not only are the edible portions of fruits and vegetables enhanced with the aforementioned component, but the leftover pieces are also sound. Vitamins, especially C and B-complex, are abundant in fruits and vegetables that are thrown out because of overproduction or aesthetic flaws (Coman et al. 2020). As byproducts of processing carrots and pumpkins, bark and peel contain carotenoids, which are precursors of vitamin A (Konrade et al. 2018; Liu et al. 2023). Vitamin A is also abundant in other waste products from fruits and vegetables since they contain the precursor to the vitamin.

B-carotene is the primary carotenoid in the seeds, while lycopene makes up 90% of the total carotenoids in tomato peel byproducts (Trombino et al. 2021). Carotenoids are found in a variety of fruit waste sources, including horticultural wastes, that was thrown out. B-carotene, lutein, zeaxanthin, and lycopene are common carotenoids in such wastes (Kultys and Kurek 2022; Liu et al. 2023).

20.3 Novel Emerging Methods for Extraction of Bioactive Compounds from Horticultural Wastes

The efficiency of traditional extraction techniques, such as the Soxhlet extraction and maceration procedures, has been demonstrated in several investigations. However, employing such methods necessitates a significant investment of time, energy, and resources (Nirmal et al. 2023). Since technology is always evolving, we cannot rely solely on traditional extraction techniques to attain more production. To get around the drawbacks of traditional extraction techniques, there are new extraction techniques (Fig. 20.2) that utilize less solvent and offer quicker extraction times, more selectivity, and higher efficiency (Nirmal et al. 2023).

20.3.1 Supercritical Fluid Extraction

Using carbon dioxide as a renewable and environmentally friendly solvent, supercritical fluid extraction has gained popularity as a green extraction method for obtaining bioactive chemicals from a variety of raw materials (Aqilah et al. 2023). Numerous industries operating in the area have accepted this strategy due to its many benefits. Because of its low cost, moderate extraction temperature, and lack of carbon dioxide toxicity, this approach is a fantastic choice for removing bioactive chemicals from agro-industrial waste (Lasunon et al. 2021; Araújo et al. 2020; Aqilah et al. 2023). Numerous studies have documented the effectiveness of SFE in removing bioactive substances derived from plant matrices (Uwineza and Waśkiewicz 2020). Alkaloids such as vinblastine (Falcão et al. 2017), colchicine (Ellington et al. 2003), caffeine (İçen and Gürü 2010), and pyrrolidine (Carrara et al. 2017) may all be extracted using SFE along with terpenes (Favareto et al.

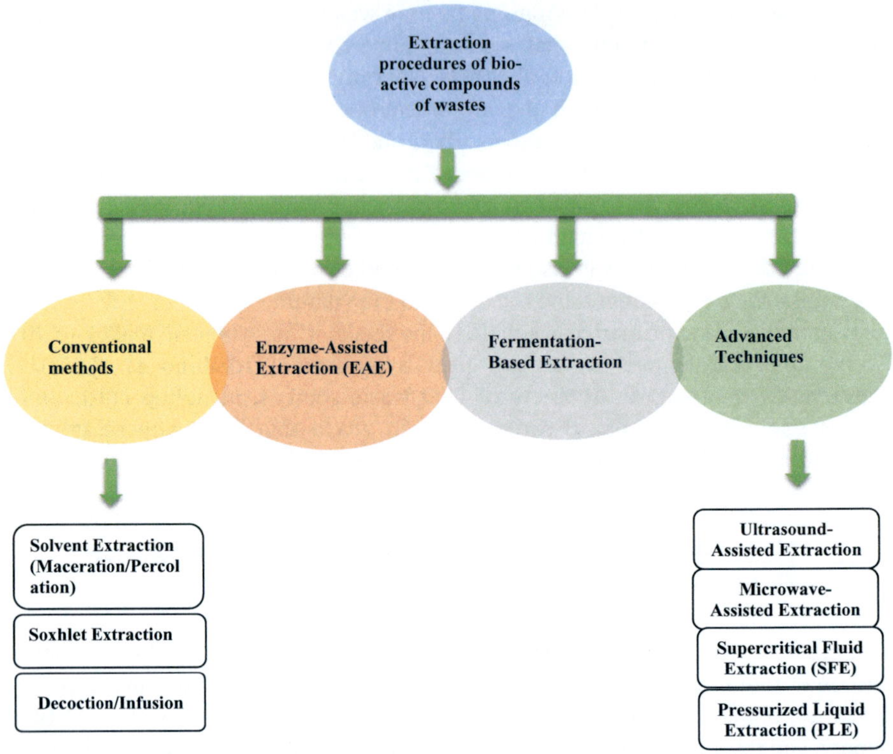

Fig. 20.2 Flow chart illustrating the extraction procedures of bioactive compounds from wastes. The process begins with a central node labeled "Extraction procedures of bio-active compounds of wastes," branching into four main categories: Conventional methods, Enzyme-Assisted Extraction (EAE), Fermentation-Based Extraction, and Advanced Techniques. Each category further details specific methods. Conventional methods include Solvent Extraction (Maceration/Percolation), Soxhlet Extraction, and Decoction/Infusion. Advanced Techniques include Ultrasound-Assisted Extraction, Microwave-Assisted Extraction, Supercritical Fluid Extraction (SFE), and Pressurized Liquid Extraction (PLE). *Note*: Sources include, Sasidharan et al. (2011), Lemes et al. (2022), Bhadange et al. (2024)

2017), flavonoids (Ouédraogo et al. 2018), phenolic compounds (Pimentel-Moral et al. 2019; Nirmal et al. 2023), and essential oils (Conde-Hernández et al. 2017). Soldan et al. (2021) used a supercritical fluid extraction technique to extract oleoresin from the industrial waste of *Capsicum annuum*. Using a supercritical CO_2 approach, Romano et al. (2022) isolated bioactive components from citrus peel (Aqilah et al. 2023). Compared to traditional extraction using acetone, Derrien et al. (2017) confirmed a better recovery of phytopigments by adopting a green extraction technique to extract lutein and chlorophyll from spinach byproducts/wastes (Sharma et al. 2021a, b).

20.3.2 Microwave-Assisted Extraction

Solid-liquid extraction using microwave heating is known as microwave-assisted extraction (MAE). The electromagnetic waves that make up microwaves range in frequency from 300 MHz to 300 GHz. The ability of the samples to absorb microwave energy and transform it into heat is what creates the microwave heating (Gomes-Araújo et al. 2021). The said technique can be used to recover a group of entities like flavonoids (Baydar et al. 2007), isoflavones (Rostagno et al. 2007), saponins (Chen et al. 2007), piperine (Raman and Gaikar 2002), carotenoids (Kiss et al. 2000), terpenes (Sánchez-Ávila et al. 2009), essential oils (Srinivas et al. 2020; Sagarika et al. 2018), polysaccharides (Lal et al. 2021), etc. (Nirmal et al. 2023). Carbone et al. (2020) used microwave-assisted extraction (MAE) to recover natural antioxidants from kiwi juice pomace. Doulabi et al. (2020) also studied the recovery of bioactive substances from eggplant peel waste by employing microwave-assisted extraction techniques (Aqilah et al. 2023). This technique has advanced to the point that it can effectively extract bioactive chemicals from a variety of horticultural wastes. Kurtulbaş et al. (2024) also extracted bioactive components from peach trash using the MAE technique. Additionally, da Rocha and Noreña (2020) used the MAE and Ultrasound-Assisted Extraction (UAE) procedures to get bioactive substances from grape pomace (Aqilah et al. 2023).

20.3.3 Pulsed Electric Field

By dissolving membrane structures, Pulsed Electric Field Extraction (PEFE) facilitates mass transfer during extraction, significantly increasing extraction yield and cutting down on extraction time (Nirmal et al. 2023). A nonthermal, environmentally friendly, and selective extraction method that uses little energy and has been shown to be successful in quick recovery times is pulse electric field (PEF) extraction (Maza et al. 2020). PEF has mostly been utilized in the food business to increase the recovery of phytochemicals from fruit and vegetable industrial waste and to decrease microbial growth/pasteurization (Pataro et al. 2020; Aqilah et al. 2023). PEFE processing is a viable method for the food, pharmaceuticals, and biotech sectors because of its energy efficiency. Proteins (Rajha et al. 2014), anthocyanins (Medina-Meza and Barbosa-Cánovas 2015), polyphenols (El Kantar et al. 2018), flavonoids (Redondo et al. 2018), and carbs can all be extracted using PEFE (Yan et al. 2017; Nirmal et al. 2023). Pulsed electric field polyphenol harvesting from *Crocus sativus*, *Sideritis scardica*, and *Vitis vinifera* was assessed by Lakka et al. (2021). The results demonstrate that PEF treatment improved all three plants' recovery in total polyphenols (Aqilah et al. 2023). When beneficial chemicals (anthocyanin monoglucosides) were extracted from grape byproducts, PEF was found to be the most effective method among those used (Corrales et al. 2008). The skin of "Merlot" grapes treated with PEF produced substantial levels of polyphenols and anthocyanins (Delsart et al. 2012; Sagar et al. 2018). The pulsed electric field approach was used by Andreou et al. (2020) to improve waste valorization and product yield in the industrial tomato process. The PEF treatment decreased peel separation work by 72.3% without compromising tomato firmness in the end, per the results (Aqilah et al. 2023). According to reports, the PEF technology helps to enhance the drying, the extraction process, dispersion, and pressing processes (Barsotti et al. 1998; Angersbach et al. 2000; Vorobiev et al. 2004; Vorobiev and Lebovka 2006; Sagar et al. 2018). Tehrani et al. (2019) used a pulsed electric field (PEF) to extract useful components from onion (*Allium cepa* L.) waste (Aqilah et al. 2023).

20.3.4 Ultrasound-Assisted Extraction (UAE)

Ultrasound is a type of sound wave that is inaudible to humans and has frequencies spanning 20 kHz to 100 MHz. As it passes through the medium, it compresses and expands it as other waves do. Cavitation, which refers to the creation, growth, and disintegration of bubbles, is a phenomenon brought on by this mechanism (Nirmal et al. 2023). The method shows promise in terms of releasing a significant quantity of energy, which is advantageous during the extraction process. Cell rupture is caused by this substantial amount of energy (Vilkhu et al. 2008). Ultrasound-assisted extraction (UAE) is a method of using powerful ultrasonic waves to extract material (Nirmal et al. 2023). An effective extraction technique that outperforms conventional extraction is ultrasound-assisted extraction (UAE). UAE has several advantages, including reduced time, easy usage, less solvent usage, lower temperatures, energy savings, and increased yield (Aqilah et al. 2023). UAE is a productive method for eliminating bioactive materials from different plant matrices, including polyphenols, flavonoids, anthocyanins, etc. (Wen et al. 2018; Kumar et al. 2021, 2022; Vilkhu et al. 2008; Nirmal et al. 2023). In their UAE of pectin from the leftover phenolic component isolated from mango peel, Guandalini et al. (2019) discovered a 50% improvement in yield (from 5.61 to 8.6). When juxtaposed with pectin extraction without ultrasound, the yield of UAE pectin from soaked mango peel rose by 31% (from 6.2 to 8.1) without compromising pectin quality (Aqilah et al. 2023). In the current environment where companies want efficiency, the method is accelerating and shows promise for extraction from multiple other horticultural wastes. Anthocyanin has been extracted from grape waste using a selected frequency of 30–40 kHz (Corrales et al. 2008), hemicellulose from grape pomace (Minjares-Fuentes et al. 2016), oil from papaya seed (Samaram et al. 2015), pectin from tomato waste (Grassino et al. 2016), and so on (Aqilah et al. 2023). Other materials that have been tested by other scientists to get the key substance from horticulture waste include the following examples. By using UAE instead of chemical extraction, de Oliveira et al. (2016) found that the pectin yield from passion fruit peel increased by 1.6 times. In a study comparing chemical and UAE extraction of pectin from grapefruit peel, Wang et al. (2016) discovered that UAE extraction yielded a higher yield (16.34%) and cut down on extraction time by 37.7% (Aqilah et al. 2023).

20.3.5 High-Pressure Extraction

According to Nirmal et al. (2023), high-pressure extraction, also called pressurized liquid extraction (PLE), expedited extraction of solvents, accelerated solvent mining, or pressurized fluid, is the method of using high pressure to retain solvents in a liquid state above boiling. Maintaining them in this form has advantages for extraction and reduces certain changes in the targeted chemicals. Because of the high pressure, solvents remain liquid above their boiling point, which causes high solvent penetration of the matrix, high lipid solubility, and high lipid solute diffusion rates (Huang et al. 2013; Nirmal et al. 2023). García et al. (2021) used pressurized liquid extraction (PLE) to study the extraction of bioactive materials from the peel of pomegranates (*Punica granatum* L.). According to the researchers' findings, the TPC and punicalagin contents were excellent (Aqilah et al. 2023). PLE provides outstanding reproducibility and drastically cuts down on extraction time and solvent usage as compared to alternative methods. Researchers have successfully extracted multiple types of bioactive components, including pectin, flavonoids, carotenoids, and phenolic compounds, using high-pressure extraction (Khan et al. 2019; Nirmal et al. 2023). The PLE approach was used by Lasta et al. (2019) to recover phenolic chemicals from beetroot waste. As per this finding, the extracts included a number of abundant phenolic components, including sinapaldehyde, vitexin, and ferulic acid (Aqilah et al. 2023). The following are some further studies that have shown promise in the realm of high-pressure extraction. Using pressure liquid extraction, Pukalskienė et al. (2021) extracted bioactive components from strawberry (Fragaria x ananassa) pomace. Quercetin-3-glucuronide, kaempferol-3-glucuronide, tiliroside, ellagic, malic, succinic, citric, and p-coumaric acids were the most common components in strawberry pomace extracts (Aqilah et al. 2023). Additionally, this study produced some novel findings that had not been discovered or observed in previous papers. For example, strawberry pomace extracts produced by PLE were shown to have cytotoxic, antiproliferative, and cellular antioxidant properties for the first time (Aqilah et al. 2023). To determine the effectiveness of high-pressure extraction compared to other procedures, some comparative experiments were performed. For instance, to recover proteins from pomegranate seed waste, Guzmán-Lorite et al. (2022) compared pressurized liquids with high-intensity focused ultrasounds. The results showed that greater quantities of proteins, peptides, and phenolics were obtained using pressurized liquid extraction techniques (Aqilah et al. 2023).

20.3.6 Membrane Technology

There are numerous advantages to pressure-driven membrane techniques like micro-, nano-, and ultrafiltration, such as low energy needs, easy scaling up, straightforward operation, high productivity, and the lack of phase transition (Gomes-Araújo et al. 2021). Such benefits of this technology make it a good alternative in bioactive compound extraction from wastes generated from horticulture. Membrane technology has been used to recover a variety of bioactive chemicals from agro-industrial wastes, including phenolic compounds, proteins, sugars, pectin, carbohydrates, and antioxidant compounds (Castro-Muñoz et al. 2016; Gomes-Araújo et al. 2021).

20.3.7 Enzyme-Assisted Extraction

Another innovative and practical method for recovering bound chemicals and boosting their yield is enzymatic pretreatment (Rosenthal et al. 1996). Enzyme-assisted cold pressing (EACP) and enzyme-assisted aqueous extraction (EAAE) are the two main methods for enzyme-assisted extraction (EAE) (Latif and Anwar 2009; Sagar et al. 2018). Immobile enzymes can lessen the complexity, cost, and limits of enzymatic processes, which include limited stability, recovery, and temperature sensitivity (Sharma et al. 2021a, b; Gomes-Araújo et al. 2021). There are not many studies on horticulture waste chemicals, nevertheless, where the extraction method has been investigated. Among the several enzymes, Celluzyme MX produced the maximum amount of overall phenolic compounds in the skin of five citrus fruits: orange, grapefruit, mandarin, Meyer lemon, and "Yen Ben" lemon (Li et al. 2006; Sagar et al. 2018). Other enzymes were employed in a different investigation to successfully extract bioactive substances such as the pectinex, celluclast, and novoferm enzyme groups. Gómez-García et al. (2012) found that the novoferm enzyme had the best effect when it came to extracting phenolic chemicals from grape waste (Sagar et al. 2018).

20.4 Impactful Properties Present in Fruits and Vegetable Waste

20.4.1 Antioxidant

Because it has the benefit of extending the storage life of food materials, the food industry is having a greater and greater influence on the growth of innovative biomaterials with antioxidant properties (Goulas et al. 2019; Aqilah et al. 2023). High concentrations of antioxidant chemicals are typically present in debris that gets thrown away from fruits and vegetables (Gómez-Estaca et al. 2015). The primary bioactive components in these waste materials are anthocyanins, flavonoids, vitamins E and C, phenolic compounds, fibers, carotenoids, and other antioxidants; as such, they are valuable sources of these substances (Sharma et al. 2021a, b). Because they can scavenge the reactive oxygen species (ROS) seen in food products (Piechowiak et al. 2021), compounds high in polyphenols, such as phenolic acids and flavonoids, have exhibited outstanding antioxidant qualities (Cadiz-Gurrea et al. 2021; Aqilah et al. 2023). Both blackberry and raspberry seed pomace waste can be potent sources of antioxidants. These oils may contain bioactive materials like phenols, tocopherols, carotenoids, and sterols, which are known antioxidants, in addition to omega-3 (a-linolenic) and omega-6 (linoleic) (ratio 1:2–4) (Radočaj et al. 2014; Sharma et al. 2021a, b). In order to improve the quality characteristics of fresh-cut apples, Venturi et al. (2019) employed potato peels, which have a stronger antioxidant potential and contain bioactive components (Aqilah et al. 2023). Compared to meat wastes, orange peel has less antioxidant potential. Although it has been demonstrated that both extracts are effective in preventing hydrogen peroxide free radicals, they caused DNA damage to human leukocytes (Park et al. 2014; Sharma et al. 2021a, b). To lower food production losses, Kurek et al. (2021) created a chitosan and pectin film combined with powdered blackcurrant pomace. This resulted from blackcurrant pomace's antioxidant qualities (Aqilah et al. 2023). Additionally, it has been noted that the natural antioxidant compounds found in tomato and potato waste make them a great protective agent for preserving vegetable oils (by preventing the oxidation of oils) (del Carmen Robles-Ramírez et al. 2016; Sharma et al. 2021a, b). To elevate the antioxidant property of disposable film as active packaging, Kam et al. (2018) used durian extract of leaves in gelatin-based film (Aqilah et al. 2023). According to Duda-Chodak and Tarko (2007), apple byproducts have a higher antioxidant potential than some other fruits. The peels of apples, oranges, Idared apple seeds, and white grapes have the highest antioxidant activity (Nirmal et al. 2023). Mango peels were used to make pectin-phenolic antioxidant films for use in food packaging (Ribeiro et al. 2021; Aqilah et al. 2023).

20.4.2 Antimicrobial Activity

Antimicrobial agents are substances that affect the destruction or prevention of the progression of harmful or spoiling bacteria (Nirmal et al. 2023). Researchers have been searching for innovative ways to stop microbial development in food without sacrificing its flavor, texture, or safety in response to recent global food-borne microbial epidemics (Aqilah et al. 2023). It is recognized that several plant parts, including fruits, leaves, peels, pods, seeds, and more, contain antimicrobial components. These components cause lysis and provide defense against harmful microorganisms by penetrating the cell membrane (Nirmal et al. 2023). There are numerous horticultural products with antimicrobial qualities that can be used to produce a variety of beneficial results. There have been reports of plum, elderberry, and Italian red grape byproducts having good inhibitory properties against possible pathogenic strains (Coman et al. 2018; Nirmal et al. 2023). The antibacterial capability of avocado peel extracts was higher than that of nisin, a naturally produced dipeptide having antimicrobial qualities. Furthermore, extracts from avocado peels demonstrated antibacterial properties against *Lactobacillus sakei*, *Weissella viridescens*, *Leuconostoc mesenteroides*, *Escherichia coli*, and *Listeria innocua* (L. innocua) (Sharma et al. 2021a, b). Extracts from elderberry skin and seeds were found to have an acceptable suppressive impact on *Bacillus cereus* growth, with a zone of inhibition of about 20 mm (Nirmal et al. 2023). Another study found that the polyphenol and tannin content of avocado seed extracts had an antimicrobial impact on *Mycobacterium avium*, *Listeria monocytogenes*, and *Staphylococcus epidermidis*.

This suggested that the tannins in avocado seed extracts could prevent urinary tract infections (Kosińska et al. 2012; Sharma et al. 2021a, b). With a maximal zone of inhibition, Gunwantrao et al. (2016) also documented the efficacy of pineapple and orange peel extracts in inhibiting the strains of harmful bacteria *Bacillus subtillis*, *Pseudomonas aeruginosa*, and *Klebsiella pneumonia*. Strong antibacterial properties were demonstrated by muscadine grape polyphenols against multiple types of food-borne pathogens, primarily *Staphylococcus aureus* (Nirmal et al. 2023). *Aggregatibacter actinomycetem* and *Porphyromonas gingivalis* were shown to be susceptible to the antibacterial activities of banana peel extracts by Kapadia et al. (2015) and Sharma et al. (2021a, b). Pomegranate peel particulates (PGP) combined with starch were recently used by Ali et al. (2019) as antimicrobial films that may also be utilized as food-grade packaging material. Both gram-positive (*S. aureus*) and gram-negative (*Salmonella*) bacteria were inhibited in their growth by PGP (Aqilah et al. 2023). Products that can act as natural antioxidants and antimicrobials in fruit juices and meat-based goods have been extracted from grape processing byproducts (Mattos et al. 2017; Sharma et al. 2021a, b). Many other fruit byproducts have been investigated for their antibacterial properties against *S. aureus* growth and activity, such as those of bananas (Mordi et al. 2016), mangoes (Mutua et al. 2017), cloudberries, and raspberries (Rauha et al. 2000; Mutua et al. 2017; Nirmal et al. 2023). Because of its antibacterial properties, beetroot pomace has been incorporated into numerous functional meals and supplements (Kushwaha et al. 2018). The antibacterial activity of cauliflower byproducts against *L. monocytogenes* was assessed (Sharma et al. 2021a, b). The skins and seeds of pomegranate have different antibacterial properties, according to Kanatt et al. (2010). While the extract from the seed showed no discernible effect against *Staphylococcus aureus* and *Bacillus cereus*, the extract taken from pomegranate peel showed remarkable antioxidant activity.

20.4.3 Anticancer Activity

It is generally known that natural pigments extracted from plants have anticancer properties. The employment of mangiferin, a naturally generated bioactive xanthonoid that is derived from the mango tree, for the treatment of cancer has been investigated by Núñez Selles et al. (2016) and Sharma et al. (2021a, b). Compounds having possible anticancer effects can be found in waste from fruits and vegetables like peels and pomace. These substances, which include carotenoids and polyphenols, have anti-inflammatory and antioxidant properties that may help lower or prevent the risk of cancer (García-González et al. 2025). Waste products from black raspberry seeds have the ability to induce apoptosis in HT-29 colon cancer cells, which inhibits the cells' ability to spread (Cho et al. 2015). Non-extractable antioxidants in freeze-dried apple wastes have stronger inhibitory effects on human cancer cells (namely, HepG2, HeLa, and HT-29) than extractable antioxidants (Tow et al. 2011; Sharma et al. 2021a, b). It has been noted that some substances, like sulforaphane, found in cole crop wastes like broccoli, can stop the formation of cancer cells

(Liang et al. 2019). Through the use of superoxide and hydroxyl anion radicals, tomato juice production waste showed anticancer effects by preventing the growth of MCF7, HeLa, and MRC-5 tumorous cells (Ćetković et al. 2012; Sharma et al. 2021a, b). Byproducts of pomegranates have anti-aging, anti-inflammatory, and anticancer properties due to the presence of bioactive components such as ellagic acid and punicalagin (Wu and Tian 2017; Nirmal et al. 2023). The strong antioxidative and anticancer activities of beetroot pomace, which demonstrated notable proliferative effects in cell lines from humans (MRC-5 and MCF7), have been attributed to betalains (Vulić et al. 2012; Sharma et al. 2021a, b). By virtue of many bioactive components, such as flavanones, flavones, and anthocyanins, citrus byproducts are also recognized for their anti-inflammatory and anti-cancerous qualities (Panwar et al. 2021). Furthermore, terpenes, citral aldehydes, and esters—the main components of citrus oil—play a significant role in surgical treatment for cancer-related conditions (Palazzolo et al. 2013; Nirmal et al. 2023).

20.4.4 Anti-inflammatory

Numerous phytochemicals, such as the natural colors found in fruits and vegetables, have drawn interest as potential medicinal agents because of their demonstrated ability to lower inflammation (Sharma et al. 2021a, b). The leftover portions of horticultural products also include significant levels of flavonoids, polyphenols, and various other chemicals that may help in reducing inflammation through a variety of ways (Ashong 2024). In order to investigate the possible anti-inflammatory properties of enzymatically processed raspberry pomace, pectinolytic preparations were utilized to extract polyphenolic chemicals and anthocyanins (Szymanowska and Baraniak 2019). These researchers discovered that the resultant anthocyanins have anti-inflammatory qualities by inhibiting the activities of lipoxygenase and cyclooxygenase 2 (Sharma et al. 2021a, b). By suppressing the expression of pro-inflammatory cytokines, the anti-inflammatory elements of pomegranate peel—punicalagin, punicalin, strictinin A, and granatin B—significantly decreased the generation of NO and PGE2 (Lee et al. 2008; Romier et al. 2008; Zhu et al. 2018). Citrus pectin can control the inflammatory response by influencing immune cells, according to numerous research. Citrus pectin has been shown in both in vitro and in vivo tests to produce endotoxins, which reduce inflammation (Ishisono et al. 2017; Sharma et al. 2021a, b). The total triterpene concentration of ten distinct pear cultivars' peel and flesh was compared by Li et al. (2014). Additionally, monomeric compounds' anti-inflammatory properties were assessed (Zhu et al. 2018). The study's conclusions included several important discoveries about the anti-inflammatory qualities of horticulture wastes. Because every chemical component in the pear peel was roughly 6–20 times more abundant than in the pear flesh. Yaguang, Hongpi, Qingpi, and Guifei peel and flesh had comparatively higher levels of total triterpenes and demonstrated more potent anti-inflammatory properties (Zhu et al. 2018). In a different study, citrus trash was used to extract limonin, a triterpenoid phytochemical (Ishisono et al. 2017). Anti-inflammatory, antibacterial, antioxidant, and anticancer properties are all exhibited by limonin (Liu et al. 2019; Sharma et al. 2021a, b).

20.4.5 Anti-obesity Activity

One of the main issues that people are dealing with nowadays is obesity, which is problematic in and of itself and causes a host of additional illnesses. It is anticipated that natural phytopigments, primarily carotenoids and anthocyanins, will significantly reduce obesity by reducing adipose tissue, or fat deposition (Sharma et al. 2021a, b). Another important factor in largely resolving this issue is the waste produced by horticulture products. For example, a number of extracts and components from mango waste have demonstrated anti-inflammatory and antioxidative qualities while also showing encouraging outcomes in lowering diabetes and obesity in people as well as animals (Sharma et al. 2021a, b). Therefore, as previously stated, because of their high quantity of healthy components, fruit and vegetable wastes—which are frequently thrown away—can have anti-obesity effects. These substances may help control weight by promoting general health, lowering inflammation, and regulating metabolism (Tariq et al. 2023). A great alternative method of preventing cancer and other conditions like diabetes, obesity, and Alzheimer's disease is betaine, which is isolated from beetroot waste (Sharma et al. 2021a, b; Zhao et al. 2018). Bioactive chemicals found in the peels, pomace, and seeds of fruits like grapes, citrus fruits, and apples, as well as the stems and cores of vegetables like broccoli, carrots, and beets, have anti-obesity properties (Bhardwaj et al. 2022). According to Devalaraja et al. (2011), bioactive proanthocyanidins found in persimmon (*Diospyros kaki* L.) fruit peel extract have antidiabetic and anti-obesity properties. Phlorizin and phloretein, which were shown to be abundant in apple pomace, were also found to decrease obesity (Shin et al. 2016; Sharma et al. 2021a, b).

20.5 Remunerative Value-Added Products from Horticultural Waste Commodities

A variety of profitable and valuable byproducts can be made from solid waste from the fruit and vegetable sectors. Wastes from various sources can be transformed both fermentatively and non-fermentatively (Joshi 2020).

20.5.1 Apple

Apple pomace can make up as much as 30% of the initial fruit, a processing byproduct produced when apple juice is made. Apple pomace, either fresh or dried, can also be fed to animals. Due to its low nitrogen content, apple pomace is not a comprehensive animal diet (Raj et al. 2016). Because apple waste is rich in nutraceutical chemicals, it can be used for a variety of reasons. Following fermentation with various yeasts and drying, apple pomace is enhanced with proteins, vitamins, minerals, and lipids, making it suitable for animal feed (Raj et al. 2016). Many companies extract dietary fibers from the pomace of apples and pears. Mechanical filtration, screening, and dehydration of the filtered materials are all steps in the extraction process. The fiber content of apple and pear pomace is 56% and 75%, respectively (Joshi 2020). Using Rhodotorula and solid-state fermentation, apple pomace can be used to produce microbial colors (Raj et al. 2016). A variety of culinary goods have been made using apple pomace. These include products made with apple pomace in different amounts, such as applesauce, jam, toffees, biscuits, etc. (Joshi 2020). Fruit waste can be fermented to produce a variety of drinks, including vinegar, wine and brandy, beer, and cider. Apple pomace can be employed to make a suitable low-alcoholic beverage with 3–5% alcohol (Raj et al. 2016). An excellent substrate for the synthesis of citric acid has been found to be apple pomace. Calcium citrate can be made from unfermented juice, and sodium citrate is created by adding a powerful solution of sodium carbonate. It is possible to filter and concentrate the precipitating calcium carbonate to create sodium citrate by crystallizing it (Hang et al. 1977; Joshi 2020). Apple pomace may be used to make cider, and adding carbonation to the beverage improves its quality (Raj et al. 2016). There are benefits to using apple pomace as an initial ingredient for making lactic acid, such as its high level of fructose and free glucose, which are great carbon sources for making lactic acid (Joshi 2020). *Trichoderma viridae* and *Aspergillus species* can be used to manufacture the enzymes cellulase and xylanase from dried apple pomace. Pomace from apple, has been demonstrated to be an effective substrate for the synthesis of numerous enzymes, including cellulose, pectinase, xylanase, and amylase. It has been discovered that apple pomace can be used for developing baker's yeast (Raj et al. 2016). Because apple pomace contains a good amount of nutrients, it works well as a medium for microbial pigment production. According to Joshi (2020), it has been used to produce microbiological colors.

20.5.2 Pineapple

On Hawaii Island, pineapple bran is much sought after as a livestock feed. In Hawaii, canary wastes are also fed to dairy calves after they have been ensiled. Fruit and vegetable processing waste has a low protein content but is high in fiber, including cellulose, hemicellulose, lignin, and silica (Raj et al. 2016). Therefore, in addition to being utilized as animal feed, pineapple waste has numerous uses for people and offers numerous benefits. By using Solid-State Fermentation (SSF) to ferment brewery waste with *Aspergillus niger*, citric acid can be produced economically (Raj et al. 2016). One of the proteolytic groups of enzymes that may be isolated from pineapples is bromelain. Its biological background and the reality that it has no documented adverse effects on human beings have made it a popular pharmacological and therapeutic substance (Maurer 2001; Meena et al. 2022). In addition to this, the enzyme offers additional advantageous properties that many sectors can utilize. According to Meena et al. (2022) bromelain is thought to be the most beneficial and extensively researched enzyme that has the potential to tenderize meat. Vinegar, a useful food addition, can be made by fermenting pineapple waste, particularly the peels (Sarangi et al. 2022). According to Pardo et al. (2014), pineapple peels (40.5%) and crown leaves (43.5%) have a high cellulose content, which makes them appropriate for cellulose derivatives that may find use in the polymer industry. Pineapple peel cellulose is transformed into carboxymethyl cellulose and utilized to create environmentally friendly packaging (Chumee and Khemmakama 2014; Meena et al. 2022). Microorganisms such as *Saccharomyces cerevisiae* may ferment pineapple waste, especially peels and cores, to produce bioethanol. Anaerobic digestion of pineapple waste can also yield biogas, an alternative source of energy (Paz-Arteaga et al. 2024).

20.5.3 Tomato

Using *Aspergillus niger*, pulp from tomatoes and the pomace from tomato waste processing can also be utilized to produce single-cell protein (Raj et al. 2016). It is possible to utilize the seeds to make edible oil. The leftovers are used to create purée, sauce, or liquid. Arotenoids can be recovered from tomato leftovers at a rate of about 37 mg/kg of waste (Sinha et al. 2024). Pomace may be dried in a number of ways, including freeze drying, drum drying, and tray drying. The resulting powder can be used to add color and flavor to soups, ketchups, and other foods (Kumari and Singh 2018). The potent antioxidant lycopene is abundant in tomato waste. For food and nutraceutical uses, lycopene can be separated via supercritical carbon dioxide extraction to produce a stable, free-flowing powder (Méndez-Carmona et al. 2022). Food packaging, medicine, and the agricultural sector can all benefit from the use of tomato waste as a raw material for the creation of microbial biomass, which includes biofertilizers and biopesticides, as well as microbial endo- and exopolymers (poly-hydroxybutyric acid) (Markam 2014). In addition, tomato waste can be utilized to improve a number of other soil-based qualities. For example, the solid, fibrous part of tomato waste that has been anaerobically digested (called compost) can be used as a soil conditioner in order to raise the soil's organic content. To provide soils with essential nutrients, organic compost made from tomato and vegetable waste can be applied as fertilizer (Markam 2014). Other bioactive substances found in tomato wastes, such as vitamins, phenolic compounds, and carotenoids, can be isolated and used in multiple sectors (Sravani et al. 2024). Additionally, tomato waste can be used to produce biogas, which lessens reliance on fuel wood while also addressing interior pollution and residual disposal issues (Markam 2014).

20.5.4 Mango

A significant amount of the stones and peels that are produced during the commercial processing of mangoes are thrown away and wind up in garbage. For growing and egg-laying chicken diets, major wastes like mango seed kernels can substitute about 20% of maize since they are an excellent source of carbohydrates (Raj et al. 2016). Mango peel and kernels can be dehydrated to create a high-quality energy meal for animal feed (Beerh et al. 1976). Pectin is abundant in apple pomace and mango peels, which are also easily obtained from processing facilities (Sinha et al. 2024). When finely crushed and defatted mango peel was soaked in a pH 2.5 sulfuric acid solution at 80 °C for 120 min, the maximum pectin output of 21.0% was achieved (Joshi 2020). In a similar manner, mango seed kernels can be used to extract glucose. Mango peel and fiber are used to make organic fertilizer and biogas (Sinha et al. 2024). Similarly, mango seed kernels may also provide starch (Raj et al. 2016). Mangoes have an average of 15% fat and 9–23% stone (Palaniswamy et al. 1974). According to one estimate, 30,000 tons of mango fat worth 20 crore might be produced each year from 0.3 million tons of dried mango seeds (Joshi 2020).

20.5.5 Grapes

After fermentation, waste from distilleries, breweries, and wineries can also be fed to animals. After cultivating bacteria on grape pomace and wine lees, these materials can also be used to make animal feed (Raj et al. 2016). According to Joshi (2020), leftover pomace and grape peel can be a good commercial supply of pectin. The skin of blue grapes also yields natural colors. By using Solid-State Fermentation (SSF) with *Aspergillus niger* on a grape pomace substrate, citric acid can be generated on an industrial scale (Raj et al. 2016). It is possible to separate and crush the seed from the pomace to produce oil, which, when purified, is edible. Ten to fifteen percent of the garbage can be turned into oil (Joshi 2020). Value-added goods made from grape pomace, such as grape pomace jam, RTS, and squash, have been developed and standardized (Gurumeenakshi et al. 2021). Ethanol can be produced when sugars found in grape pomace ferment (Zhao et al. 2023). Biochar, which may be utilized for environmental remediation and soil enhancement, can be made from grape pomace (Abreu et al. 2024).

20.5.6 Ash Gourd

The pulp and peel of sweets are removed during manufacturing and can be employed as animal feed. It is used in Asian nations and has a lot of delicious waxy compounds in its rind (Sinha et al. 2024). Fruits can be coated with wax made from ash gourd peels, which prolongs their shelf life and may have health advantages (Sreenivas et al. 2011). Chromium (VI) and other heavy metals can be eliminated from water using ash gourd peel waste as a bio-sorbent (Sreenivas et al. 2014). Byproducts from ash gourds contain useful functional substances with a variety of uses, such as terpenoids, polyphenols, and essential oils (Gupta et al. 2021). Other value-added products, including bioethanol, enzymes, and edible mushrooms, can also be made from ash gourd waste (Bhushan et al. 2025).

20.5.7 Banana

Dairy and beef cattle are frequently fed banana waste. It is also useful for growing yeast and making animal feed. For use as chicken feed, ripe banana peels—which make up roughly 22–30% of the material—are sun-dried (Raj et al. 2016). The physicochemical characteristics of starch have been investigated, and the procedure for extracting starch from pseudostems has been standardized (Subrahmanyan et al. 1957; Joshi 2020). After the banana fruit bunch is harvested, a large amount of plant biowaste is produced, of which 30–34% is pseudostem. This is a component of considerable nutritional value, nevertheless. They have catecholamines and polyphenols, which are good antioxidants. Food, pharmaceuticals, and medical medications all contain it (Srivastava 2023). It is possible to increase the amount of dietary fiber in bread, cookies, and meat by drying and grinding banana peels into flour (Zaini et al. 2022). The male banana bud can be used to make a pickle, which is a high-value product. One such fermented food (pickle) made from male inflorescence is thakku, which is delicious and keeps well at room temperature for a year (Srivastava 2023). Waste from bananas, particularly the fibers, can also be utilized to make a number of nonfood products. For instance, the fiber from the banana pseudostem is blended with cotton or silk to create handicrafts including sarees, blankets, bags, table mats, and wall hangings, as well as ropes and craft papers (Srivastava 2023). The use of banana waste is not restricted to this, though; there are numerous additional applications as well. Generally speaking, the core stem portion of the banana is used in cooking. Either manually or mechanically, juice is removed from the stem's center core. Potassium and fiber are abundant in the stem. It has the ability to dissolve kidney stones (Kumar et al. 2021; Srivastava 2023). Additionally, banana pseudostems contain roughly 5% edible starch that can be collected and used in a variety of ways (Raj et al. 2016). A highly nutritious energy drink, banana juice is made from overripe, wasted fruit and infused with basil seeds. Omega-3 and omega-6 fatty acids, as well as potassium, are abundant in the juice. The beverage lowers body heat, prevents constipation, treats acidity, lowers body weight, and regulates blood sugar levels (Srivastava 2023).

20.5.8 Potato

Potato peeling and slicing can yield starch from the leftovers, which can then be utilized as a thickening agent. The nontoxic fungus *Pleorotus ostreatus* can be utilized to produce protein from potato peels treated with 0.04% ammonium chloride (Raj et al. 2016). To reduce its detrimental impacts on the environment, improve soil nutrition, reduce the demand for artificial synthetic fertilizers, and effectively boost food production, potato peel wastes could be turned into biofertilizer (Khanal et al. 2024). Fuel-grade ethanol can be produced by fermenting and liquefying the starch found in potato peel and other "zero value" byproducts from potato processing (Hashem and Darwish 2010). In addition to being used as dietary fiber or pharmaceuticals, the value-added products produced from the processing of potato peel have a wide range of uses, including biosorbents, which are used to treat effluents and many other contaminated sources, including metals, pigments, colors, biocomposites, and materials for packaging (Khanal et al. 2024). Potato peel flour was added to cake and dough by Jeddou et al. (2017) as a source of protein and dietary fiber. They observed significant color change between the control and peel-assisted samples. When peel and flour are combined, the cake's hardness, yellowness, and lightness are all significantly reduced (Ijaz et al. 2024). Acid hydrolysis was used to create potato peel nanocrystals of cellulose and cellulose microcrystallines. Additionally, because they are biodegradable, these nanocrystals outperform metallic nanoparticles in the realm of nanocomposites (Raigond et al. 2018; Khanal et al. 2024). Different forms of dietary fiber have taken the place of wheat flour in the manufacturing of baked goods. According to a study, leftover potato peel has a higher water-holding capacity than wheat fiber and has less starch and total dietary fiber (Ijaz et al. 2024). In addition to the applications of potato waste that have already been discussed, there are some consequences for human health in the medical profession. Potato peel is an effective source of starch and can be used as an anti-ulcer and wound-healing agent. Potato peel's remarkable quality makes it suitable for use as medical bandages (Ijaz et al. 2024). These potato peel bandages offer a number of advantages and health benefits over conventional ones. The following are some advantages of using potato peel bandages: (i) a quicker return of the skin to its natural texture and color; (ii) a quicker recovery of the epithelium; (iii) a less painful bandage removal process; and (iv) comfort. The fact that these bandages do not shrink over the course of application is one of their additional benefits (Pathak et al. 2018; Ijaz et al. 2024). The effectiveness of potato peels in preserving the postharvest shelf life of baked goods was examined in a different type of study. Here, potato peel was utilized as a natural antioxidant resource to prolong biscuit shelf life by delaying lipid oxidation (Rowayshed et al. 2015; Ijaz et al. 2024).

20.5.9 Apricots

The kernels of apricot fruit can also be a great source of oil. After pulping, any leftover stone can be utilized to extract hydrocarbons. The oil has a purity level similar to almond's. The oil that is extracted can be employed to make soap, lip salve, face lotion, and other products (Sinha et al. 2024). Bioactive substances including polyphenols and carotenoids, which have uses in both food and medicine, can be extracted from apricot pulp. It is possible to extract apricot kernel oil and use it in nutraceuticals, cosmetics, and even as a biodiesel fuel (Kasapoğlu et al. 2020). The leftover press cake from apricot kernel oil extraction is a great source of protein isolate and essential oil. Additionally, "Hawan Samagri" for religious events can be made with press cake (Sinha et al. 2024). Activated carbon, which has uses in water purification and energy storage, can be made from apricot kernel shells. Additionally, they can be utilized to create carbon dots (CDs), which may find utility in fluorescence detection (Hong-Bo et al. 2022). Protein from apricot kernel press cake may be extracted and utilized in therapeutic food uses (Sharma et al. 2014). Biodiesel, a renewable and sustainable fuel source, may be made from apricot kernel oil (Kim et al. 2024). Herbal teas with a variety of health advantages can be made with apricot kernel flour (Akhone et al. 2022). Byproducts from the processing of apricot juice can yield useful ingredients such as pectin, which can be employed in food compositions and other components for a range of industrial uses (Ali et al. 2025).

20.5.10 Citrus

Essential oils can be found in abundance in citrus trash. Citrus essential oils can range in concentration from 1:5 to 1:20. Perfumes, medications, cosmetics, and soap scenting all use the oil. One important byproduct of the citrus industry is D-limonine (Raj et al. 2016). Oil, pectin, and other byproducts are abundant in citrus garbage. Around 1600 tons of orange oil, 25,000 tons of lemon oil, and 180 tons of grapefruit oil are produced worldwide from essential oils obtained from all types of fruits, particularly citrus (Wright 1995; Joshi 2020). The main byproduct of processing in the juice industry is the citrus rind. Citrus skins are the primary source of pectin globally. Citrus fruit rinds are typically used to extract pectin. But before processing the pectin, the essential oil from the peel is removed (Sinha et al. 2024). Danish manufacturers of "Genu" pectin virtually have a monopoly because they produce year-round after sourcing the necessary amount of dried citrus peels from around the globe (Joshi 2020). The flavedo contains about 70% of the carotenoids found in citrus fruits (Garzon and Hours 1992). Xanthophylls and other carotenoids are the main ingredients. By keeping the dehydrated skins in a cool silo and promptly detaching them, carotenoids are retained (Sinha et al. 2024). Lime, lemon, apple, galgal, and other materials can be used to make citric acid (Joshi 2020). Additionally, citrus waste from various processes can be successfully used to create some fermented beverages. In several European nations, orange peel is commonly consumed and used to make vinegar (Gera and Kramer 1969; Joshi 2020).

20.5.11 Cabbage

Fermenting cabbage waste from food manufacturing can yield a variety of enzymes. Waste from sauerkraut can be fermented with the aid of *Candida utilis* to produce the enzyme invertase. The food manufacturing sector makes extensive use of this enzyme (Raj et al. 2016). Bioactive substances such as phenolics and isothiocyanates can be extracted from cabbage trash and utilized as bioherbicides (Shinali et al. 2024). It is possible to use the native microflora found in cabbage trash to produce enzymes such as cellulase, amylase, and protease. *Xanthamonas compestris* can be used to turn cabbage trash into microbial gums like xanthan (Raj et al. 2016). Biochar, which can be utilized as an amendment to the soil and for carbon sequestration, is created by pyrolyzing cabbage waste (Zhang et al. 2022). Peroxidase, which can be utilized to break down contaminants in wastewater, can be obtained from cabbage waste (Joel et al. 2020). Anaerobic digestion can be utilized to create biogas or bioethanol from cabbage waste (Jameel et al. 2024). Vermicompost, which can be utilized as fertilizer, is created when earthworms break down cabbage waste (Shinali et al. 2024).

20.6 Future Prospects

Agricultural products like fruits and vegetables produce waste throughout postharvest processing, dispersion, and consumerism. The majority of byproducts are thrown away; however, some are used as fuel, building materials, or animal feed (Aqilah et al. 2023). The following discusses a few of the potential future developments in the arena of horticultural waste management.

- Making use of waste and byproducts from food manufacturing and agriculture has gained importance in recent years. The peel, skin, fruits, kernels, seeds, and leaves that make up fruit and vegetable waste have many benefits, including antibacterial, antioxidant, and anti-browning qualities, as well as bioactive compounds that can be used in many ways.
- To do this, however, a great deal of study is needed to identify more innovative methods for fully extracting the various bioactive substances found in horticultural wastes, which can then be applied in other industries to generate income.
- Developments in bioprocessing technologies, new inventions in pigment stabilization, innovative and economically viable resources, appropriate extraction techniques, the ability to apply a technique to a variety of pigments, the potential for creating pigment-enriched food ingredients and bioactives, and a trend toward creating pigment-enriched foods and bioactives are all necessary for the extraction of natural pigments.
- To lessen field loss and satisfy population hunger, businesses and academia must conduct extensive research to generate more and more value-added products from various horticultural wastes.
- More people in both urban and rural regions need to be aware that fruit and vegetable waste can be turned into some amazing value-added products rather than just being thrown out. Additionally, the general public is to be informed that eating snacks made from horticulture waste that has been dumped is just as healthy as eating them in their traditional form.
- In order for the waste management industry to see a significant increase in startups and entrepreneurship, government subsidies are necessary.

20.7 Conclusion

It is clear that a significant amount of materials are wasted from the horticultural industry every year, amounting to a staggering sum. Because horticultural products are inherently highly perishable, the loss is exacerbated by the additional items that are removed from them. Therefore, in order to address the issue, the chapter offers a critical analysis of waste management. The materials that are discarded from both industrial fruit and vegetable processing and table eating have a high functional value and are packed with bioactive chemicals. There is currently a need for more knowledge of the medicinal advantages of these chemicals, such as their antibacterial, antioxidant, anti-cancerous, and anti-obesity qualities. In order to extract these substances more and more, technologies will also need to be improved. Thus, the extracted compounds can be used in food research and the creation of functional foods, as well as in the chemical, pharmaceutical, cosmetic, and food sectors. As a result, it can be said that, in light of the annual exponential growth in the population, the need for food and nutrition must be reduced. This can be achieved by utilizing both conventional agricultural and horticultural products as well as those made from unconventional waste materials.

References

Abreu T, Sousa P, Gonçalves J, Hontman N, Teixeira J, Câmara JS, Perestrelo R (2024) Grape pomace as a renewable natural biosource of value-added compounds with potential food industrial applications. Beverages 10(2):45. https://doi.org/10.3390/beverages10020045

Afifi MM (2011) Enhancement of lactic acid production by utilizing liquid potato. Int J Biol Chem 5:91–102. https://scialert.net/abstract/?doi=ijbc.2011.91.102

Ajila CM, Rao UP (2013) Mango peel dietary fibre: composition and associated bound phenolics. J Funct Foods 5(1):444–450. https://doi.org/10.1016/j.jff.2012.11.017

Akhone MA, Bains A, Tosif MM, Chawla P, Fogarasi M, Fogarasi S (2022) Apricot kernel: bioactivity, characterization, applications, and health attributes. Foods 11(15):2184. https://doi.org/10.3390/foods11152184

Ali A, Chen Y, Liu H, Yu L, Baloch Z, Khalid S et al (2019) Starch-based antimicrobial films functionalized by pomegranate peel. Int J Biol Macromol 129:1120–1126. https://doi.org/10.1016/j.ijbiomac.2018.09.068

Ali EA, Mohammed DM, Abd El Gawad F, Orabi MA, Gupta RK, Srivastav PP (2025) Valorization of food processing waste byproducts for essential oil production and their application in food system. Waste Manag Bull 3:100200. https://doi.org/10.1016/j.wmb.2025.100200

Andreou V, Dimopoulos G, Dermesonlouoglou E, Taoukis P (2020) Application of pulsed electric fields to improve product yield and waste valorization in industrial tomato processing. J Food Eng 270:109778. https://doi.org/10.1016/j.jfoodeng.2019.109778

Angersbach A, Heinz V, Knorr D (2000) Effects of pulsed electric fields on cell membranes in real food systems. Innovative Food Sci Emerg Technol 1(2):135–149. https://doi.org/10.1016/S1466-8564(00)00010-2

Aqilah NMN, Rovina K, Felicia WXL, Vonnie JM (2023) A review on the potential bioactive components in fruits and vegetable wastes as value-added products in the food industry. Molecules 28(6):2631. https://doi.org/10.3390/molecules28062631

Araújo RG, Rodriguez-Jasso RM, Ruiz HA, Govea-Salas M, Pintado ME, Aguilar CN (2020) Process optimization of microwave-assisted extraction of bioactive molecules from avocado seeds. Ind Crop Prod 154:112623. https://doi.org/10.1016/j.indcrop.2020.112623

Ashong A (2024) Examining the role of phytochemicals in cancer prevention and treatment. Clin Nutr Hosp Diet 44(3):1–2. https://doi.org/10.12873/0211-6057.44.03.230

Balasundram N, Sundram K, Samman S (2006) Phenolic compounds in plants and agri-industrial by-products: antioxidant activity, occurrence, and potential uses. Food Chem 99(1):191–203. https://doi.org/10.1016/j.foodchem.2005.07.042

Barsotti L, Merle P, Cheftel JC (1998) Traitement des aliments par champs electriques pulses. 1. Aspects physiques. Sci Aliment 18(6):583–601

Baydar NG, Özkan G, Yaşar S (2007) Evaluation of the antiradical and antioxidant potential of grape extracts. Food Control 18(9):1131–1136. https://doi.org/10.1016/j.foodcont.2006.06.011

Beerh OP, Raghuramaiah B, Krishnamurthy GV (1976) Utilization of mango waste: peel as a source of pectin. J Food Sci Technol 13(2):96–97. http://ir.cftri.res.in/id/eprint/6196

Bener M, Shen Y, Apak R, Finley JW, Xu Z (2013) Release and degradation of anthocyanins and phenolics from blueberry pomace during thermal acid hydrolysis and dry heating. J Agric Food Chem 61(27):6643–6649. https://doi.org/10.1021/jf401983c

Bhadange YA, Carpenter J, Saharan VK (2024) A comprehensive review on advanced extraction techniques for retrieving bioactive components from natural sources. ACS Omega 9(29):31274–31297

Bhardwaj K, Najda A, Sharma R, Nurzyńska-Wierdak R, Dhanjal DS, Sharma R et al (2022) Fruit and vegetable Peel-enriched functional foods: potential avenues and health perspectives. Evid Based Complement Alternat Med 2022(1):8543881. https://doi.org/10.1155/2022/8543881

Bhushan B, Nanda V, Jindal N (2025) Strategies for processing and valorization of ash gourd byproducts: a comprehensive review. Food Chem 473:143040. https://doi.org/10.1016/j.foodchem.2025.143040

Cadiz-Gurrea MDLL, Pinto D, Delerue-Matos C, Rodrigues F (2021) Olive fruit and leaf wastes as bioactive ingredients for cosmetics—a preliminary study. Antioxidants 10(2):245. https://doi.org/10.3390/antiox10020245

Carbone K, Amoriello T, Iadecola R (2020) Exploitation of kiwi juice pomace for the recovery of natural antioxidants through microwave-assisted extraction. Agriculture 10(10):435. https://doi.org/10.3390/agriculture10100435

Carrara VS, Filho LC, Garcia VAS, Faiões VS, Cunha-Júnior EF, Torres-Santos EC, Cortez DAG (2017) Supercritical fluid extraction of pyrrolidine alkaloid from leaves of *Piper amalago* L. Evid Based Complement Alternat Med 2017(1):7401748. https://doi.org/10.1155/2017/7401748

Carvalho F, Lahlou RA, Silva LR (2025) Exploring bioactive compounds from fruit and vegetable by-products with potential for food and nutraceutical applications. Foods 14(22):3884. https://doi.org/10.3390/foods14223884

Castillejo N, Martínez-Zamora L (2024) Bioactive compounds from fruit and vegetable waste: extraction and possible utilization. Foods 13(5):775. https://doi.org/10.3390/foods13050775

Castro-Muñoz R, Yáñez-Fernández J, Fíla V (2016) Phenolic compounds recovered from agro-food by-products using membrane technologies: an overview. Food Chem 213:753–762. https://doi.org/10.1016/j.foodchem.2016.07.030

Ćetković G, Savatović S, Čanadanović-Brunet J, Djilas S, Vulić J, Mandić A, Četojević-Simin D (2012) Valorisation of phenolic composition, antioxidant and cell growth activities of tomato waste. Food Chem 133(3):938–945. https://doi.org/10.1016/j.foodchem.2012.02.007

Chang JI, Tsai JJ, Wu KH (2006) Composting of vegetable waste. Waste Manag Res 24(4):354–362. https://doi.org/10.1177/0734242X06065727

Chau CF, Chen CH, Lee MH (2004) Comparison of the characteristics, functional properties, and in vitro hypoglycemic effects of various carrot insoluble fiber-rich fractions. LWT-Food Sci Technol 37(2):155–160. https://doi.org/10.1016/j.lwt.2003.08.001

Chen Y, Xie MY, Gong XF (2007) Microwave-assisted extraction used for the isolation of total triterpenoid saponins from *Ganoderma atrum*. J Food Eng 81(1):162–170. https://doi.org/10.1016/j.jfoodeng.2006.10.018

Chitturi S, Talatam VG, Vuppu S (2013) Studies on protein content, protease activity, antioxidants potential, melanin composition, glucosinolate and pectin constitution with brief statistical analysis in some medicinally significant fruit peels. Der Pharm Lett 5:13–23

Cho H, Jung H, Lee H, Yi HC, Kwak HK, Hwang KT (2015) Chemopreventive activity of ellagitannins and their derivatives from black raspberry seeds on HT-29 colon cancer cells. Food Funct 6(5):1675–1683. https://doi.org/10.1039/c5fo00274e

Chumee J, Khemmakama P (2014) Carboxymethyl cellulose from pineapple peel: useful green bioplastic. Adv Mater Res 979:366–369. https://doi.org/10.4028/www.scientific.net/AMR.979.366

Coman MM, Oancea AM, Verdenelli MC, Cecchini C, Bahrim GE, Orpianesi C et al (2018) Polyphenol content and in vitro evaluation of antioxidant, antimicrobial and prebiotic properties of red fruit extracts. Eur Food Res Technol 244:735–745. https://doi.org/10.1007/s00217-017-2997-9

Coman V, Teleky BE, Mitrea L, Martău GA, Szabo K, Călinoiu LF, Vodnar DC (2020) Bioactive potential of fruit and vegetable wastes. Adv Food Nutr Res 91:157–225. https://doi.org/10.1016/bs.afnr.2019.07.001

Conde-Hernández LA, Espinosa-Victoria JR, Guerrero-Beltrán JÁ (2017) Supercritical extraction of essential oils of *Piper auritum* and *Porophyllum ruderale*. J Supercrit Fluids 127:97–102. https://doi.org/10.1016/j.supflu.2017.03.026

Corrales M, Toepfl S, Butz P, Knorr D, Tauscher B (2008) Extraction of anthocyanins from grape by-products assisted by ultrasonics, high hydrostatic pressure or pulsed electric fields: a comparison. Innovative Food Sci Emerg Technol 9(1):85–91. https://doi.org/10.1016/j.ifset.2007.06.002

da Rocha CB, Noreña CPZ (2020) Microwave-assisted extraction and ultrasound-assisted extraction of bioactive compounds from grape pomace. Int J Food Eng 16(1–2):20190191. https://doi.org/10.1515/ijfe-2019-0191

Das JN, Basak S, Sablania V, Negi R, Yadav N, Rustagi S, Singh S, Rai AK, Shreaz S, Yadav AN (2024) Food waste as potential bioresources for extraction of carotenoid of nutraceutical importance: current research and future challenges. J Appl Biol Biotechnol 12(5):1–15. https://doi.org/10.7324/JABB.2024.162743

de Oliveira CF, Giordani D, Lutckemier R, Gurak PD, Cladera-Olivera F, Marczak LDF (2016) Extraction of pectin from passion fruit peel assisted by ultrasound. LWT-Food Sci Technol 71:110–115. https://doi.org/10.1016/j.lwt.2016.03.027

del Carmen Robles-Ramírez M, Monterrubio-López R, Mora-Escobedo R, del Carmen Beltrán-Orozco M (2016) Evaluation of extracts from potato and tomato wastes as natural antioxidant additives. Arch Latinoam Nutr 66(1):66–73

Del Valle M, Cámara M, Torija ME (2006) Chemical characterization of tomato pomace. J Sci Food Agric 86(8):1232–1236. https://doi.org/10.1002/jsfa.2474

Delsart C, Ghidossi R, Poupot C, Cholet C, Grimi N, Vorobiev E et al (2012) Enhanced extraction of phenolic compounds from Merlot grapes by pulsed electric field treatment. Am J Enol Vitic 63(2):205–211. https://doi.org/10.5344/ajev.2012.11088

Derrien M, Badr A, Gosselin A, Desjardins Y, Angers P (2017) Optimization of a green process for the extraction of lutein and chlorophyll from spinach by-products using response surface methodology (RSM). LWT-Food Sci Technol 79:170–177. https://doi.org/10.1016/j.lwt.2017.01.010

Devalaraja S, Jain S, Yadav H (2011) Exotic fruits as therapeutic complements for diabetes, obesity and metabolic syndrome. Food Res Int 44(7):1856–1865. https://doi.org/10.1016/j.foodres.2011.04.008

Dharmaraj BM (2022) Management of agriculture waste in protecting environment. Seminar report, University of Agricultural Sciences, Bangalore. RAWE2020. https://rawe2020.in/wp-content/uploads/2022/11/Seminar-Report_pdf

Dhingra D, Michael M, Rajput H, Patil RT (2012) Dietary fibre in foods: a review. J Food Sci Technol 49:255–266. https://doi.org/10.1007/s13197-011-0365-5

Doulabi M, Golmakani MT, Ansari S (2020) Evaluation and optimization of microwave-assisted extraction of bioactive compounds from eggplant peel by-product. J Food Process Preserv 44(11):e14853. https://doi.org/10.1111/jfpp.14853

Duda-Chodak A, Tarko T (2007) Antioxidant properties of different fruit seeds and peels. Acta Sci Pol Technol Aliment 6(3):29–36

Dugo G, Di Giacomo A (2002) Citrus: the genus citrus, 1st edn. Taylor and Francis Group, Boca Raton. https://doi.org/10.1201/9780203216613

Egyankosh (2022) Unit 14: value addition of agri waste, pp 288–304. https://www.egyankosh.ac.in/bitstream/123456789/104885/1/Unit%2014.pdf

El Kantar S, Boussetta N, Lebovka N, Foucart F, Rajha HN, Maroun RG et al (2018) Pulsed electric field treatment of citrus fruits: improvement of juice and polyphenols extraction. Innov Food Sci Emerg Technol 46:153–161. https://doi.org/10.1016/j.ifset.2017.09.024

Ellington E, Bastida J, Viladomat F, Codina C (2003) Supercritical carbon dioxide extraction of colchicine and related alkaloids from seeds of *Colchicum autumnale* L. Phytochem Anal 14(3):164–169. https://doi.org/10.1002/pca.702

Falcão MA, Scopel R, Almeida RN, do Espirito Santo AT, Franceschini G, Garcez JJ et al (2017) Supercritical fluid extraction of vinblastine from *Catharanthus roseus*. J Supercrit Fluids 129:9–15. https://doi.org/10.1016/j.supflu.2017.03.018

Favareto R, Teixeira MB, Soares FAL, Belisário CM, Corazza ML, Cardozo-Filho L (2017) Study of the supercritical extraction of Pterodon fruits (Fabaceae). J Supercrit Fluids 128:159–165. https://doi.org/10.1016/j.supflu.2017.05.028

Federal TVET Agency (2020) Developing waste management strategies: module from horticulture crop production level-IV. https://dl.ojocv.gov.et/admin_/ttlm_/M18-Develop%20waste%20management.pdf

Friedman M (1997) Chemistry, biochemistry, and dietary role of potato polyphenols. A review. J Agric Food Chem 45(5):1523–1540. https://doi.org/10.1021/jf960900s

Ganesh KS, Sridhar A, Vishali S (2022) Utilization of fruit and vegetable waste to produce value-added products: conventional utilization and emerging opportunities—a review. Chemosphere 287:132221. https://doi.org/10.1016/j.chemosphere.2021.132221

García P, Fredes C, Cea I, Lozano-Sánchez J, Leyva-Jiménez FJ, Robert P et al (2021) Recovery of bioactive compounds from pomegranate (*Punica granatum* L.) peel using pressurized liquid extraction. Foods 10(2):203. https://doi.org/10.3390/foods10020203

García-González AG, Rivas-García P, Escamilla-Alvarado C, Ramírez-Cabrera MA, Paniagua-Vega D, Galván-Arzola U et al (2025) Fruit and vegetable waste as a raw material for obtaining functional antioxidants and their applications: a review of a sustainable strategy. Biofuels Bioprod Biorefin 19(1):231–249. https://doi.org/10.1002/bbb.2698

Garg N (2023) Value addition of horticultural waste. Central Institute for Subtropical Horticulture, Lucknow

Garzon GG, Hours RA (1992) Citrus waste: an alternative substrate for pectinase production in solid state culture. Bioresource Technol 39(1):93–95. https://doi.org/10.1016/0960-8524(92)90061-2

Gera IB, Kramer A (1969) The utilization of food industries waste. In: Chichester CO, Mark EM, Stewart GF (eds) Advances in food research. Academic Press, New York, p 28

Gomes-Araújo R, Martínez-Vázquez DG, Charles-Rodríguez AV, Rangel-Ortega S, Robledo-Olivo A (2021) Bioactive compounds from agricultural residues, their obtaining techniques, and the antimicrobial effect as postharvest additives. Int J Food Sci 2021(1):9936722. https://doi.org/10.1155/2021/9936722

Gómez-Estaca J, Calvo MM, Sánchez-Faure A, Montero P, Gómez-Guillén MC (2015) Development, properties, and stability of antioxidant shrimp muscle protein films incorporating carotenoid-containing extracts from food by-products. LWT-Food Sci Technol 64(1):189–196. https://doi.org/10.1016/j.lwt.2015.05.052

Gómez-García R, Martínez-Ávila GC, Aguilar CN (2012) Enzyme-assisted extraction of antioxidative phenolics from grape (*Vitis vinifera* L.) residues. *3*. Biotech 2:297–300. https://doi.org/10.1007/s13205-012-0055-7

Gorinstein S, Zachwieja Z, Folta M, Barton H, Piotrowicz J, Zemser M et al (2001) Comparative contents of dietary fiber, total phenolics, and minerals in persimmons and apples. J Agric Food Chem 49(2):952–957. https://doi.org/10.1021/jf000947k

Goulas V, Hadjivasileiou L, Primikyri A, Michael C, Botsaris G, Tzakos AG, Gerothanassis IP (2019) Valorization of carob fruit residues for the preparation of novel bi-functional polyphenolic coating for food packaging applications. Molecules 24(17):3162. https://doi.org/10.3390/molecules24173162

Gowe C (2015) Review on potential use of fruit and vegetables by-products as a valuable source of natural food additives. Food Sci Qual Manag 45(1):47–61

Grassino AN, Brnčić M, Vikić-Topić D, Roca S, Dent M, Brnčić SR (2016) Ultrasound assisted extraction and characterization of pectin from tomato waste. Food Chem 198:93–100. https://doi.org/10.1016/j.foodchem.2015.11.095

Gray ST (2006) Food waste management and value added products. J Food Sci 69(3):102–104. https://doi.org/10.1111/j.1365-2621.2004.tb13340.x

Guandalini BBV, Rodrigues NP, Marczak LDF (2019) Sequential extraction of phenolics and pectin from mango peel assisted by ultrasound. Food Res Int 119:455–461. https://doi.org/10.1016/j.foodres.2018.12.011

Gunwantrao BB, Bhausaheb SK, Ramrao BS, Subhash KS (2016) Antimicrobial activity and phytochemical analysis of orange (*Citrus aurantium* L.) and pineapple (*Ananas comosus* (L.) Merr) peel extract. Ann Phytomed 5(2):156–160. https://doi.org/10.21276/ap.2016.5.2.22

Gupta P, Chikkala S, Kundu P (2021) Ash gourd and its applications in the food, pharmacological and biomedical industries. Int J Veg Sci 27(1):44–53. https://doi.org/10.1080/19315260.2019.1699222

Gurumeenakshi G, Geetha P, Rajeswari R (2021) Utilization of grape pomace as a raw material for the development of value added products. Int J Curr Microbiol App Sci 10(03):772–779. https://doi.org/10.20546/ijcmas.2021.1003.098

Guzmán-Lorite M, Marina ML, García MC (2022) Pressurized liquids vs. high intensity focused ultrasounds for the extraction of proteins from a pomegranate seed waste. Innovative Food Sci Emerg Technol 77:102958. https://doi.org/10.1016/j.ifset.2022.102958

Hang YD, Splittstoesser DF, Woodams EE, Sherman RM (1977) Citric acid fermentation of brewery waste. J Food Sci 42(2):383–384. https://doi.org/10.1111/j.1365-2621.1977.tb01504.x

Hasan MM, Bashir T, Ghosh R, Lee SK, Bae H (2017) An overview of LEDs' effects on the production of bioactive compounds and crop quality. Molecules 22(9):1420. https://doi.org/10.3390/molecules22091420

Hashem M, Darwish SM (2010) Production of bioethanol and associated by-products from potato starch residue stream by *Saccharomyces cerevisiae*. Biomass Bioenergy 34(7):953–959. https://doi.org/10.1016/j.biombioe.2010.02.003

Hong-Bo XU, Sheng-Hai ZHOU, Ming-Yue LI, Zi-Han WANG, Yan-Mei TIAN, Xu-Qin WANG (2022) Preparation of biomass-waste-derived carbon dots from apricot shell for highly sensitive and selective detection of ascorbic acid. Chin J Anal Chem 50(12):100168. https://doi.org/10.1016/j.cjac.2022.100168

Huang HW, Hsu CP, Yang BB, Wang CY (2013) Advances in the extraction of natural ingredients by high pressure extraction technology. Trends Food Sci Technol 33(1):54–62. https://doi.org/10.1016/j.tifs.2013.07.001

Hussain H, Mamadalieva NZ, Hussain A, Hassan U, Rabnawaz A, Ahmed I, Green IR (2022) Fruit peels: food waste as a valuable source of bioactive natural products for drug discovery. Curr Issues Mol Biol 44(5):1960–1994. https://doi.org/10.3390/cimb44050134

İçen H, Gürü M (2010) Effect of ethanol content on supercritical carbon dioxide extraction of caffeine from tea stalk and fiber wastes. J Supercrit Fluids 55(1):156–160. https://doi.org/10.1016/j.supflu.2010.07.009

Ijaz N, Bashir S, Ikram A, Zafar A, Ul Ain HB, Ambreen S et al (2024) Valorization of potato peel: a sustainable eco-friendly approach. CyTA-J Food 22(1):2306951. https://doi.org/10.1080/19476337.2024.2306951

Indian Institute of Horticulture Technology (2025) Sustainable waste management in horticulture. https://iht.edu.in/sustainable-waste-management-in-horticulture/

Ishisono K, Yabe T, Kitaguchi K (2017) Citrus pectin attenuates endotoxin shock via suppression of toll-like receptor signaling in Peyer's patch myeloid cells. J Nutr Biochem 50:38–45. https://doi.org/10.1016/j.jnutbio.2017.07.016

Jameel MK, Mustafa MA, Ahmed HS, Jassim Mohammed A, Ghazy H, Shakir MN et al (2024) Biogas: production, properties, applications, economic and challenges: a review. Results Chem 7:101549. https://doi.org/10.1016/j.rechem.2024.101549

Jaime L, Moll´a E, Fern´andez A, Mart´ın-Cabrejas MA, L´opez-Andr´eu FJ, Esteban RM (2002) Structural carbohydrate differences and potential source of dietary fiber of onion (Allium cepa L.) tissues. J Agric Food Chem 50:122–8

Jeddou KB, Bouaziz F, Zouari-Ellouzi S, Chaari F, Ellouz- Chaabouni S, Ellouz-Ghorbel R, Nouri-Ellouz O (2017) Improvement of texture and sensory properties of cakes by addition of potato peel powder with high level of dietary fiber and protein. Food Chemistry 217:668–677. https://doi.org/10.1016/j.foodchem.2016.08.081

Joel EB, Mafulul SG, Adamu HE, Goje LJ, Tijjani H, Igunnu A, Malomo SO (2020) Peroxidase from waste cabbage (*Brassica oleracea capitata* L.) exhibits the potential to biodegrade phenol and synthetic dyes from wastewater. Sci Afr 10:e00608. https://doi.org/10.1016/j.sciaf.2020.e00608

Joshi VK (2020) Value addition to fruit and vegetable processing waste-an appraisal. Int J Food Ferment Technol 10(2):35–58. https://doi.org/10.30954/2277-9396.02.2020.2

Kam WYJ, Mirhosseini H, Abas F, Hussain N, Hedayatnia S, Chong HLF (2018) Antioxidant activity enhancement of biodegradable film as active packaging utilizing crude extract from durian leaf waste. Food Control 90:66–72. https://doi.org/10.1016/j.foodcont.2018.02.036

Kammerer D, Claus A, Schieber A, Carle R (2005) A novel process for the recovery of polyphenols from grape (*Vitis vinifera* L.) pomace. J Food Sci 70(2):C157–C163. https://doi.org/10.1111/j.1365-2621.2005.tb07077.x

Kanatt SR, Chander R, Sharma A (2010) Antioxidant and antimicrobial activity of pomegranate peel extract improves the shelf life of chicken products. Int J Food Sci Technol 45(2):216–222. https://doi.org/10.1111/j.1365-2621.2009.02124.x

Kapadia SP, Pudakalkatti PS, Shivanaikar S (2015) Detection of antimicrobial activity of banana peel (*Musa paradisiaca* L.) on *Porphyromonas gingivalis* and *Aggregatibacter actinomycetemcomitans*: an in vitro study. Contemp Clin Dent 6(4):496–499. https://doi.org/10.4103/0976-237X.169864

Kasapoğlu ED, Kahraman S, Törnük F (2020) Apricot juice processing byproducts as sources of value-added compounds for food industry. Eur Food Sci Eng 1(1):18–23

Khalid MU, Shabbir MA, Mustafa S, Hina S, Quddoos MY, Mahmood S et al (2021) Effect of apple peel as an antioxidant on the quality characteristics and oxidative stability of mayonnaise. Appl Food Res 1(2):100023. https://doi.org/10.1016/j.afres.2021.100023

Khan SA, Aslam R, Makroo HA (2019) High pressure extraction and its application in the extraction of bio-active compounds: a review. J Food Process Eng 42(1):e12896. https://doi.org/10.1111/jfpe.12896

Khanal S, Karimi K, Majumdar S, Kumar V, Verma R, Bhatia SK et al (2024) Sustainable utilization and valorization of potato waste: state of the art, challenges, and perspectives. Biomass Convers Biorefinery 14(19):23335–23360. https://doi.org/10.1007/s13399-023-04521-1

Kim M, Choi D, Kim JY, Jung S, Tsang YF, Lin KYA, Kwon EE (2024) Direct conversion of apricot seeds into biodiesel. Bioresour Technol 395:130339. https://doi.org/10.1016/j.biortech.2024.130339

Kiss GAC, Forgács E, Cserháti T, Mota T, Morais H, Ramos A (2000) Optimisation of the microwave-assisted extraction of pigments from paprika (*Capsicum annuum* L.) powders. J Chromatogr A 889(1–2):41–49. https://doi.org/10.1016/s0021-9673(00)00440-4

Konrade D, Kļava D, Grāmatiņa I, Kampuse S, Kince T (2018) Crispbreads with carrot and pumpkin processing by-products. Proc Latv Acad Sci 72(2):91–96. https://doi.org/10.2478/prolas-2018-0015

Kosińska A, Karamac M, Estrella I, Hernández T, Bartolomé B, Dykes GA (2012) Phenolic compound profiles and antioxidant capacity of Persea americana Mill. peels and seeds of two varieties. J Agric Food Chem 60(18):4613–4619. https://doi.org/10.1021/jf300090p

Krishnamoorthi A, Anushi, Gouthami Y, Minz V, Behera SD, Singh SK, Shilpa, Prasad L, Jain S (2023) Bioactive compounds from horticulture crops and their utilization: a comprehensive review. Int J Plant Soil Sci 35(23):268–277. https://doi.org/10.9734/ijpss/2023/v35i234240

Kultys E, Kurek MA (2022) Green extraction of carotenoids from fruit and vegetable byproducts: a review. Molecules 27(2):518. https://doi.org/10.3390/molecules27020518

Kumar H, Bhardwaj K, Sharma R, Nepovimova E, Kuča K, Dhanjal DS et al (2020) Fruit and vegetable peels: utilization of high value horticultural waste in novel industrial applications. Molecules 25(12):2812. https://doi.org/10.3390/molecules25122812

Kumar K, Srivastav S, Sharanagat VS (2021) Ultrasound assisted extraction (UAE) of bioactive compounds from fruit and vegetable processing by-products: a review. Ultrason Sonochem 70:105325. https://doi.org/10.1016/j.ultsonch.2020.105325

Kumar M, Barbhai MD, Esatbeyoglu T, Zhang B, Sheri V, Dhumal S et al (2022) Apple (*Malus domestica* Borkh.) seed: a review on health promoting bioactivities and its application as functional food ingredient. Food Biosci 50:102155. https://doi.org/10.1016/J.FBIO.2022.102155

Kumari A, Singh J (2018) Physical and chemical evaluation of tomato and its value addition. Int J Curr Microbiol App Sci 7(4):2851–2862. https://doi.org/10.20546/ijcmas.2018.704.325

Kurek M, Benbettaieb N, Ščetar M, Chaudy E, Repajić M, Klepac D et al (2021) Characterization of food packaging films with blackcurrant fruit waste as a source of antioxidant and colour sensing intelligent material. Molecules 26(9):2569. https://doi.org/10.3390/molecules26092569

Kurtulbaş E, Sevgen S, Samli R, Şahin S (2024) Microwave-assisted extraction of bioactive components from peach waste: describing the bioactivity degradation by polynomial regression. Biomass Convers Biorefinery 14(8):9609–9619. https://doi.org/10.1007/s13399-022-02909-z

Kushwaha R, Kumar V, Vyas G, Kaur J (2018) Optimization of different variable for eco-friendly extraction of betalains and phytochemicals from beetroot pomace. Waste Biomass Valoriz 9:1485–1494. https://doi.org/10.1007/s12649-017-9953-6

Lakka A, Bozinou E, Makris DP, Lalas SI (2021) Evaluation of pulsed electric field polyphenol extraction from *Vitis vinifera*, *Sideritis scardica* and *Crocus sativus*. *ChemEngineering* 5(2):25. https://doi.org/10.3390/chemengineering5020025

Lal AN, Prince MV, Kothakota A, Pandiselvam R, Thirumdas R, Mahanti NK, Sreeja R (2021) Pulsed electric field combined with microwave-assisted extraction of pectin polysaccharide from jackfruit waste. Innovative Food Sci Emerg Technol 74:102844

Lasta HFB, Lentz L, Rodrigues LGG, Mezzomo N, Vitali L, Ferreira SRS (2019) Pressurized liquid extraction applied for the recovery of phenolic compounds from beetroot waste. Biocatal Agric Biotechnol 21:101353. https://doi.org/10.1016/j.bcab.2019.101353

Lasunon P, Phonkerd N, Tettawong P, Sengkhamparn N (2021) Effect of microwave-assisted extraction on bioactive compounds from industrial tomato waste and its antioxidant activity. Food Res 5(2):468–474. https://doi.org/10.26656/fr.2017.5(2).516

Latif S, Anwar F (2009) Physicochemical studies of hemp (*Cannabis sativa*) seed oil using enzyme-assisted cold-pressing. Eur J Lipid Sci Technol 111(10):1042–1048. https://doi.org/10.1002/ejlt.200900008

Lee SI, Kim BS, Kim KS, Lee S, Shin KS, Lim JS (2008) Immune-suppressive activity of punicalagin via inhibition of NFAT activation. Biochem Biophys Res Commun 371(4):799–803. https://doi.org/10.1080/10408398.2016.1251390

Lemes AC, Egea MB, de Oliveira Filho JG, Gautério GV, Ribeiro BD, Coelho MAZ (2022) Biological approaches for extraction of bioactive compounds from agro-industrial by-products: a review. Front Bioeng Biotechnol 9:802543. https://doi.org/10.3389/fbioe.2021.802543

Li BB, Smith B, Hossain MM (2006) Extraction of phenolics from citrus peels: II. Enzyme-assisted extraction method. Sep Purif Technol 48(2):189–196. https://doi.org/10.1016/j.seppur.2005.07.019

Li X, Wang T, Zhou B, Gao W, Cao J, Huang L (2014) Chemical composition and antioxidant and anti-inflammatory potential of peels and flesh from 10 different pear varieties (*Pyrus* spp.). Food Chem 152:531–538. https://doi.org/10.1016/j.foodchem.2013.12.010

Liang J, Hänsch GM, Hübner K, Samstag Y (2019) Sulforaphane as anticancer agent: a double-edged sword? Tricky balance between effects on tumor cells and immune cells. Adv Biol Regul 71:79–87. https://doi.org/10.1016/j.jbior.2018.11.006

Liu S, Zhang S, Lv X, Lu J, Ren C, Zeng Z et al (2019) Limonin ameliorates ulcerative colitis by regulating STAT3/miR-214 signaling pathway. Int Immunopharmacol 75:105768. https://doi.org/10.1016/j.intimp.2019.105768

Liu Z, de Souza TS, Holland B, Dunshea F, Barrow C, Suleria HA (2023) Valorization of food waste to produce value-added products based on its bioactive compounds. Processes 11(3):840. https://doi.org/10.3390/pr11030840

Livadariu O, Maximilian C, Rahmanifar B, Cornea CP (2023) LED technology applied to plant development for promoting the accumulation of bioactive compounds: a review. Plants 12(5):1075. https://doi.org/10.3390/plants12051075

Lobo MG, Dorta E (2019) Utilization and management of horticultural waste. In: Postharvest technology of perishable horticultural commodities. Woodhead Publishing, pp 639–666. https://doi.org/10.1016/B978-0-12-813276-0.00019-5

Mamma D, Christakopoulos P (2014) Biotransformation of citrus by-products into value added products. Waste Biomass Valoriz 5:529–549. https://doi.org/10.1007/s12649-013-9250-y

Markam S (2014) Value addition of tomato. J Med Plant Stud 2(1):97–103

Martău GA, Teleky BE, Ranga F, Pop ID, Vodnar DC (2021) Apple pomace as a sustainable substrate in sourdough fermentation. Front Microbiol 12:742020. https://doi.org/10.3389/fmicb.2021.742020

Mattos GN, Tonon RV, Furtado AA, Cabral LM (2017) Grape by-product extracts against microbial proliferation and lipid oxidation: a review. J Sci Food Agric 97(4):1055–1064. https://doi.org/10.1002/jsfa.8062

Maurer HR (2001) Bromelain: biochemistry, pharmacology and medical use. Cell Mol Life Sci 58:1234–1245. https://doi.org/10.1007/pl00000936

Maza MA, Pereira C, Martínez JM, Camargo A, Álvarez I, Raso J (2020) PEF treatments of high specific energy permit the reduction of maceration time during vinification of *Caladoc* and *Grenache* grapes. Innov Food Sci Emerg Technol 63:102375. https://doi.org/10.1016/j.ifset.2020.102375

Medina-Meza IG, Barbosa-Cánovas GV (2015) Assisted extraction of bioactive compounds from plum and grape peels by ultrasonics and pulsed electric fields. J Food Eng 166:268–275. https://doi.org/10.1016/j.jfoodeng.2015.06.012

Meena L, Sengar AS, Neog R, Sunil CK (2022) Pineapple processing waste (PPW): bioactive compounds, their extraction, and utilisation: a review. J Food Sci Technol 59:4152–4164. https://doi.org/10.1007/s13197-021-05271-6

Méndez-Carmona JY, Ascacio-Valdes JA, Alvarez-Perez OB, Hernández-Almanza AY, Ramírez-Guzman N, Sepúlveda L et al (2022) Tomato waste as a bioresource for lycopene extraction using emerging technologies. Food Biosci 49:101966. https://doi.org/10.1016/j.fbio.2022.101966

Minjares-Fuentes R, Femenia A, Garau MC, Candelas-Cadillo MG, Simal S, Rosselló C (2016) Ultrasound-assisted extraction of hemicelluloses from grape pomace using response surface methodology. Carbohydr Polym 138:180–191. https://doi.org/10.1016/j.carbpol.2015.11.045

Mordi RC, Fadiaro AE, Owoeye TF, Olanrewaju IO, Uzoamaka GC, Olorunshola SJ (2016) Identification by GC-MS of the components of oils of banana peels extract, phytochemical and antimicrobial analyses. Res J Phytochem 10:39–44. https://doi.org/10.3923/rjphyto.2016.39.44

Moreno AD, Duque A, González A, Ballesteros I, Negro MJ (2021) Valorization of greenhouse horticulture waste from a biorefinery perspective. Foods 10(4):814. https://doi.org/10.3390/foods10040814

Mutua JK, Imathiu S, Owino W (2017) Evaluation of the proximate composition, antioxidant potential, and antimicrobial activity of mango seed kernel extracts. Food Sci Nutr 5(2):349–357. https://doi.org/10.1002/fsn3.399

National Academy of Agricultural Sciences (2019) Saving the harvest: reducing the food loss and waste. Policy Brief No. 5, New Delhi, India. http://naasindia.org/documents/Saving%20the%20Harvest.pdf. Accessed 8 Oct 2019

Ncobela CN, Kanengoni AT, Hlatini VA, Thomas RS, Chimonyo M (2017) A review of the utility of potato by-products as a feed resource for smallholder pig production. Anim Feed Sci Technol 227:107–17. https://doi.org/10.1016/j.anifeedsci.2017.02.008

Negi T, Kumar Y, Sirohi R, Singh S, Tarafdar A, Pareek S et al (2022) Advances in bioconversion of spent tea leaves to value-added products. Bioresour Technol 346:126409. https://doi.org/10.1016/j.biortech.2021.126409

Nirmal NP, Khanashyam AC, Mundanat AS, Shah K, Babu KS, Thorakkattu P et al (2023) Valorization of fruit waste for bioactive compounds and their applications in the food industry. Foods 12(3):556. https://doi.org/10.3390/foods12030556

Núñez Selles AJ, Daglia M, Rastrelli L (2016) The potential role of mangiferin in cancer treatment through its immunomodulatory, anti-angiogenic, apoptopic, and gene regulatory effects. Biofactors 42(5):475–491. https://doi.org/10.1002/biof.1299

Ouédraogo JCW, Dicko C, Kini FB, Bonzi-Coulibaly YL, Dey ES (2018) Enhanced extraction of flavonoids from Odontonema strictum leaves with antioxidant activity using supercritical carbon dioxide fluid combined with ethanol. J Supercrit Fluids 131:66–71. https://doi.org/10.1016/j.supflu.2017.08.017

Palaniswamy KP, Muthukrishnan CR, Shanamugavellu KG (1974) Physico-chemical characteristics of some varieties of mango. Indian Food Packer 28(5):12

Palazzolo E, Laudicina VA, Germanà MA (2013) Current and potential use of citrus essential oils. Curr Org Chem 17(24):3042–3049. https://doi.org/10.2174/13852728113179990122

Panwar D, Panesar PS, Chopra HK (2021) Recent trends on the valorization strategies for the management of citrus by-products. Food Rev Intl 37(1):91–120. https://doi.org/10.1080/87559129.2019.1695834

Papoutsis K, Shams Z, Yazdani M, Muneer F, Rahmatov M, Shariatipour N, Villanueva-Gutierrez EE, Sahoo M (2025) Natural preservation strategies: the potential of bioactive compounds from fruit and vegetable waste to reduce postharvest losses. Food Chem X 30:102917. https://doi.org/10.1016/j.fochx.2025.102917

Pardo MES, Cassellis MER, Escobedo RM, García EJ (2014) Chemical characterisation of the industrial residues of the pineapple (*Ananas comosus*). J Agric Chem Environ 3(2):53–56. https://doi.org/10.4236/jacen.2014.32B009

Park JH, Lee M, Park E (2014) Antioxidant activity of orange flesh and peel extracted with various solvents. Prev Nutr Food Sci 19(4):291. https://doi.org/10.3746/pnf.2014.19.4.291

Parra-Pacheco B, Cruz-Moreno BA, Aguirre-Becerra H, García-Trejo JF, Feregrino-Pérez AA (2024) Bioactive compounds from organic waste. Molecules 29(10):2243. https://doi.org/10.3390/molecules29102243

Pataro G, Carullo D, Falcone M, Ferrari G (2020) Recovery of lycopene from industrially derived tomato processing by-products by pulsed electric fields-assisted extraction. Innovative Food Sci Emerg Technol 63:102369. https://doi.org/10.1016/j.ifset.2020.102369

Patel A, Temgire S, Borah A (2021) Agro-industrial waste as source of bioactive compounds and their utilization: a review. Pharma Innov J 10(5):192–196. https://doi.org/10.22271/tpi.2021.v10.i5c.6197

Pathak PD, Mandavgane SA, Puranik NM, Jambhulkar SJ, Kulkarni BD (2018) Valorization of potato peel: a biorefinery approach. Crit Rev Biotechnol 38(2):218–230. https://doi.org/10.1080/07388551.2017.1331337

Paz-Arteaga SL, Cadena-Chamorro E, Goméz-García R, Serna-Cock L, Aguilar CN, Torres-León C (2024) Unraveling the valorization potential of pineapple waste to obtain value-added products towards a sustainable circular bioeconomy. Sustainability 16(16):7236. https://doi.org/10.3390/su16167236

Pfaltzgraff LA, Cooper EC, Budarin V, Clark JH (2013) Food waste biomass: a resource for high-value chemicals. Green Chem 15(2):307–314. https://doi.org/10.1039/C2GC36978H

Piechowiak T, Skóra B, Grzelak-Błaszczyk K, Sójka M (2021) Extraction of antioxidant compounds from blueberry fruit waste and evaluation of their in vitro biological activity in human keratinocytes (HaCaT). Food Anal Methods 14(11):2317–2327. https://doi.org/10.1007/s12161-021-02056-7

Pimentel-Moral S, Borrás-Linares I, Lozano-Sánchez J, Arráez-Román D, Martínez-Férez A, Segura-Carretero A (2019) Supercritical CO_2 extraction of bioactive compounds from *Hibiscus sabdariffa*. J Supercrit Fluids 147:213–221. https://doi.org/10.1016/j.supflu.2018.11.005

Plazzotta S, Manzocco L, Nicoli MC (2017) Fruit and vegetable waste management and the challenge of fresh-cut salad. Trends Food Sci Technol 63:51–59. https://doi.org/10.1016/j.tifs.2017.02.013

Pukalskienė M, Pukalskas A, Dienaitė L, Revinytė S, Pereira CV, Matias AA, Venskutonis PR (2021) Recovery of bioactive compounds from strawberry (Fragaria × ananassa) pomace by conventional and pressurized liquid extraction and assessment their bioactivity in human cell cultures. Foods 10(8):1780. https://doi.org/10.3390/foods10081780

Radočaj O, Vujasinović V, Dimić E, Basić Z (2014) Blackberry (*Rubus fruticosus* L.) and raspberry (*Rubus idaeus* L.) seed oils extracted from dried press pomace after longterm frozen storage of berries can be used as functional food ingredients. Eur J Lipid Sci Technol 116(8):1015–1024. https://doi.org/10.1002/ejlt.201400014

Raigond P, Raigond B, Kochhar T, Sood A, Singh B (2018) Conversion of potato starch and peel waste to high value nanocrystals. Potato Res 61:341–351. https://doi.org/10.1007/s11540-018-9381-4

Raj D, Senapati AK, Patel NL (2016) Waste management in horticulture processing industry. In: Patel NL, Chawla SL, Ahlawat TR (eds) Commercial horticulture. New India Publishing Agency, pp 391–401

Rajha HN, Boussetta N, Louka N, Maroun RG, Vorobiev E (2014) A comparative study of physical pretreatments for the extraction of polyphenols and proteins from vine shoots. Food Res Int 65:462–468. https://doi.org/10.1016/j.foodres.2014.04.024

Raman G, Gaikar VG (2002) Microwave-assisted extraction of piperine from Piper nigrum. Ind Eng Chem Res 41(10):2521–2528. https://doi.org/10.1021/ie010359b

Rauha JP, Remes S, Heinonen M, Hopia A, Kähkönen M, Kujala T et al (2000) Antimicrobial effects of Finnish plant extracts containing flavonoids and other phenolic compounds. Int J Food Microbiol 56(1):3–12. https://doi.org/10.1016/s0168-1605(00)00218-x

Reddy AS, Kasa VP, Samal B, Dubey BK, Yadav V, Pandey DS (2025) Sustainable agricultural waste management in India: innovations, challenges, and future perspectives. Biomass Bioenergy 202:108261. https://doi.org/10.1016/j.biombioe.2025.108261

Redondo D, Venturini ME, Luengo E, Raso J, Arias E (2018) Pulsed electric fields as a green technology for the extraction of bioactive compounds from thinned peach by-products. Innovative Food Sci Emerg Technol 45:335–343. https://doi.org/10.1016/j.ifset.2017.12.004

Ribeiro ACB, Cunha AP, da Silva LMR, Mattos ALA, de Brito ES, de Souza MDSM et al (2021) From mango by-product to food packaging: pectin-phenolic antioxidant films from mango peels. Int J Biol Macromol 193:1138–1150. https://doi.org/10.1016/j.ijbiomac.2021.10.131

Romano R, De Luca L, Aiello A, Rossi D, Pizzolongo F, Masi P (2022) Bioactive compounds extracted by liquid and supercritical carbon dioxide from citrus peels. Int J Food Sci Technol 57(6):3826–3837. https://doi.org/10.1111/ijfs.15712

Romier B, Van De Walle J, During A, Larondelle Y, Schneider YJ (2008) Modulation of signalling nuclear factor-κB activation pathway by polyphenols in human intestinal Caco-2 cells. Br J Nutr 100(3):542–551. https://doi.org/10.1017/s0007114508966666

Rosenthal A, Pyle DL, Niranjan K (1996) Aqueous and enzymatic processes for edible oil extraction. Enzym Microb Technol 19(6):402–420. https://doi.org/10.1016/S0141-0229(96)80004-F

Rostagno MA, Palma M, Barroso CG (2007) Microwave assisted extraction of soy isoflavones. Anal Chim Acta 588(2):274–282. https://doi.org/10.1016/j.aca.2007.02.010

Rowayshed G, Sharaf AM, El-Faham SY, Ashour M, Zaky AA (2015) Utilization of potato peels extract as source of phytochemicals in biscuits. J Basic Appl Res Int 8(3):190–201

Sagar NA, Pareek S, Sharma S, Yahia EM, Lobo MG (2018) Fruit and vegetable waste: bioactive compounds, their extraction, and possible utilization. Compr Rev Food Sci Food Saf 17(3):512–531. https://doi.org/10.1111/1541-4337.12330

Sagarika N, Prince MV, Kothakota A, Pandiselvam R, Sreeja R, Mathew SM (2018) Characterization and optimization of microwave assisted process for extraction of nutmeg (*Myristica fragrans* Houtt.) mace essential oil. J Essent Oil Bear Plants 21(4):895–904. https://doi.org/10.1080/0972060X.2018.1517613

Samaram S, Mirhosseini H, Tan CP, Ghazali HM, Bordbar S, Serjouie A (2015) Optimisation of ultrasound-assisted extraction of oil from papaya seed by response surface methodology: oil recovery, radical scavenging antioxidant activity, and oxidation stability. Food Chem 172:7–17. https://doi.org/10.1016/j.foodchem.2014.08.068

Sánchez-Ávila N, Priego-Capote F, Ruiz-Jiménez J, de Castro ML (2009) Fast and selective determination of triterpenic compounds in olive leaves by liquid chromatography–tandem mass spectrometry with multiple reaction monitoring after microwave-assisted extraction. Talanta 78(1):40–48. https://doi.org/10.1016/j.talanta.2008.10.037

Sarangi PK, Singh TA, Singh NJ, Shadangi KP, Srivastava RK, Singh AK et al (2022) Sustainable utilization of pineapple wastes for production of bioenergy, biochemicals and value-added products: a review. Bioresour Technol 351:127085. https://doi.org/10.1016/j.biortech.2022.127085

Sasidharan S, Chen Y, Saravanan D, Sundram KM, Latha LY (2011) Extraction, isolation and characterization of bioactive compounds from plants' extracts. Afr J Tradit Complement Altern Med 8(1):1–10. https://doi.org/10.4314/ajtcam.v8i1.60483

Schroeder JW (1999) By-products and regionally available alternative feedstuffs for dairy cattle. North Dakota State University. NDSU Extension Service. http://www.ext.nodak.edu/extpubs/ansci/dairy/as1180w.html

Sharma R, Gupta A, Abrol GS, Joshi VK (2014) Value addition of wild apricot fruits grown in North–West Himalayan regions—a review. J Food Sci Technol 51:2917–2924. https://doi.org/10.1007/s13197-012-0766-0

Sharma R, Oberoi HS, Dhillon GS (2016) Fruit and vegetable processing waste: renewable feed stocks for enzyme production. In: Dhillon GS, Kaur S (eds) Agro-industrial wastes as feedstock for enzyme production: apply and exploit the emerging and valuable use options of waste biomass. Academic Press Elsevier, London, pp 23–59. https://doi.org/10.1016/B978-0-12-802392-1.00002-2

Sharma M, Usmani Z, Gupta VK, Bhat R (2021a) Valorization of fruits and vegetable wastes and by-products to produce natural pigments. Crit Rev Biotechnol 41(4):535–563. https://doi.org/10.1080/07388551.2021.1873240

Sharma P, Gaur VK, Sirohi R, Varjani S, Kim SH, Wong JW (2021b) Sustainable processing of food waste for production of bio-based products for circular bioeconomy. Bioresour Technol 325:124684. https://doi.org/10.1016/j.biortech.2021.124684

Sharoba AM, Farrag MA, Abd El-Salam AMA (2013) Utilization of some fruits and vegetables wastes as a source of dietary fibers in cake making. J Food Dairy Sci 4(9):433–453. https://doi.org/10.21608/jfds.2013.72084

Shin SK, Cho SJ, Jung UJ, Ryu R, Choi MS (2016) Phlorizin supplementation attenuates obesity, inflammation, and hyperglycemia in diet-induced obese mice fed a high-fat diet. Nutrients 8(2):92. https://doi.org/10.3390/nu8020092

Shinali TS, Zhang Y, Altaf M, Nsabiyeze A, Han Z, Shi S, Shang N (2024) The valorization of wastes and byproducts from cruciferous vegetables: a review on the potential utilization of cabbage, cauliflower, and broccoli byproducts. Foods 13(8):1163. https://doi.org/10.3390/foods13081163

Sinha G, Chattopadhyay A, Maruthi B, Sinha SK, Pratyusha P (2024) Waste products of horticultural crops: an alternative for developing entrepreneurship. In: Sustainable horticulture, vol 2. Apple Academic Press, pp 27–38

Slatter B (2024) Crop loss/waste on Australian horticulture farms, 2023–24. Australian Bureau of Agricultural and Resource Economics and Sciences, Department of Agriculture, Fisheries and Forestry. https://www.agriculture.gov.au/abares/research-topics/surveys/horticulture-crop-loss-23-24

Soldan ACF, Arvelos S, Watanabe ÉO, Hori CE (2021) Supercritical fluid extraction of oleoresin from *Capsicum annuum* industrial waste. J Clean Prod 297:126593. https://doi.org/10.1016/j.jclepro.2021.126593

Someya S, Yoshiki Y, Okubo K (2002) Antioxidant compounds from bananas (*Musa cavendish*). Food Chem 79(3):351–354. https://doi.org/10.1016/S0308-8146(02)00186-3

Sravani A, Patil CR, Sharma S (2024) Value-added product development utilising the food wastes. In: Valorization of biomass wastes for environmental sustainability: green practices for the rural circular economy. Springer Nature Switzerland, Cham, pp 287–301. https://doi.org/10.1007/978-3-031-52485-1_16

Sreenivas KM, Chaudhari K, Lele SS (2011) Ash gourd peel wax: extraction, characterization, and application as an edible coat for fruits. Food Sci Biotechnol 20(2):383–387. https://doi.org/10.1007/s10068-011-0054-1

Sreenivas KÁ, Inarkar MÁ, Gokhale SV, Lele SS (2014) Re-utilization of ash gourd (*Benincasa hispida*) peel waste for chromium (VI) biosorption: equilibrium and column studies. J Environ Chem Eng 2(1):455–462. https://doi.org/10.1016/j.jece.2014.01.017

Sridhar A, Ponnuchamy M, Kumar PS, Kapoor A, Vo DVN, Prabhakar S (2021) Techniques and modeling of polyphenol extraction from food: a review, Environmental Chemistry Letters. Springer International Publishing. https://doi.org/10.1007/s10311-021-01217-8

Srinivas Y, Mathew SM, Kothakota A, Sagarika N, Pandiselvam R (2020) Microwave assisted fluidized bed drying of nutmeg mace for essential oil enriched extracts: an assessment of drying kinetics, process optimization and quality. Innovative Food Sci Emerg Technol 66:102541. https://doi.org/10.1016/j.ifset.2020.102541

Srivastava A (2023) Conversion of banana wastes into value added products in small scale industries: a review. J Glob Values 14:9–14. https://doi.org/10.31995/jgv.2023.v14iS4.003

Subrahmanyan V, Lal G, Bhatia DS, Jain NL, Bains GS, Srinath KV et al (1957) Studies on banana pseudostem starch: production, yield, physico-chemical properties and uses. J Sci Food Agric 8(5):253–261. https://doi.org/10.1002/jsfa.2740080502

Szymanowska U, Baraniak B (2019) Antioxidant and potentially anti-inflammatory activity of anthocyanin fractions from pomace obtained from enzymatically treated raspberries. Antioxidants 8(8):299. https://doi.org/10.3390/antiox8080299

Tariq A, Sahar A, Usman M, Sameen A, Azhar M, Tahir R et al (2023) Extraction of dietary fiber and polyphenols from mango peel and its therapeutic potential to improve gut health. Food Biosci 53:102669. https://doi.org/10.1016/j.fbio.2023.102669

Tehrani MG, Elhamirad AH, Azarpazhooh E, Pedramnia A, Sharayei P (2019) Natural valuable compound extraction from onion by-products using a pulsed electric field. Int J Biol Chem 12(1):171–180. https://doi.org/10.26577/ijbch-2019-1-i22

Tow WW, Premier R, Jing H, Ajlouni S (2011) Antioxidant and antiproliferation effects of extractable and nonextractable polyphenols isolated from apple waste using different extraction methods. J Food Sci 76(7):T163–T172. https://doi.org/10.1111/j.1750-3841.2011.02314.x

Trombino S, Cassano R, Procopio D, Di Gioia ML, Barone E (2021) Valorization of tomato waste as a source of carotenoids. Molecules 26(16):5062. https://doi.org/10.3390/molecules26165062

Uwineza PA, Waśkiewicz A (2020) Recent advances in supercritical fluid extraction of natural bioactive compounds from natural plant materials. Molecules 25(17):3847. https://doi.org/10.3390/molecules25173847

Venturi F, Bartolini S, Sanmartin C, Orlando M, Taglieri I, Macaluso M et al (2019) Potato peels as a source of novel green extracts suitable as antioxidant additives for fresh-cut fruits. Appl Sci 9(12):2431. https://doi.org/10.3390/app9122431

Vilkhu K, Mawson R, Simons L, Bates D (2008) Applications and opportunities for ultrasound assisted extraction in the food industry—a review. Innovative Food Sci Emerg Technol 9(2):161–169. https://doi.org/10.1016/j.ifset.2007.04.014

Vorobiev E, Lebovka NI (2006) Extraction of intercellular components by pulsed electric fields. In: Pulsed electric fields technology for the food industry: fundamentals and applications. Springer Nature, pp 153–193. https://doi.org/10.1007/978-0-387-31122-7_6

Vorobiev E, Jemai AB, Bouzrara H, Lebovka N, Bazhal M (2004) Pulsed electric field-assisted extraction of juice from food plants. In: Novel food processing technologies. CRC Press, pp 127–152. https://doi.org/10.1201/9780203997277.ch5

Vulić J, Čanadanović-Brunet J, Ćetković G, Tumbas V, Djilas S, Četojević-Simin D, Čanadanović V (2012) Antioxidant and cell growth activities of beet root pomace extracts. J Funct Foods 4(3):670–678. https://doi.org/10.1016/j.jff.2012.04.008

Wang W, Ma X, Jiang P, Hu L, Zhi Z, Chen J et al (2016) Characterization of pectin from grapefruit peel: a comparison of ultrasound-assisted and conventional heating extractions. Food Hydrocoll 61:730–739. https://doi.org/10.1016/j.foodhyd.2016.06.019

Wang M, Yang F, Yan X, Chao X, Zhang W, Yuan C, Zeng Q (2022) Anti-diabetic effect of banana peel dietary fibers on type 2 diabetic mellitus mice induced by streptozotocin and high-sugar and high-fat diet. J Food Biochem 46(10):e14275. https://doi.org/10.1111/jfbc.14275
Wen C, Zhang J, Zhang H, Dzah CS, Zandile M, Duan Y et al (2018) Advances in ultrasound assisted extraction of bioactive compounds from cash crops—a review. Ultrason Sonochem 48:538–549. https://doi.org/10.1016/j.ultsonch.2018.07.018
Wright J (1995) Essential oils. In: Ashurst PR (ed) Food flavouring. Black A & P, Elasgow, p 24
Wu S, Tian L (2017) Diverse phytochemicals and bioactivities in the ancient fruit and modern functional food pomegranate (*Punica granatum*). Molecules 22(10):1606. https://doi.org/10.3390/molecules22101606
Yadav S, Malik K, Moore JM, Kamboj BR, Malik S, Malik VK, Arya S, Singh K, Mahanta S, Bishnoi DK (2024) Valorisation of agri-food waste for bioactive compounds: recent trends and future sustainable challenges. Molecules 29(9):2055. https://doi.org/10.3390/molecules29092055
Yan LG, He L, Xi J (2017) High intensity pulsed electric field as an innovative technique for extraction of bioactive compounds—a review. Crit Rev Food Sci Nutr 57(13):2877–2888. https://doi.org/10.1080/10408398.2015.1077193
Yusuf M (2017) Agro-industrial waste materials and their recycled value-added applications. In: Handbook of ecomaterials, 1st edn. Springer, pp 1–11
Zaini HM, Roslan J, Saallah S, Munsu E, Sulaiman NS, Pindi W (2022) Banana peels as a bioactive ingredient and its potential application in the food industry. J Funct Foods 92:105054. https://doi.org/10.1016/j.jff.2022.105054
Zeyada NN, Zeitoum MAM, Barbary OM (2008) Utilization of some vegetables and fruit waste as natural antioxidants. Alex J Food Sci Technol 5:1–11. https://doi.org/10.21608/ajfs.2008.20136
Zhang Y, Cheng X, Wang Z, Tahir MH, Wang Z, Wang X, Wang C (2022) Full recycling of high-value resources from cabbage waste by multi-stage utilization. Sci Total Environ 804:149951. https://doi.org/10.1016/j.scitotenv.2021.149951
Zhao G, He F, Wu C, Li P, Li N, Deng J et al (2018) Betaine in inflammation: mechanistic aspects and applications. Front Immunol 9:1070. https://doi.org/10.3389/fimmu.2018.01070
Zhao Z, Sun L, Sha Z, Chu C, Wang Q, Zhou D, Wu S (2023) Valorisation of fresh waste grape through fermentation with different exogenous probiotic inoculants. Heliyon 9(6):e16650. https://doi.org/10.1016/j.heliyon.2023.e16650
Zhou X, Liu L, Li J, Wang L, Song X (2022) Extraction and characterization of cellulose from Jerusalem artichoke residue and its application in blueberry preservation. Foods 11(8):1065. https://doi.org/10.3390/foods11081065
Zhu F, Du B, Xu B (2018) Anti-inflammatory effects of phytochemicals from fruits, vegetables, and food legumes: a review. Crit Rev Food Sci Nutr 58(8):1260–1270

Smart Strategies in Skill-Based Remunerative Horticulture

21

Prahlad Deb and Amit Baran Sharangi

Abstract

Horticulture is rapidly transitioning from labor-intensive, traditional practices to smart, skill-oriented production systems driven by digital technologies and precision tools. This paper synthesizes recent advances in smart horticulture, highlighting the roles of Information and Communication Technology (ICT), the Internet of Things (IoT), artificial intelligence, machine learning, robotics, and protected cultivation in enhancing productivity, resource-use efficiency, and farm profitability. The integration of sensor-based irrigation, AI-enabled pest and disease diagnostics, drone-assisted field monitoring, and automated greenhouse management has enabled data-driven decision-making across the entire value chain, from planting to postharvest handling. Evidence from diverse production systems demonstrates substantial gains in yield, quality, water savings, and input optimization, alongside reductions in labor dependency and environmental footprints. Beyond technology adoption, the study emphasizes the critical importance of smart skill development to ensure effective utilization of these innovations. Targeted capacity-building initiatives, covering digital literacy, drone operation, data interpretation, and agripreneurship, are emerging as key enablers for empowering farmers, rural youth, and women to participate in technology-enabled horticulture. Institutional support through Krishi Vigyan Kendras, agri-incubators, e-learning platforms, and public-private partnerships is accelerating knowledge transfer and fostering rural agri-tech enterprises. At the

P. Deb (✉)
Department of Horticulture & Postharvest Technology, Institute of Agriculture, Visva-Bharati, Sriniketan, India
e-mail: prahlad.deb@visva-bharati.ac.in

A. B. Sharangi
Department of Plantation, Spices, Medicinal and Aromatic Crops, Faculty of Horticulture, Bidhan Chandra Krishi Viswavidyalaya, Mohanpur, West Bengal, India

A. Das, A. B. Sharangi (eds.), *Remunerative Horticulture*,
https://doi.org/10.1007/978-981-92-1007-7_21

same time, persistent challenges such as high initial investment, digital divides, infrastructure limitations, and shortages of trained personnel continue to constrain widespread adoption, particularly among smallholders. Overall, smart horticulture represents a transformative pathway toward climate-resilient, inclusive, and market-responsive agricultural systems. By aligning advanced technologies with localized skill development and supportive policy frameworks, horticulture can evolve into a dynamic engine of sustainable livelihoods, youth engagement, and economic diversification. The convergence of innovation, human capital, and institutional collaboration will be central to realizing the full potential of future-ready horticultural ecosystems.

Keywords

Smart horticulture · Precision agriculture · ICT and automation · Artificial Intelligence in farming · Skill development · Agripreneurship

21.1 Introduction

Horticulture, the science and art of cultivating fruits, vegetables, flowers, medicinal, and aromatic plants, has emerged as one of the most vibrant and remunerative sectors in agriculture. With its capacity for high productivity, diverse income streams, and strong market orientation, it plays a vital role in ensuring nutritional security, rural employment, and sustainable livelihoods (FAO 2023; NHB 2023). Particularly in peri-urban and rural areas, horticulture has opened avenues for smallholder farmers to transition from subsistence to commercial agriculture (Giger et al. 2022).

In recent years, the sector has undergone a paradigm shift from conventional production methods to smart, skill-based, and market-responsive practices. This transformation has been fueled by challenges such as labor scarcity, climate variability, resource limitations, and the demand for higher-quality produce (Mana et al. 2024). Smart horticulture incorporates a blend of digital tools, automation, and entrepreneurial skills to increase productivity while ensuring sustainability and profitability (Kumar et al. 2024; Gebresenbet et al. 2023). At the heart of this revolution is the integration of advanced technologies such as Information and Communication Technology (ICT), Artificial Intelligence (AI), Machine Learning (ML), Internet of Things (IoT), and agri-robotics. These tools enable data-driven decision-making, real-time monitoring, predictive analytics, and resource optimization (Miller et al. 2025; Sharma and Shivandu 2024). For instance, AI models are being used to detect diseases, forecast yields, and manage nutrient inputs with high precision (Delfani et al. 2024; Aijaz et al. 2025). Meanwhile, ICT platforms are enabling remote advisories, e-extension services, and market access for smallholders (Mulungu et al. 2025).

Robotic systems and drones have been successfully deployed in tasks such as weeding, spraying, and aerial monitoring, thus reducing manual labor and improving accuracy in field operations (Meesaragandla et al. 2024). Sensor-based irrigation and fertigation systems are enabling precise input delivery based on real-time soil and crop data, contributing to both water use efficiency and cost reduction (Meriç 2025; Adamo et al. 2025). These innovations, though initially capital-intensive, offer long-term gains by improving yields, lowering risks, and enhancing produce quality. Despite these advancements, a significant challenge persists—the availability of a skilled workforce that can effectively operate and manage these technologies. Traditional horticultural training often lacks modules on digital tools, automation, and business management, resulting in a skills mismatch (Poenaru et al. 2025). Addressing this gap requires targeted training that integrates both technical knowledge and hands-on practice in smart tools and digital platforms (Rikala et al. 2024).

Various government initiatives have recognized this gap and are working to build capacity in skill-based horticulture. The Skill India Mission, National Horticulture Mission, and Digital Agriculture Mission are promoting training programs that align with the evolving demands of the sector (MSD&E 2024). Institutions such as ICAR, Krishi Vigyan Kendras (KVKs), and agri-incubators are conducting short courses and field demonstrations to expose farmers to AI tools, drone operations, and digital recordkeeping (Pathak et al. 2022). Such interventions are vital in empowering rural youth and entrepreneurs to build agri-tech careers or start their own ventures.

This movement toward smart, skill-oriented horticulture also aligns with the global goals of sustainability, climate resilience, and digital inclusion. Smart technologies help mitigate the adverse effects of climate change by enabling adaptive practices, such as early disease alerts, drought-tolerant planting recommendations, and precision fertigation (Bhatnagar et al. 2024). Technologies like AI-driven soil health monitoring and remote-sensing-based weather forecasting can ensure better preparedness and lower vulnerability to environmental shocks (Rajak et al. 2023; Mmbando 2025). Moreover, smart horticulture provides a strong foundation for value chain integration, especially through digital traceability, blockchain, and online agri-marketing platforms. These not only reduce the role of middlemen but also ensure better price realization and consumer trust (Lv et al. 2023). By involving women and marginalized communities in tech-based roles such as drone pilots, e-commerce operators, and nursery entrepreneurs, the sector can become more inclusive (Sharma 2023).

Ultimately, the future of horticulture lies in redefining how we grow, who grows, and how gains are distributed across the ecosystem. By embracing technology and aligning it with localized skill-building, the sector can evolve into a dynamic platform for sustainable growth, youth engagement, and economic diversification. The pathway to resilient and profitable horticulture will be paved by a synthesis of traditional wisdom, modern innovation, and strong human capital.

21.2 Evolution of Horticulture Toward a Smart Approach

Horticulture—once grounded in traditional wisdom, seasonality, and heavy manual labor—has undergone a profound transformation. In recent years, it has shifted into a truly smart, data-driven, and resilient sector, resulting from both external pressures and technological innovations.

21.2.1 Traditional Practices and Their Limitations

Historically, horticultural techniques were deeply rooted in indigenous knowledge, such as manual planting, organic soil preparation, and instinctive pest control (Habiyaremye and Korina 2021). These practices served well in stable environments but faltered as conditions evolved. For instance, erratic weather, including unpredictable rainfall and temperature swings, undermined the timing of planting and harvest cycles (Souravi and Rajsekharan 2019). This inconsistency has continued to disrupt production outcomes. Simultaneously, labor shortages became more pronounced as rural populations migrated toward urban centers, creating seasonal instability in labor availability and inflating costs. Pest and disease outbreaks further exposed limitations, with farmers often resorting to blanket pesticide applications that yield inconsistent results and raise environmental and health concerns. Additionally, poor postharvest storage and transport infrastructures led to significant losses, in some cases up to 40% of total yield (Anand and Barua 2022). Lastly, outdated irrigation and fertilization methods not only consumed excessive resources but also negatively impacted soil and water systems (Ingrao et al. 2023). These limitations collectively motivated the need for more predictable, resource-efficient, and resilient cultivation systems (Lakhiar et al. 2024).

21.2.2 Drivers for Modernization

Three primary drivers have fueled the shift toward smart horticulture: (a) *Market demand:* Consumers today demand produce that is fresh, traceable, pesticide-safe, and nutritionally enriched (Ahmad et al. 2024a, b). Globalization and e-commerce, producing consistent, high-value horticultural goods for urban and export markets, have become critical (Han et al. 2023). (b) *Climate change:* With agriculture becoming increasingly vulnerable to extreme weather, including sudden droughts and floods, there is growing adoption of climate-smart agriculture, such as precision irrigation, greenhouse systems, and real-time weather monitoring (Balasundram et al. 2023). (c) *Labor shortages:* As rural youth pursue nonfarming jobs, a decline in agricultural labor has intensified. Automation, including drones for spraying and tack shifts, robotic seeders, and sensor networks, provides key solutions. Together, they have pushed horticulture from low-input, manual methods toward automated, connected, and intelligent systems.

21.2.3 High-Value Crops and Protected Cultivation

Responding to consumer demand and climate instability, many growers are turning to high-value crops, such as berries, off-season vegetables, floriculture, and protected cultivation systems (Table 21.1) like greenhouses, polyhouses, vertical farms, and hydroponics (Jain et al. 2023). Protected systems have yielded many notable benefits. Yield improvement is a common result in protected cultivation.

Table 21.1 Protected cultivation of horticultural crops increased yield when compared with the traditional open system cropping, along with citations

Sl. no.	Region	Yield improvement	Reported by
1.	Uttarakhand, India	Yield increased by 45–90% for most high-value vegetables	Kumar et al. (2021)
2.	Finland	Off-season cultivation of high-value vegetable crops increased 25–35% yield under protected condition	Lehikoinen et al. (2021)
3.	Guangxi Zhuang region, Nanning, China	Protected cultivation of yard long bean increased yield around 25%	Zhang et al. (2024)
4.	Taiwan	Productivity of the tomato increased by 32% when cultivated under protected conditions	Sotelo-Cardona et al. (2021)
5.	Florida	Yield of strawberries under protected cultivation increased upto 37%	Menzel (2025)

Experimental trials in Uttarakhand showed vegetable yields increasing by 40–95% under protected cultivation (Kumar et al. 2021). Another benefit of protected cultivation is the extended opportunity for growing seasons for the crops. Protected systems support year-round production, reducing sensitivity to outdoor climate variations (Lehikoinen et al. 2021). Protected cultivation systems also provide more resource-use efficiency than open cultivation systems. Hydroponic systems can reduce water use by up to 95%, and vertical farms can enhance land use up to tenfold (Ram et al. 2024). Premium quality output is also another advantage of protected cultivation compared to the conventional cultivation system. Controlled environments help reduce pest pressure, allowing safer or pesticide-free production, thereby appealing to premium markets (Nordey et al. 2017). Although initial investment is high, public-private partnerships and subsidies, especially in countries like India, have pushed protected systems into broader adoption (Fabre and Stephane 2021).

21.2.4 From Conventional to Technology-Enhanced Systems

Today, horticulture is shifting into knowledge- and technology-intensive production systems, utilizing five key pillars: (a) *ICT and data tools:* Applications like "PusaKRISHI," "Digital Green," and other mobile advisory services are providing real-time, localized information in many developing regions (Kanagaraj and Anandaraja 2021). A remarkable example is the "AgroTIC" system in Colombia, which utilizes mobile image analytics to alert citrus farmers about pest threats (Arrubla-Hoyos et al. 2022). (b) *IoT and remote sensing:* Sensor kits now measure soil moisture, temperature, light, and humidity—enabling micro-irrigation and greenhouse climate control (Rajak et al. 2023). Drones and satellite data are increasingly used to assess crop health, predict yields, and detect disease early (Guebsi et al. 2024; Rejeb et al. 2022). (c) *AI and machine learning:* Advanced AI tools are now predicting yields, identifying pests, designing irrigation scheduling, and classifying plant health metrics (Javaid et al. 2023). In one study, tomato cultivation monitored via IoT and AI demonstrated improved water efficiency and fruit grading (Galaverni et al. 2025). (d) *Robotics and mechanization:* Robots, drones, and AI-driven transplanting, sorting, and spraying systems are automating intensive labor tasks (Mana et al. 2024). In greenhouses, monitoring systems manage light, humidity, and nutrient delivery, minimizing manual intervention (Al-Qudah et al. 2025). This convergence of traditional horticulture and modern technology empowers farmers to fully monitor and optimize their operations from planting to harvest.

21.2.5 Small-Scale Farmers and Tech Democratization

Despite initial high costs, various factors have made smart techniques accessible to smallholder farmers, viz., declining prices for sensor kits and automation equipment, widespread mobile internet access, support from extension programs and livelihood initiatives (Abiri et al. 2023), etc. Small-scale tech pilots report 20–30% water saving using micro-irrigation and sensor kits, while pesticide use dropped by 30% with drone scanning in select trials (Hafeez et al. 2023). The AgroTIC project in Colombia showed a farmer-reported yield increase of 15–20% after adopting mobile diagnostics (Arrubla-Hoyos et al. 2022).

21.2.6 The Smart Horticulture Era Is Data-Driven and Profitable

Modern horticulture is fully characterized by data-informed decision-making, enabled by real-time sensors, platforms, analytics, and connectivity. Growers monitor soil moisture, weather, nutrient status, and canopy health remotely. Using this data, irrigation, pest control, temperature regimes, and harvest maturity decisions are optimized (Ahmad and Sohel 2025). There are several key benefits of smart horticulture. Data-driven smart technologies provide consistent yield improvements and better-quality produce (Kumar et al. 2024). In most cases, waste reductions from optimized use of resources are 20–40% (Papadopoulos et al. 2024). Smart technologies are also saving labor due to mechanization. Higher market value of the produce with thorough traceability and premium-grade produce (Kabir et al. 2023) is also being harvested. Resilience to climatic unpredictability and supply chain challenges is also another advantage of smart data-driven technology.

Crucially, the smart horticulture is not just for large farms. It also drives social inclusion, as women and youth increasingly occupy technology-enabled roles in managing greenhouses, operating agricultural drones, leading value-added processing units, or running e-commerce harvest hubs (Rotz et al. 2019). Governments and NGOs are deploying rural agri-tech centers, digital storefronts, incubators, and credit channels to ensure broader participation (Ajambo et al. 2022).

The greenhouse is connected to a data logger, which interfaces with various actuators and sensors (Fig. 21.1). Actuators include heaters, valves, pumps, vents, and

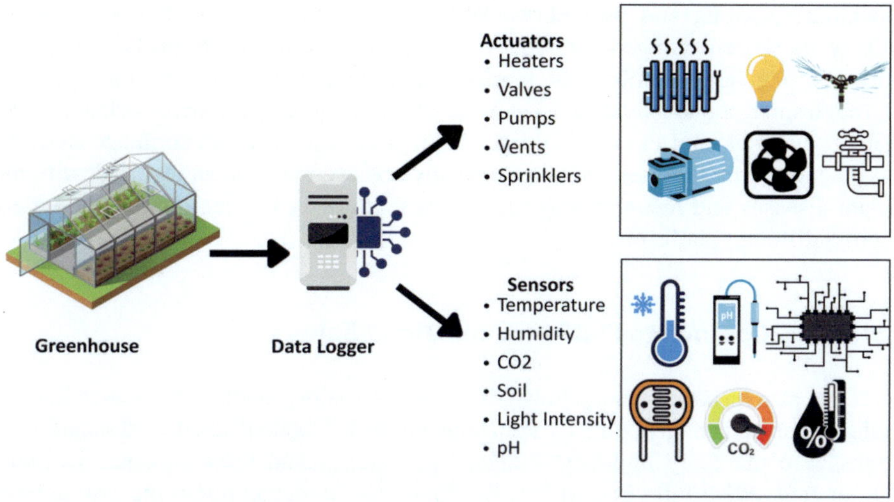

Fig. 21.1 Diagram illustrating a greenhouse system with a data logger (self-composed)

sprinklers, depicted with corresponding icons. Sensors measure temperature, humidity, CO_2, soil conditions, light intensity, and pH, also shown with relevant icons. Arrows indicate the flow of information between the greenhouse, data logger, actuators, and sensors.

21.3 Role of ICT in Modern Horticulture

Information and Communication Technology (ICT) has become an essential driver of transformation in horticulture, enabling farmers to transition from tradition-bound practices to *data-informed, market-responsive systems*. From mobile apps to remote sensing, ICT is reshaping every step of the horticultural value chain by improving access to information, reducing inefficiencies, and promoting inclusion.

21.3.1 Mobile Apps for Extension Services (MAAS) and Market Linkages

Mobile applications act as convenient conduits between farmers, extension specialists, and consumers. Mobile applications developed by ICAR for mango, guava, and aonla farmers guide postharvest processing and help to prevent losses at the producer level (Kambale et al. 2021; Verma and Garg 2021). These application tools are smart enough to automate ingredient calculations and playback of audio instructions, making cottage-scale value addition accessible. A video-based advisory on phones revealed that India's MAAS initiative enhances farmer decision-making (Kumar et al. 2022). Integrations like PusaKRISHI and Digital Green deliver localized planting and pest alerts. WeChat-based platforms in China, such as Zhe'yang'shi, encouraged adoption of soil testing and balanced fertilization practices among large numbers of farmers, suggesting better nutrient management through smartphone guidance (Li et al. 2022a, b). Applications such as PlantVillage (Ali et al. 2024) and Plantix (Siddiqua et al. 2022) use AI-powered image analysis to diagnose crop diseases. Plantix especially achieved high accuracy in identifying plant diseases and recommending treatments, although UI and app quality varied across different platforms.

21.3.2 E-Commerce Platforms for Direct Sales

ICT has also reshaped market linkages and value chains, enabling farmers to bypass middlemen and connect directly with consumers. E-Choupal and digital mandi systems have markedly improved market transparency and farmer income in India (Kaur et al. 2014; Villacis et al. 2024). Mobile-based market platforms have helped smallholders to avoid price exploitation and improve net margins in Kenya (Morepje et al. 2024). Transaction costs have also been reduced through SMS alerts and online bidding platforms in many countries. AgroTIC facilitated the listing and sale of citrus and vegetable produce in Colombia, linking over 200 farmers with agronomists and buyers (Hinojosa et al. 2023). This model led to measurable income gains and shorter value chains, making it a promising blueprint for smallholder-driven e-commerce.

21.3.3 ICT-Based Advisory Systems for Weather and Pest Alerts

Timely advisories covering weather fluctuations, pests infestation chances, nutrition, and market prices are essential in horticulture. The rise of mobile ICT tools, social media, and video platforms is nowadays providing real-time alerts to the growers (Khatri et al. 2024), which also increases the yield (Table 21.2). As per the report of FAO (2023), SMS and voice-based weather services that allowed farmers to plant and market more strategically have gained success in Kenya and Ethiopia. Smartphone-driven advisory systems, such as MAAS, that combine multimedia and push notifications can alert farmers about emerging pests and pest-resistant seed options (Khatri et al. 2024). The text alerts around market and weather conditions also lowered yield loss risks and led to better selling prices in India (Camacho and Conover 2019). On the other hand, Plant Village reduced disease incidences by providing real-time, AI-based notifications in Kenya (Shehu et al. 2025).

21.3.4 GIS and GPS in Resource Management

GIS and GPS technologies have streamlined land use planning and resource optimization in horticulture. Biradar et al. (2024) applied GIS in Karnataka to create land

Table 21.2 ICT- based digital services technology adopted in many regions, showing its benefit in yield improvement. The references have also been mentioned

Sl. no.	Region/ country	Yield improvement	Reference
1.	China	ICT and digital services for ground-level farmers improved the area coverage and yield of cereals and horticultural crops by 23%	Ding et al. (2022)
2.	India	Use of ICT by the farmers at the grass-root level increased the substantial yield as well as the return of the farmers at the grass-root level who are associated with both the cultivation of field and horticultural crops	Mukherjee et al. (2025)
3.	India	ICT interventions, including mobile apps, SMS messaging, educational videos, and voice calling services; 74% indicated higher yields, and 68% observed enhanced incomes	Mulungu et al. (2025)

capability maps, enabling horticulturalists to align crop options with soil and terrain constraints, thus optimizing land use. GPS-augmented remote sensing protocols are now being used in agroforestry to improve precision agriculture (Karada et al. 2023). Integration of GPS with sensors supports easy and precise variable application of water, nutrients, and plant protection products (Maraveas et al. 2025). Global horticulture is increasingly adopting multispectral drone imagery combined with GIS to monitor orchard health and yield (Mwinuka et al. 2022; Popescu et al. 2023). Satellite imagery can also be merged with GIS to achieve 99% land cover mapping accuracy, and it can support large-scale planning and resource monitoring (Zhang and Li 2022).

21.3.5 Integrated ICT for Smart Greenhouses

Protected cultivation, such as smart greenhouses, relies heavily on ICT-driven automation and sensor systems. Sensor networks within the greenhouse system can manage irrigation, control of temperature and relative humidity, and nutrient delivery to the plants (Fig. 21.2). Actuators, cameras, and cloud platforms enable centralized control as ICT for such monitoring and control (Hosny et al. 2025). IoT sensors and automation can reduce input waste and improve the crop health cum growth uniformity (Mowla et al. 2023). Many greenhouse systems now feature smartphone-based dashboards that allow remote monitoring of local climate and system faults (Bersani et al. 2022).

It includes elements like artificial intelligence and machine learning, represented by icons, leading to an internet cloud for data analysis and decision-making. Communication technologies are categorized by range: short (Bluetooth), medium (WiFi, Zigbee), and long (LPWAN, Sigfox, NB-IoT, LoRa, LoRaWAN), depicted with icons and a communication tower. Applications and displays are shown on a laptop with images of agricultural monitoring. Various sensors and devices, such as soil sensors, insect traps, robots, UV sensors, eco-cams, and drones, are illustrated in a field setting.

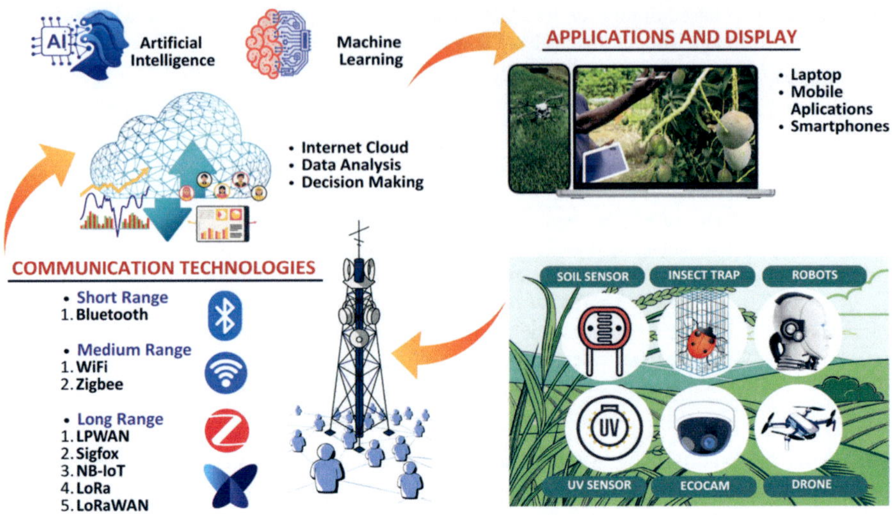

Fig. 21.2 The image is a flow chart illustrating the integration of technology in agriculture (self-composed)

21.3.6 Integrating ICT in Extension Systems

Extension services have been evolving from manual, in-field training models toward blended ICT-enabled systems. Khatri et al. (2024) provided a health check on digital extension in India, concluding that smallholder farmers benefit most when phone tools are paired with traditional peer learning and agent-led follow-up. Both digital literacy and culturally tailored content matter, suggesting that translation to local languages and simple UI are critical in such cases (Reddy et al. 2023). The increasing synergy between mobile-based tools and radio, television, and digital formats is creating inclusive farmer learning ecosystems (Ahmad et al. 2024a, b; Gouroubera et al. 2025). Many Krishi Vigyan Kendras in India have demonstrated that combining e-learning modules with field demonstrations boosts adoption of ICT-driven horticultural methods (Natarajan et al. 2024).

21.3.7 Addressing Challenges of Digital Divide and Infrastructure

Despite wide potential, several barriers remain. Unreliable rural internet, low digital skills, and limited access to smart devices are key obstacles (Khatri et al. 2024). Power outages and high cost of devices also hinder uptake in places like Nigeria (Agbeyangi and Lukose 2025). Extension agents are sometimes not well-known about the devices, and they need training in digital tools. This is the cause of not just their distribution to farmers. In response, governments and NGOs are expanding rural broadband and subsidizing smartphones. The partnerships between public extension agencies and private technical incubators are also lowering device and service costs for the reduction of the digital divide (Younus et al. 2024). In the Philippines, local ICT hubs train farmers on app usage and moderate online forums, sensitizing them to digital practices and boost tech adoption.

The future lies in deeper convergence where AI, IoT, GIS, 5G, and blockchain form integrated platforms. AI drones and edge computing will enable real-time yield forecasts, pest surveillance, and supply chain integration. Transfer learning enhanced remote sensing, and GIS will inform adaptive land use planning under climate volatility. Blockchain, combined with GPS and mobile interfaces, is beginning to appear in traceability projects. Pilot programs issuing consumers QR-coded provenance labels linked to farm and weather data, offering certified traceability. ICT integration into horticulture is yielding measurable benefits like improved productivity through precision irrigation, disease alerts, and crop diagnostics (Table 21.3). Market benefits will also be higher via better pricing and reduced wastage. Increased resource efficiency will be obtained by enabling variable water and fertilizer application based on GPS-GIS systems (Biradar et al. 2024). Better and increased inclusion of ICT will be more available, especially for women and youth, through digital learning and micro-entrepreneurship opportunities. In this context, resilience will be gained in the face of climate shocks via real-time advisories and predictive systems.

ICT serves as a catalyst that integrates innovation, inclusion, and income enhancement within horticulture. By providing mobile access to expert advisories, enabling direct and transparent marketplaces, offering precision analytics via GIS/GPS, and supporting protected cultivation, ICT is enabling farmers to transition toward sustainable, profitable, and climate-resilient practices. As digital literacy rises and connectivity spreads, the synergy of AI, blockchain, IoT, and smart analytics promises a horticultural transformation in the next decade, making horticulture smarter, fairer, and greener.

Table 21.3 Table showing technology implications across different regions, along with references

Sl. no.	Region/ country	Technology implication	Reference
1.	Puer, China	Novel spectrum splitting technology can increase the productivity of greenhouse crops, including vegetables, fruits, and flowers	Omer et al. (2025)
2.	Jeonju, South Korea	Optimization of artificial intelligence in greenhouses for the production of vegetables is increasing the yield, quality, and shelf life	Fuentes et al. (2026)
3.	Nanjing, China	Advanced environmental management of weather components of greenhouses increases the yield and price of vegetables considerably	Li et al. (2022a, b)

21.4 Robotics and Mechanization in Horticulture

Robotics and mechanization are revolutionizing horticulture, making it more efficient, consistent, and less reliant on manual labor. As labor costs rise and farm workers become scarcer, technologically driven methods are emerging as a viable long-term solution. These innovations are transforming tasks such as planting, weeding, harvesting, sorting, and greenhouse management, ushering in a new era of precision, sustainability, and quality.

21.4.1 Robotic Weed Removal, Harvesting, and Sorting

Recent years have seen a surge in the deployment of robotic weeders designed to distinguish and eliminate weeds with remarkable precision. Wu et al. (2020) described robots equipped with computer vision and GPS that target weed patches millimeter by millimeter, reducing herbicide use by 55% on experimental farms. These devices not only conserve agrochemicals but also minimize environmental damage. In parallel, robotic harvesters are being introduced across fruit and vegetable production. Apple picking robots are also capable of recognizing vine roll patterns and delicately collecting ripe fruit, maintaining consistent bruise-free quality in over 90% of picks (Anand et al. 2023). Similarly, tomato harvesters are fitted with multispectral vision that accurately detects the ripeness of the tomato fruit and picks accordingly, ensuring better color, firmness, and shelf life (Onishi et al. 2019). Automated sorters now follow harvesting robots, employing high-resolution cameras and AI algorithms to grade produce based on size, color, weight, and defects. The robotic sorting systems in strawberry greenhouses increased grading speed by 80% while reducing misclassification by 30%, delivering more uniform packaged produce (Dewi et al. 2020).

21.4.2 Automated Greenhouse Operations

Modern greenhouses increasingly rely on automated systems to maintain optimal growing conditions. Climate control units, irrigation setups, ventilation systems, and lighting adjustments are customized by AI-driven algorithms that analyze environmental data and crop growth stages. Greenhouse growing tomatoes using real-time sensors and automation where climate settings are optimized at 5-min intervals, based on external weather forecasts and internal microclimate readings, maintaining ideal conditions 96% of the time (Šalagovič et al. 2024). As a result, yield increased by 20%, while water use decreased by 40%. In cucumber and pepper greenhouses, Çeliktopuz et al. (2025) reported centralized irrigation and fertigation systems equipped with flow meters that adjust nutrient delivery by analyzing plant water uptake. This method reduced fertilizer use by 25% and produced more uniform fruit size, which is an essential attribute for domestic consumers. Moreover, automated ventilation has become essential for guarding against disease. Greenhouse fans and vents automatically respond to humidity and temperature thresholds, significantly cutting the incidence of fungal and bacterial infections by 60% (Qiu et al. 2022).

21.4.3 Labor-Saving Devices for Precision Input Application

Efficient application of seeds, planting materials, nutrients, and protection chemicals is a growing focus in horticulture. Traditional broadcast methods often result in uneven coverage and product overuse or gaps that decrease yield. Seed-planting robots equipped with GPS and cameras can determine optimal seeding depth and spacing, ensuring 95% germination uniformity in leafy vegetables and transplant trays (Lu et al. 2024). Similarly, nutrient applicator robots detect plant nutrient deficiency and spray microdoses at variable rates directly at the root zone, cutting fertilizer usage by up to 35% (Nadafzadeh et al. 2024). Targeted spray drones, equipped with multispectral sensors, are also gaining momentum. A 70% reduction in fungicide use on strawberries and grapes is owing to drone-identified hotspots (Aljawasim et al. 2023). These drones apply precise chemical amounts only where necessary, reducing pesticide loads and environmental impact.

21.4.4 Impact on Labor Efficiency and Crop Quality

Automation resulted in the requirement of a single-handed operator instead of the earlier tasks that used to require 4–6 laborers. In a greenhouse trial, robotic weeding and harvesting led to a 60% reduction in labor hours per acre (Upadhyay et al. 2024). In conventional fields, robotic weeders handled 10 acres per day with a single driver, compared to 2 acres by a four-person manual crew. Similarly, autonomous harvesters accomplished 3–4 full tomato harvest cycles daily, exceeding the capacity of 20 pickers (Yaguchi et al. 2016). These efficiency gains translate into reduced operational costs. Apple harvest robot usage cut annual labor expenses by 45%, while simultaneously improving the fruit's cosmetic quality (Yi and Mhamed 2024). Automation also enhances postharvest integrity. AI-assisted sorting reduced bruising during strawberry packing by 40%, which resulted in longer storage life and better market prices (Pugazhendi et al. 2023).

21.4.5 Modular Technology for Small and Midsized Farms

Concerns about high startup costs have been a barrier to using robotics in mid-scale horticulture. However, modular and customizable systems are making automation more accessible. Small-frame autonomous robots specifically for peppers and cucumbers can be rented monthly, making them cost-effective for farms with 5–10 acres (Henten et al. 2002). Cutulle and Maja (2021) developed a tractor-mounted weeder that can be hooked up to small utility vehicles, requiring minimal investment. The spread of these scalable systems allows growers to gradually build toward larger investments in sustainable mechanization.

21.4.6 Long-Term Benefits of Robotics Investment

Despite initial expenses, the benefits of mechanization are increasingly clear, particularly with sustained operation. A 3-year trial shows that cost reductions of 25–30% per acre in leafy lettuce systems are due to labor savings and reduced chemical use (Kaur et al. 2023). The robotic tomato pickers consistently delivered premium-grade produce, boosting market returns by 18% (Jun et al. 2021). Automation also reduces risk related to labor. The greenhouse automation ensured timely harvesting even during labor shortages, preventing yield loss of up to 35% during peak seasons.

21.4.7 Integration with AI and IoT

The future lies in interlinking robotics, AI, and IoT for real-time automation, diagnostics, and optimization (Fig. 21.3). The systems where robotic harvesters transmit yield and fruit quality data back to cloud servers are now in use (Miller et al. 2025). This data is then used by AI to forecast harvest schedules and pricing, driving dynamic marketing and logistics strategies. Meanwhile, closed-loop nutrient systems where soil-plant sensors feed data to robots that update nutrient formulations daily are in use (Zhai et al. 2020). This ensures near-optimal nutrition and reduces waste. Besides, autonomous creep robots equipped with edge computing can execute complex tasks such as grafting and fruit thinning under open-sky environments (Chen et al. 2024).

The advent of robotics is shifting horticulture into a skilled domain. Operators and technicians will need training in programming and maintenance of robots in the

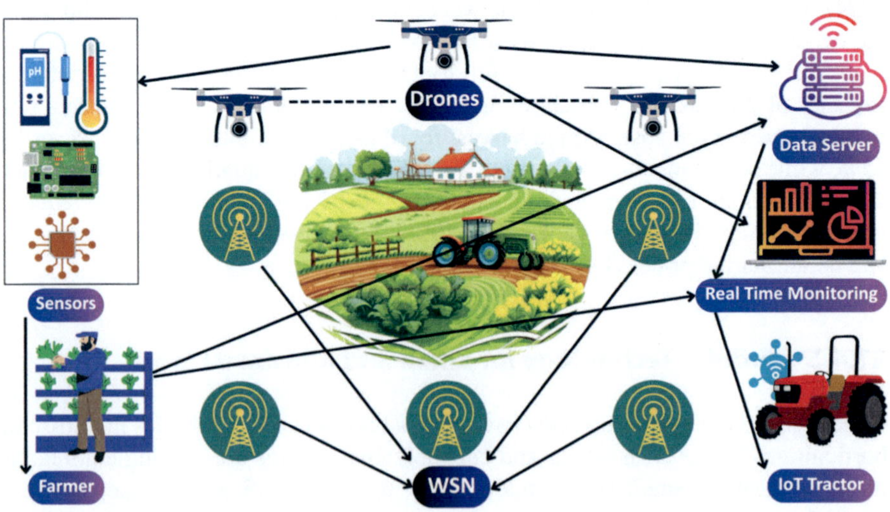

Fig. 21.3 Diagram illustrating a smart agriculture system. Central image of a farm with a tractor, surrounded by interconnected elements: drones, sensors, a farmer, a data server, real-time monitoring, and an IoT tractor. Wireless sensor networks (WSN) connect these components, indicating data flow and communication for efficient farm management (self-composed)

field of horticulture. The creation of the concept of "HortiBots" and their training programs to prepare the farm laborer for managing robot-assisted greenhouses is going on (Jørgensen et al. 2006). Such initiatives, like university-robotics laboratories, support workforce transformation and attract intelligent youth into agriculture, freeing them from monotonous manual labor while commanding higher wages. Despite the numerous benefits, robot integration faces several barriers, such as high capital costs, which mean that many small farms lack the upfront funds, and technical complexity that requires a specialized workforce for repair and troubleshooting. Additionally, regulatory uncertainty delays approval of robots in public spaces, and data security risks arise from farm system data being managed by third-party IoT platforms, which are also some of the major barriers. To overcome these challenges, the economic models of Javaid et al. (2024) share that farmers can jointly purchase and rotate robotic equipment. Rental schemes, subsidies, and training programs to expand the adoption of robotics in the field of horticulture can be introduced. Regulatory frameworks like the EU's "Agri-Robotics Safety Standards" are being finalized to ensure inspection, operator safety, and product quality (Martins et al. 2023). Meanwhile, efforts have shown that blockchain could secure data on yield, pesticide usage, and harvest timing, protecting farmer privacy. Robotics and mechanization are powering the next wave of horticultural modernization. They are enhancing efficiency, reducing dependency on labor, and delivering precise interventions that heighten productivity and crop quality. From robotic weeders and pickers to intelligent greenhouse systems, the future of horticulture is fast becoming automated and interconnected. Modular robotics models are bringing automation within reach of smallholders. Combined with AI and IoT-enabled analytics, these systems promise sustainable, resilient, and profitable farming.

21.5 Artificial Intelligence (AI) and Machine Learning (ML) Applications in Horticulture

Artificial Intelligence (AI) and Machine Learning (ML) are ushering in a new era in horticulture, one defined by precision, prediction, and data-driven decision-making. By processing vast datasets from sensors, aerial imagery, climate stations, and on-farm records, these systems empower growers, enhance yields, reduce risk, and improve economic resilience.

21.5.1 AI-Based Pest and Disease Detection

One of the most dramatic applications of AI in horticulture is early detection of pest infestation and plant disease through image recognition (Table 21.4). Convolutional neural networks (CNNs) integrated into smartphone apps and drone platforms can identify minute foliar changes long before human eyes can see those (Chakrabarty et al. 2025). In strawberries, a vision system with 94% accuracy in spotting gray mold at the pre-symptomatic stage was described by Giakoumoglou et al. (2023).

Table 21.4 Artificial Intelligence-based detection of pests and diseases in horticultural crops, along with citations

Sl. no.	Application of AI-based pest and disease detection technology	Reference
1.	Crop pest detection by a three-scale convolutional neural network with attention in horticultural crops	Wang et al. (2023)
2.	Algorithm for convolutional neural networks in the detection of pests using Googlenet	Yulita et al. (2023)
3.	The achievement of the deep convolutional neural network proposed a method of identifying various types of trapped insect species by making predictions based on available images using a database for horticultural crops	Nam and Hung (2018)

Similarly, tomato leaf detection models that achieved 92% precision under field trials have also been reported by Islam et al. (2025). With AI-enabled diagnostic applications like PlantX and SmartCrop, pesticide application becomes targeted and minimal, reducing agrochemical use by 30–60% (Javaid et al. 2023). Farmers can respond to localized disease hotspots rather than treating entire fields, curbing costs and environmental impact.

21.5.2 Machine Learning (ML) Models for Yield Prediction and Resource Optimization

Machine learning models are also excelling in yield forecasting and resource management. ML systems can predict production volumes with impressive precision using real-time metrics, like rainfall, temperature, soil moisture, and plant growth, by analyzing historical yield data alongside them (Waqas et al. 2025). A seasonal case study by Morales and Villalobos (2023) on apple orchards used random-forest regression and achieved a yield forecast within 5% error margin. ML model forecasting shows 93% accuracy in weekly harvests in tea plantations (Elbasi et al. 2023). This also enables better resource allocation and logistics planning. ML aids in optimized irrigation and fertilization beyond yield. Implementation of gradient-boosted trees in a cucumber greenhouse reduced water and fertilizer use by 25% and 18%, respectively (Emmanuel et al. 2022). Elsewhere, ML-driven water scheduling reduced the irrigation frequency of sugarcane fields by 40%, reducing stress on groundwater reserves (Kingsly et al. 2024).

21.5.3 Decision Support Systems for Planting, Inputs, and Harvesting

AI-powered decision support systems (DSS) are helping growers to make informed choices about planting, inputs, and harvest timing. Tools like AgriSense integrate weather forecasts, pest risk analysis, and phenology models into dashboards and mobile alerts (Soussi et al. 2024). AI systems in olive orchards can suggest pruning dates based on bloom cluster predictions (Sharma and Shivandu 2024). Sensor-driven models implemented in grape vineyards can recommend optimal harvest windows, leading to a 12% increase in wine quality scores (Mizik 2023). The DSS tools helped to reduce fruit cracking losses in banana plantations by 20% and elevated market value (Das and Ahmed 2023). These systems also advise variety selection, suggesting cultivars best suited for current soil, climate, and market conditions, blending local knowledge with data insights.

21.5.4 Automated Sorting, Grading, and Packaging

AI has significantly transformed postharvest processing through automated grading and packaging systems. Modern robots equipped with deep learning can examine fruit size, shape, color uniformity, and surface blemishes during high-speed conveyor movement (Kondo 2009; Tan et al. 2025). Yu et al. (2024) reported a 35% increase in throughput using vision-based sorting, with fewer grade errors in citrus packing lines. Similarly, mango sorting robots can achieve 98% accuracy in identifying underripe fruits, improving shelf presentation and reducing customer returns. These systems uphold food safety by removing damaged produce, ensuring only premium quality is shipped. They also reduce postharvest waste and labor costs, crucial in high-volume operations.

21.5.5 AI in Supply Chain and Logistics

AI is now playing a vital role in horticulture supply chains, optimizing packing schedules, refrigeration logistics, and market delivery. An AI-based cold-chain monitoring tool can track temperature, humidity, and transit delays, triggering alerts to minimize spoilage (Jackson et al. 2024). Logistics firms use Machine Learning pipelines to match logistics needs based on destination, crop type, and harvest volume (Odimarha et al. 2024). A case study by Talaviya et al. (2020) described an AI routing engine that reduced transit time by 18% and crop loss by 10% in green-grocery delivery networks. Moreover, consumer-facing digital traceability is often powered by blockchain-stored AI data, and it allows buyers to verify the origin, handling, and freshness of produce. The traceable produce via QR codes fetched 10–15% higher prices in urban markets due to consumer trust (Dey et al. 2021).

21.5.6 Democratizing AI for Smallholders

While many of these innovations emerged from large farms, AI platforms are increasingly accessible to small and marginal farmers. Mobile-based AI diagnosis tools like PlantXLite and AgriAI have reached remote regions in India and Sri Lanka (Siddiqua et al. 2022). The smallholders in Punjab, India, are using ML-driven weather forecasts that saved 25% on irrigation costs (Bhatia and Singh 2024). The open-source iMap tool allows extension agents to use drones and ML for disease mapping in East Africa, improving regional responses (Appiah et al. 2024). Open-source ML libraries like TensorFlow and PyTorch are being localized by universities and NGOs to train farmers and data-village technicians (Li et al. 2024), helping bridge the AI knowledge gap.

21.5.7 Integration of AI, IoT, and Robotics

AI is being tightly integrated with IoT and robotics to enable real-time automation and learning systems in the future. In greenhouse trials, robots are now collecting yield, disease, and environmental data and feeding it into AI models to fine-tune conditions autonomously (Kumari et al. 2025). Connected robotic arms can thin fruits and perform selective pruning based on AI assessment. Drone scouts feed leaf-spot data to AI systems and trigger ground robots for the treatment of infected trees within hours in orange orchards (Partel et al. 2019). This is the essence of smart fields in horticulture, where AI, sensing, and robotics work together in feedback-driven harmony.

21.5.8 Economic, Quality, and Sustainability Impacts

The implications of artificial intelligence and machine learning in horticulture are profound, like yield gains of 10–20% have been reported in multiple trial systems (Opara et al. 2024). Input efficiency has been improved by 25–40%, limiting water and nutrient overuse (Botero-Valencia et al. 2025). Crop quality showed consistent gains as reported by Gatou et al. (2024) in grapes. Environmental impact was minimized through reduced agrochemical application, with a 30–60% drop in pesticide use and 40% lower nitrogen runoff. The income growth of 10–25% was observed when AI insights helped time market entry and VOC control at critical harvest stages (Aijaz et al. 2025).

Despite these benefits, several barriers remain in the integration of AI and ML in the field of horticulture. Data availability is limited in some smallholder contexts, and connectivity limitations slow AI real-time deployment in remote landscapes. Moreover, model transparency hampers adoption, and farmers resist "black-box" AI systems as they cannot understand (Chourlias et al. 2025). Costs of AI and ML integration and technological infrastructure are also prohibitive, underscoring the need for leasing, cooperatives, and public funding. Promising solutions include lightweight models, edge-based AI, regional data cooperatives, and joint funding schemes. AI and ML are reshaping horticulture from reactive to predictive and automated systems. With emerging tools like transfer learning, federated learning, and explainable AI, ethical, transparent, and evolving systems are within reach. In the coming years, we can expect full-chamber AI scouting, sensory robotic sprayers, harvest prediction dashboards, and digital marketplaces accurately orchestrated by ML pipelines. Such systems not only boost farming resilience but also entice digitally fluent youth to adopt agriculture as a modern, technology-driven profession.

21.6 Smart Skill Developments for a Future-Ready Workforce in Horticulture

As horticulture adopts cutting-edge technologies, developing a future-ready workforce has become essential in this field. Growers can no longer rely solely on hands-on experience; instead, modern horticulture requires capabilities in digital tools, precision farming, greenhouse automation, and data-driven decision-making. This transformative shift requires substantial upskilling, which in turn depends on comprehensive training ecosystems and inclusive entrepreneurship initiatives.

21.6.1 The Need for Digital and Technical Upskilling

The horticultural landscape is becoming increasingly digitalized, with IoT sensors, drones, AI analytics, and mobile apps transforming farm management. Digital proficiency now determines a grower's ability to maximize resource efficiency, responsiveness to pest alerts, and market alignment (Papadopoulos et al. 2024). Without understanding sensor-readouts or app-based advisories, lessons from old-school training may be lost in translation (Food and Agriculture Organization [FAO] 2023). Theoretical knowledge alone is insufficient, and practical skills are of the utmost vital importance. Precision irrigation systems demand both digital literacy and technical understanding (Ben Farah and Amara 2025). Field experiments showed that even seasoned farmers misconfigure automated systems when digital competencies are narrow. Meanwhile, Rajak et al. (2023) found that smallholder farmers who received drone-usage training had 30% higher monitoring accuracy and quicker responses to crop stress. The challenge is that digital literacy in rural areas remains uneven. A survey by Sharma et al. (2021) in Maharashtra reported that only 40% of farmers could reliably navigate smartphone apps. Hence, upskilling initiatives must begin with foundational training, basic ICT, language-agnostic platforms, and responsive user interfaces to connect human capital with technology effectively.

21.6.2 Training Modules for ICT, Drones, and Data Analytics

Skill development curricula are evolving to match technological complexity, such as ICT for agriculture. Modules typically begin with smartphone use, mobile-based advisory apps, and digital recordkeeping. Paliwal and Kumari (2024) designed a curriculum for the farmers to learn, install, and operate extension apps; interpret market signals; and upload photo logs of crops. 70% of trainees actively used these apps within 3 months, during cultivation seasons, reporting cost savings of 15% in inputs. Similarly, drone operation training emphasizes flight-safe regulations, payload delivery, pest-spraying safety, and image interpretation. Hands-on drone clinics led to more effective field scouting. Even trainees could detect early-stage blight symptoms with 90% accuracy within 6 weeks (Marinus et al. 2021). The modules, in the case of data analytics and AI adoption, cover basic data cleaning, interpretation of sensor readings, and feeding datasets into AI dashboards. Through the hands-on workshops, farmers are able to input temperature, plant biomass, and humidity data into ML models that help 80% of participants comprehend yield predictions and irrigation planning (Elbasi et al. 2023). Modular training reduces complexity, allowing learners to master components sequentially, such as digital basics, drone proficiency, and data analytics. This scaffolded approach enhances competency and builds confidence within the growers.

21.6.3 Role of Institutions

Krishi Vigyan Kendras (KVKs) have evolved into local training hubs. Recently, KVKs offering AI-integrated greenhouse demos trained over 1200 farmers in five agricultural seasons. After receiving training, 65% adopted sensor-enabled drip irrigation systems, and 55% incorporated integrated pest management (IPM). Agricultural universities are partnering with such training centers to develop ICT-skilled farm technicians. The university–KVK incubation models facilitated 6-month certification programs, culminating in internships with technology providers and micro-enterprises (Malhotra et al. 2022). Agri-tech startups are also playing a catalytic role. Companies like FarmSense and AgroEdge are collaborating with universities to offer micro-Massive Open Online Courses (MOOCs) focused on drone piloting, automated greenhouse setups, and smartphone-based pest diagnostics (Zelenić et al. 2024). These platforms often feature interactive chatbots for on-demand clarifications, helping learners overcome literacy barriers. Moreover, partnerships between NGOs, banks, and start-ups are launching mobile training units, essentially ICT labs on wheels that travel to rural areas, creating democratizing access. Within a year of deployment, these units had trained over 5000 farmers in drone flight basics and ICT tool usage for crop planning.

21.6.4 Skill Incubation for Rural Agri-Entrepreneurship

Upskilling is useful, but its potential multiplies when paired with entrepreneurship. Skill incubation centers offer a complete ecosystem from prototyping to market access. Sumalan et al. (2020) describe incubation hubs with hardware labs for assembling nutrient sensor kits and greenhouses. Trainees learn electronics, firmware loading, and troubleshooting. Several graduates from these hubs have launched start-ups selling low-cost IoT kits and drone services in their districts. Entrepreneurship modules often run as bootcamps that introduce business planning, microfinance access, pricing strategies, and customer acquisition (Vaz et al. 2023). Virtually all participants developed proposals for agri-service businesses, such as drone mapping rentals or greenhouse monitoring services, and eight teams had secured seed funding within 6 months. A longitudinal study by Pan et al. (2024) found that rural entrepreneurs specialized in horticultural services increased household income by 25–40% within 2 years. Such independent agri-entrepreneurship reduces rural outmigration and enhances local technological resilience.

21.6.5 E-learning and Hybrid Learning: Scaling Access

While offline training is key, e-learning platforms provide continuity and scalability. "AgriTech Online Academy," a popular MOOCs in India, offers modules on IoT, AI, and greenhouse systems. Moonlighting-friendly webinar series allows farmers to learn outside harvest hours, reinforcing hands-on training previously received. Language adaptability is another important issue in this regard. Content localized in Hindi, Tamil, and Telugu has doubled participation compared to English-only courses (Groff 2010). Khan et al. (2023) found that video-based learning with voice-over instructions led to 80% retention among learners with minimal formal education. Blended learning models mixing in-person sessions with online follow-up provide strong outcomes. Participants in such hybrid courses adopted an average of three new practices (like drone scouting and sensor-driven irrigation) within two growing cycles (Belt and Lowenthal 2021).

21.6.6 Supporting Infrastructure: Digital Literacy and Connectivity

Liu et al. (2025) have reported that villages with more than 70% smartphone penetration and broadband speeds of more than 5 Mbps supported significantly faster adoption of ICT and drone modules. In contrast, areas with patchy 2G connectivity failed to maintain participation, affecting 40% of e-learning cohorts. Policy efforts, like digital broadband expansion and digital literacy missions, are also critical. Subsidies for smartphones and data plans boost participation in low-income communities. Digital libraries in local languages, offline app versions, and kiosk-based training have mitigated internet constraints.

21.6.7 Measuring Returns: Benefits to Farmers and Communities

Smart skill development delivers tangible benefits. Drone-monitored, sensor-irrigated plots reported 15–25% yield increases compared to control fields (Hafeez et al. 2023). Input reduction studies showed a 10–20% drop in fertilizer and pesticide costs post-training (Paliwal and Kumari 2024). Incubated agri-entrepreneurs achieved 25–40% growth in annual income. Trainings led to new agri-tech service roles, drone operators, data analysts, greenhouse supervisors, and startup founders. Shifting of labor from urban migration to rural agri-science generated social value (Anand and Raj 2019). These metrics make a compelling case for sustained investment.

21.6.8 Challenges and Enablers

Major enablers include public-private collaboration, donor funding, policy incentives, and strong training infrastructure. But challenges have remained, like trainer shortage being a critical issue, and only 25% of district KVK staff having strong ICT backbones. Institutionalizing trainer training-of-trainers (ToT) models can address this. Cost barriers limit access to drones, sensors, and software subscriptions. Agri-tech start-ups are exploring shared-asset models and pay-as-you-go services. Moreover, cultural resistance can stall uptake as some farmers equate digital tools with risk. Community-led demonstration plots show 90% adoption compared with pilot spread to non-participants (Chacko 2023). Similarly, regulatory support is needed, and licensing for drones, data policies, and drone-negative areas requires clarification.

21.6.9 Roadmap to a Tech-Savvy Horticulture Workforce

Better foundation skills are to be developed through basic digital literacy training in vernacular languages, smartphone use, and app navigation. Technical upskilling can be attained through focused modules on drones, sensors, greenhouse automation, and AI dashboards in local centers. Entrepreneurship training is required for better agribusiness planning, finance, service design, and market research. More hybrid learning platforms, like blended in-person modules with MOOCs, webinars, and mobile refreshers, are to be arranged. More infrastructure support is to be given by expanding rural broadband, arranging subsidized devices, and developing mobile delivery units. Public-private partnerships are to be aligned with government, NGOs, universities, and start-ups toward shared training goals. Monitoring and certification are needed to track learner progress and offer accredited certificates to boost credibility. Local incubation hubs are to be established for hardware labs, DIY sensor kits, and business planning spaces.

This roadmap can catalyze sustained change toward smart horticulture, inclusive growth, and climate resilience. In sum, smart skill development is no longer optional but essential for horticulture's next generation. By integrating digital literacy, drone operation, AI comprehension, and entrepreneurship training, institutions empower rural youth and farmers to compete in a future-focused sector. These initiatives benefit individuals and uplift communities—enhancing livelihoods, preserving tradition, and creating sustainable agri-tech ecosystems. As knowledge gaps close, adoption spreads. As sustainability demands grow, skill-driven horticulture becomes market-ready, climate-smart, and socially inclusive. For horticulture to thrive tomorrow, equipping people today with the right skills is mission-critical.

21.7 Challenges and Opportunities in Smart Skill Development for Horticulture

As horticulture undergoes a transformation driven by digital tools, precision farming, and automation, the integration of smart strategies also introduces a set of challenges. At the same time, it creates vast opportunities for inclusive growth, employment generation, and global competitiveness. Navigating this duality requires a clear understanding of the systemic barriers that hinder technological adoption and the enablers that can unlock horticulture's full potential in a technology-driven future.

21.7.1 The Digital Divide and Unequal Access

One of the foremost challenges facing smart horticulture is the digital divide, which is the unequal access to technology, digital infrastructure, and digital literacy across regions and socioeconomic groups. While urban and peri-urban farmers increasingly embrace mobile apps, drones, and AI tools, large swathes of small and marginal farmers in rural areas remain excluded from these advancements and advantages. Limited access to high-speed internet and affordable smartphones constrains real-time access to information on weather forecasts and pest alerts as well as market trends. The broadband penetration in rural areas of India remains significantly lower than in cities, which hinders the widespread deployment of app-based advisory systems and e-learning platforms (Sindakis and Showkat 2024). Additionally, the digital illiteracy of village-level growers further marginalizes older or less-educated farmers who may struggle to interact with interfaces in English or even in their local languages. Localization of digital content and language-agnostic interfaces are essential to reduce such digital illiteracy. Digital training must begin with foundational ICT skills before moving to more complex tasks such as AI-powered decision-making or drone operation. Public digital literacy programs like the PMGDISHA (Pradhan Mantri Gramin Digital Saksharta Abhiyan) offer hope, but scaling and contextualizing these for horticultural applications remains to be progressed.

21.7.2 High Initial Investment and Skill Requirements

Another major barrier is the high initial cost of acquiring smart horticultural tools. Drones, IoT-based irrigation systems, greenhouse automation kits, and sensor networks require substantial upfront capital, which is often beyond the reach of smallholder farmers. Even when equipment is made available through subsidies or cooperatives, its effective use hinges on technical training, which itself requires resources, time, and appropriate delivery mechanisms. Moreover, using drones or interpreting AI dashboards requires specific competencies. These investments risk becoming underutilized or misused without continuous skilling and field-level demonstrations. The shortage of trained facilitators compounds this challenge, especially in remote areas. Although Krishi Vigyan Kendras (KVKs), agri-polytechnics, and private start-ups are beginning to offer training, their reach and standardization need improvement. A promising solution is the development of shared service models, such as drone-on-hire centers or community resource hubs that pool infrastructure for collective access. These must be coupled with modular, multilingual, and hands-on training to ensure uptake and sustainability.

21.7.3 Opportunities for Inclusive Growth and Employment

Despite these challenges, smart horticulture offers tremendous opportunities for inclusive growth, especially when skills are democratized. The benefits cascade beyond higher productivity, particularly when farmers are equipped with the ability to leverage digital tools. There are measurable improvements in resource-use efficiency, product quality, climate resilience, and postharvest value addition. Smart horticulture enables diversified employment pathways beyond traditional farm labor. Trained individuals can become drone operators, greenhouse technicians, mobile advisory agents, or AI-supported agri-consultants. Smart skill incubation centers helped rural youth to launch over 300 micro-enterprises in drone mapping, app-based pest diagnosis, and sensor installations across three states in India. Thus, the sector can absorb not only farmers but also aspiring entrepreneurs and technology graduates. Moreover, India has the potential to expand its export base, especially in fruits, spices, floriculture, and organic produce, by ensuring quality through technology-enabled systems with increasing global demand. Blockchain-based traceability, AI-powered grading, and digital certification mechanisms can enhance the credibility and open avenues to premium international markets.

21.7.4 Government Schemes and Public-Private Partnerships

Recognizing both the barriers and opportunities, government policies and schemes have begun to focus on mainstreaming smart horticulture through strategic investments and partnerships. Schemes like the National Mission on Sustainable Agriculture (NMSA), National Horticulture Mission (NHM), etc., incorporate components for promoting ICT tools, polyhouses, precision farming, and skill development. The Sub-Mission on Agricultural Mechanization (SMAM) provides financial assistance for purchasing modern equipment, while initiatives like the Agri-Clinics and Agri-Business Centres (ACABC) promote entrepreneurship and training through Public-Private Partnerships (PPP) models. Additionally, RKVY-RAFTAAR and the Start-up India campaign support technology-based agribusiness incubation, many of which focus on horticulture. Public-private partnerships (PPPs) play a vital role in amplifying these schemes. Companies such as AgNext, CropIn, and FarmSense collaborate with universities and KVKs to deliver mobile-based training, AI analytics, and customized digital dashboards. These firms act as innovation carriers, rapidly scaling solutions while adapting them to field realities. For example, AgroEdge Technologies partnered with rural Industrial Technical Institutes (ITIs) and FPOs (Farmer Producer Organizations) to train over 10,000 farmers in sensor installation and data logging. These models illustrate that smart horticulture can become an engine for equitable rural transformation when supported by coherent policies, targeted subsidies, and cross-sector collaboration.

The convergence of smart tools and skill development in horticulture represents a critical inflection point for the sector. While the road ahead is not without obstacles, chief among them are the digital divide, investment barriers, and training needs. On the other hand, the opportunities for employment, sustainability, entrepreneurship, and global competitiveness are equally profound. Effective collaboration among stakeholders, government, academia, private firms, and farming communities to design inclusive, accessible, and scalable systems will only show success. Smart horticulture becomes not just a technological vision but a viable pathway to secure and sustainable livelihoods for millions when digital tools are matched with human capacity, and innovation is paired with equity.

21.8 Conclusion and Future Outlook

As the world confronts the intertwined challenges of food insecurity, climate change, labor shortages, and shrinking arable land, horticulture is emerging not only as a solution but also as a strategic driver of agricultural transformation. The role of smart strategies that leverage digital innovation, automation, and skill-based practices has become increasingly vital in this rapidly evolving landscape. These technologies are not just about boosting productivity from automated greenhouses to AI-driven decision support systems; they are fundamentally reshaping the way that horticulture is practiced, taught, and scaled for the future. At the heart of this transformation lies a critical insight that the future of horticulture will be defined not only by tools and technologies but also by the people who wield them. While innovations such as drones, IoT sensors, robotics, and machine learning offer immense potential, their true value is only realized when farmers, entrepreneurs, and extension professionals are equipped with the skills to adopt and adapt them. Bridging the gap between advanced technology and grassroots skill readiness is, therefore, a nonnegotiable step toward achieving sustainability, profitability, and resilience in the sector.

This skill-technology nexus is especially relevant in the context of India and other developing economies, where the horticulture sector supports millions of small and marginal farmers. Digital illiteracy, cost barriers, and limited access to training remain major hurdles. However, many encouraging developments, such as mobile-based learning platforms and incubation programs backed by government and private-public partnerships, in many cases demonstrate that the digital divide can be narrowed with the right policy support and community engagement. Initiatives from Krishi Vigyan Kendras, agri-tech start-ups, and universities are proving that localized, modular, and hands-on training programs can transform rural youth into tech-savvy agri-entrepreneurs capable of navigating a modernized value chain.

Moreover, smart horticulture is inherently inclusive. It not only optimizes resource use and enhances environmental sustainability but also creates diverse employment opportunities across the value chain from drone pilots and greenhouse managers to AI model trainers and digital extension agents. This diversification is critical in reversing rural outmigration and fostering local economic development. The trained individuals in protected cultivation and digital tools have significantly increased productivity, income, and confidence levels among farming communities. Looking ahead, the future of horticulture will be increasingly data-driven, automated, and knowledge-intensive. Farms will rely on predictive analytics, real-time diagnostics, and precision input delivery to manage every aspect of crop production. AI-powered algorithms will forecast yield, optimize planting windows, and detect pest threats with minimal human intervention. Robots will harvest fruits with surgical accuracy, while blockchain-based systems will ensure traceability from farm to fork. These technologies will be accessible not only to large agribusinesses but, through shared infrastructure models and cooperative ownership, also to smallholders and rural collectives.

However, this vision will remain aspirational unless accompanied by continued investment in human capital. The development of a future-ready workforce that is adaptable, digitally literate, and innovation-oriented is the foundation of a smart horticultural ecosystem. This invites a massive shift in how modern agricultural education is delivered, from theory-heavy curricula to hands-on, tech-enabled, problem-solving modules; from instructor-led lectures to peer-to-peer, field-based, and e-learning environments. Curriculum reform, industry collaborations, and the inclusion of soft skills such as entrepreneurship, communication, and risk management are equally vital. Governments must sustain and expand their role as enablers. Smart subsidies, digital infrastructure investments, and inclusive policy frameworks can accelerate the diffusion of innovation to the grassroots. Supporting start-ups, funding research on climate-resilient and tech-integrated practices, and facilitating global market access for high-value horticultural exports will further strengthen the sector's growth trajectory. Data governance, cybersecurity, and ethical AI use have equally important roles. As farms generate and rely on more data, frameworks must ensure that information remains secure, farmers retain ownership, and AI tools are trained on diverse, localized datasets to avoid biases or failures in varied conditions.

In conclusion, smart horticulture is not a distant ideal but an emerging reality that promises sustainability, resilience, and prosperity for millions. The transformation will not be led by technology alone but by skillful minds, adaptive institutions, and inclusive policies. A future-ready horticulture sector must be founded on innovation, empowered by skill, and sustained by smart practices. As the agriculture sector continues to evolve, the convergence of knowledge, technology, and people will ensure that horticulture remains not only a source of food and income but also a beacon of opportunity in the twenty-first-century economy.

References

Abiri R, Rizan N, Balasundram SK, BayatShahbazi A, Abdul-Hamid H (2023) Application of digital technologies for ensuring agricultural productivity. Heliyon 9(12):e22601. https://doi.org/10.1016/j.heliyon.2023.e22601

Adamo T, Caivano D, Colizzi L, Dimauro G, Guerriero E (2025) Optimization of irrigation and fertigation in smart agriculture: an IoT-based micro-services framework. Smart Agric Technol 11:100885. https://doi.org/10.1016/j.atech.2025.100885

Agbeyangi AO, Lukose JM (2025) Telemedicine adoption and prospects in sub-Sahara Africa: a systematic review with a focus on South Africa, Kenya, and Nigeria. Healthcare (Basel) 13(7):762. https://doi.org/10.3390/healthcare13070762

Ahmad U, Sohel F (2025) Evaluating decision support systems for precision irrigation and water use efficiency. Digit Eng 4:100038. https://doi.org/10.1016/j.dte.2025.100038

Ahmad B, Sarkar MAR, Khanom F, Lucky RY, Sarker MR, Rabbani MG, Ray SRR, Rahman MN, Sarker MNI (2024a) Experience of farmers using mobile phone for farming information flow in Boro rice production: a case of Eastern Gangetic plain. Soc Sci Humanit Open 9:100811. https://doi.org/10.1016/j.ssaho.2024.100811

Ahmad MF, Ahmad FA, Alsayegh AA, Zeyaullah M, AlShahrani AM, Muzammil K, Saati AA, Wahab S, Elbendary EY, Kambal N, Abdelrahman MH, Hussain S (2024b) Pesticides impacts on human health and the environment with their mechanisms of action and possible countermeasures. Heliyon 10(7):e29128. https://doi.org/10.1016/j.heliyon.2024.e29128

Aijaz N, Lan H, Raza T, Yaqub M, Iqbal R, Pathan MS (2025) Artificial intelligence in agriculture: advancing crop productivity and sustainability. J Agric Food Res 20:101762. https://doi.org/10.1016/j.jafr.2025.101762

Ajambo S, Ogutu S, Birachi E, Kikulwe E (2022) Digital agriculture platforms: understanding innovations in rural finance and logistics in Uganda's agrifood sector. International Food Policy Research Institute, pp 1–49

Ali AH, Youssef A, Abdelal M, Raja MA (2024) An ensemble of deep learning architectures for accurate plant disease classification. Eco Inform 81:102618. https://doi.org/10.1016/j.ecoinf.2024.102618

Aljawasim BD, Samtani JB, Rahman M (2023) New insights in the detection and management of anthracnose diseases in strawberries. Plants 12(21):3704. https://doi.org/10.3390/plants12213704

Al-Qudah R, Almuhajri M, Suen CY (2025) Unveiling the potential of sustainable agriculture: a comprehensive survey on the advancement of AI and sensory data for smart greenhouses. Comput Electron Agric 229:109721. https://doi.org/10.1016/j.compag.2024.109721

Anand S, Barua MK (2022) Modeling the key factors leading to post-harvest loss and waste of fruits and vegetables in the agri-fresh produce supply chain. Comput Electron Agric 198:106936. https://doi.org/10.1016/j.compag.2022.106936

Anand A, Raj S (2019) Agritech startups: the ray of hope in Indian agriculture (discussion paper 10). MANAGE–Centre for Agricultural Extension Innovations, Reforms, and Agripreneurship (CAEIRA). https://www.manage.gov.in/publications/discussion%20papers/MANAGE-Discussion%20Paper-10.pdf

Anand R, Malkani P, Rathod S, Dharmender D, Swain S, Kumar S, Kurmi R (2023) Robotic harvesters for fruits and vegetables. https://doi.org/10.22271/ed.book.2337

Appiah O, Hackman KO, Diallo BAA, Ogunjobi KO, Diakalia S, Valentin O, Abdoul-Karim D, Dabire G (2024) PlanteSaine: an artificial intelligent empowered mobile application for pests and disease management for maize, tomato, and onion farmers in Burkina Faso. Agriculture 14(8):1252. https://doi.org/10.3390/agriculture14081252

Arrubla-Hoyos W, Ojeda-Beltrán A, Solano-Barliza A, Rambauth-Ibarra G, Barrios-Ulloa A, Cama-Pinto D, Arrabal-Campos FM, Martínez-Lao JA, Cama-Pinto A, Manzano-Agugliaro F (2022) Precision agriculture and sensor systems applications in Colombia through 5G networks. Sensors 22(19):7295. https://doi.org/10.3390/s22197295

Balasundram SK, Shamshiri RR, Sridhara S, Rizan N (2023) The role of digital agriculture in mitigating climate change and ensuring food security: an overview. Sustainability 15(6):5325. https://doi.org/10.3390/su15065325

Belt E, Lowenthal P (2021) Video use in online and blended courses: a qualitative synthesis. Distance Educ 42(1):1–31. https://doi.org/10.1080/01587919.2021.1954882

Ben Farah S, Amara N (2025) Lab to farm: mapping knowledge transfer channels and determinants from researchers' perspective – a systematic literature review. J Innov Knowl 10(1):100650. https://doi.org/10.1016/j.jik.2025.100650

Bersani C, Ruggiero C, Sacile R, Soussi A, Zero E (2022) Internet of things approaches for monitoring and control of smart greenhouses in industry 4.0. Energies 15(10):3834. https://doi.org/10.3390/en15103834

Bhatia S, Singh SP (2024) Assessing groundwater use efficiency and productivity across Punjab agriculture: district and farm size perspectives. Agriculture 14(8):1299. https://doi.org/10.3390/agriculture14081299

Bhatnagar S, Chaudhary R, Sharma S, Janjhua Y, Thakur P, Sharma P, Keprate A (2024) Exploring the dynamics of climate-smart agricultural practices for sustainable resilience in a changing climate. Environ Sustain Indic 24:100535. https://doi.org/10.1016/j.indic.2024.100535

Biradar IB, Sreenatha A, Reddi S, Prasanna S, Tuppad G, Hiremath R, Naragund R, Maheswarappa HP (2024) Land use planning by using GIS techniques for land resource inventory (LRI)–Northern Dry Zone of Karnataka, India. Asian J Soil Sci Plant Nutr 10:27–36. https://doi.org/10.9734/ajsspn/2024/v10i2257

Botero-Valencia J, García-Pineda V, Valencia-Arias A, Valencia J, Reyes-Vera E, Mejia-Herrera M, Hernández-García R (2025) Machine learning in sustainable agriculture: systematic review and research perspectives. Agriculture 15(4):377. https://doi.org/10.3390/agriculture15040377

Camacho A, Conover E (2019) The impact of receiving SMS price and weather information on small scale farmers in Colombia. World Dev 123:104596. https://doi.org/10.1016/j.worlddev.2019.06.020

Çeliktopuz E, Can M, Eriş H (2025) Optimizing cucumber growth: integrating smart irrigation with various fertilization strategies in greenhouse conditions. Not Bot Horti Agrobot Cluj-Na 53:14332. https://doi.org/10.15835/nbha53114332

Chacko J (2023) Behavioural barriers to technology adoption in agriculture (Master's seminar report). Department of Agricultural Extension, College of Agriculture, Kerala Agricultural University

Chakrabarty S, Deb CK, Marwaha S, Haque MA, Kamil D, Bheemanahalli R, Dhillon MK, Shashank PR (2025) Application of artificial intelligence in insect pest identification – a review. Artif Intell Agric 6:100XXX. https://doi.org/10.1016/j.aiia.2025.06.005

Chen Z, Lei X, Yuan Q, Qi Y, Ma Z, Qian S, Lyu X (2024) Key technologies for autonomous fruit- and vegetable-picking robots: a review. Agronomy 14(10):2233. https://doi.org/10.3390/agronomy14102233

Chourlias A, Violos J, Leivadeas A (2025) Virtual sensors for smart farming: an IoT- and AI-enabled approach. IoT 32:101611. https://doi.org/10.1016/j.iot.2025.101611

Cutulle MA, Maja JM (2021) Determining the utility of an unmanned ground vehicle for weed control in specialty crop systems. Ital J Agron 16(4):1865. https://doi.org/10.4081/ija.2021.1865

Das T, Ahmed A (2023) Evaluation of postharvest losses of banana growers in North East India: an ordered probit analysis. Curr Agric Res J 11(2):484–498. https://doi.org/10.12944/CARJ.11.2.10

Delfani P, Thuraga V, Banerjee B, Chawade A (2024) Integrative approaches in modern agriculture: IoT, ML and AI for disease forecasting amidst climate change. Precis Agric 25:2589–2613. https://doi.org/10.1007/s11119-024-10164-7

Dewi T, Pola R, Oktarina Y (2020) Fruit sorting robot based on color and size for an agricultural product packaging system. Bull Electr Eng Inform 9(4):1438–1445. https://doi.org/10.11591/eei.v9i4.2353

Dey S, Saha S, Singh A, McDonald-Maier K (2021) FoodSQRBlock: digitizing food production and the supply chain with blockchain and QR code in the cloud. Sustainability 13(6):3486. https://doi.org/10.3390/su13063486

Ding JP, Li J, Liu J, Zhang W, Jia X (2022) ICT-based agricultural advisory services and nitrogen management practices: a case study of wheat production in China. J Integr Agric 21(6):1799–1811. https://doi.org/10.1016/S2095-3119(21)63859-5

Elbasi E, Zaki C, Topcu AE, Abdelbaki W, Zreikat AI, Cina E, Shdefat A, Saker L (2023) Crop prediction model using machine learning algorithms. Appl Sci 13(16):9288. https://doi.org/10.3390/app13169288. Preprints, 202305.1519.v1. https://www.preprints.org/manuscript/202305.1519/v1

Emmanuel A, Hensel O, Esau T, Elijah O, ZainalAbidin MS, Ajibade S, Yerima O, Nasirahmadi A (2022) Precision irrigation management using machine learning and digital farming solutions. AgriEng 4(1):70–103. https://doi.org/10.3390/agriengineering4010006

Fabre A, Stephane S (2021) The impact of public-private partnerships (PPPs) in infrastructure, health and education (working paper). Toulouse School of Economics, University of Toulouse Capitol

FAO (2023) The state of food security and nutrition in the world 2023: urbanization, agrifood systems transformation and healthy diets across the rural–urban continuum. Food and Agriculture Organization of the United Nations. https://doi.org/10.4060/cc3017en

Food and Agriculture Organization of the United Nations (FAO) (2023) AU Digital Agriculture Strategy (DAS) and implementation plan 2024–2030, pp 1–139. https://faolex.fao.org/docs/pdf/au225158.pdf

Fuentes A, Yoon S, Park DS (2026) AI-driven decisions for smarter, sustainable greenhouse vegetable production. Int J Veg Sci 32(1):1–7. https://doi.org/10.1080/19315260.2025.2564423

Galaverni M, Oddi G, Preite L, Belli L, Davoli L, Marchioni I, Rodolfi M, Solari F, Beghè D, Ganino T, Vignali G, Ferrari G (2025) An IoT-based data analysis system: a case study on tomato cultivation under different irrigation regimes. Comput Electron Agric 229:109660. https://doi.org/10.1016/j.compag.2024.109660

Gatou P, Tsiara X, Spitalas A, Sioutas S, Vonitsanos G (2024) Artificial intelligence techniques in grapevine research: a comparative study with an extensive review of datasets, diseases, and techniques evaluation. Sensors 24(19):6211. https://doi.org/10.3390/s24196211

Gebresenbet G, Bosona T, Patterson D, Persson H, Fischer B, Mandaluniz N, Chirici G, Zacepins A, Komasilovs V, Pitulac T, Nasirahmadi A (2023) A concept for application of integrated digital technologies to enhance future smart agricultural systems. Smart Agric Technol 5:100255. https://doi.org/10.1016/j.atech.2023.100255

Giakoumoglou N, Pechlivani EM, Sakelliou A, Klaridopoulos C, Frangakis N, Tzovaras D (2023) Deep learning-based multi-spectral identification of grey mould. Smart Agric Technol 4:100174. https://doi.org/10.1016/j.atech.2023.100174

Giger M, Reys A, Anseeuw W, Mutea E, Kiteme B (2022) Smallholders' livelihoods in the presence of commercial farms in Central Kenya. J Rural Stud 96:343–357. https://doi.org/10.1016/j.jrurstud.2022.11.004

Gouroubera MW, Akinhola AS, Segnon AC, Ismail M-M, Zougmoré RB (2025) Drivers and impacts of mobile phone-mediated scaling of agricultural technologies: a meta-analysis. Front Sustain Food Syst 8:1514546. https://doi.org/10.3389/fsufs.2024.1514546

Groff C (2010) Language, education, and empowerment: voices of Kumauni young women in multilingual India. Doctoral dissertation, University of Pennsylvania. Publicly Accessible Penn Dissertations. https://repository.upenn.edu/entities/publication/48be9ffd-1e66-4052-bc76-fa083c5ec639

Guebsi R, Mami S, Chokmani K (2024) Drones in precision agriculture: a comprehensive review of applications, technologies, and challenges. Drones 8(11):686. https://doi.org/10.3390/drones8110686

Habiyaremye A, Korina L (2021) Indigenous knowledge systems in ecological pest control and post-harvest rice conservation techniques: sustainability lessons from Baduy communities. Sustainability 13(16):9148. https://doi.org/10.3390/su13169148

Hafeez A, Husain MA, Singh SP, Chauhan A, Khan MT, Kumar N, Chauhan A, Soni SK (2023) Implementation of drone technology for farm monitoring and pesticide spraying: a review. Inf Process Agric 10(2):192–203. https://doi.org/10.1016/j.inpa.2022.02.002

Han A, Liu P, Wang B, Zhu A (2023) E-commerce development and its contribution to agricultural non-point source pollution control: evidence from 283 cities in China. J Environ Manag 344:118613. https://doi.org/10.1016/j.jenvman.2023.118613

Henten VEJ, Hemming J, van Tuijl BAJ, Kornet JG, Meuleman J, Bontsema J, van Os EA (2002) An autonomous robot for harvesting cucumbers in greenhouses. Auton Robot 13(3):241–258. https://doi.org/10.1023/A:1020568125418

Hinojosa C, Sanchez K, Camacho A, Arguello H (2023) AgroTIC: bridging the gap between farmers, agronomists, and merchants through smartphones and machine learning. arXiv. https://doi.org/10.48550/arXiv.2305.12418

Hosny KM, El-Hady WM, Samy FM (2025) Technologies, protocols, and applications of internet of things in greenhouse farming: a survey of recent advances. Inf Process Agric 12(1):91–111. https://doi.org/10.1016/j.inpa.2024.04.002

Ingrao C, Strippoli R, Lagioia G, Huisingh D (2023) Water scarcity in agriculture: an overview of causes, impacts and approaches for reducing the risks. Heliyon 9(8):e18507. https://doi.org/10.1016/j.heliyon.2023.e18507

Islam SU, Ferraioli G, Pascazio V (2025) Tomato leaf detection, segmentation, and extraction in real-time environment for accurate disease detection. Agri Eng 7(4):120. https://doi.org/10.3390/agriengineering7040120

Jackson I, Namdar J, Saénz MJ, Elmquist RA III, DávilaNovoa LR (2024) Revolutionize cold chain: an AI/ML driven approach to overcome capacity shortages. Int J Prod Res 63(6):2190–2212. https://doi.org/10.1080/00207543.2024.2398583

Jain S, Kore D, Kishorkumar K, Mohapatra A, Baksh H, Kumar V, Mohanty S, Haokip S (2023) A comprehensive review on protected cultivation of horticultural crops: present status and future prospects. Int J Environ Climate Chang 13:3521–3531. https://doi.org/10.9734/IJECC/2023/v13i113528

Javaid M, Haleem A, Khan IH, Suman R (2023) Understanding the potential applications of artificial intelligence in agriculture sector. Adv Agrochem 2(1):15–30. https://doi.org/10.1016/j.aac.2022.10.001

Javaid M, Haleem A, Singh RP, Sinha AK (2024) Digital economy to improve the culture of industry 4.0: a study on features, implementation and challenges. Green Technol Sustain 2(2):100083. https://doi.org/10.1016/j.grets.2024.100083

Jørgensen R, Sørensen C, Maagaard J, Havn I, Jensen K, Søgaard HT, Sørensen L (2006) HortiBot: a system design of a robotic tool carrier for high-tech plant nursing. CIGR J Sci Res Dev 9:13. https://cigrjournal.org/index.php/Ejounral/article/view/937/931

Jun J, Kim J, Seol J, Kim J, Son H (2021) Towards an efficient tomato harvesting robot: 3D perception, manipulation, and end-effector. IEEE Access 9:17631–17640. https://doi.org/10.1109/ACCESS.2021.3052240

Kabir ANMF, Alam MJ, Begum IA, McKenzie AM (2023) Consumers' interest and willingness to pay for traceable vegetables: an empirical evidence from Bangladesh. Future Foods 7:100214. https://doi.org/10.1016/j.fufo.2022.100214

Kambale P, Chowhan V, Bhattad R, Ragunath S (2021) Mobile applications in agriculture. J Agric Ext Manag 22(2):49

Kanagaraj M, Anandaraja N (2021) Mobile-based agricultural apps and portals for farmers' welfare in India. Karn J Agric Sci 34:562–566

Karada M, Bajpai R, Singh M, Singh A, Agnihotri D, Singh B (2023) A review on advances in agriculture and agroforestry with GPS and GIS. Int J Plant Soil Sci 35:150–160. https://doi.org/10.9734/ijpss/2023/v35i62849

Kaur S, Jha S, Mandal R (2014) Information, efficiency, and sustainability in Indian agricultural markets: E-Choupal, ITC's private initiative. SSRN Electron J. https://doi.org/10.2139/ssrn.2526061

Kaur B, Dimri S, Singh J, Mishra S, Chauhan N, Kukreti T, Sharma B, Singh SP, Arora S, Uniyal D, Agrawal Y (2023) Insights into the harvesting tools and equipment's for horticultural crops: from then to now. J Agric Food Res 14:100814. https://doi.org/10.1016/j.jafr.2023.100814

Khan A, Saeed M, Anwar MN, Kanwal L (2023) Unleashing the Potential: A Study of the Effectiveness and Impact of YouTube Educational Content on Student Learning Outcomes. Journal of Mass Communication & Journalism 13(5):538

Khatri A, Lallawmkimi M, Rana P, Panigrahi C, Minj A, Koushal S, Ali M (2024) Integration of ICT in agricultural extension services: a review. J Exp Agric Int 46:394–410. https://doi.org/10.9734/jeai/2024/v46i123146

Kingsly I, Balasubramanian M, Arivelarasan T, Kumar S (2024) Water withdrawal and sustainability concerns in sugarcane cultivation in Western India: an application of the SURE model. Irrig Drain 74(1):1–15. https://doi.org/10.1002/ird.3014

Kondo N (2009) Robotization in fruit grading system. Sens Instrumen Food Qual 3(2):81–87. https://doi.org/10.1007/s11694-008-9065-x

Kumar P, Kar A, Singh DR, Perumal A, Shivamurthy SGC, Reddy KV, Badal PS, Kamble AL, Kamalvanshi V, Jha GK, Nain MS, Pachiyappan P, Alataway A, Dewidar A, Elansary HO (2021) Protected cultivation of horticultural crops in Uttarakhand: an economic analysis. Agronomy 11(4):692. https://doi.org/10.3390/agronomy11040692

Kumar A, Singh S, Yadav M, Bhuj B, Dhar S, Pruthi N, Kumar R, Bajpai V, Rizwan M, Jyoti K, Thapa R, Kumar V, Kumar H, Mishra B, Kumar V, Rajput A, Singh A, Kumar R (2022) Artificial intelligence, internet of things (IoT) and smart agriculture for sustainable farming: a review. Arch Plant Sci 11(11):6. https://doi.org/10.21746/aps.2022.11.11.6

Kumar V, Sharma K, Kedam N, Patel A, Kate TR, Rathnayake U (2024) A comprehensive review on smart and sustainable agriculture using IoT technologies. Smart Agr Technol 8:100487. https://doi.org/10.1016/j.atech.2024.100487

Kumari K, Mirzakhani Nafchi A, Mirzaee S, Abdalla A (2025) AI-driven future farming: achieving climate-smart and sustainable agriculture. Agric Eng 7(3):89. https://doi.org/10.3390/agriengineering7030089

Lakhiar IA, Yan H, Zhang C, Wang G, He B, Hao B, Han Y, Wang B, Bao R, Syed TN, Chauhdary JN, Rakibuzzaman M (2024) A review of precision irrigation water-saving technology under changing climate for enhancing water use efficiency, crop yield, and environmental footprints. Agriculture 14(7):1141. https://doi.org/10.3390/agriculture14071141

Lehikoinen P, Tiusanen M, Santangeli A, Rajasärkkä A, Jaatinen K, Valkama J, Virkkala R, Lehikoinen A (2021) Increasing protected area coverage mitigates climate-driven community changes. Biol Conserv 253:108892. https://doi.org/10.1016/j.biocon.2020.108892

Li B, Zhuo N, Ji C, Zhu Q (2022a) Influence of smartphone-based digital extension service on farmers' sustainable agricultural technology adoption in China. Int J Environ Res Public Health 19(15):9639. https://doi.org/10.3390/ijerph19159639

Li X, Hu X, Song S, Sun D (2022b) Greenhouse management for better vegetable quality, higher nutrient use efficiency, and healthier soil. Horticulturae 8(12):1192. https://doi.org/10.3390/horticulturae8121192

Li H, Rajbahadur GK, Bezemer C-P (2024) Studying the impact of TensorFlow and PyTorch bindings on machine learning software quality. arXiv. https://doi.org/10.48550/arXiv.2407.05466

Liu L, Cheng W, Kuo H-W (2025) A narrative review on smart sensors and IoT solutions for sustainable agriculture and aquaculture practices. Sustainability 17(12):5256. https://doi.org/10.3390/su17125256

Lu J, Liu S, Wang Q, Liao M (2024) Research on device and sensing technology for precision seeding of potato. Agriculture 14(12):2146. https://doi.org/10.3390/agriculture14122146

Lv G, Song C, Xu P, Qi Z, Song H, Liu Y (2023) Blockchain-based traceability for agricultural products: a systematic literature review. Agriculture 13(9):1757. https://doi.org/10.3390/agriculture13091757

Malhotra SK, Verma R, Bhasin P, Horam Y, Gupta KB (2022) National education policy, 2020, agricultural education. ICAR, Agri Rise 1(1):1–89

Mana AA, Allouhi A, Hamrani A, Rehman S, El Jamaoui I, Jayachandran K (2024) Sustainable AI-based production agriculture: exploring AI applications and implications in agricultural practices. Smart Agric Technol 7:100416. https://doi.org/10.1016/j.atech.2024.100416

Maraveas C, Arvanitis KG, Bartzanas T, Loukatos D (2025) Potential applications of quantum sensors in agriculture: a review. Comput Electron Agric 235:110420. https://doi.org/10.1016/j.compag.2025.110420

Marinus W, Descheemaeker KKE, van de Ven GWJ, Waswa W, Mukalama J, Vanlauwe B, Giller KE (2021) "That is my farm" – an integrated co-learning approach for whole-farm sustainable intensification in smallholder farming. Agric Syst 188:103041. https://doi.org/10.1016/j.agsy.2020.103041

Martins JJ, Silva M, Santos F (2023) Safety standards for collision avoidance systems in agricultural robots – a review. In: Tardioli D, Matellán V, Heredia G, Silva MF, Marques L (eds) ROBOT2022: fifth Iberian robotics conference, vol 589. Springer. https://doi.org/10.1007/978-3-031-21065-5_11

Meesaragandla S, Jagtap MP, Khatri N, Madan H, Vadduri AA (2024) Herbicide spraying and weed identification using drone technology in modern farms: a comprehensive review. Results Eng 21:101870. https://doi.org/10.1016/j.rineng.2024.101870

Menzel CM (2025) A review of strawberry under protected cultivation: yields are higher under tunnels than in the open field. J Hortic Sci Biotechnol 100(3):286–313. https://doi.org/10.1080/14620316.2024.2449026

Meriç MK (2025) Implementation of a wireless sensor network for irrigation management in drip irrigation systems. Sci Rep 15:14157. https://doi.org/10.1038/s41598-025-97303-w

Miller T, Mikiciuk G, Durlik I, Mikiciuk M, Łobodzińska A, Śnieg M (2025) The IoT and AI in agriculture: the time is now—a systematic review of smart sensing technologies. Sensors 25(12):3583. https://doi.org/10.3390/s25123583

Ministry of Skill Development and Entrepreneurship (2024) Skill India Digital and To Improve Employability of Youth, Specific steps to align the training programs to market needs and improve the employability of trainees (Press Release). https://www.pib.gov.in/PressReleasePage.aspx?PRID=2034983

Mizik T (2023) How can proximal sensors help decision-making in grape production? Heliyon 9(5):e16322. https://doi.org/10.1016/j.heliyon.2023.e16322

Mmbando GS (2025) Harnessing artificial intelligence and remote sensing in climate-smart agriculture: the current strategies needed for enhancing global food security. Cogent Food Agric 11(1):2454354. https://doi.org/10.1080/23311932.2025.2454354

Morales A, Villalobos FJ (2023) Using machine learning for crop yield prediction in the past or the future. Front Plant Sci 14:1128388. https://doi.org/10.3389/fpls.2023.1128388

Morepje MT, Sithole M, Msweli N, Agholor I (2024) The influence of e-commerce platforms on sustainable agriculture practices among smallholder farmers in sub-Saharan Africa. Sustainability 16(15):6496. https://doi.org/10.3390/su16156496

Mowla MN, Mowla N, Shahen AFMS, Rabie K, Shongwe T (2023) Internet of things and wireless sensor networks for smart agriculture applications: a survey. IEEE Access 11:145813–145852. https://doi.org/10.1109/ACCESS.2023.3346299

Mukherjee S, Padaria RN, Burman RR, Velayudhan PK, Mahra GS, Aditya K, Sahu S, Saini, Mallick S, Quader SW, Shravani K, Ghosh B, Bhat AG (2025) Global trends in ICT-based extension and advisory services in agriculture: a bibliometric analysis. Front Sustain Food Syst 9(2025):1430336. https://doi.org/10.3389/fsufs.2025.1430336

Mulungu K, Kassie M, Tschopp M (2025) The role of information and communication technologies-based extension in agriculture: application, opportunities and challenges. Inf Technol Dev 31(4):1117–1146. https://doi.org/10.1080/02681102.2025.2456232

Mwinuka PR, Mourice SK, Mbungu WB, Mbilinyi BP, Tumbo SD, Schmitter P (2022) UAV-based multispectral vegetation indices for assessing the interactive effects of water and nitrogen in irrigated horticultural crops production under tropical sub-humid conditions: a case of African eggplant. Agric Water Manag 266:107516. https://doi.org/10.1016/j.agwat.2022.107516

Nadafzadeh M, Banakar A, AbdananMehdizadeh S, ZareBavani M, Minaei S, Hoogenboom G (2024) Design, fabrication and evaluation of a robot for plant nutrient monitoring in greenhouse (case study: iron nutrient in spinach). Comput Electron Agric 217:108579. https://doi.org/10.1016/j.compag.2023.108579

Nam NT, Hung PD (2018) Pest detection on traps using deep convolutional neural networks. Pub:Association for Computing Machinery, New York, NY. ISBN: 9781450364706. https://doi.org/10.1145/3232651.3232661

Natarajan K, Senguttuvan K, Murugan PP (2024) Impact of training programmes of KrishiVigyan Kendra (KVK) towards socio-economic development of weaker section farmers in Cuddalore district. Madras Agric J 111(4–6):17(1–10). https://doi.org/10.29321/MAJ.10.300017

NHB (2023) Annual reports of National Horticulture Board, Gurgaon. https://www.nhb.gov.in/annual_report.aspx

Nordey T, Basset-Mens C, De Bon H, Martin T, Déletré E, Simon S, Parrot L, Despretz H, Huat J, Biard Y, Dubois T, Malézieux E (2017) Protected cultivation of vegetable crops in sub-Saharan Africa: limits and prospects for smallholders. a review. Agron Sustain Dev 37:53. https://doi.org/10.1007/s13593-017-0460-8

Odimarha A, Ayodeji S, Abaku E (2024) Machine learning's influence on supply chain and logistics optimization in the oil and gas sector: a comprehensive analysis. Comp Sci IT Res J 5(3):725–740. https://doi.org/10.51594/csitrj.v5i3.976

Omer AAA, Li M, Zhang F, Kolaly WE, Zhang X, Saad AG, Liu W (2025) Enhancing cultivation environment and increasing productivity in greenhouse rooftop with novel spectrum-splitting technology. Heliyon 11(13):e43740. https://doi.org/10.1016/j.heliyon.2025.e43740

Onishi Y, Yoshida T, Kurita H, Fukao T, Arihara H, Iwai A (2019) An automated fruit harvesting robot by using deep learning. Robomech J 6:13. https://doi.org/10.1186/s40648-019-0141-2

Opara IK, Opara UL, Okolie JA, Fawole OA (2024) Machine learning application in horticulture and prospects for predicting fresh produce losses and waste: a review. Plants 13(9):1200. https://doi.org/10.3390/plants13091200

Paliwal H, Kumari J (2024) Transforming agriculture: a comprehensive review of mobile-based ICTs and innovations in India's agricultural extension services, with a focus on the multimedia agricultural advisory system (MAAS). Int J Agric Extension Social Dev 7:88–92. https://doi.org/10.33545/26180723.2024.v7.i2b.304

Pan Y, Zhang S, Zhang M (2024) The impact of entrepreneurship of farmers on agriculture and rural economic growth: innovation-driven perspective. Innov Green Dev 3(1):100093. https://doi.org/10.1016/j.igd.2023.100093

Papadopoulos G, Arduini S, Uyar H, Psiroukis V, Kasimati A, Fountas S (2024) Economic and environmental benefits of digital agricultural technologies in crop production: a review. Smart Agric Technol 8:100441. https://doi.org/10.1016/j.atech.2024.100441

Partel V, de Morais Nunes L, Stansly P, Ampatzidis Y (2019) Automated vision-based system for monitoring Asian citrus psyllid in orchards utilizing artificial intelligence. Comput Electron Agric 162:328–336. https://doi.org/10.1016/j.compag.2019.04.022

Pathak H, Mishra JP, Mohapatra T (2022) Indian agriculture after independence. Indian Council of Agricultural Research. http://www.icar.org.in/sites/default/files/2023-02

Poenaru MM, Manta LF, Gherțescu C, Manta AG (2025) Shaping the future of horticulture: innovative technologies, artificial intelligence, and robotic automation through a bibliometric lens. Horticulturae 11(5):449. https://doi.org/10.3390/horticulturae11050449

Popescu D, Ichim L, Stoican F (2023) Orchard monitoring based on unmanned aerial vehicles and image processing by artificial neural networks: a systematic review. Front Plant Sci 14:1237695. https://doi.org/10.3389/fpls.2023.1237695

Pugazhendi P, Gnanavel BK, Sundaram A, Sathiyamurthy S (2023) Advancing post-harvest fruit handling through AI-based thermal imaging: applications, challenges, and future trends. Discov Food 3. https://doi.org/10.1007/s44187-023-00068-2

Qiu Y, Zhou Y, Chang Y, Liang X, Zhang H, Lin X, Qing K, Zhou X, Luo Z (2022) The effects of ventilation, humidity, and temperature on bacterial growth and bacterial genera distribution. Int J Environ Res Public Health 19(22):15345. https://doi.org/10.3390/ijerph192215345

Rajak P, Ganguly A, Adhikary S, Bhattacharya S (2023) Internet of things and smart sensors in agriculture: scopes and challenges. J Agric Food Res 14:100776. https://doi.org/10.1016/j.jafr.2023.100776

Ram N, Sagar J, Beese S, Singh M (2024) Role of hydroponics in improving water-use efficiency and food security. Int J Environ Clim Chang 14:608–633. https://doi.org/10.9734/IJECC/2024/v14i23976

Reddy P, Chaudhary K, Hussein S (2023) A digital literacy model to narrow the digital literacy skills gap. Heliyon 9(4):e14878. https://doi.org/10.1016/j.heliyon.2023.e14878

Rejeb A, Abdollahi A, Rejeb K, Treiblmaier H (2022) Drones in agriculture: a review and bibliometric analysis. Comput Electron Agric 198:107017. https://doi.org/10.1016/j.compag.2022.107017

Rikala P, Braun G, Järvinen M, Stahre J, Hämäläinen R (2024) Understanding and measuring skill gaps in industry 4.0 — a review. Technol Forecast Soc Chang 201:123206. https://doi.org/10.1016/j.techfore.2024.123206

Rotz S, Gravely E, Mosby I, Duncan E, Finnis E, Horgan M, LeBlanc J, Martin R, Tait Neufeld H, Nixon A, Pant L, Shalla V, Fraser E (2019) Automated pastures and the digital divide: how agricultural technologies are shaping labour and rural communities. J Rural Stud 68:112–122. https://doi.org/10.1016/j.jrurstud.2019.01.023

Šalagovič J, Vanhees D, Verboven P, Holsteens K, Verlinden B, Huysmans M, Van de Poel B, Nicolaï B (2024) Microclimate monitoring in commercial tomato (*Solanumlycopersicum* L.) greenhouse production and its effect on plant growth, yield and fruit quality. Front Hortic 3:1425285. https://doi.org/10.3389/fhort.2024.1425285

Sharma R (2023) Role of technology in empowering women: a pathway to inclusive development. J Electr Syst 19(4):482–487

Sharma K, Shivandu SK (2024) Integrating artificial intelligence and internet of things (IoT) for enhanced crop monitoring and management in precision agriculture. Sens Int 5:100292. https://doi.org/10.1016/j.sintl.2024.100292

Sharma U, Chetri P, Minocha S, Roy A, Holker T, Patt A, Joerin J (2021) Do phone-based short message services improve the uptake of agri-met advice by farmers? A case study in Haryana, India. Clim Risk Manag 33:100321. https://doi.org/10.1016/j.crm.2021.100321

Shehu HA, Ackley A, Mark M, Eteng OE (2025) Artificial intelligence for early detection and management of *Tutaabsoluta*-induced tomato leaf diseases: a systematic review. Eur J Agron 170:127669. https://doi.org/10.1016/j.eja.2025.127669

Siddiqua A, Kabir MA, Ferdous T, Ali IB, Weston LA (2022) Evaluating plant disease detection mobile applications: quality and limitations. Agronomy 12(8):1869. https://doi.org/10.3390/agronomy12081869

Sindakis S, Showkat G (2024) The digital revolution in India: bridging the gap in rural technology adoption. J Innov Entrep 13:29. https://doi.org/10.1186/s13731-024-00380-w

Sotelo-Cardona P, Lin MY, Srinivasan R (2021) Growing tomato under protected cultivation conditions: overall effects on productivity, nutritional yield, and pest incidences. Crops 1(2):97–110. https://doi.org/10.3390/crops1020010

Souravi K, Rajsekharan PE (2019) Traditional knowledge in horticulture. In: Rajsekharan PE, Rao VR (eds) Conservation and utilization of horticultural genetic resources. Springer, pp 621–636. https://doi.org/10.1007/978-981-13-3669-0_21

Soussi A, Zero E, Sacile R, Trinchero D, Fossa M (2024) Smart sensors and smart data for precision agriculture: A review. Sensors 24(8): 2647. https://doi.org/10.3390/s24082647

Sumalan RL, Stroia N, Moga D, Muresan V, Lodin A, Vintila T, Popescu CA (2020) A cost-effective embedded platform for greenhouse environment control and remote monitoring. Agronomy 10(7):936. https://doi.org/10.3390/agronomy10070936

Talaviya T, Shah D, Patel N, Yagnik H, Shah M (2020) Implementation of artificial intelligence in agriculture for optimisation of irrigation and application of pesticides and herbicides. Artif Intell Agric 4:58–73. https://doi.org/10.1016/j.aiia.2020.04.002

Tan Y, Liu X, Zhang J, Wang Y, Hu Y (2025) A review of research on fruit and vegetable picking robots based on deep learning. Sensors 25(12):3677. https://doi.org/10.3390/s25123677

Upadhyay A, Zhang Y, Koparan C, Rai N, Howatt K, Bajwa S, Sun X (2024) Advances in ground robotic technologies for site-specific weed management in precision agriculture: a review. Comput Electron Agric 225:109363. https://doi.org/10.1016/j.compag.2024.109363

Vaz R, de Carvalho JV, Teixeira SF (2023) Developing a digital business incubator model to foster entrepreneurship, business growth, and academia–industry connections. Sustainability 15(9):7209. https://doi.org/10.3390/su15097209

Verma H, Garg N (2021) Mobile apps: potential ICT tools to disseminate technologies on value addition of mango, guava and aonla fruits. J AgriSearch 8(1):67–71. https://doi.org/10.21921/jas.v8i01.19568

Villacis AH, Kopp T, Mishra AK (2024) Agricultural marketing channels and market prices: evidence from high-value crop producers in India. Econ Anal Policy 81:1308–1321. https://doi.org/10.1016/j.eap.2024.02.004

Wang X, Zhang S, Wang X, Xu C (2023) Crop pest detection by three-scale convolutional neural network with attention. PLoS One 18(6):e0276456. https://doi.org/10.1371/journal.pone.0276456

Wu X, Aravecchia S, Lottes P, Stachniss C, Pradalier C (2020) Robotic weed control using automated weed and crop classification. J Field Robot 37:322–340. https://doi.org/10.1002/rob.21938

Yaguchi H, Nagahama K, Hasegawa T, Inaba M (2016) Development of an autonomous tomato harvesting robot with rotational plucking gripper. In: 2016 IEEE/RSJ international conference on intelligent robots and systems (IROS), pp 652–657. https://doi.org/10.1109/IROS.2016.7759122

Yi T, Mhamed M (2024) Developments in automated harvesting equipment for the apple in the orchard: review. Smart Agr Technol 9:100491. https://doi.org/10.1016/j.atech.2024.100491

Younus M, Zaenuri M, Suswanta. (2024) Public-private collaboration to overcome the digital divide in digital transformation of government. Digital Zone J Teknol Inf Komun 15(1). https://doi.org/10.31849/digitalzone.v15i1.17027

Yu Y, An X, Lin J, Li S, Chen Y (2024) A vision system based on CNN-LSTM for robotic citrus sorting. Inf Process Agric 11(1):14–25. https://doi.org/10.1016/j.inpa.2022.06.002

Yulita IN, Rambe MFR, Sholahuddin A, Prabuwono AS (2023) A Convolutional Neural Network Algorithm for Pest Detection Using GoogleNet. AgriEngineering. 5(4):2366–2380. https://doi.org/10.3390/agriengineering504014

Waqas M, Naseem A, Humphries UW, Hlaing PT, Shoaib M, Hashim S (2025) A comprehensive review of the impacts of climate change on agriculture in Thailand. Farming System 3(1):100114. https://doi.org/10.1016/j.farsys.2024.100114

Zelenić A, Işıtan A, Ionescu C, Sevillano C, Gok C, Vasileiou K, Sulak M, Vurchio P, Kutlubay R, Alessandroni V, Hedžet Z (2024) Digital traineeship in agriculture. Digital Farmer, Erasmus Cooperation, Istanbul, pp 1–239

Zhai Z, Martínez JF, Beltran V, Martínez NL (2020) Decision support systems for agriculture 4.0: survey and challenges. Comput Electron Agric 170:105256. https://doi.org/10.1016/j.compag.2020.105256

Zhang C, Li X (2022) Land use and land cover mapping in the era of big data. Land 11(10):1692. https://doi.org/10.3390/land11101692

Zhang N, Liu L, Li H, Wei W, Liang G, Tang Y, Zhao Y, Wei O, Yang Q (2024) Effects of protected cultivation on agronomic, yield, and quality traits of yard-long bean (*Vigna unguiculata* ssp. *unguiculata* cv.-gr. *sesquipedalis*). Horticulturae 10(11):1167. https://doi.org/10.3390/horticulturae10111167

22 Cropping System Approach to Horticultural Crop Production for Agricultural Sustainability: A Climate-Smart Production Strategy

Sagar Maitra, Sumit Ray, Masina Sairam, Lalichetti Sagar, Shaik Rishitha, Dinkar J. Gaikwad, D. T. Santosh, Nihal Ravindranath, Chinmaya Jena, Satya Prasad Nanda, Sarthak Pattanayak, Anjana J. Atapattu, Tharindu D. Nuwarapaksha, and Akbar Hossain

Abstract

The world's agricultural sector is currently at a crossroads of climatic anomalies, food insecurity, decreasing arable land, and a growing population. Hence, they force a paradigm shift towards sustainable and resilient agricultural practices. Traditional methods of agricultural production flourished in intensive agriculture, especially after the fossil fuel-driven Green Revolution, which has raised questions about agricultural sustainability over time. In this context, the proper adoption of sustainable cropping systems emerged as an ecologically rational, strategically managed, practically relevant and economically viable option. Such systems are aligned with ecosystem services, the UN's Sustainable Development

S. Maitra · S. Ray · M. Sairam · L. Sagar · S. Rishitha · N. Ravindranath · C. Jena · S. P. Nanda
M.S. Swaminathan School of Agriculture, Centurion University of Technology and Management, Paralakhemundi, Odisha, India

D. J. Gaikwad · D. T. Santosh
Research Centre of Excellence for Smart Agriculture, Centurion University of Technology and Management, Paralakhemundi, Odisha, India

S. Pattanayak
Odisha University of Agriculture and Technology, Bhubaneswar, Odisha, India

A. J. Atapattu (✉) · T. D. Nuwarapaksha
Agronomy Division, Coconut Research Institute of Sri Lanka, Lunuwila, Sri Lanka

A. Hossain
Soil Science Division, Bangladesh Wheat and Maize Research Institute, Dinajpur, Bangladesh

A. Das, A. B. Sharangi (eds.), *Remunerative Horticulture*,
https://doi.org/10.1007/978-981-92-1007-7_22

Goals (SDGs) and climate-smart agriculture. In a horticultural system, crops are of high economic value and are perishable and sensitive to climate change; however, integrating a polyculture can increase resource utilisation while providing better monetary benefits. The diverse canopy structure and architecture of horticultural crops provide a conducive environment for intercropping by regulating the microclimate, which mitigates climatic extremes, namely, wind speed, temperature, moisture depletion and erosion. The integration of intercrops, such as legumes, into a monocultured horticultural system not only fixes atmospheric nitrogen biologically, which minimises the dependency on synthetic fertiliser and enhances soil fertility, but also disrupts the pest and disease cycle while decreasing the need for chemical inputs. The different root depths of horticultural crops provide access to soil moisture and nutrients and maintain their compatibility with intraspecific competition. Intercropping of horticultural crops helps mitigate the risk of economic losses due to climate change. The cropping system substantially contributed to ecosystem stability, year-round carbon sequestration, and soil health; minimised nutrient losses; and provided vegetative soil cover. Multistrata intercropping can be considered in the directions of crop combination, mechanisation and yield assessment tools, which can be bridged through targeted research and development, capacity building, extension services, and policy support. Hence, climate-smart strategies can focus on agricultural complexity under a changing climate scenario to develop a sustainable climate-smart horticultural cropping system.

Keywords

Climate change · Crop productivity · Intercropping systems · Multistrata cropping · Sequential cropping · Sustainable Development Goals

Abbreviations

4R	Right time, Right rate, Right source and Right place
AI	Artificial Intelligence
ATER	Area-time equivalence ratio
B:C	Benefit-to-cost ratio
CAS	CRISPR-associated proteins
CO_2	Carbon dioxide
CRISPR	Clustered Regularly Interspaced Short Palindromic Repeats
CSA	Climate-smart agriculture
CSHCS	Climate-smart horticultural cropping system
DHCS	Diversity horticultural cropping system
GHG	Greenhouse gas
HCS	Horticultural cropping system

IS	Intercropping system
LER	Land equivalent ratio
MEY	Maize equivalent yield
PCT	Protected cultivation technology
pH	Power of hydrogen
SCS	Sequential cropping system
SDG	Sustainable Development Goals
SOC	Soil organic carbon
SOM	Soil organic matter
SPI	System productivity index
USD	United States dollar
WUE	Water use efficiency

22.1 Introduction

Operating within the planet's ecological boundaries while ensuring food and nutritional security for the ever-increasing population and challenged by a changing climate scenario has brought the global agricultural landscape to a critical juncture (Santosh et al. 2023a; Sairam et al. 2023a, b; Gouda et al. 2024; Maity et al. 2025). Food security was achieved using fossil fuel-based inputs and high-yielding crop cultivars, such as those of the Green Revolution, which catered to the caloric requirements of the human body without considering nutritional requirements or agricultural sustainability. Although horticultural crops, namely, fruits, nuts, vegetables, spices and condiments, and ornamental plants contribute only to caloric needs, they are nutritional powerhouses that contribute more than 50% of essential vitamins, minerals and antioxidants, combating hidden hunger and chronic diseases, especially in developing countries (Ahmed et al. 2023, 2024). However, the resilience of horticultural systems is under siege because of increasing temperatures, erratic rains, extreme weather conditions, degrading arable land and depleting groundwater levels. Additionally, conventional horticultural systems, often characterised by monoculture and intensive input regimes, directly and indirectly contribute to soil degradation, biodiversity loss and imbalances in ecosystem services and increase vulnerability to climatic perturbations (Kaur et al. 2024). Moreover, the role of horticulture extends beyond nutrition to economic viability and environmental stewardship. The market value for horticultural products is 1.5 trillion USD, driven by investments in precision technology exceeding 30 billion USD, which can revolutionise production efficiency (Crawford et al. 2024). Berries, leafy greens, and tropical fruits not only diversify diets but also generate rural livelihoods (Knez et al. 2024). Moreover, speciality crops exhibit plasticity because of their indeterminate growth pattern and shallow root systems, which increase their resilience to environmental cues, ensuring high-quality harvests. However, the perishable nature of these crops leads to postharvest losses averaging approximately 30–40%, increasing the stakes for sustainable intensification (Anand and Barua 2022; National Economic Forum 2025).

The cropping system addresses climatic vulnerabilities while evolving from rudimentary farmer practices to science-backed frameworks, which integrate agronomic, physiological and edaphic principles (Ghosh et al. 2021; Maitra et al. 2025a). Cropping systems delineate the temporal and spatial arrangement of crops for maximising yields and synergies (Maitra and Gitari 2020). In horticulture, monocultures are characterised by soil nutrient depletion, pest and disease buildup and plateauing yields, which can be strategically managed by diverse assemblages that mimic natural polyculture ecosystems through cropping systems (Belete and Yadete 2023; Kavita et al. 2025). For example, sequential cropping not only exploits temporal niches by relay planting but also increases the land equivalent ratio to >1 (Ruillé et al. 2026). Such approaches increase annual productivity, capitalise on residual moisture and organic matter, stimulate root exudates, enhance nutrient cycling and provide economic stability to small farmers (Hariharan et al. 2022). Other cropping systems, such as intercropping systems or mixed stands, further amplify the multifaceted ingenuity of horticultural cropping. The juxtaposing or complementary effect of legumes in a horticultural system fixes atmospheric nitrogen and enhances light interception, water uptake, and below-ground resource partitioning, increasing yield increases from 20% to 50% (Maitra et al. 2024a; Luo et al. 2024). Alley cropping, particularly with speciality crops such as medicinal herbs (lemongrass, *tulsi*, vetiver) in hedges, not only provides protection and produce but also increases soil organic carbon because of the decaying residues (Grebner et al. 2013; Maitra et al. 2025b). Moreover, multi-storied systems in tropical agroforestry, with fruits at the top canopy layer, midstratum nuts and ground-level herbs, mimic a forest ecosystem and capture vertical gradients in irradiance and humidity (Ali et al. 2024). Additionally, degraded soils, affected by compaction, salinisation and biodiversity losses due to monoculture and monocentric tillage, can be countered by a diverse cropping system through rhizosphere engineering. Improved soil structure not only increases water infiltration and retention but also buffers against drought stress, enriching the soil food web by increasing nutrient use efficiency and suppressing soil-borne pathogens (Pan et al. 2018). Hence, the use of cropping system management allows the regulation of living ecosystems both above and below the soil. The multifaceted approach of the cropping system aligns with ecosystem services and the UN's Sustainable Development Goal (SDG), such as 'SDG 1 (No poverty), SDG 2 (Zero hunger), SDG 3 (Good health and well-being), SDG 12 (Responsible consumption and production), SDG 13 (Climate action) and SDG 15 (Life on land)' (Ray et al. 2025a). Furthermore, a cropping system approach for horticultural crop production can be considered a climate-resilient practice for achieving the SDGs (Fig. 22.1).

In the current era of climate uncertainty, a cropping system that enhances resilience by increasing biological and economic risk, modifying and recreating a soothing microclimate, reducing soil temperature and evaporative losses, and tapping deep soil moisture acts as a windbreak and shelter belt while enhancing productivity. Soil health is the cornerstone of climate-smart horticulture (Zhang et al. 2023). However, the future of horticultural cropping systems is promising and demanding; translational science, which moves beyond generic recommendations to

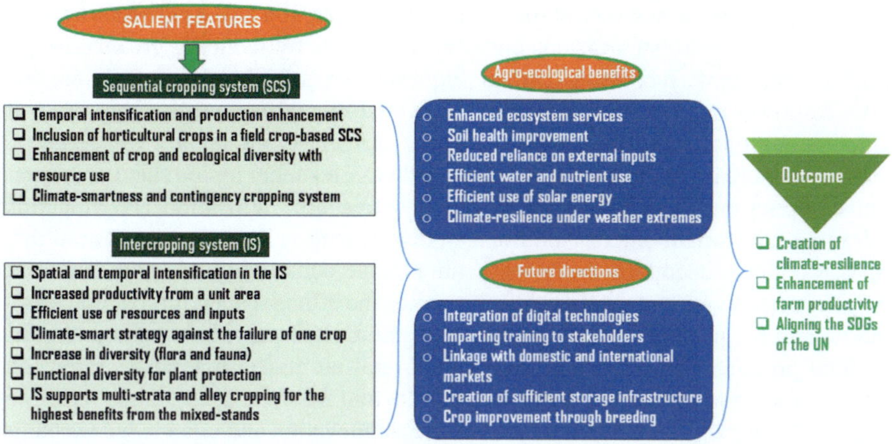

Fig. 22.1 Horticulture-based cropping systems aligned with climate resilience and Sustainable Development Goals

context-specific, predictive designs, locations and market-specific recommendations, is needed (Sashika et al. 2024). The leveraging of technologies such as remote sensing, real-time input management, molecular tools for below-ground interactions, and crop–crop and crop–environmental interaction modelling can increase system efficiency. Moreover, economic models for valuing ecosystem services, policy frameworks, incentivised frameworks, producer and consumer awareness, proper postharvest management and sustainable production practices are crucial drivers of system sustainability. This chapter focuses on the pros and cons of the system approach in horticultural cropping systems and future directions for the transformation of horticultural production as a remunerative venture.

22.2 Multifaceted Benefits of Cropping Systems

Horticultural intensification through a cropping system can be a transformative approach towards climate-smart horticulture. Such approaches can address climate variability, resource scarcity and food security. The strategic modification of the spatiotemporal arrangement of multiple crops in cropping systems harnesses resources, enhances ecological resilience, and distributes risk. The Diversified Horticultural Cropping System (DHCS) spans socioeconomic and social dimensions (Ghosh et al. 2023). Varying root patterns in DHCS enhance soil structure, increase soil organic matter (SOM) and facilitate nutrient cycling (Fan et al. 2025). Diversity in the horticultural cropping system (HCS) creates habitat heterogeneity, which increases beneficial insect populations and disrupts the life cycle of pathogens (Meena et al. 2023). Additionally, compared with monocropping, the DHCS has better adaptability to various climatic constraints because of its portfolio effect, where the impaired performance of one crop component is compensated by the other, improving the stability of the system. Moreover, soil cover enhances biomass production and mitigates various climatic constraints by reducing greenhouse gas (GHG) emissions, increasing carbon sequestration and decreasing soil erosion. DHCS spreads market risks, scattering the availability of harvested output and considering seasonal variability, which supports farmers economically (Meena et al. 2023). An increase in the land equivalence ratio (LER) under mixed stand or sequential cropping enhances efficiency in terms of resource utilisation and economic advantages for smallholders. Additionally, diversification in produce opens multiple market options, decreases dependency on a single commodity price and enhances bargaining power. Crop diversity also promotes the efficient allocation of household labour and financial resources. Furthermore, the DHCS provides year-round horticultural products and consistent access to several nutrients, including vitamins, to address food and nutritional security (Mihrete and Mihretu 2025). The DHCS not only fosters experimentation but also enhances innovation and knowledge exchange among farming communities. Hence, the DHCS not only caters to nutritional and food security but also strengthens the social status of farmers. Therefore, HCSs have multiple benefits, such as an efficient configuration and design with a proper understanding of context-approach farming, which should be environmentally sound, economically viable and socially acceptable.

22.3 Sequential Cropping System with Horticultural Crops

A sequential cropping system (SCS) is defined as a system in which two or more crops are grown on the same piece of land in a year or season in sequence. The SCS is specifically known as a temporal intensification strategy, where the growing degree days are captured efficiently throughout the growing seasons of crop life cycles and integrated with crop phenology under favourable environmental conditions (Hariharan et al. 2022). Compared with monocropping, the SCS helps increase land utilisation, solar radiation interception, and the efficient use of nutrients and water across the season, all of which improve yearly production. The short duration of vegetables such as okra, radish, spinach and *Amaranthus* creates an opportunity for multiple cropping in an HCS on the basis of market demand, labour availability, and seasonal windows (Shekhar 2024). The strategic selection of crops in such systems depends on crop duration, input requirements, pest and disease dynamics, soil fertility, labour demand and market needs. However, as a climate-smart strategy, the SCS has several advantages. Several horticultural crops, namely, garlic, onion, cucurbitaceous crops, cole crops and legumes, are commonly considered for the SCS (Das et al. 2021). The diversified sowing and harvesting times spread the climatic risk across the season, reducing the vulnerability to localised weather extremes (Ranjan 2025). The changing climatic pattern may affect the first crop, but the subsequent cop may perform well and cover the losses under favourable conditions. Short-duration leafy and nonleafy vegetables may function as contingency crops under weather extremes and crop failure (Saurabh et al. 2023). Cropping can be adjusted according to the needs of the farmer and the seasonal weather forecast. However, the SCS can be adopted on the basis of agroecological concerns, adaptability, local needs and market demand to ensure the viability of the cropping system. Earlier research conducted on the same topic has shown some pathbreaking findings for the adoption of a sustainable SCS (Table 22.1).

Diversity and continuous cropping in the SCS contribute to nutrient cycling and disease suppression, increase soil microbial populations and improve soil structure (Pramanick et al. 2023). However, intensive cropping poses a risk of soil degradation if adequate soil organic carbon (SOC) does not remain in the soil. Multiple cropping methods extract primary and micronutrients from the soil, which are essential for horticultural crops (Ahmed et al. 2023). The strategic addition of sequential crops, such as nitrogen-fixing legume vegetables and cover crops and adjustments in the timing and formulation of fertilisers with organic amendments can be adapted for efficient nutrient management in the SCS. Additionally, in the SCS, crop rotation disrupts various pest and disease cycles. Horticultural pests and diseases are host-specific and can be managed through temporal diversification.

The sequenced crop can also utilise the residual moisture available from the previous crop and season (Kumar et al. 2020a). The temporal matching of cropping sequences with water demand is an important climate adaptation strategy in water-scarce environments. For instance, a 10-year field study on sequential vegetation cropping on sandy soil under drip irrigation revealed an approximately 25–35% reduction in total water application compared with that under monocropping, with yields of tomato increasing by 30–40 t/ha, followed by those of cucumber (20–25 t/ha) (Stanley et al. 1991). In a field study of rice followed by onion, which utilised residual soil moisture and reduced the irrigation requirement by 25–35% compared with that under sole onion cultivation (Kumar et al. 2020b).

Furthermore, the continuous presence of roots in the soil also promotes soil aggregation, decreases soil erosion and increases the microbial population (Witzgall et al. 2024). Swapping a fallow period in a cropping system with continuous soil cover increases C sequestration and reduces GHG emissions (Ghosh et al. 2021; Mwadalu et al. 2022; Yang et al. 2024). Although GHG emissions largely depend on tillage operations, fertiliser application and cropping intensification, continuous soil cover collectively decreases GHG emissions while enhancing SOC through residual decomposition (Mehra et al. 2018). However, the implementation of sequential cropping requires sophisticated planning and management capabilities. Inputs, land preparation, sowing, harvesting and processing within a tight temporal window can be properly processed to enhance the efficiency of the SCS. Additionally, farmers should have crop-specific knowledge, market dynamics, details of seasonal climatic conditions and proper access to inputs to optimise cropping sequences and management practices.

Table 22.1 Benefits of sequential cropping systems with horticultural crops under various options

Cropping system	Salient findings	References
Rice–potato–onion	Rice–potato–onion recorded highest rice equivalent yield (26.94 t/ha) compared to other rice-based cropping systems	Sharma et al. (2004)
Green gram–radish–potato	Green gram–radish–potato recorded the highest potato equivalent yield (50.63 t/ha) and production efficiency (200.89 kg/ha/day), compared to other SCSs studied	Kushwah et al. (2011)
Rice–potato–onion	Rice–potato–onion mint recorded the highest potato equivalent yield (53.40 t/ha) and land use efficiency (95%), compared to other cropping sequences. Furthermore, the system noted a 138% yield increment over rice–wheat SCS	Singh et al. (2017)
Jute–rice–vegetables	Jute–rice–potato and jute–rice–okra recorded higher crop yields with 43.1 t/ha and 30.2 t/ha, which were superior to jute–rice–wheat cropping system (12.7 t/ha) Higher phosphate-solubilising bacteria and phosphate-solubilising capacity were noted with the abovementioned cropping systems	Manna et al. (2017)
Maize + green gram – green gram + maize – tomato	Maize + green gram (1:2) – green gram + maize (1:1) – tomato recorded the highest system productivity (17.55 t/ha) and BC ratio (4.40) than other intensive cropping systems	Mokidul et al. (2018)
Rice–vegetables	Rice–pumpkin cropping system recorded rice equivalent yield (21.89 t/ha), with a B:C ratio 3.84 and rice–tomato yielded 34.42 t/ha with a B:C ratio of 2.86. Both cropping systems registered higher rice equivalent yield and B:C ratio than rice–fallow, rice–rapeseed and rice–lentil cropping systems	Kalita et al. (2018)
Sponge gourd–rice–broccoli	Sponge gourd–rice–broccoli cropping recorded highest rice equivalent yield of 14.44 t/ha with a B:C ratio of 3.26, compared to other SCSs with vegetable as a component	Ekka et al. (2019)
Rice–rice–cucurbitaceous vegetables	Rice–rice–watermelon cropping system recorded highest yield (26.10 t/ha) and B:C ratio (2.66) than other cucurbitaceous vegetables grown as the third crop in the intensive cropping system	Punitha et al. (2023)
Cabbage–maize–cabbage	Cabbage–maize–cabbage showed highest soil water storage during different growth stages and economic benefits	Wang et al. (2024)
Rice–garlic	Rice–garlic SCS recorded highest economic yield of 48.46 t/ha against Rice–wheat cropping system, which yielded 4.99 t/ha Rice–garlic cropping recorded highest rice equivalent yield of 33.70 t/ha, which was 185% higher than rice–wheat cropping system (11.82 t/ha) Rice–garlic SCS recorded the highest benefit-to-cost (B:C) ratio of 3.17	Kamini et al. (2025)

22.4 Intercropping System with Annual Horticultural Crops

In intercropping systems, two or more crops are grown on the same piece of land simultaneously (Maitra et al. 2021, 2024b; Maheswari et al. 2025). The mixed stand harnesses resources from temporal and spatial arrangements. The integration of various species of horticultural crops simultaneously in the same space creates complex interactions above and below the soil. These interactions facilitate enhanced microbial interactions, nutrient cycling, water use efficiency, efficient harnessing of sunlight, and soil health (Li et al. 2023). The coexistence of diversified species also enhances productivity, stability and ecological sustainability (Table 22.2). In plant–plant interactions, a conducive microenvironment is created because of varying canopy cover, root systems and biochemical environmental interactions. A conducive environment supports better growth of flora and fauna above- and below-ground.

The microbial networks are strengthened, enhancing nutrient cycling in an intercropping system. Moreover, in mixed stands, the arrangement of species with a specific spacing or area affects intercropping or strip cropping, whereas the duration of sowing and overlapping of growing periods affects relay cropping (Ray et al. 2025b). The mixed stand of crops extends from agronomical factors related to ecological, economic and social dimensions. A mixed stand enhances the LER by more than 1.3, indicating 30% or more productivity per unit land compared with monocultures (Glaze-Corcoran et al. 2020). Such an increase in LER is essential in the current era, when the world is facing the issue of arable land shrinkage due to industrialisation and urbanisation (Gaikwad et al. 2022; Santosh et al. 2023b; Maitra et al. 2023; Ray et al. 2024). Temporal insurance is created for the species because of the portfolio effect, which ensures compensation for decreased productivity. Additionally, the continuous soil cover during the crop periods reduces soil moisture evaporation losses, which decreases the water need for the crop. These spatial modifications are important for the cultivation of horticultural crops in arid and semiarid regions. The integration of heat-tolerant species, such as drumstick, guava, and amla, with heat-sensitive species, namely, strawberry, spinach and lettuce, in a mixed stand mitigates heat stress, ensuring productivity across variable conditions. In addition to moisture conservation, the continuous ground cover and biomass production in a mixed stand increase carbon sequestration, decrease GHG emissions and soil erosion and reduce the dependency on fertilisers, especially in the case of mixed stands with legumes.

22.4.1 Intercropping Field Crops and Vegetables

The mixed stand combining field crops with vegetables aligns with the polyculture principle of DHCS. In field crops and vegetable-based intercropping systems, field crops such as cereals, pulses, and oilseeds coexist with horticultural vegetables, which facilitates a multifunctional production system (Kumari et al. 2022). The mix stands between field crops and vegetables, ensuring food and nutritional security and diversity while fulfilling the economic objectives of the farmers. Agronomic

Table 22.2 Benefits of intercropping systems with horticultural crops

Cropping system	Salient findings	References
Cotton + beet and cotton + chilli	Cotton + beet and cotton + chilli registered a higher seed cotton equivalent yield of 1766 and 1746 kg/ha, respectively, over seed cotton yield of uniform-row–planted cotton (1618 kg/ha)	Maitra et al. (2001)
Maize + pea	Highest green cob equivalent yield (17.6 t/ha) was recorded for sweet corn + pea (2:1) over sole corn productivity (13.8 t/ha). The highest monetary benefit as B:C ratio (2.7) was recorded in sweet corn + pea intercropping system (2:1)	Sahoo (2009)
Maize + bush bean	Highest productivity of maize (10.57 t/ha) and bush bean (6.27 t/ha) were registered with maize + bush bean (1:1) intercropping system; however, sole maize yielded (10.9 t/ha) of maize grain, indicating a higher gross productivity from a unit area. The same treatment also recorded the Maize Equivalent Yield (MEY) (14.04 t/ha) and B:C ratio (3.54) over the pure stand of maize and other maize + vegetables intercropping systems	Ali et al. (2015)
Maize + eggplant/ hot pepper	Maize + eggplant recorded the highest LER (2.08) in northern region of Ghana, which was superior to their pure stands. However, for maize + pepper, the row ratio of 1:2 established its superiority over other spatial arrangements of both crops	Tignegre et al. (2018)
Maize + vegetable cowpea	Maize + vegetable cowpea at 2:2 ratio recorded the highest MEY of 9.67 t/ha, compared to other maize-based intercropping systems Furthermore, the highest LER (1.53), ATER (1.23), System Productivity Index (SPI) (10.25) and BC ratio (3.34) were recorded for maize + vegetable cowpea (2:2)	Parimaladevi et al. (2019)
Maize + cabbage	Maize + cabbage (1:1) intercropping resulted in the highest maize equivalent yield (25.5 t/ha) and BC ratio (4.54), proving the advantage of intercropping system over sole maize	Khanum et al. (2020)
Amaranthus + jute	*Amaranthus* + jute intercropping recorded highest LER (1.89) and B:C ratio of 1.51 over sole *Amaranthus* and sole jute cropping systems	Ayodeji et al. (2021)
Maize + jute (as leafy vegetable)	Maize + jute (as leafy vegetable) when broadcast, recorded the highest MEY of 19.28 t/ha and 114.7% increase in MEY over sole maize in Gazipur of Bangladesh	Begum et al. (2022)
Maize + cluster bean	Maize + vegetable cluster bean (2:3) recorded maize yield of 6285 kg/ha and cluster bean yield of 5035 kg/ha; however, sole maize yielded 6756 kg grains/ha Maize + cluster bean (2:3) noted higher values of LER (1.89) and ATER (1.40), which indicated the spatial and temporal advantage of the maize + cluster bean intercropping system.	Nandi et al. (2022)
Maize + broccoli	The disease severity and incidence of *Exserohilum turcicum* were recorded to be the lowest in maize + broccoli intercropping compared to sole maize. Brassicas in the mixed stand released glucosinolates and other sulphur-containing volatiles that were responsible for lowering the pathogen load.	Vaishnavi et al. (2025)

crops sown with wider spacing can accommodate vegetables that are more closely spaced. Such arrangements allow complete intermingling of plants while creating an intimate association with different species of plants. When plants with different canopy cover types are grown on the same land, they can harvest sunlight from different heights and extract nutrients and water from different soil depths because of their varying root architectures (Dechant et al. 2020). The canopy cover of field crops such as maize, jute and sorghum captures light from above, and short vegetables, especially shade-loving leafy crops such as spinach and mint, grow luxuriantly. Furthermore, the temporal dimension is advantageous when short-duration vegetables mature earlier than field crops do, as they reach their canopy cover and reduce competitive interference during the critical growth phase (Sethi et al. 2025). The combined growth of crops on the same land also increased the LER above unity. A field study of maize intercropped with okra revealed that LER ratios were greater than 1 in intercropping treatments, and spatial crop arrangements also minimised competition between the two crops while maintaining comparable outputs (maize yield of 3.58 t/ha, okra productivity of 11.98 t/ha) and crop diversity (Ajayi et al. 2024). An experiment on maize and tomato intercropping recorded LER values and increased the yield of both crops (El-Mehy and Mohamed 2018).

22.4.2 Intercropping with Vegetables

The polycultures of vegetable crops require an understanding of complementary growth habits, resource requirements and temporal development patterns (Adamczewska-Sowińska and Sowiński 2019). Vegetables in vegetable-based intercropping systems, crops with varying growth habits, durations and species, are considered for enhancing the efficiency of such systems. The biochemical compatibilities of vegetables are crucial in these mixed stands, as vegetables such as fennel, sunflower and *brassicas* (broccoli, cabbage, cauliflower, etc.) exhibit allelopathic effects on companion crops (Ghosh 2021). In a mixed stand of leafy vegetables, such as lettuce or spinach, combined with root vegetables such as carrot, radish or beet, the space below and above ground is used most efficiently. Additionally, leafy vegetables are shallow-rooted and are harvested early, whereas root vegetables generally have access to the deeper layer of soil, which helps them to bulk and are harvested late (Williams 2025). Hence, the varying harvesting times and root and shoot architecture in the mixed stand enhance input use efficiency while providing continuous economic support to the farmer. For example, an experiment on cucumber–lettuce intercropping revealed LER >1 in all intercropping treatments, but the highest LER, 1.55–1.71, was recorded in the 'Vanda' lettuce cultivar intercropping treatment (Ribas et al. 2020). In a field trial of intercropping cauliflower with knolkhol, spinach beet, Swiss chard, lettuce, coriander, fenugreek and radish, the highest LER (1.35) and a benefit–cost ratio (5.19) were recorded in the cauliflower + coriander intercropping system (Shah et al. 2025). Summer savoury (*Satureja hortensis* L.) is a potential herb suitable for mixed stands with leaves, beans and carrots (Jankowska and Wojciechowicz-Żytko 2016; Namazi et al. 2022; Rezaei-Chiyaneh et al. 2023). Moreover, several medicinal and aromatic crops, such as *Coleus forskohlii*, *Portulaca oleracea* and *Andrographis paniculata*, are also intercropped with a variety of vegetables (Aćimović et al. 2025).

22.4.3 Intercropping in Fruit Orchards

Fruit orchards consider multiple dimensions, such as tree architecture, phenological patterns, environmental requirements and market demand (Lauri et al. 2008). Combining tree species with varying growth habits creates vertical and horizontal structural diversity, which optimises space and light interception. Mixing deciduous species, such as apple, pear and crab apple, with evergreen species, such as rosemary and lavender, attracts pollinators, increases their population, and supports fruit setting in fruit species (Zhou and Guo 2025). In a fruit orchard, tree species of fruit crops trap resources from the subsoil and surface soil, whereas shallow-rooted crops remove nutrients from the surface layer. The perennial nature of fruit trees supports drought tolerance, and fruit crops such as ber, aonla and custard apple are known for their resistance to drought. Moreover, the allelopathic effects of trees such as black walnut and citrus should be considered. The integration of legume crops at the beginning of orchard establishment supports the need for nutrients. Citrus requires abundant nitrogen, whereas figs need less nitrogen; hence, the integration of such trees along with legumes can increase the productivity of mixed-fruit orchards. The LER of mixed-fruit orchards is also greater than that of monocultures. Several species of fruit crops in the orchard also present more market opportunities for farmers (Deb et al. 2022). A field study on mango + pineapple and dragon fruit + ber intercropping revealed a maximum LER of 1.68, a water productivity of 16.54 kg/m^3 with a partial factor productivity of 32.87, and a crop productivity of 27.55 Mg/ha, which were 3.3 times greater than those of a monoculture of fruits (Kishore et al. 2025). The perennial nature of these trees makes them amenable to large amounts of stored carbon, hence promoting carbon sequestration. The residues of the crops further add SOM while enriching the soil microorganism population (Maitra et al. 2026). In another experiment on strawberry-based intercropping, the productivity increased by 18.1% in French bean and by 26.9% in zucchini, with a combined LER of 1.2 for a strawberry + zucchini intercropping system (Badawy and Abdalla 2022).

Deep-rooted fruit crops, such as mango and guava, access moisture from the subsoil, whereas shallow-rooted crops, such as onion, garlic and lettuce, utilise surface moisture in the mixed stand. Temporal changes can also be matched with seasonal needs, as water-demanding crops can be cultivated during the rainy season, whereas dry-season crops, such as garlic, onion and winter vegetables (peas, cauliflower and cabbage), can be grown in sequence during the winter and summer. The temporal matching of cropping systems with water demand is an important climate adaptation strategy in water-scarce environments.

22.4.4 Alley Cropping and Speciality Crops

Alley cropping is a mixed stand of agroforestry in which multiple crop species are accommodated in the alley (Maitra et al. 2025a). High-value horticultural crops, also known as speciality crops such as herbs, medicinal plants, spices, nurseries and ornamentals, are integrated with alley cropping to make it an economically profitable and ecologically intensive production system (Yadav 2025). Multiple spatial arrangements of trees, as alternatives to cropped alleys in parallel strips, are common practices. The diversity of tree species during plantation should consider factors such as shade tolerance, growth habits, management practices and market demand. Shade-tolerant speciality herbs such as mint, basil or coriander; medicinal plants (ashwagandha or aloe vera); and spices such as black pepper or ginger are grown under the partial shade of alley crops. Temporal diversity is managed by pruning and adjusting the growing periods. The alley crops are grown first, followed by shade-loving plants. The canopy cover by the tree crops not only provides shade but also acts as a natural windbreak for the other crops. Windbreaks reduce evapotranspiration by 20–30% while improving water use efficiency as a form of drought resilience (Wolz et al. 2017). Moreover, during the growing period, the crop residue, leaf litter and pruned biomass from the alley crops increase the SOM. The addition of organic carbon to the soil helps increase carbon sequestration. In alley cropping, if leguminous species such as *Gliricidia sepium*, *Leucaena leucocephala* and *Sesbania grandiflora* are considered, the entire cropping system and all crops grown receive benefits through nitrogen fixed by legumes and shared among adjacent crops (Handayanto 2025). Additionally, the canopy covers of the alley crops not only provide shade but also decrease evaporation losses of soil moisture from the ground, reduce erosion by slowing the runoff velocity, and ensure that extensive root systems firmly hold the soil (Kokila et al. 2024). The diversity maintained in the cropping system provides diverse habitats for the proliferation of natural enemies and pollinators (Gawdiya et al. 2025). The quality and quantity of the output of speciality crops are enhanced, opening new markets for farmers. Moreover, the limited tillage operations in the perennial cropping system reduce GHG emissions while increasing the soil microbial population.

22.4.5 Multi-Storied Cropping

In multi-storied cropping, crops of different heights are grown temporally on the same land, which mimics natural forest architectures. Multi-storied systems integrate tall trees, medium-height shrubs and low-statured herbaceous crops in the same area, creating a three-dimensional polyculture system. The three-dimensional coverage harnesses maximum solar radiation and creates shade for the shade-loving plants growing below (Jiang et al. 2025). Minimal competition while maximising resource use, especially vertically, is the fundamental principle of multi-storied cropping. The tallest plant needs to be light-demanding and capable of developing a relatively large canopy, such as coconut or arecanut, whereas the middle-height crop should be partially shade-loving, such as papaya or guava; the low-height plants should be completely shade-loving crops, such as pineapple, leafy vegetables or ginger, and the ground layer can have creepers such as black pepper or vanilla. The varying levels of tree height act as shelter belts and windbreaks, protecting other plants from heatwaves and high-velocity winds. The altered microclimate is generally cooler than the open field by 2–5 °C (Rajgopal 2023). The microclimate created under this system is conducive to growing high-value or off-season crops (Sharma et al. 2020). Furthermore, the perennial nature of the crops under this system results in high levels of carbon storage while reducing GHG emissions. Moreover, the residue increases soil organic matter, and the extensive root architecture reduces soil erosion. The simultaneous presence of drought-tolerant and moisture-demanding species on the same land provides insurance against erratic rainfall patterns. Among the various plantation crops, coconut has versatile importance and offers ample scope for multistrata cropping in intensive and remunerative horticulture.

22.4.5.1 Coconut Monocropping: Limitations and Land Use Efficiency

Coconut monocropping has clear drawbacks in how it uses land and canopy space, which limits farmers' productivity and income. Traditional coconut farms use only approximately 22% of the land and 30% of the canopy for growing crops (Nuwarapaksha et al. 2022). As shown in Fig. 22.2, monocrop coconut cultivation has a long initial loss period, with cumulative returns turning positive approximately 14–15 years later, indicating a delayed payback period. Because coconut palms have shallow roots, much of the soil is left unused. This, along with the long wait before coconuts start producing, makes it difficult for farmers to earn a steady income (Thomas et al. 2018).

These challenges explain why better, more sustainable farming methods are needed to help farmers improve productivity. Switching from monocultures, which

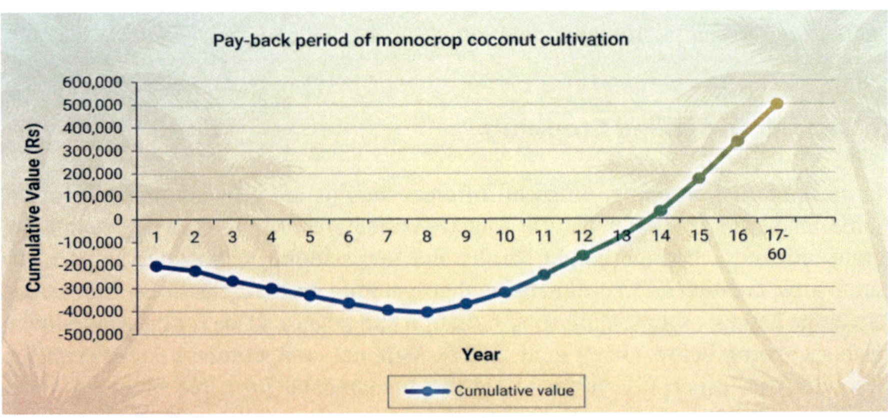

Fig. 22.2 Payback period of coconut monocultures

can result in the loss of biodiversity and soil, to multi-storied systems can improve both the environment and the economic stability of farmers (Dissanayaka et al. 2023). Diverse farming systems fit with climate-smart agriculture and nature-based solutions, helping address social and environmental challenges while supporting people and biodiversity. These systems make coconut farming more dynamic, help manage natural resources, and improve ecosystem services.

22.4.5.2 Conceptual Framework of Multi-Storied Cropping Systems

Multi-storied cropping systems are defined as the intentional cultivation of multiple crop species in vertical layers to increase both productivity and resilience (Atapattu et al. 2025a). These systems, frequently described as nature-based solutions, involve the cultivation of diverse farming crops within various vertical layers, promoting both economic benefits and improved crop productivity. This multi-storied approach increases resource utilisation by differentiating niches for various species, consequently enhancing the overall efficiency and toughness of the farming system (Kumar and Kunhamu 2022). Coconut palms form the upper canopy, whereas intercrops such as fruits, tubers, spices, beverages, legumes and fodder species occupy the middle and lower layers (Fig. 22.3).

Such traditional land use systems represent accumulated indigenous knowledge, often developed over thousands of years without external inputs, offering time-tested examples of sustainable land management (Dissanayaka et al. 2025). Within these intricate agroecosystems, coconut palms often serve as a principal component, around which various other economically valuable species are intimately associated, forming a managed community of plants.

Fig. 22.3 Different cropping systems cultivated under coconut: (**a**) banana (fruit), (**b**) cassava (tuber), (**c**) pepper (spice), (**d**) cocoa (beverage), (**e**) *gliricidia* (legume crop) and (**f**) CO-3 (fodder crop)

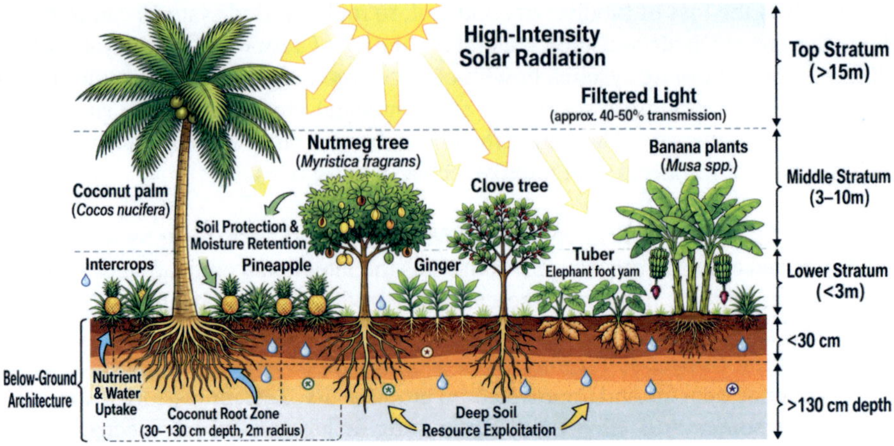

Fig. 22.4 Vertical stratification and resource use in coconut-based multi-storey cropping systems

22.4.5.3 Potential of Coconut-Based Multi-Storied Systems

The purposeful layering of vegetation takes advantage of vertical space utilisation, where different plant species inhabit distinct strata on the basis of their light, water and nutrient requirements, thereby fostering resource complementarity and limiting interspecific competition (Fig. 22.4). This finely detailed arrangement permits enhanced biodiversity and environmental equilibrium, mimicking natural forest ecosystems while simultaneously providing a continuous income stream through diversified outputs (Chauhan et al. 2025). The deliberate selection and spatial arrangement of compatible species inside these systems are intended to optimise positive interactions, such as nutrient cycling and pest regulation, while lessening negative interactions, such as competition for resources (Atapattu et al. 2025b).

22.4.5.4 Biological and Agronomic Potential

Coconut-based multistate systems demonstrate strong biological and agronomic advantages because they integrate diverse species for maximum productivity and ecosystem function (Atapattu et al. 2024). This diversity enables vertical stratification, with the use of the coconut canopy to support multiple plant communities. Vertical arrangements allow intercrops to capture sunlight at different heights and access resources from various soil depths, increasing productivity over monocultures (Nuwarapaksha et al. 2022). Niche complementarity suggests that diversity enhances resource use, productivity and stability. Multiple species can optimise the use of sunlight, carbon dioxide, water and nutrients. Diversification also increases functional biodiversity, supporting beneficial insects and microorganisms that are essential for ecosystem services. Interactions among plant components promote nutrient cycling and biological nitrogen fixation, reducing reliance on synthetic fertilisers and minimising environmental impacts (Atapattu et al. 2026).

22.4.5.5 Social and Economic Potential

The economic viability of coconut-based multistate systems lies in their capacity to increase income, reduce risks and improve farm livelihoods (Nuwarapaksha et al. 2023a). Diverse systems provide economic benefits through improved farm performance and food security, leading to better livelihoods for farming communities. These systems generate income even during the typically unproductive establishment phase of coconut palms, improving benefit–cost ratios and land productivity (Nuwarapaksha et al. 2022). Diversified income streams reduce financial risks from the price fluctuations of a single crop, offering greater economic security. Reduced reliance on external inputs further lowers production costs, increasing profitability and making these systems attractive alternatives to monocultures. Integrating livestock, such as sheep, under rubber trees has proven profitable, suggesting similar benefits for coconut-based systems (Nuwarapaksha et al. 2023b). Integrated models with various intercrops and allied enterprises have shown productivity increases of up to tenfold compared with those with solely coconut cropping. This approach not only increases output but also buffers against financial volatility and crop failure, supporting more stable agricultural economies.

22.4.5.6 Environmental and Ecological Potential

Coconut-based multistate systems strongly support biodiversity, ecosystem health, and climate mitigation through their multifaceted environmental contributions (Dissanayaka et al. 2022). These multifunctional systems support biodiversity conservation by providing diverse habitats and resources for different plant and animal species (Udumann et al. 2025). The existence of several species and complex interactions fosters ongoing ecological enrichment. This diversity supports more beneficial insects, birds and microorganisms, resulting in resilient and stable agroecosystems. These systems also increase carbon sequestration through multiple perennial species and greater biomass, contributing to climate change mitigation (Atapattu et al. 2025a, b, c). Measuring these contributions is important for understanding agroforestry's environmental value in terms of global mitigation efforts. Beyond carbon, diversified systems enhance disaster resilience and climate adaptation by supporting robust ecosystems that better withstand environmental shocks. This resilience comes from functional redundancy and diversity, allowing continued productivity even when some components are affected.

22.5 Legume Effect in Mixed Stands and Sequential Cropping

The integration of legumes in the various HCSs provides transformative yet sustainable multifunctional benefits. The benefits include soil health regeneration; nitrogen enrichment; increased SOC, pest, disease and weed management; and the ability to meet the dietary needs of the household. The symbiotic relationship of legumes, resulting in atmospheric nitrogen fixation, helps in restoring the nitrogen content in the soil, which helps companion or succeeding crops in their growth and development while decreasing the need for chemical fertilisers (Jena et al. 2022; Maitra et al. 2024c; Nori et al. 2025; Ray et al. 2025c). Additionally, the residues or litter from the plant decompose in the soil when organic carbon is added. Such advantages increase soil fertility over time. The root exudates and nodule decomposition with time not only add nitrogen but also increase microbial colonisation in the soil, which helps in nutrient cycling. In mixed stands, fine roots, which have a limited lifespan, provide localised nitrogen-rich zones to the roots of companion crops when carbon is added to the soil (Panchal et al. 2022; Mirriam et al. 2022; Hemasree et al. 2025a, b). Hence, the nodules provide a continuous supply of nitrogen in the cropping system during and after the growth of legumes. The root architecture of legumes creates an extensive network that improves infiltration, drainage and aeration. Root exudates, mucilage, root-associated microbes and organic matter added because of senescence increase and stabilise soil aggregates through physical enmeshment and biochemical binding, which continue even after legumes are harvested (Maitra et al. 2021). The nitrogen-sparing effect, coupled with soil health improvement, has a complementary effect on other crops in the system.

Intercropping of broccoli and faba bean resulted in LER >1, with total nitrogen concentrations ranging from 0.11% to 0.13%, phosphorus concentrations ranging from 11 to 39 mg/kg and exchangeable potassium concentrations ranging from approximately 277 to 364 mg/kg (Marcos-Pérez et al. 2023). In another field trial of broccoli intercropped with faba bean or cowpea, higher biological nitrogen fixation under 1:1 intercropping and curtailment in the use of synthetic fertiliser of approximately 30% were recorded (Gómez 2024). Hence, the economic advantages of legume integration in the cropping system are not limited to LER and productivity but also promote the efficient use of inputs. The selection of legumes in an HCS needs to consider several factors, such as soil pH, companion effects, allelopathic effects, phosphorus and potassium needs of horticultural and legume crops, growth habits and market demand.

22.6 Benefits of a Diversified Horticultural System for Plant Protection

Diversification of crop species, either in sequence or mixed form, supports crop intensification in multiple ways, even for crop protection. Moreover, the multidimensional movement of diverse root systems helps to disrupt the pest, disease and weed cycle. The root exudates of diverse crops further strengthen the soil microbiome. Additionally, the soothing environment helps increase the population of natural enemies because of habitat complexity, diverse food resources and diverse microhabitats. The associational resistance of multiple species suppresses pest populations, disease cycles and weed infestations. An earlier study revealed that mixed stands of mango + guava + lemon reduced fruit fly and aphid infestations, whereas citrus + marigold + legume suppressed the nematode load (Chadfield et al. 2022). For instance, a 4-year study on onion-potato SCS revealed an improved microbial population with an increase in organic matter (4.3–5.1%) and a reduction in stem

Table 22.3 Pest population dynamics in intercropping systems with horticultural crops

Intercropping systems	Pest population dynamics	References
Eggplant + radish (4:1)	The mixed stand reduced the shoot damage of eggplant by fruit and shoot borer (*Leucinodes orbonalis* Guenee) for approximately 50%	Navasero and Calumpang (2013)
Mustard + onion	Aphid (*Lipaphis erysimi* Kalt.)–infested branch/plant was less (1.47) in the mixed stand, compared to sole mustard	Afrin et al. (2017)
Eggplant + cluster bean (4:1)	The maximum fruit damage was noted in sole-cropped plot 35.81%, while fruit damage was 28.03% in cluster bean intercropped plot	Chinnadurai et al. (2019)
Cabbage + carrot	Reduced population of diamondback moth	Singh (2021)
Broccoli + *faba* bean	Reduced foraging of flea beetle	Singh (2021)
Broccoli + vetch/bean between rows	Decrease in the foraging period. Decreasing the population of *Phyllotreta cruciferae*	Mir et al. (2022)
Groundnut + common beans	Common beans' sticky tendrils kept aphids away A reduction in aphids by decreasing foraging time of *Aphis craccivora*	Mir et al. (2022)

canker severity (0.4% vs 0.91%) and black scurf incidence (7% vs. 54%) in potato and pink root severity in onions (8.2% vs. 15.4%) (Wright et al. 2014). In a 3-year field trial on a broccoli–lettuce cropping system, up to 50% reductions in lettuce leaf-drop incidence and infection by Sclerotia, ranging from 40% to 60%, were noted compared with lettuce monocropping, with enhanced activity of beneficial microbes in the soil (Hao et al. 2003).

Compared with monocultures, mixed stands of vegetable crops have better ecological interactions, enhancing resource use efficiency and suppressing pest, disease and weed populations while enhancing productivity (Kelty 1992; Avasiloaiei et al. 2022). In a mixed stand, synergistic effects of crops to combat pest attacks are observed. In terms of intercropping among horticultural crops and in combination with field crops, the pest population decreased (Table 22.3). The diverse aromas of crops such as cabbage or kale mixed with onions or leeks confuse pests. Furthermore, the inclusion of legumes in the cropping system provides diverse habitats for natural enemies and pollinators. Secondary metabolites with repellent or antifeedant properties, such as volatile organic compounds, phenolic acids and allelochemicals, protect companion crops from pests. Legumes such as mung bean and soybean also act as trap crops for pests, such as *Helicoverpa armigera* (cotton bollworm) and *Spodoptera litura* (armyworm) (Kumawat et al. 2022). Additionally, crop rotation or intercropping with legumes breaks the disease cycle of non-leguminous crops. The fast-growing habit of legumes covers the ground, acts as a live mulch and decreases the weed population while enhancing WUE, which is important in arid and semiarid areas.

22.7 Climate Smartness of Horticultural Cropping Systems

The climate-smart approaches of HCSs move beyond traditional horticultural and agronomic practices. The climate-smart Horticultural Cropping System (CSHCS) aligns with the three pillars of climate-smart agriculture (CSA), i.e., productivity, adaptation and mitigation, and moves beyond it towards various SDGs and ecosystem services (Bhatnagar et al. 2024). The CSHCS interrelates and integrates various mechanisms, namely, functional biodiversity, biogeochemical cycling, microclimate engineering and hydrological regulation (Ali 2024). This approach also focuses on a paradigm shift from carbon sources and climate change to carbon sinks and climate-resilient, sustainable strategies while considering socioeconomic viability as the centre (Raza et al. 2025). The inclusion of various species and genetic diversity in the system provides resistance to climate change through varying stress tolerance mechanisms, structural diversity and microclimate modifications. Moreover, the water management approach is most important because of erratic rainfall patterns, increasing drought and depleting groundwater levels. Climate-smart water strategies focus on enhancing WUE and conservation (Kumar et al. 2023; Santosh et al. 2023a). The approach includes atmospheric moisture collaring, rainwater harvesting, the use of micro-irrigation and decreasing water losses. SOC enhancement through continuous residue retention and decomposition increases the water-holding capacity. Continuous soil cover in various HCSs, where plants act as live mulch, enhances soil cover and decreases evaporation losses. In cropping systems where the canopy cover of horticultural crops provides shade and protection from heatwaves, it facilitates water conservation, decreasing evapotranspiration losses (Scavo et al. 2022). The use of micro-irrigation methods such as drip and micro-sprinkler systems enhances water use efficiency up to 90%, which is important for crops with high water requirements, such as banana or citrus (Santosh et al. 2021, 2023c). The integration of soil moisture monitoring equipment and systems, such as tensiometers, capacitance sensors or thermal imaging, helps in better irrigation scheduling. Appropriate soil moisture management in cropping systems not only increases productivity but also the shelf life and quality of horticultural products (Al-Shammary et al. 2024).

A well-planned HCS should also focus on carbon-smart approaches. The perennial nature of horticultural crops makes them favourable for increasing carbon sequestration (Ray et al. 2025a). Additionally, the perennial nature of some horticultural crops limits tillage in an HCS, which decreases GHG emissions. The residues, litterfall, and pruned parts of horticultural crops add carbon to the soil when decomposition is allowed. Proper HCSs create a long-term carbon sink, resulting in the accumulation of 20–30 tonnes of CO_2 per hectare (Ilakiya et al. 2024). An increase in SOC enhances water retention, the microbial population and nutrient cycling. Nutrient management plays a critical role in the CSHCS. Diverse plant stands have varying nutrient requirements, which need to be catered to for better and higher-quality produce. The excessive use of synthetic fertilisers leads to GHG emissions, which can be managed through precise nutrient management (Sagar et al. 2023, 2025). Precision nutrient management can be adopted through soil testing, plant tissue analysis, and site-specific and sensor-based real-time management, which can overcome the over- or underapplication of fertilisers in HCSs (Sairam et al. 2023a, b, 2025a; Nori et al. 2025). The 4R framework, which includes the right time, right rate, right source and right place, should be considered (Johnston and Bruulsema 2014). Additionally, integrated nutrient management is a superior approach for maintaining sustainability without hindering productivity. The integration of legumes and nitrogen-fixing crops such as soybean, cowpea, green gram or groundnut in HCSs curtails the need for synthetic fertilisers (Akchaya et al. 2025). The diverse root architecture in the HCS taps nutrients from different soil strata. Continuous soil cover also decreases nutrient losses through volatilisation, leaching, and runoff. Such approaches progressively improve soil fertility, maintaining the proportion of soil carbon and nitrogen.

The global temperature increase is estimated to be 1.5 °C by 2050, and the consequences warrant the need for smart approaches in remunerative horticultural cropping systems. One such smart technology is protected cultivation technology (PCT) or controlled environment agriculture (Cermak 2024). Protected cultivation technology (PCT) with greenhouses, shade houses, polytunnels and rain shelters enables the production of in- or off-season crops on the basis of market needs, despite the climatic conditions in the open field (Sairam et al. 2025b). The climate inside PCT structures can be manipulated on the basis of crop needs, facilitating a congenial environment with better product quality and quantity. These modifications also create a microclimate that facilitates the growth of speciality crops or other enterprises. In Odisha, farmers cultivate paddy straw mushrooms under bamboo and tree shade, producing 8–12 kg of mushrooms per cycle with an annual income of Rs. 50,000–1.5 lakhs (Mohapatra et al. 2018). A farmer of Cuttack, Odisha, cultivated mushrooms within a low-cost bamboo structure unit and earned almost Rs. 2 lakhs monthly from selling mushrooms and mushroom-based value-added products (Misra 2026). The temporal diversity in crops in the HSC, such as agroforestry, multi-storied cropping and alley cropping, acts as a physical barrier to heatwaves and decreases evapotranspiration. In areas with waterlogging, strategic selection of crops such as *Eucalyptus*, *Casuarina* or *Syzygium cumini* can mitigate this situation through bio-drainage (Manik et al. 2019). For instance, a field experiment of bio-drainage using high-transpiring plants such as different species of *Lucerne*, Sesbania or eucalyptus recorded LERs of 1.2–1.5, fixed nitrogen concentrations of approximately 50–200 kg/ha and reduced excess moisture concentrations by 500–1000 mm/year over 2–5 years (Maitra and Zaman 2017). Furthermore, digital agriculture tools, such as real-time decision-making, predictive analysis, automation and data integration, can help mitigate climate change. The major losses noted in the HCS are due to the perishability of the produce, which can be managed by blockchain-based traceability and a circular economy (Bisht and Singh 2024). The perishable products can be further processed to increase their shelf life. Additionally, nutrient recycling can be further refined from the waste produced after food processing. These residues can be converted into animal feed, compost and bioactive compounds. This circular economy can further reshape the value chain, decrease losses and provide economic stability to farmers.

22.8 Future Prospective of Climate-Smart Horticultural Cropping Systems

The future of HCSs is directly dependent upon the changing climate scenario, which needs technological breakthroughs. Genomic technologies can lead to a paradigm shift for the development of new and climate-resilient crop varieties. The use of technologies such as CRISPR/CAS or speed breeding can increase the tolerance of new varieties to heat, drought and biotic stresses, whereas conventional breeding should also be continued simultaneously. However, ensuring equitable access to these technologies for small farmers needs to be considered. Moreover, AI and digital agriculture can transform the predictive analytics and decision-making capacity of farmers. The integration of weather data, soil sensors and crop monitoring can further aid in resource management. In areas with labour scarcity, autonomous robots can be used for intercultural and harvest management. The controlled environment and vertical farming will further expand with increasing urbanisation and meeting the modified market demands; specific strategies are required in HCSs. HCSs consume more energy, which can be mitigated by renewable energy sources. The use of underutilised indigenous crops also needs to be integrated with HCSs to address climatic risk. Technological advancements need to focus on regenerative horticulture, where ecosystem restoration should also be considered. HCSs can be further designed for improving carbon sequestration, enhancing SOC and restoring soil health while being aligned with the SDGs and ecosystem services. Climate-smart approaches often overlook existing and indigenous knowledge systems, capacity building and policy support. The major gap in innovation and scientific understanding is the transfer of knowledge and skills, capacity building and policy support to farmers. Furthermore, the creation of efficient and sufficient storage infrastructure may ensure the profitability of perishable crops. These gaps should be bridged by proper policy support, awareness among farmers and appropriate extension services. Hence, the future of HCSs will be determined by the integration of technologies and collective commitment towards achieving sustainable, productive and socioeconomic scenarios, which can nourish the global population with improved livelihoods and the planet.

22.9 Conclusion

The cropping system approach has not been comprehensively studied earlier with respect to horticultural crops; therefore, an initiative has been taken to gain deeper insights into the benefits of cropping systems for remunerative horticulture. The cropping system with horticultural crops is a blend of crops cultivated under varied temporal and spatial distributions, covering multiple strata, where crops should interact with one another, targeting sustainable and remunerative horticultural production. This chapter highlights all possible directions of system approaches for sustainable horticultural crop production aligned with Sustainable Development Goals. Future research should focus on the latest technology-enabled precision horticulture, which has already flourished for field crops and can further show the way forward to agricultural sustainability. However, the integration of horticultural crops, included in cropping systems, is potentially sound because of the agroecosystem, suitability, farming facilities and all components. In other words, horticultural crop production is an integral part of agriculture, where the time warrants the production of more output from a unit area to achieve food and nutritional security. Moreover, the growth of horticulture-based cropping systems depends on targeted research and development activities, capacity building, proper extension services and policy support wherever needed.

Author Contributions Conceptualisation and original draft: Sagar Maitra (SM), Sumit Ray (SR), Anjana J. Atapattu (AJA), Akbar Hossain (AH); Original draft: SM, SR, Masina Sairam (MS), Lalichetti Sagar (LS), AJA, Tharindu D. Nuwarapaksha (TDN), Shaik Rishitha (SRI); Review and Editing: SM, Santosh D.T. (SDT), Dinkar J. Gaikwad (DJG), Satya Prasad Nanda (SPN), Sarthak Pattanayak (SP), Nihal Ravindranath (NR), Chinmaya Jena (CJ), AJA, AH; Infographics/visualisation: SM, SR, SRI, ADA, DTS, SPN, TDN; Resources and supervision: SM, SPN, CJ, NR, DJG.

Declaration Authors declare that they have read the final manuscript and there is no conflict of interest.

References

Aćimović M, Navarro Rocha J, Ibraliu A, Červenski J, Sikora V, Winter S, Lončar B, Pezo L, Salamon I (2025) Intercropping medicinal and aromatic plants with other crops: insights from a review of sustainable farming practices. Agronomy 15(12):2692. https://doi.org/10.3390/agronomy15122692

Adamczewska-Sowińska K, Sowiński J (2019) Polyculture management: a crucial system for sustainable agriculture development. In: Meena R (ed) Soil health restoration and management. Springer, Singapore, pp 279–319. https://doi.org/10.1007/978-981-13-8570-4_8

Afrin S, Latif A, Banu NM, Kabir MM, Haque SS, Ahmed ME, Tonu NN, Ali MP (2017) Intercropping empowers, reduces insect pests and increases biodiversity in agro-ecosystem. Agric Sci 8(10):1120. https://doi.org/10.4236/as.2017.810082

Ahmed N, Zhang B, Chachar Z, Li J, Xiao G, Wang Q, Hayat F, Deng L, Narejo M, Bozdar B, Tu P (2023) Micronutrients and their effects on horticultural crop quality, productivity and sustainability. Sci Hortic 323:112–512. https://doi.org/10.1016/j.scienta.2023.112512

Ahmed M, Babayola M, Bake ID (2024) Role of horticultural crops in food and nutritional security: a review. J Food Process 8:1–6. https://doi.org/10.31579/2637-8914/226

Ajayi AJ, Fasinmirin JT, Faloye OT (2024) Growth and yield of maize/okra intercrop under different plant spacings and cropping systems. Ecol Front 45(2):328–336. https://doi.org/10.1016/j.ecofro.2024.10.005

Akchaya K, Parasuraman P, Pandian K, Vijayakumar S, Thirukumaran K, Mustaffa MRAF, Rajpoot SK, Choudhary AK (2025) Boosting resource use efficiency, soil fertility, food security, ecosystem services, and climate resilience with legume intercropping: a review. Front Sustain Food Syst 9:1527256. https://doi.org/10.3389/fsufs.2025.1527256

Ali U (2024) Climate-smart horticultural practices: building resilience in a changing environment: a scoping review. Agric For Fish 13(5):202–209. https://doi.org/10.11648/j.aff.20241305.19

Ali MR, Rahman MS, Asaduzzaman M, Hossain MM, Mannan MA (2015) Intercropping maize with different vegetables. Bangladesh Agron J 18(1):49–52. https://doi.org/10.3329/baj.v18i1.25566

Ali M, Bari, Rahman M, Sarmin I (2024) Multi storied woodlot-based agroforestry system for improved resource utilisation and incomes for farmer. Heliyon 10(16):e36096. https://doi.org/10.1016/j.heliyon.2024.e36096

Al-Shammary AAG, Al-Shihmani LSS, Fernández-Gálvez J, Caballero-Calvo A (2024) Optimising sustainable agriculture: a comprehensive review of agronomic practices and their impacts on soil attributes. J Environ Manag 364:121487. https://doi.org/10.1016/j.jenvman.2024.121487

Anand S, Barua M (2022) Modelling the key factors leading to postharvest loss and waste of fruits and vegetables in the agri-fresh produce supply chain. Comput Electron Agric 198:106936. https://doi.org/10.1016/j.compag.2022.106936

Atapattu AJ, Nuwarapaksha TD, Udumann SS, Dissanayaka NS (2024) Integrated farming systems. In: Babu S, Singh R, Rathore SS, Das A, Singh VK (eds) Agricultural diversification for sustainable food production, Sustainability sciences in Asia and Africa. Springer, Singapore, pp 89–127. https://doi.org/10.1007/978-981-97-7517-0_4

Atapattu AJ, Dissanayaka NS, Udumann SS, Nuwarapaksha TD (2025a) Carbon farming: making agroforestry fit for the future. In: Raj A, Jhariya MK, Meena RS (eds) Agroforestry for monetising carbon credits. Springer Nature, Cham, pp 147–179. https://doi.org/10.1007/978-3-031-97151-8_7

Atapattu AJ, Nuwarapaksha TD, Dissanayaka DMNS, Udumann SS (2025b) Unravelling the role of plant–microbe interactions in sustainable agriculture. In: Babalola OO, Amund OO, Akanmu AO (eds) Ecofriendly frontiers. Springer Nature, Cham, pp 3–38. https://doi.org/10.1007/978-3-031-98700-7_1

Atapattu AJ, Nuwarapaksha TD, Udumann SS, Dissanayaka NS (2025c) Agroforestry system: a game changer for sustainable land use practices. In: Raj A, Jhariya MK, Meena RS (eds) Agroforestry for monetising carbon credits. Springer Nature, Cham, pp 197–232. https://doi.org/10.1007/978-3-031-97151-8_9

Atapattu AJ, Nuwarapaksha TD, Dissanayaka NS, Udumann SS (2026) Contributing to global food security through efficient nitrogen management. In: Tonni AK, Abdelkader A (eds) Point source nitrogen pollution. Elsevier, pp 421–451. https://doi.org/10.1016/B978-0-443-27352-0.00027-6

Avasiloaiei DI, Calara M, Benchea MC, Cristea TO (2022) Review on the positive influence of intercropping systems for organic vegetable growing. Sci Pap Ser B Hortic 66(1):397–408

Ayodeji OO, Atungwu JJ, Fadeyi JO, Ifezue DU, Okoye HO, Akinwale AM (2021) Profitability of dry season farming of Jute Mallow (*Corchorus olitorius*) and Amaranth (*Amaranthus cruentus*) inter and sole cropping systems in Abeokuta, Nigeria. J Agric Ext 25(3):49–59. https://doi.org/10.4314/jae.v25i3.5

Badawy I, Abdalla R (2022) Yield, quality, land equivalent ratio, and economic viability of intercropping combinations of summer squash, faba bean and strawberry. Assiut J Agric Sci 53(1):74–89. https://doi.org/10.21608/ajas.2022.120373.1093

Begum A, Karim Kakon S, Mian M, Rahman J, Chowdhury J (2022) Intercropping of jute as leafy vegetable with hybrid maize under different planting systems. Bangladesh J Agric Res 47(3):269–279. https://doi.org/10.3329/bjar.v47i3.68288

Belete T, Yadete E (2023) Effect of mono cropping on soil health and fertility management for sustainable agriculture practices: a review. Plant Sci 11:192–197. https://doi.org/10.11648/j.jps.20231106.13

Bhatnagar S, Chaudhary R, Sharma S, Janjhua Y, Thakur P, Sharma P, Keprate A (2024) Exploring the dynamics of climate-smart agricultural practices for sustainable resilience in a changing climate. Environ Sustain Indic 24:100535. https://doi.org/10.1016/j.indic.2024.100535

Bisht A, Singh SP (2024) Postharvest losses and management of horticultural produce: a review. J Sci Res Rep 30(3):305–320. https://doi.org/10.9734/jsrr/2024/v30i31881

Cermak A (2024) The effects of climate change. NASA Science. Available at: https://science.nasa.gov/climate-change/effects/

Chadfield VGA, Hartley SE, Redeker KR (2022) Associational resistance through intercropping reduces yield losses to soil-borne pests and diseases. New Phytol 235(6):2393–2405. https://doi.org/10.1111/nph.18302

Chauhan S, Kengoo N, Kishore K, Haksinhbhai MR, Rana P (2025) Carbon dynamics in agroforestry systems: implications for climate change mitigation and adaptation. Int J Environ Climate Chang 15(8):109–133. https://doi.org/10.9734/ijecc/2025/v15i84960

Chinnadurai S, Chandrasekaran M, Yasodha P, Kumanan K (2019) Influence of clusterbean as an intercrop in the management of eggplant shoot and fruit borer, *Leucinodes orbonalis* Guenee (Crambidae: *Lepidoptera*). Pharma Innov J 8(6):400–403

Crawford D, Wang J, Singh R (2024) AI's trillion-dollar opportunity. Bain. Available at: https://www.bain.com/insights/ais-trillion-dollar-opportunity-tech-report-2024/

Das S, Mitra B, Saha A, Mandal S, Paul PK, El-Sharnouby M, Hassan MM, Maitra S, Hossain A (2021) Evaluation of quality parameters of seven processing type potato (*Solanum tuberosum* L.) cultivars in the eastern sub-Himalayan plains. Foods 10(5):1138. https://doi.org/10.3390/foods10051138

Deb D, Dutta S, Erickson R (2022) The robustness of land equivalent ratio as a measure of yield advantage of multicrop systems over monocultures. Exp Results 3:e2

Dechant B, Ryu Y, Badgley G, Zeng Y, Berry JA, Zhang Y, Goulas Y, Li Z, Zhang Q, Kang M, Li J, Moya I (2020) Canopy structure explains the relationship between photosynthesis and sun induced chlorophyll fluorescence in crops. Remote Sens Environ 241:111733. https://doi.org/10.1016/j.rse.2020.111733

Dissanayaka D, Nuwarapaksha T, Udumann S, Dissanayake D, Atapattu AJ (2022) A sustainable way of increasing productivity of coconut cultivation using cover crops: a review. Circ Agric Syst 2(1):1–9. https://doi.org/10.48130/CAS-2022-0007

Dissanayaka DMNS, Dissanayake DKRPL, Udumann SS, Nuwarapaksha TD, Atapattu AJ (2023) Agroforestry a key tool in the climate-smart agriculture context: a review on coconut cultivation in Sri Lanka. Front Agron 5:1162750. https://doi.org/10.3389/fagro.2023.1162750

Dissanayaka NS, Nimanthika N, Nuwarapaksha TD, Udumann SS, Atapattu AJ (2025) Agroforestry and climate-smart agriculture. In: Atapattu AJ (ed) Agroforestry and climate-smart agriculture. IGI Global Scientific Publishing, pp 173–200. https://doi.org/10.4018/979-8-3693-8282-0.ch006

Ekka AB, Jha KK, Kispotta LM (2019) Identification of suitable vegetable-based cropping sequence in relation to economic yield and profitability. J Pharmacogn Phytochem SP5:37–41

El-Mehy AA, Mohamed MHM (2018) Yield and economic evaluation of maize and tomato as affected by cropping systems and some growth stimulants. Middle East J Appl Sci 8(1):209–222

Fan H, Miao R, Guo C, Bao X, He W, Sun Y, Zhao C (2025) Research on the effect of diversified cropping on crop quality: a review. Agriculture 15(5):456. https://doi.org/10.3390/agriculture15050456

Gaikwad DJ, Ubale NB, Pal A, Singh S, Ali MA, Maitra S (2022) Abiotic stresses impact on major cereals and adaptation options – a review. Res Crop 23:896–915. https://doi.org/10.31830/2348-7542.2022.ROC-913

Gawdiya S, Sharma RK, Singh H, Kumar D (2025) Crop diversification as a cornerstone for sustainable agroecosystems: tackling biodiversity loss and global food system challenges. Discov Appl Sci 7(5):373. https://doi.org/10.1007/s42452-025-06855-z

Ghosh D (2021) Studies on the changes of biochemical, microbiological and sensory parameters of sauerkraut and fermented mixed vegetables. J Homepage 5:78–83. https://doi.org/10.26656/fr.2017.5(1).193

Ghosh D, Brahmachari K, Das A, Hassan MM, Mukherjee PK, Sarkar S, Dinda NK, Pramanick B, Moulick D, Maitra S, Hossain A (2021) Assessment of energy budgeting and its indicator for sustainable nutrient and weed management in a rice–maize–green gram cropping system. Agronomy 11(1):166. https://doi.org/10.3390/agronomy11010166

Ghosh PK, Hazra KK, Roy A (2023) Contributions of diversified cropping systems to plant nutrition, soil health, and environmental services. Indian J Fertil 19:1108–1117

Glaze-Corcoran S, Hashemi M, Sadeghpour A, Jahanzad E, Afshar RK, Liu X, Herbert SJ (2020) Understanding intercropping to improve agricultural resiliency and environmental sustainability. Adv Agron 162:199–256. https://doi.org/10.1016/bs.agron.2020.02.004

Gómez M (2024) Biological nitrogen fixation and intra- and interspecific competition in two vegetable/legume intercropping systems and their relationship to crop yield. Sci Hortic 338:113659. https://doi.org/10.1016/j.scienta.2024.113502

Gouda HS, Singh YV, Shivay YS, Biswas DR, Bana RS, Poornima S, Manu SM, Maitra S, Sairam M, Salmen SH, Alharbi SA (2024) Root parameters and water productivity of rice and wheat in a rice–wheat cropping system as influenced by enriched compost and crop establishment methods. J Agric Food Res 18:101317. https://doi.org/10.1016/j.jafr.2024.101317

Grebner DL, Bettinger P, Siry JP (2013) Common forestry practices. In: Introduction to forestry and natural resources. Elsevier eBooks, pp 255–285. https://doi.org/10.1016/b978-0-12-386901-2.00011-7

Handayanto E (2025) Knowledge. CIFOR-ICRAF. Available at: https://www.cifor-icraf.org/knowledge/publication/30444/

Hao J, Subbarao KV, Koike ST (2003) Effects of broccoli rotation on lettuce drop caused by *Sclerotinia minor* and on the population density of sclerotia in soil. Plant Dis 87(2):159–166. https://doi.org/10.1094/pdis.2003.87.2.159

Hariharan G, Abhayapala R, Karunarathna B, Zakeel MCM (2022) Efficient utilisation of rice fallow through pulse cultivation. In: Meena RS, Kumar S (eds) Advances in legumes for sustainable intensification. Elsevier eBooks, pp 71–92. https://doi.org/10.1016/b978-0-323-85797-0.00028-8

Hemasree KR, Sairam M, Maitra S, Ray S, Maheswari N, Gaikwad DJ, Santosh DT (2025a) Effects of sulphur and zinc on growth and productivity of summer cowpea under conditions of southern Odisha. Crop Res 60(1&2):28–34. https://doi.org/10.31830/2454-1761.2025.CR-1012

Hemasree K, Maitra S, Ray S, Sairam M, Gudla SL, Amith S (2025b) Impact of sulphur and zinc fertilisation on yield, quality, nutrient uptake and economics of cowpea (*Vigna unguiculata* L.) cultivation. Farm Manag 10(2):115–123. https://doi.org/10.31830/2456-8724.2025.FM-179

Ilakiya T, Parameswari E, Swarnapriya R, Yazhini G, Kalaiselvi P, Davamani V, Singh S, Vinothini N, Dharani C, Garnepudi SL, Ajaykumar R (2024) Unlocking the carbon sequestration potential of horticultural crops. J Carbon Res 10(3):65. https://doi.org/10.3390/c10030065

Jankowska B, Wojciechowicz-Żytko E (2016) Effect of intercropping carrot (*Daucus carota* L.) with two aromatic plants, coriander (*Coriandrum sativum* L.) and summer savory (*Satureja hortensis* L.), on the population density of select carrot pests. Folia Hortic 28:13–18. https://doi.org/10.1515/fhort-2016-0002

Jena J, Maitra S, Hossain A, Pramanick B, Gitari HI, Praharaj S, Shankar T, Palai JB, Rathore A, Mandal TK, Jatav HS (2022) Role of legumes in cropping system for soil ecosystem improvement. In: Hanuman SJ (ed) Ecosystem services: types, management and benefits. Nova Science Publishers, Inc, pp 1–21

Jiang N, Li F, Tan M, Yu H, Qu A, Feng H (2025) Three-dimensional shading models for estimating global radiation on photovoltaic module. Renew Energy 242:122333. https://doi.org/10.1016/j.renene.2024.122333

Johnston A, Bruulsema T (2014) 4R nutrient stewardship for improved nutrient use efficiency. Procedia Eng 83:365–370. https://doi.org/10.1016/j.proeng.2014.09.029

Kalita J, Deka B, Kalita DN (2018) Assessment of rice-based cropping systems for maximising productivity and profitability in Kamrup district of Assam. Int J Agric Sci 10(18):7209–7211

Kamini Y, Muskan R, Maurya BM (2025) Assessment of rice-based multiple cropping system at Bansagar area of Kymore–Satpura region. Int J Res Agron 8(1):497–500. https://doi.org/10.33545/2618060X.2025.v8.i1g.2448

Kaur S, Bedi M, Singh S, Kour N, Bhatti SS, Bhatia A, Kumar M, Kumar R (2024) Monoculture of crops: a challenge in attaining food security. In: Advances in food security and sustainability: environmental challenges in attaining food security. Elsevier, London, pp 197–213. https://doi.org/10.1016/bs.af2s.2024.07.008

Kavita, Sankhyan NK, Singh D, Amita, Mandar RP, Mehta T (2025) Soil health degradation in monoculture: challenges and remediation strategies. Vigyan Varta 6(5):230–232

Kelty MJ (1992) Comparative productivity of monocultures and mixed-species stands. In: Kelty MJ, Larson BC, Oliver CD (eds) The ecology and silviculture of mixed-species forests, Forestry sciences, vol 40. Springer, pp 125–141. https://doi.org/10.1007/978-94-015-8052-6_8

Khanum M, Bazzaz M, Ahmed B, Huda HM (2020) Intercropping of cabbage with maize. Bangladesh Agron J 22(1):115–120. https://doi.org/10.3329/baj.v22i1.44943

Kishore K, Sangeetha G, Rupa T, Ganeshmurthy A, Samant D, Acharya G, Srinivas P (2025) Assessment of mango-based intercropping systems for productivity, resource use efficiency and environmental sustainability in tropical region of India. Next Sustain 5:100121. https://doi.org/10.1016/j.nxsust.2025.100121

Knez M, Mattas K, Gurinovic M, Gkotzamani A, Koukounaras A (2024) Revealing the power of green leafy vegetables: cultivating diversity for health, environmental benefits, and sustainability. Glob Food Sec 43:100816. https://doi.org/10.1016/j.gfs.2024.100816

Kokila A, Nagarajaiah C, Hanumanthappa CD, Shivanna B, Sathish K, Mahadevamurthy M (2024) Effect of tree canopy cover on soil moisture dynamics in different agroforestry systems under semiarid condition. Int J Environ Climate Chang 14:485–495. https://doi.org/10.9734/ijecc/2024/v14i94432

Kumar BM, Kunhamu TK (2022) Nature-based solutions in agriculture: a review of the coconut (*Cocos nucifera* L.)-based farming systems in Kerala, “The Land of Coconut Trees”. Nat Based Solut 2:100012. https://doi.org/10.1016/j.nbsj.2022.100012

Kumar R, Mishra JS, Rao KK, Mondal S, Hazra KK, Choudhary JS, Hans H, Bhatt BP (2020a) Crop rotation and tillage management options for sustainable intensification of rice-fallow agro-ecosystem in eastern India. Sci Rep 10(1):11146. https://doi.org/10.1038/s41598-020-67973-9

Kumar S, Singh R, Mishra JS, Kumar P (2020b) Enhancing water productivity through crop diversification and utilisation of residual soil moisture in rice-based cropping systems. Agric Water Manag 240:106290. https://doi.org/10.5772/intechopen.76904

Kumar S, Yadav A, Kumar A, Hasanain M, Shankar K, Karan S, Dayal P (2023) Climate smart irrigation practices for improving water productivity in India: a comprehensive review. Int J Environ Climate Chang 12:333–348. https://doi.org/10.9734/ijecc/2023/v13i123689

Kumari M, Pandey S, Giri VP, Chauhan P, Mishra N, Verma P, Tripathi A, Singh SP, Bajpai R, Mishra A (2022) Integrated approach for technology transfer awareness of traditional knowledge for upliftment of circular bioeconomy. In: Sunita V, Ashok P, Bhaskar T, Mohan SV, Daniel CW, Tsang (eds) Biomass, biofuels, biochemicals. Elsevier eBooks, pp 613–636. https://doi.org/10.1016/b978-0-323-89855-3.00005-4

Kumawat A, Bamboriya SD, Meena RS, Yadav D, Kumar A, Kumar S, Raj A, Pradhan G (2022) Legume-based intercropping to achieve the crop, soil, and environmental health security. In: Meena RS, Kumar S (eds) Advances in legumes for sustainable intensification. Elsevier eBooks, pp 307–328. https://doi.org/10.1016/b978-0-323-85797-0.00005-7

Kushwah SS, Singh OP, Gupta BS (2011) Evaluation of potato-based crop sequence for crop diversification in Malwa region of Madhya Pradesh. J Hortic Sci 6(2):166–168. https://doi.org/10.24154/jhs.v6i2.429

Lauri PE, Costes E, Regnard JL, Brun L, Simon S, Monney P, Sinoquet H (2008) Does knowledge on fruit tree architecture and its implications for orchard management improve horticultural sustainability? An overview of recent advances in the apple. Int Symp Hortic Eur 817:243–225. https://doi.org/10.13140/2.1.3307.6486

Li W, Liu Q, Xie L, Yin C (2023) Interspecific plant–plant interactions increase the soil microbial network stability, shift keystone microbial taxa, and enhance their functions in mixed stands. For Ecol Manag 533:120851. https://doi.org/10.1016/j.foreco.2023.120851

Luo CL, Duan HX, Wang YL, Liu HJ, Xu SX (2024) Complementarity and competitive trade-offs enhance forage productivity, nutritive balance, land and water use, and economics in legume–grass intercropping. Field Crop Res 319:109642. https://doi.org/10.1016/j.fcr.2024.109642

Maheswari N, Maitra S, Sairam M, Ray S, Sagar L, Santosh DT, Gaikwad DJ (2025) Impact of real-time nitrogen management on the performance of maize–cowpea intercropping system. Crop Res 60:139–146. https://doi.org/10.31830/2454-1761.2025.CR-1018

Maitra S, Gitari H (2020) Scope for adoption of intercropping in organic agriculture. Indian J Nat Sci 11(36):28624–28631

Maitra S, Zaman A (2017) Biodrainage in waterlogged areas for agricultural sustainability. Int J Bioresour Sci 4(2):79–84. https://doi.org/10.5958/2454-9541.2017.00013.5

Maitra S, Samul SK, Roy DK, Mondal AK (2001) Effect of cotton-based intercropping system under rainfed conditions in Sunderban region of West Bengal. Indian Agric J 45(3&4):157–162

Maitra S, Hossain A, Brestic M, Skalicky M, Ondrisik P, Gitari H, Brahmachari K, Shankar T, Bhadra P, Palai JB, Jena J, Sairam M (2021) Intercropping—a low input agricultural strategy for food and environmental security. Agronomy 11(2):343. https://doi.org/10.3390/agronomy11020343

Maitra S, Praharaj S, Brestic M, Sahoo RK, Sagar L, Shankar T, Palai JB, Sahoo U, Sairam M, Pramanick B, Nath S, Venugopalan VK, Skalický M, Hossain A (2023) Rhizobium as biotechnological tools for green solutions: an environmentally friendly approach for sustainable crop production in the modern era of climate change. Curr Microbiol 80:219. https://doi.org/10.1007/s00284-023-03317-w

Maitra S, Sain S, Sairam M, Shankar T, Ray S, Banerjee M, Malik GC (2024a) Impact of Rabi maize–legume intercropping system on growth, yield and crop competition in maize. Res Crop 25(3):409–415. https://doi.org/10.31830/2348-7542.2024.roc-1093

Maitra S, Sairam M, Ray S, Santosh DT, Gaikwad DJ, Gitari HI, Mukesh G, Maheswari N (2024b) Applauding soil quality improvement through legume-based intercropping system: a review. Int J Bioresour Sci 11(1):59–69. https://doi.org/10.30954/2347-9655.01.2024.7

Maitra S, Sairam M, Santosh DT (2024c) Evaluation of the growth, flower quality, yield and vase life of gerbera (*Gerbera jamesonii* L.) cultivars under open ventilated saw-tooth type polyhouse. Res Crop 25(1):104–109. https://doi.org/10.31830/2348-7542.2024.ROC-1055

Maitra S, Ray S, Sagar L, Sairam M, Pramanick B, Gitari H, Santosh DT, Jagannath GD, Atapattu A, Chappa LR (2025a) Cropping system approach in climate-resilient food crop production. In: Pramanick B, Singh SV, Maitra S, Celletti S, Hossain A (eds) Sustainability sciences in Asia and Africa. Springer, Singapore, pp 327–356. https://doi.org/10.1007/978-981-96-7699-6_15

Maitra S, Sairam M, Sagar L, Ray S, Santosh DT, Gaikwad DJ, Gitari HI (2025b) Alley cropping: a sustainable agroforestry system for enhancing productivity, soil health and biodiversity. In: Plant biotechnology and sustainable agriculture: bridging the gap for global food security. CABI, pp 235–248. https://doi.org/10.1079/9781800627659.0016

Maitra S, Santosh DT, Jena J, Behera SD, Ray S, Sairam M, Palai JB, Gaikwad DJ, Rishitha S, Gitari HI, Atapattu AJ (2026) Understanding plant responses to abiotic stress through systems biology. In: Tariq A (ed) Systems biology in crop improvement. Academic Press, pp 59–100

Maity B, Sairam M, Ray S, Maitra S, Dash B, Guchhait D (2025) Evaluation of polyhalite as a potassium and sulphur source for improving the growth and productivity of summer groundnut (*Arachis hypogaea* L.). Crop Res 60:357–362. https://doi.org/10.31830/2454-1761.2025.CR-1069

Manik SMN, Pengilley G, Dean G, Field B, Shabala S, Zhou M (2019) Soil and crop management practices to minimise the impact of waterlogging on crop productivity. Front Plant Sci 10:140. https://doi.org/10.3389/fpls.2019.00140

Manna US, Biswas S, Mazumdar SP, Majumdar B, Sasmal S (2017) Effect of different jute (*Corchorus olitorius* L.) based cropping systems on soil quality under farmers' field condition in the eastern Indo-Gangetic plain. Int J Curr Microbiol Appl Sci 6(9):3324–3334. https://doi.org/10.20546/ijcmas.2017.609.411

Marcos-Pérez M, Sánchez-Navarro V, Zornoza R (2023) Intercropping systems between broccoli and fava bean can enhance overall crop production and improve soil fertility. Sci Hortic 312:111834. https://doi.org/10.1016/j.scienta.2023.111834

Meena A, Kumar A, Meena A (2023) Patterns, determinants and challenges of horticulture diversification in India. Int Res J Bus Stud 16(1):99–110. https://doi.org/10.21632/irjbs.16.1.99-110

Mehra P, Baker J, Sojka RE, Bolan N, Desbiolles J, Kirkham MB, Ross C, Gupta R (2018) A review of tillage practices and their potential to impact the soil carbon dynamics. Adv Agron 150:185–230. https://doi.org/10.1016/bs.agron.2018.03.002

Mihrete TB, Mihretu FB (2025) Crop diversification for ensuring sustainable agriculture, risk management and food security. Glob Chall 9(2):2400267. https://doi.org/10.1002/gch2.202400267

Mir MS, Saxena A, Kanth RH, Raja W, Dar KA, Mahdi SS, Bhat TA, Naikoo NB, Nazir A, Amin Z, Mansoor T (2022) Role of intercropping in sustainable insect-pest management: a review. Int J Environ Climate Chang 12(11):3390–3403. https://doi.org/10.9734/ijecc/2022/v12i111390

Mirriam A, Mugwe J, Raza MA, Seleiman MF, Maitra S, Gitari HH (2022) Aggrandising soybean yield, phosphorus use efficiency and economic returns under phosphatic fertiliser application and inoculation with *Bradyrhizobium*. J Soil Sci Plant Nutr 22(4):5086–5098. https://doi.org/10.1007/s42729-022-00985-8

Misra NR (2026) Odisha farmer builds low-cost mushroom unit; clocks Rs 2 lakh monthly. Available at: https://30stades.com/farming/odisha-farmer-builds-low-cost-mushroom-unit-clocks-rs-2-lakh-monthly-10990375

Mohapatra KB, Sahoo AK, Chinara N, Pani BK (2018) Status and constraints of straw mushroom cultivation in Odisha. Available at: https://epubs.icar.org.in/index.php/MR/article/view/70644

Mokidul I, Nath LK, Samajdar T (2018) Sustainable diversification of maize (*Zea mays* L.)–legumes cropping systems for productivity, profitability and resource-use efficiency in West Garo Hills of Meghalaya, India. Legum Res 43(3):427–431. https://doi.org/10.18805/LR-3970

Mwadalu R, Mochoge B, Mwangi M, Maitra S, Gitari H (2022) Response of Gadam sorghum (*Sorghum bicolor*) to farmyard manure and inorganic fertiliser application. Int J Agric Environ Biotechnol 15(1):51–60. https://doi.org/10.30954/0974-1712.01.2022.6

Namazi Y, Rezaei-Chiyaneh E, Moghaddam SS, Battaglia ML (2022) The effects of microbial inoculation and intercropping on yield and active ingredients of savory (*Satureja hortensis* L.) intercropped with common bean (*Phaseolus vulgaris* L.). Int J Environ Sci Technol 19:8273–8288. https://doi.org/10.1007/s13762-022-04024-y

Nandi S, Maitra S, Shankar T, Panda M, Sairam M (2022) Impact of intercropping of vegetable legumes in summer maize on productivity and competitive ability of crops. Crop Res 57(3):122–127. https://doi.org/10.31830/2454-1761.2022.017

National Economic Forum (2025) Addressing postharvest losses in India: a silent crisis in agriculture. National Economic Forum. Available at: https://nationaleconomicforum.org/nef_articles/addressing-post-harvest-losses-in-india-a-silent-crisis-in-agriculture/

Navasero MV, Calumpang SMF (2013) Influence of intercropping vegetable crops on the population of some arthropods associated with eggplant (*Solanum melongena* L.) and damage potential of eggplant fruit and shoot borer (*Leucinodes orbonalis* Guenee) in the presence of radish (*Raphanus sativus* L.). Philipp Agric Sci 96(3):247–255

Nori M, Masina S, Sagar M, Sumit R, Hemasree KR, Roman Kumar M, Dinkar JG, Rajesh Shriram K (2025) Influence of precision nitrogen management in Rabi maize and cow pea intercropping system under varied planting proportions in south Odisha. Plant Sci Today 12(3):1–9. https://doi.org/10.14719/pst.8194

Nuwarapaksha TD, Udumann SS, Dissanayaka DMNS, Dissanayake DKRPL, Atapattu AJ (2022) Coconut based multiple cropping systems: an analytical review in Sri Lankan coconut cultivations. Circ Agric Syst 2:8. https://doi.org/10.48130/CAS-2022-0008

Nuwarapaksha TD, Rajapaksha UN, Ekanayake J, Weerasooriya SA, Atapattu AJ (2023a) Exploring the economic viability of integrating Jamnapari goat into underutilised pastures under coconut cultivations in Coconut Research Institute, Sri Lanka. Biol Life Sci Forum 27:27. https://doi.org/10.3390/IECAG2023-14990

Nuwarapaksha TD, Udumann SS, Dissanayaka NS, Atapattu AJ (2023b) Coconut-based livestock farming: a sustainable approach to enhancing food security in Sri Lanka. In: Kiba DI (ed) Transitioning to zero hunger. MDPI, Basel, pp 197–213. https://doi.org/10.3390/books978-3-03897-863-3-8

Pan R, Martinez A, Brito T, Seidel E (2018) Processes of soil infiltration and water retention and strategies to increase their capacity. J Exp Agric Int 20(2):1–14. https://doi.org/10.9734/JEAI/2018/39132

Panchal P, Preece C, Peñuelas J, Giri J (2022) Soil carbon sequestration by root exudates. Trends Plant Sci 27(8):749–757. https://doi.org/10.1016/j.tplants.2022.04.009

Parimaladevi C, Ramanathan SP, Kumar SN, Suresh S (2019) Evaluation of maize-based intercropping system in Thamirabarani basin of Tamil Nadu. J Pharmacogn Phytochem 8(3):4051–4056

Pramanick B, Kumar M, Naik BM, Singh SK, Singh SK, Kumar M, Singh SV, Singh SV (2023) Soil carbon–nutrient cycling, energetics, and carbon footprint in calcareous soils with adoption of long-term conservation tillage practices and cropping systems diversification. Sci Total Environ 912:169421. https://doi.org/10.1016/j.scitotenv.2023.169421

Punitha A, Prasad AS, Sivagamy K (2023) Evaluation of cucurbits under rice-based cropping sequence in Tiruvallur district. Madras Agric J 110(1–3):129–133. https://doi.org/10.29321/MAJ.10.000764

Rajgopal K (2023) Advances in multitier cropping systems for HRT 604 course. Studocu. Available at: https://www.studocu.com/in/document/university-of-agricultural-sciences-bangalore/advances-in-production-technology-of-plantation-medicinal-aromatic-and-spice-crops/advances-in-multi-story-cropping-sysytem/68401768

Ranjan P (2025) Crop diversification and modern cropping systems in Indian agriculture. Available at: https://licchavilyceum.com/crop-diversification-and-modern-cropping-systems-in-indian-agriculture/

Ray S, Maitra S, Sairam M, Sravya M, Priyadarshini A, Shubhadarshi S, Padhi DP (2024) An unravelled potential of foliar application of micro and beneficial nutrients in cereals for ensuring food and nutritional security. Int J Exp Res Rev 41(Spl Vol):19–42. https://doi.org/10.52756/ijerr.2024.v41spl.003

Ray S, Maitra S, Sairam M, Sameer S, Sagar L, Divya BS, Gitari HI (2025a) The nexus between intercropping systems, ecosystem services and sustainable agriculture: a review. Res Crop 26(1):1–11. https://doi.org/10.31830/2348-7542.2025.ROC-1166

Ray S, Maitra S, Sairam M, Lalichetti S (2025b) Impact of maize–groundnut intercropping system on growth and yield as influenced by phosphorus, potassium and sulphur fertilisation. Crop Res 60:308–316. https://doi.org/10.31830/2454-1761.2025.CR-1054

Ray S, Maitra S, Masina S, Maity B (2025c) Effect of precision nitrogen management and plant stimulants on the growth and yield in a Rabi maize–summer groundnut cropping system. Crop Res 60:299–307. https://doi.org/10.31830/2454-1761.2025.CR-1047

Raza A, Khare T, Zhang X, Rahman MM, Hussain M, Gill SS, Chen Z, Zhou M, Hu Z, Varshney RK (2025) Novel strategies for designing climate-smart crops to ensure sustainable agriculture and future food security. J Sustain Agric Environ 4(2):e70048. https://doi.org/10.1002/sae2.70048

Rezaei-Chiyaneh E, Mahdavikia H, Alipour H, Dolatabadian A, Battaglia ML, Maitra S, Harrison MT (2023) Biostimulants alleviate water deficit stress and enhance essential oil productivity: a case study with savory. Sci Rep 13(1):720. https://doi.org/10.1038/s41598-022-27338-w

Ribas RGT, Filho ABC, Dutra AF, Barbosa JC, De Souza Rolim G (2020) Land equivalent ratio in the intercropping of cucumber with lettuce as a function of cucumber population density. Agriculture 10(3):88. https://doi.org/10.3390/agriculture10030088

Ruillé M, Beillouin D, Prudhomme R (2026) Ecological drivers of intercropping performance for enhanced global crop production. NPJ Sustain Agric 4(1):1–9. https://doi.org/10.1038/s44264-025-00110-z

Sagar L, Maitra S, Singh S, Sairam M (2023) Impact of precision nutrient management on rice growth and productivity in southern Odisha. Agric Sci Dig 43(6):812–816. https://doi.org/10.18805/ag.D-5824

Sagar L, Maitra S, Masina S, Ray S (2025) Influence of different nutrient management strategies on growth, yield and greenhouse gas emissions of irrigated rice. Int J Exp Res Rev 49:1–9. https://doi.org/10.52756/ijerr.2025.v49.001

Sahoo SC (2009) Yield and economics of sweet corn–vegetable intercropping system. Range Manag Agrofor 30(2):155–158

Sairam M, Gaikwad D, Pramanick B, Maitra S (2023a) Conservation agriculture impacts on soil organic carbon and soil properties. In: Maitra S, Gaikwad DJ, Santosh DT (eds) Advances in agricultural technology. Griffon

Sairam M, Maitra S, Sahoo U, Sagar L, Krishna TG (2023b) Evaluation of precision nutrient tools and nutrient optimisation in maize (*Zea mays* L.) for enhancement of growth, productivity and nutrient use efficiency. Res Crop 24(4):666–677. https://doi.org/10.31830/2348-7542.2023.ROC-1016

Sairam M, Maitra S, Sagar L, Biswas T, Bárek V, Brestic M, Hossain A (2025a) Application of precision nutrient tools for the optimisation of fertiliser requirements and assessment of the growth and productivity of maize (*Zea mays* L.) in the northeastern Ghat of India. J Agric Food Res 21:101958. https://doi.org/10.1016/j.jafr.2025.101958

Sairam M, Ray S, Padhi DP, Dayal PAS, Dt S, Maitra S (2025b) Growth, productivity and quality of gerbera cultivars under forced- and open-ventilated polyhouse conditions. Res Crop 26(4):675–681. https://doi.org/10.31830/2348-7542.2025.ROC-1256

Santosh DT, Tiwari KN, Maitra S (2021) Influence of drip irrigation and plastic mulch on young cashew (*Anacardium occidentale* L.). Int J Bioresour Sci 8(2):87–92

Santosh DT, Maitra S, Tiwari KN (2023a). Optimizing nutrients through drip fertigation and plastic mulch on the growth and yield in banana (*Musa acuminata* L.). Res Crop 24(2):330–336. https://doi.org/10.31830/2348-7542.2023.ROC-11163

Santosh DT, Debnath S, Maitra S, Sairam M, Sagar L, Hossain A, Moulick D (2023b) Alleviation of climate catastrophe in agriculture through adoption of climate-smart technologies. In: Chatterjee U, Shaw R, Kumar S, Raj AD, Das S (eds) Climate crisis: adaptive approaches and sustainability. Springer Nature, Cham, pp 307–332. https://doi.org/10.1007/978-3-031-44397-8_17

Santosh DT, Maitra S, Tiwari KN (2023c) Optimising nutrients through drip fertigation and plastic mulch on the growth and yield in banana (*Musa acuminata* L.). Res Crop 24(2):330–336. https://doi.org/10.31830/2348-7542.2023.roc-11163

Sashika MN, Gammanpila H, Priyadarshani S (2024) Exploring the evolving landscape: urban horticulture cropping systems–trends and challenges. Sci Hortic 327:112870. https://doi.org/10.1016/j.scienta.2024.112870

Saurabh, Kumar T, Kumar M (2023) Vegetable production under challenging conditions—abiotic stress: a review. Int J Adv Biochem Res SP-7(2):13–19

Scavo A, Fontanazza S, Restuccia A, Pesce GR, Abbate C, Mauromicale G (2022) The role of cover crops in improving soil fertility and plant nutritional status in temperate climates: a review. Agron Sustain Dev 42(5):93. https://doi.org/10.1007/s13593-022-00825-0

Sethi D, Sood S, Kumar V, Sharma K (2025) Spatiotemporal changes in area, production and productivity of vegetables: a review. Agric Rev 46(5):757–765

Shah TA, Bhat MA, Shah ST (2025) Land equivalent ratio as influenced by intercrops in cauliflower (*Brassica oleracea* var. *botrytis*). Adv Appl Therm Chem Sci 1(1):1–8. https://doi.org/10.21276/AATCCReview.2025.13.01.158

Sharma RP, Pathal SK, Haque M, Raman KR (2004) Diversification of traditional rice (*Oryza sativa*)-based cropping system for sustainable production in south Bihar alluvial plains. Indian J Agron 49(4):218–222. https://doi.org/10.59797/ija.v49i4.5202

Sharma A, Chaudhary N, Sharma R, Anjanawe S (2020) Multistoried cropping system in horticulture: an approach for more crop per unit area of land. J Pharmacogn Phytochem 9:25–28

Shekhar C (2024) Impact of high-yielding and short-duration varieties on shifts in cropping patterns. AgriInsights. Available at: https://www.agriinsights.in/2024/09/impact-of-high-yielding-and-short-duration-varieties.html

Singh H (2021) Integrated pest management of vegetable crops. Int J Agric Plant Sci 3(2):4–7

Singh RK, Tiwari JK, Singh RK, Singh AK (2017) Rice (*Oryza sativa*)–potato (*Solanum tuberosum*) based cropping sequences in relation to production potential and economic returns under irrigated ecosystem of central plains. Indian J Agric Sci 87(10):1277–1282. https://doi.org/10.56093/ijas.v87i10.74793

Stanley CD, Csizinszky AA, Clark GA, Prevatt JW (1991) Sequential cropping for vegetable production using microirrigation on sandy soils in southwestern Florida. HortTechnology 1(1):72–76. https://doi.org/10.21273/HORTTECH.1.1.72

Thomas GV, Krishnakumar V, Dhanapal R, Srinivasa Reddy DV (2018) Agro-management practices for sustainable coconut production. In: The coconut palm (*Cocos nucifera* L.) – research and development perspectives. Springer, Singapore, pp 227–322. https://doi.org/10.1007/978-981-13-2754-4_7

Tignegre JBS, Larbi A, Tenkouano A, Nurudeen A, Asante M et al (2018) Optimisation of maize–vegetable (African eggplant and pepper) intercrops in northern, upper west and upper east regions of Ghana. JOJ Hortic Arboric 1(2):555–557. https://doi.org/10.19080/JOJHA.2018.01.555557

Udumann SS, Dissanayaka NS, Nuwarapaksha TD, Atapattu AJ (2025) Agroforestry education and training for the next generation. In: Atapattu AJ (ed) Agroforestry education and training for the next generation. IGI Global Scientific Publishing, pp 449–472. https://doi.org/10.4018/979-8-3693-8282-0.ch015

Vaishnavi K, Raina Saha A, Roy A, Bhattacharya PM, Chowdhury AK, Mandal R, Mandal S (2025) Performance of maize (*Zea mays* L.) with intercropping of different vegetable crops during Rabi season in West Bengal, India. Int J Plant Soil Sci 37(9):596–603. https://doi.org/10.9734/ijpss/2025/v37i95738

Wang X, Wang T, Wang L, Liu E (2024) The effects of different rotations of beans, maize, and cabbage on soil moisture and economic benefits. Agronomy 14(3):479. https://doi.org/10.3390/agronomy14030479

Williams J (2025) 20 Vegetables that have shallow roots. Farming Shelter. Available at: https://farmingshelter.com/vegetables-that-have-shallow-roots/

Witzgall K, Steiner F, Hesse B, Riveras-Muñoz N, Rodríguez V, Teixeira P, Li M, Oses R, Seguel O, Seitz S, Wagner D, Scholten T, Buegger F, Angst G, Mueller C (2024) Living and decaying roots as regulators of soil aggregation and organic matter formation from the rhizosphere to the detritusphere. Soil Biol Biochem 197:109503. https://doi.org/10.1016/j.soilbio.2024.109503

Wolz KJ, Lovell ST, Branham BE, Eddy WC, Keeley K, Revord RS, Wander MM, Yang WH, DeLucia EH (2017) Frontiers in alley cropping: transformative solutions for temperate agriculture. Glob Chang Biol 24(3):883–894. https://doi.org/10.1111/gcb.13986

Wright P, Falloon R, Hedderley D (2014) Different vegetable crop rotations affect soil microbial communities and soilborne diseases of potato and onion: literature review and a long-term field evaluation. N Z J Crop Hortic Sci 43(2):85–110. https://doi.org/10.1080/01140671.2014.979839

Yadav G (2025) High value crops – significance & challenges – explained pointwise. Free UPSC IAS Preparation Syllabus and Materials for Aspirants. Available at: https://forumias.com/blog/high-value-crops-significance-challenges-explained-pointwise/

Yang X, Xiong J, Du T, Ju X, Gan Y, Li S, Xia L, Shen Y, Pacenka S, Steenhuis TS, Siddique KHM, Kang S, Butterbach-Bahl K (2024) Diversifying crop rotation increases food production, reduces net greenhouse gas emissions and improves soil health. Nat Commun 15(1):198. https://doi.org/10.1038/s41467-023-44464-9

Zhang M, Han Y, Li D, Xu S, Huang Y (2023) Smart horticulture as an emerging interdisciplinary field combining novel solutions: past development, current challenges, and future perspectives. Hortic Plant J 10(6):1257–1273. https://doi.org/10.1016/j.hpj.2023.03.015

Zhou S, Guo J (2025) Mixed evergreen and deciduous forests boost local climate resilience, nutrient dynamics, and photosynthetic performance assessed by remote sensing. For Ecol Manag 586:122677. https://doi.org/10.1016/j.foreco.2025.122677

MIX
Papier aus verantwortungsvollen Quellen
Paper from responsible sources
FSC® C105338

FSC
www.fsc.org

If you have any concerns about our products,
you can contact us on
ProductSafety@springernature.com

In case Publisher is established outside the EU,
the EU authorized representative is:
Springer Nature Customer Service Center GmbH
Europaplatz 3, 69115 Heidelberg, Germany

Printed by Libri Plureos GmbH
in Hamburg, Germany